PRESENTI

ARCHAEOLOGIST'S TOOLKIT

SERIES EDITORS: LARRY J. ZIMMERMAN AND WILLIAM GREEN

The Archaeologist's Toolkit is an integrated set of seven volumes designed to teach novice archaeologists and students the basics of doing archaeological fieldwork, analysis, and presentation. Students are led through the process of designing a study, doing survey work, excavating, properly working with artifacts and biological remains, curating their materials, and presenting findings to various audiences. The volumes—written by experienced field archaeologists—are full of practical advice, tips, case studies, and illustrations to help the reader. All of this is done with careful attention to promoting a conservation ethic and an understanding of the legal and practical environment of contemporary American cultural resource laws and regulations. The Toolkit is an essential resource for anyone working in the field and ideal for training archaeology students in classrooms and field schools.

Volume 1: *Archaeology by Design*
By Stephen L. Black and Kevin Jolly

Volume 2: *Archaeological Survey*
By James M. Collins and Brian Leigh Molyneaux

Volume 3: *Excavation*
By David L. Carmichael and Robert Lafferty

Volume 4: *Artifacts*
By Charles R. Ewen

Volume 5: *Archaeobiology*
By Kristin D. Sobolik

Volume 6: *Curating Archaeological Collections:*
  From the Field to the Repository
By Lynne P. Sullivan and S. Terry Childs

Volume 7: *Presenting the Past*
By Larry J. Zimmerman

PRESENTING THE PAST

Larry J. Zimmerman

ARCHAEOLOGIST'S TOOLKIT
VOLUME 7

ALTAMIRA
PRESS

A Division of Rowman & Littlefield Publishers, Inc.
Walnut Creek • Lanham • New York • Oxford

For Karen Zimmerman, who more than thirty years ago pointed out that archaeology is mostly science fiction. She was right.

ALTAMIRA PRESS
A Division of Rowman & Littlefield Publishers, Inc.
1630 North Main Street, #367
Walnut Creek, CA 94596
www.altamirapress.com

Rowman & Littlefield Publishers, Inc.
A Division of Rowman & Littlefield Publishers, Inc.
4501 Forbes Boulevard, Suite 200
Lanham, MD 20706

PO Box 317
Oxford
OX2 9RU, UK

Copyright © 2003 by ALTAMIRA PRESS

All rights reserved. No part of this publication may be reproduced, stored in a retrieval system, or transmitted in any form or by any means, electronic, mechanical, photocopying, recording, or otherwise, without the prior permission of the publisher.

British Library Cataloguing in Publication Information Available

Library of Congress Cataloging-in-Publication Data
Zimmerman, Larry J. 1947–
  Presenting the past / Larry J. Zimmerman
  p. cm.—(The archaeologist's toolkit ; v. 7)
  Includes bibliographical references (p.   ) and index.
  ISBN 0-7591-0403-4 (cloth : alk. paper) — ISBN 0-7591-0025-X (pbk. : alk. paper)
  1. Archaeology—Methodology. 2. Archaeology—Public relations. 3. Communication of technical information—Handbooks, manuals, etc. 4. Communication of technical information—Audio-visual aids. 5. Public speaking—Handbooks, manuals, etc. 6. Report writing—Handbooks, manuals, etc. I. Title. II. Series.

CC75.7 .Z56 2003
808'.06693—dc21   2002152407

Printed in the United States of America

∞™ The paper used in this publication meets the minimum requirements of American National Standard for Information Sciences—Permanence of Paper for Printed Library Materials, ANSI/NISO Z39.48–1992.

CONTENTS

Series Editors' Foreword   ix

Acknowledgments   xiii

1 Out of Site, Out of Mind?   1
Making Excuses • Plan of the Book

2 Recognizing Our Audiences   7
Ethics and Accountability • Educating Our Students • Public Accountability: What Do Different Audiences Want and Need? • Archaeological Style and the Real World • Jargon

3 Choosing the Right Medium   15
Who Is My Intended Audience? • What Are My Contractual or Other Obligations to These Audiences? • What Kind of Budget Do I Have to Prepare My Work? • Given These Factors, What Media Are Available to Me? • What Will It Cost to Distribute My Work to the Intended Audiences? • Conclusion

4 Developing Needed Skills and Tools   21
Basic Writing Skills • Beyond the Basics of Writing • Dealing with References • Conclusion

5 Computers and Presenting the Past — 41
Basic Computer Skills You'll Need • Advanced Skills You'll Need • Computer Hardware • Software • Learning Curves • Conclusion

6 Visual Archaeology — 47
The Cultural Life of Images • Photographs versus Drawings • Tools

7 Lone Ranger or Team Player? — 59
Developing the Team • Outsourcing Work or Hiring In-House

8 Publish or Perish? Communicating with Colleagues — 65
Abstracts • Conference Papers and Public Talks • Poster Papers? How to Decide and How to Deliver • Electronic Sessions • Organizing and Chairing a Session • Conclusion

9 From Presented to Printed — 91
CRM Reports • Journal Articles and Book Chapters • Books and Monographs • Publishing and Career Development

10 Bringing the Past to Life and Presenting It with Style — 109
Hands-on Learning Promotes Understanding • Action Is Better Than Description • Games and Toys • Archaeological Fiction • Cartoons • Movies and Videos: Live Action versus Documentary • Public Archaeology • Events • Exhibits and Museums

11 Media Method or Media Madness? — 121
With Friends Like the Media, Who Needs Enemies? • Attitudes • Learn Something about Media • Finding a Hook • Deciding When to Go Public • Getting Noticed • Other Advice and Cautions • Ethics, Contexts, and Professional Responsibility • Conclusion

12 The Future of Presenting the Past — 135
Archaeology on the Web • CD-ROMs and DVDs • Storyboarding: A Crucial Concept for Websites and

CDs/DVDs • Interactivity: Bringing the Past to Life • Conclusion

Appendix: Some Archaeology Journals on the Web   145

References   147

Index   153

About the Author and Series Editors   161

SERIES EDITORS' FOREWORD

The Archaeologist's Toolkit is a series of books on how to plan, design, carry out, and use the results of archaeological research. The series contains seven books written by acknowledged experts in their fields. Each book is a self-contained treatment of an important element of modern archaeology. Therefore, each book can stand alone as a reference work for archaeologists in public agencies, private firms, and museums, as well as a textbook and guidebook for classrooms and field settings. The books function even better as a set, because they are integrated through cross-references and complementary subject matter.

Archaeology is a rapidly growing field, one that is no longer the exclusive province of academia. Today, archaeology is a part of daily life in both the public and private sectors. Thousands of archaeologists apply their knowledge and skills every day to understand the human past. Recent explosive growth in archaeology has heightened the need for clear and succinct guidance on professional practice. Therefore, this series supplies ready reference to the latest information on methods and techniques—the tools of the trade that serve as handy guides for longtime practitioners and essential resources for archaeologists in training.

Archaeologists help solve modern problems: They find, assess, recover, preserve, and interpret the evidence of the human past in light of public interest and in the face of multiple land use and development interests. Most of North American archaeology is devoted to cultural resource management (CRM), so the Archaeologist's Toolkit focuses on practical approaches to solving real problems in CRM and

public archaeology. The books contain numerous case studies from all parts of the continent, illustrating the range and diversity of applications. The series emphasizes the importance of such realistic considerations as budgeting, scheduling, and team coordination. In addition, accountability to the public as well as to the profession is a common theme throughout the series.

Volume 1, *Archaeology by Design*, stresses the importance of research design in all phases and at all scales of archaeology. It shows how and why you should develop, apply, and refine research designs. Whether you are surveying quarter-acre cell tower sites or excavating stratified villages with millions of artifacts, your work will be more productive, efficient, and useful if you pay close and continuous attention to your research design.

Volume 2, *Archaeological Survey*, recognizes that most fieldwork in North America is devoted to survey: finding and evaluating archaeological resources. It covers prefield and field strategies to help you maximize the effectiveness and efficiency of archaeological survey. It shows how to choose appropriate strategies and methods ranging from landowner negotiations, surface reconnaissance, and shovel testing to geophysical survey, aerial photography, and report writing.

Volume 3, *Excavation*, covers the fundamentals of dirt archaeology in diverse settings, while emphasizing the importance of ethics during the controlled recovery—and destruction—of the archaeological record. This book shows how to select and apply excavation methods appropriate to specific needs and circumstances and how to maximize useful results while minimizing loss of data.

Volume 4, *Artifacts*, provides students as well as experienced archaeologists with useful guidance on preparing and analyzing artifacts. Both prehistoric and historic-era artifacts are covered in detail. The discussion and case studies range from processing and cataloging through classification, data manipulation, and specialized analyses of a wide range of artifact forms.

Volume 5, *Archaeobiology*, covers the analysis and interpretation of biological remains from archaeological sites. The book shows how to recover, sample, analyze, and interpret the plant and animal remains most frequently excavated from archaeological sites in North America. Case studies from CRM and other archaeological research illustrate strategies for effective and meaningful use of biological data.

Volume 6, *Curating Archaeological Collections*, addresses a crucial but often ignored aspect of archaeology: proper care of the specimens

and records generated in the field and the lab. This book covers strategies for effective short- and long-term collections management. Case studies illustrate the do's and don'ts that you need to know in order to make the best use of existing collections and to make your own work useful for others.

Volume 7, *Presenting the Past,* covers another area that has not received sufficient attention: communication of archaeology to a variety of audiences. Different tools are needed to present archaeology to other archaeologists, to sponsoring agencies, and to the interested public. This book shows how to choose the approaches and methods to take when presenting technical and nontechnical information through various means to various audiences.

Each of these books and the series as a whole are designed to be equally useful to practicing archaeologists and to archaeology students. Practicing archaeologists in CRM firms, agencies, academia, and museums will find the books useful as reference tools and as brush-up guides on current concerns and approaches. Instructors and students in field schools, lab classes, and short courses of various types will find the series valuable because of each book's practical orientation to problem solving.

As the series editors, we have enjoyed bringing these books together and working with the authors. We thank all of the authors—Steve Black, Dave Carmichael, Terry Childs, Jim Collins, Charlie Ewen, Kevin Jolly, Robert Lafferty, Brian Molyneaux, Kris Sobolik, and Lynne Sullivan—for their hard work and patience. We also offer sincere thanks to Mitch Allen of AltaMira Press and a special acknowledgment to Brian Fagan.

LARRY J. ZIMMERMAN
WILLIAM GREEN

ACKNOWLEDGMENTS

This book derives from many discussions and arguments with colleagues about how best to tell people what we find and do. Larry Bradley, Rich Fox, Brian Molyneaux, and other staff members of the University of South Dakota Archaeology Laboratory have always been good sounding boards and full of good ideas, and always willing to tell each other in good-natured ways that we were "full of it." Colleagues on the SAA Committee on Ethics provided a wide range of ideas, and the insights gained from the committee's 2000 and 2001 sessions on the media were especially useful. John Doershuk, Claire Smith, Tristine Smart, Joe Watkins, Alison Wylie, Randy McGuire, Kris Hirst, Shirley Schermer, Lynn Alex, and many others shared lots of insight and experiences to help form many of my ideas. My coeditor for the Archaeologist's Toolkit series, Bill Green, has been an extraordinary colleague and friend. Mitch Allen is a profoundly patient man who exemplifies what a good editor should be. Because this book is partly about publishing, his insights proved to be extremely valuable. Thanks to you all.

Finally, I'd like to thank Karen Zimmerman, who has put up with my whining for years.

1
OUT OF SITE, OUT OF MIND?

Once upon a time in archaeology, grizzled, field-hardened professors told their students that "you aren't a real archaeologist unless you die with at least one unfinished site report." The press of salvage archaeology and the heady, money-dripping days of early CRM projects hardly left time to do more than get artifacts out of the ground and dash on to survey the next sewage lagoon or bridge replacement. An awful truth exacerbated these time pressures: Many of us who got into archaeology did so because it was the fieldwork and its discoveries that excited us more than anything else. The lab was always a distant second choice for us and mostly boring. The rare exception came when some new technology became available to allow us bit of analytical or interpretive wizardry.

When it came to actually writing up the project, the pain really started. Writing was a cold bucket of water dousing the flames of passion for the field. To be sure, some proved bold and daring enough to throw together a conference paper, usually outlined on the plane on the way to the conference or in the hotel room the night before a "yawner" of an 8 A.M. session. A few brazenly went so far as to turn their work into a monograph read by a few dozen colleagues or a journal article skimmed by a few hundred, provoking envy from many and establishing celebrity for some.

Accountability to a range of publics changed all that. Pesky contract managers who paid the bills for some federal or state agency started to hound us for reports so that they could jump through legislative and regulatory hoops. Deep down, we knew that the reports just lined a bookshelf or engorged a file drawer in some minion's office. Were the deadlines really all that important? To make matters worse, somebody came up with the bright idea that the folks who really paid the bills—taxpayers or shareholders—might actually be interested in what we were finding with the contract dollars their congressional delegates or CEOs always seemed so hesitant to spend. They wanted us to write public reports and articles for popular magazines, set up traveling exhibits, and present agency-sponsored projects in a way the public could understand.

To top it all off, modern archaeology, with its emphasis on multidisciplinary approaches, brought with it the complexities of collaborative projects. Single-authored, jargon-laden, mostly descriptive reports were no longer acceptable even to colleagues, let alone CRM bureaucrats. Our professional organizations went so far as to codify the idea that we ought to write up our fieldwork.

Unfortunately, the truth of the matter is that many archaeological projects do go unreported, for a variety of reasons, including a lack of time and funding, difficulties interpreting complex cultural remains, and fear of professional criticism. British archaeologist Peter Addyman claims that up to 60 percent of modern excavations go unpublished after ten years, and only 10 percent of excavations funded by the National Science Foundation since 1950 ever reached print. In Israel the problems are worse, with about 39 percent of the excavations from the 1960s, 75 percent in the 1970s, and 87 percent in the 1980s going unreported (Renfrew and Bahn 1996:535–36).

Most of us also learned along the way that archaeology is a destructive process. Our excavations can wreck a site as completely as a bulldozer or a chisel plow. That's why most of us fully realize the need to report our excavations. If we don't, and the artifacts just sit on a shelf in our labs, we've just contributed to site destruction. If we think carefully about it, we should fully understand the problem and have a profoundly guilty conscience about any sites we've dug that we haven't reported.

If the past is a public heritage, as many archaeologists consider it to be, when we don't report a site we've dug, we commit a double theft. We've first stolen from the people paying the bills and then from the public of whose heritage we say we are the stewards. We become little different from the looters we condemn who dig up artifacts to sell for profit in the lucrative antiquities market.

MAKING EXCUSES

The reasons we don't get the work written up can be many, starting with our own attitudes toward writing, but often, we just don't allow ourselves time or budget to get the job done. We can make excuses about how tough it is to convince a contract manager whose eye is always on the bottom line that it really does take about triple or greater the amount of time in the lab as it does in the field to get analysis, interpretation, and write-up done. We constantly tell our students this in textbooks and lectures, but why don't we make the same effort to educate the bean counters?

We also make excuses about how our labs don't have the best equipment or the latest version of our favorite word-processing software, but we ought to be ashamed of ourselves. Ludmila Koryakova of the Laboratory of Archaeological Researches at Ural State University told me of how Russian archaeologists often struggle to stay ahead of rampant site destruction, with relatively few resources at their disposal. Site reports often get handwritten in pencil, with only one copy of the report in existence because they don't have copiers! She lives in fear that the single copies will somehow be lost and all records of important sites destroyed forever. When I gave her an old laptop computer, she was ecstatic!

If we put aside our own lethargy and excuse making, we may notice something we might truly lack, and that is adequate training to prepare our reports. Although many archaeology texts describe the need to report our work, they rarely give any clue as to how difficult this task is or to the tools or strategies for actually doing it. Few of us can recall any courses specifically geared to writing our reports or presenting our materials. Although we probably took part in field and lab methods classes, studied topical issues or culture areas, and lived and slept theory, almost no one was taught anything about presenting the past. We were somehow to learn this by reading other reports, or by listening to colleagues give conference papers, or by figuring it all out by osmosis. This may be a major failing of our educational system in archaeology.

Another problem for us is that at least since the 1950s, archaeology very much has been a multidisciplinary team effort. For our excavations and analyses, we require many specialists, as noted in the other books in this series. We need our geomorphologists and archaeobiologists in the field with us. We need them in the lab, too, where they might handle analysis on everything from sediment particle size to gas chromatography on charred residue in pottery. Although we know full well that the days of a professor, three graduate students, and a cloud of dust are over, many of us still see what we do as a more or less solitary venture. Some of us work in very small CRM firms, and we rely on contracting out much of our specialty labor. We might get a report from our consultant, but it's up to us to massage it into our final reports. Others of us work in a university system where joint publications are unfortunately not given the same weight toward tenure and promotion that single-authored books, monographs, and papers might be. The result is that we tend to see production of our reports as a mostly solitary chore when it should often be a team production.

Finally, and paradoxically, archaeologists are prodigious borrowers. We can bend almost any theoretical approach to our wishes, and we easily latch onto new technologies for field and laboratory work. However, when it comes to preparing and presenting our reports, we have been slow to adjust to new media. Many of us are still firmly hooked on hard copy when more appropriate media are readily available for everything from site reports to books like this one!

PLAN OF THE BOOK

The intent of this book is to provide you with basic tools you'll need to present the past. Topics will be wide-ranging, sometimes reading like a primer but also providing resources if you have already mastered the basics. Chapter 2 considers archaeology's audiences and how to recognize them. Chapter 3 is geared toward helping you decide on the media you'll use to meet audience, personal, and budgetary needs for your presentation. Chapter 4 looks at basic writing skills and how to develop them, and it also considers some of the complexities of writing such as style, jargon, and dealing with references. Chapter 5 takes a brief look at computers and software, more about their use than the specifics of hardware and software. Chapter 6 considers visual archaeology, the creation and use of images in our presentations, from drawings to video. Chapter 7 examines team approaches to presenting the past. Chapters 8 and 9 essentially start with presentations of various kinds, from conference papers and luncheon talks, and move to the world of publishing, from peer review to working with editors. Chapter 10 looks at alternative ways to bring the past to life, from exhibits and events to cartoons and movies. Chapter 11 shows how you can work with the media to publicize the past. Chapter 12 deals with new technologies and how we will present the past in the future.

There is structure to this. In one sense, the book follows what might be considered the processes of presenting the past; that is, the text moves from looking at audiences to selecting materials for them, to preparing and delivering the materials. Another thread linking chapters is the kinds of presentations or media available to us, from presentations to digital technologies. As complex as this approach might seem, much derives from a few key issues.

The starting point for the rest of this book is a single question: For whom do we do archaeology? However you answer the question, your answer(s) dictate how you present the past. Archaeology has a wide range of constituencies, including both colleagues and the public, so effective presentation of the past needs to begin with the differences in audiences.

2
RECOGNIZING OUR AUDIENCES

The single greatest problem in presenting the past may be figuring out who our audience is. Many of us never even think about the problem that can be. Professionally we are an odd lot. As archaeologist and essayist Loren Eisley (1971:81) observed in his book *The Night Country*, "A man [sic] who has looked with the archaeological eye will never quite see normally again." To a degree, we pride ourselves in our peculiar viewpoint. Those of us who consider ourselves first to be anthropologists should recognize that anthropologists are often marginal to their own societies. We never really quite fit in. Sociocultural anthropologists, at least according to the popular stereotype, study the Other, that exotic, non-Western member of a small-scale society living in one of the world's remote locales. At least anthropologists study living people, but we archaeologists study cultures long gone, where we don't even have to deal with the living, who just tend to get in the way. To paraphrase L. P. Hartley (1953:xvi) in *The Go Between*, the past becomes our foreign country.

To a degree we revel in our marginality and isolation, and we certainly get along best with our own kind. I don't mean this in a negative or harshly critical way; both actually have their utility for our field- and lab work. Still, these characteristics can get in our way when it comes to communicating what we do. Although our relative isolation as a profession can provide a modicum of comfort, most archaeologists realize that they do not do their work in a vacuum. During the past three decades, with the rise of cultural resources management (CRM) and the advocacy of a wide range of publics for some level of control of what they consider to be their own pasts, archaeologists have come to understand the very public nature of what we do. In recognition, the Society for American Archaeology (SAA) spent several years in the early 1990s developing an ethics code, many segments of which are geared toward communication with colleagues and various publics, as well as how to educate our students to do so.

ETHICS AND ACCOUNTABILITY

During the early 1990s, the SAA began a series of deliberations, small conferences, and open-member comment on how the society might revise its ethics code (see Lynott and Wylie 1995). The Executive Committee of the SAA approved the code in 1996 (see Principles of Archaeological Ethics at www.saa.org/Aboutsaa/Ethics/prethic.html to read the code). The principle of stewardship of the past in both practice and promotion is at the core of the code, and there is some level of recognition of the discipline's publics in virtually every principle. Certainly Principle No. 2, "Accountability," directly recognizes our links to other communities, as does Principle No. 4, "Public Education and Outreach." Principle No. 6, "Public Reporting and Publication," brings it all home in terms of this book. As part of the principle states, "Within a reasonable time, the knowledge archaeologists gain from investigation of the archaeological record must be presented in accessible form (through publication or other means) to as wide a range of interested publics as possible."

EDUCATING OUR STUDENTS

Soon thereafter, the SAA's Education Committee began a lengthy examination of the ways in which archaeologists educate their students, culminating in a substantial report, *Teaching Archaeology in the 21st Century* (Bender and Smith 2000), in which they outlined several "Principles for Curriculum Reform." Presentation of the past is clearly mentioned in two of them, along with possible topics for consideration in student training:

- *"Written and Oral Communication."* Archaeology depends on the understanding and support of the public. For this to occur, archaeologists must communicate their goals, results, and recommendations clearly and effectively. Archaeology training must incorporate

training and frequent practice in logical thinking as well as written and oral presentation. Possible topics: Clear writing (implied clear thinking), clear speaking, public speaking, computer literacy.
- *"Basic Archaeological Skills."* Students planning on a career in archaeology must have mastered a set of basic cognitive and methodological skills that enable them to operate effectively in the field and laboratory contexts. These skills must span the range of basic professional responsibility: excavation, analysis, report writing, and long term curation. Possible topics: Observation skills, inferential skills, basic map skills, organizing and assessing data, knowledge of the law, technical writing.

The "Written and Oral Communication" principle is fairly direct about the importance of communicating well to our publics. As important, the principle states that students should be trained in how to do it. The "Basic Archaeological Skills" principle specifically mentions writing for colleagues. Between the SAA Ethics Code and the principles regarding archaeology education, there can be no doubt that communicating the past well and to a range of publics is extremely important.

PUBLIC ACCOUNTABILITY: WHAT DO DIFFERENT AUDIENCES WANT AND NEED?

Few archaeologists would disagree that the past is important and that archaeology can be a powerful tool for understanding it. Most would probably agree that the scientific method behind archaeology is probably the greatest contribution of the Western tradition to world heritage. At the same time, "science is a very human form of knowledge" (Bronowski 1973:374). Scientists want to "proselytize" the public as much as any other belief system does in order to make them think like we do. We archaeologists tend to be offended when others don't think the past is as important as we think it is, when what they do might "destroy" a past we hold so dear. We work hard in our public education programs to show our publics how important it is and why they should support preservation, or at least conservation, of the past. However, do we really know what our publics think?

A coalition of archaeological organizations, including the SAA, the Archaeological Conservancy, the Archaeological Institute of America, the Bureau of Land Management, the Fish and Wildlife Service, the

Forest Service, the National Park Service, and the Society for Historical Archaeology, used a Harris poll to determine how Americans view archaeology. A random sample of 1,016 adults across the continental United States answered questions about the public's grasp of, and participation in, archaeology (SAA 2000). The SAA's press release summary of the poll shows that "a majority (60%) of the public believes in the value to society of archaeological research and education."

The devil is in the details, however. Although nearly everyone felt there should be laws protecting archaeological resources, when it came to laws about resources on private land, there was a 16 percent drop, from 96 percent to 80 percent in support. When it came to removing archaeological objects from a foreign country without permission, 64 percent said this should not happen. There is some confusion about the role of archaeologists, with a tendency to link archaeology to recent discoveries. When it came to how people learned about archaeology:

> The majority of respondents learned about archaeology through television (56%) and books, encyclopedias, and magazines (33%), followed by newspapers (24%). Learning about archaeology in school accounted for 23% of respondents at the college level, 20% at the secondary level, and 10% at the primary level, although the vast majority (90%) believed that students should learn about archaeology as part of the school curriculum from their earliest years. Most of the public (88%) have visited a museum exhibiting archaeological material, while 1 in 3 people (37%) have visited an archaeological site. (SAA 2000)

For most of the public, however, the primary use of archaeology is probably for entertainment. Certainly people think the past is important, until it gets in the way of their ownership, their job, or their pet project. The fact that the two largest media for learning about it are television and museum visits should tell us something important: If we want to get our messages across to the public, we need to find ways to teach that are entertaining *and* intellectually enlightening. If you doubt that others feel this way, read Brian Fagan's (2002) diatribe "I Am So Tired of Jargon and Narrow Teaching . . ." in the *SAA Archaeological Record*.

Think about what got you into archaeology in the first place. More than likely, you read or saw on television some popularized story about a discovery. For those from an era when television was limited, it might have been a book such as C. W. Ceram's *Gods, Graves, and Scholars*; if you are younger, you might have gotten pulled in by the Ice Man in the Alps or even a movie like *Raiders of the Lost Ark*.

When you took your first academic classes, you began to understand that learning academic archaeology was work; entertainment was no longer so important. If your first dig was in a hot Iowa cornfield working on a "culture of poverty" lithic scatter, you learned that the romance of archaeology flies very quickly. As you became a professional, something deeper kept you going, perhaps the chance of discovery or some intangible quality of field- and lab work. Fun, yes, but hardly the sort of fun most of the public likes!

On a more serious level, we also need to learn that archaeology can be a very cruel discipline. Like history, archaeology can literally undercut a people's belief system. Little wonder that many tradition-oriented American Indians despise an archaeology that sometimes seeks to debunk their origin stories. The problem, of course, is how to deal with archaeological interpretations about the past and our kind of "truth" versus a concern for people's feelings. That is no small matter!

We shouldn't have to dumb down everything we do so that it "sells" or makes people "feel good" about themselves. Rather, we had better figure out that just doing archaeological reports doesn't cut it if we want our publics to learn about, let alone buy into, our disciplinary views about the past. Worse, it may even alienate them if done without sensitivity to their concerns or needs. Except for all of us liking the exciting elements of discovery—the newest, the oldest, and the greatest—archaeology's is an exceedingly difficult message. By its very nature, archaeology has problems in dealing with the individual in the past (except, of course, in historical archaeology where there may be documentary evidence). We deal with the material results of cultural norms. We are forced to objectify, and that often takes the "humanity" out of our work. With the human side gone, what we write is often boring. Even most archaeologists have a tough time reading the 105th description of a potsherd in a report or looking at the eighth chart of a factor analysis. If we can figure out ways to put people back into our work, we have a better chance of success. There *is* fascinating archaeology out there, and it sells!

As an archaeologist, you need to make sure that you understand this point. Yes, you must write reports for your contracting agency in CRM and for your colleagues. Be aware, however, that if the discipline is truly accountable to the public and ever hopes to accomplish anything remotely related to public education, we must recognize the need for different kinds of presentations of the past. We have to work diligently to find ways of presenting the past that are sensitive to the wants and needs of our publics.

ARCHAEOLOGICAL STYLE AND THE REAL WORLD

To do this sort of presentation of the past, we need to consider our own attitudes. Although the following may seem like "preachy" statements from a bully pulpit, some reflection on who we are is vital to our communication skills. In the public stereotypes of archaeologists, we are myopic, dedicated to the quest for the past. No small number of archaeologists see themselves as Indiana Jones types: adventurous, attending to social convention only as necessary to pay the bills, something of a loner, and dedicated more than anything to protection of the past. Some of this brashness and insensitivity translates into our work, with one archaeologist (Mason 1997) even going so far as to declare that our job is to challenge peoples' oral traditions!

Most archaeologists are not so insensitive, but our approaches often alienate as much as educate. We are not a discipline that has normally valued writing for the public; in fact, in academia doing so has often been frowned upon by promotion and tenure committees. We are also a field that is vastly more critical of the work of peers than are scholars in many other fields, and it does not serve us well with our audiences. Our public bloodlettings more often make us appear to be jealous and petty. In terms of communicating with our publics and sometimes even our colleagues, we seem not to care whether our audiences understand us at all. Some seem to have a "let them eat cake" attitude: If they can't understand my terminology and my syntax, they are nothing but uneducated louts, so screw 'em!

If you are entering the field of archaeology or are a relatively new professional, some good advice suggests that you can be both a decent human being *and* a good archaeologist. In no way can this book change your basic attitudes, but perhaps it can make you think differently about the way you communicate with both colleagues and various archaeological publics. Nowhere is our insensitivity more apparent than in our use of jargon, so let's start there.

JARGON

Some years ago, I got into a discussion with a historian colleague, both of us Plains specialists, about why American Indians—and, for that matter, the general public—didn't pay much attention to archaeologists. He proposed a couple of possibilities, both on target. He maintained that ordinary people, including Indians, pretty much wanted to know about tribes with familiar names, and he asked why we archaeologists were so reluctant to assign tribal names to prehistoric people. I pointed out that population movement and culture histories were rather more complex than simply assigning tribal names, most of those designated by white people, anyway. After all, as a profession we had countless arguments about taxonomy. It was no small feat to document the culture history of any group archaeologically, and we didn't want to fall victim to the simplistic schemes that some historians often proposed. Then he leveled an accusation: "How the hell is someone supposed to understand the subtleties of what the Terminal variant of the Initial Middle Missouri phase means, anyway?"

He had a good point. We archaeologists do love our jargon. Being raised as a processual archaeologist in the late 1960s and early 1970s, I recall the jargon flowing freely. There were new paradigms galore, and terms like *deviation amplification* brought an intellectual orgasm. A student had to be up on the latest usage to be "with it." The analytical archaeology course I took in 1971 was a veritable jargon orgy!

Jargon is not just made up, empty, meaningless words. It is specialized language and can promote extremely efficient communication if everyone involved knows it. However, jargon can also be empty and meaningless, not to mention boring, to someone who doesn't know the code. The problem for archaeology is that when we use jargon, we can rarely be assured that even our colleagues know what it means, let alone the nonspecialist. It's fine to talk Terminal Middle Missouri variant to a colleague who is also a specialist in Plains Village cultures of the northern Plains, but one might want to avoid doing so to a colleague who might specialize in the Plains Village cultures of the southern Plains. Certainly you wouldn't want to use it with an avocational archaeologist. You would probably be better off saying "ancestral Mandan" so she at least has an inkling of what you are talking about, in spite of the risk of possible misunderstandings on technical issues about whether Terminal Middle Missouri might have associations with other Siouan-speaking groups.

The point here is simple: Let the jargon match the audience. Jargon is fine so long as it has relatively wide acceptance. Use jargon carefully, usually only with those groups you are relatively certain know the meanings. Writing about Terminal Middle Missouri variant is fine for *Plains Anthropologist*, a professional journal, but it wasn't so good for my *Peoples of Prehistoric South Dakota*, a popular book on South

Dakota archaeology. If using jargon is crucial to accuracy, then it might be important to contextualize it by taking the time to explain what it means. One need not write down to an audience, but one should certainly respect the limitations of their knowledge and training and give them the opportunity to understand. The real problem for many of us is that we are so used to our jargon, we don't even really recognize that it is jargon. We use *projectile point* so often we rarely realize that most people use the more specific *arrowhead* or *spearhead* (even though they might be wrong about the kind of projectile point it is!).

Jargon is especially difficult in CRM reports because the audience is not always clear. Colleagues who review your report, and hopefully the archaeologists who work for the agency for which the report is being done, will at least have some clue about the jargon. CRM reports do have other audiences, including contract managers, planners, and other cultural heritage administrators, who haven't a clue about the jargon. Along with protecting site-sensitive information, aversion to our jargon is one reason that many agencies demand public versions of many CRM reports.

Jargon is only one element in working with audiences. So many other issues about how we present the past stem from who we think comprises our audience. The media we choose to communicate with them are also crucial, which is the subject of the next chapter.

3
CHOOSING THE RIGHT MEDIUM

With so many media available today, from standard print sources to a range of new electronic media, how do you choose those that are best for presenting your work? The answers are not as simple as one might think. For those working in the academic sphere, most still treasure the printed media of books and journals. CRM specialists may have to consider publication in multiple ways given stipulations of a project's scope of work. Independent research projects may allow substantial freedom to choose any medium you wish. However, additional complications may be related to intended audiences.

Certainly, professional books, journals, and CRM reports are not the best way to reach a general audience or perhaps even a professional audience who does not specialize in your area of interest but who may benefit by knowing something of your work. Given increased efforts at public outreach, many agency CRM managers discovered a long time ago that public versions of reports were a good idea, making preparation of such documents part of scopes of work. Even with reports prepared at more than one level, delivery of the documents to the intended audiences can be a problem. Books, journals, and even glossy public reports are costly to produce, and purchasing them is no small matter for many of us, let alone for a general public with a less than fanatic interest in archaeology.

Essentially, you need to ask the following questions in deciding on a medium:

1. Who is my intended audience?
2. What are my contractual or other obligations to these audiences?

3. What kind of budget do I have to prepare my work?
4. Given these factors, what media are available to me?
5. What will it cost to distribute my work to the intended audiences?

Let's look briefly at each of these questions.

WHO IS MY INTENDED AUDIENCE?

Chapter 2 provided a discussion of audiences that might have led you to think that audience is easy to determine. If so, you are probably thinking too narrowly. Avocational archaeologists, for example, often read articles written for state archaeological journals. Professional colleagues might also use your World Wide Web site developed for a lay audience to prepare lectures for their introductory classes. Consider a range of possibilities when you prepare materials. This doesn't mean that you necessarily have to "dumb down" presentations or that you have to continually worry about what colleagues might think. What it does mean is that with relatively little planning and effort, you should be able to provide satisfactory materials for several audiences at once.

WHAT ARE MY CONTRACTUAL OR OTHER OBLIGATIONS TO THESE AUDIENCES?

You may have contractual obligations that instruct you to gear your presentations toward particular audiences. If you are working on a CRM project, check the scope of work carefully to see who intended audiences will be. If it is just for a state or federal agency CRM specialist, you may only need an administrative report, one that can be loaded with CRM jargon for efficient communication aimed at best protecting the resource. You and the agency manager both know what a Phase I Reconnaissance Survey is. If the scope says the report will be sent out for review, however, your problem is more complex. Is it a report to be reviewed by other CRM agencies, or are professional specialist colleagues with intimate knowledge of an area also to be reviewers? If the latter, you may need to provide more detailed information on theory and culture history than the CRM manager might find understandable, so you'll need to find a middle ground or provide adequate, clear explanation of jargon in the text. Many larger CRM projects now require as part of the scope of work that the contractor prepare a shorter version of the project's findings for the general public.

You may thus find yourself having to prepare a multilevel report, but don't assume that you can just abstract the longer report into the shorter report. Jargon is jargon, and a nonprofessional audience still might not understand it, nor will they likely care about your justification for your sampling method or detailed descriptions of each flake found by your surveyors. Regular contact with your contracting officer might help you find an appropriate level for each report.

For non-CRM reports, you will be writing for a specific audience from the start. If you suspect multilevel audience concerns might apply, and even though you won't have contractual obligations for multiple-level reporting, you might discuss the issues with your editor. If you can provide for different audiences, you will likely increase the size of your audiences.

WHAT KIND OF BUDGET DO I HAVE TO PREPARE MY WORK?

Of course, this question is one you should have asked before you took on a project! If you are doing a CRM report, there may be particular requirements for distribution of the report. For example, the United States National Park Service has recently started requiring that contractors prepare both a hard copy of the report and also an electronic copy in a .pdf format (portable document file) for easy insertion on the World Wide Web or on a CD-ROM. Although the Adobe Acrobat software itself is inexpensive, there will be some time, and possibly substantial labor costs for converting your report to .pdf files. When you prepare a cost proposal or bid for a contract, you might wish to factor in some of these cost elements.

Your manuscripts, reports, or other projects may require lots of detailed graphics. Web and CD-ROM presentations may also require skill in web markup, design, or even instructional design to be effective for an intended audience. If you are like most archaeologists, you probably aren't trained as a graphic artist or design specialist, let alone a computer programmer or education specialist. You may need to hire additional staff or outsource some of the work. Hiring work done by others adds to your costs.

A CAUTION

Everyone wants to do work of the highest quality, although we sometimes find it less than practical to do so due to the many constraints of time, personnel, budget, and, frankly, our own (lack of?) talent and that of our staff. Most of us have been trained to consider what we do to be part of a scientific process of archaeology and hope to add something, however small, to the culture history or key research topics of the areas in which we work. This often motivates us to do more than is necessary for what the people paying the bills really need.

This point is especially true for small CRM projects. Why prepare a ten-page report when an agency will allow a letter report? More than one state historical preservation review and compliance officer have received lengthy reports with "boilerplate" culture histories and major methodological descriptions that report no sites and contain only a few paragraphs about survey methods. Your time and that of your staff is valuable. Why waste it trying to do "science" when what the entire agency wants is a report to help its staff manage the cultural resources for which they are responsible?

GIVEN THESE FACTORS, WHAT MEDIA ARE AVAILABLE TO ME?

This is not usually a tough call. Most of us will probably fall back on the standard hard-copy paper report, but given the rapidly increasing number of options, perhaps we should be more selective. If we need only prepare a letter report or a report of a few pages, paper is just fine. If we have to prepare a report of some length, with the bulk of it done on a word processor, then why not consider publishing it in some electronic format, if the contracting agency is willing or if the audience is appropriate? The savings can be substantial, and the amount of included material provided to readers can be expanded.

WHAT WILL IT COST TO DISTRIBUTE MY WORK TO THE INTENDED AUDIENCES?

This may not always be your problem, but it may be that of the agency or publisher. Its staff will also be thinking about how to best present materials to a particular audience. Someone has to think about it, so it might be best if you think of it as you prepare a report, a paper, or a book. It may make the difference in getting published!

Lots of questions regarding the report seem to appear almost immediately. Should you include photographs? Would color illustrations be useful? What sort of figures do you need, how many of them, and how big? Do you have lots of tabular data that you would like to include but that only a few readers might find useful? If you can plan ahead a bit while you prepare a document, you'll be better off in the long run when it comes to producing the final product.

CONCLUSION

When it comes to all the questions in chapters 3 and 4 about audience and media, one thing becomes clear very quickly: Presenting the past is complicated! One shouldn't think that it is a matter of just sitting down and writing. Differing audiences require you to think about the ways in which you can present the past. Differing media require different kinds of skills. Most archaeologists haven't thought much about the skills they need to present the past or how to go about acquiring them. The next chapter aims to remedy that situation.

4
DEVELOPING NEEDED SKILLS AND TOOLS

You wouldn't go to the field without a reasonably well-developed research design to guide your survey or excavations, and you certainly couldn't do the work without the right tools. The same should apply to presenting the past. You might develop a strategy for preparing manuscripts, drawings, or photographs whether you are doing a CRM report, a book, or an article.

BASIC WRITING SKILLS

Just as we don't like to have untrained people working in the field or lab, we should not assume that it's okay to have untrained people preparing reports. During the many years that I ran a CRM operation for the University of South Dakota, I had the opportunity to read the many reports prepared by our lab staff. Some were excellent, but many had to be rewritten before they went out. Bright people wrote most of them, but they obviously didn't have writing skills. These days, managers in the business world often like to complain that fresh college graduates come to them without basic communication skills, and I'm afraid to admit that they are mostly right. Usually, on-the-job training can help people overcome the problem, but it takes close supervision, something that may be a luxury in a small CRM operation.

When colleagues have called me for references for colleagues or former students who are seeking employment with some CRM firm, one of the most commonly asked questions has been either "Can they write?" or "Can you get reports out of them?" For CRM firms, this is no small matter. Projects are done to deadlines, and if the project staff can't produce the report, people might not get paid, and the next contract might be impossible to get.

If you are hiring new people, don't hesitate to ask them for writing samples. If they can't produce one, or they give you one that's not well written, don't hire them. If you have staff who write poorly and you'd like to keep them, you can provide training. If, heaven forbid, you are yourself not so good at writing, you may wish to take some training. This book is not meant to be a writing handbook; thousands of "how to write" books are available. However, a few concepts might help you understand some of the problems of writing and get you thinking about developing and shaping your own or some staff member's basic writing skills.

WRITING IS A LINEAR MEDIUM MADE UP OF CIRCULAR THINKING

When we put words on paper, we tend to think of the process in terms of X leading to Y leading to Z, a nice straight line sequence. Even our thinking as archaeologists is posed in this way in terms of time where culture X came before culture Y and culture Z. The problem is that even our analyses of archaeological cultures don't proceed that way. They just end up that way on a culture history chart or in our article about a site. The actual process is more meandering. We sort and measure sherds hoping to do a ceramic seriation, we attempt to correlate certain styles with certain strata to give us a sequence, and so forth. In the end, we impose a kind of linearity on our materials. The process is difficult, without question!

Writing is also like that, and we'd be better off if we just recognized this fact from the start. I've had more problems with introductions than I care to think about. Even in this book, as I write this chapter, I've only begun to think about writing the introduction! I'm also certain that it will be rewritten about five times before I actually send it in. I've also had more students fail to submit papers because they couldn't get past the introduction. I've also written the sections on preparing abstracts, giving conference papers, and dealing with the media before I tackled this section.

Few of us are capable of enough organizational skill to just sit down and write something from beginning to end. We don't think that way, so why should we force ourselves to write that way? We think in jumbled ways. We go over and over things in our heads. We try to puzzle out things. There's nothing wrong with writing parts of a book, paper, or anything else "out of order." Ultimately these things do need to be put in a linear sequence. Readers can't follow the jumble in our heads. So how do you get there?

OUTLINES HELP IMPOSE WRITING'S LINEAR STRUCTURE

Writing an outline is something most of us were taught in elementary school, but with the press of writing papers against deadlines, most of us quickly forget how valuable they actually are. Outlines are really nothing but an overview of where you intend to go with your writing. They need not be cast in stone and, in fact, should be malleable as "great thoughts" surface about what you are writing.

For some kinds of books, though, outlines must be detailed and essentially allow little deviation. A colleague and I worked on a book for a publisher who demanded a detailed outline ahead of time so his design staff could get busy with layout. We had to think in terms of two-page "spreads" that would have 600 words, with a box that would have 150 words. This was tough, but when it came time to actually write, it was unbelievably easy. In fact, the outline took longer than the writing, the former about three months and the latter only a month!

Many people actually do an outline in their heads, so why not put it on paper? Doing so allows you to write on various parts at different times. In this way, you'll see the writing as connected, but you won't be locked into linear writing. An added benefit on large projects is that outlines allow you to break the project into pieces of manageable size so you don't feel overwhelmed.

TELL 'EM WHAT YOU'RE GOING TO TELL 'EM, TELL 'EM, AND TELL 'EM WHAT YOU TOLD 'EM

In the U.S. Air Force, we were taught this adage about being direct and concise, in both writing and speaking, especially when we were presenting information to higher-ranking officers who supposedly had little time. The adage is a lesson in linearity, but it also makes a point about structure and reinforcement of points as you move through a work. In essence, it is the old writing structure about introduction, body, and conclusion, which you might have learned in your first composition classes. The introduction contains your primary theses and statements about how you will demonstrate your contentions. The body contains data, examples, and thinking about both, but organized in a logical and replicable fashion. Your conclusion contains the assessment of your data and argumentation, as well inferences about it.

In classes I usually tell students how to do well on my essay exams or papers. I liken the writing process to that of sands in an hourglass. At the top of the hourglass, the glass is wide and there is a clutter of ideas. You need to tell the reader generally about the body of information you have and how you are going to organize it. As the sand passes through the constriction, it flows through a grain or two at a time. This is the detailed presentation of data and examples, with the constriction being the logic you impose on it. These data and examples collect in the bottom of the hourglass, but they are hopefully now in a slightly more ordered state than they were. This more organized collection is your conclusion, where you tell the reader what it all means.

I tell students that this structure also should apply to each unit, chapter, or paragraph in the work. For the latter, one uses the general topic sentence to start a paragraph, followed by the body of the paragraph, which contains data and logic, and the transition sentence, which leads the reader to the next topic sentence. Certainly, other analogies might suffice and the approach may not work for all forms of writing, but after you've read a few student essays, you understand how few people really understand the linear structure of writing. In archaeology, you can write using a stream of consciousness, but you'll probably get the report sent back to you by your professor, boss, or the contracting agency!

UNDERSTAND THAT YOUR WORDS ARE NOT ALL THAT PRECIOUS

Sportswriter Red Smith commented more or less, "Writing is easy. You sit down at the typewriter and open a vein." After you've slaved away for hours, days, or months on a manuscript, it's really tough to have somebody critique your work. If you take writing seriously, you put some of your lifeblood into it, so every word seems to be one of those drops of blood, but you know that losing a few drops won't kill you. If you give it some serious thought, you will recognize that most of what you write as a professional archaeologist is not poetry, in which well-chosen words may actually be precious. Your writing can almost become formulaic or "boilerplate" in the CRM world, where you need to crank out reports relatively quickly. Even in your other work, your quest is usually for easily understood, rapidly written prose. As one who has edited several books and professional journals, I can tell you straightforwardly that nothing is more of a pain than an author who thinks his or her words are "golden." (More on working with editors later.) You'll write more rapidly, be a more likable person to editors, and generally be a happier soul if you understand that letting others see and comment on your work can improve it.

THE MORE YOU WRITE, THE BETTER YOU GET

Writing is like anything else. The more you practice, the better you get. If you start an exercise program, your muscles sometimes hurt if you aren't used to it. Writing is the same way. If you haven't written much, it's tough at the beginning but gets easier.

Writers tend to develop various strategies that work for them. Some insist that you have to write some every day, but as with exercise, that takes great discipline, perhaps even obsession. At the same time, if you can write a page or two a day, you've got a solid book at the end of each and every year.

Most people will never be that disciplined. If you are like some, you may "write" every day, though not always on paper. You may be one to mull over a subject in your head, then "download" it in nearly finished form to your word processor. Some like to worry about word choice from the beginning, slaving like Flaubert to find *le mot juste*. Others are more concerned with content and meaning, spewing a James Joyce–like stream of consciousness, with any wordsmithing saved for the end.

You need to develop your own strategies for writing. If you haven't thought about it much, look at how you tend to go about things and what works for you. However, don't make excuses or lie to yourself about your real approach. Once you've got a strategy figured out, stick to it, and practice, practice, practice.

READ LOTS, BUT PAY ATTENTION TO THE WRITING, NOT JUST THE CONTENT

You really can't be a good writer unless you pay attention to good writing. You can learn a great deal by reading others. Certainly, archaeologists have a style. One simple example is our use of the passive voice. We accept passive voice more than most other fields because it's really difficult to know who acted on what in the past in order to allow the active voice. Mostly, we just don't pay attention. (If you don't know what I'm talking about here, it's the difference between the following sentences: "The excavation grid was laid out by the crew" [passive] versus "The crew laid out the excavation grid" [active].) At the same time, you also see approaches you like and don't like, so you can model your approach to what you like.

FIND ONE OR TWO GOOD WRITING REFERENCES AND USE THEM

It is really important that you have a basic writing book by your desk or an electronic version on your computer and use it. If your style is more conservative, this might be a book such as Strunk and White's *The Elements of Style*, from Allyn and Bacon (2000). If you are younger and more in tune with the information age, you might like Verve Press's (1999) *Adios, Strunk and White: A Handbook for the New Academic Essay* by Gary and Glynis Hoffman. Another favorite is *Line by Line: How to Improve Your Own Writing* by Claire Kehrwald Cook, published by Houghton Mifflin (1986) for the Modern Language Association. Its strength is that the author has been a copyeditor and provides techniques used by editors. She gives insights into how editors think about text when they work over a manuscript. The standard for all writers, however, is *The Chicago Manual of Style*, used by everyone in the publishing process. Its primary strength comes from its currency. Constant revision keeps it up-to-date, and recent editions have emphasized the process of publishing as much as the elements of writing. Which book you have is probably not all that important; you just need to have some basic writing guides at your disposal.

Besides having these guides, having access to a good dictionary, thesaurus, and other basic writing tools should go without saying, but oddly enough, most people don't have them nearby. Good, inexpensive dictionaries and other reference tools are available for your computer. Some word processors have simple tools built in, where all you need to do is to highlight a word and click on the tool(s). There are also many such tools available on line. *Your Dictionary* (www.yourdictionary.com/) is one good example, providing a wide range of English and other language tools, but there are many others. There's not much excuse for not having access to these resources.

SOME FINAL POINTS ABOUT WRITING BASICS

You can follow all the advice offered thus far. You can buy books galore on writing and study them in depth. You can fantasize about writing all you like. What it comes down to is sitting your rear end in a chair, putting your fingers on the keyboard, and grinding away. That's the single hardest part. Even after thirty years of writing I still have to force myself into the chair, and frankly, it doesn't get all that much easier. However, no matter what you may think about the prose in this book, after thirty years, the writing *has* gotten better!

BEYOND THE BASICS OF WRITING

Assuming you have some level of mastery over basic writing and some level of self-discipline about doing it, you can work toward acquiring additional tools. As presented in the following paragraphs, they don't necessarily come with any priority but are tied to the concept of making writing easier and you more capable of providing a product, whether it be a paper for a graduate seminar, a full-blown CRM report, or a scholarly monograph.

STYLES AND HOW TO DEAL WITH THEM

Figuring out writing styles is one of the most difficult elements in writing, especially when you have to write for a wide range of audiences. As I'm writing this, I'm glancing over at a copy of *Dig* magazine for kids. I'm contemplating writing a piece on "fringe" archaeology that one of the contributing editors asked me to think about. Frankly, the task is daunting at best. Writing for kids can be a bear when you have been writing academic archaeology for thirty years. We archaeologists have our own style, with its own jargon, sentence structure, and linear presentation, and I have to wonder whether I can do it.

Even in considering the style of this series and volume, the editors and publisher had several discussions about what we could and couldn't do. We didn't want lots of in-text citations or footnotes common to academic writing because they tend to disrupt the flow of text for a reader. We wanted more of a textbook style, but even there we had concerns. We wanted to keep the text "light," with relatively little jargon and few overly complex sentences. To do this, we needed to keep things more conversational in tone, so we allowed authors to use contractions and other writing styles usually anathema to academic writing, such as using arcane words such as *anathema* and *arcane*.

So, how can you figure out what style is for what audience? The first thing to do is to read widely within archaeology to get an idea of the range of styles that we use. If you can assess a target audience, you might be able to figure out what the style might be. If it's for a popular audience, vocabulary and sentence structures should be less complex. Jargon should be fully explained or absent. For more educated lay audiences, sentence structures can be complex, as can vocabulary, but jargon should still be minimized. For professional audiences, in which scholars know the jargon and are used to the style, you can write for that audience, but the style should not be convoluted or the prose turgid.

ARCHAEOLOGICAL STYLES AND HIDDEN AUDIENCES

Once you've figured out what style is for what audience, you then need to craft the prose for that group. Mitch Allen (2002) has taken a close look at what he calls archaeology's "hidden audience"—nonspecialists—and how to write for them in particular. He provides an analysis of who they are and gives ten tips for writing for them. Of course, you must also write for professional colleagues, so let's take look at how this can play out using some examples from my own writing (I choose them not because I think they are particularly good, but so that I don't have to pick on anyone else!). I've chosen selections from writings about the Crow Creek massacre and the associated issue of repatriation, mostly because I've done lots of pieces on each for a range of audiences.

Let's start with a recent piece for an educated audience, but not just professional colleagues (Zimmerman 2001:169). In considering my audience, I understood from the editor that readers would be educated, concerned with human rights issues, and at least moderately aware of repatriation debates. Here is part of the second paragraph:

> But do archaeologists export the data and move the creative process and their results away from Native American nations? The answer is unequivocally yes. American Archaeology is an edifice of scientific colonialism, and this has crippled its relationships with Native Americans. The crucible of the repatriation and reburial issues has made it painfully

clear that archaeological interactions with Indians often have been inept and torturous. Many Native Americans go so far as to call archaeology irrelevant or inaccurate.

Notice that the sentence structure is somewhat complex, but with only one compound sentence, and there are lots of difficult words (e.g., *edifice, crucible, inept, torturous*). There is one concept, scientific colonialism, that is jargon, which I actually define in the first sentence of the piece.

Work for a fully professional audience of archaeologists can be more data-dense, as in this paragraph taken from an article about the Crow Creek massacre for *Plains Anthropologist*, a professional journal (Zimmerman and Bradley 1993:217). A few nonspecialists might read it, but there was no expectation of that. Midway through the piece is a discussion of malnutrition in the remains:

> In preliminary evaluation of the skeletal materials, investigators found numerous transverse lines in radiographs of long bones. Examination also revealed evidence suggesting iron deficiency anemia in 28 skulls, 18 with orbital cribra, four with orbital cribra associated with other skull lesions, and six with porotic hyperostisis. Both localized and generalized periostial reactions were very prevalent in the Crow Creek bones. Subsequent investigations have uncovered additional supporting evidence indicating repetitious and prolonged malnutrition in many Crow Creek Skeletons.

Ouch! This paragraph is actually one that requires not only archaeology jargon but also knowledge of osteology and paleopathology. Do you know what a periostial reaction is? These terms never get defined in the text. Sentence structure is relatively complex, with several clauses, and the paragraph is data-dense.

Let's now look at a paragraph from my *Peoples of Prehistoric South Dakota* geared toward an audience of nonspecialists. Certainly, some colleagues would use the book, but South Dakota citizens were the target audience. They would range from junior high school students through adults. The paragraph again describes Crow Creek (Zimmerman 1985:51):

> Archaeologists know from skeletons found at the site that a pattern of warfare had been going on for some time. At least two individuals had been scalped earlier and had survived, only to be killed during the Crow Creek massacre. Others had been wounded earlier by arrows, the points embedded in bone, which had grown over them.

This paragraph has some complexity, with several clauses and one compound sentence, but words are relatively short, with no real jargon. There is one instance of passive voice ("Others had been . . ."), which I would now change. Sentence length is longer than one might write for children, and vocabulary is more complex.

I wrote the following paragraph for a Crow Creek Massacre website (www.usd.edu/anth/crow/ccwhat.html) specifically aimed at middle school children:

> Some archaeologists believe that the attack was carried out by the Middle Missouri villagers who came down from the north. They might have been unhappy that the Initial Coalescent people had moved into the area and had taken their land. Other archaeologists do not believe this is a good explanation. These scientists believe that the cause had to do with the environment and overpopulation.

The vocabulary is more limited and sentence structure a bit simpler. I notice a pesky passive voice, which I would now change. The content is more emotional, from the point of view of both the reasons for the massacre and the difference of opinion among the archaeologists. Children are probably the most difficult to write for because their abilities change so quickly. For children, ways to present the past need to be more action oriented and emotional.

What do kids like to read about archaeology? To some degree, they like to read about other kids. In looking at each piece in *Dig*, the first thing I noticed was that all but two of the articles and many of the news items started with a question within the first line or two. I suppose this approach is to get the child thinking about the subject, but it also helps set a kind of imaginary stage. Here's an example:

> About 20,000–30,000 years ago, an 8- to 10-year-old boy skidded barefoot through a muddy cave in what is now southern France. Think that kid would get a kick out of knowing that his tootsie marks would become the oldest footprints ever found in Europe?

You should notice that the first sentence sets a temporal and spatial stage, but, more important, it sets the action. The question is a hook, drawing the child into the scene. You might also notice that the tone is conversational and that the question is not even a proper sentence.

Many of you might scoff at such approaches if you found them in a student's paper or a colleague's report, but consider that the approach is successful. *Dig*'s subscription base is growing dramatically, with an even higher readership, and questions e-mailed to "Dr. Dig" are surprisingly numerous and sophisticated.

Finally, let's look briefly at fiction with a paragraph from my only published short story (Zimmerman 1986), "Redwing," a pair of vignettes, one set in the past and one in the present, linked by a skeleton and a bird:

> He seldom thought about the people represented by the village debris. He couldn't know their individual lives, only the general patterns of their daily existence. When he worked with burials, however, they made it more personal. Knowing that he was working on a young female caused him to see Sarah. He still didn't understand why she had to die. She was only 37.

The style is looser, close to that for children's writing, with relatively short sentences. The paragraph actually contains an important theoretical idea about how archaeologists see individuals, but in the emotional context of Sarah's death. Fiction allows a writer to push the edges of grammar a bit, with sentence structure that may not be standard. You can play with point of view in ways not available to you in nonfiction, and active voice is crucial. In fiction, a style is more difficult to pin down because it is so open and variable.

For the most part, archaeology students learn style through osmosis. Professors rarely teach anything about it, even if a department happens to have a writing course. In a way, that's too bad, because we would be much better communicators if we did. The problem with hoping students will learn by absorbing what they read is that much of what students, especially graduate students, read is the cutting-edge material in a profession. Obviously, they need to do that, but cutting-edge materials are often the most logically complex and jargon-laden prose in a profession. Theory is supposed to be simple, but explaining theory usually isn't. Because their professors praise their students' supposed understanding of complex articles on theory, students read more and more of it, mouthing the jargon but sometimes not really understanding it. We often use it to impress each other rather than to communicate. In the end, very impressed people can be talking away, not really understanding each other! The same is true of our writing.

You may be tempted to say that you shouldn't have to worry about whether the masses can understand your prose. Remember, however, that you've probably often thought of the contract manager for some CRM project as having the intelligence of a charcoal sample. Do you really want your report rejected because the stupid bureaucrat can't understand what you've written? Just tone it down. Table 4.1 summarizes some tips for keeping your writing clean and simple, and sidebar 4.1 describes a useful tool for helping in that regard: readability indexes.

DEALING WITH REFERENCES

As if style wasn't enough of a problem, writing has a number of technical aspects to complicate matters. If you are reading this book, you have already probably written at least a few papers and know some of the problems. References seem to be a problem for lots of writers. The need for them is fundamentally simple: Readers need to know how you arrived at your conclusions. All scholarship is not done in a vacuum but is cumulative, built on the work of others.

Our readers need to be able to connect what we've concluded to the work of others, the data we've used, and what our work implies. The reader also has a right to know who provided what elements of your thinking. They might like to know whether you actually quoted or interpreted another scholar correctly. You might have taken some

Table 4.1. Tips to Help Simplify Your Writing

1. Figure out who your audience is.
2. Write shorter sentences by eliminating compound sentences (you know, the ones with conjunctions).
3. Try eliminating clauses.
4. Count multiple-syllable words, which are often more abstract (think jargon here).
5. Use shorter words wherever possible, lowering the average number of syllables per word.
6. Don't be afraid to write in first person.
7. Use active voice wherever possible.
8. Use anecdotes and stories to raise the emotional content.
9. Every so often, calculate a readability index to see whether your writing meets what you think the appropriate reading level is for your audience. If it doesn't, then rewrite.
10. Read the work aloud. Normally, people don't speak in as complicated a way as they write. If the prose sounds complicated, it probably is.

> ### 4.1. READABILITY INDEXES
>
> Can you learn to write at several different levels and really know what you are doing? One approach is to use readability indexes. These provide a grade-level estimate of skills required to understand written material. Although there are ongoing debates about what this really means, the tools can still be used to estimate relative difficulty of materials for readers.
>   Most indexes tend to:
>
> - count the number of words in a sentence;
> - count the words with three or more syllables;
> - look at prepositions, such as *of, from, with,* and *by,* which tend to make sentences too long and destroy sentence rhythm; and
> - look at "lazy words" such as *and, it, this,* and *there* or "personal" words or sentences including pronouns.
>
> This approach does provide subjectivity. If you want more information, look at Dr. Jay's *Power Writing Home Page* (www.csun.edu/~vcecn006/read1.htm), which provides an excellent summary of many such tools.
>   Do readability indexes work? They can help you learn basic skills, but you can't use them to check everything. You are better off learning the problem areas of writing that add complexity and teaching yourself different styles of writing.

idea out of context or misunderstood what another author wrote. Much of this has to do with the idea that science is supposed to be replicable, so being able to follow the reasoning of another scholar is important. Most writers know this, and there really is no reason to fight against it, as I've seen many students do.

DO YOU REALLY NEED THAT CITATION?

One of the biggest debates in writing is whether to use citations or at what level you should cite your sources. Avocational archaeologists often hate citations, as do nonacademics in general. Whether superscripted numbers for footnotes or in-text citations, they certainly disrupt the flow of text for the reader. Certainly, if one is doing academic writing for colleagues, citation is imperative! However, most of us who've written for a long time tend to get warped a bit in graduate school. As professors read our papers, the query "Source?" commonly appeared in the margin or next to some idea. In addition, as students learn the process of academic writing, they learn citation overkill, where they string together several citations about one general idea. One doesn't have to read much to see this in archaeological writing. If an idea is common knowledge, you may not need to cite another source at all. (Check out sidebar 4.2 for a quick discussion of copyrights and plagiarism, which are another big reason why we need to mess with references in the first place.)

An important consideration is the intended audience. For popular works, avoid lots of citations. Audiences neither like nor need them. They accept your reasoning as an authority (although most of us would advise them not to) or the argumentation itself (the wiser approach). When writing for young people, avoid citations altogether. A better approach for anything popular is to have a section of sources and suggested readings, kept as brief as possible. Even for some professional publications, lots of citations may be unnecessary. Textbooks are an example, on the border between popular and professional. You'll

4.2. PLAGIARISM AND COPYRIGHT

Plagiarism is using the ideas or words of others without giving them credit. Plagiarism has become more common, especially with the World Wide Web and other digital media providing easy means of finding and directly copying text from published works. Many students claim not to know that plagiarism is wrong, and several scholars have blamed careless research for not taking notes that some words are direct quotes.

Paraphrasing the work of others is generally okay, but the idea should be attributed to them with a citation. A good rule of thumb is "When in doubt, cite!" Some kinds of popular writing don't allow citation in the text, so be certain to paraphrase and then to list the work in your sources or acknowledgments.

Copyright is legal ownership of words, ideas, images, and sounds. Use requires permission. Generally academic works allow more flexibility than popular works. Attributed quotations of several lines are considered "fair use," "but seek permission for use of very long quotations, especially if there are lots of them. (This includes illustrations or music.) Generally you seek permission from the holder of the copyright. This may be the publisher or author, and it may involve a fee for use or a particular credit line they would like you to use. The work usually lists the holder of copyright. The general rule: When in doubt, ask permission!

see that publishers use different approaches to them. In this series, the series editors and the publisher made a conscious decision to keep citations to a minimum. We rely instead on suggested readings, largely because The Archaeologist's Toolkit books are like textbooks or handbooks, not academic treatises.

The real reason references upset writers is that their research is poorly organized. You've often been in a hurry doing research, with a class or contract report deadline looming. The written materials you've found excite you. The source you've got is excellent, and you are scribbling notes like crazy. In the rush, you don't note the page number of the source, or the volume number, or the first names of the editor. When it comes time to put in the reference, you simply don't have the unrecorded element in your notes. Nothing—that's right, *nothing*—is more frustrating than to be held up by such piddling details! If you've got the source in your own personal library, this might not seem like a big deal. However, a trek to the university library is far less fun, and you've got a special headache if the volume you just took back there three days ago was one that took weeks for your interlibrary loan librarian to track down!

If you can't have a source right in front of you when you write, which is usually the case on a large project, you may need a system to help you keep careful track of sources and information. I use an approach that can be useful, but I assure you that I preach it far better than I practice it!

Normally, as I read something that I am not using immediately for a writing project, I have a notebook or computer with me. The first element that I put on a page is an absolutely complete bibliographic citation including the book's ISBN and library call number in the place I found it. As I'll get to soon, I know from experience that several different bibliographic styles are favored by archaeologists and other anthropologists, so I try to cover all the bases, everything from publisher location to full author names exactly as they are on the copyright or title page. I start through the work, taking notes along the way. As I take a note, I leave space in the left-hand margin for the page number from which I take an idea or about which I write a comment. These are normally paraphrases or snippets of the text. If I think a quotation is cool, I copy it down, with the page number in the left-hand margin. I put quotation marks around the text! Then I double-check the quotation to see that I have it down correctly, even to the capitalization and punctuation. This step is important: If you don't get it down correctly, you can do substantial damage to your own work, let alone misrepresent the work of someone else. Besides, tracking down a quotation later is often more difficult than getting publication information on a source.

As I have time to digest the work, I often think about how it connects to my own projects (common), or I'll come up with some brilliant, synthetic thoughts of my own (rare). For both, I put "P.N." in the margin for "Personal Note" to let myself know later that these are my thoughts about the work. I proceed through the book or article this way, often generating many pages of notes (see table 4.2 for a sample page).

I hear students and colleagues talk about taking marginal notes in books, even to the point of having strata of notes on second or third readings. Others highlight key elements with markers. These approaches assume that you own the volume—God forbid, and shame on you, that you do this in a library copy! The problem with both approaches is that you have to go back to the original to get material you need. Besides, writing notes helps you remember the material better.

What note taking allows is full bibliographic control of your research. This approach is really no different from what you were prob-

Table 4.2. Sample Notebook Page for Bibliographic and Research Control

Source: Feder, Kenneth L.
2002 *Frauds, Myths, and Mysteries: Science and Pseudoscience in Archaeology.* McGraw-Hill Mayfield, Boston. 4th edition.
ISBN: 0-7674-2722-X, UI Library Call Number: CC140.F43.2001

p. 2	Survey among tv viewers showed 25% believe in fortune telling, 12% in astrology, 22% in clairvoyance, 3% in fortune cookies.
P.N.	This corresponds highly with the numbers in the "gullibility" index I give my Lost Tribes, etc., class.
p. 4	Feder notes that his own college classes show beliefs similar to the public; see his charts on p. 5.
p. 6	"'Science,' after all, is merely a process of understanding the world around us through the application of logical thought."
P.N.	. . . is science perfect? "Of course not." Science has rejected seemingly outrageous claims (plate tectonics) only to have them proved right.

ably taught about using 3 × 5–inch index cards for taking notes and organizing research. Computers have changed all that, of course. I still like to have a notebook for note taking, but I do often take my laptop to the library for research. I still take notes in more or less the same way, but in a database or in my word processor. The query functions of most database programs and find functions of a word processor allow fast searching for terms or strings of text. There are also stand-alone text analysis programs to do this for you. You can easily search your whole database for related terms or concepts. Mostly, they allow you to get things organized and keep track of tons of materials easily, even over a whole career.

BUILDING A PERSONAL OR COMPANY ANNOTATED BIBLIOGRAPHY

By the time you hit graduate school, you know that you have read and will continue to read books, monographs, CRM reports, and articles as a regular part of your scholarship and pleasure. If I could give only one piece of advice to graduate students or young professional archaeologists, I would suggest that you start as soon as possible to build a lifetime annotated bibliography of all you read. I certainly wish that I had done so. As adept as you might be at remembering what you read, you will come to a point at which you tell yourself, "Now I know that I read somewhere that. . . ." If you have a bit of discipline and can use a database program, you can save lots of frustration because you can't remember where you read it.

I have one colleague who actually was disciplined enough to do this and has used her bibliography effectively. Since grad school, she has put together a bibliography of more than five thousand items, and she did a good job of keeping track of all the publication information. She uses it regularly when she writes for filling in her citations. She did make one mistake, however. What she acknowledges that she did wrong was not to make it an *annotated* bibliography. An annotated bibliography includes any notes you might have taken on the work. It can be a series of phrases, full sentences, or even abstracts. My colleague admits that as she reads more, she tends to forget more. She may recall that she read a particular work, but not necessarily what it was about in any detail. She wishes she had included annotations.

If you regularly take notes on what you read, you are already more than halfway there on compiling an annotated bibliography.

Certainly, you can keep your notes in file folders or notebooks, but a better bet is to enter the information on some form of database program. Doing so will help immensely when you write, especially as you build your career. The sooner you start, the better, but it's never too late!

BIBLIOGRAPHIC SOFTWARE

If you are adept at building relational databases, you could design your own bibliographic database. These are powerful, and you can include lots of information about each source. But why go to the trouble?

Almost since the beginning of word processing on desktops, a number of bibliographic software programs have been available for relatively little money. Many are good and constantly improving. Each scholar probably has unique needs, but some general features ought to be considered. Several offer bibliographic styles for archaeology and anthropology, including those associated with the journals *American Antiquity* and *American Anthropologist*. The most useful have a user interface linked directly to your word processor as well as web connectivity and searching. For what features should you look?

Start with the size of database the product will hold. Certainly, for a personal bibliography, many products mentioned will easily handle most needs, holding many thousand references. On the other hand, it might be useful to build a lab or office annotated bibliography, which might require a product that will handle substantially more references. Look for a product that can handle a relatively large comment field or text field in addition to the basic reference information. This will allow you to put in all your notes on each source. If you are limited in the number of characters of text you can enter, avoid the product!

You probably already know this, but *data fields* are the pieces of information you enter, such as publisher location, author last name, or publication date. Find a product that will allow you to alter the names of data fields, delete certain fields, or add new ones. Look for a product that has lots of bibliographic styles, particularly for archaeology or anthropology. These templates are already formatted in the style of the journal and allow you quickly to set up bibliographies/references cited sections of your paper just by entering the author name/date in your text. Find a product that will allow you to import files from another bibliographic product easily. If you decide to change programs, you won't have to do lots of work to change products.

STYLES GUIDES AND WHERE TO FIND THEM

Few technical matters cause professors and students, or contractors and contract managers, more grief than reference and citation styles, but there are other concerns for style as well, such as how illustrations get treated, measurement systems, and use of numbers. Stylistic differences may seem minimal, often just punctuation or capitalization. However, if you pay attention to these elements up front, you'll be the darling of copyeditors and save a lot of time.

The single-best approach is to model your manuscript according to what the journal, publisher, agency, or professor wants. Usually journals will have a style guide, sometimes published in the back of every issue, sometimes published as a somewhat lengthy document periodically, and, more frequently these days, on websites for the journal. For American archaeology, the de facto standard is *American Antiquity* (used for this volume and series), published in *American Antiquity* (1992, 57:749–70) and on the Web at www.saa.org/Publications/StyleGuide/styframe.html. However, other journals do use other approaches, so you are well advised to look for the style guide for the journal before submitting a manuscript.

Certain book publishers prefer certain styles. Many, for example, will tell authors to use the *Chicago Manual of Style* for a guide or to follow MLA (Modern Language Association) or APA (American Psychological Association), of which most anthropological styles are variations. Certain agencies, such as the United States Department of Defense, may also have preferred styles for reports done under contract with them.

The simple rule of thumb before you write is to find out whether there is a preferred style and follow it. This will save you lots of time and grief later. If you cannot find a style guide, or if it doesn't cover the kind of source you have cited, look at a recent issue of the publication to see whether you can find how similar sources have been treated (this has been a problem for Internet and other electronic sources), and follow that approach. In fact, if you can't figure out *any* source from the style guide, do the same thing. You'll usually come out fine.

CONCLUSION

So many of the research and writing tools discussed in this chapter have changed because of computers, especially with word processing, and these days only the rare archaeologist doesn't use them. Still, some remain intimidated by the technology. In chapter 5, we'll consider core issues about archaeological computing and presenting the past.

5
COMPUTERS AND PRESENTING THE PAST

Tips on writing fill books. As tough as writing can be, for some archaeologists the physical act of putting words to paper is more daunting. Many have to do their first drafts with pen and paper, handwriting giving them time to think and create. Others prefer writing's buggy whip, the typewriter. Some prefer dictation, usually to their secretary's chagrin. Most of us use computers and word processors these days, not to mention lots of other associated software and hardware. There is no good reason here to cover much about computers; plenty of sources are available, from good computer magazines loaded with tips to online sources that cover just about everything you could want to know about computers. More important, rapid technological change dates all these subjects very quickly.

Few sources are directed specifically at archaeological computing, but a good recent one is McPherron and Dibble's (2002) *Using Computers in Archaeology: A Practical Guide.* Much of the book is about computers in field archaeology, from total stations to GPS, but one chapter considers digital publication, looking at such topics as platforms and file formats and issues surrounding them. For specific computer applications, the *Dummies* and *Idiot's Guide* series are actually pretty good starting places, but you won't find many references to archaeology in them. Some of the best computer-related material specific to archaeology has appeared in the Society for American Archaeology's *Bulletin,* now the *SAA Archaeological Record.*

Some archaeologists seem to have strange attitudes about computers, but computers are neither sorcerer's apprentice nor savior. Computers are really nothing all that special, but from the way many of us talk, you would believe they are magic. Most of us seem to have a love–hate relationship with technology that has caused us to lust after the latest and best computers at the same time that many of us are afraid of them.

We'd be far better off if we could get past the awe of computers. To help, let's try to pin down some key ideas about using them. These aren't hard and fast, but they may help you when it comes to figuring out what you need to use computers effectively in your writing or in other tasks of presenting the past.

BASIC COMPUTER SKILLS YOU'LL NEED

Although these may seem a bit silly as basic skills, the number of people who have problems with them is surprising. To use computers well, you will need to pick up these skills:

- Learn to type reasonably well.
- Become adept at using the mouse, touch pad, trackball, or other method for moving the cursor around the screen.
- Learn to RTFM (geek for "read the f*#$ing manual).
- Learn something about file structure, directories, and pathways used by your computer.

These skills will take you a long way toward mastering computers and the software you will need to accomplish what you wish.

ADVANCED SKILLS YOU'LL NEED

People who are mystified by computers think that every new piece of equipment or software is something unique that has to be learned with a great expenditure of energy and four-letter words. The truth is that much of what you need to learn is similar to what you have already learned. Macs are similar in operation to PCs running on Windows; Word is similar to other word processors; Photoshop is similar to many other image programs. So if you want to be skilled at computer use, you'll need to do the following:

- Find out what is similar in a computer or application to what you already have used, and what is different.

- Learn how to follow digital or hard-copy tutorial materials available from the manufacturer or through third parties.
- Learn enough to call technical support or a geek friend if you hit a real snag.

When it actually comes to learning new programs, many think that a class will help them. A class may give basics, but what learning software or hardware takes is mostly just sitting down and "playing" until you reach a comfort level or until you can make it do what you want.

COMPUTER HARDWARE

The odds are good that you will work in a situation where you will use computer hardware that is already on your desk. If you are buying your own computer or are afforded the opportunity to get new equipment, you may find yourself in a quandary about what to get. Should that be the case, think about the following issues:

- *What will your primary uses for the computer be?* In archaeology we still have to word-process, use e-mail, develop databases, and budget on spreadsheets. You will want a computer that handles these tasks with reasonable speed, which is just about any newer computer. If you want to do anything with graphics, you'll want a faster computer with a larger monitor, a big hard drive, and probably a rewritable CD or DVD drive.
- *Will your primary usage be in the office, home, or field?* If you travel much or are in the field a great deal, you might consider buying a laptop instead of a desktop. Archaeological fieldwork is not the best place for any computer unless it is specifically protected against harsh conditions, especially dust and rain. Having both is the best of both worlds, but if you decide on a laptop and use it in the field, protect it!
- *What kinds of computers are used in your office and at home?* You'll find that compatibility is useful, so you may want to be sure that platforms/operating systems are the same in both places.
- *What is your budget for buying hardware?* Buy the fastest processor, biggest hard drive, and largest monitor you can afford. Remember that the most expensive equipment is the fastest and biggest, so if one or two steps down in technology will do, you

probably won't notice the difference and will save money. Also remember that the equipment will be cheaper and out-of-date tomorrow.
- *What peripherals might be useful?* We archaeologists like our toys, and computers are great playgrounds. Depending on needs and budget, you can put just about everything you want into and onto your computer. Computer toys can also be outrageous money and time sinks, but some peripherals such as printers are necessities. If you will work with images, you may need large-capacity, rewritable storage devices such as CD and/or DVD drives, scanners, and a range of other hardware. When you consider these devices, also factor in costs of media, printer cartridges, and other expendable supplies. Much depends on what you are going to do, and specific software may require specific hardware to operate and to output results.

TACTICS FOR BUYING HARDWARE

If you've answered these questions and have a good notion of what you need and what you want, then it's time for some detective work. Frankly, word of mouth is not such a bad way to find out about the quality of products, service, and support. Ask around. If the product is specifically for archaeology, ask other archaeologists who might know the product. If information technologists are on staff, ask them what they like to work with (and fix!).

Read reviews. You can find these in numerous computer magazines and online. Many of them will also give you a good idea of what you might find for "street price" as opposed to retail.

Finally, a bit of advice: Once you've bought, stop looking! You'll always find a cheaper price tomorrow.

SOFTWARE

The latest hardware means nothing unless you have the software to make it do what you want. In some ways, software is much more difficult to discuss because of the vast variety, much greater than the hardware. Some of the same questions about hardware apply to software, especially those related to your uses and compatibility within an office or home.

TRY BEFORE YOU BUY

Although software is generally cheaper than hardware, most of it is still expensive, especially when it comes to specialty software such as statistics, mapping, graphics, or Geographic Information Systems (GIS). You owe it to yourself and/or your organization to investigate before you buy. Control the urge to buy software on a whim just because it looks cool. Explore these alternatives:

- Borrow a copy from a friend, but only to try. If you like it, buy a legal copy.
- Try shareware. You can often download a program, see whether it works for you, and then register the copy by paying a fee.
- Try "lite" and demo versions of the software that you download and try out for a period of time, often thirty days, or a number of uses before you have to buy a copy. Go to the company website to see whether it offers demo copies for download and trial.
- Try freeware, programs produced and distributed for free, but sometimes paid for by advertisements that appear on your screen. You can sometimes get rid of ads if you pay a fee for the "pro" version. Some freeware also may not be full-featured.
- If you are on a college or university campus, the local bookstore or campus computer center may offer academic pricing or may have purchased a site license that can save many dollars.
- Read reviews of software before you buy anything.

LEARNING CURVES

A *learning curve* is the amount of time and energy required to learn a piece of hardware or software. If you are not at all familiar with a type of software, learning curves can be extremely steep. Once you've learned the basics, the curve flattens a bit, so learning becomes easier. If you buy new hardware or software, plan on spending time learning it. There really are no shortcuts. Classes, tutorials, *Dummies* and *Idiots* books, or software can be helpful, but in the end you just have to go through the product systematically. Remember, however, that most of your skills and knowledge from one piece of hardware or software are often transferable to another.

Ultimately if you need help, don't be afraid to ask for it, but your local computer guru will be more inclined to help if you've really tried first. You might consider three *T*'s with software: *try*, *toy*, and *trouble*. When you get new software, install it and just *try* it. Look for similar constructions to like products you know. See what's different. Then *toy* with it—just play with it! This gives you some familiarity with moving around in the system. Take a document or image created with a prior version or a similar product, and see how compatible the old version is with the new or how the new product handles it. Then *trouble* your way through the product. Use it on something you need to create or do. You may need to use the help screens or worry your way through the tutorial or manual. Sometimes even third-party manuals can help. *Try* and *toy* are fun; *trouble* isn't. When you've done this, you'll probably have figured things out. Then, practice what you've learned regularly and often—use it or lose it!

CONCLUSION

More than anything else in presenting the past, computers have changed the ways in which we deal with images, the subject of the next chapter. Computers show up again in a discussion of digital imaging. You just can't escape them!

6
VISUAL ARCHAEOLOGY

Visual archaeology is any use of illustrative materials in support of archaeological presentation. Given that archaeological writing is about as dry as the dirt we usually dig, most of us won't quibble about the importance of images to our work. We rely heavily on slides for our conference or public talks, we zing our colleagues about the quality of artifact illustrations in their site reports, and we fiddle with new ways to provide better images to present our work. By and large, we do visual archaeology badly. Unless you've been very fortunate in your training, the odds are slim that you've had a course or even part of one on creating and using illustrations or other graphics with your work. Even more of a problem is that most of us don't really give much thought to our visuals until we are getting ready for our final report, a conference paper, or a Rotary luncheon talk.

As with writing and public speaking, the visual part of archaeology was something else you were apparently supposed to learn by osmosis. Some field methods volumes have chapters on field photography (see Heizer and Graham 1967:148–57 or Hester, Shafer, and Feder 1997:159–76 for good examples), but the coverage is minimal. Mostly it is geared toward field recording, where the emphasis is on data ordination, not presentation. Other works on both photography and drawing have been specific to artifact classes or techniques. Most of us have had to pick up skills by taking outside classes or just by doing them. As with many other skills used in archaeology, we should recognize that being an illustrator or photographer could be a full-time occupation. Sadly, we also know that most project budgets can't afford a site photographer, and most labs can't put a full-time artist on staff. As with many other parts of our jobs, we end up doing most of the work by our poorly trained selves.

If that's where you find yourself, you should at least know some basics and some sources. The intent here is in no way to cover all there is to know about visual archaeology. Rather, we'll look into some core theoretical and practical aspects of the use of images and provide some basic tools to get you started. The chapter will touch on photography, drawing, computer graphics, and video, but not in much detail, and mostly to provide source materials. Each topic could be a book and probably several!

THE CULTURAL LIFE OF IMAGES

Images are vastly more complex than most of us think. As Molyneaux (1997:1) notes, "Understanding a picture requires the education and direction of attention to its meaningful aspects." As he goes on to consider, images in scientific illustration by way of appearing to be exactly what they represent—the way we think we use illustration and photography in archaeology—claim truth to nature and science but are nonetheless saturated with meaning (Molyneaux 1997:3). The pictures and drawings we create and select are not value-neutral; they reflect our theoretical approaches, our aesthetic sense, and even our budgets. We should pay more attention to just what it is that images say about us as archaeologists, our profession, and our concern for our audiences, not just what we think they say about the archaeological record.

To do this is not all that easy, but we at least need to think about how we create and use images. At very least we should keep the following points in mind:

- Each image should be chosen to support or amplify a point made by the text of a report or your presentation.
- Images should not be "filler."
- As with the text, images need to be chosen with the audience in mind.
- Consider the values connoted by each image.

Illustrations of both sites and artifacts have a history at least as ancient as their first appearance in medieval manuscripts, but generally the development of archaeological illustration parallels the development of archaeology, and both are tied to the development of technology to produce the illustrations. Piggott's (1979) *Antiquity Depicted: Aspects of Archaeological Illustration* provides a good history, but the general trend has been from topographical recording, to drawing of excavations, to drawings and engravings of artifacts, to photography of both, to digital recording and manipulation. Photography also has an early beginning, traceable back to the photographs of W. H. Fox Talbot of manuscripts, engravings, and busts in England. By the 1850s, photography was regarded as a panacea for problems of documenting materials, but it was followed by disillusion when scholars realized that photos could distort evidence (Dorrell 1989:1). Whatever the problems with photography, it certainly became part of the standard techniques for recording artifacts and features in the field and laboratory by the late 1800s.

Illustration of representative artifacts has always been an important element in good communication about the past. Certainly photographic documentation is useful and necessary, but for detail about the skill of the flint knapper or pot maker, nothing is as good as excellent illustration. Some might say that photos are more "authentic" or real, but all graphics are in some way representation of three dimensions in two. In both drawings and photographs, for example, certain elements can be (de)emphasized by the angle of light, shadowing, and distance from the object. As Piggott (1978:7) notes, "The form of illustration cannot be divorced from its purpose and the requirements of the society in which the given visual language gains currency."

In discussing Piggott's statement, Adkins and Adkins (1989:1) say that illustration needs to be tailored to the audience and to the purpose of the illustration before a "level of technology" used to execute the illustration can be decided. Upon deciding on these elements, the illustrator has little room for "maneuver" except in how the variables change. Reading the Adkinses' history of archaeological illustration is instructive, especially in their discussion of the impact of photography. They clearly demonstrate that "archaeological illustrations are interpretive diagrams rather than attempts at realistic or artistic portrayal" (Adkins and Adkins 1989:7–8).

PHOTOGRAPHS VERSUS DRAWINGS

Photographs are deceptive. They seem to be realistic in that they record all that is visible to the lens of the camera and capable of being recorded on film or digital media. They can be manipulated, especially if they are digital, but they are essentially not very selective. They do not convey all of reality and, in fact, miss lots of important information and include irrelevant information. Their weakness is that they present all information "equally."

Drawings, on the other hand are selective from the start, emphasizing details the illustrator and archaeologist want the reader to see. They are interpretive, meant to convey key information about an artifact or feature. Drawings can also be an aid to study (Dillon 1985:6) because they force close observation, with constant reference to the object and lots of handling, rechecking, and measurements.

As it does with text, the intended audience dictates the kind of drawings used. Some of the conventions used in archaeological drawings, like technical jargon, may not be meaningful to nonspecialists. Use of drawings or photographs may also be a function of cost. Generally the cost of producing both has gone down. In terms of production, drawings are probably more expensive in terms of production time, but they can convey substantially greater amounts of information. Photographs are relatively easy to take, especially if they are digital, and manipulation of images has gotten easier. Layout and printing costs of photographs used to be prohibitive, which is why some archaeologists used line drawings for artifacts, but current technologies, particularly using desktop publishing tools, allow relatively inexpensive layout and printing of both. Color drawings and photographs are still costly to print.

Archaeologists have a substantially greater number of ways to present information by using illustrations than they did a few decades ago. Choice of graphic media should be made on the basis of both intended audience and the sort of information to be conveyed to the audience, not ease of production or cost. Consider these issues carefully as you decide.

TOOLS

Having the right tools is important to just about everything we do in archaeology, and that's no different with visual archaeology. One can get by with few and inexpensive tools, but quality does show in illustrations. Good tools do not make good illustrations, but they help!

SOURCES FOR ARCHAEOLOGICAL DRAWING

Several books provide a good coverage of archaeological drawing. Adkins and Adkins' (1989) *Archaeological Illustration* is perhaps the most thorough. Their volume discusses a bit of history and theory of archaeological illustration, as well as considering equipment, drawing under a range of conditions including the field and lab, and drawing different kinds of artifacts and features. Their next to last chapter is especially relevant to this volume with coverage of drawing for a wide range of reproduction types, from print to other media such as slides, microfiche, and overheads. They also look at matters of copyright, but more current material about it is online.

Still useful is *The Student's Guide to Archaeological Illustrating* (Dillon 1985) in the UCLA Institute of Archaeology's Archaeological Research Tools series. The articles focus on particular types of illustrations or artifacts. Especially useful is Armstrong's (1985) chapter on drawing maps.

The most complete coverage of drawing any artifact type is Addington's (1986) *Lithic Illustration: Drawing Flaked Stone Artifacts for Publication*. Her last chapter on the working relationship of the archaeologist and the illustrator should perhaps be read first. Its two pages provide a quick study on how to work with a specialist.

A little book by Brodribb (1971) called *Drawing Archaeological Finds* may be the best small book on illustration. The book is organized around the process of drawing, and that helps one to understand its complexities. In fact, more than anything else from all these books, the most useful element novices may get from them is to understand the *process* of drawing.

Support is also available from a professional organization of archaeological illustrators, the Association of Archaeological Illustrators and Surveyors, an international body for those engaged in all aspects of professional archaeological illustration and survey. They produce a journal, *Graphic Archaeology*, newsletter, and website (www.aais.org.uk/). They have also developed a series of technical papers for drawing lithics, pottery, wooden artifacts, and a number of different artifacts classes.

EQUIPMENT FOR DRAWING

All the books list equipment that you might need for drawing. As might be expected, different tools may be needed for different types of drawings, and there are a lot of tools. Adkins and Adkins (1989:11–39) not only list the equipment but also detail uses and varieties. Much of the equipment is relatively inexpensive, but special drawing tools can be expensive. If you intend to produce in-house drawings, you can build your collection of equipment over time. Certainly some of the equipment listed has been overtaken by technological change.

SOURCES FOR ARCHAEOLOGICAL PHOTOGRAPHY

Very basic information on photography in the field is available in chapters in the widely used field methods guides (Heizer and Graham 1967:148–57; Hester, Shafer, and Feder 1997:159–76). Of use in them is description of different kinds of field photos (site views, features, underwater, etc.) and the use of visible scales in photos, as well as some issues with special types of photos, such as with petroglyphs.

More detailed coverage is available in Dorrell's (1989) *Photography in Archaeology and Conservation*. He includes a good history and then looks at specific issues of photography in the field and lab, with a chapter on preparing materials for publication. Though dated, this chapter raises important issues regarding the use of 35mm transparencies in publication, especially color balance between slides. What is apparent is that anyone should seek the advice of a potential publisher regarding the type and quality of photos to be used.

A Practical Guide to Archaeological Photography by Howell and Blanc (1992) may be the best place to start if you haven't done much photography. After a fair coverage of camera function and ways to use light, the guide looks at photography in both the field and laboratory. The last chapter, "Review," is something you should look at even if you are experienced. It has "index cards" that are worth the trouble to copy if you can't afford a copy of the book itself. The seventeen cards contain topically organized, bulleted lists of key elements about everything from how to shoot artifacts to f-stops and depth of field. Appendix D contains a somewhat useful series of points about preparing slide presentations.

In truth, if you are a novice to photography, starting with the guides listed here might not be as effective as very basic guides on photography such as *Photography for Dummies* (Hart and Richards 1998). If you are more skilled, you might like to see what professional photographers say about archaeological site and artifact photography. An article in *American Archaeology* (Anonymous 2001) features limited advice from Eldon Leiter, Jerry Jacka, David Whitley, and Steven Wall. The advice is interesting, and the article makes the point that photography is an art that takes years of experience to do well.

DIGITAL PHOTOGRAPHY

Rapid improvements in digital imaging have dramatically increased the quality and utility of digital photography in both the field and lab. Combined with digital scanning and a range of software to manipulate the images you generate, there is not much you can't do with images, and the gap between drawing and photography can be readily bridged.

Little in basic source materials for archaeology discusses digital photography. Although the seventh edition of *Field Methods in Archaeology* (Hester, Shafer, and Feder 1997:166) mentions digital photography, the material is minimal and dated. A more current overview appears in McPherron and Dibble's (2002:160–77) *Using Computers in Archaeology: A Practical Guide.* Most useful discussion of what can be done is in the fringe literature of notes and newsletters. An article by Rick (1999) provides a recent survey of digital cameras in archaeology, with good coverage of the technology and several key issues in their use. It's now true that for almost every purpose in archaeological presentation except slides and enlargements above 11 × 14 inches, digital photography equals or surpasses film photography.

DIGITAL IMAGING

Digital cameras derive their strength from the connection to the computer and its use in manipulating images. On a superficial level, computer graphics—that is, digital imaging—is really easy. You can quickly learn to scan or use a digital camera or grab an image off the web, but doing more with them requires substantially greater knowledge. The learning curve is as steep as any in computing, short of learning to program. Graphics may be tougher in that using them effectively also requires a level of artistic skill most of us either don't have or haven't developed. There is no way to cover the topic thoroughly here, so we'll just touch on the basics, looking at what you might need for digital imaging. A digital camera, a scanner, and a graphics program are the core of a digital imaging system.

DIGITAL CAMERAS

A digital camera, as the name implies, stores images digitally rather than using film. Once you take a picture, you can download it to a computer and manipulate it with your graphics program. Digital cameras are now relatively commonplace, with quality improving and prices dropping. The resolution of digital cameras is limited by camera memory, the optical resolution of the mechanism that digitizes the image, and the output device used to show the picture.

At present, the best digital cameras are not capable of producing film-quality resolution, although this is changing dramatically. The greatest advantage of digital cameras is that you avoid the film process altogether. For fieldwork, this has huge advantages because you can see almost instantly whether you've got the picture you want and retake if you need to. Because there is no film or processing cost, you can take lots of pictures at extremely low cost, amortizing the higher cost of the camera and media very quickly. For presentation, you can get reasonably high-quality images and easily download, manipulate, and incorporate them into websites, CD-ROMs, and publications.

SCANNERS

A scanner is a must for anyone, and prices are so low that some package computer deals even come with a scanner. Scanners are the computer equivalent of a copy machine, but the items being copied are digitized so that your computer can read them. Scanners provide a convenient way for entering text and graphics on your computer. Even most low-end scanners have software that will allow you to scan and manipulate graphics, as well as optical character recognition (OCR) software that will convert scanned text into text your word processor will read.

Scanners come in a range of forms. Pen and other hand-held scanners can read a line or more of text at a time as you move over it. Their benefit is portability, but their drawback is speed. Some scanners require that a sheet of paper be pulled though the mechanism. They offer portability, but if you need to scan from a book, they are impractical. Flatbed scanners are the best bet for most purposes. They come in a range of sizes and have a flat glass surface on which you lay a document or picture. A moving light source, much like that in a copy machine, captures the image.

Digital cameras and scanners have other important uses. They can actually be used to help create "drawings" of three-dimensional objects. One can scan or photograph artifacts such as pottery sherds or stone tools and then use a graphics program to eliminate or enhance certain elements of the object, much as one might do with a drawing. Houk and Moses (1998) discuss the process using a scanner, but it is similar with a digital camera.

OTHER TOOLS

You might also find the following pieces of equipment useful:

- *Slide or negative scanner.* Most scanners do not allow 35mm color transparencies (slides) or negatives to be scanned well. Specialized scanners do a better job.
- *Web camera.* Attached directly to your computer, web cams allow you to photograph and input images from the real world to the World Wide Web.
- *Graphics tablets.* You can draw with a pencillike stylus on a touch-sensitive pad and have what you create put into your computer for modification in a graphics program, bypassing paper drawings altogether.
- *Graphics programs.* These allow you to take the images generated by a scanner or camera and manipulate them with your computer. Simplified programs for graphics and OCR (which allows you to scan text) usually come bundled with the scanner or camera. These are usually a limited or "light" edition (LE), meaning that the program is not full featured or may not be as capable of handling the data. Most bundled graphics software allows you to resize, rotate, crop, and do minor adjustments to the images. Some even allow you to compress the file size of image files for e-mailing and web use. However, if you want to do more with images or work with large images, even isolate elements of images, a higher-end graphics program is best. Most such programs will probably do more than you can use, but in addition to the features of the LE software, you will want to have some level of color and brightness/contrast adjustment, the ability to put text and draw directly onto the image, and the ability to reduce the size of the image file for easy use on the World Wide Web or for e-mail. Without trying to explain it in detail here, you may also wish to be able to move back and forth between the two main systems of dealing with images, vector and

raster graphics, particularly if you do work in GIS. Finally, you'll want a program that allows you to prepare both print and electronic images, so find one that allows you to use and convert between a range of image formats. There is a wide range of image formats, with three standards in common use on the web, but many in use for print media. For print, the formats are more variable, and certain publishers prefer certain formats. Most high-end programs handle all the common formats.

SPECIALTY GRAPHICS PROGRAMS

Beyond these graphics basics, even discussing graphics gets complicated, especially with issues such as raster versus vector images, file formats, and file compression (see McPherron and Dibble 2002:160–70 for a brief discussion). Some issues are program-specific.

Preparing images for the web brings even more presentation choices. One element that many people like to put on web pages is animation or movement. You can make simple animated images with free and shareware programs, and some graphics programs also include animation. High-end specialty programs allow for creation of extremely sophisticated animations without dramatically increasing file size, though some require special plug-ins or players (downloadable for free) to see the animation.

Other programs allow you to sketch an object, and then have it turned into a real image. Computer-aided design and drafting (CADD) programs may become very useful if you are working with architects and engineers on a CRM project, but good ones are very expensive, with steep learning curves. If you need CADD, perhaps some other part of your organization has a useful program. GIS programs also contain specialty applications that allow you to produce images to better comprehend the patterning present in data. In the realm of the web, some programs allow you to build three-dimensional images (actually, they are quasi-3D; see Rick and Hart 1997 for a discussion of processes and possibilities). Some programs are set up specifically to allow you to build buttons and navigation tools for the web.

If computer graphics interest you, just go to a computer store and look around, go online, or look at any of several magazines having to do with graphics. Most specialty graphics programs are relatively expensive, so get reviews, talk to other users, and use trial software downloadable from the web before you buy.

VIDEO

Videos and films are just about the closest approximation to real life. They are also a very good way to educate at the same time that we entertain, and they are media our publics generally prefer over others. At the same time, they are probably the media archaeologists have the least experience producing. They have been relatively expensive, film especially so. Many archaeologists have used video in the field to document activities and sites, but they have rarely taken more than "home movies" because of the expense and skill required. However, changes in technology, especially digital camcorders and relatively easy-to-use editing software, now allow archaeologists to produce video near professional quality.

Because video seems to be a recording of real life, audiences might think that what they see is closer to reality or somehow more authentic than other kinds of presentation. The truth is, however, that video is subject to the same biases as other images, a limitation we should remember and of which we should inform our audiences. Although little has been written about archaeological video and film, there is a solid body of literature about ethnographic film from which to draw. Heider's (1976; 2001:426–33) work provides a good starting point regarding theoretical issues, many of them applying to archaeological film.

STORYBOARDING

Among the more difficult issues is the fact that text is linear, but pictures are not. In video, nonlinear pictures must be crunched into what is essentially a linear text, a script. Storyboarding can help the process. In storyboarding, the point of view of the camera and the action are linked by putting them onto storyboards, or drawings of how the story should be told. The storyboarding process is also very useful in developing multimedia product, a subject that will be discussed in chapter 12.

GETTING YOUR PRODUCTION NOTICED

Even if you do video on a shoestring budget, distributing the product is not cheap. Sponsored by Archaeological Legacy Institute, the Archaeology Channel (TAC) at www.archaeologychannel.org partners with producers of archaeological video or film and broadcasts the films using streaming media on the web (it does the same for audio materials). If nothing else, TAC is a good place to find out what is going on in archaeology video, and it can be a relatively inexpensive way to distribute your productions.

Although archaeologists have tried artifact drawings and most have done photographs, making a quality video without help is probably well outside our abilities. Because we recognize video as a useful medium for communicating with the public, we will usually have such work done by professionals. When it comes to presenting the past, perhaps it is time to consider what we do to be as much a product of teamwork with specialists and consultants as it is in our field- and lab work, the subject of chapter 7.

7
LONE RANGER OR TEAM PLAYER?

If all the tasks archaeologists are called upon to do haven't scared you by now, you must be a better person than most! The problem for archaeologists is that our field is so vast that we tend to become jacks-of-all-trades and masters of none. A few of us may have had time to develop a specialization, but we have to know something about almost every aspect of our discipline, from how to do surveying to how to analyze faunal remains collected from a site. Many of us can know something about many of these things, but often just enough to be dangerous! The same is true when it comes to presenting the past. Many of the tasks, such as basic write-up, are relatively straightforward, but do we really have the time and intellectual capacity to learn photography, graphics, CD-ROM development, and broadband web technologies? True, one can know about all these things and maybe even do a few of them reasonably well, but the end product usually will not be as good as you might like.

We have gotten to the point in our field- and lab work that we are comfortable with hiring specialists as consultants, even to the point of building them into our bids for projects and our grant budgets. At the same time, most of us rarely consider the similar situation when we come to preparing the final report on our project, even for the illustrations. We try to do it alone, and because of that, our product is often less than it could be, and sometimes less than it should be. If you are fortunate enough to work in a large operation, your overhead costs may include hiring illustrators, photographers, and web designers, but if you are like many, you just can't afford these specialists and still be competitive in the CRM world. If you want to improve the product and can consider a team approach to presenting the past, what kinds of people should you get, and how do you find them?

DEVELOPING THE TEAM

As you hire staff for projects, you almost always consider a range of qualifications. You need to hire people who are good in the field at a variety of tasks, from personnel management to excavation. You also want to know that they can handle themselves in the lab. As already noted, you probably will want people who are capable of writing the reports on the projects to which they are assigned. At the same time, you might consider putting a qualification in a job ad for someone who can also illustrate or do web design or some other task related to presenting the past. If you get lucky, you might be able to assemble a pretty good team.

If you work in a university, you may have a plethora of low-cost but fairly skilled student labor at your disposal, people who need to build their résumés or portfolios for future employment. Most universities also have people who "hang around" long after getting their degrees, just because they like the university atmosphere. They survive by working at temporary assignments or low-paying hourly jobs, but they are nonetheless highly skilled.

If you plan on building a CRM or academic laboratory, you should do some longer-term planning and work toward hiring a specialized staff. Some of us have built report preparation into budgets for many years; even a small survey project has a line item for the task. Target all or some portion of the report preparation money from all the projects toward hiring an illustrator, computer specialist, or whatever you most need at the time. Also consider building in some of each contract's overhead or indirect costs to include hiring these people. If you happen to be in a situation where you can request permanent employees as budget lines, then build toward adding staff according to your presentation needs, just as you would for field- or laboratory work.

Figuring out what kinds of people you will need can be difficult, and much depends on your office or program's goals. You should spend some time planning how to develop such a team. What do you most need? Consider adding the following team members:

- *Editor/copyeditor.* As much as most of us might like to think we are great writers, everyone can be improved by editing. Good editors have many skills, but they pay more attention to detail than most people. They must know spelling and grammar but also need to be able to work within archaeological styles and standards. They should also be good at proofreading. With enough experience in archaeology, they might even be able to do indexing well.
- *Illustrator/graphics specialist.* People with an ability to draw, and in ways that you want them to, are important to any team. The trick is in finding someone willing to learn what is important to show in his or her illustrations, based on collaboration with an archaeologist. These team members need to understand that what they are doing is not "art" in the sense that they must follow reasonable standards of archaeological illustration. These days, they should also have some level of experience with computer graphics, not just drawing.
- *Designer.* You may be fortunate enough to find an illustrator who also has skill at design. Design is what some might call the layout of a report, posters, your website, or any graphics product that comes from your office. Designers help set a style for the products and your whole office. They also must have facility with computer graphics.
- *Digital media specialist.* If you need to develop an office or company website, or if you wish to present materials electronically, you may wish to hire an individual capable of working with digital media. Such a person does not necessarily need a degree in computer science. Most computer science programs don't teach the kinds of skills needed to do work with the web or CD-ROMs. As important are people skills; digital media specialists must listen to be able to carry out the media goals of those in your office. They can't be the stereotypical computer geek and do well at the job. Skills with specific computer programs for the web or graphics are crucial, but they don't need to be able to field-strip a computer!
- *Photographer.* Many archaeologists consider photography to be just an extension of note taking. We are often not very good at taking pictures, which shows at lots of conference presentations. Basic skills are not too hard to learn, but much beyond that takes training and experience. High-budget projects may be able to hire photographers for the field, but most forget about it for the lab. Consider what bad photos might tell a client about your operation. These days, a photographer should be adept at both film and digital

photography to be useful to your organization. You might also want this team member to know about, or learn, video.
- *Instructional designer/educational specialist.* Many of us have spent more time than we like to remember in academic settings. We've taken lots of formal courses and may even have taken additional training in workshops. We've developed a feeling for when instruction is good or bad. That doesn't mean, however, that we have the skill to design and execute a good course. We may need help. Many archaeology organizations do some level of instruction, from teaching skills to staff members to working on certification programs for avocational archaeologists. For instruction delivered at a distance electronically or through correspondence, design becomes even more crucial. Individuals who develop such materials must have training or substantial experience in constructing, packaging, and delivering educational products. As important, they have to learn what it is they are producing, so they must have some level of experience with archaeology.
- *Public relations/marketing specialist.* How do you let your publics know what you are producing or what events you are planning unless you have someone doing advertising for your products? This topic is covered in detail in chapter 11.

OUTSOURCING WORK OR HIRING IN-HOUSE

You can probably think of some other specialists to add to the list of people you could use in your office. If you just consider the positions listed here, you'll quickly realize that the annual costs would be well in excess of $300,000, a figure most CRM firms or university labs just can't afford when most can't even pay for core archaeological specialists like geomorphologists! This means that you either must have your archaeology staff develop the skills as a sideline to their regular chores or must outsource the work or hire temporary staff to do the tasks in-house. You will probably have to outsource for specific tasks on projects, anyway, so why not do so with much of your presentation work?

If you are in this situation, consider the costs and benefits. Costs for hiring short-term consultants or specialists are generally higher per hour than having specialists on your staff. At the same time, paying salaried staff also includes benefit costs, and you pay for any downtime when you don't have work for them. As you may already know from your fieldwork when you hire consultants, these folks are outsiders. They know that they are contributing to a product, but they may not feel they are part of the team and may not fully understand your approaches and goals.

If you outsource work, you need to pay particular attention to making these consultants understand your project's goals. While the scientific specialists we bring in have often set themselves up to work specifically with archaeologists, most of the consultants you bring in to help with presentation probably have a wide range of clients for whom they work. Good consultants also realize this problem and work hard to understand your needs and goals, sometimes to the point of being annoying (this is actually a good trait in a consultant). If they don't ask lots of questions, they probably aren't going to be a good consultant. Achieving good relationships and understanding does take your time and work, but if consultants work for you regularly, they eventually develop a feel for your needs.

Finding such people can often be difficult. If your are part of a university, consider yourself lucky in that most universities have a wide range of offices specifically designed to provide consulting services in every presentation specialty listed earlier. In some cases, you buy their services at an hourly rate, and they can even produce the final products you need, such as 1,500 copies of a CD-ROM or 30 copies of a contract report. If you work in the private sector but near a university, you can often use university services, probably at rates competitive with, or even a bit lower than, private business. If you need to hire services from the private sector, you can try a number of approaches to finding quality consultants. You can ask colleagues or competitors whom they use. Competitors may be hesitant to tell you, but that's not very likely. You can ask each potential service provider for references and samples of his or her work. You can ask for competitive bids from several providers, and probably should. You may even enjoy being on the other side of the bid process for a change!

Whatever the case, you will need to be clear about your needs, and you need to make your consultants part of a team if you can. If you find good consultants who figure out your needs and how best to work within your style and structure, you'll obviously want to go back to them. The better the situation for them in terms of working conditions and clarity of goals, the better work you'll get. If you go back to them often, you may find that all your costs go down, in both time and money, especially if you treat them with respect. This should go without saying, but consultants have skills you need, and they are professionals who deserve as much respect for their skills and knowledge as anyone else associated with your project.

8
PUBLISH OR PERISH? COMMUNICATING WITH COLLEAGUES

Communicating your work to colleagues is a crucial element of your work as an archaeologist. Doing good archaeology demands a dependence on other archaeologists who provide feedback about your work. However, colleagues may also be your most critical audience, some of them even your direct competitors, competing for contracts, positions, and reputations. For those in academic positions, communicating with professional colleagues is a key to earning promotion and tenure. For CRM archaeologists, if you don't produce reports, you don't eat!

Certainly, one can communicate with colleagues, as we often do, over a beer at a conference, but ultimately, more formal communication is necessary. This means presenting papers at conferences, preparing articles for a range of publications, and, for many, writing lengthy reports, monographs, and books. These tasks sometimes seem daunting, especially if you are at the beginning stages of a career. You might have seen the curriculum vitae (vita or c.v.) of your professor listing all of her or his publications and presented papers, thinking the whole time how much you struggled to get that ten-page paper done for the class you just took. Don't be too impressed; generally speaking, your professional life is long. That vita didn't appear overnight. If you develop certain skills and persevere, you'll find that your vita will look like that some day—and perhaps even better than your prof's!

A good place to start is by giving conference papers, the traditional way to inform colleagues about your work. Presenting papers also gives you a chance to discuss your ideas with other archaeologists before you put the work into print. However, even a paper starts with some writing: the abstract you send in to the conference to tell its organizers about your paper. Sadly but truly, an abstract is one of the most difficult kinds of writing there is.

ABSTRACTS

If you've been around archaeology very long, you'll know that learning how to construct an abstract is an essential skill. Everything seems to require them. Most conferences want an abstract, usually at some excruciatingly terse length. The conference program committee may use it to evaluate your paper for possible presentation. If they accept it, the abstract is put into the meeting program for colleagues to read to see whether they want to hear your paper or, barring that, to contact you later for a copy of the full paper.

Just about every CRM report short of the letter report seems to demand an abstract. If one might be a bit harsh, one starts to suspect that CRM bureaucrats don't like to read long CRM reports, so they have their contract managers demand abstracts of lengthier documents. Cynicism aside, abstracts can serve a more important function, even in the CRM world. Depending on the agency or state, abstracts of reports sometimes get published so that colleagues or other organizations can see key results from their research areas or jurisdictions. Most quality journals also demand abstracts of papers. In truth, there is such a glut of archaeological and other information being created that most scholars have learned how to skim or read abstracts to find out whether an article is worth their time or fits their research interests. Many online databases and abstract journals are now available, so abstracts are often published as stand-alone documents to aid researchers.

Abstracts are important! If you want someone to hear your paper or read your work, you need to pay careful attention to writing your abstract as you do to your full manuscript. Still, the whole idea of an abstract is tricky. You want to provide enough material to entice readers to hear your paper at a meeting or read the whole paper in a journal. At the same time, you don't want to give so much that they feel they have all your data, logic, and ideas fully encapsulated in the abstract so that they don't need to attend your presentation or read the full paper. Finding the right balance is often a dilemma.

Wherever the abstract is printed or whatever its length, its purpose is singular: to provide a summary of your work that can be quickly accessed and assessed by colleagues. As simple as it may seem, writing an abstract is no easy task, especially when you have to cram your whole project into as few as one hundred words! Sometimes simply knowing what an abstract is can help. An abstract should be a stand-alone statement that briefly conveys the essential information of a paper, article, CRM report, monograph, or book. In a short, nonrepetitive style, it presents the work's objectives, methods, results, and conclusions.

Abstracts are deceiving. They tend to appear in the first part of a paper, or they are usually sent to a conference program committee months before you give your paper. Even though they seem to come before anything else, the best abstracts are written after actually completing the full project. To do the best job, you really need to have finished your work before you can choose and summarize the essential information. Self-contained compactness is crucial, but if you try to write the abstract before the work is concluded, it can be a profoundly difficult task.

WHAT DO YOU PUT IN AN ABSTRACT?

STYLE GUIDES

What goes into an abstract may fully depend on the guidelines provided by the conference to which you are submitting the paper or the journal, publisher, or agency to which you are sending the manuscript. Some publications may have a required style for abstracts; guidelines for authors provided by a publisher or by organizers of specific conferences may provide specific instructions. The Society for American Archaeology, for example, asks for a two-hundred-word factual summary of the contents and conclusions of a submitted paper, with specific references to new information being presented and some indication of its relevance. The SAA specifically notes that the abstract should not be an introduction to the paper or an outline of it, with each section being reduced to a sentence. It also expressly asks authors to avoid passive voice.

TYPES OF ABSTRACTS

There are generally two types of abstracts, and they are not necessarily mutually exclusive. *Informative* abstracts summarize the entire report and give the reader an overview of the facts to be presented in detail in the paper. This type of abstract is normally of the sort used for conference papers, short reports, or relatively brief manuscripts such as that for a journal article. *Descriptive* abstracts are useful for longer documents such as monographs or CRM reports in which one cannot easily summarize all the findings of the project. It's a bit like the scope of work for a CRM project or a statement of purpose, describing a work's organization but not all of its content.

ABSTRACT STYLE

An abstract is no place for fancy style and convoluted prose. Writing needs to be straightforward and sparse. It doesn't need to have the same sentence structure or flow of the paper or report. Certainly, good grammar remains essential, but concise presentation of content is crucial.

WRITING THE ABSTRACT

Remember, in an abstract you are making a claim to knowledge. Thus, you succinctly need to tell the reader the boundaries of that knowledge, including time and space, as well as the limits to which you are willing to generalize from that knowledge. For archaeology this point is crucial, and many abstracts fail to make such boundaries clear. The abstract should precisely describe objectives, methods, results, and conclusions. Avoid any background information, don't make reference to the literature at all, and dump any thought of giving details of your methods or theoretical position.

December and Katz (1996) suggest that "writing an abstract involves boiling down the essence of a whole paper into a single paragraph that conveys as much new information as possible." They offer that

> one way of writing an effective abstract is to start with a draft of the complete paper and do the following:
> - Highlight the objective and the conclusions that are in the paper's introduction and the discussion.
> - Bracket information in the methods section of the paper that contains keyword information.
> - Highlight the results from the discussion or results section of the paper.
> - Compile the above highlighted and bracketed information into a single paragraph.

- Condense the bracketed information into the key words and phrases that identify but do not explain the methods used.
- Delete extra words and phrases.
- Delete any background information.
- Rephrase the first sentence so that it starts off with the new information contained in the paper, rather than with the general topic. One way of doing this is to begin the first sentence with the phrase "this paper" or "this study."
- Revise the paragraph so that the abstract conveys the essential information.

Following this advice is not as easy as it might seem, but one does get better with practice. Remember, finally, that the abstract and the paper, monograph, or report are essentially parallel documents, but each must stand on its own.

A DOSE OF REALITY

Despite all the good advice out there about writing abstracts, most of you have gone to hear a particular paper at a meeting and found that it didn't sound much like the abstract published in the conference's abstracts volume. Many of you know exactly why that happens. Deadlines for submitting a presentation at a meeting often fall months before the conference. You might just have gotten out of the field or only be nearing completion of a complex bit of analysis. You haven't gotten anywhere close to being done with the manuscript. Still, you feel a need to give a paper on your work. What are you to do but make a guess about what you are actually going to find? You've probably heard all the good suggestions discussed already, but the odds are that you probably didn't have time to use them. Certainly, it's easier to write an abstract following this process when you've actually completed a manuscript, but when it comes to preparing an abstract for a conference paper, just about all of this advice goes out the window. If the abstract is due tomorrow, what should you do?

- Pay attention to some of the advice, at least.
- Give the boundaries of your work, temporally and spatially, unless the work is more generalizing.
- Be scrupulously honest, stating exactly how far along you are in your work, but it is fair to speculate what data analysis at this time seems to indicate.

- Avoid using jargon, even if you think such BS might snow the people reviewing your abstract or a reader; inappropriately used jargon just makes you look silly.
- Don't overextend your data, even if you think it might be supportive of certain hypotheses you propose.

Abstracts are vastly more important tools than most scholars realize. Most of us tend to think of them as a bother, but given their importance in convincing other scholars to read our work, we really should pay more attention to the way in which we construct them.

CONFERENCE PAPERS AND PUBLIC TALKS

One's first professional paper seems like such a rite of passage, remembered for the rest of your career. (If you're lucky, you'll be the only one who remembers it!) Giving professional presentations and papers is a professional obligation, used to inform colleagues about our research and giving them a chance to provide criticism. Publication in print and on the web provides other mechanisms for feedback, but having a group of colleagues together at one time provides a compelling dynamic. Giving talks is part of our accountability to the public, an ethical obligation.

WHY WE GIVE PAPERS AND PRESENTATIONS

The real reasons we give conference papers or public talks are easy to forget. In the crush of academic life, we get pushed to present papers in order to have our way to a conference paid or to build our credentials. In the CRM world, an agency's scope of work may require you to give a professional presentation to showcase its project. Sometimes in university life, you are asked to give presentations as part of a well-defined service obligation. These mask the real reason for presentations.

More than anything, conference papers are for the exchange of information with colleagues. It's important to have your peers know about your work and to swap ideas with you based on their own experiences on similar subjects. Presentations to groups not made up of colleagues are service, a way to pay archaeology's bills, so to speak, by letting them know what we learn from the tax or foundation dollars we spend. For all the energy spent on preparing for the papers and talks, most of us tend to forget this. Whatever the politics of conferences, whatever the possibilities of getting shot down by a colleague or a heckler, remember why you give a paper. The idea is to get your data, thoughts, and logic into a larger arena than that between your ears or around the company coffeepot!

GLOSSOPHOBIA

If you've never given a paper before, the process might seem daunting or the fear overwhelming. Your anxiety probably starts with *glossophobia*, the fear of speaking in front of an audience. Public speaking is a fear shared by many people (more than 40 percent). In fact, according to the *People's Almanac Book of Lists*, the fear of speaking in public is the number one fear of all fears, while the fear of dying is a mere number seven! The truth, however, is that speaking in public probably won't kill you. You need to learn to deal with the fear, or you simply will not do all that well in archaeology. Hiding with your nose in an excavation unit or at the end of a caliper in a lab just won't work.

You undoubtedly have had a number of opportunities to learn and practice public speaking, but if you are like most, you probably found ways to avoid it. As an undergraduate, you probably had to take a required speech class but, like lots of students, put it off until the last semester of your final year (along with the math requirement?). In graduate training, you might have had ample opportunity to report your findings in a seminar class, but perhaps you avoided certain seminars entirely because you were afraid to present findings in class. If your glossophobia is that great, you probably need some professional help to overcome the fear, but if your fear is average, there are a lot of things you can do to control anxiety.

The single most important piece of advice is that you prepare yourself well. This tip might seem obvious, but in the crunch of time, most of us don't take into account what preparation means and leave scant time for it. You need to do more than just read your paper over a few times. Good advice on overcoming your anxiety and on how to prepare can be found all over the Internet and on many bookshelves, but some of the best comes from Toastmasters International, a fine organization dedicated to helping people become better public speakers. They have chapters in most areas, many of them on college and university campuses. You can get tips on speaking and experience, as well as "safe" criticism from them. Their website (www.toastmasters.org/tips.htm) has excellent tips for preparing yourself.

TIPS AND REALITIES

The Toastmasters International tips and others like them are all fine, but the reality of public presentation is a different thing altogether. Most tips are good and will help you, but some are crucial to success as a speaker.

KNOW THE ROOM

During a two-year stint as a National Lecturer for Sigma Xi, the Scientific Research Society, I gave more than thirty lectures. I spoke in everything from a bare classroom with a slide projector on the arm of a student desk to fancy lecture halls where the podium was as well equipped as the bridge of the *Starship Enterprise*. Usually I was afforded a chance to check out the room, something I'd advise every time and something I do at every conference before I give a paper. You just don't know what you will find!

In some ways, I'd say that you also owe this level of acquaintance with the room to your audience. They have a right to see and hear you, or at least to expect you to be prepared (although some problems may just be out of your control). Consider the following:

- If possible, ask someone else to come with you, and have them listen to you speak in the room.
- If the room is so large that a microphone is needed, be sure that you know how to use it. How far from the mike can you lean or move and still have it pick up your voice? If you have a lavaliere microphone, how far can you move around on the cable without strangling yourself or tripping over it? If you have a radio microphone, consider yourself lucky!
- How do your visuals look at a distance? Check their size from different spots in the room, so that you can describe them better if graphics are too small.
- Learn how the slide projector works, and run through your slides if there is time.
- See whether there is a pointer, water, or other things you might need.

If nothing else, by checking these things out you'll at least feel comfortable, but doing so may also help you avoid disaster during your talk.

KNOW YOUR AUDIENCE

Before you speak somewhere, always find out the nature of the audience. If you haven't been told, ask! Based on your assessment of audience background, your entire tone, style of presentation, and parts of the content may need to be varied. For an audience of professional colleagues, you have a right to expect that they will at least know most of the jargon. Certainly, your professional paper can and should be different from that given to the local archaeological society. You should be smart enough to recognize audience differences and prepare for them. If the presentation is something you might give several times, you can actually prepare talks with different slides and text for different audiences. Though doing so takes a bit of extra work, it's worth it.

KNOW YOUR MATERIAL

Nothing should give you confidence so much as the fact that you know your material better than anyone else, but that's not all there is to it. The real trick is practice. Having your logic of presentation and any visual materials well prepared is the starting place, but real confidence in your material comes from practice. You can only feel confident if you've gone through the presentation enough that you don't really need to worry about the material. You know what argument is next, you know what slide follows the one on the screen and that it is not upside down, and you know how an audience might react to what you show or say. Nothing can throw you.

Knowing your material will help you succeed in a way that more archaeologists giving conference papers should remember: Don't read your paper; give it! Nothing is more deadly than a person reading a paper. Attend any conference, and you'll hear many read papers. Unless the presenter is very skilled, has used a speaking style, and has practiced carefully, reading a paper can be a disaster. Most people don't write like they speak; they go long, and the audience can't follow complex spoken arguments because they don't have time to think about them like they have when they read. Usually, you are better off having an outline to speak from or, as many do, using your illustrations as your outline. If you are nervous and need a "crutch," you might have a full copy with you in case you get lost, but do your best to avoid using it.

When you first start giving talks or lectures, being nervous is natural, but you can even prepare for that. I require that my teaching assistants give a lecture in my classes so that they can get a feel for what that situation is like. Most are justifiably terrified at their first lecture. I always tell them to prepare just about twice as much material as they think they'll need because with their nervousness, they'll rush through it. After you calm down, you can always cut material if you are going long, but if you don't have enough, it can be pretty difficult to wing your way through another fifteen minutes.

The more you practice, the more you can pay attention to other elements of your presentation. You can get to a point where it's almost like an out-of-body experience. You know the material so well and are so practiced that you can almost step outside yourself to objectively assess how you are doing. You can pay attention to the "acting" part of public speaking. Most important, you can pay attention to your audience and how they are responding to what you're saying, using the feedback to adjust pace, explain terms better, and assess their understanding. Believe it or not, you can actually enjoy lecturing to large groups.

Obviously, delivering conference papers is different but nonetheless requires detailed preparation. If you are a student, ask your professor whether there are opportunities for you to give a paper. You might be able to tailor it to an audience of advanced students, filling in for your professor for a class period. Perhaps a department seminar is being held where you can try out your paper. Wouldn't you rather bomb in front of your fellow graduate students and faculty members than at a professional meeting? If there aren't adequate departmental opportunities, you can organize graduate student brown-bag lunches where you can try out your talk. If nothing else, find a friend or two so that you can practice; do ask for their honest feedback. If you are a young professional in a nonacademic setting, perhaps you can arrange a small Friday afternoon seminar for your CRM firm. Again, if nothing else, try to arrange for a few friends or colleagues to hear you give the talk aloud.

ANSWERING QUESTIONS

Most talks you give will probably generate questions. At conferences, questions are usually limited by the amount of time available or the structure of the session. Luncheon talks may be limited by the amount of time the audience has. Other talks may be open-ended for questions. You rarely know what kinds of questions will come out of a talk, but if you've given a talk often, you can usually predict some of them. If you are speaking on something controversial, be prepared for challenging questions, but don't be afraid of them. After all, you wouldn't be speaking on the topic if you didn't have evidence or some level of confidence. If you are speaking to a class or audience of non-specialists, the questions can be lots of fun (though eventually you will have a heckler who wants you to justify the cost of a dig or some such silliness). When your talk affords an opportunity for questions, remember these tips:

- Relax and listen carefully to the question.
- Allow questions during the talk itself if you are comfortable with that structure.
- Ask for clarification if you couldn't hear or didn't understand the question.
- If the question is long and complicated, try to break it up into parts.
- Some people ask questions that are actually statements, not requiring an answer, so don't worry about the answer, and don't be afraid to challenge them gently if they are off target.
- If someone tries to dominate or consistently challenge you, tell him or her you will be happy to deal with the question afterward, allowing others a chance to ask questions.
- Show respect for every person who asks a question, even if the question is poorly phrased. Really stupid questions are rare, but many are badly asked.
- If you don't know the answer, don't be afraid to say you don't know!

Don't be afraid to have some fun with questions. In fact, a general rule suggests that the more questions asked, the better your talk was. The more practice you get, the easier it is to handle questions. Believe it or not, you'll actually start to enjoy the tough ones.

GAINING EXPERIENCE: FINDING OPPORTUNITIES TO SPEAK

If you haven't done much public speaking, start small with your talks. Undergraduates can give papers and presentations at low-key meetings such as at a chapter or state meeting for an archaeological society or the anthropology section of the state Academy of Sciences at its annual meeting. These venues usually have supportive people, sometimes with classmates or people from the archaeology lab where you work. You'll be nervous but fine.

Then move up. Find a larger but friendly place to give a paper and gain experience. Some regional meetings such as the Plains Conference or the Midwest Archaeological Conference have reputations as being places where graduate students give their first professional papers. If you are confident, then why not go for one of the national meetings, such as those of the SAA or the American Anthropological Association? (See sidebar 8.1 for some tips on presenting at the SAA annual conference and sidebar 8.2 for pointers on using graphics.) You might consider teaming yourself with another student or one of your professors on a joint paper where you only give part of it and are supported by a more experienced speaker.

There are also lots of local places to get experience. Many service organizations are constantly in need of luncheon or banquet speakers.

8.1. HELPFUL HINTS ON PRESENTING A PAPER AT THE ANNUAL MEETING OF THE SOCIETY FOR AMERICAN ARCHAEOLOGY (1997, REVISED 1999)

A. Before the Meeting

1. Read your paper a number of times aloud to make sure it is within the 15 minute maximum. (Printed out in 12 pt. font, a 15-minute presentation consumes 5 double-spaced pages. No more than 14–15 slides should accompany your presentation.)
2. Once you are happy with your paper, do a "dress" rehearsal with slides for an audience. Make sure that you are still within the 15-minute time limit. Ask for criticism and revise if necessary. Ideally, your presentation should be 14 minutes long.
3. If your illustrations contain text or graphics, make sure that they can be read from the back of a large room. Captions on figures and graphs should be in a larger font than you would use for publication, i.e., 21 pt. or larger. A slide should only remain on the screen for as long as it is relevant to what you are saying. You can leave gaps in the slides so that the audience concentrates on you and not an irrelevant illustration.
4. Print your paper in a large font (e.g. 16 pt.), so that you can easily read it in dim light. Mark clearly on the paper when a slide is to be shown. (If you intend to discuss the illustration, make sure that you have included

The sandwich buffet and rubber chicken circuit for Lions Club, Optimists, and assorted other service groups is excellent. One way to put yourself on this circuit is to register with your university's speakers bureau if it has one. Schoolteachers also look for specialists to speak to their classes. If you have a local anthropology or natural history museum, you might inquire whether it needs docents or tour guides. This role can get monotonous after your fifteenth tour, but it will give your a chance to get comfortable in front of different sorts of groups and used to answering insightful and inane questions.

Giving these talks gives you an early start in returning something of our profession to the general public. The situation is not too difficult, and audiences truly are grateful to you. Speaking to the public or your peers is an important part of what an archaeologist does, but something that rarely comes automatically to anyone. To become a proficient speaker requires preparation and lots of practice.

---

the discussion in the timing of your paper. To be sure about this, write down what you want to say about the illustration.)
5. Put your slides in a tray to bring to the meeting (SAA does not supply slide trays). Keep your paper, illustrations, and presentation attire with you on the airplane. Don't allow your presentation to suffer just because your luggage was lost!
6. You will be asked by interested scholars for a copy of your paper. Prepare distribution copies, which might be copies of the paper from which your presentation was abstracted.

B. At the Meeting

1. As soon as you register, check the date and time of your paper. It may have changed from the schedule published in the preliminary program.
2. Make sure you know in which room you will be presenting your paper.
3. Visit the room before your symposium or session begins. Check that your slides will show up on the screen properly. Make sure you know how the projector controls work.
4. Arrive at your symposium 15 minutes early. Introduce yourself to the chair. Talk to the volunteer running the projector and lights. Explain what sort of lighting you would like.
5. Smile!

8.2. SOCIETY FOR AMERICAN ARCHAEOLOGY
GRAPHIC PRESENTATION TIPS

Dos

Use landscape orientation; screens at most presentation forums are designed for landscape format slides.

Enhance the readability of text by using LARGE (24 pt or better) font size.

Use lower and upper case letters (your eye has more "topographic" information to use than with block letters).

Use line weight, style, symbol, etc. to convey important information.

Keep it simple.

Choose fonts with serif.

Check for misspellings.

Maintain consistency in images, legends, colors, between slides.

Check graphic margins for slides; most graphics packages have a media format default set for the monitor or for paper.

Don'ts

Use profile [portrait] orientation; the tops or bottoms of slides will likely not be projected onto the screens.

Use small (less than 24 pt) font size; there is no excuse for "You probably can't read this, but. . . ."

Overuse of block letters; use them with the maximum size font on the graphic.

Use red and green on same slide (40% of men cannot distinguish between the two).

Use complex slides (break down ideas).

Use a sans serif font.

Substitute paper graphic for slide graphic; paper and slide presentations have different audiences and environments, and therefore require different designs.

Use 3D graphics when 2D will suffice.*

Presentation Tip

Slide Presentations: Figure on using one (1) slide per minute of presentation. The human mind cannot take in complex graphical Information at a rate faster than this.

Reference

Tufte, Edward R.
1983 *The Visual Display of Quantitative Information.* Graphics Press, Cheshire, CT.

*Edward Tufte and other experts in data communication argue that our eyes are very good at interpreting 3D forms and often do so in unintended ways. Therefore, the possibility of miscommunication is greater here. You should closely inspect 3D graphs; they are not what they seem.

DEVELOP SOME "CANNED" TALKS

If you spend time developing a good talk or lecture for the general public, or in some cases even a professional lecture, spend time to package or "can" it. Think carefully about what worked and what didn't, then make appropriate changes. Be certain to keep any slides together along with any other support materials you need. You'd be surprised how many people are just so relieved to have a talk done that they don't even think about the possibility of giving it again. They take the slides out of the slide tray and forget about the talk, and then they later regret that they didn't just leave the slides in the tray.

If the talk was successful and the slides are laboratory or office property, get permission to have them copied so that you have a set. The cost of a slide tray and copies of even thirty or so slides is rather small compared to the time you might need to reconstruct the whole talk. You can even package it to give to different kinds or levels of audiences. If the opportunity to give a lecture a second or third time comes up, you'll be very glad you canned the lecture.

POSTER PAPERS?
HOW TO DECIDE AND HOW TO DELIVER

For academic conferences, poster papers have almost become a necessity to allow the conference program committee to squeeze in several more presentations. Cynicism suggests that allowing for more presenters is the primary reason poster papers came into existence, but posters actually are a fine idea. The format is straightforward. With posters grouped in common themes (methodological, topical, geographic area, etc.), presenters come to a room set aside for the poster papers. The themes change on a schedule like that for regular paper sessions, but poster sessions usually allow at least an hour or more per session for posters to be viewed. The presenter sets up the poster and is available for questions as viewers come by. The posters appear on standard-size poster boards, often four by eight feet. Sometimes the boards will be on a table so that a presenter can lay out printed copies of the paper text, business cards, artifacts, and other objects as they need them. At other times, the poster boards are stand-alone, with no table space.

The benefits of poster papers are many. The most important benefits are that viewers can take more time to digest the paper and to ask questions. The fifteen minutes allowed for most oral papers is usually not enough time, and presenters tend to rush through their talk or simply read a written paper. Graphics are sometimes ill chosen for viewers, with almost no time allowed for them to digest the content. Unless the session is well planned to include questions and well chaired to keep things on schedule, there is almost never time for them until after the session ends. Poster sessions usually allow for more interchange between presenter and viewer. An added benefit for some is that there is no public performance and little need to worry about speaking in public. Most interaction is one-on-one or in very small groups.

Poster sessions do have some drawbacks. For some strange reason, many colleagues think that poster papers are less important than oral papers. Depending on your situation, you may need to forego posters for traditional presentation, especially if you are on a tenure and promotion track. You may not get as large an audience as traditional sessions provide. Much of this depends on where organizers place the poster session. At a recent Plains/Midwest joint conference, the posters were in the lobby area of the conference center, a terrific location through which everyone passed several times a day. At the recent American Anthropological Association (AAA) meeting, posters were right by the book exhibit area, also a good location. If posters get shoved into a room out of the way, it can be the kiss of death for having your work seen. The very best location may be near where foods for session breaks get placed. (Conference organizers, take note of all this!)

All the problems aside, with the poster paper format you may get a more attentive, interested audience able to spend more time looking at your work. You can answer individual questions and make useful contacts, usually in a less time-stressed setting. Although one-on-one contact is terrific, one drawback is that you will constantly repeat the same information as your audience changes.

HOW DO YOU PUT TOGETHER A GOOD POSTER PAPER?

What makes for a good poster paper? Just as with a traditional delivery, you first need a good paper! However, poster papers are essentially show-and-tell. Limit the amount of information directly on the poster. As with museum exhibits, your poster will compete for attention with other posters in the room. Thus, you cannot—no, *must not*—try to put the whole paper up on the poster. People usually won't spend the time reading all your text. A recent trend in poster sessions has been toward doing just that. Using Adobe PageMaker or some other layout program, many have taken to putting all their text and graphics on a single large, printed poster. These look really slick but are almost impossible to work your way through as a reader, especially if the poster session is crowded. Too many try to cram thirty pages of text onto a poster along with all their graphics. This utterly defeats the purpose of poster papers, whose strength is in providing conclusions and graphics. In terms of presenting information, some of the most effective posters look the least slick.

You can still be slick, but don't overwhelm the viewer with lots of complex text or graphics. One effective approach is to do the paper as a series of Power Point or other presentation software slides. With these, you can print out color pages and simply tack them on the board. They also allow for a great deal of flexibility with different poster board setups. Woe to the archaeologist who has gone to great effort and expense to print out a three-foot by four-foot color poster only to find that he doesn't have that much space or has a trifold poster setup! One paper by a senior archaeologist at the recent AAA was little more than photographs of people, artifacts, and excavations relating to the historical Colorado Coal Wars along with small amounts of text, all thumbtacked to the poster board. The poster was stark and minimalist, almost antislick, but still very effective. That archaeologist knew something special about how to do posters!

Posters work best with graphical information, so you should try to determine whether the core of your paper could best be presented graphically. Then present the core, not all the information and logic of the paper. Put up only key points of logic and conclusions as text, but emphasize the graphics. You can have copies of the whole paper with you or on the table to give to those who are really interested. If you have lots of statistics and charts that wouldn't show well in a traditional paper and whose presentation is vital to your conclusions, a poster may be ideal. On the other hand, if you have a largely theoretical paper, with few graphics, stick with traditional papers. If you have lots of artifact photos or actual artifacts to lay out, a poster is again ideal. If you have table space for actual artifacts, presenting them there can be effective. Sadly, however, if you display artifacts, you do need to worry about theft, even among colleagues. If you use them, be there with them all the time, and pay attention.

If you are showing a specialized computer program, such as a website or CD-ROM, poster sessions are also ideal, with one drawback. If you draw a crowd, no more than a few people at a time can see the screen. On the other hand, people can actually see or use the product with fewer constraints on time. However, you need to be sure there are electrical outlets, Internet connections, and access to other needed computer connections.

For a good poster paper, do the following:

1. Write a good paper.
2. Decide whether to do the poster. To do so:
    - Decide whether the information can be well presented graphically. If it would be difficult, do a traditional paper.
    - Find out from the organizers how long the posters will be up and where they will be shown at the conference.
3. Select four or so key points to emphasize on your poster if they can presented graphically.
4. Choose graphics wisely to demonstrate the points but also to catch the eye of viewers.
5. Find out ahead of time what the organizers will provide. This includes:
    - the size of poster space;
    - the nature of the poster board (a wall, a flat bulletin board, a tri-fold, table space);
    - pins, tape, and other accessories including computer capabilities.

    If you wish to bring your own setup, let the organizers know so they can tell whether it is feasible. Conference facilities may not allow it.
6. Organize your information within the constraints of the poster setup.
7. You may also wish to consider having business cards, a summary handout with contact information or website addresses where people can see the whole paper, and a few copies of the whole paper for those who might like more information or who wish to contact you.

If you are considering a poster paper, two articles in the *SAA Bulletin* have addressed how to construct good poster papers and offer a range of tips (Neiman 1994; Baxter 1996). Neiman provides some detailed how-to advice. Baxter's (1996:9,31) article is especially useful. She provides a list of criteria for judging poster paper quality at the

SAA meetings (see sidebar 8.3) developed by David Anderson, a former SAA Annual Meeting program chair and a judge for poster sessions at several SAA meeting. She also relates some tips from Alanah Woody who won the SAA Student Poster Award in 1994. Many of these are the same as discussed earlier but are worth repeating. She notes:

- Preparing a poster is an expensive venture. Matte board, photo enlargements, and other graphic materials can add up quickly!
- Have hard copies of a "real" publishable paper on hand to go with your poster, complete with chaining arguments, citations, and a bibliography. Separate bibliographies and business cards are also good ideas.
- Posters are not the same as paper. You do not have space to support every argument. Use your poster to spark enough interest so people will ask for a copy of your paper.
- Read how-to articles about creating an effective poster, but also add your own touches to make your poster unique and noteworthy.

Baxter makes another excellent suggestion. If you've gone to all the trouble and expense of preparing a poster, she suggests that you recycle it. In one sense, a poster is like a small museum exhibit. If the poster is well done, it's probably worth showing again. Perhaps it can

---

8.3. JUDGING A GOOD POSTER

Subject matter: A great site or topic makes a great poster!
Production values: Legible text, colorful artwork, and effective arrangement are essential for a poster.
Clarity: Is the technical argument well presented? Is the flow of information logical and easy to follow?
Absorbable level of detail: Do not try to squeeze in too much information by using small type or presenting too many graphics.
First impressions are critical: Effective posters have a "hook" either in the subject matter or in its presentation.
Cleverness and originality of presentation: Posters that are new or unique and effective in presenting information will stand out.
(From David Anderson, in Baxter 1996:31.)

be put up in your lab area, much as the Office of the State Archaeologist of Iowa does. If the subject matter is not too esoteric, the poster might go well as a temporary exhibit in a small museum on your campus, town, or county.

The AAA website (www.aaanet.org/mtgs/poster.htm) also provides a substantial set of suggestions for preparing poster papers. Although the materials give AAA-specific guidelines, the page asks excellent questions everyone giving a poster paper should consider. It also contains a short bibliography of materials from other fields about poster papers.

ELECTRONIC SESSIONS

A recent trend allowed by the World Wide Web is to put whole conference sessions on the Internet. As with poster papers, part of the reason for doing this is to allow for more papers in conferences, but the trend also has other benefits. One problem with traditionally delivered papers, and even poster papers, is that the presenters must summarize what they've done to meet time or space constraints. Likewise, viewers rarely have much time to consider what they've heard or seen. With electronic sessions, full papers with a wide range of illustrative materials can be put on the web well ahead of the conference. All session participants and possible attendees can read the papers and have time to think about them. At the actual conference, the papers are briefly summarized, and all the remaining time can be spent on discussing them among the participants and attendees. This approach may entail a more efficient use of time, but it has some drawbacks.

Participants do have to provide a full draft version of their paper ahead of time. If procrastinators are participating in the session, as they always seem to, papers may still come in late or in poor shape. If you are a procrastinator, don't offer to participate in an electronic session! Someone also has to be able to put the papers up on the web. This is really not big problem, but it does take time, and one does need to have access to web space.

Some participants are paranoid about having their work on the web, either because they feel embarrassed at others seeing their work or because they are worried about people stealing their ideas. Session materials can be password protected, but this may limit the audience. If you fit these definitions of paranoid, the electronic session is probably not for you! The conference organizers must also list the session as part of the regular conference sessions and provide the web address to possible attendees well ahead of time. If they don't, no one but participants will have a chance to preview the papers and may not wish to attend the session.

On the whole, the electronic approach is a new one, with organizations such as the SAA only just trying it out. Having organized my first such session for the 2001 SAA meetings, I can now make several observations about how it went. On the plus side, the discussion was fantastic. One audience member e-mailed me about how she enjoyed not being read or lectured to and having a real opportunity for substantive discussion. Many of the attendees participated, and when asked how they liked the session, they were uniform in praise of the format. The presenters agreed that the discussion was excellent and that their own anxieties about being there were dramatically lower than when giving a paper. On the minus side, when I asked the audience how many had read the papers ahead, only five had. When I asked how many even knew of the session before getting to the meetings, only six said they had. Clearly, the format is useful, but there are lots of bugs to work out—and if you are a session organizer, you'll find there are sometimes more than you'd like.

ORGANIZING AND CHAIRING A SESSION

Sometime in your career you will find a compelling subject on which you are doing research and know that others are doing similar research. You may also be part of a collaborative CRM project team, perhaps even its director, and will realize that your team needs to present its findings. On a sad occasion—when a colleague or mentor has died, for example—you may wish to assemble former students and colleagues to memorialize the individual. For a more joyful occasion, you may wish to assemble colleagues and students to honor the contributions and retirement of a colleague or mentor in a session that sometimes results in a written collection called a *festschrift*. Whatever the situation, you will need to organize the colleagues around a central theme and propose the theme and session for a meeting that eventually will get turned into an edited volume.

Doing so may be no small feat, or it may be little effort at all, mostly depending on your colleagues. Usually colleagues feel honored to be invited to present a paper in a session, although they may eventually complain about deadlines or your heavy-handed nagging. Memorials and festschrifts may have no central theme except the person around whom the session gets organized. Thematic sessions often require more organization.

Planning for any session needs to begin well ahead of the meeting dates, in fact, well ahead of the submission deadlines for meetings. Most regional meetings have deadlines closer to the actual meetings than do national meetings, but in both cases submission of a complete session package may be eight to twelve months ahead of the meetings. This allows organizers time to arrange sessions and rooms. Good session planning might allow at least two to three months ahead of the deadline, but even that may be pushing it.

Once you have a theme for a meeting, you should choose an appropriate venue. Festschrift and memorial sessions are often more appropriate for regional or small national meetings unless the honoree is/was a nationally recognized figure in archaeology. In some ways, so are sessions surrounding a single site unless the site has national or international significance. Thematic sessions will do for most meetings. If you have an idea for a thematic meeting but don't always know who might be working on the topic, you may wish to include a note about the planned session in regional or national newsletters; some even have "cooperation" columns. You can also send out e-mails to colleagues you know who are working on the topic or who may have students or other colleagues who are doing so. With contacts made, you can begin to assemble the session.

Most organizations have very strict structures for submission of sessions. Be sure to get all the necessary forms for the participants and distribute them in plenty of time for submission by the deadlines. Forms usually come in printed flyers mailed out to organization members or are now available on many organizations' websites. National organizations usually demand submission of an entire session packet. The packet usually consists of a session abstract, the forms of each of the participants, and, more and more often, advanced registration materials and fees for all session participants. All this may sound easy, but getting participants to get materials in on time, submitting their abstracts in the proper form (assuming there is a standard form), or paying their registration fees in advance often takes more nagging than you might imagine!

As you prepare to submit the session, read the abstracts over carefully. Try to find commonalties that will allow you to order the papers for the session. Look for common ideas that may link papers, and then play with different orderings. Oddly enough, this is almost like doing the lineup for a baseball game. You need something to gain the audience's attention early, so you may want to put "bigger names" near the start, followed by the more esoteric or weaker papers in the middle, and then, near the end, the more theoretical or synthetic papers, even another "bigger name." This can be followed by discussants who are chosen for their expertise or ability to summarize and critique.

After you submit the session, and assuming the organization accepts it, you need to inform all session members that the session is moving ahead. You will need to set a deadline for submission of the session papers to you, especially if you have one or more discussants, individuals who critique the papers or provide a synthesis of the session. These individuals need to see the papers ahead of time to make appropriate comments. In truth, all the papers making it to the discussant in time to digest them before the session is a rarity, and one of great joy for the discussant. This accomplishment often takes genuine harassment of participants, but it will make for a better session. Having been a discussant many times, I know how difficult it is to listen to a paper and have to invent sage comments just from listening to a hastily prepared and badly delivered paper! One of my more jovial colleagues has even taken to giving cute prizes for people who get papers in on time and for those who don't. Believe me, you don't want one of his prizes for those who don't get papers in on time.

A bit before the session, you would be wise to contact the presenters to see whether any unusual problems or needs have popped up. Even though abstract and session submission forms ask what presentation equipment, such as slide projectors, are needed, some presenters insist on special equipment, such as a computer with Internet connections and a projector. These may or may not be possible for the organizations, but you can do the presenter the courtesy of checking. You should also tell the presenters to assemble at a particular time and place before the session to introduce themselves to you. If this isn't possible, try to leave them your hotel number at the information center many conferences have, and ask them at least to call to say they have arrived. Nothing is more nerve-wracking than waiting to see whether your presenters have arrived when the session is about to start. If you can't meet before the session, ask them to be certain they introduce themselves to you before the session starts, particularly if you don't know them well.

If you organized the session, plan on the first paper being your own, or plan for a short introduction of the session topic to an audience that may not be fully aware of what the session is about. During the session, your job will be to introduce the speakers, so you may wish to ask them to submit a very brief biography for your introduction.

The biggest problem at sessions is time. Many speakers do very well at staying within allotted times, but others are a disaster! You certainly need to bring a watch or timer of some sort. You may wish to instruct the speakers on how you will notify them of the time. Some session chairs hold up small sheets of paper with the amount of time left: "ten minutes," "five minutes," "one minute," "stop." Be judicious in the use of such papers because they can interrupt a speaker who is well rehearsed and organized to the extent that the talk takes more time should she lose her place. On the other hand, unless you have had a speaker drop out, you may need to be strict about cutting off a speaker. Don't be embarrassed about it; just do it! A speaker going over time limits takes time from other speakers, limits any time for discussion, and may throw off a session following yours, if scheduling is tight.

FROM GIVING A PAPER TO DISCUSSING A SESSION

If you are a session organizer, you may wish to find an individual who can help you tie the content of all the papers together. If your session is a memorial or festschrift, you may wish to have someone present an overview of the honoree's contributions. This should usually be someone who has been a colleague or a former student. You might ask for some reminiscences as well as a few comments about each paper. For a thematic or site-centered session, you should choose discussants who are extremely knowledgeable about the topic, whose work is respected, and who are capable of synthesizing information quickly.

To be asked to be a discussant is both an honor and a chore. If you agree to be a discussant, your own sins will come back to haunt you! You'll curse the presenters who never get papers to you and recall every time you never got your paper to the discussant ahead of time. There are several approaches to being a discussant. If you are a rare individual who can "wing it" upon hearing a group of papers, you can simply listen and then get up and give cogent comments. Most of us are not that adept, so a better approach is to take what papers have come in, read them a couple of times, and then devise a general set of comments into which additional papers will probably fit.

As you sit in the session, listen for changes that speakers who submitted advanced papers might have made. Listen carefully to those you've not read. Take notes on each, but not too many to pull together. As you listen to the last paper, pull things together, and pray that if there is more than one discussant, you are not the first! Being last can have drawbacks, however, in that others may say all the things you wanted to say. In any event, thank the session organizers and paper presenters briefly; then open with general comments, followed by something about each paper. You don't have much time, so you must hit only key points, both good and bad, about each. You do not necessarily need to address each paper in order, though doing so may be easier. You can end with more general comments about the subject or any general issues you'd like to raise regarding the subject. This can serve as a spark to urge further research or address particular questions or concerns about the subject.

As a discussant, you have a certain degree of power. If everyone in the group is senior and the subject is of import to archaeology, you may have been chosen because the organizer(s) wanted you to be controversial. Senior people are very capable of defending themselves, so you need not worry about offending. However, in my own experience, I can't say that I've heard a paper yet without some redemptive qualities. If the participants are early in their careers or even in the middle, what you say can have a profound influence on their self-image. You need to be honest in your criticisms, and you don't need to give false praise, but you should avoid attacking the speaker. Some individuals let power and reputation go to their heads, even "counting coup" on graduate students when discussing a paper. No one likes to be humiliated!

Such behavior is unfortunate, at best. If you can find good qualities in the paper, try to balance gentle criticism with faint praise, or don't say anything. If you find serious fault, you can approach the individual after the session, buy him a cup of coffee or a beer, and offer more direct criticism and even assistance in fixing problems.

Another problem for discussants is to heap praise on some papers while saying nothing about others, which sends a message that the latter are not worth your time. If you can, try to say something about each paper unless it is so bad that it needs more private discussion. If there is more than one discussant, you might arrange to chat with your counterparts ahead of time and agree to concentrate on certain papers so each gets some level of attention. If you must use being a discussant as a bully pulpit, steer clear of comments that can be seen as ad hominem attacks or as self-serving. The true role of a discussant is to help improve the papers by providing feedback with suggestions for research, clarifying logic or arguments, and providing a synthesis of key ideas from the session. In a sense, it is "instant" peer review, aimed at improving the product.

CONCLUSION

Papers presented at conferences are an important way of disseminating information to colleagues, but the audience is relatively small, even compared to the number of readers of state or regional journals. Giving a conference paper should not be considered an end in itself, unless of course, you are just trying to pry travel money out of the hands of your company's CEO or your dean! Even then, conferences are so cluttered with papers that if getting travel money is the only reason to give a paper, you should do yourself and colleagues a favor and at least make it a poster or electronic symposium presentation.

The real reasons for conference papers are to let colleagues know what you are doing, to exchange ideas on research, and to get feedback on your work. Putting the paper into writing should be a more important goal than giving conference papers. Your work can be read by more people and will have a greater chance of having an impact on the profession. This is the subject of chapter 9.

9
FROM PRESENTED TO PRINTED

Archaeologists have to do lots of writing. We have our day-to-day chores, from the e-mails and memos we send colleagues to proposals to agencies we hope will fund our next project. Some of these find their way into our presentations of the past, so we need to spend time doing them well. You may not worry about an e-mail sent to a colleague explaining your views about Woodland tradition taxonomy, but you might find your colleague asking for permission to quote your statement in an article. Your outline of a research design in a proposal may end up as part of a chapter in a CRM report.

The materials in this chapter, however, are geared toward the kinds of formal presentations of the past that might end up in print. Because much of this series is aimed at CRM practitioners, we'll start with CRM reports, which are specific to projects prepared for agencies, the major product of the vast majority of research money spent on archaeology. They are often heavily peer reviewed but are called a "gray literature" because they usually don't have a wide distribution. Relatively short works, journal articles, and newsletters are periodical or serial literature produced by professional organizations or other publishers. Books and monographs are longer treatments of a subject produced by a variety of publishers and may be collections of shorter chapters. These printed forms are linked by common issues of peer review and editing, which we'll consider at some length.

CHAPTER 9
CRM REPORTS

Cultural resource management reports have become the core of data production in archaeology. Many projects are not CRM based, but the majority of basic fieldwork no longer comes from private or academic research programs. CRM reports are so variable in form that it is difficult to discuss them in a general sense, but some issues are common in the preparation of most reports.

The initial element of concern is that the report should meet the requirements of the organization contracting with you for the work. After all, it's paying for the work. Some agencies, especially if they are federal, may have particular requirements in terms of formats and perhaps even style guides. You should be certain before you start any writing that you know and understand these requirements. Doing so may save you lots of rewriting time! Don't waste time trying to convince a contracting agency to do things your way instead of theirs, unless what they ask violates professional and personal ethics. Some things, such as the spelling of *archeology* without the *a*, instead of *archaeology* with it, are just not worth worrying about!

As important as agency contract requirements are any standards for reports developed by organizations such as the State Historic Preservation Office (SHPO), state archaeologist, or professional associations. These may be fairly detailed, even to the kind and size of maps or illustrations to be included. Some SHPOs allow letter reports for negative findings so long as there is enough detail for them to make an assessment of the quality of work. Others may require a more complete report, even when you find nothing.

The main elements of any CRM report these:

- A statement of the project area, in terms of both planned land disturbance and archaeological coverage
- A section on the culture history of the project area in terms of what might be expected concerning site types and cultures and how what might be found is connected to the state plan for cultural resources
- Field methods used
- A section on the geomorphology and soils of an area and how your project assessed possible buried sites
- Sites and materials recovered
- An assessment of the archaeological materials in terms of significance regarding the National Register of Historic Places

- Any recommendations you have regarding cultural resources within the project area

Besides meeting these elements of content or format, you should also pay careful attention to the scope of work under which you were acting, as well as any proposal you submitted regarding the project. Be certain that you address every important aspect of both documents. Doing so will provide both the contracting agency and the SHPO with ways of seeing that the required work got done.

THE PROBLEM OF BOILERPLATES

Boilerplates are documents or parts of documents prepared ahead of time and then worked into a report. The idea is to save you from constantly having to rewrite the same part of a document many times. For example, many organizations have prepared boilerplate sections about the culture history of an area or about the geomorphology of a region. Boilerplates are meant to be altered to meet the unique situation of each project, but overuse without alteration becomes report padding. Try to make the boilerplate sensitive to any unique situations in an area.

FROM DRAFT TO FINAL REPORT

If the CRM project is small, the only readers may be the contracting agency and the SHPO's review and compliance department. Longer reports, usually from testing and mitigation projects, get read by more people. Many federal agencies tend to have deadlines for various draft versions of the report, so when it comes to differences of opinion about the content and format of a report, most elements are negotiable. If you've maintained good relationships with the contracting officer, agency, and SHPO, minor problems can quickly be dealt with. More difficult matters, such as whether your project actually met the scope or your proposal, may not be easily handled, but if you've kept the various parties informed, even they can be overcome.

When it comes to reports, do your best to meet the needs of all parties, particularly if you hope to have other contracts with them in the future. Certainly, someone does read all your reports, which may include many of your peers and competitors. Make suggested changes where they are reasonable, but be sure the contracting agency knows why you didn't make other changes. In the highly stressful world of CRM, you'll live longer if you don't get bent out about every little thing!

GETTING YOUR CRM REPORT NOTICED

Getting your report noticed beyond an agency or region is often difficult. However, you should make an effort to be certain some notice of your report is available beyond just a line on your résumé or vita. Some state archaeological society newsletters or journals contain sections set aside specifically to list recent CRM reports and abstracts. In the United States, the Reports module of the National Park Service's National Archeological Database

> is an expanded bibliographic inventory of approximately 240,000 reports on archeological investigation and planning, mostly of limited circulation. This 'gray literature' represents a large portion of the primary information available on archeological sites in the U.S. NADB-Reports can be searched by state, county, worktype, cultural affiliation, keyword, material, year of publication, title, and author. (National Park Service 2000)

The database is profoundly useful and an excellent way of being sure people know about CRM projects in each state. Reports from federal agencies increasingly are being made available on line in full text using portable document files (.pdf).

JOURNAL ARTICLES AND BOOK CHAPTERS

Professional journal articles and book chapters are the heart and soul of the theoretical side of both CRM and academic archaeology. Producing them is more difficult than CRM reports because they are less "formulaic," though the deadlines usually are more flexible and there may be fewer peer reviewers.

JOURNAL ARTICLES

Journals are widely variable in topic and scope within archaeology, so your initial task is finding a journal appropriate for your manuscript. If you are near a university campus with a good library, you'll find it worth your time to visit regularly to go through the journals to which the library subscribes, or you can look for the journal on the web. (See the appendix for a listing of the URLs for some archaeology journals with websites.)

STATE JOURNALS AND NEWSLETTERS

Certainly, if your manuscript is mostly a descriptive report on an archaeological site or artifact with limited interest outside the area, the best fit may be in one of the many good local or state archaeological society journals. If you've got a short, localized contribution, perhaps the newsletter of the society would be even better.

State and local journals or newsletters contain a lot of terrific archaeology, often of more use to scholars than articles in major journals. They also usually have a need for material, and time from submission to publication is often substantially less than that for major journals. One reason is that they are rarely peer reviewed.

REGIONAL JOURNALS

Regional journals cover a wide range of material, from site reports to methodological to theoretical as they relate specifically to an area or region. For archaeology in the United States, these regions are often roughly congruent with culture areas. In existence since the 1950s, *Plains Anthropologist,* for example, publishes archaeology and some Plains ethnography. Articles in regional journals tend to be peer reviewed; most usually deal with subjects that have utility and implications for the broader region rather than more limited locales. Regional journals work well ahead of deadlines, so you can probably plan on a minimum of six to eight months before your article comes out.

TOPICAL JOURNALS

Topical journals tend to consider issues surrounding a subject matter defined by a methodological or theoretical concern that crosscuts regional or national boundaries. This might be something like field methods, as is the case in the *Journal of Field Archaeology,* or issues surrounding archaeological and heritage management in the public sphere, as are considered in *Public Archaeology*.

If you have a manuscript that is topical, be sure that the subject matter fits the types of materials covered by the journal. If in doubt, you might send an inquiry to the editor asking whether he or she will look at the manuscript or an abstract before sending it out for peer review.

"BIG-TIME" JOURNALS

Big-time journals are those that tend to be recognized by most archaeologists as the major and prestigious journals in the field. Most are published by professional organizations. For example, most archaeologists in the Americas and the rest of the world recognize *American Antiquity* and *Latin American Antiquity*, published by the Society for American Archaeology, as key journals. On a world basis, most regard *Antiquity* as important. Many also acknowledge general anthropology journals such as *American Anthropologist* or *Current Anthropology* as publishers of important archaeological articles, although they publish articles in all subfields of anthropology.

As you might imagine, these journals are the most difficult ones in which to publish, not entirely because they always contain the best pieces but because of the competition. Quality is generally better than many journals; the number of submissions is high. Although some attention is given to fair coverage of archaeological time periods or regions, articles with theoretical, methodological, and topical content useful to a wider audience tend to be published before others. In truth, unless your article has broad implications for the field, you may find a wider readership for it in a topical or regional journal. However, if you are an academic, worried about earning promotion and tenure, there is enough elitism in most institutions to warrant the effort of submitting articles to the more prestigious journals. Be prepared for turnaround time from first submission to publication of a year or more.

EDITED VOLUMES AND BOOK CHAPTERS

Edited volumes contain a series of papers or chapters compiled around a certain theme. Topics are as variable as all the possible ideas or themes in archaeology. Edited volumes are sometimes the spawn of a conference session from which all or the best papers are drawn; sometimes extra, solicited papers get added to the mix.

Whatever the origin, writing book chapters is much like writing journal articles. Book editors have a difficult job, especially when it comes to writing style. Edited volumes, just like journals, are notoriously uneven in the content of articles and writing. Good editors generally try to make your piece fit the whole. If an editor suggests changes in your writing in order to bring your chapter into the volume's writing style, just do it. The only time you should fight this is when the suggested change significantly alters your meaning.

Book chapters are carefully reviewed by the volume editor(s). In addition, sometimes the publisher will send the whole volume out for review to anonymous reviewers.

NOTES, REPORTS, AND NEWSLETTERS

Most organizations have some kind of newsletter for keeping members informed. These contain everything from editorials, letters to the editor, committee reports, obituaries, job ads, conference notices, cooperation columns, and shorter articles of a narrow focus. Some journals publish notes or reports, which tend to be very focused, and short articles, usually fewer than five published pages, describing a new discovery or discussing a new methodological or interpretive technique. Turnaround time on these is usually more rapid than for articles, and some journals do not submit them to extensive peer review, if they are reviewed outside the editorial staff at all. For younger scholars, submitting pieces as notes or reports may be less stressful than trying to publish articles in the prestigious journals right away.

BOOK REVIEWS

Most journals also publish book reviews. Reviews help readers allocate their time (and money) by giving clues as to what may or may not be a useful volume. For younger scholars, being contacted to write a review is relatively rare, so don't wait for it to happen. Some journals ask potential reviewers to identify themselves. Another approach is a bit more aggressive, but if you know of a book that has just been published, you might contact the review editor and ask to do the review.

Book reviews may be a good place for younger scholars to start publishing. However, book reviews are not as easy to write as some might think because of limited space. Consider yourself fortunate if you are allowed more than two hundred to four hundred words.

As a reviewer, you generally get to keep the book after the review. If you agree to do a review, be certain to do it! When I served as book review editor for *Plains Anthropologist,* about a quarter of the books I sent out never got reviewed, even after several reminders. Here are some other matters to consider:

- Don't agree to do a review unless you can be completely fair.
- Do the review by the deadline, or make arrangements for returning the book or for an extension.
- Stay close to the word limit.
- Don't focus on petty issues of production unless they truly detract from the book's content.
- Pay special attention to the arguments made by the book. Do they make sense?
- Don't simply repeat the dust jacket blurbs that promote the book. Analyze critically.
- Remember that just about every book has both good and bad points.
- Remember that what goes around, comes around. The author whose book you review today may review yours tomorrow.

One variation is the review essay, in which a reviewer examines, compares, and contrasts a number of books centering on a theme, published at about the same time. These are sometimes more difficult to write but do provide for interesting comparisons and are almost articles in themselves. They allow a reviewer to include his or her own analytical constructs in addition to reviewing the books. In recent years, more journals are including reviews of videos, computer programs, museum exhibits, and multimedia. If you are asked or request to do such a review, remember that these media are different from books and may require different expertise.

BOOKS AND MONOGRAPHS

Monographs tend to be specialized treatises on very narrow subjects, such as a lengthy discussion of a particular class of objects or an extended version of a site report. Books tend to be more general in coverage, perhaps an analysis of a particular theoretical approach or theme, using examples from several archaeological sites or regions.

A monograph often represents the culmination of a relatively narrow analytical project resulting from something such as a grant or contract, while a book may begin out of your interest in a particular research topic, more comprehensive and synthetic than most monographs tend to be. Monographs also tend to be published by more specialized presses, such as in a museum or university press series. Most monographs also tend to have a relatively smaller number of pages than books. The press run (the number of copies printed) also tends to be small, usually a few hundred. If you intend to publish a monograph, you'll probably need to have the full manuscript completed before you contact anyone about publishing the work. University presses and museums are likely monograph publishers, because they are not in business to make a profit. Some commercial scientific publishers also will have monograph series.

Books tend to be longer and have larger print runs, but this depends mostly on the publisher and the topic of the volume. Books often begin life as a prospectus rather than a finished manuscript. A *prospectus* is essentially a proposal to a publisher to produce a manuscript on a particular topic. The format of a prospectus is fairly standard across the industry, but with enough variation that you may wish to write a publisher for its requirements. Most ask for these items:

- A description of the volume
- An assessment of the intended audience
- An assessment of the "competition"—that is, other books similar to the one you are proposing
- A table of contents
- A completion schedule

Some publishers also ask for a writing sample, especially if the book is to be single authored. They may request names of possible reviewers, too.

The press will usually send the prospectus to reviewers to ask their opinion, and depending on what they write, the press may offer a contract. The process may take some time, so be prepared for a wait of several months depending on the reviewers and the publisher's decision-making process.

Pick your possible publisher wisely. Find out which publishers are publishing in your area or topic. This may be as easy as looking on your bookshelf. Looking online or at a few catalogs from a publisher may give you more information, or you might also ask a well-published colleague. You should also consider the amount of advertising a publisher does to market books.

If you've done much publishing, you've probably already made lots of mistakes. For example, when it came to publishing my dissertation a quarter-century ago, I accepted the first offer that came along to publish it, in a solid but not well-recognized or well-marketed state series. After I heard that the monograph was being used in several classes and it had gotten some good reviews, I realized that I probably could have gotten it published in the "hot" New Archaeology press of that era. That might have done more for the early years of my career. Most recently, a colleague and I did a trade book (a book for general audiences). The book is beautiful, with lots of color illustrations, and inexpensive. We accepted a generous one-time payment up front. However, after the U.S. publisher let it go out of print after selling ten thousand copies, another U.S. publisher picked it up, marketed it better, and in just over six months sold thousands more. We've also learned that our book was translated into other languages. We would have done much better if we had bargained a bit on contractual matters. Most academics are so eager to see their names in print anywhere that they often don't even consider the question of royalties, republication rights, and the like.

Monographs usually don't pay and are often financial losers for the publishers. Books and monographs published by university presses and commercial scholarly presses will often pay royalties, but the chances of making much money are slight. The structure is widely variable, usually a percentage of sales, sometimes incrementally increased as sales move up. Trade books are a different ball game altogether, with substantially higher sales numbers and more generous royalties.

If a publisher offers you a contract, do your best to honor its terms. If you suspect that there will be problems, make a point of keeping in regular contact with your publisher.

THE MADDENING TRIVIA: PROOF PAGES AND INDEXES

As you work through the book publication process, you'll face two issues that can make you crazy: proofs and indexes.

The proofs are the final step of your work as an author. They are actual mock-ups of the pages of your book or article. The publisher sends them to you for a final look before the manuscript gets printed.

Read proofs carefully! They are ultimately your responsibility. More embarrassing screw-ups happen here than anywhere else in the publishing process. Sometimes you do a diligent job proofreading the text, but forget the big stuff, such as the cover. In my own case, my middle initial (*J*) has been *A* once (my own fault) and *W* (my coeditor's fault) on covers.

There are lots of tried and true techniques for reading proofs. Some suggest reading the text backward; other say that they need to be read aloud to another person who follows with the last submitted version before proofs. Why these techniques? You are too close to the material and know what it's supposed to say. If you silently read the proofs, you tend to fill in with what is correct instead of recognizing errors.

Finally, by the time you are at the proof stage, the only changes you can make are minor. Most publishers allow minor changes of a word or two, but anything that changes pagination, especially for journals, may not be allowed or may be allowed but with a charge. Most publishers only give a few days for reading proofs.

Most book contracts require you as author or editor to provide an index of the volume based on the page proofs. Resist the temptation to rely solely on electronic indexing using your word processor files. They will index proper nouns but not necessarily choose the words you want or order them as you wish. Go through a manuscript and enter the terms and subterms alphabetically on your word processor, listing the page numbers as you go. If you are working with several authors and time allows, you can also ask them to highlight the terms in the proofs that they find to be important.

Some publishers will provide indexes done by professional indexers, but because they don't know what's important in your field, they may not do as good a job as you. Publishers will usually charge you for a professionally prepared index. Remember how often you use indexes, and you'll know how important the task is.

PUBLISHING AND CAREER DEVELOPMENT

If you are a young or soon-to-be professional archaeologist, you'll get lots of advice from people on how to develop your publishing career: Only books really count; publish only journal articles; publish both, but only with the best publishers or journals; avoid popular media; don't do edited volumes; and so on. The truth is that in many cases, opportunity is what guides you, not your own plan. Let's say you've just finished a dissertation, just read by a big name in the field who invites you to submit a manuscript for a chapter in a book she's editing. Do you turn her down because your mentor told you to do only books? Frankly, to do so would be nuts! Opportunity doesn't knock all that often, so sometimes you have to allow for the serendipitous.

One good way is to devise a "lifetime" research plan. Look realistically at your career goals and your interests, then figure out how to reach them. Do you hope to find an academic job at a top-flight university by the time you are thirty-five? Do you like CRM enough to stay in it your whole career as the owner of your own firm? Your publications need to be geared to allow you to accomplish your research needs and your career ambitions, but it won't happen overnight. Plan a realistic and relatively loose schedule, allowing for some changes in goals as your interests change and opportunities shift, but still fulfill your ethical obligation to be accountable to the public.

If you are disciplined, you can really do a lot of writing. If you write only five hundred finished words (about two double-spaced pages) daily, that's seven hundred manuscript pages a year, the equivalent of two to three books or twenty articles every year! That will make you one of the most published people in archaeology. Be even half that productive, and there will be time to meet most needs for individual research plans, obligations to the public, and career ambitions, as well as to have a little fun! To make this happen, you need to learn to do the following:

1. Do research that will be usable for several works. You may need to finish a CRM report right now, but the research about the culture history of the river drainage in which you are working can be used for your future book on the archaeology of your state.
2. Consider multiple outlets for the same piece of work. The people who won't read that methodological article on Clovis point manufacture in *American Antiquity* might be interested in that unique point from central South Dakota you discuss in *South Dakota Archaeology*.
3. Mix the level of publication types. A good mix of professional and popular articles and varying lengths of publications make you well rounded and someone meeting both professional and public obligations.
4. Start small and work up. You don't need to hit the big-time journals right away. Graduate students publishing in the state journal or newsletter will be well ahead of most of their cohort. As a

young professional, you can work toward publishing in journals at the national level as well as writing books.
5. Don't be afraid to collaborate. As an archaeologist, you often work as a member of a team, so why not publish as a team member? Pick reliable collaborators.

Lots of people can help you with your career—mentors, colleagues, friends, and spouses—so don't be afraid to ask them what they think or to work with you on some project. Another who will sometimes seem like an enemy, but who is really one of your best friends in the whole process, is the editor.

EDITING AND THE PUBLICATION PROCESS

There always seems to be a level of tension between writers and editors, and it's probably good that there is. In college and graduate school, you wrote papers at deadlines, allowing little time for feedback. Most of us didn't face an editor until we submitted a manuscript for publication and didn't quite know how to react to real criticism.

In fact, many of us see criticism of our work as bad, not something that can help us improve. Few actually *like* to see their work criticized. At the same time, constructive criticism is extremely useful, especially for writing. As you write, you are so close to the process and product that you have difficulty seeing flaws of structure, poor word choice, and even typographical errors. As an exercise, find an old paper you've written for a class, an old CRM report, or something else you've written that you were pleased with at the time. Read it. After you are finished, try to honestly assess your feelings about it. Unless you have extraordinary writing skill (or an extremely strong ego!), you'll probably not like what you read. Time away from the piece has given you distance and a better perspective.

SELF-EDITING

As I age, I find that I'm my own worst editor. That has its strengths but sometimes can be debilitating. A major strength is that I don't send any of my writing out of my office until I've rewritten it several times. I've revised the manuscript for this book four or five times before I let my series coeditor, Bill Green, even glance at it. Using Word, I'll Flesch-

Kincaid Index (a readability index with my word processor) a few paragraphs at random to see whether I'm hitting about the right audience level (grade level 8.5 for this paragraph, by the way). I've learned to use my word processor's grammar checker, and you should, too. I'll throw out entire sections because they just plain suck (but keep them in a discard file in case I can use a revised version of what I throw away). Most of the time, I feel relatively confident and just make minor changes, such as moving a paragraph around, changing words here and there, and sometimes expanding an idea.

For self-editing, at least give yourself a few days before you reread something editorially. Your best editing happens after some time has elapsed. Another good piece of advice is to read aloud what you've written. If it sounds stilted or if sentence structure sounds convoluted, it probably is. Read *exactly* what is written on the page or screen because your mind sometimes rewrites what you see on paper. Editing on-screen is difficult. To save paper, do an editorial read on screen, but eventually print a rough copy and edit that. You'll often find more problems.

COLLEAGUES AND FRIENDS AS EDITORS

Once you've done what you can to clean up your work, you need to move on to the hard part: letting someone else read what you've written. Most people find this hard to do. Letting someone else read what you've done brings a fear of criticism, but it's really more than that. Writing is a display of your logical and reasoning skills, your scholarly abilities, and, at some level, your "soul." Who wouldn't be afraid?

Get over the fear by having classmates, friends, or close colleagues read your work before putting it into the peer review system. Be aware that people close to you can be harsh critics, so you might prepare yourself for that. In my Ph.D. grad school classes, even though all of us in seminars were pretty good friends, we used to delight in criticizing each other's work. We joked about razor-sharp Marshalltown trowels flashing in the light like sacrificial knives ready to rip the heart out of each other's papers over typos and poor word choices. If you understand that this is not meant to be hurtful, you can deal with it pretty easily and get used to criticism meant as it should be: a way to improve your thinking and writing. If you're uncomfortable with criticism, start with friends because the peer review system *can* be worse.

PEER REVIEW

Peer review—examination of your work by colleagues who know something about your subject area—can be one of the most traumatic parts of professional writing, even for seasoned professionals with lots of publications. Most professional journals and most reputable academic publishers send your manuscript out to a "jury" of your peers. In the CRM world, an agency might send your report out to several colleagues from the area to see whether your colleagues think you did the project and report at professional standards.

The trauma of peer review is exacerbated by the fact that peer review is often anonymous; that is, the reviewers don't get identified unless they choose to tell you who they are. In theory, the idea of anonymity is important because it allows reviewers to really write what they think about a manuscript, without fear of retribution from the author.

If all parties act professionally, anonymous peer review serves an important purpose by assuring that what gets published is of high quality. There is more to peer review than publish/don't publish recommendations. Most journals or publishers have a set series of questions, which, if answered faithfully by a reviewer, can help an author solve problems of logic, writing, and scholarship, as well as to suggest additional questions or interpretations. The problems for authors arise when reviewers don't do a good job or allow their personal biases to overshadow their professional responsibility to the author.

Some people take delight in skewering colleagues over their work, a tendency anonymous peer review makes easy. I make a point of signing all my reviews of the work of others, especially when the author is someone who is not my favorite person. I do this because I believe that I need to be responsible for my opinions. Manuscripts that don't have some redeeming quality are rare. If you can't say something nice, at least offer some suggestions on how to improve the work.

Sometimes there are also legitimate differences of opinion, and sometimes the result can be almost amusing, even if frustrating, when, for example, two reviewers like or dislike exactly the opposite papers in an edited volume or recommend changes to a manuscript that would do exactly the opposite of what the other suggests. Tom

Biolsi and I coedited a series of conference papers, ultimately published in 1997 by the University of Arizona Press (UAP) as *Indians and Anthropologists: Vine Deloria, Jr. and the Critique of Anthropology*. We sent the manuscript to another press first, which used two peer reviewers. Both reviews came back negative, each really disliking *exactly* the opposite group of papers. None of them suggested any changes except dumping the papers they didn't like. On that basis, the press rejected the book. We then sent the manuscript to UAP along with the reviews from the other press. With the earlier reviews in hand, UAP sent the collection out to two more reviewers, and virtually the same thing happened, but with this, UAP realized that the reason was not that the papers were bad but that the subject was controversial. They have reprinted the book after the first run sold out. For us, the process was traumatic and frustrating, but not without humor.

FINAL STEPS IN THE PROCESS

The last editorial barriers involve the editor, copyeditor, publisher, or contracting agency. An editor takes peer review comments, reads the work, and then may demand, ask for, or negotiate any substantive changes. After you've agreed on changes and resubmitted the manuscript, a copyeditor usually goes over it in detail looking for contradictions, missing elements, and spelling, typographical, grammatical, or stylistic errors. The copyeditor will be certain that your manuscript meets their style guide, to which you already should have paid attention. Sometimes, especially if a manuscript has few changes to make, both the editor and copyeditor do their work before asking you to resubmit a clean version. After making these changes and resubmitting, your work is nearly done. You may get page proofs to examine one last time for errors or to incorporate very small additional changes you'd like to make. Books may involve the added step of indexing.

These stages may seem straightforward, but there can be problems. The greatest danger comes when you get reviewer and editorial comments back and simply can't face them. You've already spent so much time on a manuscript and are so close to it that making changes is just about the last thing you want to do. If you don't get to the changes quickly, you might not get to them at all. Knowing this, many editors give you time limits at various stages of the process. I rather like the maxim a colleague uses to motivate himself: "If I don't get on top of it, it gets on top of me." Get the work done as soon as possible.

The role of the editor in the publishing process is crucial. His job is to present a work to a reader that is clear, consistent, and clean. You will rarely run into an editor who wants to change your work dramatically, and good editors want to keep your style as much as possible. When Mitch Allen at AltaMira Press worked with Bill Green and me on the Archaeologist's Toolkit series, we agreed beforehand that we needed to maintain as much as possible a general style and tone in the books. They were not to be academic treatises but more like primers. We wanted them to be as lively as possible. We understood that with nearly a dozen authors and coauthors, we absolutely would not be able to control this, and we knew we would get a range of writing styles. We agreed on certain conventions of presentation such as bulleted lists, cross-references, and boxes.

The editor's job largely deals with content and style, while the copyeditor's is in the details. Copyeditors are the real workhorses of the publishing world. Pay attention: They usually know more about grammar and style than you do! Still, they can make errors because they are usually not experts in the content of a field—but I've not met a copyeditor yet who has not improved my work.

Having worked as a university press director, an editor, and an author, I'd say that the author causes more problems than editors ever do. Usually troubles stem from an author's attitude. Many authors, especially those less seasoned but sometimes even experienced writers, think that everything they write is wonderful. Younger writers usually just have worked so hard on something that they can't imagine it could be improved by editing. To reiterate: I've never read a manuscript that could not be improved by a good editor. Whatever the case, I urge authors to be humble enough to be flexible. Work with your editors, not against them!

BEING AN EDITOR

Sometimes you may end up on the other side, editing a manuscript. Doing CRM reports today is often a team project, with specialists in everything from ethnobotany to remote sensing. For a final report, you may have to serve as an editor for the other people's sections. You may also find yourself being an editor of a set of chapters sent to a publisher. You may be selected as a journal editor. As an editor, you also need to be humble enough to be flexible. Certainly, you do have power. Don't abuse it; to do so is unethical and will not serve archaeology well.

You have a right to respect and an obligation to be certain that the best possible work goes into the journal or volume. You will face the arrogant or stubborn author. I once had an excellent paper submitted to me as editor of *Plains Anthropologist*, but it needed a couple of important changes. I would have been embarrassed if the paper had appeared without them. The author was adamant about not making the changes, so I agreed to publish the paper as it was, but only if I could add a statement at the bottom of the first page that certain changes had been recommend but not made by the author. The author made the changes before it went to print.

You'll face the other problems of materials getting back to you on time or not getting back at all. The peer review process is the worst of it. Establish a time frame and stick to it. If a reviewer doesn't respond, find another quickly.

Despite the problems, journal and volume editors do an important job in archaeology by evaluating and distributing important works. Editing reinforces the importance of collaboration and, believe it or not, builds respect for colleagues. Editing brings other rewards as well. The contacts they allow can be important for building your career. Caring, competent, and patient editors with a solid sense of humor make far more friends than enemies. Editing is a tough job, but something every archaeologist needs to try, if only once.

10

BRINGING THE PAST TO LIFE AND PRESENTING IT WITH STYLE

For all the fine writing we may do as professional archaeologists, the public mostly finds our writing boring, or at least overly complex. For people who have never seen or tried flint knapping, no written explanation is sufficient to help them understand the process, even if an archaeologist is masterful at creating images from words. The minute they see and try real flint knapping, the better they understand both the process and our explanations of it. The same applies to our discussions of how people in the past lived. Mostly our work describes the norms of life in the past; individuals and their feelings are usually absent from what we present. Sometimes we forget that objects aren't people, a major complaint about archaeology from members of descendent communities.

We know why we do this: Our data are limited, and we feel that they don't justify saying more. Should we even try to get past the limits? As scientists, we don't like to speculate too much. After all, if we do, we'll probably face the criticism and the disdain of our peers. That's what makes many of us shy away from popularizing archaeology. Archaeologists seem to have a love–hate relationship with our popularizers. As in other sciences, some of us have smeared colleagues who popularize science (e.g., Carl Sagan). Are we jealous of their skill, celebrity, or earnings? At the same time, most of us also know what a useful service they perform for a public that believes most of us "suck the life" out of what can be really interesting stuff. I was recently shocked when three of my colleagues who teach introductory prehistory proclaimed how boring the class was.

If the professor thinks the class is boring, pity the poor students! Maybe our critics are right: We must find solutions to the deadening effect we have on archaeology. Fortunately, some of our colleagues have found ways to put life back into the dusty old bones. The first step is to understand that hands-on learning works well.

HANDS-ON LEARNING PROMOTES UNDERSTANDING

The old saw that seeing is believing contains a major truth. I've tried and tried to explain a spear thrower to students. There is never any real understanding until I bring atlatls and darts for the students to use. After the initial fear of making fools of themselves dissipates, their eyes light up when they start to get the hang of the instrument and see the increased range and impact of a spear. This is the same reason, of course, that we organize field schools. All the book descriptions and drawings of stratigraphy, for example, mean nothing compared to having to read soil layers in a cut bank. If there is a way to use hands-on techniques in any public venue, do it!

Hands-on experience may be as simple as passing around a few artifacts or reproductions when you give that service organization lunch talk discussed in chapter 8 or having a small display of objects people can touch or handle before or after your talk. If there is a way for people to do or handle things safely, work this in. This is no big secret—learning sometimes comes better with doing than with reading or listening.

ACTION IS BETTER THAN DESCRIPTION

Dig magazine does a masterful job at drawing the reader into thinking about life in the past, not archaeology. For example, the short news piece on a child's footprints found in Chauvet Cave in France that I discussed in chapter 4 about writing styles is a fine example of much of the magazine's writing. Let's look at the statement again:

> About 20,000–30,000 years ago, an 8- to 10-year-old boy skidded barefoot through a muddy cave in what is now southern France. Think that kid would get a kick out of knowing that his tootsie marks would become the oldest footprints ever found in Europe?

The first sentence sets the action, and the second draws the reader directly into the thoughts of that child from millennia ago. Of course, we don't write this way for our colleagues or advanced students, but perhaps we ought to do more of it for our entry-level students and the general public.

Our public talks should try to pull the listeners into the action. Often this might be no more difficult than the sentence from *Dig:* "Think that kid would . . . ?" You might be able to enliven archaeology by making connections directly to the current lives or memories of listeners and readers. You can draw links to objects they are familiar with in their daily lives, or even do something as simple as ask them what they are reminded of when they see an artifact. This might be introduced by a question such as "Did you ever wonder . . . ?" This approach involves the audience directly in the substance of your presentation, with choices to make about what you say. Action makes archaeology interactive; the reader or listener becomes a partner in the product (this is a crucial concept for new media, as will be discussed in chapter 12).

GAMES AND TOYS

The popularity of hands-on activities makes games and toys almost ideal media for archaeology, but we rarely think of them as ways to bring life to the past. Some involve simulated excavation. One good example is Running Press's 1994 *Lost Civilizations Explorer's Kit* in which you use a Popsicle stick "trowel" to excavate a miniature Mayan ruin from Tikal. The accompanying booklet tells how real excavations get done and provides other activities such as making a *Patolli* board game played by the Aztec and Maya. Another activity is writing in Mayan hieroglyphs. Similarly, *Our Amazing Ancestors Science Kit*, published by Educational Designs, Inc. (1997), deals with a much earlier period. Twenty-four activities range from putting Amazing Ancestor stamps on a six-foot-long time chart to making and painting small plaster casts of fossil skulls and cave painting. There is also instruction on how to make and play "Out of Africa," a board game about (pre)human migration.

Older children and some adults enjoy role-playing games. One that's fun is *Tribes: It's 50,000 B.C.—Where Are Your Children?* published by Steve Jackson Games (1998) and written by science fiction writer David Brin and Steve Jackson. Each player represents a tribesperson. The object of the game is biological success—that is, to reproduce and have many children reach adulthood. The game's rules teach many concepts of early cultural evolution.

Computer games are sometimes variations of role-playing games. Microsoft's *Age of Empire* series is a terrific strategy game with several modules in which players build and maintain empires. For all its silliness about Atlantis, the game segments of GTE Entertainment's *Timelapse* are pretty good graphically and factually for Egypt, the Maya, the Anasazi, and Easter Island. McGraw-Hill Home Interactive's *Pyramid: Challenge of the Pharoah's Dream* is funny and surprisingly accurate regarding Egyptian culture, tools, pyramid building, and daily life along the Nile. Players use math, science, and logic to solve problems. The Lara Croft action–adventure series is more stereotypical than useful, but the games are fun.

With all games and toys, most archaeologists are likely to worry about accuracy. If so, we might consider suggesting games or serving as technical advisers for them. Creating a few may help you develop interesting activities for the public, and you may even find a market for them.

ARCHAEOLOGICAL FICTION

Some archaeologists and other authors have taken the approach that fiction can be used to teach as well as entertain. After all, we rarely get glimpses of individuals from the past and often wonder what they must have been like. We might even fantasize about having a time machine to allow more intimate glimpses into the past and how people acted, thought, and felt. We know they have a story. Archaeologist John Whittaker (1992:56–58), for example, discusses the archaeology of the Sinagua, a prehistoric complex of the American Southwest that may be related to the Anasazi. He writes, "Few of us would bother with archaeology if we weren't emotionally involved with the past. We don't dig for dry bones and dusty potsherds, but for people. . . . I care about them and want to tell their story." Most of us agree, but while we crave that knowledge, we understand there is no real way to know the past. Fortunately, several writers are willing to try to help us understand what life may have been like. They write short stories and novels, and by doing so, they help us make connections between the humanity of past lives and our own.

Writing fiction is absorbing, exhilarating, and mostly humbling. I can't say much about the process, except that I've published one short story about archaeology (Zimmerman 1986) and have several failed starts of a novel. Novels require a delicate balance between character and plot, connected by vast amounts of hard work and creativity. Although there are numerous bad novels about archaeology, there are also many good ones (see sidebar 10.1). Authors of the good ones do solid research before each book, and while scholars might nay-say some elements, most of each story rings true.

Plenty of books are available on how to write fiction. If you want to try it, read a few books and jump in. I'm glad that some authors are willing to give archaeological fiction a shot. We need them.

10.1. RECOMMENDED ARCHAEOLOGICAL NOVELS

- W. Michael and Kathleen O'Neal Gear
  Two CRM archaeologists have turned their imagination to the North American past in their extremely popular First North Americans series (titles contain *People of . . .*) and the Anasazi Mysteries series, both published by Tom Doherty Associates.
- William Sarabande
  The First Americans (Bantam Books) series details lives of some of America's first inhabitants.
- Kathleen King
  Cricket Sings: A Novel of Pre-Columbian Cahokia (Ohio University Press, 1983).
- Jean Auel
  Clan of the Cave Bear, the first and best of her The Earth's Children series (Crown, 1980).
- Piers Anthony
  Tatham Mound. Fantasy author turns his attention to the impact of the DeSoto expedition on Florida's Native populations (Morrow, 1991).
- Sharman Apt Russell
  Essayist turned some of her research for the nonfiction *When the Land Was Young: Reflections on American Archaeology* (Addison-Wesley, 1996) into *The Last Matriarch* (New Mexico, 2000), a novel of the American Southwest eleven thousand years ago.

CARTOONS

Most archaeologists probably don't consider cartoons to be a medium for communicating with the public, but cartoons can be an effective way of communicating a point. Most of us have pinned a Gary Larson cartoon from *The Far Side* on our bulletin board at one time or another. Bill Tidy's cartoons for *Archaeology* magazine or Robert Humphrey's in *AnthroNotes* bring a chuckle. Tidy has teamed with archaeologist Paul Bahn for a longer, raunchy, and hysterically funny treatment of human history in *Disgraceful Archaeology* (Bahn and Tidy 1999). Segments of Larry Gonick's (1990) *The Cartoon History of the Universe* provide a fair summary of human cultural evolution in comic form. The best and most hilariously to the point in terms of demonstrating archaeological thinking is David Macaulay's (1979) *Motel of the Mysteries* with an excavation of the Toot-N-Come-On Motel in the year 4022.

Cartoons allow an easy connection between the past and contemporary life through the use of anachronisms and humor. They make us think in slightly different ways, usually with a good laugh. To do cartooning requires relatively modest artistic talent and an ability to quickly find a core message. Some archaeologists can do this well as part of their "doodling." If it comes easily, you might consider adding cartoons to your repertoire of ways to present the past.

MOVIES AND VIDEOS: LIVE ACTION VERSUS DOCUMENTARY

Just as most archaeologists don't consider writing fiction, most also don't consider making movies. Certainly, most of us understand the impact of the movies on our profession. The astounding success of *Raiders of the Lost Ark* and its sequels showed the influence of films in bringing archaeology to public attention. Generations of moviegoers have seen archaeologists as eccentric pith- or fedora-helmeted treasure hunters. Portrayals of archaeologists differ little from the opening scenes of Boris Karloff's 1933 *The Mummy* through today's *Tomb Raider* video game, spin-off movie, and Lara Croft action figures. (At least more women archaeologists are being portrayed.)

We might not always like the way our profession or our methods get portrayed in theatrical films, but the only way to fight this is to work with the motion picture industry—a fight we won't win because stereotypes sell. On the other hand, by turning the subject of our work into screenplay, we have a better chance. Even though it was a box office flop, consider *Rapa Nui*, the movie about Easter Island. It did a respectable job of showing a possible way of building and transporting the *moai*, the huge stone heads raised by the people of the island, and it put life into oral tradition, environmental reconstructions, and a wide range of good archaeological issues. Archaeologists can serve as technical consultants on such movies. Perhaps more of us should also suggest topics to the movie industry, and a few of us might even try writing a screenplay. For those interested in doing screenplays, books and even software on how to do them can easily be found by doing a web search using the terms *writing* and *screenplay* together.

Archaeologists should consider making documentaries about our work. Easter Island provides another good example of how this can be done. The 1998 Nova documentary *The Secrets of Easter Island* shows the other side of the *Rapa Nui* coin. The video focuses on the work of Jo Anne Van Tilburg and others, often using experimental archaeology. The spin-off website for the program at www.pbs.org/wgbh/nova/easter/ provides a wealth of print and interactive material, including a small video game simulating life on the island, use of natural resources, and building moai. With the American public getting most of its information about archaeology from television, and with cable channels such as Discovery Channel, the Learning Channel, and the History Channel each needing well over one hundred hours of programming a week, there is a definite market for archaeology documentaries.

The problem, as with writing novels or screenplays, is that few of us have any training in preparing documentaries. Filmmaking is a full-time occupation requiring years of training. At the same time, most of us can be extremely useful in providing story ideas to documentary producers or acting as technical consultants (see chapter 11). And now, with affordable digital video cameras and quality video-editing software, creating your own documentary is not difficult. You won't likely be able to prepare a broadcast-quality documentary, but you might be able to prepare good, compelling work for your classes, a museum, or some other local venue.

Seek out quality sources for help. A terrific one is Barry Hampe's *Making Documentary Films and Reality Videos: A Practical Guide to Planning, Filming, and Editing Documentaries of Real Events* (published by Owlet, 1997). Hampe discusses preproduction (developing ideas, writing at various stages, scheduling, etc.), production (recording picture and sound, directing, interviewing, etc.), and postproduction (preparation, editing, and the wrap) phases of filmmaking. He writes about selecting equipment, hiring necessary crew, directing actors and nonactors, and conducting on-camera interviews. Although the process might seem daunting, Hampe shows how to plan, recognize, and record visual evidence and how to organize it into a visual argument. I'll admit that I've not yet tried anything lengthy, but I have worked on a couple of commercials for a powwow, and I'd like to do more.

Working as a consultant or interviewee on a video can be a bit intimidating. I've now consulted on several programs, and I've been on camera several times. The most difficult work was for the BBC Horizon program *Bones of Contention*. The taping had a separate audio track, a video track, the traditional movie clapboard, and just about everything else needed for a movie. A thoroughly professional crew and director can often help you along, and after a couple of takes, you don't feel so strange, though I'd not like to do it as a regular part of my job.

The beauty of video is that you can reach a large audience and not have to be there. Seeing a video can bring life to the past, but it still can't substitute for being a part of the past or, more appropriately, interacting with the past. There is no substitute for a real hands-on activity like what one can get in a school classroom, a museum, or "archaeology days" activities.

PUBLIC ARCHAEOLOGY

For eight years, the University of South Dakota (USD) Archaeology Laboratory worked with the Vermillion, South Dakota, Middle School on a project we called "Archaeology Days" (Zimmerman et al. 1994) Collaborating with the sixth-grade teachers, we literally took over the curriculum for about two weeks. Teachers primed the kids with readings in their literature books, some videos, and a large lunchroom display about archaeology, but the core was hands-on activities. One full day was devoted to ancient technologies and another to fieldwork in the plow zone of a heavily disturbed site. The project had an impact. The kids saw the connections between their areas of learning. The teachers were full of stories about changed interests and behaviors.

The USD project is but one example of many. Several excellent books provide educational contexts and ideas for doing similar projects. *The Archaeology Education Handbook: Sharing the Past with Kids* (Smardz and Smith 2000) is a guidebook aimed at educating children in archaeology. *Presenting Archaeology to the Public: Digging for Truths* (Jameson 1997) provides case studies of ways to translate archaeological information for broader public audiences. The SAA provides a wide range of public education materials on its website at www.saa.org/education/edumat.html. Of special note is the *Archaeology and Public Education Newsletter* available on the web at www.saa.org/PubEdu/a&pe. The SAA Public Education Committee also has a Network of State and Provincial Archaeology Education Coordinators, a good place to make contacts for information and planning (www.saa.org/Aboutsaa/Committees/n-penet.html).

EVENTS

Among the more impressive recent developments in archaeology are events specifically aimed at bringing archaeology to the public. Many states now have "Archaeology Months" or "Archaeology Weeks" full of public-oriented events. These might be as simple as lectures or as complicated as participation in experimental archaeology. Favorites include hands-on activities such as letting people try spear throwing, pottery making, or simulated digs. Some include one-day field schools. Iowa, for example, promotes a wide range of activities, usually hosted by local communities but attended by at least one professional archaeologist. The Iowa Archaeology Month website (www.uiowa.edu/~osa/focus/public/iam) provides lots of detail. Wyoming has an "Archaeology Awareness Month" and also promotes archaeology-related events during other times of the year (wyoshpo.state.wy.us/waamindx.htm). Many states have similar activities, and several produce stunning posters to promote the events. At its annual meeting, the SAA hosts a contest for the best poster, and several of the posters have become collector's items.

Archaeological events with a longer history are field schools or volunteer opportunities for lay people. Some archaeologists have raised objections to programs, arguing that we might be training people to loot sites or allowing untrained individuals to dig in potentially important sites. To deal with these issues, many archaeologists include consideration of ethics and request that volunteers spend a minimum number of days on a project. Training takes time, and usually just about the time volunteers have learned the basic skills, they leave.

Archaeology events take money, but you can find grants from state humanities councils and private donors. Events also take staff time and operations resources. On the whole, however, the growing number of these events suggests that archaeologists consider them to be worth the effort.

EXHIBITS AND MUSEUMS

From curiosity cabinets to full-blown permanent galleries, exhibits and museums provide the public with a major point of contact with archaeology. According to the Harris Interactive Poll discussed in chapter 1, fully 88 percent of the public has visited a museum with archaeological content. Museums are a major tourist destination; even small roadside exhibits can pull in visitors. Museums and exhibit design are part of a large and complex discipline, but one in which many archaeologists are employed or involved; truly, saying much in this book about professional museum studies is impossible. At the same time, many of us find ourselves involved in putting up temporary or permanent small exhibits, sometimes related to our CRM or research projects, often on little or no budget, so a few comments may be useful.

Several sources can help if you are faced with building such an exhibit, and you usually don't have time for learning the ins and outs of museum theory. *Museum News,* produced by the American Association of Museums (AAM), is a magazine for professional museum personnel covering a wide range of issues. Similarly, *Curator: The Museum Journal* is mostly for museum professionals, but it provides solid information about exhibition development, studies of visitors, and education. Its "Technical Notes" section can be useful for problem solving.

Hundreds of books are also available. A good place to start is at the AAM Online Bookstore Catalogue (secure.aam-us.org:80/nonmembers/web_store.cgi), which boasts that it contains "the most comprehensive list of titles in the museum profession." Annotations about the content of each volume listed are especially useful.

The web is replete with museum websites, and a search for terms related to building small, temporary exhibits turns up lots of useful sources. From them one can digest several key questions about small exhibits:

- Can a small exhibit give enough information?
- What is the intended life of the exhibit?
- Who has authority over the exhibit and its contents?
- Who is responsible for maintaining the exhibit?
- What are the provisions for security of artifacts and other materials?

As in other aspects of archaeology, making exhibits may benefit from the use of specialists if budgets allow. If a CRM project for a government or large private client has a public component, solicit the agency's or client's advice regarding an exhibit; there may be professional staff who can help.

Whether it's an event or exhibit, all the time, skill, energy, and money you put into it don't mean a thing if no one comes or sees it. Nothing is more disheartening than having no audience. How do you let the public know about archaeology? Archaeology makes a good story, but you have to figure out effective ways to get it to them. That's where the media come in.

11
MEDIA METHOD OR MEDIA MADNESS?

Whether it's a CRM project, a grant from the National Science Foundation, or even a gift from a private donor, most funding for archaeology comes from the public. We just can't afford to fund our own research. As a discipline, we've pondered the best strategy for finding ways to get across the importance of what we do, not only so the public will keep the money coming but also to show them that the past needs to be protected. It's really simple: You have to *work* with the public. Some of us are active supporters of avocational archaeology, whether it's with the local archaeological society, a project allowing volunteer participation, a popular book, or an archaeological website. But working with the public is probably one of the most difficult things we do. How do we find ways to get the public interested at all, let alone educate them about quality archaeological practice?

Thankfully, most people have some level of interest in the past. There is mystery, adventure, and romance in what we do. Archaeology "sells." Many archaeologists who regularly involve the public in their work know this. They also know that media are among the best ways to distribute information. Whether it's a simple local television or radio news spot about an ongoing excavation, a newspaper article, or a full-fledged television documentary, the public does pay attention to what we do.

At the same time, many of us are suspicious of the media. We've probably not had any training in the media, so we really don't know how to handle them very well. Almost every archaeologist has fallen victim to bad reporting. Flawed news stories can easily make you "gun-shy" about all media.

CHAPTER 11

WITH FRIENDS LIKE THE MEDIA, WHO NEEDS ENEMIES?

David Pendergast (1998:15) tells an amusing story from his first press conference about his work at the Mayan Altun Ha site in Belize. After he answered a reporter's question about the use of a string of pearls in front of the face producing crossed eyes in infancy, a headline appeared saying, "Mayas Liked Their Women Cross-eyed." His only comfort was that they spelled his name correctly! My own experience with the Crow Creek Massacre site along the Missouri River in central South Dakota was similar. A *New York Times* reporter asked who had committed the massacre. I hypothesized that nomadic bands had done it, but fearing that he might not know what *nomadic* meant, I used the word *roaming*. In the story, *roaming* unfortunately became *Roman*, with a very uppercase *R*!

Pendergast (1998:15) points out that reporters today are better prepared than their predecessors. They come to interviews with questions better than those most undergraduates ask. Many now come to the story trying to stimulate what he calls fact-based interest. He suggests that the conditions in which we operate with media are greatly improved. The reasons are many but have to do in part with an increased need for news given the expansion of both numbers and types of media outlets. In spite of improvements, he warns that we need to take care in how we present our work and its results. We also need to change our attitudes about media and learn more about them.

ATTITUDES

Archaeologists seem to have a confused relationship with media. We know that we benefit by taking our work to the public, even at the risks we face from misrepresentation of our work, not to mention straightforward reporting errors. Many of us naively used to believe, and probably still do, that we worked in concert with media to bring our projects to the public in an effort to generate greater interest, which in turn would increase support for funding. We were and are grateful that media show interest in us, almost to the degree that we've had an inferiority complex about them. After all, how could anyone be interested in this exotic old stuff?

If anything, the recent Harris poll (Society for American Archaeology 2000) should tell us that such attitudes are very wrong. When 96 percent say there should be laws protecting sites, when 80 percent agree that public funds should be used for archaeology, and when 88 percent have visited a museum with an archeology exhibit, we need to recognize that archaeology has intrinsic value. The reporters recognize this, so why don't we? Archaeology is hardly likely to get to pop culture status, but even a dig in a local area easily raises public curiosity about what is being discovered. Perhaps we need to change our attitudes and strategies about working with media.

LEARN SOMETHING ABOUT MEDIA

The Harris poll also asked respondents about the sources of information from which they learned about archaeology. Although many visit them, fewer than 9 percent chose museums. Television was the most important source at 56 percent, with 33 percent mentioning books and magazines, and 23 percent listing newspapers (Anonymous 1999).

Many archaeologists think they are somewhat sophisticated in terms of media; after all, most probably have done several newspaper and television interviews. Many archaeologists also seem to have at least a slight disdain for media because media are part of pop culture, not something an intellectual elite should worry about. They sometimes don't trust the media because of bad experiences. Thus, the idea of using the media—things like video games, advertising/marketing, broadcast news, television and radio (other than news), movies, newspapers and magazines, music, and the Internet—would probably cause some to choke. Media, however, are a major source for the public's ideas about archaeology, so a wise archaeologist should become truly media-literate.

Media literacy is concerned with developing an informed and critical understanding of the nature of mass media, the techniques they use, and the impact of these techniques. The aim is to increase understanding and enjoyment of how the media work, how they produce meaning, how they are organized, and how they construct reality (see www.media-awareness.ca/eng/).

All media, even popular culture such as archaeological reproductions and trinkets sold to tourists at museum gift shops at archaeological sites, embody points of view, whether or not they are consciously intended. They are based on decisions as broad ranging as what story is to be told and what camera angles might be used, but they lead to questions about how media are organized and how they construct reality. As with images discussed in chapter 6 on visual archaeology, media materials are *not* just reflections of reality. They are preconstructed "products" containing values and ideologies—in our case, about the value of the past and its conservation and preservation. However, most of us are at least aware of this because we understand that we need the media to promote these values. Many archaeologists even recognize that media are mostly a commercial venture, geared to make a profit and controlled by a relatively small number of people, people who are "gatekeepers."

How do these issues play out in media that cover archaeology? As Marshall McLuhan noted years ago, each medium has its own grammar and codifies reality in its own particular way. Different media will report the same event but create different impressions and messages. We need to learn how to work with each to develop the messages we would like to convey regarding the past.

There are lots of opinions and no easy answers as to how to do this, but we do have those on whose experience we can draw. The Society for American Archaeology sponsored a session on media at its 2000 annual meeting, during which four media specialists in archaeology spoke and answered questions. Evan Hadingham represented the PBS science series *Nova*, and Steve Burns represented the Discovery Channel; both are major producers of archaeology television documentaries. Peter Young, the editor of *Archaeology Magazine*, and John Noble Wilford, science writer for the *New York Times*, represented print media. As one might expect, opinions varied, but they agreed on many ideas. One of the important questions was how to know when the story is important enough to "go public," or when does archaeology become news?

FINDING A HOOK

At the SAA seminar, Wilford addressed the issue most directly. One of his most important pieces of advice: Every story needs a "hook," some idea that would first draw the attention of the media and the public. Discoveries that are earlier than we knew or suspected, or new interpretations about them, can be important hooks where a connection can be made to broader issues. New applications of technology or methods can also be a hook, as can something in a discovery that challenges an existing theory or hypothesis or that makes a significant step in solving a question about the past. He noted that almost anything about origins usually provides a good hook. However, he cautioned that almost every story also has to be part of a bigger picture, with meaning for human culture, or part of an active professional debate about the subject. There is nothing wrong with developing hooks that may not be at the core of the ideas that we as scholars think are important, provided they help educate the public about our work

DECIDING WHEN TO GO PUBLIC

When are you ready to go public with a story? This is a difficult call. The story may be so compelling that it becomes public knowledge whether you like it or not, so if your topic is spectacular or controversial, you should be ready at any time. At the SAA seminar, Young discussed the remarkable story of the New York City slave burials and the controversy surrounding them, which raised all sorts of ethical questions about when to go public and what to do once the story breaks. In other cases, you may be able to control the release of the story, such as when an article is about to be released in a peer-reviewed journal or on a website. In such cases, taking a story public may be a matter of when you are ready to face your peers and the media.

As Young noted, having preexisting relationships with the media helps. Although local media reporters are often young and inexperienced, they are usually not "out to get you" and may be your first point of contact to the media outside your area. Hadingham added that in video, it's best if you can make yourself and the project known to producers as early as possible. Many of the seminar presenters noted that as journalists, they do tend to have favorite topics they follow and usually will report.

GETTING NOTICED

If you are preparing to go to the field, are in the field, or are doing something that is likely to be of interest to the public, how do you get noticed? If you figured out what the hook might be and have carefully considered the story in terms of linkages to controversy, new discoveries, and the human side of the story, you are really only part way there.

PUBLIC RELATIONS OFFICES

If the organization for which you work has a public relations (PR) office, as most universities and many larger CRM offices might, media professionals in that office can work with you on presenting your story and getting media attention. Use them if at all possible because they do have experience. They can help you find or clarify the hook if you haven't done so. Experienced public relations people can be worth a great deal to you, so if you find yourself in an organization with a PR office, make yourself known to the staff as soon as you can. Let them know your areas of expertise and the kinds of projects you do.

This can work for you in several ways. If they know you are active and that the things you do are interesting, they'll often contact you on a regular basis. Many PR offices, especially in universities, get calls from media organizations that are looking for experts in areas on which they are doing stories. If you have identified yourself as having a level of expertise in a subject, they'll direct the media to you for comment. If you work well with journalists, you become part of the reporter's contact list, and they may contact you directly. The PR office maintains a list of media contacts, from local to national outlets, and this can be of substantial benefit.

PRESS RELEASES

Press releases are short statements sent to a wide range of media to attract attention to your story. Press releases are not easy to write, so get help from your PR office if you can. If you are on your own, you can still write a good press release. The American Anthropological Association has put together advice on writing a good press release (see sidebar 11.1). Notice that it agrees with the advice from the media professionals at the SAA Ethics meeting.

CHOOSING THE RIGHT LEVEL

Archaeological work often seems very mundane, and thus in your mind it becomes unimportant. For the people of Podunk, however, you as the first archaeologist coming to their town might be big news. You might be amazed to find the editor of the local weekly waiting for you some morning at the survey area. At the other extreme, we

> **11.1. THE AMERICAN ANTHROPOLOGICAL ASSOCIATION'S *GUIDE TO THE PERFECT PRESS RELEASE***
>
> - The best release is no more than a single page, approximately two hundred words arranged in a maximum of three paragraphs.
> - Choose a simple yet compelling title (maximum five words is ideal) to catch the reader's eye. This should capture the most interesting aspect of what you will discuss. As an exercise, imagine yourself in the supermarket checkout line scanning the newspapers and tabloids. What causes you to stop and look twice at one of the headlines? It's probably not the article that begins, "Scientists Find Evidence for Extraterrestrial Life." Instead, you'll probably perk up at the one boldly announcing, "Man in the Moon Waves Back."
> - Before beginning to write, ask yourself what is the single most interesting result or aspect of your research—what does this study tell us about ourselves? If your paper is about research in Indonesia, how does it relate to our lives in the United States? Does it tell us something about being human?
> - Lead with your best line, rather than building up to it. This is the opposite of writing for an academic audience. Don't expect the reader to bear with you as you first clarify all the whys and wherefores. If details are absolutely necessary in this format, cover them briefly after you have pitched your most appealing sentence.
> - Use active tense; avoid the passive and impersonal ("We found . . ." instead of "It was found that . . .").
> - Avoid jargon. Visualize explaining the topic to your Aunt Mabel or twelve-year-old nephew. If possible, try the draft on someone outside the field.
> - Use at least one concrete example to illustrate your point.
> - Give the reporter a reason to ask for more. Although this is a teaser, it is critical to deliver a substantive bottom line. Instead of making vague promises that you will discuss the implications of child rearing in Africa, tell the reader up front that you have observed African mothers to be more nurturing/fulfilled/better prepared than their American sisters—and why.
>
> Adapted from www.aaanet.org/press/relguide.htm. This guide is offered as a public service by the American Anthropological Association and is posted on its website at www.aaanet.org.

sometimes do make interesting discoveries but for all our efforts can't get anyone interested. The problem is one of being too close to what we do to see it in terms of public interest. Another problem is presenting the project in a way that is only of interest to an archaeologist. Thus, one of the main skills to learn is how to pitch the story at the right level.

LOCAL PAPERS

If you are going to do a project in a small town, or even a medium-sized one, making contact with the local paper makes good sense for a lot of reasons. A story in the local paper lets people in the area know who you are and what you are doing. Small-town folks are terribly curious. One of our crews doing a highway relocation survey ventured into a small northwest Iowa town for supplies. The residents are all white, and most are very Dutch Reformed; everybody knows everybody else, and many share the same last name. One of our crew members was African American, and several had beards, a few had tattoos and piercing, and they weren't all that clean. Even in a state car, within minutes of driving into town, they had two police cars following them around and finally "chatting" with them when they stopped at the store.

An added benefit of a story in the paper is that people will often seek you out, many providing you with information about sites in the area. You get to see collections, hear the local stories, and have a chance to interact with folks. For some, especially landowners, such a story may smooth the way for permissions (see Toolkit, volume 2).

Consider writing a series of stories on regional or state archaeologies and releasing them to a wide range of local weekly or daily papers. Klesert (1998) provides excellent resources for how to go about writing for local papers. Set up the stories in ways that allow for quick editing so that a paragraph or two can be quickly eliminated or reduced without dramatically changing the meaning. The odds are that the papers will print the story in some form. If you provide contact information, you may find local collectors at your door or on the phone, ready to give you information on sites in areas where you might do future surveys.

LOCAL TELEVISION AND RADIO

The same benefits can derive from television and radio coverage of a story, but these media outlets usually appear in larger markets and cover a broader region than most local papers. They also tend to be more interested in stories of a bit more importance than some archaeologists looking for sites at the proposed city park shelter. They like controversy, which they see as more newsworthy, because it has that hook. Still, they also like to do stories on discoveries about the past of a region, so if you find some interesting sites, they might do a story, and if you are doing excavations, they are almost bound to cover the project.

Mostly, getting a story done takes a phone call to the station newsroom or news editor. On the downside, they often don't have much background in a subject or the good sense to get the information they need. Radio formats are often different, usually interviews done by phone or in a studio. They are usually not limited to a two- or three-minute story, so they may take the time with you to explore the subject a bit better.

LARGER MARKET MEDIA

Larger markets are more difficult in that they tend to have more news, just for the simple fact of having more people. They generally have better quality in reporting but less time to spend on stories. Here the hook becomes critical to drawing attention. If the market includes a number of communities, the stations may feel an obligation to report something from each of them as news arises, so you may be able to get coverage if your work happens to be the biggest news from that place on a particular day. A better approach is for you to show the importance of a project to the whole market. If you can begin to meet the conditions outlined by the media experts at the SAA panel, you have a better chance of hitting the larger market.

Certainly, the newsworthy stories containing controversy or new, exciting discoveries are an immediate hook, but most archaeology is more mundane. With the right hook, the work might become the subject of a feature. Here is where you need to play on the importance of the work in broader human themes or the application of new technologies or approaches or theories.

For television, the major expansion of cable stations in recent years plays in our favor. For example, at the SAA session, Steve Burns of the Discovery Channel noted that Discovery broadcasts one hundred or more hours of science features and news a week; it needs materials. Because of archaeology's scientific, humanistic, and historical content, archaeology is of possible use to a wide range of specialty cable networks. Burns added that these cable networks often become the first points of contact after a news or feature story appears in print. If they see a story in print, with a hook, the network may actually contact you.

OTHER ADVICE AND CAUTIONS

There is much more to working with the media than getting them to notice you. Most of us just aren't trained to be aware of many of these problems, and learning them through experience is absolutely the hard way.

MASTER THE SOUND BITE

Sound bite is a term you've no doubt heard, where a person says one very short thing, and that becomes the focus of the story. Sound bites work the same way for television, radio, and print media. They are essentially the hook, but in a very direct form, meant to catch the attention of the audience. At a reburial ceremony at Wounded Knee on the Pine Ridge Reservation, Bob Dotson of NBC interviewed me for *Nightly News.* His crew did more than an hour of video on me, and when the segment was aired, I was on for an entire eight seconds of a four-minute piece! Dotson took one pithy statement from me. Since that time, I've worked hard to think about what to say to reporters about stories well ahead of interviews. If I come up with something good, I boil it down into as few catchy words as possible.

To give you an idea of sound bites, my favorites have been "The Crow Creek Massacre changed the way archaeologists viewed Plains Indian warfare" (which is actually a bit too long) and "Indians are from now, not just back then." The latter has been used in papers more often than I can probably count. As silly as sound bites really are, if you consider them ahead of time, you can get a point across quickly.

RESIST TEMPTATION TO CRITICIZE COLLEAGUES

Steve Burns made this point in his discussion at the SAA session, but I've seen colleagues criticize others too often, and I've done so myself a few times. Openly criticizing colleagues in the media puts a question in the mind of the reader/viewer about you. Are you stuck-up—that is, do you think too highly of your own work? Are you being too defensive and using your attack on a colleague to cover weaknesses in your own ideas? You can fairly question ideas of others, but stay away from attacking the person.

PREEMPT WRONG INFORMATION WHERE POSSIBLE

More than once I've given an interview about a field project and had basic information come out all wrong about the site, the age, the cultural complex, and just about anything else a reporter can mess up. One way to handle the problem is to provide basic information about the project ahead of time by doing up a fact sheet about the project. If you are on a CRM project, you can describe the project, the agency funding it, the kind of site or sites found, the cultural complexes, dates associated with them, and artifact types found, with brief explanations of each. In that way your lanceolate projectile point won't become an arrowhead, or its 8,000 B.P. date won't become 8,000 B.C. You may want to include phonetic spellings of difficult names or words to avoid mispronunciations.

AVOID DENIABILITY

If there is an error in reporting, or if you misstated, or if your remarks get taken out of context, there is nothing you can do to make it right. For the record, you should probably report necessary corrections, but whatever you do, your corrections will not get the same coverage as the original story. Sometimes, attempts at corrections can even make things worse.

DEVELOP WAYS TO PRESENT MATERIALS VISUALLY

Steve Burns in the SAA session gave the good advice that archaeologists need to learn how to speak visually—that is, to develop visual imagery through words. For television, he also suggested that archaeologists use available resources to develop real imagery, especially 3D graphics. Pictures do sell a story.

TAKE CHANCES

Burns also noted that we archaeologists are often too cautious. We worry too much about our colleagues or that the media will make us look bad. He didn't mean that we should make statements beyond what our data are capable of supporting. Rather, he meant that conjecture is not always bad, especially if presented that way.

STAY ON POINT

We archeologists tend to be afraid of what colleagues will think, so as we speak to interviewers, we tend to drift away from the story trying to cover our rear ends and get lost in parenthetical asides. Stay on the point, and try to give relatively concise answers. Save your full logic for your articles and books!

ETHICS, CONTEXTS, AND PROFESSIONAL RESPONSIBILITY

The darker side to all of this goes beyond knowing how to present yourself and your work to the media. Archaeologists mostly love what they do; as we sometimes say, archaeology is one of the best things you can do out of bed. We get enthused about what we find and what it means. At the same time, we rarely think about the contemporary repercussions of what we find on the people who are the readers, viewers, and, most important, subjects of our work. We like to think that we are scientists, thoroughly objective in our interpretations. We tout academic freedom; we do not want to feel limited about what we can and cannot say. We need to be cautious. A brief story from Iowa might be instructive (and others could be told):

There has always been controversy about whether syphilis started in the Old World or in the Americas and in which direction it was transmitted. A sexually transmitted disease, syphilis is one of a number of treponemal diseases similar to yaws. Each is difficult to distinguish from the others without specialized tests or solid contexts. After examining some remains from an eight-hundred-year-old burial mound, an Iowa paleopathologist proclaimed to the media that he found evidence of syphilis in Iowa before the Europeans arrived. This made the front page of a statewide paper, and the next day in some Iowa schools, classmates harassed Indian children by telling them, "Your mothers are whores!" That also made front page. The scholar gave his best assessment of the data without thought to its impact; the story is not dissimilar to more recent controversies about evidence of cannibalism at Anasazi sites in the Southwest.

Some in our profession say that we need not be responsible for it—indeed, that it is our job to challenge people's beliefs about themselves and their cultures (see Clark 1996; Mason 1997). Being so naive in the rightness of our own judgments about the past serves no good purpose. We do not need to make archaeology into a cruel discipline. If we reach conclusions that are likely to be harmful to descendant communities, we need to figure out ways to present that information that might make it less damaging. This is not just "political correctness" but respect. We need to carefully contextualize our words to mute the repercussions. In the instance of the syphilis story, for example, the paleopathologist might have mentioned how hard it is to tell syphilis from similar diseases in bone and asked the reporter to emphasize the debate over its origins and the limitations of his findings. We can't control what comes out, but we need to be aware and respectful; that's also part of accountability to the public.

CONCLUSION

Dealing with the media is no easy matter, but it is necessary to what we do. Issues are complex and add another dimension to an already difficult job. With a bit of education in media literacy, you can make the job far easier on yourself. As you have experiences with the media, begin to catalog the incidents so that you can improve your skills. Pay attention to what works well and what doesn't. As with writing and speaking, practice doing some of the skills discussed in this chapter, such as writing a press release or giving an interview. Have a colleague or friends critique your performance. Prepare a handout for reporters who might appear at your next project. Working with the media can actually be fun and rewarding, but if you aren't ready, you may find yourself in a small public relations fiasco or, at very least, embarrassed. So much happens very quickly in the media, so the bad that happens is usually forgotten quickly.

Other forces, affecting both the media with which we deal and all the ways we present the past, will change much of what we do in the future. Most of these forces have to do with new technologies as ways of delivering content to our publics. That's our final topic in chapter 12.

12
THE FUTURE OF PRESENTING THE PAST

The shift in ways to think about presenting the past over the past decade has been rapid and remarkable. Early adopters of the World Wide Web will probably recall looking at information about archaeology on the web late in 1994, a mere decade ago. At last estimate, some twenty million new websites pop up a year, but there's no real way of knowing. A surprising number of them, with a wide range of quality, contain information about archaeology.

CD-ROMs preceded the web a bit, but they were primarily for games and a few educational products, mostly commercial ventures. Archaeology has taken a long time to start using them. Still, at the exhibit room and in publishers' flyers at the 2002 SAA meeting, I counted about a dozen that used archaeological sites and data as a primary component of the disk.

DVDs have also become readily available, allowing vastly more data storage for inclusion of large numbers of images, video, and sound. Development of new media seems to come quickly, posing new challenges.

New technologies always seem to fascinate archaeologists in other areas of their work, so I've been a bit surprised that we have not picked up on the use of these new ways of presenting material to both colleagues and the public as quickly as I thought we would. One reason we've been slow is that using new technologies requires learning new hardware and software. Another, however, is that many of us are trained to think in the very linear ways that writing promotes. The web and CD-ROMs promote a different way of seeing and presenting materials, one that is multilinear. Although there is an underlying structure to what has been called hypermedia, a user can move around in the medium easily, and it is vastly more dependent on sound, images, and interactivity than any medium we have yet produced in archaeology.

To use the new technologies, we are faced with problems of design, computer programming, and production of audio and video. Most of us just don't have the training and certainly don't have the time to deal with substantial learning curves in each element of the production. We are comfortable with the idea of controlling content, but not how to design or produce it for the new media.

What is necessary to produce the highest-quality hypermedia is a team of people, with each member a specialist in one or two elements. As discussed in chapter 7, one may need to add a graphic designer, a video specialist, an audio specialist, and a computer programmer—more! Every team member adds greater costs to the project. However, producing a website, CD-ROM, or DVD can be done with reasonably skilled nonspecialists. The basics are really pretty simple.

ARCHAEOLOGY ON THE WEB

If you haven't used the World Wide Web to look up information about archaeology, you are probably really unusual, a practicing Luddite, or a real technophobe. Most who have used the web understand its potential for providing information to both your colleagues and archaeology's other publics:

- It allows you to quickly combine images, text, sound, and video.
- It is certainly cheaper than printing a book or multiple copies of a report.
- It lets anyone in the world with web access see your materials quickly.

WEB BASICS

The fundamentals of website construction are easy. To use the web, you need a computer with some level of access to the Internet, a web browser, a web server, and a way to do HTML (hypertext markup language). These are explained in lots of books, and building websites can be accomplished with lots of different tools, including your own word processor, which probably offers an option of saving text files as a web page. To do exactly what you want may be more complicated and so may require learning a WYSIWYG (what you see is what you get) web-editing program or learning to do HTML "on the fly."

DEVELOPING WEB APPEAL

You can't use the web for very long without seeing something you'd like to have on your own website. Many fancy animations, sounds, and image placement require special programs for you to construct and may require a web browser plug-in. Plug-ins are usually free, but the programs to design what the plug-in is used to see are usually not. Some browsers come with many of the plug-ins already installed; download and install new ones as you need them. Some of these specialized programs produce what is essentially eye candy to draw in your audience, but others provide essential services.

One good example of an essential plug-in is Adobe Acrobat Reader because many organizations, especially government agencies, put documents on the web that way, including CRM reports, requests for proposals, and the like. Acrobat Reader reads .pdf (portable document files), which are facsimile copies of documents you produce with your word processor. To create .pdf files may require the installation of the full Adobe Acrobat program, which works in tandem with common word processors. Many fancy elements on the web can be done with high-end HTML editors. Forms, for example, are easy to create, but they require your web server to have special programs. These programs have been written in Perl, C++, or some other language and get called into and used by your web page through cgi (common gateway interface). Other fancy stuff is programmed in Java, a language developed especially for the web; Javascript is similar. Some high-end HTML editors will help you by writing the program for what you want to do.

Most of us don't want to get involved in such detailed programming, but we don't mind fiddling with code someone else has developed. One easy approach is to see a program you like on a website, "borrow" the code, and then modify it to your needs. That takes minimal programming experience. For some eye candy, you will need to buy a program, learn it, and create what you need.

As the web has proliferated, so have the tools. If you are going to use them, consider their costs and benefits in both time and money for what you hope the user will get out of it. Another concern is that until broadband Internet is available to more users, some of your creations will only be accessible to relatively narrow audiences. There are a few tools that archaeologists can use effectively.

VRML (VIRTUAL REALITY MODELING LANGUAGE)

Virtual reality is common in video games, but a number of archaeologists have written about it or applied it to websites and CD-ROMs. *Virtual reality* is the creation of a facsimile of the real thing, sometimes with three dimensions but as often with two dimensions on the video or computer screen. To build virtual reality is complex but certainly holds promise for archaeology.

A VRML document takes the form of a human-readable text file describing a three-dimensional scene. This text file is a list of commands that tell the computer to place objects of given sizes and colors at specific locations within a virtual world. The benefit of VRML is that lots of data can be represented visually in ways that allow you to see patterns that might not be apparent in a spreadsheet of numbers or measurements. In analysis of data, this feature can help spot problems with data shortage or weakness, or it may provide insight to organization or linkages of data. For presentation of some data, virtual reality is a major plus to help those not familiar with the data "see" what you are writing or talking about.

CRM will find VRML-generated data output to be a useful tool for predictive modeling. GIS 3D projections are already a form of virtual reality, and "fly-through" can make site predictive models seem much more real to project managers. VRML is also a way to make data meaningful to nonspecialists; for example, a recent article by Gummerman and Dean (2000) discusses the application for Anasazi archaeology. Vince and Garside-Neville (1997) provide a summary as well, but they give a somewhat less positive view due to speed and file sizes. The electronic journal *Internet Archaeology* (intarch/york.ac.uk/) presents several excellent examples of VRML use, so peruse the issue's table of contents.

One of the early models on the web was that of Chetro Ketl (Kanter 2001). If you've not seen VRML, this is a good place to start, but you will need to download a few plug-ins onto your computer if you don't already have them.

DIGITAL AUDIO AND DIGITAL VIDEO

If you've used the web, you've probably already encountered audio. Digital audio is relatively simple. You can prepare digital audio if you can connect a microphone to your computer's sound card. For short audio files, where great clarity is not all that important (some music and most spoken voice), you literally can speak or play right into the microphone. Most computers come with a program to allow you to do simple sound editing to help clean the sound up a bit, but for music, where clarity is important, or for mixing, you may want to use a specialized program that handles the growing range of audio file types.

As already discussed in chapter 6, video can easily be adapted to use on the web. The core issues in digital video use on the web are file size and bandwidth. Video file size is extremely large, so compression of files is necessary (which is why most video doesn't fill the whole screen), as is sampling, which decreases the number of frames per second (fps). The lower the fps, the more jerky the video looks. *Bandwidth* is essentially the amount of information that can move from computer to computer in a certain amount of time. Slow telephone modems have a much smaller bandwidth than cable modems, for example, and this affects the speed at which the user gets to see the material from your website. For people with low bandwidth, the World Wide Web becomes the World Wide Wait!

Streaming media uses specialized file formats and compression to allow the video and audio to begin playing as soon as they start downloading. This lets the user see/hear immediately, and compression doesn't have to be so severe. Bandwidth still has an impact.

SOME PROBLEMS WITH USING THE WEB (AND A FEW FIXES)

For all the benefits the web brings for relatively low-cost presentations of archaeological materials and easy access to them, it also can create problems. All of them are issues you need to consider if you decide to use the web to deliver archaeological materials to professional colleagues or the public.

- *Problem 1: The web is more ephemeral than printed materials.* Websites tend to go away more quickly than most of us like. The medium is constantly changing, so that what a site had on it yesterday may not be on it today. If sites are maintained, content that

was once featured may now be buried or removed from the site. If you find material on a website that is very useful to you, you may want to save the particular page or site on disk or perhaps even print it. If you want to save the whole website, that's also easy using purchased programs, shareware, or freeware.
- *Problem 2: Materials on the web are not peer reviewed.* Anyone with resources can put up a website to say just about anything. The web contains a wealth of junk, but you might remember that most libraries do as well. Many journals and even some monographs on the web are starting to be peer reviewed, which raises their credibility. Even if not peer reviewed, they may well be put up by a competent archaeologist and edited by competent editors. As with any material you use as a scholar, you need to know how to evaluate websites. You'll find lots of advice about this on the web itself and perhaps need to consider how this advice needs to be modified to suit archaeological information. Alexander and Tate (1999) offer excellent suggestions for both evaluating websites and providing users with ways to evaluate your own materials.
- *Problem 3: Flash sometimes takes precedence over content.* Websites can now do so much with graphics, sound, and video that site layout, art, and action sometimes take precedence over content. If you are on a slow connection, this becomes profoundly annoying. At the same time, be aware that web users tend to be younger and that many have grown up with the eye candy of television, MTV, and video games. "Flash" can bring attention to your website, but too much becomes a problem. One good way to figure out what you should put on your site is to look at others you find well organized and attractive, and then design a site like them. In truth, much beyond simple web layout might best be left to specialists in graphics and web design.

Remember, though, that on the web, *content is king*, no matter what might be possible in terms of the software and eye candy. Less is probably best in terms of the fancy design, but at the same time, websites with text alone defeat the purpose of the medium. A good mix of media is fine, but don't go overboard.
- *Problem 4: Some websites are very hard to navigate.* If you have visited some websites with lots of materials, you can sometimes quickly get lost in the many pages of information provided. Good sites provide good navigation tools.
- *Problem 5: The real cost of the web comes in maintaining the website!* If your website contains any content that can change outside of

your control, your single biggest cost will be in maintaining the site. The costs mostly come in time, not money, trying to keep content current. This will especially be the case if your website contains lots of links to other sites. You'll be vastly happier if you provide original content and as few links to other sites as possible. If your site contains active data such as databases, these can also take a great deal of time to maintain. Whatever the complexity of your website, count on the fact that you'll need to provide regular maintenance.

CD-ROMs AND DVDs

CD-ROMs actually appeared shortly before the web, but they are fundamentally the same. They both use hypermedia, linking a range of audio, graphics, video, and text about a subject. The major difference is that a CD-ROM resides in your own computer. Up to seven hundred megabytes (MB) of files get stored on a CD-ROM disk rather than on a web server. In contrast, DVD-ROMs hold a minimum of 4.7 gigabytes (GB) of information. This huge capacity can hold a full-length movie and then some. The way they work is the same, but they sometimes require different players.

Designing a CD or DVD is much the same as designing a website, with a few added complexities because they are stand-alone products. The web is mostly platform-independent, but CDs and DVDs are not. You must decide whether to make your product playable by Windows, Macintosh, or both. Public schools tend to use Macintosh systems heavily, so you'll probably want to provide for both. Fortunately, this is mostly a matter of adding easily acquired software to make your program run on each platform.

One simple way around this is to use HTML, which is platform-independent, to set up a website, and then transfer the whole thing to the CD/DVD. To make the product work, you need to instruct users to open a web browser and which file name to type in the location/URL/address line of the browser. Once they've done that, they are on their way. If they do this with a live Internet connection, you can even include links to external websites.

You might ask why you'd want to put the product on a CD/DVD instead of the web. The answers are simple:

- There will be no problems with down servers.
- Slow web connections are nonexistent.

- Bandwidth is not a concern.
- Vastly more material can be included than on most websites.

If you choose not to use an HTML-based approach, you'll want to acquire a multimedia authoring program to build your CD/DVD. Most have a steep learning curve, but they are worth the trouble for the amount of control you get in combining and sequencing your materials.

STORYBOARDING: A CRUCIAL CONCEPT FOR WEBSITES AND CDs/DVDs

When you learned how to write, your teachers probably taught you the use of outlines, a way of organizing your thoughts before you put them onto paper. In a sense, the outline keeps you from getting lost in a maze of ideas, citations, illustrations, and other items you'd like in your document. Storyboarding is a similar concept but borrowed from filmmaking. In the case of the web, CD-ROMs, and DVDs, storyboards help you keep your files, images, and direction in place. Guthrie (2000) provides the following storyboarding steps:

1. Define the purpose of your page.
2. Identify the audience for this page (primary, secondary, or other).
3. Choose the main heading for the page.
4. Identify major categories of information that will go on your page, and place related information in these categories.
5. Determine how you will link information. Identify which links are relative and which are absolute.
6. Decide what graphics and images you will use (your .jpg and .gif files).
7. Plan how you will provide navigation through the pages (e.g., buttons, bars, text, bottom of the page, side of the page, or top of the page).
8. Draw a simple sketch depicting how the pages or topics link to one another. (For additional information, see the storyboarding bibliography compiled by Marie Wallace at www.llrx.com/columns/sbbiblio.htm.)

A growing body of professional computer software can help with storyboarding, but it tends to be expensive. You might do just as well with a notepad. Take it from lots of bad experience: Storyboarding can really help your product!

INTERACTIVITY: BRINGING THE PAST TO LIFE

The strength of using new media for presenting the past is that they can bring the past to life by allowing the user to interact in ways that text and a few pictures cannot do. Hypermedia provide ways for an individual to make choices about where to go, what to see, and in what order. Words alone can fire imaginations, but not in the same way. Interactivity helps the user understand the complexities of ancient cultural systems and many of the processes at work in them. Interactivity is definitely nonlinear.

The problem with being trained in a discipline is that you sometimes have trouble thinking outside the box. For archaeologists trained in linear approaches to discovering and interpreting the past, this is a problem. If you intend to design new media, you'll need to train yourself to think in different ways. One good way to get a feel for this is to try lots of video games and educational programs just to understand the approach. There are even some featuring archaeology (Watrall 2002). Certainly, you need to try to see things the way your audiences might. See what's out there, and consider the possibilities.

Let me go out on what is a rather secure limb: Print media are not going to go away, but the web and its descendants will be the future for delivering archaeological materials to colleagues and our publics. There is plenty of room for CDs/DVDs, and in the near term, you'll see a great deal of experimentation using virtual reality, streaming audio and video, and a host of other applications. With wider distribution of broadband web access, especially Internet 2, much more will be possible, so much so that predicting what you'll see on the web in a decade is like trying to predict what you'll find when you put a test pit in an archaeological site. The key for your success in using the web or any other new media will be your own creativity and your budget.

We shouldn't be afraid of these developments or let our tradition of emphasizing print media stand in the way of experimentation and adoption of new technologies. Thankfully, there are always a few who push the edges. Over the past decade, several challenging articles have appeared in the *SAA Bulletin* as part of a "Net Works" column, helping expand the horizon for presenting the past. New technologies bring new issues and problems, and colleagues are addressing them. For example, the SAA recently published a collection of papers, *Delivering Archaeological Information Electronically* (Carroll 2002), that addresses a range of issues from peer review to archiving of materials.

CONCLUSION

This book began with a discussion of archaeologists' ethical obligations to the past, and it will end that way. Stewardship, however it gets defined, is complex and underpins everything we do. It is not just a matter of protecting the physical remains of the past or creating an intellectual record of the past. If we agree that it is a public heritage, then we have to make the past *public;* that is, we have to make what we do "transparent" in terms of the ways we do and think about things. We have to let people know about what we find but also, as important, what we believe to be its impacts for their lives. This is a profound responsibility. If the past is a public heritage, then archaeologists are partners with the public as stewards of that heritage. Presenting the past is part of that partnership. Do it well, do it creatively, and do it responsibly.

APPENDIX

SOME ARCHAEOLOGY JOURNALS ON THE WEB

Many Archaeology journals have associated websites. Some list tables of contents, and some have the same articles as the print editions. URLs do change, so if the links listed here do not work, use a search engine to try to find the site.

Aerial Archaeology Newsletter: www.nmia.com/~jaybird/AANewsletter/
African American Archaeology Newsletter: www.mindspring.com/~wheaton/NSA.html
African Archaeology Review: www.plenum.com
American Antiquity: www.saa.org/Publications/AmAntiq/amantiq.html
Andean Past: kramer.ume.maine.edu/~anthrop/AndeanP.html
Anthropology Today (some archaeology): lucy.ukc.ac.uk/rai/
Antiquity: intarch.ac.uk/antiquity
Archaeoastronomy & Ethnoastronomy News: www.wam.umd.edu/~tlaloc/archastro/
Archaeological Dialogues: archweb.LeidenUniv.nl/ad/home_ad.html
Archaeology Ireland Magazine: slarti.ucd.ie/pilots/archaeology/
Archaeology Magazine: www.he.net/~archaeol/index.html
Assemblage (The Sheffield Graduate Journal of Archaeology): www.shef.ac.uk/uni/union/susoc/assem/index.html
At the Edge: www.gmtnet.co.uk/indigo/edge/atehome.htm
Berkeley Archaeology: www.qal.berkeley.edu/arf/
British Archaeology: britac3.britac.ac.uk/cba/ba/ba.html
Bulletin of Information on Computing in Anthropology: lucy.ukc.ac.uk/bicaindex.html

CRM: www.cr.nps.gov/crm
Cultural Anthropology Methods (CAM): www.lawrence.edu/~bradleyc/cam.html
Current Anthropology (some archaeology): www.artsci.wustl.edu/~anthro/ca
Current Archaeology: www.archaeology.co.uk
Forum Archaeologiae (Austrian Journal of Archaeology): allergy.hno.akh-wien.ac.at/forum/
Glasgow University Archaeological Research Division (GUARD) Digest Reports: www.gla.ac.uk/Acad/Archaeology/guard/guard
Internet Archaeology: intarch.york.ac.uk
Journal of Field Archaeology: jfa-www.bu.edu
Journal of Material Culture: www.sagepub.co.uk/journals/details/mcu.html
KMT: A Modern Journal of Ancient Egypt: www.sirius.com/~reeder/kmt.html
Latin American Antiquity: www.saa.org/Publications/LatAmAnt/latamant.html
Midcontinental Journal of Archaeology: www.uiowa.edu/~osa/publica/mcja/mcja.htm
Online Archaeology: avebury.arch.soton.ac.uk/Journal/journal.html
SAA Bulletin: www.sscf.ucsb.edu/SAABulletin

REFERENCES

Addington, Lucile R.
    1986   *Lithic Illustration: Drawing Flaked Stone Artifacts for Publication.* University of Chicago Press, Chicago.

Adkins, Lasley, and Roy A. Adkins
    1989   *Archaeological Illustration.* Cambridge University Press, Cambridge.

Alexander, Janet E., and Marsha Ann Tate
    1999   *Web Wisdom: How to Evaluate and Create Information Quality on the Web.* Erlbaum, Mahwah, N.J.

Allen, Mitch
    2002   Reaching the Hidden Audience: Ten Rules for the Archaeological Writer. In *Public Benefits of Archaeology*, edited by Barbara J. Little, pp. 244–51. University Press of Florida, Gainesville.

Anonymous
    1999   Tuning in on Public Opinion. *Common Ground*, Winter 1999:5.
    2001   Photographing Archaeological Sites. *American Archaeology* 5(1):21–28.

Armstrong, Douglas V.
    1985   Maps. In *The Student's Guide to Archaeological Illustrating*, edited by B. Dillon, pp. 27–42. University of California–Los Angeles Institute of Archaeology Archaeological Research Tools, Volume 1, Los Angeles.
    2001   *Evaluating Web Resources.* www2.widener.edu/Wolfgram-Memorial-Library/webevaluation/webeval.htm. Viewed April 20, 2001.

REFERENCES

Bahn, Paul, and Bill Tidy
    1999   *Disgraceful Archaeology or Things You Shouldn't Know about the History of Mankind.* Tempus Publishing, Charleston, S.C.

Baxter, Jane E.
    1996   Getting Graphic! Making an Effective Poster. *SAA Bulletin* 14(5):9, 31.

Becker, Howard S.
    1986   *Writing for Social Scientists: How to Start and Finish Your Thesis, Book, or Article.* University of Chicago Press, Chicago.

Bender, Susan J., and George S. Smith (eds.)
    2000   *Teaching Archaeology in the 21st Century.* Society for American Archaeology, Washington, D.C.

Brodribb, Conant
    1971   *Drawing Archaeological Finds.* Association Press, New York.

Bronowski, Jacob
    1973   *The Ascent of Man.* Little, Brown, Boston.

Carroll, Mary S. (ed.)
    2002   *Delivering Archaeological Information Electronically.* Society for American Archaeology, Washington, D.C.

Clark, G. A.
    1996   NAGPRA and the Demon-Haunted World. *SAA Bulletin* 14(5):3.

December, John, and Susan Katz
    1996   Abstracts. Rensselaer Polytechnic Institute: The Writing Center. www.rpi.edu/llc/writecenter/web/text/abstracts.html. Viewed April 16, 2001.

Deloria, Vine
    1995   *Red Earth, White Lies: Native Americans and the Myth of Scientific Fact.* Scribner's, New York.

Dillon, Brian
    1985   Introduction. In *The Student's Guide to Archaeological Illustrating,* edited by B. Dillon, pp. 3–8. Archaeological Research Tools, Volume 1. University of California at Los Angeles, Institute of Archaeology.

Dorrell, Peter
    1989   *Photography in Archaeology and Conservation.* Cambridge University Press, Cambridge.

Ducette, Martin
    1999   *Digital Video for Dummies.* Wiley, New York.

Dvorak, John
 2001 Closing in on Perfection. *PC Magazine* 20(6):79.
Eiseley, Loren
 1971 *The Night Country*. Scribner's, New York.
Fagan, Brian
 2002 I Am So Tired of Jargon and Narrow Teaching. . . . *SAA Archaeological Record* 2(2):5–7.
Gonick, Larry
 1990 *The Cartoon History of the Universe*, Volumes 1–7, *From the Big Bang to Alexander the Great*. Doubleday, New York.
Gummerman, George J., and Jeffrey Dean
 2000 Artificial Anasazi. *Discovering Archaeology* 2(2):44–51.
Guthrie, Sally
 2000 *Storyboarding*. web.utk.edu/~sguthrie/j416/storyboard.htm. Viewed April 23, 2001.
Halsey, John R.
 1991 "State Secrets": The Protection and Management of Archaeological Site Information in Michigan. In *Ethics and Professional Anthropology*, edited by C. Fluehr-Lobban, pp. 115–19. University of Pennsylvania Press, Philadelphia.
Hart, Russell, and Dan Richards
 1998 *Photography for Dummies*. Wiley, New York.
Hartley, L. P.
 1953 *The Go Between*. Hamilton, London.
Heider, Karl
 1976 *Ethnographic Film*. Austin, University of Texas Press.
 2001 *Seeing Anthropology: Cultural Anthropology Through Film*. Allyn and Bacon, Boston.
Heizer, Robert F., and John A. Graham
 1967 *A Guide to Field Methods in Archaeology: Approaches to the Anthropology of the Dead*. National Press, Palo Alto, Calif.
Hester, Thomas R., Harry J. Shafer, and Kenneth L. Feder
 1997 *Fields Methods in Archaeology*. Mayfield, Mountain View, Calif.
Houk, Brett A., and Bruce K. Moses
 1998 Scanning Artifacts: Using a Flatbed Scanner to Image Three-Dimensional Objects. *SAA Bulletin* 16(3):36–39.

Howell, Carol L., and Warren Blanc
    1992   *A Practical Guide to Archaeological Photography*. University of California at Los Angeles, Institute of Archaeology.

Jameson, John H., Jr. (ed.)
    1997   *Presenting Archaeology to the Public: Digging for Truths*. AltaMira, Walnut Creek, Calif.

Kantner, John
    2001   Sipapu-Anasazi Great House Model. sipapu.gsu.edu/kinshow/kin.tliish.html. Viewed April 16, 2001.

King, Julie Adair
    1999   *Digital Photography for Dummies*. Wiley, New York.

Klesert, Anthony L.
    1998   You Too Can Write Good: Writing about Archaeology for Local Newspapers. *SAA Bulletin* 16(3):40–41.

Lynott, Mark, and Alison Wylie (eds.)
    1995   *Ethics in American Archaeology: Challenges for the 1990s*. Society for American Archaeology, Washington, D.C.

Macaulay, David
    1979   *Motel of the Mysteries*. Houghton Mifflin, New York.

Mason, Ronald J.
    1997   Letter to the Editor, *SAA Bulletin* 15(1):3.

McPherron, Shannon P., and Harold L. Dibble
    2002   *Using Computers in Archaeology: A Practical Guide*. McGraw-Hill, Mayfield, Boston.

Molyneaux, Brian L.
    1997   Introduction: The Cultural Life of Images. In *The Cultural Life of Images: Visual Representation in Archaeology*, edited by B. Molyneaux, pp. 1–10. Routledge, London.

National Park Service
    2000   National Archeological Database Reports. www.cast.uark.edu/products/NADB/nadb.mul.html. Viewed April 4, 2001.

Neiman, Fraser D.
    1994   A Poster Primer: A Few Tips for Planning Your Poster Session. *SAA Bulletin* 12(1):13–14.

Pendegast, David
    1998   The Bedfellows Are Less Strange These Days: The Changing Relationship between Archaeologists and the Media. *SAA Bulletin* 16(5):15–16.

Piggott, Stuart
  1979 *Antiquity Depicted: Aspects of Archaeological Illustration.* Thames and Hudson, London.

Renfrew, Colin, and Paul Bahn
  1996 *Archaeology: Theories, Methods and Practice.* Thames and Hudson, London.

Rick, John
  1999 Digital Still Cameras and Archaeology. *SAA Bulletin* 17(1):37–41.

Rick, John, and Dakin Hart
  1997 Panoramic Virtual Reality and Archaeology. *SAA Bulletin* 15(5):14–19.

Smardz, Karolyn E., and Shelley J. Smith (eds.)
  2000 *The Archaeology Education Handbook: Sharing the Past with Kids.* AltaMira, Walnut Creek, Calif.

Society for American Archaeology
  1992 Editorial Policy, Information for Authors, and Style Guide for *American Antiquity* and *Latin American Antiquity*. *American Antiquity* 57:749–70 and www.saa.org/Publications/StyleGuide/saaguide.pdf. Viewed May 28, 2002.
  1996 Principles of Archaeological Ethics. www.saa.org/Aboutsaa/Ethics/prethic.html. Viewed April 23, 2002.
  1999 *Teaching Archaeology in the 21st Century: Promoting a National Dialogue.* www.saa.org/Education/Curriculum/principles1.html. Viewed May 28, 2002.
  2000 Poll Finds Public Support of Archaeology. www.saa.org/Pubrel/publiced-poll.html. Viewed May 28, 2002.

Vince, Alan, and Sandra Garside-Neville
  1997 Publishing multimedia in archaeology. *Internet Archaeology.* intarch.york.ac.uk/news/eva97.html. Viewed April 23, 2001.

Watrall, Ethan
  2002 Interactive Entertainment as Public Archaeology. *SAA Archaeological Record* 2(2):37–39.

Whittaker, John
  1992 Hard Times at Lizard Man. *Archaeology* 45(4):56–58.

Zimmerman, Larry J.
  1985 *Peoples of Prehistoric South Dakota.* University of Nebraska Press, Lincoln.
  1986 Redwing. *Anthropology and Humanism Quarterly* 11(1):8–9.

1995 Regaining Our Nerve: Ethics, Values and Transforming Archaeology. In *Ethics in American Archaeology: Challenges for the 1990s*, edited by M. Lynott and A. Wylie, pp. 64–67. Society for American Archaeology, Washington, D.C.

2001 Usurping Native American Voice. In *The Future of the Past: Archaeologists, Native Americans and Repatriation*, edited by Tamara L. Bray, pp. 169–84. Garland, New York.

Zimmerman, Larry J., and Lawrence E. Bradley
1993 The Crow Creek Massacre, Initial Coalescent Warfare and Speculations about the Genesis of Extended Coalescent. *Plains Anthropologist*, 38(145):215–26.

Zimmerman, Larry J., Steve Dasovich, Mary Engstrom, and Lawrence Bradley
1994 Listening to the Teachers: Warnings about the Use of the Archaeological Agenda in the Classroom. In *The Presented Past: Archaeology, Museums and Public Education*, edited by P. Stone and B. Molyneaux, pp. 359–74. Routledge, London.

INDEX

AAA. *See* American Association of Archaeology
AAM Online Bookstore Catalogue website, 118
abstract: for CRM report, 66–67, 68; for public talk, 66–67
abstract, for publication, 65–67; descriptive, 68; informative, 67–68; meeting deadline for, 69–70; specifications for, 67; style for, 68; writing, 68–69
accountability, 1–2, 8, 9–11
Addington, Lucile R., 51
Adkins, Lasley, 49, 51, 52
Adkins, Roy A., 49, 51, 52
Adobe Acrobat, 17
Age of Empires (computer game), 112
Alexander, Janet E., 140
Allen, Mitch, 28, 107
American Anthropological Association, press release guide of, 127
American Anthropologist, 38, 96
American Antiquity, 96, 102; style guide of, 38, 39; website for, 145
American Archaeology, 52–53
American Association of Archaeology (AAA), 84

American Indian, 11, 12–13, 30
Anderson, David, 83
annotated bibliography, 37–38
anonymous peer review, 105
Anthony, Piers, 113
AnthroNotes, 114
Antiquity Depicted: Aspects of Archaeological Illustration (Piggott), 49
APA (American Psychological Association) style guide, 39
Archaeological Channel (TAC), 57–58
Archaeological Conservancy, 9–10
Archaeological Illustration (Adkins & Adkins), 51
Archaeological Institute of America, 9–10
Archaeological Legacy Institute, 57
archaeological project, prevalence of unreported, 2
Archaeological Record, 82–83
archaeologist, stereotype of, 12, 114
Archaeology, 114
Archaeology and Public Education Newsletter (SAA), 117
"Archaeology Days" project, 116–17
Archaeology Education Handbook (Smardz & Smith), 117

INDEX

Armstrong, Douglas V., 51
Association of Archaeological Illustrators and Surveyors, 51
audience: child, 27, 30–31; contractual obligations to, 16–17; determining, 16; educated, 28–29; hidden (nonspecialist), 16, 17, 28, 29–30; professional archaeologist, 16, 28, 29; in public speaking, 73; using jargon in communicating with, 12–14
Auel, Jean, 113

Bahn, Paul, 114
bandwidth, 139
Baxter, Jane E., 82–84
Bender, Susan J., 8–9
bibliography, annotated, 37–38
Biolsi, Tom, 106
Blanc, Warren, 52
boilerplate, for CRM report, 93
Bones of Contention (TV program), 116
book chapter, writing, 96–97
book/monograph: book vs. monograph, 98–99; index for, 61, 101, 106; proof pages for, 100–101; prospectus for, 99; publishing, 99–100; trade, 100
book review, 97–98
Brin, David, 112
Bulletin. See *SAA Archaeological Record*
Bureau of Land Management, 9–10
Burns, Steve, 124, 129, 130, 131

CADD. See computer-aided design and drafting
Carroll, Mary S., 144
cartoon, as presentation medium, 114
The Cartoon History of the Universe (Gonick), 114

CD-ROM/DVD presentation medium: creating, 135, 141–42; using storyboarding in creating, 142–43
CD-ROM presentation, cost of, 17
Chicago Manual of Style, 26, 39
citation, reference: bibliographic/research control (sample), 36; bibliographic software for, 38–39; building annotated bibliography, 37–38; style guide for, 39–40; tracking through bibliographic/research control, 35–37; uses for, 32–33; when to use, 32–34
computer, for presenting archaeology: advanced needed skills, 42–43; basic needed skills, 42; hardware for, 43–44; learning curves for using, 45–46; reference materials for, 41; software for, 17, 18, 38–39, 44–45, 81
computer-aided design and drafting (CADD), 56
conference paper: reasons to present, 70–71. See also abstract; public speaking
copyediting, 61, 106, 107
copyright, 34
CRM (cultural resource management) report, 91: abstract for, 66–67, 68; availability of, 94; boilerplate for, 93; main elements of, 92–93; meeting client needs, 93–94; peer review of, 105
Crow Creek Massacre website, 30
cultural resource management report. See CRM report
Current Anthropology, 96

data field, 38
Dean, Jeffrey, 138
December, John, 68–69

Delivering Archaeological Information Electronically (Carroll), 144
Department of Defense, report style guide of, 39
descriptive abstract, 68
designer, 61
Dibble, Harold L., 41, 53
Dig magazine, 27, 30–31
digital audio/video, for website development, 139
digital camera, 54, 55
digital media specialist, 61
Dillon, Brian, 51
Disgraceful Archaeology (Bahn & Tidy), 114
Dorrell, Peter, 52
Dotson, Bob, 130

editorial process: colleague/friend as editor, 104–5; copyediting, 61, 106, 107; for CRM report, 61, 107–8; editor role in, 107–8; indexing, 61, 101, 106; page proof review, 106–7; peer review, 105–6, 108; self-editing, 103–4
Eisley, Loren, 7
electronic conference session, 84–85; chairing, 88; discussant for, 87, 88–90; festschrift, 85, 86, 88; memorial, 85, 86, 88; organizing, 85–87, 88; thematic, 86, 88
ethics, 8, 132–33
event, as presentation medium, 117–18
exhibit/museum, as presentation medium, 118–19

Fagan, Brian, 10
Feder, Kenneth L., 53
festschrift electronic conference session, 85, 86, 88
fiction, archaeological, 112–13

Field Methods in Archaeology (Hester, Shafer, & Feder), 53
field photography, 47
Fish and Wildlife Service, 9–10
Forest Service, 10
freeware, 45

game/toy, as presentation medium, 111–12
Garside-Neville, Sandra, 138
Gear, Kathleen O'Neal, 113
Gear, W. Michael, 113
geographic information system (GIS), 56, 138
glossophobia, 71–72
The Go Between (Hartley), 7
Gonick, Larry, 114
Graphic Archaeology, 51
graphic tablet, for visual archaeology, 55
graphics program, for visual archaeology, 55–56
gray literature, 91
Green, Bill, 107
Gummerman, George J., 138
Guthrie, Sally, 142

Hadingham, Evan, 124
Hampe, Barry, 115–16
hands-on learning, 27, 30–31, 110–11
hardware, computer: buying tactics for, 44; for presenting archaeology, 43–44
Hart, Russell, 52
Hartley, L. P., 7
Heider, Karl, 57
Hester, Thomas R., 53
Houk, Brett A., 55
Howell, Carol L., 52

illustration, for visual archaeology, 48–49
illustrator/graphics specialist, 61

index, for book/monograph, 61, 101, 106
informative abstract, 67–68
instructional designer/educational specialist, 62
Internet Archaeology, 138, 146
Iowa Archaeology Month website, 117
Israel, prevalence of unreported projects in, 2

Jacka, Jerry, 53
Jackson, Steve, 112
Jameson, John H., Jr., 117
jargon, 12–14, 17, 28, 30, 31, 70
journal: article for, 91, 94–95; book review for, 97–98; notes/reports for, 97; peer review of, 95; peer review of article, 95, 105; prestigious, 96, 102; regional, 95; state, 95, 102–3; topical, 95–96; website for, 138, 145–46
Journal of Field Archaeology, 95, 146

Katz, Susan, 68–69
Ketl, Chetro, 138
King, Kathleen, 113
Klesert, Anthony L., 128
Koryakova, Ludmila, 3

Larson, Gary, 114
learning curve, definition of, 45
Leiter, Eldon, 53
Lithic Illustration: Drawing Flaked Stone Artifacts for Publication (Addington), 51
Lost Civilizations Explorer's Kit, 111

Macaulay, David, 114
Making Documentary Films and Reality Videos (Hampe), 115–16
marketing specialist/public relations, 62

McLuhan, Marshall, 124
McPherron, Shannon P., 41, 53
media, gaining attention of, 125; in larger market, 129; in local paper, 128; in local television/radio, 128–29; through press release, 126, 127; through public relations office, 126
media, working with, 121–22; attitude for, 122–23; cautions concerning, 130–32; going public with story, 125; learning about, 123–24; story hook, 124–25, 129
media literacy, 123
memorial electronic conference session, 85, 86, 88
Molyneaux, Brian L., 48
monograph. *See* book/monograph
Moses, Bruce K., 55
Motel of the Mysteries (Macaulay), 114
movie/video, as presentation medium, 114–16
multidisciplinary nature, of archaeology, 2, 3–4
Museum Curator: The Museum Journal, 118
museum/exhibit, as presentation medium, 118–19
Museum News, 118

National Archaeological Database (NADB), 94
National Park Service, 17, 94
National Science Foundation, unreported projects funded by, 2
National Wildlife Service, 10
Neiman, Fraser D., 82
Network of State and Provincial Archaeology Education Coordinators website (SAA), 117
newsletter, writing for, 95, 97, 102–3
The Night Country (Eisley), 7

optical character recognition (OCR), 54, 55
Our Amazing Ancestors Science Kit, 111
outline, writing, 23

passive voice, 25–26, 30
peer review, 105–6, 108; anonymous, 105; of CRM report, 91, 105; of journal, 95; of journal article, 95, 105; lack on World Wide Web, 140
Pendergast, David, 122
Peoples of Prehistoric South Dakota (Zimmerman), 13–14
photographer, 61–62
photography: archaeological, 52–53; digital, 53; field, 47; photograph vs. drawing, 49–50; resources for, 52–53
Photography for Dummies (Hart & Richards), 52
Photography in Archaeology and Conservation (Dorrell), 52
Piggott, Stuart, 49
plagiarism, 34
Plains Archaeologist, 13, 29, 95
poster paper, for conference: benefits of/drawbacks to, 79–80, 84; judging, 83; recycling, 83–84; reference sources for, 82–83; simplicity of, 80–81; tips for, 82, 83; vs. traditional paper, 81–82
A Practical Guide to Archaeological Photography (Howell & Blanc), 52
presentation medium: action *vs.* description in, 110–11; archaeological fiction, 112–13; cartoon, 114; choosing, 18; cost of distributing, 18–19; event, 117–18; exhibit/museum, 118–19; game/toy, 111–12; hands-on activity/learning, 27, 30–31, 110–11, 117; movie/video, 114–16; public archaeology education, 116–17
Presenting Archaeology to the Public (Jameson), 117
press release, 126, 127
proof pages, for book/monograph, 100–101, 106–7
prospectus, for book/monograph, 99
Public Archaeology, 96
public archaeology education, 116–17
public relations/marketing specialist, 62
public speaking: abstract for, 66–67; fear of, 71–72; finding opportunity for, 75–77; packaging a good talk, 79; reasons to present, 70–71. See also public speaking, tips for
public speaking, tips for: answering questions, 74–75; audience, 73; material, 73–74; presenting graphics at SAA, 78; presenting paper at SAA, 76, 77; room, 72–73
public view/beliefs, concerning archaeology, 10, 122–23
publishing: and career development, 101–3. See also book/monograph; editorial process; journal
Pyramid: Challenge of the Pharaoh's Dream (computer game), 112

Raiders of the Lost Ark (film), 114
Rapa Nui (film), 115
readability index, 33
reference. See citation, reference
reference tool, for writing, 26
regional journal, 95
report: lack on archaeological project, 2–4. See also CRM report
review essay, 98

Richards, Dan, 52
Rick, John, 53
Russell, Sharman Apt, 113
Russia, archaeology in, 3

SAA. *See* Society for American Archaeology
SAA Archaeological Record, 10, 41, 143, 146
SAA Bulletin. *See SAA Archaeological Record*
Sarabande, William, 113
scanner, 54–55
scope of work, CRM, 15, 16, 17
The Secrets of Easter Island (documentary), 115
self-editing, 103–4
Shafer, Harry J., 53
shareware, 45
slide/negative camera, 55
Smardz, Karolyn E., 117
Smith, George S., 8–9
Smith, Red, 24
Smith, Shelley J., 117
Society for American Archaeology (SAA): abstract specifications of, 67; ethics code of, 8, 145–46; graphic presentation tips of, 78; paper presentation tips of, 76, 77; poll on public view of archaeology by, 9–10; public education resources of, 117; on student training, 8–9
Society for Historical Archaeology, 10
software: bibliographic, 38–39; freeware/shareware, 45; for layout, 81; for .pdf file, 17, 18
sound bite, 130
South Dakota Archaeology, 102
staff: outsource vs. in-house, 17, 62–63; team development, 60–62
state journal/newsletter, 95, 102–3
storyboarding, 57, 142–43

The Student's Guide to Archaeological Illustrating (Dillon), 51
style guide, 39, 67

TAC. *See* Archaeological Channel
Talbot, W. H. Fox, 49
Tate, Marsha Ann, 140
Teaching Archaeology in the 21st Century (Bender & Smith), 8–9
technology, for presenting the past: CD-ROM/DVD, 141–42; increase in use of, 135; interactivity of, 143. *See also* World Wide Web
television, as presentation medium, 115
thematic electronic conference session, 86, 88
Tidy, Bill, 114
Timelapse (computer game), 112
Toastmasters International, 71–72
topical journal, 95–96
toy/game, as presentation medium, 111–12
trade book, 100
training, in archaeology: basic archaeological skills, 9; written/oral communication, 8–9
Tribes: It's 50,000 BC—Where Are Your Children? (computer game), 111–12

Using Computers in Archaeology: A Practical Guide (McPherron & Dibble), 41, 53

Van Tilburg, Jo Anne, 115
video/digital audio, for website development, 139
video/movie, as presentation medium, 57, 114–16
Vince, Alan, 138
virtual reality modeling language (VRML), 138

visual archaeology: artist for, 47–48; definition of, 47; illustrations, 48–49; photographs vs. drawings, 49–50; photography, 47, 49; reference source for, 47, 49
visual archaeology, tools for, 50; archaeological photography, 52–53; digital imaging, 53–56; digital photography, 53; digital photography resources, 53; drawing equipment, 51–52; drawing resources, 51; specialty graphics programs, 56; video, 57–58
VRML. *See* virtual reality modeling language

Wall, Steven, 53
web camera, 55
web presentation, cost of, 17
website: AAM Online Bookstore Catalogue, 118; American Association of Archaeology, 84; for Archaeology Channel, 57; for archaeology journal, 138, 145–46; for Crow Creek Massacre, 30; for dictionary, 26; for electronic journal, 138; Iowa Archaeology Month, 117; for poster presentation, 84; SAA ethics code, 8, 145–46; storyboarding for, 142–43; Wyoming "Archaeology Awareness Month," 117
website development: digital audio/video for, 139; plug-ins for, 137; virtual reality modeling language for, 138
Whitley, David, 53
Whittaker, John, 112
Wilford, John Noble, 124
Woody, Alanah, 83
World Wide Web: basics of, 136–37; common problems using, 139–41
writing, as circular process, 22–23
writing, developing basic skills in, 21–22; accept editing, 24–25; find strategy, 25; outlines, 23; practice, 25; by reading, 25–26; structure, 23–24; using references to aid in, 26
writing style: for child audience, 27, 30–31; for educated audience, 28–29; for fiction, 31; for hidden audience (nonspecialist), 17, 28, 29–30; for professional audience, 28, 29; simplicity in, 31–32; tips for simplifying, 32
Wyoming "Archaeology Awareness Month" website, 117

Young, Peter, 124, 125
Your Dictionary, 26

Zimmerman, Larry, 13–14

ABOUT THE AUTHOR AND SERIES EDITORS

Larry J. Zimmerman is the head of the Archaeology Department of the Minnesota Historical Society. He served as an adjunct professor of anthropology and visiting professor of American Indian and Native Studies at the University of Iowa from 1996 to 2002 and as chair of the American Indian and Native Studies Program from 1998 to 2001. He earned his Ph.D. in anthropology at the University of Kansas in 1976. Teaching at the University of South Dakota for twenty-two years, he left there in 1996 as Distinguished Regents Professor of Anthropology.

While in South Dakota, he developed a major CRM program and the University of South Dakota Archaeology Laboratory, where he is still a research associate. He was named the University of South Dakota Student Association Teacher of the Year in 1980, given the Burlington Northern Foundation Faculty Achievement Award for Outstanding Teaching in 1986, and granted the Burlington Northern Faculty Achievement Award for Research in 1990. He was selected by Sigma Xi, the Scientific Research Society, as a national lecturer from 1991 to 1993, and he served as executive secretary of the World Archaeological Congress from 1990 to 1994. He has published more than three hundred articles, CRM reports, and reviews and is the author, editor, or coeditor of fifteen books, including *Native North America* (with Brian Molyneaux, University of Oklahoma Press, 2000) and *Indians and Anthropologists: Vine Deloria, Jr., and the Critique of Anthropology* (with Tom Biolsi, University of Arizona Press, 1997). He has served as the editor of *Plains Anthropologist* and the *World Archaeological Bulletin* and as the associate editor of

American Antiquity. He has done archaeology in the Great Plains of the United States and in Mexico, England, Venezuela, and Australia. He has also worked closely with a wide range of American Indian nations and groups.

William Green is the director of the Logan Museum of Anthropology and an adjunct professor of anthropology at Beloit College, Beloit, Wisconsin. He has been active in archaeology since 1970. Having grown up on the south side of Chicago, he attributes his interest in archaeology and anthropology to the allure of the exotic (i.e., rural) and a driving urge to learn the unwritten past, abetted by the opportunities available at the city's museums and universities. His first fieldwork was on the Mississippi River bluffs in western Illinois. Although he also worked in Israel and England, he returned to Illinois for several years of survey and excavation. His interests in settlement patterns, ceramics, and archaeobotany developed there. He received his master's degree from the University of Wisconsin at Madison and then served as Wisconsin SHPO staff archaeologist for eight years. After obtaining his Ph.D. from the University of Wisconsin at Madison in 1987, he served as state archaeologist of Iowa from 1988 to 2001, directing statewide research and service programs including burial site protection, geographic information, publications, contract services, public outreach, and curation. His main research interests focus on the development and spread of native agriculture. He has served as editor of the *Midcontinental Journal of Archaeology* and *The Wisconsin Archeologist;* has published articles in *American Antiquity, Journal of Archaeological Research,* and other journals; and has received grants and contracts from the National Science Foundation, National Park Service, Iowa Humanities Board, and many other agencies and organizations.

CURATING ARCHAEOLOGICAL COLLECTIONS

ARCHAEOLOGIST'S TOOLKIT

SERIES EDITORS: LARRY J. ZIMMERMAN AND WILLIAM GREEN

The Archaeologist's Toolkit is an integrated set of seven volumes designed to teach novice archaeologists and students the basics of doing archaeological fieldwork, analysis, and presentation. Students are led through the process of designing a study, doing survey work, excavating, properly working with artifacts and biological remains, curating their materials, and presenting findings to various audiences. The volumes—written by experienced field archaeologists—are full of practical advice, tips, case studies, and illustrations to help the reader. All of this is done with careful attention to promoting a conservation ethic and an understanding of the legal and practical environment of contemporary American cultural resource laws and regulations. The Toolkit is an essential resource for anyone working in the field and ideal for training archaeology students in classrooms and field schools.

Volume 1: *Archaeology by Design*
By Stephen L. Black and Kevin Jolly

Volume 2: *Archaeological Survey*
By James M. Collins and Brian Leigh Molyneaux

Volume 3: *Excavation*
By David L. Carmichael and Robert Lafferty

Volume 4: *Artifacts*
By Charles R. Ewen

Volume 5: *Archaeobiology*
By Kristin D. Sobolik

Volume 6: *Curating Archaeological Collections:
From the Field to the Repository*
By Lynne P. Sullivan and S. Terry Childs

Volume 7: *Presenting the Past*
By Larry J. Zimmerman

CURATING ARCHAEOLOGICAL COLLECTIONS

FROM THE FIELD TO THE REPOSITORY

Lynne P. Sullivan
S. Terry Childs

ARCHAEOLOGIST'S TOOLKIT
VOLUME 6

AltaMira
PRESS

A Division of Rowman & Littlefield Publishers, Inc.
Walnut Creek • Lanham • New York • Oxford

ALTAMIRA PRESS
A Division of Rowman & Littlefield Publishers, Inc.
1630 North Main Street, #367
Walnut Creek, CA 94596
www.altamirapress.com

Rowman & Littlefield Publishers, Inc.
4501 Forbes Boulevard, Suite 200
Lanham, MD 20706

PO Box 317
Oxford
OX2 9RU, UK

Copyright © 2003 by ALTAMIRA PRESS

All rights reserved. No part of this publication may be reproduced, stored in a retrieval system, or transmitted in any form or by any means, electronic, mechanical, photocopying, recording, or otherwise, without the prior permission of the publisher.

British Library Cataloguing in Publication Information Available

Library of Congress Cataloging-in-Publication Data

Sullivan, Lynne P.
  Curating archaeological collections : from the field to the repository / Lynne P. Sullivan and S. Terry Childs.
     p. cm.— (Archaeologist's toolkit)
  ISBN 0-7591-0402-6 (cloth : alk. paper) — ISBN 0-7591-0024-1 (pbk. : alk. paper)
 1. Archaeological museums and collections. 2. Archaeological museums and collections—Management. 3. Antiquities—Collection and preservation. 4. Museum technique. I. Childs, S. Terry. II. Title. III. Series.

  CC55 .S85 2003
  930.1'074'73—dc21
                                                                          2002015057

Printed in the United States of America

∞™ The paper used in this publication meets the minimum requirements of American National Standard for Information Sciences—Permanence of Paper for Printed Library Materials, ANSI/NISO Z39.48–1992.

CONTENTS

Series Editors' Foreword	vii
Acknowledgments	xi
1 Introduction	1
2 **A Brief History of Archaeological Curation in the United States**	5
The Museum Era of Archaeology: Nineteenth Century to the 1930s • Early Federal Archaeology Programs: 1930s and 1940s • The Postwar Construction Boom and the "New Archaeology": 1945 to 1970 • Making versus Caring for Collections: The 1970s and Beyond • Conclusion	
3 **The Current Status of Archaeological Collections**	23
Federal Legislation and Policy • Key Elements of the Curation Crisis • The Bright Side	
4 **Repositories: What Are They, and What Do They Do?**	45
Kinds of Repositories • What a Repository Does and Why • Responsibilities and Training of Repository Staff • Conclusion	
5 **Managing Curated Collections: The Basics**	59
Acquisitions Policies and Practices • Accessioning • Cataloging • Collections Preparation: Labeling and	

Conservation • Storage • Inventory Control and Data Management • Deaccessioning • Public Access and Use • Conclusion

6 **Making a Collection: Fieldwork Practices and Curation Considerations**   79
Before the Field: Project Design • In the Field: Sampling and Conservation • In the Laboratory: Applying the Sampling Strategy and More Conservation • In Your Office after the Field Project: Records Management • Conclusion

7 **Working with a Repository**   91
Arranging for Long-term Curation • Using Curated Collections • Conclusion

8 **The Future of Archaeological Collections Curation**   103
Access: Collections in the Computer Age • Use of Curated Collections • The "Big Picture": Curated Collections as Samples of the Archaeological Record • Encouraging Repositories to Curate Representative Samples of the Archaeological Record • Coordinated Regional Planning • Support for Curating Archaeological Collections • Is It All Worth It?

Appendix: Useful Internet Sites Related to Curating Archaeological Collections   121

References   127

Index   141

About the Authors and Series Editors   147

SERIES EDITORS' FOREWORD

The Archaeologist's Toolkit is a series of books on how to plan, design, carry out, and use the results of archaeological research. The series contains seven books written by acknowledged experts in their fields. Each book is a self-contained treatment of an important element of modern archaeology. Therefore, each book can stand alone as a reference work for archaeologists in public agencies, private firms, and museums, as well as a textbook and guidebook for classrooms and field settings. The books function even better as a set, because they are integrated through cross-references and complementary subject matter.

Archaeology is a rapidly growing field, one that is no longer the exclusive province of academia. Today, archaeology is a part of daily life in both the public and private sectors. Thousands of archaeologists apply their knowledge and skills every day to understand the human past. Recent explosive growth in archaeology has heightened the need for clear and succinct guidance on professional practice. Therefore, this series supplies ready reference to the latest information on methods and techniques—the tools of the trade that serve as handy guides for longtime practitioners and essential resources for archaeologists in training.

Archaeologists help solve modern problems: They find, assess, recover, preserve, and interpret the evidence of the human past in light of public interest and in the face of multiple land use and development interests. Most of North American archaeology is devoted to cultural resource management (CRM), so the Archaeologist's Toolkit focuses on practical approaches to solving real problems in CRM and public archaeology. The books contain numerous case studies from all parts of the continent, illustrating the range and diversity of applications. The series emphasizes the importance of such realistic considerations as budgeting, scheduling, and team coordination. In addition, accountability to the public as well as to the profession is a common theme throughout the series.

Volume 1, *Archaeology by Design*, stresses the importance of research design in all phases and at all scales of archaeology. It shows how and why you should develop, apply, and refine research designs. Whether you are surveying quarter-acre cell tower sites or excavating stratified villages with millions of artifacts, your work will be more productive, efficient, and useful if you pay close and continuous attention to your research design.

Volume 2, *Archaeological Survey*, recognizes that most fieldwork in North America is devoted to survey: finding and evaluating archaeological resources. It covers prefield and field strategies to help you maximize the effectiveness and efficiency of archaeological survey. It shows how to choose appropriate strategies and methods ranging from landowner negotiations, surface reconnaissance, and shovel testing to geophysical survey, aerial photography, and report writing.

Volume 3, *Excavation*, covers the fundamentals of dirt archaeology in diverse settings, while emphasizing the importance of ethics during the controlled recovery—and destruction—of the archaeological record. This book shows how to select and apply excavation methods appropriate to specific needs and circumstances and how to maximize useful results while minimizing loss of data.

Volume 4, *Artifacts*, provides students as well as experienced archaeologists with useful guidance on preparing and analyzing artifacts. Both prehistoric and historic-era artifacts are covered in detail. The discussion and case studies range from processing and cataloging through classification, data manipulation, and specialized analyses of a wide range of artifact forms.

Volume 5, *Archaeobiology*, covers the analysis and interpretation of biological remains from archaeological sites. The book shows how to recover, sample, analyze, and interpret the plant and animal remains most frequently excavated from archaeological sites in North America. Case studies from CRM and other archaeological research illustrate strategies for effective and meaningful use of biological data.

Volume 6, *Curating Archaeological Collections*, addresses a crucial but often ignored aspect of archaeology: proper care of the specimens

and records generated in the field and the lab. This book covers strategies for effective short- and long-term collections management. Case studies illustrate the do's and don'ts that you need to know in order to make the best use of existing collections and to make your own work useful for others.

Volume 7, *Presenting the Past*, covers another area that has not received sufficient attention: communication of archaeology to a variety of audiences. Different tools are needed to present archaeology to other archaeologists, to sponsoring agencies, and to the interested public. This book shows how to choose the approaches and methods to take when presenting technical and nontechnical information through various means to various audiences.

Each of these books and the series as a whole are designed to be equally useful to practicing archaeologists and to archaeology students. Practicing archaeologists in CRM firms, agencies, academia, and museums will find the books useful as reference tools and as brush-up guides on current concerns and approaches. Instructors and students in field schools, lab classes, and short courses of various types will find the series valuable because of each book's practical orientation to problem solving.

As the series editors, we have enjoyed bringing these books together and working with the authors. We thank all of the authors—Steve Black, Dave Carmichael, Terry Childs, Jim Collins, Charlie Ewen, Kevin Jolly, Robert Lafferty, Brian Molyneaux, Kris Sobolik, and Lynne Sullivan—for their hard work and patience. We also offer sincere thanks to Mitch Allen of AltaMira Press and a special acknowledgment to Brian Fagan.

LARRY J. ZIMMERMAN
WILLIAM GREEN

ACKNOWLEDGMENTS

No book is solely the product of its authors, and this one is no exception. We can't possibly thank everyone individually, but we wish to recognize several persons for their help and for their tolerance of this project. First, we thank Bill Green and Larry Zimmerman for inviting us to participate in the Toolkit series. We appreciate Bill's patient editorial work that helped make our manuscript more readable. We also thank Mitch Allen at AltaMira Press for his comments and for keeping the series on track, and we thank the members of the production staff at AltaMira Press for turning our computer files and stacks of paper into an actual book.

Several people read sections of our draft manuscript and provided useful information and helpful suggestions, including Harriet Beaubien, Cindy Stankowski, and Eileen Corcoran. Discussions at the National Archaeological Collections Management Conference in San Diego and with members of the Society for American Archaeology's new Committee on Curation also provided needed feedback on ideas as well as confirmed the widespread nature of curatorial problems and the many permutations.

Jeff Chapman and the staff of the McClung Museum patiently tolerated Lynne's time spent on the manuscript in the midst of developing major new exhibits on Tennessee archaeology. Lindsay Kromer, photographer at the McClung Museum, provided us with the photos, and Jennifer Barber, graduate assistant, helped with proofreading. Terry attributes her extensive exposure to and involvement with federal curation to the guiding interest of Frank McManamon and

Veletta Canouts in the National Park Service's Archeology and Ethnography Program, where she works by day.

We also thank our many teachers of the value of curation and curated collections—including some archaeologists we never personally met who kept meticulous records and well-organized materials. We also learned from those at the other end of the spectrum, but we prefer to thank those who provided the positive experiences!

Our colleagues fully in the museum world also deserve thanks for putting up with the somewhat warped perspectives of two archaeologists who know enough about collections care (especially conservation) to be dangerous.

Last but not least, we thank *you*, the reader, for wanting to learn about archaeological curation. The more people who care enough to learn about this topic, the more likely it is that archaeological collections will receive the care they need to be useful to future generations.

1
INTRODUCTION

Many archaeologists believe curation is something that happens only *after* fieldwork. This notion, along with a related set of historical circumstances, has contributed to a crisis in curation of archaeological collections. We hope to show that curation is a *process* that begins in the field and continues in the repository. As such, curation is the responsibility of many individuals and institutions, from field archaeologists and their sponsors and employers, to collections managers and curators and the repositories that employ them.

The curation crisis did not happen overnight, nor will it quickly be resolved. One thing is for certain: The crisis will be perpetual unless there are coordination and shared goals among all the players to assure adequate care for archaeological collections. If this book helps archaeologists take steps in that direction, it will have served its purpose.

Archaeologists typically do not learn about curatorial issues and practices in school. Graduate programs rarely offer such classes, nor is discussion of curation usually a part of the material covered in the relatively few classes on cultural resource management. Even museum classes do not offer the kinds of information archaeologists need to deal with curation. These classes usually focus on exhibits, not on collections management concerns or collections issues in the field. This critical omission in basic archaeological training still is not fully recognized by the profession, even in the context of ongoing discussions of curricular reform (Bender and Smith 2000). In this book, we aim to help archaeologists understand the history behind the curation crisis, where we are today as a profession and the legal aspects, and where we seem to be headed. We also consider basic technical information about making collections with curatorial considerations in mind, and the hows and whys of working with repositories to ensure adequate long-term curation. We summarize salient points and procedures in tables for easy reference, and we include case studies and examples to help readers relate our discussions to real-world situations.

Before discussing curation itself, we need to define the kinds of materials that comprise archaeological collections. Many archaeologists, museum professionals, and archivists tend to think of archaeological collections as artifacts and specimens—that is, the pottery sherds, animal bones, chert flakes, projectile points, and so forth that archaeologists collect in the field. These objects are only part of an archaeological collection. Artifacts and specimens are worthless for research or interpretation unless they are accompanied by documentation that records, often very precisely, where these things were found—that is, the provenience. The integration of these records with the objects and specimens is what separates a systematic archaeological collection from an antiquarian collection. *Archaeological* collections are made to collect and preserve information that is useful for research and interpretative purposes, while *antiquarian* collections are made solely for the intrinsic value of the artifacts. Thus, without the accompanying records, a collection of "old things" is just that; it is not an archaeological collection.

The records that form significant elements of archaeological collections include field notes, photographs, maps, and suites of forms (level, feature, photo log, etc.), as well as other paperwork and electronic data made to record and preserve contextual and analytical information. Other kinds of records also may be important components of archaeological collections. In today's world of cultural resource management projects, scopes of work, proposals, and limited-distribution reports often are the only documents that contain essential contextual information on where, how, and why fieldwork was done. So, in short, when we use the term *archaeological collections* in this book, we mean both the collected materials and the associated records.

Now that we've defined what we curate, we can define curation. What is archaeological curation, and why is it important? As noted earlier, we view curation as a process, one that is never ending and that requires constant vigilance. Archaeological curation is the process of managing and interpreting archaeological collections over the long term. By *managing*, we mean making and caring for collections, and rendering them accessible for multiple purposes. By *long term*, we mean as long as the constituent materials of the objects and records can be preserved. *Interpretation* in a curatorial sense usually refers to using collections for public education. We only touch on this latter aspect of archaeological curation (see Toolkit, volume 7); our primary focus is on the managerial aspects of curation because that is where the most serious problems lie.

Regarding the importance of curation, we hope that the research significance of collections is self-evident, at least to archaeologists. Archaeological research relies on material objects and their archaeological contexts (i.e., proveniences), so the collections resulting from research become invaluable and irreplaceable. Quite simply, without collections we cannot do our work. They are as much our tools as trowels and shovels. Yet, from our collective behavior in regard to curation, one would think that we value our trowels and shovels much more than we do the collections! Furthermore, archaeological collections are important not only to archaeologists. Peoples whose cultural heritage is intimately intertwined with collections—often the only tangible evidence of their past—also care a great deal about these materials.

Another equally important aspect of curation is accountability. Many, if not most, archaeological collections are in public trust. When curatorial practices are poor or nonexistent, everyone loses: Archaeologists suffer loss of irreplaceable research data, the general public suffers loss of an expensive and valuable educational resource, and those whose heritage may be linked to the collections lose that part of themselves. As the makers of archaeological collections, archaeologists are among the persons and institutions accountable for ensuring proper care of these materials. Why should the public spend money to preserve information from archaeological resources if the basis of that information, the collections, are not preserved for the long term? The credibility of archaeologists, and indeed archaeology, quickly becomes called into question when collections are lost, mismanaged, or destroyed because the field archaeologist did not consider curation needs.

Before discussing the organization of this book, we need to clarify our use of the terms *repository* and *museum*. A repository simply is a place to care for collections. It does not necessarily have the public educational functions of a museum, such as exhibits and public programs. The heart of museums, on the other hand, traditionally has been the collections and their care, yet these institutions also exist to educate the general public through displays and programming. Today, repositories that are museums, and those that are not, care for archaeological collections. We discuss the types of repositories in chapter 4, but suffice it to say here that our use of the term *museum* means an institution with collections and exhibits, while the term *repository* refers more broadly to a range of institutions that care for collections.

Because we are archaeologists ourselves, it is perhaps not surprising that the organization of this book proceeds from the past to the future. We begin in chapter 2 with a history lesson that shows how the making of collections gradually became separated from the care of collections. In that context, we provide a case study of a large collection, made in connection with a New Deal–era federal project, as an example of the kinds of materials that now are being curated, the kinds of issues and circumstances that surround their creation and care, and their research potential. Then we move to the present day, with an overview of the laws and regulations that pertain to collections and a consideration of the many pressing issues related to curation. The next three chapters provide basic technical information on curatorial practice, common procedures followed by repositories, and what archaeologists need to know to arrange effective agreements for long-term collections care. Chapter 6 outlines the kinds of curatorial concerns archaeologists and managers need to consider in designing and implementing field projects. Chapter 7 focuses on how to work with a repository both before a collection is made and once it is ready to be delivered after fieldwork and analysis. The final chapter discusses collections accessibility and use issues and the "big picture" in terms of what is being curated and the cost of doing so. An appendix provides a list of useful websites that deal with aspects of archaeological curation.

As more archaeologists become familiar with the background, procedures, and issues of curating archaeological collections, from the field to the repository, we will be able to move forward to resolve the current crisis. All archaeologists who make collections must help plan for the long-term care of those collections. The very hard lessons we've already learned in terms of loss of irreplaceable materials need to be communicated to students so that as younger generations enter the profession we don't repeat past mistakes. Learning from archaeology's own past and changing old attitudes and habits are critical challenges for curating archaeological collections.

2
A BRIEF HISTORY OF ARCHAEOLOGICAL CURATION IN THE UNITED STATES

A general trend in American archaeology seems to be increased separation between the making of archaeological collections (i.e., fieldwork) and the caring for archaeological collections (i.e., curation). In the nineteenth and early twentieth centuries, these activities more often were accomplished under the direction of single individuals or institutions. Now, multiple individuals and institutions often are involved in various stages of archaeological fieldwork, analysis, reporting, and curation.

A brief review of the history of archaeological curation in the United States is useful for understanding the current situation. As part of this review, we note some changes in archaeological collecting and collections, and we look in particular at one large collection as a case study of the history, contents, uses, and care of older but significant archaeological materials.

THE MUSEUM ERA OF ARCHAEOLOGY: NINETEENTH CENTURY TO THE 1930S

Museums are the oldest and perhaps most common kind of public repository for archaeological collections. The histories of museums and archaeology are closely intertwined. In fact, the beginnings of archaeological chronology are rooted in museology. In Denmark in 1819, J. C. Thomsen developed the "three-age" system (Stone, Bronze, and Iron) for European prehistory to organize the displays of the national museum. In 1873, Otis Mason used Lewis Henry Morgan's now-antiquated model of social evolution with its stages of savagery, barbarism, and civilization to organize the collections at the Smithsonian Institution (Conn 1998; Willey and Sabloff 1977). This close association meant that making, caring for, and interpreting archaeological collections all were accomplished under one roof. By the end of that era, however, museums no longer were at the forefront of developments in archaeology, and the making of collections began to become separate from collections care. The early histories of museums and archaeology in the United States chronicle this process.

The nature of museums has changed through time. A few museums were created in the United States toward the end of the eighteenth century, including the Charleston Museum in South Carolina that opened in the 1770s and Peale's Museum in Philadelphia (1872). These early museums were "cabinets of curiosities," featuring shrunken heads, Egyptian mummies, two-headed beasts, and other items intended to entertain the public (Alexander et al. 1992). The antiquarians who collected archaeological objects at this time also typically saw them as curiosities. Methodical collection, study, and arrangement of objects, as well as subject specialization, including the differentiation between art and natural history museums, came about a century later. It was not until the second half of the nineteenth century that the concept of the public educational museum became established (Burcaw 1997; Conn 1998). This is also the same era when archaeology emerged as a formal professional discipline. The late nineteenth and early twentieth centuries have even been dubbed "the museum era" of American anthropology (Jonaitis 1992; Willey and Sabloff 1977).

The maintenance of elite European society was a major impetus for the establishment of public museums. The notion that museums should be places for public education was influenced by revolutions in Europe, which made aristocrats nervous enough to enact government-subsidized programs in an effort to raise the lot of the common people. Englishman James Smithson apparently saw a similar need in the United States and gave funds to establish a national public museum, the Smithsonian Institution, established by Congress in 1846 (Conn 1998).

The world's fairs of that era also were efforts by elites on both continents to "uplift" (and pacify) the "common people" and to teach the values and ideas of the ruling class. World's fairs of the late nineteenth century, in both Europe and the United States, often included anthropologically oriented exhibits that also boosted the growth of museums. In the United States, the 1876 Centennial Exposition in

Philadelphia helped the development of the American Museum of Natural History in New York City and the Smithsonian's National Museum of Natural History. Archaeological collections from North America became associated with such natural, rather than cultural, history museums for a reason: Prejudices and class structures often relegated studies and objects of minority groups to the natural world rather than to the world celebrated by the dominant Euro-American culture (Burcaw 1997; Conn 1998).

By the end of the nineteenth century, the United States had numerous well-established general and natural history museums, most of which were in the east. The rise of American museums was influenced by, but did not exactly mimic, European museums. Unlike many European countries, the United States did not organize a national museum with a regional structure. Americans also advanced a populist perspective that museums were places for public education, "public service," rather than strictly research. In fact, because the administration of the American Museum of Natural History embraced this attitude to the virtual exclusion of support for basic research, Franz Boas, the "father of anthropology," left his post there in the early 1900s for a faculty position at Columbia University. The pendulum on this issue continues to swing back and forth in many U.S. museums (Conn 1998; Jonaitis 1992). The emphasis away from research by museums undoubtedly helped foster the separation between archaeological collecting and curation.

Before this separation began, the making of archaeological collections changed from antiquarian fascination to systematic observation, recording, and collection. The emphasis on systematic recording by late nineteenth-century investigators resulted in the first significant holdings of curated archaeological, as opposed to antiquarian, collections in the United States (Willey and Sabloff 1977).

Both archaeology and museology were becoming professionalized as well. Archaeologists employed by museums did most of the early fieldwork in the United States, and the collections were curated by those institutions. Museums began to use the collections in public displays to interpret the research of their staffs, who were directly involved in making, curating, and interpreting the collections.

The most influential institutions in archaeology during the late nineteenth and early twentieth centuries were the Smithsonian Institution, the American Museum of Natural History, and the Peabody Museum of Archaeology and Ethnography at Harvard. The Museum of the American Indian, or Heye Foundation, in New York City also was active during this period and amassed large collections. Its focus, however, was more on collecting than on the intellectual development of the discipline. Notables such as Cyrus Thomas, Frederick Ward Putnam, Arthur C. Parker, M. R. Harrington, William C. Mills, and Warren K. Moorehead built substantial archaeological collections in these and other institutions (Willey and Sabloff 1977).

From the mid–nineteenth century through the 1920s, the Bureau of American Ethnology (BAE) of the Smithsonian Institution employed many research associates in an active archaeological fieldwork program throughout the United States. The dominant question of the period concerned the origin of American Indians and resolution of the Moundbuilder Myth, which posited that a race of people other than the American Indians built the earthen mounds of the Americas. The BAE sponsored the first major archaeological survey of the period, resulting in E. G. Squier and E. H. Davis's *Ancient Monuments of the Mississippi Valley* (1848). Cyrus Thomas became director of mound exploration within the BAE in 1882. His publication in 1894 of the results of his extensive field investigations into the Moundbuilder issue settled, once and for all, the debate in favor of an American Indian origin. Subsequent work by William Henry Holmes and Aleš Hrdlička at the Smithsonian established that humans were a relatively recent addition to the New World, as compared with the Old (Willey and Sabloff 1977).

At the American Museum, Boas had worked to link archaeology and anthropology as early as 1899, and he set up a joint training program with Columbia University. He and other museum staff, including Nels C. Nelson, who also had worked for the Museum of Anthropology at the University of California, were instrumental in bringing stratigraphic excavation techniques from Europe to the Americas (Browman and Givens 1996).

Frederick Ward Putnam and his student Arthur C. Parker were influential museologists as well as museum archaeologists. Putnam was a leading figure of the period as an excavator, sponsor, administrator, and founder of museums and departments of anthropology, and he has been termed by Willey and Sabloff (1977:52) the "professionalizer of American archaeology." He became the curator of the Harvard Peabody Museum in 1874, bringing that institution to national attention. Putnam also helped found the Field Museum of Natural History in Chicago, the Anthropology Department at the University of California (Berkeley), and the Anthropology Department at the American Museum of Natural History. He trained a cadre of aspiring archaeologists, mainly through apprenticeships rather than formal academic methods, in the basics of careful fieldwork and recording.

Parker became the first salaried archaeologist hired by New York state. He worked for the State Museum from 1904 to 1924. He went on to become director of the Rochester (New York) Museum and, in 1935, the first president of the Society for American Archaeology. Like his mentor, Parker stressed professionalism. He changed the State Museum's policy for obtaining collections from purchasing archaeological specimens from collectors to financing professional excavations by museum staff. He also pioneered popular interpretation of archaeological research in museum exhibits (Sullivan 1992b).

The close connections between museums and archaeology also appear to have influenced the first federal law regarding protection and preservation of archaeological sites and collections. The Antiquities Act of 1906 (16 U.S.C. 431–33) stipulates that objects collected under the act are "for permanent preservation in public museums" and that "every collection made under the authority of the act and of these rules and regulations shall be preserved in the public museum designated by the permit and shall be accessible to the public." The act further specifies that if the designated public museum ceases to exist, the collections revert to the national "depository" (sic). Implicit in this law is the notion that arrangements for collections care will be made before a permit for fieldwork is granted. It was not until 1960 (see a later section) that mention is made again of where collections are to be curated (but without explicit mention of the kind of institution to be used), and it was not until 1979 that the law required making such arrangements prior to beginning fieldwork on federal land.

By the 1920s, archaeological research programs were established in many states, commonly as part of anthropology departments in universities. Museums became more concerned with direct public service. As universities trained aspiring archaeologists and museums turned toward increased public outreach, the heyday of museum-based research was over by the end of the 1920s. Some universities maintained museums with missions to educate the general public, but many universities did not maintain or develop museums. This shift of archaeology from museums to universities meant that the curation of archaeological collections was becoming removed from a museum context where they were used for research and to teach the lay public. Instead, collections were being placed in an academic context where they continued to be used for basic research, as well as for professional training of archaeologists. This change meant that archaeological collections generally became less accessible to the general public. Perhaps public accountability for curation also became deemphasized as part of this change.

The use of archaeological collections as research tools rather than for exhibits is exemplified by the establishment of the Ceramic Repository for the Eastern United States at the Museum of Anthropology at the University of Michigan. The Ceramic Repository was organized in 1927 to develop and curate a collection of pottery sherds that would facilitate comparative knowledge of the distribution of artifacts in the eastern United States and to serve as a kind of "clearinghouse" for pottery studies. Carl E. Guthe, a Harvard graduate, was appointed director. Sherds submitted to the repository had to have detailed provenience information, and Guthe stressed that the emphasis of the repository was not on beautiful objects for museum displays but on research specimens (Lyon 1996).

Guthe continued in the museum field as director of the Rochester Museum and the New York State Museum, and he became highly influential in the development of curatorial techniques after World War II. His publications, *So You Want a Good Museum* (1957) and *The Management of Small History Museums* (1959), were standard references for many years.

Another development in the making of archaeological collections was that field schools, such as those taught by Fay-Cooper Cole at the University of Chicago, became a venue for archaeological fieldwork. Anthropology departments thus became responsible for the curation of the archaeological collections they made, although this responsibility often fell to the individual faculty member—a situation that continues today in many academic institutions.

Cole had worked at the Field Museum before going to the university. As a faculty member, he developed a program of archaeological research in Illinois as a training ground for graduate students in anthropology. A major part of Cole's program was summer field schools, which he ran for nearly a decade. These field schools were instrumental in developing modern archaeological field techniques, including the development of systems of horizontal and vertical control, techniques for excavating mounds and villages, and systematic record keeping on detailed forms (M. Fowler 1985). Cole's emphasis on scientific data collection procedures, including systematic excavation and recording techniques, also apparently instilled in his students a respect and preservation ethic for the collections they made (Sullivan 1999). However, archaeological fieldwork and postfieldwork curation continued to diverge.

EARLY FEDERAL ARCHAEOLOGY PROGRAMS: 1930S AND 1940S

The New Deal's public works projects of the 1930s profoundly affected archaeological fieldwork and curation. In many regions, university anthropology departments, sometimes in conjunction with university museums, conducted large field projects that provided work for the unemployed. These federal projects had significant impacts on U.S. archaeology, both in creating extremely large collections that now serve as major data banks and in formulating new techniques, methods, and theories. A few of these projects actually spawned museums and filled them with large collections, but long-term curation was not part of the New Deal–era programs.

The curatorial fate of collections from these projects runs the full spectrum of possibilities. At one extreme, some collections still are unwashed. At the other, the work of vigilant and determined individuals resulted in construction of repositories to care for these materials. Regardless of condition, the New Deal–era collections and records contain a wealth of information pertinent to current archaeological problems as well as to historical research about the development of the discipline. They can be highly useful, if they are accessible.

A major concern of archaeology during the 1930s was formulation of regional syntheses. These were based on experimental chronologies and culture classification schemes without the benefit of absolute dating techniques. The antiquity of the archaeological record in the Americas was not yet known, nor was the existence of the eastern Archaic or Paleoindian fully recognized. Classification schemes were developed to define archaeological complexes based on associated traits in an attempt to gain some spatial-temporal control over the data. For example, pottery typology became an important tool in these classification schemes. Different suites of pottery styles became associated with (and often definitive of) archaeological complexes that many researchers viewed as actual "cultures." A widely used system was the Midwestern Taxonomic Method, developed and advocated by W. C. McKern of the Milwaukee Public Museum. The 1930s also saw experiments with new field methods and techniques to improve recovery of contextual and functional information (Fairbanks 1970).

The federal work programs of the Great Depression fostered the attitude that fieldwork was more valuable than curation because the programs' goal was to put many people to work, and archaeological fieldwork is labor-intensive. While laboratory work, data analysis, and publication initially were not viewed as acceptable costs, long-term curation was not even considered to be a part of these programs. Another rather insidious aspect of the devaluation of archaeological curation during the Depression had to do with sexist cultural mores. Curatorial duties and lab work often fell to women in the federal work programs, reinforcing marginalization of both the work and the workers (White et al. 1999).

The movement of field projects from museums to universities also may have influenced the curation prejudice. While universities often have tended to see field training of students as an integral part of their missions, the maintenance of research collections has been seen as peripheral. Such a view reinforces differential valuing of making versus caring for collections.

The historic preservation movement, which began in the late nineteenth century and became intertwined with archaeology in the 1930s, also contributed to the curation prejudice. The goal of the movement initially was to preserve historic buildings. The architects and architectural historians who led the historic preservation movement and who drafted the historic preservation laws that pertained to archaeological sites did not adequately address the curation of collections such as those produced by archaeological research. Mention of collections in the Historic Sites Act of 1935 was relegated to a general empowerment of the secretary of the interior, through the National Park Service, to "secure, collate, and preserve drawings, plans, photographs, and other data of historic and archaeologic sites, buildings and objects." The fact that artifacts and specimens were not mentioned reflects the lack of involvement of archaeologists in crafting this law. More significantly, where and how these "data" were to be secured, collated, and preserved was not specified (T. King 1998; T. King et al. 1977).

A CASE STUDY: THE CHICKAMAUGA BASIN PROJECT

A case study of one New Deal project helps us appreciate early federal archaeology programs in terms of curation-related problems in program administration, the kinds of information that are available from this era, and the kinds of materials that currently are being curated. The Chickamauga Basin project in southeastern Tennessee (Lewis et al. 1995) not only laid the groundwork for much subsequent research in the Upper Tennessee Valley, so much so that the archaeological phases in the region are named for sites in the Chickamauga Basin, but also helped build an infrastructure for archaeological research in the state. The collected information is the only systematic documentation of major sites that now are inundated or destroyed. These collections thus exemplify the need to ensure that archaeological materials are preserved and accessible for future scholars.

The Chickamauga Basin project was one of several archaeological projects conducted in conjunction with the construction of reservoirs by the Tennessee Valley Authority (TVA). Archaeological work for the Chickamauga reservoir on the Tennessee River near Chattanooga began in early 1936 and continued through 1939. A survey of the basin, conducted under the direction of William Webb of the University of Kentucky, located some seventy sites. Thirteen multiple-component sites were chosen for excavation. Thomas M. N. Lewis, who had been hired by the University of Tennessee in 1934 as its first archaeologist, ran the project under the newly established Division of Anthropology, a section of the History Department. Lewis was from Wisconsin and had worked with W. C. McKern at the Milwaukee Public Museum (Sullivan 1999).

A new federal relief program, the Works Progress Administration (WPA), provided the labor. Lewis hired young archaeologists from several universities as field supervisors, including several of Cole's University of Chicago students who would go on to become well-known archaeologists. Jesse Jennings was the main supervisor. The large WPA crews (fig. 2.1) made it possible to excavate huge areas, and as Lewis notes in the project's manual of field and laboratory procedures, the field supervisors had to expend "much shoe leather" to keep up with the excavations. The excavated sites included five Mississippian platform mounds, eight Late Woodland burial mounds, sixteen Late Woodland shell middens, a multicomponent Woodland midden, and portions of nine Mississippian villages and one historic Cherokee village. Nearly 2,000 burials and 165 structures were excavated, mapped, and photographed (Sullivan 1999; Lewis et al. 1995).

Dealing with the collections was an issue from the start of the TVA-WPA work because multiple institutions and personalities were involved and suitable laboratories and repositories were lacking. William Webb, a physicist interested but not formally trained in archaeology, ran the first TVA reservoir project in the Norris Basin. At the end of that project, TVA gave the artifacts to the University of

Figure 2.1. One of the WPA field crews that worked in the Chickamauga Basin. The field crew members were almost always male. Stuart Neitzel and Charles Fairbanks (at left) were the University of Chicago–trained site supervisors.

Tennessee, which at the time did not have a repository. Webb took the osteological collections and field records to the University of Kentucky, which had established an archaeology program and lab in the 1920s (Lyon 1996). This probably was not the first example of an unfortunate split of materials that should have been curated together. It certainly was not the last.

Webb became the TVA archaeological consultant and resisted spending money on laboratories to process collections. He did so because the government was not legally required to do archaeology and because he feared that requests for large sums of money would be an impetus to end the projects. It was not until 1938 that the WPA archaeological program was restructured to support laboratories, analysis, and publication. By the time this could be done in Tennessee, the Chickamauga laboratory work was three years behind the fieldwork. Boxes of artifacts were piled up in the laboratory, and there was a large backlog of uncataloged and unanalyzed materials.

In 1938, near the end of the Chickamauga fieldwork, Lewis hired Madeline Kneberg, a student of Cole, to be lab director. Her goal for the laboratory work was to catch up with the fieldwork. She oversaw materials preparation, restoration, and cataloging, as well as analysis.

The Tennessee lab had forty workers and six supervisors (mostly Chicago students) at its peak in 1939–1941.

Artifacts were cleaned in the field and shipped to the Knoxville laboratory. The lab then cataloged and analyzed the collected materials with the assistance of several specialists from other institutions. Under Kneberg, the lab also developed an innovative attribute-based system for artifact classification, a technique for pottery vessel reconstruction, and numerous card files for analytical purposes and collections management. For the Chickamauga project alone, the lab classified over 360,000 pottery sherds and some 100,000 stone, bone, shell, and copper artifacts; identified almost 7,000 animal bones to species; reconstructed several hundred pottery vessels; and examined all of the nearly 2,000 recovered skeletons for age, sex, and pathologies.

Lewis and Kneberg wrote a manual that described the field and laboratory methods used by the Tennessee projects. The field methods were largely derived from systems developed by Cole's field schools but also included techniques developed under Lewis's supervision and experimentation. The manual also explained Kneberg's complicated catalog and file system that was used to preserve provenience and contextual information and to prepare data for publication.

The entry of the United States into World War II interfered with project publication goals, as New Deal projects were shut down and many of the staff were drafted for the war effort. In 1946, the University of Tennessee Press published a report by Lewis and Kneberg on one excavated site, Hiwassee Island, but the report for the entire project remained in incomplete draft form until 1995. The curated project records made it possible to complete and publish this important document (Lewis et al. 1995).

Although the New Deal archaeology programs eventually did provide for laboratory processing and analysis of collections, they did not provide for long-term curation. This was a particular problem in Tennessee where no state or university museum or repository existed. To house the vast collections resulting from the University of Tennessee WPA work, and to provide the growing anthropology program with a home, Lewis and Kneberg convinced the university to build a museum. Lewis had been an advocate for a museum supported by the State of Tennessee ever since the university hired him. The Frank H. McClung Museum finally was built in 1960 with private funding donated to the university, some two decades after the end of the New Deal archaeology projects (Sullivan 1999) (fig. 2.2). The McClung Museum continues to curate the

Figure 2.2. Madeline D. Kneberg at work in the laboratory of the Frank H. McClung Museum. Many of the pots on the shelves behind her are from the WPA excavations in the Chickamauga Basin. The museum was established and built in the 1960s, some thirty years after the WPA investigations. Kneberg's and Thomas M. N. Lewis's long-term and persistent advocacy were essential factors in getting an appropriate repository for these collections.

Chickamauga Basin collections and remains the major repository of archaeological collections from Tennessee.

The Chickamauga collections are suitable for a wide range of research because of the systematic and standardized collecting and recording techniques, preservation of within-site provenience information, and large samples of various materials (see Cantwell et al. 1981) (fig. 2.3). The collections are especially precious because the opportunity for continued excavation no longer exists at most of the sites. Examples of the kinds of data available from the Chickamauga Basin include: documentation of large portions of villages showing spatial relationships between structures and other features, plan drawings of many structures, large and detailed data sets for studies of mortuary practices and human biology, and large numbers of intact artifacts for technological and stylistic comparative studies

Figure 2.3. Examples of the records made by the Chickamauga Basin Project, including a plan map of the Dallas site (a Mississippian village), a feature form for a structure on the Dallas site mound, 3.5 × 5–inch photographs of the mound structure (the original negatives are silver nitrate), and a ceramic analysis card showing the pottery counts from the mound structure (written in the abbreviation code developed in the WPA lab). The records are now invaluable components of the Chickamauga Basin collections and must be curated in perpetuity.

(Lewis et al. 1995). A number of dissertations and theses have relied on data from the Chickamauga Basin collections but certainly have not exhausted their potential. One wonders what would have happened to these significant and irreplaceable collections if the McClung Museum had not been built.

THE POSTWAR CONSTRUCTION BOOM AND THE "NEW ARCHAEOLOGY": 1945 TO 1970

The postwar construction boom had major impacts on American archaeology. Sites were being lost at unprecedented rates. Dams began to be built throughout the country immediately after the war. Clearing and leveling of agricultural lands were extensive during the 1950s and 1960s under federal cost-sharing programs. The Eisenhower administration began building interstate highways in the 1950s. "Urban renewal" of the early 1960s began leveling older sections of cities. Archaeologists scrambled to save sites and to get federal legislation passed to stop the rampant destruction (T. King 1998; T. King et al. 1977). Site salvage was the battle cry of the day; no one had time to think about what was to happen to the materials once they were saved from the bulldozers.

The River Basin Salvage Program was one federal archaeology program developed in response to these massive construction projects. It began in 1945 at the Smithsonian Institution. Through a memorandum of understanding between the Smithsonian and the National Park Service (NPS), institutions in several states were contracted to do surveys for archaeological sites that would be affected by Bureau of Land Reclamation and Army Corps of Engineers reservoir projects (Friedman and Storey 1999). The program was modeled largely after the WPA programs, was authorized under the Historic Sites Act of 1945 (16 U.S.C. 461–67) (T. King 1998; T. King et al. 1977), and, not surprisingly, lacked any specific provisions for long-term curation of the collections.

The NPS gradually became the lead agency in the reservoir salvage effort, and its Interagency Archeological Salvage Program assumed much of the administration of the reservoir projects. The federal Reservoir Salvage Act of 1960 (16 U.S.C. 469) formalized this role (T. King 1998). This law mentions, almost as an afterthought, "The Secretary [of the Interior] shall consult with any interested Federal and State agencies, educational and scientific organizations, and private institutions and qualified individuals, with a view to determining the ownership of and the most appropriate repository for any relics and specimens recovered as a result of any work performed as provided for in this section." Unlike the 1906 Antiquities Act, the kinds of institutions to serve as repositories were not named, the public educational value of collections and accessibility issues were not mentioned, and the critical need to make curation arrangements prior to fieldwork was ignored. Curation definitely was not a front-burner issue as salvage archaeology became a substantial part of the fieldwork being done in the United States.

A number of laws passed in the 1960s provided for archaeological assessments before construction projects (see Toolkit, volume 1). The most comprehensive was the National Historic Preservation Act of 1966, which required federal agencies to consider the effects of their undertakings on historic properties, including archaeological sites. Other laws, including the National Environmental Policy Act of 1969, and portions of laws pertaining to the Departments of Transportation and of Housing and Urban Development also required such considerations. None of these laws, however, provided specific guidance on the curation of the resulting archaeological collections.

Not only was the amount of archaeological fieldwork increasing in the 1960s, but also the discipline of archaeology was changing in ways that resulted in the collection of more materials. Archaeologists began to ask new kinds of questions of the archaeological record and to devise techniques for collecting new kinds of samples and materials to answer these queries. The "New Archaeology" investigated aspects of human behavior and processes of culture change. It asked questions about environmental and social relationships, and it used systematically collected data to answer them. This approach contrasted with more descriptive and classificatory approaches that sought to characterize past cultures through regional syntheses based on artifact types (Willey and Sabloff 1977). For example, the new emphasis on human relationships with the natural environment stimulated detailed analysis of the uses of wild versus domesticated plants. A technique new to archaeology, flotation, allowed recovery of carbonized plant materials from soil samples. New emphasis was placed on sieving archaeological deposits through fine-mesh screens to enable recovery of small-scale specimens. Such new techniques greatly increased the kinds of information that could be gleaned from archaeological sites (see Toolkit, volume 5). They also produced more and different kinds of materials to be curated. By the early 1970s, the fact that there were significant problems with curating archaeological collections was becoming widely apparent (Ford 1977).

MAKING VERSUS CARING FOR COLLECTIONS: THE 1970S AND BEYOND

The segregation of making and caring for collections and lack of consideration of long-term curation issues continued to be problems in the 1970s, in relation both to federal preservation laws and to the profession of archaeology (Lindsay et al. 1980). Chapter 3 provides an in-depth look at the history of the laws and how they influenced the current curation situation. For now, we can say that these laws dramatically increased the amount of archaeological fieldwork done in the United States, so the amount of material to be curated also dramatically grew. Repositories, including museums, saw their collection storage areas fill to capacity with no means to properly store, catalog, or inventory the materials. These increases, on top of the previous three decades of large federal programs with no support for curation, were leading to a major crisis.

Attitudes and assumptions among professional archaeologists and curators also affected collections care in the 1970s. Academic archaeologists in particular often assumed that storage areas could be found for collections. Usually, it did not cross their minds to arrange for care ahead of time or that collections required more than just storage. These attitudes and practices were, and still are, prevalent because curation was rarely discussed at professional meetings or taught in graduate schools. Furthermore, university anthropology departments often were unable to secure curatorial space because collections care typically was (and is) not an institutional priority. Important collections, therefore, were relegated to dank, dark basements; asbestos-coated surplus warehouses; old and unrenovated army barracks; and mice-infested and leaky barns (Trimble and Meyers 1991). After saving collections of artifacts from destruction in the field, archaeologists were permitting the loss and degradation of their research value out of the field.

Another development that has contributed to the segregation of making collections and curating them relates to the fact that private consulting firms, instead of university anthropology departments, began to do much of the field archaeology in the United States in the late 1970s. Because such firms are not directly tied to museums, their field projects often are done without the direct involvement, support, approval, or supervision of the repository that will be the custodian of the collection. The priority of cultural resource management is largely the sites in the field, not the collections that are the legacies of those sites. This focus is understandable due to the realities of site loss, but severe problems arise when the ultimate destiny of the resulting collections is not considered as part of the entire project.

Further separation between collecting and curating is occurring because collections management is becoming a specialized profession. The museum curator who commonly served as all-purpose researcher, collections caretaker, conservator, and exhibit designer in earlier decades still survives in small museums. But these tasks cannot be done by one person in the larger museums that are usually the primary repositories for research and federally owned collections (even though budget cutting may force a return to the days, and problems, of the do-it-all curator). Furthermore, collections managers today are not necessarily trained archaeological researchers. Even when they are, they cannot be expected to have expertise in the multiple research specialties of modern archaeology (Sullivan 1992a). Also, each one of the many kinds of repositories has a different mandate, different audiences, and different collections policies (see chapter 4).

CONCLUSION

Attitudes and practices in field collecting and collections care have changed over the last hundred years. The changes are not inherently bad. One positive aspect is that collections are now beginning to be cared for by professionals who are trained for this work. Many people and institutions often are involved in activities that once were done by single individuals or institutions. The problems of earlier federal programs can be avoided if all parties coordinate their efforts and plan specifically for collections care. Successful long-term curation should not have to depend solely on the persistent efforts of determined individuals, as was the case with the New Deal–era collections in Tennessee.

Recent efforts are bringing us to a different place today, as we shall see in chapter 3. Many collections are housed in improved storage conditions, and we have better information on the contents of many collections. We also have increased public accountability for collections care and use. However, archaeologists cannot afford to proclaim the issues resolved, given that collections continue to grow and require sustained support. A major question for the twenty-first century is, How do we manage this growth so that collections care does not mean loss of significant research materials or a return to substandard basements, barns, and warehouses? We've been there, done that. What can we learn from these experiences?

Archaeologists and repositories have learned two important lessons over a long period of time, and the hard way. One lesson is that proper care of collections does not just happen. Collections must be managed by knowledgeable staff with adequate facilities and support at their disposal. The other lesson is that good curation costs real money, a resource that is chronically scarce in archaeology (Marquardt et al. 1982; McManamon 1996). In the next chapter, we will learn more about how a national curation crisis developed and about the efforts that are being made to correct this situation.

3

THE CURRENT STATUS OF ARCHAEOLOGICAL COLLECTIONS

The vast majority of archaeological collections around the United States are now under federal or state ownership and are in a state of crisis. As we saw in chapter 2, significant problems began during the New Deal era. As we shall explore here, problems accelerated in the 1970s and early 1980s, influenced by at least three factors.

One factor was the historic preservation movement. As museums and archaeology were developed and professionalized, support slowly increased for historic preservation. The movement emphasized many tangible and valuable products of past cultural systems that deserved to be preserved for the benefit of future generations (T. King et al. 1977). The focus was primarily on historic properties, including archaeological sites and the data they contained. Unfortunately, the logical relationships between the preservation and conservation of historic properties and the associated collections, records, and reports preserved in museums and repositories were not well made.

Another factor more specific to the discipline of archaeology was the ethic of conservation (Lipe 1974; McGimsey and Davis 1977). The earlier ethic of excavation had encouraged the destruction of sites through research and salvage excavations in order to learn as much as possible about past cultures. This information was then supposed to be shared within the archaeological community. The conservation ethic taught that archaeological sites are nonrenewable and irreplaceable, that the maximum amount of information must be retrieved from each project, and that some sites must be conserved and preserved for future use. It encouraged careful long-term management of those resources so that well-informed decisions could be made about their significance, survival, and best use. The conservation ethic also encouraged more communication between archaeologists and the public about the results of excavations and their relevance to the public. Although understated, there was some realization that the collections and records resulting from archaeological projects would become increasingly valuable and useful as sites were destroyed for both research and compliance (Christenson 1979; Marquardt 1977a; McGimsey and Davis 1977:63).

Native American concerns are a third factor that will affect archaeological sites and collections well into the future. Native peoples protested against excavation of their sacred and ancient places, insensitive treatment of their ancestors' skeletal remains in museum storage boxes and exhibits, and the poor care of Native American archaeological and ethnographic museum collections in general. While it was recognized that Indian cultural centers did not meet minimum museum requirements for storage and display, Native Americans complained that neither did many of the museums and repositories holding archaeological collections (McGimsey and Davis 1977:94).

In this chapter, we examine the federal and state legislation enacted in response to these concerns, including their intended effect on the care and maintenance of archaeological collections, records, and reports. We then outline the urgent issues affecting archaeological collections today that have emerged from legislative efforts and archaeological practice. Finally, we discuss the progress that has been made.

FEDERAL LEGISLATION AND POLICY

The Archeological and Historic Preservation Act (AHPA) of 1974 (the Moss-Bennett Act) called for "the preservation of historical and archaeological data (including relics and specimens) which might otherwise be irreparably lost or destroyed as the result of . . . any Federal construction project or federally licensed activity or program." AHPA stipulated that up to 1 percent of the costs of a federal project could be spent on the recovery, protection, or preservation of endangered data. This act also stated that the secretary of the interior must consult with designated groups or individuals "with a view to determining the ownership of and the most appropriate repository for any relics and specimens recovered as a result of any work performed." AHPA called for the secretary of the interior to issue regulations for the long-term curation of significant artifacts and associated collections, thus beginning action to rectify the lack of curation planning for federal archaeology discussed in chapter 2.

Five years later, the Archeological Resources Protection Act of 1979 (ARPA) was passed. ARPA acknowledged the national problems involving the preservation and care of huge numbers of sites and collections, and it supplied a modicum of help for collections. First, it clearly stipulated that sites and objects from public lands were the property of the United States. Second, it called for the preservation of objects and associated records in a "suitable" institution, implying that an institution had to meet some sort of minimum curation standards. Third, it permitted the secretary of the interior to issue regulations about the exchange and ultimate disposition of archaeological collections.

Unfortunately, the regulations entitled *Curation of Federally-Owned and Administered Archeological Collections* (36 CFR Part 79) were not issued until 1990. In the meantime, a Government Accounting Office (GAO) report entitled *Cultural Resources—Problems Protecting and Preserving Federal Archeological Resources* (GAO 1987) clearly showed that the curation crisis was already in full swing: repositories were understaffed, overstuffed, and underfunded. The report strongly urged the promulgation of standards and guidelines to rectify this situation.

When finally issued, the regulations clearly assigned responsibilities to the federal agencies who owned archaeological collections and provided considerably more guidance than had been previously available. These guidelines have served as a model that agencies could use to develop more specific policies, and some agencies have done so (e.g., the U.S. Army Corps of Engineers, several bureaus of the Department of the Interior, and the U.S. Air Force). Table 3.1 summarizes key components of the regulations.

Within a month of the promulgation of the curation regulations, the Native American Graves Protection and Repatriation Act (NAGPRA) was passed. NAGPRA responds to American Indian concerns regarding the treatment of their ancestors' remains and significant related collections. It provides a procedure for the repatriation of Native American, Native Hawaiian, and Native Alaskan human skeletal remains, funerary objects, sacred objects, and objects of cultural patrimony that are currently held by federal agencies or by museums that receive federal funds (McKeown et al. 1998). If cultural affiliation can be identified between a particular set of prehistoric or historic remains

Table 3.1. Key Components of 36 CFR Part 79

1. Its legal authorities (Reservoir Salvage Act, Section 110 of NHPA, and ARPA)
2. Guidelines for the long-term management and preservation of preexisting and new collections
3. Methods to obtain curatorial services, such as
   - placing a collection in a federal repository
   - entering into a cooperative arrangement through a formal agreement with a non-federal repository
   - transferring a collection to another federal agency
   - requiring permitted projects on public or Indian lands under the Antiquities Act or ARPA to provide for curatorial services as a condition of the permit
4. Methods to fund curatorial services by federal agencies, such as
   - using funds appropriated by Congress for specific activities at a federal, state, or nonfederal repository
   - charging licensees and permittees reasonable costs for curatorial activities as a condition of the license or permit
   - placing collections in a repository that agrees to provide curatorial services at no cost to the U.S. government
5. Terms and conditions for curatorial services to include in contracts, memoranda, and agreements
6. Standards to determine if a repository can provide adequate long-term curatorial services, such as
   - professional museum and archival practices to accession, label, catalog, store, inventory, and conserve collections
   - maintenance of complete and accurate records on collections, as well as the
   - protection of documents from fire and theft
   - dedicated space to properly store, study, and conserve collections
   - secure storage, laboratory, study, and exhibit areas for collections
   - qualified museum professionals to manage and preserve a collection
   - collections inspections and inventories
   - access to collections
7. Scientific, educational, and religious uses of collections, including some conditions and restrictions
8. Procedures for conducting inventories and inspections

and a present-day Native American tribe, Native Hawaiian organization, or Native Alaskan corporation, or if the lineal descendants request the return of those remains, the federal agency or museum must do so as quickly as possible.

In many ways, NAGPRA has had a more positive impact on the current and future status of archaeological collections than the regulations in 36 CFR Part 79. This is because NAGPRA demanded that every federal agency and federally funded museum conduct an inventory to determine the contents of its collections and to address the ownership of specific items. Institutions had to meet specific deadlines for compliance or be subject to penalties. NAGPRA also provided some much needed funding to assist museums and tribes in the mandated process. Finally, NAGPRA has forced archaeologists and curators to cope with a dreaded and contested process: deaccessioning of items, in this case through repatriation.

So there is both good news and not-so-good news concerning the current status of archaeological collections and curation. The not-so-good news first: Early laws promoting archaeology did not provide guidance or direction on long-term collections management or how to make collections publicly accessible and useful. As discussed in chapter 2, early federal archaeology programs had no curation plans, and large collections were inadequately cared for. Most federal agencies with collections relied on the good graces of nonfederal repositories to manage their collections for free. In return, the repositories had little interaction with or interference from the agencies, and their staff had immediate access to the materials for research and interpretive projects.

This trend has continued. For example, the U.S. Army Corps of Engineers spent approximately $165 million on archaeological projects between 1975 and 1990 and virtually nothing on curation (Trimble and Meyers 1991:3). Few people anticipated the quantities of new collections that would be created, the storage space that would be required, or the curatorial effort necessary due to the National Historic Preservation Act and the Archaeological and Historic Preservation Act.

The good news is that 36 CFR Part 79, as well as NAGPRA and its implementing regulations (43 CFR Part 10), directed more attention to collections. Specifically, 36 CFR Part 79 provided guidance on establishing formal relationships between agencies and repositories for long-term care. It also acknowledged the costs involved, including the obligations of federal agencies to pay for curation. Because NAGPRA required an inventory of human skeletal remains, funerary objects, sacred objects, and objects of cultural patrimony, federal agencies had to figure out what they owned and where it was all located. Repositories were compelled to identify the owners of the collections under their care. The creation of better inventories has yielded knowledge of the collections' contents, storage locations, and conditions, as well as better access to them for research and interpretation. Finally, both 36 CFR Part 79 and NAGPRA have reinforced the need for collections policies, repository mission statements, and records management systems for both the owners of collections and the repositories.

Let us now examine some of the specific effects of the legislative efforts and the related issues that have been raised within the archaeological and museum professions over the last twenty-five years.

KEY ELEMENTS OF THE CURATION CRISIS

A few voices began to identify trouble concerning the storage and care of archaeological collections, records, and reports as early as the mid-1970s (Ford 1977; Marquardt 1977a). Concerns included inadequate care and visible deterioration of existing collections, inaccessibility to collections for research due to poor or nonexistent inventories or catalogs, and little or no security to protect the collections (Ford 1977; Lindsay et al. 1979). Some collections, such as some from WPA projects of the 1930s, were uncleaned and unanalyzed, and they had no associated project reports (T. King et al. 1977:24).

Several other concerns were related to archaeological practice. A critical problem, which continues today, was that archaeologists were not taking responsibility for the collections they generated (Christenson 1979; Marquardt 1977a; Marquardt et al. 1982). Excavation activities and research interests almost always had higher priority than curation of collections (Lindsay et al. 1980:93). This situation was exacerbated by the fact that field projects were largely run by universities without collections facilities or mandates to care for collections, as noted in chapter 2. The primary focus of graduate programs was on research accomplished through excavation or survey rather than collections. Virtually no courses in archaeological curation or collections management existed, nor do they today. The subliminal message from many graduate programs was and still is that the long-term preservation and care of collections are not the concerns of archaeologists but are the responsibility of curators or collections managers as soon as a collection is delivered to a repository.

Also in the late 1960s, new classes of materials, such as soil and botanical samples, were collected to help reconstruct the environmental contexts of past cultures, as noted in chapter 2. In the past, collections often were culled. For example, WPA researchers counted and then discarded all the plain, shell-tempered body sherds from the Chickamauga Basin collections (see chapter 2) because the analytical value of these materials was viewed as negligible and curatorial space was limited. Over time this practice became regarded as unethical. Instead of discarding large amounts of material after analysis, "everything" was saved for maximum information yield and to build up a data bank of materials in anticipation of changes in intellectual foci and new analytical techniques (Dunnell 1984; Sonderman 1996; Woosley 1992).

During the 1970s, increasing numbers of projects yielded only unpublished technical reports with little interpretation (McGimsey and Davis 1977). Few results of salvage and cultural resources management work were accessible through widely distributed, formal publications. Instead, information resided in reports of very limited circulation that were not cataloged or available in most libraries (see Toolkit, volume 7). Because project results and reports are poorly known, they rarely contribute to the goals of archaeology in general or interpretive syntheses for particular regions and themes. Also, new projects may duplicate previous work. However, these limited-distribution reports often provide the only information about projects, including critical contextual and locational data needed for proper use of the resulting collections. Therefore, timely production, widespread awareness, and access to copies of this "gray" literature are of considerable concern (Lipe and Redmond 1996).

By the early 1980s, new collections from cultural resource management (CRM) work began to pour into repositories, and cries for help focused on the new materials. Some repositories were already full and could not take in any new collections. For other repositories, the rate of acquisition became overwhelming because they did not have the staff or the funding base to do the required cataloging, information management, conservation, and storage work (Bell 1990; Marquardt et al. 1982; McGimsey and Davis 1977). Some of the issues mentioned earlier became exacerbated, while other significant concerns began to appear: ownership, standards, professional responsibilities, costs, information management, management of associated records, NAGPRA and consultation, deaccessioning, illicit trade, and repository certification. A national initiative on curation also has been proposed. Each of these issues is addressed next.

OWNERSHIP

Ownership of artifact and document collections is a significant issue in light of the wide range of collection owners, differences in their responsibilities, and confusion over who owns what over time. Owners of archaeological collections include federal, state, and local agencies, federally recognized tribes, CRM firms, museums and repositories, university departments, individual researchers, and private landowners. In many cases, dependency relationships have formed between groups, blurring the ownership of specific collections.

Federal agencies, federally recognized tribes, museums that have received federal funding, and some state agencies have specific legal responsibilities for the collections they own. For federal and many state and local agencies, these responsibilities include the following:

- Identifying repositories that meet minimum curation standards, negotiating conditions of long-term care, and depositing collections in those repositories
- Funding long-term care of both new and existing collections
- Inventorying collections
- Providing access to the collections in order to meet demands for public accountability and use

Federal agencies and institutions that have received federal funding must comply with NAGPRA. Although most agencies and institutions submitted their NAGPRA collection summaries in 1993 and inventories in 1995, they are still involved with consultation and repatriation processes that require staff expertise and funding. Federally recognized tribes must meet the requirements of laws and regulations such as ARPA and 36 CFR Part 79, but requirements for tribes are less well defined than those for federal agencies. Other organizations, such as CRM firms working on nonfederal and nonstate projects, many private and university museums, individual researchers, and private landowners, control the destiny of their collections, often with few policies or standards for long-term care. When compliance and mitigation work is done by a federal or state agency on private land, the landowner has full rights to the collection. The landowner may waive these rights and formally donate the collection to a repository. Archaeologists and agency staff who work with private landowners should encourage them to donate their collections to a repository that meets the standards of 36 CFR Part 79.

Given the number of owners of archaeological collections, the increasing costs involved in their care, and the inconsistent controls over long-term collections management and care, ownership sometimes has become contentious. For many decades prior to 36 CFR Part 79 and NAGPRA, museums and repositories curated federal collections for free in return for immediate access to them for research and exhibit. Now, in times of shrinking budgets, increased sizes and numbers of incoming collections, and stricter curation legislation, many repositories want federal agencies to acknowledge ownership and be fully accountable for funding collections care. Many repositories also want agencies that cannot do so to give up ownership. Agencies should know what they own and where it is located, while repositories should be able to identify and locate a specific collection and know its owner. Unfortunately, government agencies often have very poor records on their collections (Ferguson and Giesen 1999; GAO 1987). Repository inventories, when available, have been primarily designed to enable research, not to identify ownership (Bell 1990:9), so many repositories have not been able to determine who owns the collections under their care (Meyers and Trimble 1993; Trimble and Meyers 1991). Ownership remains unclear in such cases.

Federal and state agencies have had to work with each other to determine ownership where, for example:

- land ownership changed between agencies;
- several agencies involved in one project had different responsibilities for permitting, funding, land management, and project work;
- multiple funding sources supported a project; or
- agencies disagreed over their involvement in a project.

Indian tribes have ownership rights to whole collections excavated on Indian lands, as well as specific items in federal collections, since the passage of ARPA and NAGPRA. Controversy over rightful ownership can occur between tribes during the NAGPRA consultation and repatriation process or due to privately owned holdings of land on many reservations. Federal or state agencies may share ownership of an archaeological collection with private landowners because it is within the rights of a private landowner to keep a portion of a collection and donate the remainder to a repository.

Ownership of collections will remain a problem in U.S. archaeology due to the responsibilities and costs involved. Given the costs of long-term care, agencies may fight to *lose* ownership rights! Because the laws and regulations are now more explicit about ownership, some controversy may dissipate as experience is gained, formal partnerships and cooperative agreements are created, viable models of cooperation are made public, and inventory and records management systems are improved.

STANDARDS FOR COLLECTIONS MANAGEMENT AND CARE

By the mid-1970s, considerable variation existed in the quality of short-term and long-term care that repositories provided for the collections they owned and those they managed for others. These disparities grew as new collections accumulated. To minimize the variability and stabilize the collections for better access and use, a few archaeologists advocated establishing criteria or standards to evaluate and improve the quality of collections care, repository facilities, curatorial services, and management controls (Lindsay et al. 1979, 1980; Marquardt 1977a; McGimsey and Davis 1977). The early criteria were general, noting the needs to retain systematic collections for research purposes, have storage facilities with conservation facilities and climate control for different materials, have record keeping for accessibility and security, and provide educational services (McGimsey and Davis 1977). Explicit standards for initial collections processing and long-term collections management were subsequently advocated (Marquardt et al. 1982). Proposed standards for archaeologists depositing a collection for curation included submission of:

- a complete catalog of objects;
- original or clear copies of all associated field records, photographs, maps, and related documents; and
- objects in a stabilized and/or conserved condition.

Repositories that accepted new collections were expected to manage accessioned collections using an information management system that, at a minimum, provided locational information for research and management (computerization was not seen as mandatory in the early 1980s). The repositories also should package, label, and store collections in a pest-free, secure, and appropriately climate-controlled area. For long-term collections management, proposed repository standards included:

- good physical integrity of the building with special-use space and health, fire, and security measures;
- an information management system to search and retrieve items from all collections and to update records; and
- staffing expertise based on the size of the repository.

Many of the collections care and management needs recognized by these proposed standards were reiterated in a later analysis of the substandard conditions of federal collections held in several southwestern repositories (GAO 1987).

The issuance of 36 CFR Part 79 in 1990 finally provided a needed minimum set of curation standards and practices to be met by all federal agencies and the repositories with which they partnered. The regulations cover the issues presented earlier, along with the need for a repository mission statement, collections policy, and records management policy. Unfortunately, however, the regulations do not establish standards for inventorying collections and their associated records to improve their identification and accessibility. The regulations do not require periodic inventories or deadlines for compliance by repositories. Nor do they establish a certification process to help agencies identify repositories that meet the minimum standards. Some felt that these additional requirements would slow down the promulgation of the regulations and add considerable burden and expense to those affected by the regulations (Ferguson and Giesen 1999).

Some federal agencies (e.g., U.S. Army Corps of Engineers, U.S. Air Force) have used the minimum set of standards in 36 CFR Part 79 as a foundation on which to develop more specific standards. As a result, some curators discovered that agencies that share space in the same repository have different sets of requirements the repository must meet (Bell 1990). Similar problems exist for the agencies that have collections in many different repositories, each with its own collection management requirements.

PROFESSIONAL RESPONSIBILITIES

"Curation is a professional responsibility; we must argue for it, insist on it, teach it, believe in it, and practice it" (Marquardt 1977a:39). Unfortunately, many archaeologists still do not consider the long-term care of collections to be as important as the sites from which the objects were recovered. Farnsworth and Struever (1977:13) bluntly note that "deep down inside many of us is the belief that the first analysis is equivalent to the last analysis of any body of archaeological data. When the analysis is done, why curate?" As a consequence, archaeologists often regard curation as a storage problem, not as a process of banking irreplaceable data, objects and documents, as a basis for research and education, and as a foundation of cultural heritage (Sullivan 1992a, 2001).

Three things need to happen to rectify professional attitudes. The first involves education and training. Basic archaeological collections management must be a required course in graduate archaeology programs so archaeologists learn their responsibilities and understand the long-term impact of their actions (see the appendix for information on an online course on archaeological curation). Such course work should encourage some students with a good grounding in archaeology to pursue a career in archaeological curation and collections management. Training in the conservation of objects and documents, information management and databases, and archiving also must be facilitated and valued by the archaeological profession. Recent discussions of new curriculum reforms for graduate archaeology education do not include collections curation (Bender and Smith 2000; Lynott et al. 1999). On the positive side, the Society for Historical Archaeology (SHA) has a collections management committee and a set of curation standards for SHA members (SHA 1993), and the Society for American Archaeology established a curation committee in 1999 and approved new ethical guidelines regarding curation in 2002 (Childs 2002).

The second need is to increase the number of curatorial jobs in repositories and in large cultural resource management firms to handle the significant workload (McGimsey and Davis 1977). A curation needs assessment for the U.S. Army Corps of Engineers found that only five of the sixteen repositories surveyed had a full-time curator for the archaeological collections they housed (Bade and Lueck 1994). In the other eleven, the responsibilities for written policies and procedures; updated, complete, and accessible management records; collections security; a computerized inventory process; artifact accessibility; and periodic inspections either were not done or were divided among staff.

Third, there needs to be much more interaction between archaeologists and repository staff. Archaeologists need to be involved in developing and updating collection management policies because these policies can significantly affect research efforts (Sullivan 1992a, 2001). At the same time, curators need to be involved in the development of archaeological research designs and collecting strategies.

CURATION COSTS

"Until the American public becomes aware that excavation makes up only about a fourth of archaeological research, laboratory analyses and curation the other three quarters, it will be extremely difficult to acquire the needed support on a continuing basis" (Hester 1977:10). Yes, the public should know how costly curation is since they are increasingly paying for it with their tax dollars. More important, however, archaeologists, CRM contractors, and state and federal agencies must learn the real costs of managing the collections they either generate or own. The reality is that excavation or survey involves one-time costs; long-term collections care involves continued costs (Woosley 1992). Yet it is well known that archaeological research and mitigation work have received considerably more dollars than curation (Society for American Archaeology [SAA] Curation Task Force 1993).

The costs of curation have been a major concern for as long as anyone has spoken up about archaeological collections management in general. The tremendous influx of new collections in the late 1970s elevated this concern (Marquardt 1977a; Marquardt et al. 1982; McGimsey and Davis 1977), further heightened by the curation standards in 36 CFR Part 79. By one count, the federal standards added twenty-one new costs to the curation process (Harris 1993), yet most agencies did not budget for curation, and most repositories did not have money to increase staff, space, or facilities (Acuff 1993; Bell 1990; Moore 1994). Nevertheless, the significant costs of curating and making accessible millions of archaeological artifacts, records, and reports are here to stay.

INFORMATION MANAGEMENT SYSTEMS FOR CATALOGS AND INVENTORIES

Accountability of and accessibility to collections are critical to their long-term care and their use for research, public interpretation, and exhibit. If we do not know what is in a collection, where its contents are located, when it was last inventoried or inspected, and where the documents and reports containing essential contextual information about the collection are located, then it is all rather useless until found. In fact, why keep a collection at all if it is not known about or accessible?

Several surveys over the years have assessed the information management capabilities of U.S. repositories. They found that basic catalogs and inventories with locational information have been seriously deficient or lacking, particularly for federal collections (Ford 1977;

GAO 1987; Lindsay et al. 1980; Meyers and Trimble 1993; Trimble and Meyers 1991). Furthermore, the records of an agency rarely agree with the records of the caretaking repository (Ferguson and Giesen 1999; Moore 1994).

Substandard collections information management persists in many settings even as archaeologists have become computer-savvy. Database technology allows access to a huge range of data, efficient searching and data retrieval, and identification of gaps, biases, and research potential in the data (Sullivan 1992a, 2001; Woosley 1992). Appropriate hardware and staff expertise are not hard to find, although often expensive. There are several software applications for cataloging collections and archives, as well as bibliographical systems for reports (see Toolkit, volume 7). Archaeologists and curators need to work together to maximize the effective use of information management technology.

MANAGEMENT OF ASSOCIATED RECORDS

The condition of artifact collections is widely deficient, but the management of the records pertaining to archaeological investigations and collections is abysmal. Poor records management has led to loss of primary archaeological research data and is leading to loss of data on the history of the discipline.

Records unite the artifacts and specimens with their proveniences, thereby making primary data useful for basic research. Field and laboratory records and project correspondence also contain epistemological information. They show how innovations have been applied to regional archaeology and shaped our current understanding of regional prehistory. Records contain copious information about the interpersonal relations and networks that often lead to use (or dismissal) of certain methods or techniques. They also provide insights into various personalities and their effects on the discipline (Sullivan 1991, 1995).

Four principal problems have been identified in records management. The first is that some archaeologists do not submit complete sets of field records with a collection (Trimble and Meyers 1991). Such collections lack contextual information and are virtually useless.

The second problem lies in recognizing who owns and is responsible for records. The 36 CFR Part 79 regulations have clarified this issue:

- Records related to federal projects are highly valuable and must be kept with an artifact collection.
- A federal agency is responsible for the administrative records regarding disposition and care of their collections.
- A repository is responsible for the long-term care of all the associated records for an artifact collection it curates.

The third problem involves storage practices. Not only are records collections rarely stored in fireproof cabinets, but there has been a tendency to separate the records from the associated artifact collection. Sometimes the documents and artifacts are stored in different buildings or towns (Meyers and Trimble 1993), and sometimes the documents have been destroyed (GAO 1987). In these cases, accessibility to and optimal use of a collection may be severely compromised.

A related matter is the condition of documents as they are stored, both in the field and in the repository. Often, various media of paper, ink, film, and video used to capture key information deteriorate due to unstable or inappropriate environmental conditions. Basic practices must be implemented to save the records:

- Completely inventory the records in a collection.
- Use archival quality paper for copies of documents.
- Separate record copies from originals to ensure preservation while maintaining access.
- Create an archives management plan to set priorities for rehabilitating damaged records.

The final problem involves the increasing creation, use, and reliance on electronic data by archaeologists through word-processed documents, digital photographs, Geographic Information System (GIS) or computer-aided drawing (CAD) data layers, database files, and websites (Carroll 2002). These types of associated records are in many ways more fragile than paper and film. There are some very basic practices that should be applied to ensure the long-term preservation of electronic data. First, at the beginning of a project, archaeologists must carefully plan the types of software and file formats that will be used to ensure compatibility and long-term utility. They also must decide whether it is practical to make paper copies of electronic information. Then, for each electronic file, basic information must be documented, such as format, version of software used, date of creation, creator's name, file relations, and database structure, and related scripts or macros. Finally, all electronic data must be migrated to up-to-date file formats and software as old software and hardware become obsolete and unusable. The migrations also must be documented. If these basic practices are not followed, important information representing considerable time, costs, and expertise will be lost forever.

NAGPRA AND CONSULTATION WITH AMERICAN INDIANS

The need for guidelines, special considerations, and sensitivity regarding the curation and use of American Indian skeletal remains, associated funerary objects, and sacred objects has been recognized for some time (Lindsay et al. 1980; McGimsey and Davis 1977). It was not until the enactment of NAGPRA in 1990, however, that federal agencies and many repositories were required to do something about it (McKeown et al. 1998). Under the threat of deadlines and fines, the primary needs in recent years have been to

- inventory collections to determine whether they contain human skeletal remains, funerary objects, sacred objects, and objects of cultural patrimony;
- resolve ownership and responsibility for related costs;
- identify and then retrieve skeletal remains on loan;
- consult with tribes about the cultural affiliation of the remains and about repatriation; and
- repatriate requested remains and materials to affiliated tribes.

The NAGPRA inventory process has been time-consuming and costly. For some agencies and repositories, it was the primary focus of staff effort for years and, in some cases, still is.

The net effect of NAGPRA has been positive, although there is still a long way to go. Tribal representatives are now in dialogues with archaeologists, curators, administrators, land managers, state historic preservation officers and staff, and many others who tended not to interact in previous decades. Some tribes have designated their own tribal historic preservation officers. Compromises and agreements are being developed on several curation issues:

- Archaeological research interests versus tribal concerns that may vary by tribe
- Appropriate analytical procedures for different types of objects
- Collections access and use procedures for skeletal remains and for objects related to traditional values and uses
- Collections access and use for traditional rites and ceremonies
- Long-term effects of conservation treatments (Odegaard 2000)

DEACCESSIONING

Deaccessioning is "the process used to remove permanently an object from a museum's collection" (Malaro 1998:217). When 36 CFR Part 79 was issued in 1990, an important section was published "as proposed" only. It concerned the authority and criteria to deaccession individual items or portions of collections. The proposed regulation met opposition from archaeologists who foresaw the development of new analytical techniques to extract important, unanticipated information from existing collections and who emphasize the finite nature of the resources held in repositories (Childs 1999; Sonderman 1996). Also, at a time when very positive steps in collections management were beginning to be made, it did not make sense to appear to promote deaccessioning. Thus, this regulation was not promulgated.

However, the need to be able to deaccession portions of collections is becoming urgent for many archaeologists and curators. Huge amounts of bulk materials such as shell, fire-cracked rock, and unanalyzed soil samples result from archaeological projects, and real limitations exist on storage space and staff to process the materials (Acuff 1993; Bell 1990; Sonderman 1996). Some archaeologists question whether these materials will ever be analyzed. Many are asking, "Why keep all of them?"

Three steps can be taken. First, key professional societies need to make deaccessioning an issue, educate their members about the subject, and become involved in finding and implementing solutions. Many archaeologists do not know what accessioning is, so how can they be expected to know about deaccessioning (Sonderman 1996)?

Second, the National Park Service should revive its effort to publish a deaccessioning regulation under 36 CFR Part 79. It should consider the appropriate conditions under which archaeological collections may be deaccessioned, such as loss, theft or involuntary destruction, abandonment and voluntary destruction, nonconformity to a repository's scope of collection, destructive analysis, return to rightful owner, and repatriation under NAGPRA. It must also consider particular types of materials, such as those that were inadvertently collected and later determined to lack archaeological or historical significance, are highly redundant and nondiagnostic, are without good research potential, are deemed to lack archaeological interest under ARPA, or are a hazard to human safety and health.

A third step is to involve more archaeologists in writing the deaccessioning policy for the repository with which they work. This interaction may help prevent inappropriate deaccessioning by a repository and facilitate appropriate deaccessioning in consultation with archaeologists. For example, inactive, relatively inaccessible storage may be appropriate for highly redundant, infrequently analyzed materials.

LOOTING OF SITES AND THE ILLICIT TRADE IN ARCHAEOLOGICAL OBJECTS

The looting and destruction of archaeological sites is ultimately about finding collectible objects that can be traded and sold for considerable prices. Many archaeologists focus on rightly condemning the act of looting and the visible effects on sites rather than on the looter's desire for the objects buried in the sites. Professional attention needs to be directed to the types of objects looted from sites, the effects of their loss on the use and interpretation of archaeological collections, and the need for better public education about the critical interrelationships between sites and the objects buried in them. Repositories are increasingly doing their part by accessioning archaeological objects with clear provenience information and improving security to prevent theft of valuable objects.

CERTIFICATION OF REPOSITORIES

A federally funded study promulgated a proposal to certify repositories that meet well-defined standards accompanied recognition of the need for guidelines and standards for long-term collections care (Lindsay et al. 1980), especially for federal collections. The proposal advocated establishment of a task force to set criteria for certification and a coordinating office to authorize accreditation and to monitor the practices of accredited repositories. It also proposed a grant program to help repositories improve their facilities and practices to achieve certification.

The regulations in 36 CFR Part 79 did not include any mechanism to certify repositories that met its standards. Although state and federal agencies and private CRM companies often want assurance about the quality of curation services for which they now pay, the issue of an accreditation program at the federal level is in suspension. Such a program is often seen to be too costly and perhaps redundant, especially because the American Association of Museums (AAM) provides accreditation for museums in general. Still, not all repositories are museums, so AAM accreditation is not the answer. On the state level, Texas was the first to initiate a program specifically to accredit repositories curating archaeological collections. The Accreditation and Review Council, a unit of the Council of Texas Archeologists, developed this program (see the appendix for its website address). Perhaps other states will follow this lead if the Texas program proves successful.

NATIONAL INITIATIVE FOR CURATION

The need for a national initiative or system to preserve and protect archaeological collections, records, and reports has been an issue since the 1970s. The focus would be the huge collections owned by federal agencies for the American public and curated by federal and nonfederal repositories. The goals of such an effort would include the following:

- Providing uniform guidance on the deposition, long-term care, and access to collections (partially achieved with the issuance of 36 CFR Part 79) and encouraging state and private groups to use the national standards
- Coordinating policy, guaranteeing consistency in policy implementation, and acting as a clearinghouse for information sharing on archaeological collection management issues
- Conducting a full assessment of the nation's collections by identifying existing repositories and evaluating the conditions of the facilities, information management systems, collections, records, and reports
- Accrediting repositories to participate in the national program and providing oversight and review of repository activities
- Creating a grants program to rehabilitate old collections, upgrade repositories, and improve information management systems

- Coordinating efforts to inventory and index archaeological reports, store and disseminate digital data, and maintain an up-to-date national database of the unpublished, gray literature (Ford 1977; Lindsay et al. 1980; SAA Curation Task Force 1993; Scoville 1977)

This effort could take several forms. It could be fully funded and managed by one federal agency on behalf of all others (Scoville 1977), coordinated by a federal agency with the state historic preservation offices (Lindsay et al. 1980), or managed by an independent, national organization with federal and nonfederal advisers (Lindsay et al. 1980). Each of these has potential problems in terms of organization and infrastructure, funding sources, budgeting priorities, implementation and monitoring responsibilities, and accountability. As a result, relatively little has been done at the national level, although further consideration of such an initiative is important at this time.

THE BRIGHT SIDE

Despite the problems surrounding archaeological collections, records, and reports that still need to be systematically confronted, some notable progress has been made in recent years.

The publication of 36 CFR Part 79 stimulated development and implementation of standards by several federal agencies and many states, including New York (New York Archaeological Council 1994), Maryland (Shaffer and Cole 1994), West Virginia, and Vermont, among many others. Cities with archaeology programs, such as Alexandria, Virginia (Magid 1991), and state/federal research programs, such as the Savannah River Archaeological Research Program (Crass 1991), also have developed curation policies and procedures.

Several important surveys were conducted in the late 1970s and 1980s to determine the status of university, state, and federal collections and repositories across the United States (Ford 1977; GAO 1987; Jelks 1989; Lindsay et al. 1979, 1980). These surveys identified and outlined the curation crisis. More recently, the U.S. Army Corps of Engineers established a Mandatory Center of Expertise in Archaeological Curation and Collections Management, and its assessments of specific collections and facilities have provided a more detailed and vivid picture of the existing deficiencies and problems of federal collections (Bade and Lueck 1994; Meyers and Trimble 1993; Trimble and Meyers 1991). The center assesses groups of collections, sets priorities for resolving the problems, funds required activities, pursues the establishment of state and regional repositories to control costs and maximize accessibility to and use of collections, writes guidelines for field collections and collections management (Griset and Kodack 1999), and helps focus professional attention on curation.

The National Science Foundation's Systematic Anthropological Collections program once provided financial assistance for improving the care and management of significant non-federal collections (Greene 1985), but the program no longer exists. The National Endowment for the Humanities now has two helpful grant programs: "Preserving and Creating Intellectual Access to Collections" provides funding for collections stabilization, storage, and the like; the "Preservation Assistance Grants" provide funding to assess the needs of a repository.

A 1994 survey by the Interagency Federal Collections Working Group helped some federal agencies determine the present locations of their collections in nonfederal repositories and promoted the establishment of formal partnerships between federal agencies and repositories (Childs 1995; Wilson 1999). Three national conferences focused on partnerships for federally associated collections have been held between 1996 and 2000 building on new understandings and issues that resulted from the survey.

Although regional repositories are not popular for organizational reasons, and because many small museums fear that the primary collections under their care will be taken away, some regional repositories have been established for specific government agency collections. These repositories were created either in anticipation of a huge influx of materials or to consolidate collections within a region for better care, access, and use. Examples include the Anasazi Heritage Center in Colorado for the Bureau of Land Management, the Portland, Oregon, district repository and the Illinois State Museum cooperative agreement with the St. Louis District of the U.S. Army Corps of Engineers, and Fort Vancouver in Washington for the National Park Service (Bush 1996; Woosley 1992).

Recent initiatives have heightened interest in the preservation of and accessibility to archaeological records and reports, including field records in all media, laboratory reports and data, oral histories, administrative documents, and correspondence. These records are essential to interpret the archaeological record and to record the history of the discipline (D. Fowler et al. 1996; Silverman and Parezo 1995). A group of anthropologists, archivists, and information managers recently formed a nonprofit organization called the Council for the Preservation of Anthropological Records (CoPAR). Its goals are to heighten professional awareness of the need to archive personal collections of documents; create research and indexing aids to increase accessibility to archived collections; facilitate repository support; and educate professionals about issues concerning preservation, digitization and information management, and ethics (Fowler et al. 1996). CoPAR has a website to help anthropologists find research materials in archives and learn how to archive their own records (see the appendix).

The National Archeological Database (NADB) has grown in recent years to provide an Internet-accessible, searchable system of information about key documents relevant to archaeologists, curators, and many others (Canouts 1992). The principal module, NADB-Reports, is a bibliographical inventory of hundreds of thousands of archaeological reports, mostly gray literature. The institution that holds a copy of a report is identified during a search of the database. NADB-NAGPRA is a compilation of key NAGPRA documents. Another module, NADB-MAPS, contains GIS maps that show national distributions of cultural and environmental resources. NADB-Permits, an inventory of federal archaeological permits issued between 1908 and 1984, should be available soon.

Another promising development is that in 1999, the Society for American Archaeology (SAA) established a standing committee on the curation of archaeological collections (Childs 2000). Curation issues now have continuing, focused attention from the archaeological profession beyond the Society for Historical Archaeology. The new ethical guidelines for curation (Childs 2002), mentioned earlier, are a product of this committee working in concert with the SAA's Committee on Ethics.

We are still in a curation crisis, but we have made progress. Better education about archaeological collections management, stronger commitment by archaeologists and professional societies to both the making and care of collections, increased numbers of cooperative efforts, and better inventory and cataloging systems will foster even greater progress.

4

REPOSITORIES: WHAT ARE THEY, AND WHAT DO THEY DO?

A repository, "a place to care for collections," is a key concept in this book. This is a facility that holds archaeological collections, associated records, and reports. If properly set up and maintained by its staff, the repository plays the primary role in the long-term stability and usefulness of the materials it holds for research, interpretation, cultural heritage, and public exhibition.

Federal regulations, 36 CFR Part 79, provide another commonly used definition of a repository:

> a facility such as a museum, archeological center, laboratory or storage facility managed by a university, college, museum, other educational or scientific institution, a Federal, State or local Government agency or Indian tribe that can provide professional, systematic and accountable curatorial services on a long-term basis. (National Park Service 1990:37632)

This broad definition recognizes several important aspects of a repository. First, there are several types of repositories across the United States that differ in significant ways. Second, the words *accountable* and *long-term* emphasize the fact that specific curatorial services must be accountable to someone over a long period of time, whether it is a board of directors, contributors to a nonprofit institution, or a federal agency and the taxpayers who support it. Finally, the term *professional* indicates that curatorial work requires professional training and expertise in order to carry out the responsibilities of managing a repository and its contents.

KINDS OF REPOSITORIES

We have grown up hearing the word *museum*, not *repository*. We visit museums, not repositories. Yet archaeological curation centers on repositories. There are several types of repositories across the United States: museums, academic repositories, tribal museums and cultural centers, government repositories, historical societies, and archives. Each type often has a different configuration of missions and goals and an emphasis on a different array of activities. Archaeologists need to understand how repositories differ in order to know how to relate to them (see chapter 7). All repositories should meet a number of common standards in order to provide the best possible long-term care for and management of the collections they contain (see chapter 5).

MUSEUMS

A useful definition of a museum, adapted from the Museum Services Act, is "a permanent, nonprofit organization, essentially educational and often aesthetic in purpose, which, utilizing professional staff, acquires tangible objects, interprets them, cares for them, and exhibits them to the public on a regular basis" (Malaro 1994:81). The key words are *nonprofit* and *educational*. A nonprofit institution benefits the public, usually by providing service(s) to the public (Butler 1989; Malaro 1994). Because benefiting the public is seen as charity, nonprofit museums are exempt from taxes. Museums also can accept tax-deductible contributions from individuals and organizations.

A nonprofit museum can be private or public (Nichols 1989). Private nonprofit museums are founded because one or more individuals donate a collection or collections, or because one or more individuals organize a nonprofit museum around their own collection(s). A museum's mission may or may not involve collecting more objects after it is established. Public nonprofit museums are established by federal, state, or local legislative action when someone donates one or more collections to a public agency or a public agency owns and creates collections. The nature of the collection and the services provided can further define a museum. For example, a natural history museum, regional museum, historical society, and historical museum all may hold archaeological collections.

The primary service a nonprofit museum provides to the public is education, and this service generally distinguishes it from a repository. Usually a museum's statement of purpose clearly states the intended results of educational service, while its mission statement provides the methods, including exhibits and public programs. Education requires an infrastructure with types of expertise other than those required to manage a repository and care for the collections it holds. For example, planning and creating an exhibit requires considerable research on the context of the collection(s) and the individual objects put on display, plus expertise on exhibit construction, a firm understanding of the exhibit's intended audience, and full consideration of security, preservation, and housekeeping issues. Public programs require working with diverse publics, including different age groups, schools, Native American and other cultural groups, and the disabled. It also involves teacher training and expertise in evaluating program effectiveness.

ACADEMIC REPOSITORIES

Secondary schools, institutes, colleges, and universities, as educational institutions, are also nonprofit organizations. Many, too, have repositories under their institutional umbrella. One type of repository is the academic museum. The Peabody Museum of Archaeology and Ethnology at Harvard University, a well-known and well-established academic museum, comes to mind, as does the nearly eponymous R. S. Peabody Museum at Andover Academy, a secondary school. They serve the public by offering exhibits and educational programs yet also are attuned to the research needs of their faculty and students. An important characteristic of this type of museum is accessibility by professionals, students, and other interested individuals, such as Native Americans, to the vast majority of the museum's collections that are behind-the-scenes. Effective accessibility requires up-to-date management of the objects and the associated documentation, a regular inventory system, adequate research space, conservation, security, and fire protection.

Another type of repository in some academic institutions is a place for long-term storage of collections generated by its faculty and students, as well as for research on those collections. Its mission, when stated, does not include public education through exhibits and programs. It fulfills its basic educational goal when faculty use the repository for teaching and research. This type of repository requires all the features of an academic museum but without public outreach: up-to-date collections management, an inventory system, adequate research space, conservation, security, and fire protection. Unfortunately, some of these features are often not available due to lack of funding and inadequate staffing.

TRIBAL MUSEUMS AND CULTURAL CENTERS

Heightened tribal interests in issues of heritage, identity and community, along with the passage of NAGPRA, have influenced the recent growth of tribal museums and cultural centers. Depending on the tribe, such a facility may be an independent, nonprofit museum with trained museum staff, an institution under the umbrella of a parent organization such as a tribal government, a cultural center, or a mixture of these (e.g., Nason 1999; Warren 1991).

Cultural centers and tribal museums have several primary functions. One is to provide heritage and language education for tribal members, as well as others, from the perspective of the tribe. This often involves using archaeological and ethnographic objects in exhibits and programs including dances, feasts, and craft demonstrations. Another is to appropriately house, care for, and respect human skeletal remains, funerary objects, sacred objects, and objects of cultural patrimony from archaeological projects that are repatriated under NAGPRA, as well as any other stored objects (Jennings 1996). Some outcomes of this function are innovative research and community-based collaboration on new methods to store, handle, and provide access to Native American collections (Nason 1999).

HISTORICAL SOCIETIES

Museums associated with historical societies tend to curate a hodgepodge of collections from a particular geographic area, usually a state or county. Many have a long history, sometimes preceding the well-established public museums. Often they started with gifts of assorted items from local residents in the spirit of documenting and preserving local history. Although well-trained staff support some historical society museums, many others depend on the good intentions of volunteers.

GOVERNMENT REPOSITORIES

Many public agencies generate and own archaeological collections. Recent data from several sources indicate that the federal government alone owns well over sixty million archaeological objects, thousands of cubic feet of objects (this group is counted differently), and thousands more linear feet of associated documentation (Ferguson and Giesen 1999; Wilson 1999). We do not know of any effort to quantify the number or cubic feet of objects owned by states, local governments, or tribes.

Many government-owned collections have been placed in the care of nongovernment museums, both private and academic, for many decades. The latest count of nonfederal museums and repositories holding federal collections for Department of Interior agencies is almost five hundred (Wilson 1999). Many federal and state agencies, however, also have chosen to care for their collections in their own facilities. While the Bureau of Land Management manages only three repositories and relies on 189 nonfederal museums and repositories to care for its collections (Damadio 1999), the National Park Service has over three hundred park and curatorial repositories to hold its museum property, the majority of which is archaeological (NPS 1998).

Like academic repositories, there are primarily two kinds of government repositories for archaeological collections. The first is a museum with a strong emphasis on public education and outreach. Examples are the Florida Museum of Natural History, the Anasazi Heritage Center operated by the Bureau of Land Management, and the visitors' center at Colonial National Historical Park. Because many agencies own and manage archaeological sites, some museums offer both artifact exhibits and visits to the archaeological sites from which they came. These museums also offer access to stored collections. Access usually is regulated to minimize risk to the objects and documents. Given the mission to provide both educational programs and access to the collections, a museum must have staff who can research and build exhibits; teach; catalog, accession, and inventory the collections; ensure security and fire prevention; and provide conservation for objects. It also must have space for researchers, students, individuals whose heritage is tied to the collections, and the interested public to examine and sometimes use the material. Native American groups, in particular, periodically visit public museums that care for objects of particular ritual or spiritual significance in order to perform ceremonies.

The other type of government repository is one that does not have active public outreach and educational activities. Its primary missions are to care for the agency's collections and to provide access to those collections for the same diverse public as the public museums. Some government collections, however, are placed in dead or low-maintenance storage that does not allow for ready access. When objects in this kind of storage are requested, repository staff must retrieve the items and make them available in a location amenable to research or other specified uses. Clearly, if collections are cared for in different storage locations under the management of one repository, the cataloging and accession information must be up-to-date and accurate.

ARCHIVES

Some materials related to archaeology might be cared for in an *archive,* a depository of records (Ellis 1993; Vogt-O'Connor 1999; Wilsted and Nolte 1991). It can hold and conserve key documents that outline the history of an individual archaeologist, the discipline, or a state, federal, or private archaeology program or society. Although many archaeologists deposit their correspondence, miscellaneous field and laboratory records, diaries, draft articles, and other documents in the same repository as the collections they produced, many others do not. The records of government archaeology programs are sent to government archives on a regular basis.

Many archives now manage digitized records, such as computer files of letters, draft articles, GIS maps, and correspondence. E-mail often is never transferred to paper, yet it may contain important facts and observations about an archaeological project or logistical details not recorded elsewhere. Archivists and document conservators must deal with constant, rapid, technological change in order to preserve documents in useable forms (Vogt-O'Connor 1999; Puglia 1999).

WHAT A REPOSITORY DOES AND WHY

Of the various kinds of repositories, museums tend to have the most diverse array of programs, usually geared to a wide audience. Museum programs can include collections care and management, basic and applied research, and public education, including exhibits. Other types of repositories sponsor a more limited range of programs. What follows is a brief overview of the various kinds of programs in which repositories may be engaged and the rationale or benefits of such programs.

COLLECTIONS MANAGEMENT

All repositories have collections care and management programs. These programs are the heart of the institutions and are what make them repositories. Collections programs tend to be the activities archaeologists most directly equate with curation. As we note, however, a museum curator may be primarily responsible for research and educational programs, while collections care may be the duty of other professional staff. In larger repositories, collections care can be the responsibility of several staff members with different training. In a small institution, one staff member may be required to be a jack-of-all-trades in regard to collections.

Managing and caring for collections in accordance with a repository's mission require work on the physical well-being, overall coherence and composition, and safety of collection items as well as their accessibility. *Conservation* is the care of the physical condition of collections. Environmental conditions, storage furniture and containers, and treatments to stabilize or repair deteriorating or damaged materials all are aspects of collections conservation.

Accessibility of collections refers to how readily collections and information can be found and used. Access to collections in repositories may be necessary for research, programming, heritage or religious purposes, or collections conservation, such as monitoring fragile materials.

Computerized inventories are nearly indispensable for keeping track of items, their storage locations, and relevant documentation. Electronic databases eliminate the need for the card catalogs that used to be the mainstays of library and museum collections management. Managing the content of a repository's collections is an intellectual endeavor with the goal of matching objects to the repository's overall mission and subject matter theme(s). For example, an institution that curates archaeological materials most likely has a defined geographic area(s) from which it will acquire such materials.

Different kinds of repositories can have differing goals for their collections programs, depending on their institutional contexts and missions. The collections of a museum, for example, are intended for public benefit and may include more items suitable for exhibition and public education than collections curated by a university anthropology department. The latter repository likely houses the products of faculty research, which may be used for teaching and research, but not often for exhibition.

A repository must have policies, procedures, and plans that are broad enough to cover all of the kinds of materials it curates. Archaeological materials may be only one category of collections managed by a repository. Natural history or general museums, for example, often curate geological and paleontological collections including gems and minerals, fossils, and core samples; decorative arts and historical items such as costumes, furniture, toys, photographs, and paintings; biological specimens including herbaria, animal skins, feathers, and taxidermy collections; and ethnographic items such as baskets, masks, spears, and clothing. The wider the range of items a repository curates, the more complex are the issues of collections care and management. More specialized knowledge is necessary to care for the collections. For example, environmental controls for herbarium collections can be quite different from those for gems and minerals. It is unrealistic to expect a professional staff member such as a conservator, who is trained to care for the physical well-being of many different kinds of materials and objects, to be well versed in the subject matter and research issues of a particular discipline.

EDUCATIONAL PROGRAMS AND EXHIBITS

Although the purpose of all repositories is at least indirectly related to education and learning through the collections they curate, not all repositories sponsor direct programming in the form of exhibits, lectures, workshops, field trips, and the like. The content of educational programs usually is connected to the repository's mission and to the subject matter of its collections. A repository can share the educational value and significance of its collections with a broad audience through public programs. Although this is a laudable and worthy goal, especially for publicly owned repositories, there is a danger that public programming can become the cart that drives the horse: collections care can become a secondary activity in terms of funding, visibility, and stature within the institution. The use of collections objects for exhibition and other public programming can present a risk to the objects, a risk that must be managed by the repository. Collections care often must compete with high-profile programs, such as blockbuster exhibits, for institutional resources and support. In some cases, however, such public programming is beneficial to collections care because it can generate funding.

RESEARCH

Research in the context of a repository may take several forms and may be basic or applied. Basic research generally is oriented toward creating new knowledge purely for the sake of learning, while applied research generally takes existing information and uses it to solve a problem or create a new application such as an exhibit or other program. Repositories such as large natural history museums may have their own staff members who are involved in basic research and who do fieldwork that adds to the repository's collections. In-house and other research staff members do basic research involving existing collections. Smaller repositories that do public programs may have staff members who do applied research, but the staff members of repositories not involved in public programming may have no research duties whatsoever in the disciplines of the collections. They may, however, be involved in research oriented toward collections care, such as discovering new ways to treat and preserve certain material types.

Like the kinds of collections it acquires, the kinds of research a repository conducts are tied to its mission and to the size and training level of its staff. In the tight-budget environments of many museums, staff members who care for and manage collections often are viewed as infrastructure, while those who do basic research more likely are considered to be a luxury. As noted in chapter 2, this was not the case in the nineteenth century when museum research staffs were needed to build collections, before academia took primary responsibility for basic research from museums. Today, with most archaeological collections made by archaeologists not associated with repositories, the need for repositories to have their own field research staff is becoming less crucial to many repository administrators.

RESPONSIBILITIES AND TRAINING OF REPOSITORY STAFF

Staffing of a repository largely is a function of its size, mission, and funding. Not surprisingly, larger repositories generally have broader missions and larger funding bases, so they tend to have the largest and most diverse staffs. Curatorial staff tend to be specialized and to have more focused responsibilities in large institutions. Staff in smaller institutions tend to be the "general practitioners" of curation, calling in specialists as consultants for specific needs. Archaeologists often use the term *curator* to refer to all professional staff members that are involved with collections in a repository. The responsibilities of curators may be quite different depending on the institutional context.

CURATORS, CONSERVATORS, REGISTRARS, AND COLLECTIONS MANAGERS: WHAT ARE THEY?

The management and care of collections has become more professionalized and more complex. The greater amounts of information, legalities, and public accountability associated with curatorial work all have contributed to the needs for increased specialization and training of repository staff. Museum studies programs at some universities (e.g., Texas Tech and George Washington University) train students specifically for jobs in collections care and management.

Table 4.1 summarizes characteristic positions, responsibilities, and training for the various kinds of curatorial staff. In some large museums or repositories, collections staff may belong to an administratively separate division distinct from research staff. In such settings, research staff members may have the title of curator yet may have little responsibility for collections management or care (the Peabody Museum of Archaeology and Ethnography at Harvard has this kind of administrative structure). Archaeologists working with repositories through curation agreements or collections research most typically deal with either collections managers or curators (M. E. King 1980; Terrell 1979). A registrar (Buck and Gilmore 1998) also may be involved if collections are to be loaned or exhibited because registrars handle risk management and insurance considerations.

A collections manager typically holds a master's degree and has experience in the discipline of the collection. An archaeology collections manager, for example, typically will have some archaeological field experience and will have taken museum studies courses in school or classes offered by institutes such as the Campbell Center in Mount Carroll, Illinois. A collections manager usually has some knowledge of the discipline but typically is not engaged in a program of basic research. Collections managers generally are trained to know

Table 4.1. Curation-Related Jobs

Job Title	Duties	Training
Registrar	Generally responsible for collections inventory, tracking, and risk management. Coordinates overall planning for collections needs.	Master's degree in museum studies or related field; computer expertise (especially database programs)
Collections manager	Responsible for day-to-day care of collections, and collections access. Advises on use of collections in exhibits, arranges loans, monitors storage conditions, routine maintenance of collections. May be responsible for certain aspects of cataloging. May write grants for collections care.	Master's degree in museum studies or field of collection; specific training and experience in collections management. Computer experience useful.
Curator	Largely dependent on size of institution. In small institutions, the curator often performs a combination of registrar, collections manager, and curator duties. In large institutions, the curator is a professional in a specific academic discipline and may be responsible for overall administration of a specific collection, research in the discipline of the collection, and/or providing expertise in that discipline for other museum programs. Writes grant applications for research, collections care, and/or exhibitions.	In a small institution, a master's degree in museum studies or a discipline relevant to the museum's collections and museum experience. In a large institution or one with large, specialized collections, a Ph.D. in the specialty for the collection being curated. Museum experience is helpful. Otherwise similar credentials as for academic jobs, although teaching experience is not critical.
Conservator	Cares for the physical condition of objects in museum collections. May write grants for collections care.	Master's degree in conservation. Requires special training in care and restoration of specific kinds of materials. Involves artistic skills and knowledge of materials analysis/chemistry.
Technician	Prepares collections for study and storage (washes, cleans, numbers, packs, etc.). Assists with field collecting.	B.A. in field of collection. Laboratory and field experience. Organizational skills.
Archivist	Appraises usefulness of records, organizes and culls accordingly so as to preserve accessibility, and preserves physical integrity. Provides for outreach through publications, exhibits, and other visible displays and applications for records.	Master's degree typically in history or library science, but degree in subject matter of records specialization is preferred, as well as specific training in archival techniques. May have Ph.D. in large, specialized institution.

more about the issues, problems, and procedures of managing and caring for collections than most archaeologists.

Repositories having only one staff member who is responsible for the collections typically call this person the curator. In larger institutions with full-time collections managers, curators may be research staff, administrators, or a combination. Such curators are experts in their disciplines and typically hold a Ph.D. Curators' research duties may include applied as well as basic research. Applied research usually involves providing content information and expertise for exhibitions ("curating" exhibits) and other public programs. A curator also may oversee a collection in his or her discipline and may supervise a collections manager who provides the day-to-day collections care and access for users. If a curator is administratively in charge of a collection, she or he also may have considerable influence on the kinds of materials that the repository acquires, but approval of acquisitions usually requires higher-level administrative sanction, such as from a board of trustees.

Many repositories do not have in-house conservators and instead hire highly trained professionals as consultants on an as-needed basis. Yet without in-house conservation expertise, it is difficult to maintain a program of physical collections care. Instead, conservation becomes a special project. Many archaeologists do not realize that the need for conservation expertise is not limited to the repository. Advice from a professional conservator on handling items in the field can be essential for preserving and stabilizing fragile items. Conservators tend to specialize in specific types of materials.

Even fewer repositories have professional archivists on staff, and communication between archaeologists and archivists has been minimal at best. Nonetheless, the importance of records in archaeological collections indicates, minimally, the need for advice from archivists for preserving these materials. Archivists have standards for their training developed by the Society of American Archivists, and, like conservators, archivists tend to specialize in certain subject areas and types of records. More and more archivists are developing expertise in the long-term management of electronic records.

Curatorial support staff may have many different titles; technician is a common one. They do the grunt work of collections care—the time-consuming tasks of cleaning, cataloging, and packing materials, similar to basic archaeological laboratory work. In a university setting, students may provide the labor pool for such duties. In other contexts, such staff may be full-time or temporary and funded by either "hard" or "soft" money.

CONCLUSION

Differing management contexts and structures affect repository activities, their perceived audiences, and their realms of accountability. The duties and responsibilities of a repository's staff relate to the institution's mission, size, and institutional context. The management structure in which curatorial staff members work differs among repositories and affects how decisions about collections issues relate to decisions concerning archaeological research. The staff members who care for archaeological collections may not themselves be trained archaeologists. Nonetheless, curatorial staff may have professional training and expertise useful to archaeologists and important for the proper care and preservation of collections. Training and expertise in both archaeology and collections management are essential for the effective curation of archaeological collections.

5
MANAGING CURATED COLLECTIONS: THE BASICS

Few archaeologists have formal training in managing collections. Courses in collections management more typically are taught in museum studies programs than in university anthropology departments. Even when anthropology departments offer courses in museum work, these classes tend to focus on developing exhibits rather than on managing collections. Although archaeologists do not need to become experts in collections management, a working knowledge of basic principles, issues, and terminology is useful for enjoying effective communication with repository staffs and for being sensitive to their concerns, policies, and procedures.

This chapter offers an introduction to the mechanics of collections management. It begins with the entry of items into the repository and ends with use of curated collections by repository patrons. We discuss the policies and procedures repositories use to acquire collections, conduct initial processing and storage, preserve the physical integrity of the objects, maintain accessibility and inventory control, deaccession objects, and enable public use. In the museum world, these policies and procedures generally fall under the rubric of *registration methods* and may be handled by a staff member called a *registrar* (Buck and Gilmore 1998). Repositories differ in the specifics of managing collections, but all face similar challenges and thus have collections management programs that generally are similar.

CHAPTER 5

ACQUISITIONS POLICIES AND PRACTICES

Repositories use the term *acquisition* to refer to the process of obtaining legal title to an object or collection of objects and formally adding it (or them) to the collections. Legal title is obtained through a *transfer of title*, which is the formal process of a change of ownership of an object from one person or organization to another (Malaro 1994). To transfer title, a *deed of gift* often is the contractual statement that must be signed by the appropriate parties, typically the owner of the object(s) and a repository representative. Examples of typical deed of gift forms can be found on several websites (see the appendix). Repositories do not like to acquire objects if there is any doubt about the vendor's or donor's rights of ownership because investment in the care of an object is not prudent if the object's owner can take it from the repository at any time. Questionable acquisition also may violate laws, such as the Archaeological Resources Protection Act (ARPA), that deal with the illicit trade of antiquities, as well as professional museum and archaeological ethics. When public money is used to care for privately owned items, these issues especially are problematic. Archaeologists need to make every effort to ascertain the ownership of collections and to transfer legal title to the repository or appropriate public agency.

A written transfer of title from a private landowner is preferred because a verbal agreement is difficult to document. Repositories often can assist with obtaining title from a recalcitrant private landowner because the public generally views repositories (especially museums) as desirable institutions. On the other hand, the contract archaeologist may be unsuccessful in obtaining title to collections because she or he may be associated with a project that the landowner did not want to occur. ARPA permits should specify repositories for federally owned collections. The contract archaeologist will need to consult with repository staff about policies relevant to curating federal collections because the repository will not actually acquire such materials. Federal and other government collections remain the property of the public entity.

Repositories have policies covering the acquisition of objects. An *acquisition policy* typically includes the following:

- A collecting policy governing what is acquired, aligning the composition of the collections with the institutional mission
- An authorization policy for agreeing to acquisitions, specifying who has the authority to accept materials on behalf of the repository

- A statement of the terms and conditions under which objects will normally be acquired
- Guidelines for determining costs for additional processing, storage, transport, and conservation (what these may cost the repository if the collection is acquired), as well as to what extent the repository will provide or arrange for such services

The governing board of a repository most often approves acquisitions, because this board is legally responsible for the repository's activities (Malaro 1998). Staff members usually are responsible for recommending acquisitions to the board. Beyond the "fit" of a particular collection with a repository's collecting policy, factors that can influence a repository's decision to acquire collections can include the amount of time and money that will be required to process, conserve, and store the objects. An ethic of the museum profession is that collections should not be acquired unless the institution can properly care for them.

ACCESSIONING

Once a repository acquires an object or collection, its formal inclusion into the institution's collections is called *accessioning* (Carnell and Buck 1998). This process follows the transfer of title and includes assigning an accession number and entry of information about the acquired object or collection into the accessions register. An accession file that contains records about the acquisition also typically is developed as part of the accessioning procedure.

A repository that agrees to care for a federal or state collection in perpetuity does not gain title of ownership to that collection unless a transfer of title is completed. Nevertheless, a repository will usually go through the accessioning procedure for those collections to optimize their long-term care.

The typical steps for accessioning a collection or object are these:

1. Evaluate and authorize an acquisition according to agreed museum policy and retain written documentation of this process.
2. Ensure that the receipt of the object is properly planned for and that appropriate long-term storage or display space is available for the object(s) to be acquired.
3. Complete a *condition report* for the object(s) to be acquired (a condition report is a brief description of the physical condition of the

objects in order to highlight any hidden costs of conservation that can be planned for or avoided).
4. Obtain unambiguous evidence of title to the object.
5. Describe the method of acquisition (e.g., bequest, field collection, gift).
6. Assign a unique number to the object or collection.
7. Record the information about the acquisition in the accessions register.

Each repository keeps an *accessions register* as a permanent record of all objects that are or have been part of the institution's collections. Entries in the accessions register include the accession number, a brief description of the object or collection, and the date of accession. An accessions register may be handwritten, generated from computer records, or a computer database. Repositories ideally should keep a copy of their accessions register in a secure place, such as microfiche or digital copies held at an outside location.

All of the documents compiled for acquisition of objects and associated records during the accessioning process are called *accession records*. These records provide written evidence of the original title to an object and the transfer of the title to the acquiring institution, and they contain a unique number (accession number) that is physically associated with all objects in an acquisition. Because they describe all acquisitions and list them by number, accession records ensure that an accessions register is maintained and information about the acquisition process is retained. Accession records also are the central place where all subsequent information about a collection's history may be maintained or cross-referenced, such as loans and conservation treatments.

Repositories generally use one of two systems for *accession numbers*. The first assigns a separate number to each item, while the second assigns the same number to all items in one accession group or individual collection and then a suffix to create a unique identity number. The second system works best for archaeological collections, which typically include many small objects. Accession numbers may take the form of a simple running sequence (e.g., 14603; 14604; 14605) or the year of accession followed by a running number (e.g., 1991.3; 1991.4; 1991.5). Identity numbers follow the accession number and may be assigned during the cataloging process. Actual placement of numbers on objects is discussed later in the section on collections preparation.

CATALOGING

Cataloging is the assembly of all primary information about each item in the collection. Archaeologists sometimes confuse cataloging with analysis, although if an object is analyzed as a single item, it usually should be cataloged as a single item. For an archaeological collection, the most important aspect of cataloging is permanently associating a specimen with its archaeological context or provenience. This correlation usually is done as part of typical archaeological laboratory procedures and involves keying a number that is associated with the specimen to a written record of its provenience. Cataloging should be done by, or under the supervision of, someone with subject knowledge as well as familiarity with the cataloging system used by the repository that will curate the collection.

The main purpose of cataloging is to identify and document an object or group of like objects, especially to collect information that is likely to be valuable as an index heading or computerized search term. Such information may include an object's identification number or code and its provenience. While the purpose of cataloging is not to record attributes or collect data for research purposes, catalog information is essential and useful for the research process. Cataloging allows the researcher to know what objects are in a collection, and the identifications made for cataloging purposes often can be used for very general analyses (e.g., sherd counts). Until it is cataloged, an item cannot be indexed properly and will not be easily accessible to museum staff or the public. A repository may also catalog the records associated with an archaeological collection, although repositories handle these records in different ways. A common way is to link the accession number to both the specimens and the records that comprise an individual archaeological collection.

There are no standards for cataloging archaeological collections used by repositories across the United States. Some federal and state repositories have clear policies and guidelines for cataloging (e.g., Department of the Interior 1997). Most repositories commonly collect the following types of information: accession number, catalog number, object name or description, material type, form, quantity, measurements, weight, cataloger name, cataloging date, location in repository, site number, state site number, provenience or collection unit, state, county, Universal Transverse Mercator (UTM) coordinates or township/range/section, field collector, date collected, and conservation and treatment comments (Griset and Kodack 1999).

Catalog or *identity numbers* provide a code for uniquely identifying objects and for linking archaeological objects to their provenience. The form they take varies among repositories, depending on individual needs and past practice. The accession number may serve as the identity number if different accession numbers are given to each item. Otherwise, an item suffix must be added to create a unique identity number. For example, a group of four projectile points brought into the museum together might be accessioned as 1991.24. Each individual point would then be numbered 1991.24.1, 1991.24.2, 1991.24.3, and 1991.24.4. Another common archaeological numbering practice is to use the tripartite Smithsonian site number (i.e., state number/county abbreviation/site number) in lieu of an accession number, but in combination with identity numbers—for example:

$$\frac{11Ms38}{15} \text{ or } \frac{40Mg31}{123}$$

A problem with this system arises when multiple investigations are conducted at the same site because there is no unique designation for the year of investigation. Collections made in different years and by different investigators thus could have objects with identical numbers. This situation should be avoided because it can cause confusion between discrete collections and lead to loss of provenience data. Therefore, a unique set of catalog numbers should be used for each phase of an investigation at a particular site.

Sometimes repositories have to work with numbering systems that they have not created. For example, a collection generated by an amateur archaeologist may already have its own numbering system. These numbers must be preserved as a link to the original records, but they can play havoc with computerized data management systems if they use a different format from the numbering system used by the repository.

COLLECTIONS PREPARATION: LABELING AND CONSERVATION

After objects are cataloged, they must be prepared for storage and for possible research, exhibit, and loans. The primary objectives of collections preparation are to preserve objects and associated records in a stable condition for the long term and to maintain their research, heritage, and educational values (Jacobson 1998). The latter requires that any action applied to a collection

- be reversible;
- be well documented;
- be respectful of the object's integrity and of the culture from which it originated and is affiliated;
- utilize nonreactive materials; and
- always maintain the connection between the object and its documentation.

A critical step during collections preparation is to *label* or mark each accessioned item or, where appropriate, group of items with its permanent identity number (Segal 1998). Labeling should use methods and materials that do not damage the object and its surface yet ensure that the labels cannot be removed accidentally. Label placement should not negatively affect the appearance of an object or its detail yet must be sufficiently visible to minimize the need to handle the object. It should not be placed in areas that might receive wear or friction, including the bottom of objects. The materials used for labeling should be reversible so they can be removed intentionally with minimal trace, even after fifty to one hundred years. They should have good aging properties and be as chemically stable as possible yet be safe for staff use without posing health risks.

Many factors influence the choice of the most appropriate labeling technique for a specific object. Most factors are closely related to the object's constituent materials and to the physical stability and roughness of the object's surface, the object's porosity, physical strength, and flexibility (Segal 1998; Sease 1994).

Taking into account the basic principles and limitations of labeling, most repositories choose to use semipermanent methods and materials to label their collections. Procedures generally involve thorough examination of the object, cleaning the surface where it is to be marked, applying a base coat of clear, reversible varnish, writing the number on the base coat with a permanent pigment-based ink or acrylic paint, and placing a thin coat of clear varnish over the number. Sufficient drying time must be allowed for each coat. For paper and photographs, writing the number in the same location in pencil is optimal. Repositories may use temporary labeling methods and materials (e.g., attachable tags, outside labels on storage boxes, loose labels) for archaeological objects or records. These are used only for highly unstable items, temporary deposits, loans, tiny objects, and objects stored outdoors.

For archaeological objects whose research value depends on their link to provenience, some repositories choose to use more permanent marking methods and materials. These procedures may help ensure that the label is not damaged or lost over time, although the semipermanent techniques are often adequate. It is important to consult with a conservator if there is any doubt about appropriate methods and materials. It also is critical to know what materials are *not* good to use for labels, primarily due to their effects over time as summarized in the list below (Segal 1998; Sease 1994):

- Typewriter correction fluids can flake and may resist solvents.
- Nail polish made of cellulose nitrate can yellow, shrink, peel, and become brittle.
- Nail polish remover often contains contaminants other than the solvent.
- Rubber cement adhesive can deteriorate, and it can stain organics.
- Pressure-sensitive tape or label adhesive can deteriorate.
- Paper labels moistened by water are often hard to remove and can stain.
- Ballpoint ink may fade, smear, and resist solvents.
- Metal fasteners and metal-edged tags can corrode, stain, abrade, and cause cracking of some materials.

The other task that is critical to collections preparation involves *condition assessment* and *conservation*. As we discuss in chapter 6, the condition of objects and conservation must be considered during project design, in the field, and in the laboratory. Once in the repository, an assessment is necessary to document the current condition of an object or record and to recommend necessary conservation treatment. The goal of conservation, then, is to maintain the preservation and survival of an object or record in its original state, to the extent possible, often through the use of interventions such as physical strengthening or chemical stabilization.

Conservation may be costly, particularly for submerged or underwater archaeology, and should be carefully budgeted. Conservation also should be a collaborative process that involves archaeologists, collections managers or registrars, curators, and conservators (Cronyn 1990). However, treatment requires the expertise and experience of a conservator who is familiar with the best interventions or treatment available for the particular constituent materials of an object. All decisions and interventions must be well documented.

STORAGE

Integral to the long-term care and preservation of collections is how they are stored (Ford 1980). *Storage* refers to both the overall conditions of the spaces where collections are kept and the safekeeping of individual objects and associated records (Swain 1998; Department of the Interior 1997).

Collections storage within a repository requires attention to spatial layout, environmental controls for different types of collections, security, fire protection, and disaster planning (table 5.1). Many repositories have areas for long-term storage that are separate from other key activities, such as exhibits, exhibit preparation, research, temporary storage, object preparation, and administration. Such physical separation enhances security as well as protection from fire and other disasters. Some repositories have off-site storage, but this alternative can hinder access, security, and monitoring.

Maintaining environmental standards minimizes the rate of deterioration, extends the lives of objects and records, and reduces the need for conservation treatment. Extreme levels and significant fluctuations in temperature and humidity may damage archaeological objects, especially those made of unstable materials such as metal or wood, bone, and other organics. They also can severely affect documents, photographs, electronic media, and audiovisual recordings. Therefore, temperature and humidity should be monitored and recorded and kept within acceptable ranges. Establishing an appropriate range of relative humidity (RH) involves consideration of local climate, the collection materials and their condition, mold growth prevention, repository structure and layout, and the RH levels to which the collection had been adapted.

Other environmental conditions that require consideration are visible light, ultraviolet radiation (i.e., sunlight), pollutants, and pests. In general, no visible light should be present in a storage area except for short periods of time. Ultraviolet light (UV) should be monitored periodically and controlled with filters if found to be excessive. UV can be particularly harmful to photographs. Gaseous and particulate pollution (e.g., dust, chemical off-gassing from unsealed wooden shelving, solvents, and acidic paper) should be monitored and controlled

Table 5.1. Basic Standards of Repositories

Environmental Controls
- Temperature and humidity
- Level and duration of visible light
- Ultraviolet radiation
- Pests
- Air pollution

Security
- Mechanical and/or electrical system for detecting and deterring intruders
- Policy on access to collections and associated documents, including systems for visitor and researcher registration, opening and closing storage and exhibition areas, and control of keys to particular areas of the repository

Fire Protection
- Fire detection and suppression equipment appropriate for the collections housed in storage and exhibition areas
- Storage of repository and collection records in appropriate fire-resistant container that is also locked when not in use
- Fire plan for the needs of the collections to prevent, detect, and suppress fire

Housekeeping
- Regular cleaning of storage and exhibit spaces based on established procedures and policy
- Maintenance and calibration of monitoring equipment

Physical Examination and Inventory
- Regular examination to detect deterioration of collections' contents
- Inventory policy to regularly confirm locations of collections and prevent loss or theft

Conservation
- Maintenance of objects in stable condition using professional conservation standards and practices

Disaster Planning
- Procedures to protect collections in the event of a natural or human-inflicted disaster

Exhibition
- Consideration of how to best preserve, protect, and minimize risk to objects when planning an exhibit
- Design and use exhibit cases and areas to promote security, housekeeping, and preservation of objects

when appropriate. A rigorous pest management program should be developed because rodents and insects can severely damage organic objects and archival materials. Many repositories have an integrated pest management program to monitor and treat pest infiltration (Jessup 1995).

Individual objects and associated records are packed and housed in an *artifact container* (e.g., bag or box) and then a *storage container* (e.g., box, larger bag, cabinet, or drawer). Two critical principles must be followed in all storage decision making. First, *provenience information must be maintained at all times.* Provenience labels must be placed on all containers so that their contents can be reunited if they become separated. Second, *all storage materials must be archival quality for long-term use.*

Several factors must be considered when deciding how to store objects and records. One is the frequency of use of the items. If an object is a type specimen that may be regularly studied or used, it is more appropriate to place it in a drawer rather than a box on a high shelf. Another factor is the diversity of materials in the collection. In general, objects and associated records are stored in artifact containers by material class: All the shells from one provenience should be in one bag, all the ceramics in another, and all the photographs in acid-free or inert plastic photographic sleeves. Objects and associated records, particularly those that are fragile or made of unstable materials that require special environmental conditions, should be separated by both artifact container and storage container. Another consideration is the size and weight of the objects. Obviously, a large mortar and pestle should not be placed in the same bag as obsidian flakes, nor should a bag containing a large mortar and pestle be placed in the same storage container as a bag of thin-walled pottery sherds. Some oddly shaped or fragile items may need special support mechanisms to prevent them from further deterioration or to enhance their visibility so as to decrease handling.

Storage of human remains and objects held sacred by various groups require additional considerations. Human remains must be handled with respect and stored in a separate area. The Native American Graves Protection and Repatriation Act of 1990 (NAGPRA) also requires that culturally affiliated, federally recognized American Indian tribes be consulted about the disposition of any remains, as well as sacred objects and objects of cultural patrimony. Therefore, these items should be accessible for consultation. Tribal members also may want to conduct ceremonies in the storage area for various purposes (McKeown et al. 1998).

The most common and suitable artifact container for most objects is the polyethylene zip-top plastic bag. It is easy to handle, lightweight, and economical and can be directly labeled. Bag thickness, no less than two millimeters, is important for long-term curation of the objects inside. Most bags should be perforated with small holes to promote air circulation and prevent the creation of microenvironments that promote mold. Bags containing objects that are sensitive to the environment, such as iron, should not be perforated (Vogt O'Connor 1996).

Care of records even from a single archaeological project is not a simple task due to the wide variety of formats and media used. The Chickamauga Basin case study we discussed in chapter 2 provides an example. Records from this project include handwritten field notes on notebook paper, standardized mimeographed field and laboratory forms that are filled out with typing and handwriting in ink and pencil, artifact catalog cards on 3 × 5–inch card stock, analytical notes and tallies that are either typed or handwritten on ledger paper, photographs and negatives (the latter of which are typical of the time period with self-destructing silver nitrate), artifact illustrations on vellum, large excavation plats on graph paper, progress reports and preliminary site reports, and correspondence (both typed and handwritten) between the various investigators and with the WPA. To maintain the rich research potential of the collections, these critical and often fragile documents must be carefully preserved.

Appropriate document containers vary by document format and size. Paper records may be contained in acid-free or buffered folders or files or in polypropylene, polyethylene, or polyester sleeves. Each map should be stored in an acid-free folder, although a divider sheet of acid-free tissue between multiple maps is minimally acceptable when expense is a significant issue. Photographs (print, negative, and transparency) should be placed in individual envelopes or sleeves made of acid-free paper or an inert plastic of polypropylene, polyethylene, or polyester. Older nitrate-based negatives can be considered hazardous materials as they deteriorate. Freezing can deter total degradation. The long-term solution is to make copies of all of these negatives. Audio- and videotapes should be stored in acid-free boxes of suitable sizes, while electronic records can be placed in appropriate plastic containers. Tapes and disks should be stored in areas free from harmful electromagnetic fields.

Unacceptable containers for objects include lightweight sandwich and food storage bags, brown paper bags, cigar boxes, plastic (not polyethylene) film vials, and glass containers that are not well insulated and may break. It is also unacceptable to use rubber bands, twist ties, adhesive tape, string, staples, or heat sealing to close artifact containers.

Storage containers, the larger boxes or polyethylene bags that contain one or more smaller artifact containers, should be made of archivally stable materials. They should also be of standard sizes for optimal stacking on shelves. Repositories vary in their box size requirements, although boxes of approximately one cubic foot are most common for objects.

INVENTORY CONTROL AND DATA MANAGEMENT

An *inventory* document is a permanent record of the physical location of all accessioned and cataloged objects and associated records. It is periodically updated during the process of *inventorying* when the physical locations of items are checked against the list of objects. The inventory process also enables an inspection of the condition of each object and document, and it identifies items that may be lost, missing, or stolen.

An inventory thus is an important means of accounting for and managing data on collections (Malaro 1998; Cowan 1998). At its most basic level, it documents what a repository has and where to find it. It can also relate to other important information critical to the mission of a repository. For federal and state agencies that depend on nonfederal and nonstate repositories to care for their collections, inventories provide a critical check on their property, helping fulfill ethical and legal obligations. It is for these reasons that 36 CFR Part 79.11 specifies periodic inventories and inspections of federal collections. For bureaus with the Department of the Interior, for example, a 100 percent inventory and inspection of all museum property above a certain value and a random sample inventory of all other cataloged items must be done on an annual basis (Department of Interior 1997). A well-documented inventory also is essential for complying with requirements of NAGPRA (McKeown et al. 1998).

A repository benefits from inventory control and good data management. Good location and description information not only enhances security, it also facilitates tracing lost, stolen, or missing objects. When such information is cross-referenced to title, it may further enhance retrieval and help resolve disputes about legal ownership. Accounting for collections increases the repository's credibility in the eyes of funding agencies, potential donors, and the general public. Also, inventories facilitate access to specific items by repository staff, researchers, and educators; access to information about the contents of collections; and data needed for planning and budgeting exhibits and educational programs.

DEACCESSIONING

Despite the claims that all archaeological collections should be preserved in perpetuity for their present and future research value and potential (see chapter 3), there are times when a collection or specific objects may legitimately be deaccessioned (Childs 1999). *Deaccessioning* is "the process used to remove permanently an object from a museum's collection" (Malaro 1998:217). The decision-making process is usually lengthy and well documented because deaccessions often have been controversial. Critics include local communities who felt their cultural heritage was being discarded, donors offended that their gifts were no longer worthwhile, and various support groups who disapproved of how the repository spent auction proceeds (e.g., *not* for new acquisitions). Once a decision is made to deaccession one or more objects, the subsequent action of removal from the repository is *disposal* (Morris 1998).

The primary reasons to deaccession and dispose of archaeological objects are when they are

- outside the repository's scope of collections (see chapter 4),
- subject to repatriation under NAGPRA (see chapter 3),
- physically deteriorated beyond viable research or educational value,
- hazardous,
- incapable of being adequately cared for by the repository, or
- determined to have been acquired illegally or unethically.

The decision to deaccession involves a number of steps (Morris 1998):

- Initiation of action with a written justification that lays out the reason to deaccession in relation to the repository's mission, collecting plan, scope of collections, and any relevant federal or state laws
- Confirmation that the object(s) was accessioned, cataloged, and well documented—a good deaccessioning policy is intimately related to a good accessions policy (Malaro 1998)
- Physical inspection by a conservator to help identify the best method of disposal
- Confirmation of title and a check of the records for any donor restrictions
- When applicable, an outside appraisal of the monetary and research value

- Internal review to ensure full knowledge of the deaccessioning plan
- Approval by the repository director and governing board, and review committee, if established
- Assignment of a deaccession number to each item to be disposed of
- Public relations, particularly with local community groups whose cultural heritage may be the subject of deaccessioning and disposal

The final act of the deaccessioning process is disposal. There are several options for archaeological objects, although every effort should be made to transfer, through donation or exchange, the objects to another repository in the vicinity for research and educational use. The primary means of disposition are as follows:

- Donation and transfer of title to another repository to maintain educational use.
- Repatriation (the restoration of control over) of human remains, funerary objects, sacred objects, and objects of cultural patrimony to affiliated Native American tribes, other appropriate culture groups, or countries of origin. Packing and delivery of repatriated objects should be done in consultation with the recipient group in order to respect traditional practices and beliefs (McKeown et al. 1998).
- Exchange with other repositories, usually of items of equal value and significance. This involves transfer of title.
- Physical destruction. This disposal technique may be applied to hazardous items, severely deteriorated items, and counterfeits. The exact method chosen should be permanent, irreversible, and well documented. Destructive analysis for research purposes, when total, is often placed in this category.
- Return to rightful owner if it is determined that the donor was not the legal owner.
- Public auction, usually to raise funds for future acquisitions. This option usually is not appropriate for archaeological items recovered in a research or CRM context for ethical reasons.

PUBLIC ACCESS AND USE

Use of collections is both a bane and a blessing for repositories. The purpose of having collections is to use them. However, their use puts them at risk because individual items may accidentally become damaged, proveniences can get lost if objects are not labeled, and a range of other problems can occur. Policies and procedures for use of collections are designed to allow use in ways that minimize the risks to collections. Chapter 7 discusses how to make arrangements with a repository to use a collection physically, and chapter 8 presents access issues in the computer age. Here, we discuss what a repository does to enable access and use of collections including issues of accessibility and consultation, on-site use and loans, and publications.

To make collections accessible and useable, a repository must have good systems of inventory control and data management. These systems should allow repository staff to provide access to those materials that are needed by the user. Repository staff members have to try to anticipate the kinds of questions users will ask about the collections. They must then structure their data and information management systems in ways that are likely to provide answers to the most common kinds of questions or, at a minimum, to furnish leads to information sources that can be used to answer the questions. For example, it should be easy for a repository to answer the question "Do you have archaeological artifacts from Illinois?" On the other hand, answering the question "Do you have a Madison point from Feature 315 at the Goody Site that was excavated in 1902 by Hiram D. Igers?" could entail a records search for a number of similar objects unless the repository has a very sophisticated computer database or a very small collection. The inquirer may be invited to conduct the records search, depending on the number of objects and volume of records. In any case, a repository has to decide just how accessible information about its collections will be based on its individual circumstances, including the size and nature of its collections, funding, and staff.

Repositories also must decide the circumstances under which physical access to collections will be allowed. Not just anyone can walk into a repository and ask to use the collections; "curiosity seekers" almost never are granted physical access to collections. Because of the potential risks, repositories almost always limit physical access to collections to those persons who have needs linked to legitimate scholarly research or public education purposes such as exhibits. Even then, access usually is limited to those items pertinent to the user's needs. Users typically are not let loose in storage areas. Secure exhibitions, not storage areas, are the appropriate venue for the browser.

NAGPRA affects collections access and use at repositories that hold Native American human remains, funerary objects, sacred ob-

MANAGING CURATED COLLECTIONS: THE BASICS 75

jects, and objects of cultural patrimony (McKeown et al. 1998). The process of inventory for NAGPRA compliance entails determining cultural affiliation through consultation with tribes. Consultation may involve tours through collections, careful consideration of appropriate handling of objects, and the need to accommodate traditional rites such as purification rituals.

Loans of collections to other institutions may be made for exhibitions or for research and teaching when in-house use is not feasible. Even small repositories often process a large volume of outgoing loans for exhibition and research. Some repositories do not have the facilities for scholars to use collections or do not have exhibits, so loaning collections is a way to provide access to both scholars and the general public. Procedures for loans made for both purposes are often identical and are intended to meet the needs of both the repository and the borrower.

Repositories have *loan policies* governing outgoing and incoming loans and the conditions under which they are made. *Loan agreements* are contracts that spell out the conditions of a specific loan, which must be signed by representatives of the borrowing and loaning institutions. Loan conditions may differ depending on the nature and intended use of the items. Borrowers almost always are responsible for the items they borrow, so repositories should make loans to institutions, not individuals. At large repositories, loan requests may be considered by a loan review committee and may need approval by the director or other administrator.

Other typical stipulations of loan policies include the following:

- Only properly accessioned and cataloged material are loaned unless there is a specific agreement that the borrower will do some of this work for the repository.
- Loans generally are made for the period of one year or less, subject to annual recall. The term *permanent loan*, although in common use, is a misnomer. *Permanent* implies that the borrower will always have the loaned item and thus may as well have title to it. The term *continuing loan* is preferable if an object will be loaned for an indefinite or lengthy period of time. Some repositories will not renew loans for more than five years.
- Unless alternate arrangements are agreed on in writing, all loaned items must be returned to the repository in the condition in which they leave.
- For research loans, collections returned must be accompanied by a report of findings, whether or not formal publication results. If for-

mally published, the repository receives a complimentary copy of the report.
- Borrowers may be subject to fees or costs involved in arranging the loan.

Procedures for setting up a loan usually include the following steps:

- Requests must be submitted in writing and addressed to the appropriate curator, collections manager, registrar, or director. Information on the intended use of collections, the requested duration of the loan period, and any other details pertinent to the request are necessary. Requests must stipulate the nature of collections use or type of research to be conducted, the accession and catalogue numbers of the objects requested, and a time framework. Research requests should also discuss expected results.
- A completed *facility report* must accompany exhibition loan requests. Facility reports describe the borrowing institution's security system, climate control, and so forth, as pertains to exhibitions. The American Association of Museums has standard forms that many institutions use.
- Each object or group of objects lent is documented on a loan agreement and may be photographed for record-keeping purposes.
- Specification of loan return is made at the time of the initial loan negotiation. Notification of loan return must be made in advance.

Repositories often receive requests for photographs of objects to be used in publications. Publicly owned repositories usually do not hold copyright to the images of the objects they curate, but private institutions may. Policies on use of photographs differ widely among repositories, but almost all institutions ask to receive a credit line on the published photograph, and users usually must pay the costs of photography. Many larger repositories prefer to have their own in-house photographer make photographs for publication purposes so they can ensure the quality of images of their objects and safety in handling.

Some repositories also publish their own series of research reports, catalogs, or other collections-related materials. As discussed in chapter 8, digital publications and catalogs, especially on the World Wide Web, provide new frontiers of public access to curated collections (see Toolkit, volume 7). In general, publications make collections more accessible because they provide detailed information in a format that can be widely circulated. Publication policies and procedures differ considerably among institutions, depending on the kinds of series and available support.

CONCLUSION

Repositories have standard sets of policies and procedures for managing collections. Acquisition, accession, and cataloging procedures document the ownership status of objects and insure that an object stays linked to its documentation. Inventory control and data management procedures establish an institution's accountability for its holdings and facilitate finding objects in the repository. Conservation, collections preparation, and storage procedures help maintain the physical integrity of curated objects. The intent of all of these procedures is to enable controlled use of collections in order to maintain the safety and long-term preservation of the objects and associated records.

6

MAKING A COLLECTION: FIELDWORK PRACTICES AND CURATION CONSIDERATIONS

Overcrowded repositories and increasing demands for public accountability of stored archaeological collections force the field archaeologist to carefully consider what will go to a repository after a field project is completed. Future access to well-managed collections for research, interpretation, and exhibit can be assured only if thought is put into how to make collections *before* going to the field, while in the field, in the laboratory processing and analyzing the artifacts, and back in the office after the project is finished. The long-term management of archaeological collections, records, and reports is not just the job of a curator but a key responsibility of the field archaeologist.

The research design should help the field archaeologist make essential curatorial decisions even before going to the field (see Toolkit, volume 1). While in the field, the collecting strategy laid out in the project design should be followed, and the archaeologist must always be aware of changes that may have to be made due to unexpected circumstances. Basic principles of conservation must be followed to ensure that the collected objects are stabilized for further appropriate care and use. In the laboratory, decisions must be made about what parts of the collection will be saved for long-term care at its designated repository, including whether any highly redundant object types need to be sampled (i.e., discarded). These decisions have implications for how to process the various objects in the collection for long-term conservation and preservation. Finally, after the project is complete and the report and articles are written, the archaeologist must consider the ethical and professional responsibilities inherent in the long-term care and preservation of field notes, maps, photographs, draft reports, laboratory notes, and the like. These documents hold valuable information for future scholars and should not get lost or destroyed.

BEFORE THE FIELD: PROJECT DESIGN

Two significant parts of a project design often are not given adequate attention: identification of a repository where the resulting collections will be curated and a well-articulated collecting strategy.

The Archaeological Resources Protection Act (ARPA) and its regulations (36 CFR Part 7.6[5]) require identification of a repository in advance of issuing a permit for any archaeological project on federal lands. Many states have similar requirements for compliance activities on state lands. However, university-based projects, often not conducted on federal or state lands, are rarely constrained by such requirements. Graduate students often learn by example to assume they can find room for the results of their fieldwork in a university museum, an anthropology department's storerooms, or their office after the project is over.

The archaeologist must select a repository before naming one in a project design. Chapter 7 provides more details on appropriate repositories for curating archaeological collections, but some basic guidelines are explained here. First, choosing a repository that is near the excavated site or surveyed land is a consideration, especially if the repositoty curates similar types of collections, thus enhancing the material's future research potential. The repository also must apply standard curation practices that permit well-managed collection growth (Sullivan 1992a; see also chapters 3, 4, and 8). It must have a mission statement, long-range goals for collections care and research, and a strong scope of collections statement. The scope of collections specifies the range of object types the repository will accept, along with other specifications such as regional or temporal foci. (Many state repositories only accept archaeological collections from their state or a few adjoining states.) The archaeologist should carefully examine this document and be aware that a repository can change its scope of collections statement and requirements. The repository may then be in a position to justify deaccessioning objects that do not fit the new scope and split up an archaeological collection.

Although no system exists to accredit repositories that meet the standards in 36 CFR Part 79, there are some other basic principles and questions to consider when choosing a repository. The archaeologist must explore the methods the repository employs to accession, catalog (when not adequately done by the archaeologist), and inventory collections, as discussed in chapter 5. Does the repository use a computerized search and retrieval system to identify the items in each collection it curates and to easily find their storage location? The physical plant of the repository is also critical. Key questions that follow the standards set forth in 36 CFR Part 79 include these: Does it have appropriate temperature, humidity, and lighting controls for different types of materials? Does it meet local fire, building, electrical, health, and safety codes? Does it have an appropriate and functioning fire detection and suppression system? Does it have an emergency management plan with procedures to respond to flood, fire, and other natural disasters, as well as acts of violence or structural and mechanical failures within the repository itself?

The archaeologist also needs to consider whether adequate funding exists to fully document, preserve, and house a collection of objects and records. Does the repository charge a fee for curation? If so, is it based on a one-time-only or an annual fee structure? What services does the fee cover? Whether or not a repository charges a fee, does the project budget cover the expenses of preparing the collection for long-term curation in compliance with the repository's standards for acceptance (i.e., nondegradable and acid-free bags, tags, and boxes; archival-quality paper for field records; archival-quality preservers for slides and photo negatives, or basic conservation treatments)? The project design and funding proposals for an archaeological project should carefully budget for these needs. In some cases, particularly for work contracted by federal agencies such as the National Park Service, there are some established guidelines for budgeting curation (NPS 2000).

After a repository has been identified, the archaeologist must set up a curation agreement with the repository staff for accessioning and curating the final collection of objects. The agreement should clearly state the expected condition of the collection upon delivery to the repository and the responsibilities of the repository for the collection's long-term care and access. It should also clearly identify the owner of the collection. If any serious conservation needs are anticipated, these should be discussed in advance to determine whether and how the repository can provide appropriate conservation services. Because a repository may change its policies, it is critical that project archaeologists, repository staff, and any other significant parties (i.e., federal or state agency staff involved in project support) discuss how any policy changes might affect the project collection over time.

A well-developed project design also includes a sound collecting strategy. The collecting strategy must be based on the theoretical or compliance focus of the work, the phase of work (i.e., survey, testing, excavation), and, whenever possible, the long-term research plans for a region (Sullivan 1992a:4–5, 2001). Incorporating a collecting strategy into the project design forces thoughtful consideration of what artifacts and other archaeological materials (i.e., radiocarbon, soil, flotation samples) should be collected, retained for preliminary analysis, kept for long-term use, and added to a repository. Table 6.1 lists the critical components of a collecting strategy.

For testing and excavation projects, identification of the time periods of interest, as well as the primary classes of objects and noncultural materials expected to be recovered, is usually not difficult. They can be based on previous research conducted in the area and the phase of work planned. It is especially important to state these expectations for archaeological projects on federal land because the Archaeological Resources Protection Act (ARPA) allows for consideration of later claims that certain artifacts or noncultural materials were "inadvertently discovered." After due process, objects ruled to be inadvertently discovered can be discarded or deaccessioned. ARPA and its regulations, 36 CFR Part 79.5(d), also recognize only sites and objects over one hundred years old as archaeological and subject to curation. Without explicit statements regarding expected artifact types and their associated dates, portions of a collection made on federal land could be discarded or not accessioned if deemed to be "historical" and younger than one hundred years.

Table 6.1. Key Elements of a Collecting Strategy

1. The principal types of objects to be recovered (e.g., pottery, shell, metal, organics, construction materials) and their expected range of variation
2. The principal time periods of interest for the project A field and/or laboratory sampling regime for those objects that either do not fit within the project design or are highly redundant The principal types of noncultural materials to be collected (e.g., soil samples, radiocarbon samples, etc.)
3. A field sampling regime for each type of noncultural material to be collected (e.g., size and number of flotation samples per five cubic meters of soil excavated) in order to meet project goals
4. A justification and appropriate sampling strategy for excavating through deposits that are not relevant to the research design
5. A provision that allows for modification of the research design and collecting strategy, as well as renegotiation of the curation agreement, if significant amounts of unexpected material are recovered

As Black and Jolly emphasize in *Archaeological Research Design* (Toolkit, volume 1), even the best project design cannot always anticipate problems that will occur or unanticipated finds that will be recovered. The project design should always permit modification of the collecting strategy and the curation agreement when large numbers of unexpected artifacts are recovered or more noncultural samples must be taken than originally planned. Such modifications will recognize and record the research and interpretative value of the additional objects or samples.

IN THE FIELD: SAMPLING AND CONSERVATION

An archaeologist faces three key issues in the field that may have ramifications for the long-term condition of and access to the resulting collection: (1) sampling classes of objects that fall outside the project design or are highly redundant; (2) sampling noncultural materials; and (3) the impact of field methodology on the long-term preservation and conservation of the objects recovered, especially those whose condition and interpretive value may be jeopardized when removed from their archaeological context.

At least one class of cultural objects recovered during testing or excavation often is highly redundant, such as fire-cracked rock, shell, slag, or window glass. Occasionally, a class of objects may be both redundant and unrelated to the project design, such as body sherds from a time period that is not the focus of research. Despite the traditional practice of archaeologists to save everything for the sake of the discipline and in anticipation of new analytical techniques, these materials can take up considerable space in a repository and, after initial analysis, are never reexamined. Archaeologists must now be sensitive to the costs of curation and the possible need for repositories to deaccession highly redundant materials in order to provide access to the rest of a collection and room for new collections. Archaeologists, therefore, need to apply consistent procedures in the field and laboratory to sample classes of objects that may not fit the project goal or are highly redundant (Sonderman 1996). During fieldwork, it is critical to confirm the need to sample a particular class of objects as presented in the collecting strategy. Do this by monitoring the amount of material that is excavated and processed in the laboratory during the project. Then change the collecting strategy if necessary to reflect any unanticipated developments.

Soil samples and palynological soil columns have often been taken in large numbers without considering whether they will all be analyzed, what their long-term storage requirements are, and whether their impacts on storage space are manageable. The collecting strategy for such noncultural samples should obtain only the number of samples that will actually be used to achieve your project goals. You should process these samples quickly and efficiently so they do not become part of the permanent, accessioned collection of your project.

Field methods and on-the-spot decision making can influence the preservation and conservation of the excavated materials. A buried object is either in a state of equilibrium with its environment (soil, biological agents, and local climatic conditions) or in the process of adapting to it (Sease 1994). It begins to change when you expose it during excavation. Although changes to most objects are not visible to the naked eye, some objects become unstable and may begin to deteriorate in a matter of minutes or over several days. You must think and act like a conservator: what "first aid" interventions should be used as objects are uncovered in order to prevent partial loss or complete disintegration? Remember that an intervention is any sort of action applied to an object, whether deliberate or accidental.

Several basic principles of field conservation should be applied as objects are exposed, lifted from the ground, and packed for transport from the excavation unit (table 6.2.) Sease (1994:xi) reminds us that

Table 6.2. Key Conservation Principles for the Field (Cronyn 1990; Sease 1994

1. Always think about the excavation tools you are using. How might they affect the constituent materials and original shape of an object?
2. If you excavate a large object or a group of objects that are clearly unstable, quickly cover them up again and contact a conservator for advice on in situ treatment and appropriate removal methods.
3. Remember that water, temperature, humidity, and sunlight, among other factors, can affect the stability of an object as it sits in the ground while you excavate around it.
4. Lifting fragile objects out of the ground and carefully packing them for transport requires specific materials, such as consolidants, and considerable preparation. It must be done with great care. It is not recommended to glue a facing material (e.g., cheesecloth, tissue, toilet paper) directly to the surface of an object to facilitate lifting, because this is an intervention that usually causes problems in the long run.
5. Some objects can be seriously damaged when they are bagged at the site with other, inappropriate objects from the excavation unit. For example, do not mix bones with large, heavy lithics.

the "temporary" treatment an object receives in the field and laboratory is often its only treatment, so application of these principles may dictate the usability and survival of the collection.

IN THE LABORATORY: APPLYING THE SAMPLING STRATEGY AND MORE CONSERVATION

Whether under a tent near the excavation or in the repository prior to accessioning, the archaeologist must make several decisions and take actions to retain the short-term and long-term research potential of the collection. One step involves deciding what to save for the long term. Another is how to maximize object preservation by processing and conducting more "first aid" conservation on particular items.

Decisions about what objects to save may influence what conservation interventions are applied. Certain actions taken in the field to facilitate sampling and analysis can affect the long-term preservation and conservation of the kept objects, especially washing and drying. If soft potsherds are washed with hard bristled brushes, for example, their surface finish and decoration may disappear. If sherds or lithics are washed that have organic residues on or in them, information may be lost or altered permanently (see Toolkit, volume 5). If cleaning agents are added to the wash water, the introduced chemical contaminants may be damaging. Drying some objects in direct, hot sun may encourage partial or full disintegration.

Some culling of materials always occurs in the laboratory. Objects that are determined to be noncultural are usually tossed in the nearest garbage can. A carefully planned process of sampling and culling can be applied to highly redundant cultural materials or objects that do not fit the project design. From the beginning, this process must involve the assistance of an archaeologist specializing in the analysis of the material(s) to be sampled. It entails several steps, each of which must be fully completed and documented before the next one is taken.

The first step is to complete *all* the preliminary analyses of the entire collection in order to fully understand the range of objects it contains. This task involves counting and weighing classes of artifacts based on common attributes and writing general descriptions of them (see Toolkit, volume 5). The specialist in the targeted redundant artifact class(es) is especially important at this time. She or he should determine the range of variation and the artifact types present. For example, hundreds of kilograms of iron slag look pretty much the same to archaeologists untrained in metallurgy. A specialist, however, might determine that a particular collection has several types of slag, each with particular and informative attributes.

After the types have been isolated, properly recorded, and described, the specialist must answer a very important question: *Now that the range of variation of this artifact class has been determined and the necessary analyses completed, should all of it be kept, or should samples be taken from the artifact types for accessioning in the designated repository?* To make this decision, the specialist must evaluate the following:

- The current state of knowledge of the artifacts and specimens
- The relative quantities of objects in each designated artifact type
- The range of variation within each type
- How the artifact types were distributed over a site or survey area
- The range of scientific methods that might be used to study the objects in the future
- The number and variety of objects needed for those studies

Once the decision is made to sample and cull, the second step is to define the appropriate sample size. Although the sample size should have been discussed in the project design and collecting strategy, it must be confirmed by the specialist in light of the known size of the project collection. Generally, the sample size needed for any future collections studies depends on the exact question(s) being investigated and the methods to be used. The goal should be to estimate the reasonable sample sizes that may be needed for future research, realizing that one can never completely ascertain what will be needed until a series of scholars try to use the collection. If the research methods are noninvasive, problems with the sample size should be minimal. If they are destructive, such as some chemical analyses, petrography, or metallography, the percentage of objects saved will decrease after each analysis.

The archaeologist does not have to take the same sample size for each artifact type. Some relatively rare object types might be sampled at 100 percent (all pieces are kept), while other object types with enormous numbers of objects can be randomly sampled at 10 or 20 percent. All decisions involved in this process and any significant observations made while removing the samples must be documented. This documentation must be kept with the final collection in the repository so future researchers understand exactly how the collection was created.

The next step involves handling the discarded materials. A place must be selected that, in the distant future, can never be mistaken as having a function different from artifact disposal. Appropriate places may be the middle of a fully excavated unit at your site or a municipal waste facility. Again, it is necessary to record where the culled objects were deposited in case someone wants to find them in the future or unknowingly stumbles upon them.

The final step is for the archaeologist to advise the curatorial staff at the designated repository about any employed sampling procedures. The same procedures then can be applied if the collection needs to be further sampled prior to accessioning in the repository. Such further sampling would only occur, however, under unusual circumstances and after prior consultation with the project director and object type or class specialist. Consultation with an official of the agency or bureau that owns the material also might be advisable if further sampling occurs at this stage of the process.

Whether sampling is done or not, other processing and conservation actions necessary to prepare the collection for long-term care and management must be considered (see Toolkit, volume 5). After cleaning and drying, these may include labeling, consolidation, reconstruction, and packing for transport to a repository. Table 6.3 lists important principles to follow when processing and providing first-aid conservation to objects. Try to ensure that it will be possible to reverse any intervention used over the long term without affecting either the object's constituent materials or future interventions that may become necessary at the repository.

IN YOUR OFFICE AFTER THE FIELD PROJECT: RECORDS MANAGEMENT

Archaeologists create a large amount of significant documentation before, during, and after a project: field and laboratory forms and notes, electronic data (e.g., databases, GPS, GIS), maps, photographs, video- and audiotapes, results of scientific analyses, administrative and legal paperwork, drawings of objects, preliminary and final reports, lecture notes, and published articles. One can never predict how these materials will be used in the future as archaeological methods and interests change over time, but one can be assured that they all have considerable value. Like the objects, they are the irreplaceable legacy of a project and its results (D. Fowler and Givens 1995). Every effort must be

Table 6.3. Key Principles of Object Processing and Conservation in the Laboratory (Cronyn 1990; Sease 1994)

1. The best conservation approach is usually to do as little as possible. Minimize the use of interventive materials (e.g., glues, strengtheners, acids), and handle the object as little as possible.
2. Always think about the tools being used to clean, process, and conserve objects. Can they damage the object in any way?
3. Do not use any treatments (e.g., glues, consolidants) that may contaminate or scar an object for scientific analysis, such as chemical analysis or microscopy. If you are uncertain about the effects of an intervention yet believe that future analyses may be conducted, retain a sizable, unadulterated sample.
4. All conservation treatment must be reversible over long periods of time. Glues or consolidants, for example, must be removable without any negative effect to the object and its constituent materials. Remember that acids to remove concretions permanently alter an object.
5. To ensure reversibility, only use interventions that have been well tested over time and whose aging properties are well known. Note: Tests on Elmer's glue show that it becomes insoluble in water over time and cannot be reversed.
6. All interventions must be fully and accurately documented, especially the materials used. Years from now, a conservator may need to do additional interventions or a researcher may select your objects for scientific analysis. They must know the conservation history of the objects.
7. Make sure the treatment records become part of the permanent record of the project.
8. Conservation interventions often involve chemicals that must be properly and carefully transported, stored, and disposed of.

made to prevent their discard or separation into disassociated parts as archaeologists move between jobs or into retirement, or their complete loss after a death. Documentation associated with federally sponsored projects is considered an integral part of the overall project collection, as well as the property of the government.

It is understandable that some archaeologists may not want to give up some of these records during their archaeological careers. They may need access to them for lecturing, writing, or further research. Also, one can never be sure how records might be used by others in isolation from the original work and aspirations. Archaeologists can take three critical steps to maximize the long-term care and management of, as well as access to, the records and reports they make.

The first task is to designate a repository or archive to curate the materials. This step needs to be done in a legal document, such as a will or a formal agreement with the repository. An archaeologist should spend some time determining the best place to deposit her or his records (Parezo and Person 1995). There should be a logical association between the individual and the repository or archive. It can be the university where the archaeologist taught for many years, the repository that holds most of the artifact collections she or he generated, or a repository with an excellent reputation in the state or region of the work or in the disciplinary specialty. Once several potential repositories or archives are identified, the records collection should be discussed with the staffs. Does the repository have a full-time archivist? What is the scope of collections of the repository? How does it handle the particular storage needs of paper records, film negatives, printed photographs, audio- and videotapes, and electronic media? How will your materials be made accessible to the scholarly and general public? Does the repository have any requirements of its potential donors, including monetary conditions?

The second step is always to be aware of the need to preserve documents and records as they are created and while they are in the archaeologist's possession (Kenworthy et al. 1985). There may be paper records, 35mm film, videos, audiotapes, maps, and computer files, so concern about the different, long-term storage requirements of each format is necessary (Griset and Kodack 1999). The most important principle to follow is to use quality materials. Even a "poor" graduate student doing dissertation research should budget for the necessary archival-quality materials to preserve the records over the next forty to fifty years of her or his career and well beyond.

The third step is to retain an electronic copy of all reports, including limited distribution CRM reports. The discipline values the contents of gray literature reports (Lipe and Redmond 1996), but it is often difficult to obtain copies once the limited number of printed copies is distributed. Some reports are available on the Internet, and in the future it may be possible to make full electronic copies of most reports broadly accessible. The archaeologists should make sure that appropriate, centralized facilities (e.g., State Historic Preservation Offices) and regional libraries receive a paper copy at least. Also, report citations need to be entered into appropriate state, regional, and national bibliographic databases (e.g., the National Archeological Database, Reports module), which help researchers find the results of previous work. Table 6.4 provides some other key principles to follow concerning the care and long-term preservation of records and reports.

> **Table 6.4. Key Principles for the Preservation of Project Records (Griset and Kodack 1999; Vogt-O'Connor 1996, 1997, 1999; Vogt-O'Connor and van der Reyden 1996)**
>
> 1. Make at least one copy of all written records, including field notes and forms, laboratory notebooks, and administrative documents, on archival paper.
> 2. Send a copy of your field notes, including descriptions of field conservation and artifact sampling procedures, to the repository that is curating your artifact collection.
> 3. Make a copy of all computer-generated documents and data files on archival paper.
> 4. Make a backup copy of all computer generated documents and data files. Label them fully with information about the associated project, date created, and software used. Keep a copy of the software application used to generate the data or documents because it may quickly become obsolete.
> 5. Recopy all computer files and applications onto fresh media every five to ten years.
> 6. Process all film at laboratories that test for destabilizing chemicals.
> 7. Use appropriate containers for proper archiving and long-term preservation of 35mm film negatives, printed photographs, slides, 8mm movie film, and videotapes.
> 8. Label all photographs, maps, video- and audiotapes, and drawings minimally with the name of the creator, date created, and description of the item.
> 9. Store records in acid-free folders and boxes.
> 10. Enter bibliographic citations of your unpublished and published reports in regional and national databases, and, whenever possible, submit electronic copies of your full reports to databases that can store large documents.

CONCLUSION

Decisions made prior to, during, and after fieldwork have significant ramifications for the long-term care and viability of the archaeological collections. Therefore, archaeologists are directly responsible for the collections they make. The importance of considering collections care and management early in the project design, and the need for a collecting strategy, cannot be underestimated. Conservation and sampling issues that arise during fieldwork and soon thereafter in the lab also need consideration. Finally, it is very important for archaeologists to preserve and provide for the care of the documents and records they generate over their careers.

7
WORKING WITH A REPOSITORY

The intent of this chapter is to acquaint archaeologists with typical policies and procedures related to making curation arrangements with repositories and using archaeological collections that are curated by a repository. Of course, no two repositories are alike. Specific policies and procedures will differ between institutions, but the general framework discussed here will provide an idea of what you can expect.

ARRANGING FOR LONG-TERM CURATION

One of the major points of this book is that curation begins *before* a collection is made. Identifying a repository and becoming acquainted with its policies and procedures is an initial step in planning a field project (see chapter 6). An archaeological collection should not be made until and unless there is a repository that can and will care for it. The search for a repository typically begins within the general geographic region (e.g., the state) from which the collection originates. State museums usually can offer assistance and may themselves be appropriate repositories. Federally owned and administered collections may have designated repositories with curation agreements in place. This arrangement alleviates the need for the archaeologist who is making a collection to find a repository, but not the need to follow pertinent policies and procedures.

CURATION FEES

Many repositories charge curation fees, particularly for collections resulting from CRM projects. Fees typically are tied to the size of a collection (e.g., number of boxes of a certain size or total cubic feet). Artifacts, other specimens, and records are often included in size calculations. Fees vary widely but tend to be similar within regions (Childs 1998; see chapter 8). The kinds of services repositories offer on a fee basis also vary. Some institutions will do cataloging, while others will not. Charging fees to funded projects is simply one way repositories are dealing with the curation crisis. Curation fees rarely, however, reflect the true total costs of curating a collection "in perpetuity." Many institutions charge fees that reflect only the costs of actually getting a collection installed in their facility—for example, storage furniture, additional staff time for placing collections in storage, and inputting data to a computerized inventory.

WHICH REPOSITORY IS APPROPRIATE?

We briefly discussed some of these issues in chapter 6, and in chapter 4 we described kinds of repositories and their governing structures. A repository that is appropriate for long-term curation of archaeological collections may be organized in any one of the ways discussed in chapter 4, but not all repositories are appropriate for the care of archaeological collections or for specific kinds of archaeological collections. Making informed judgments regarding a proper curatorial institution often is a responsibility of the archaeologist making a collection. Some institutions would quite willingly take an archaeological collection but are inappropriate repositories.

A primary concern is whether a repository is a legitimate long-term institution. Is the institution configured to exist and continue curating collections after the individuals currently responsible are no longer involved? Evidence of such long-term interest may take the form of a charter or incorporation, depending on the laws of a particular state. To curate archaeological collections "in perpetuity," the repository itself must be configured legally to exist in perpetuity or minimally to have a legal agreement as to what will become of the collections if the repository closes. Collections in the public trust usually should not pass into private ownership. If a potential repository is privately rather than publicly owned and the collection in question is (or will be) in the public trust, it will be especially important to make legal provisions for protecting the collection.

A second question to ask a repository is whether it curates research collections. As explained in chapter 5, an institution's mission statement and scope of collections or acquisition policies define the kinds of collections a repository acquires and the uses it makes of them. Institutions that curate collections mainly for exhibition or for teaching are not likely to be appropriate curatorial facilities for archaeological materials of research value. Related to this concern is whether the institution already is curating professionally generated archaeological collections with significant research value. Many local historical museums curate prehistoric or historic objects that originated in archaeological deposits, but these collections typically are antiquarian in nature, not archaeological, as we defined in chapter 1. Such institutions may be very interested in having an archaeological collection to add to their displays and may even offer to take it for "free"—that is, to charge no curation fees, especially if the collection originated in their geographic area. Such institutions may not have staffs that understand the importance of records and provenience to an archaeological collection and, although well intentioned, may inadvertently destroy the research value of the collection. This problem certainly does not pertain to all local historical societies or small museums. Many smaller institutions have very capable staffs and well-curated collections. The archaeologist who is making curation arrangements must make sure the repository is indeed an appropriate one for archaeological collections (see chapter 6). It is always prudent to seek advice from the staff of a reputable repository that curates archaeological collections in a nearby region.

Once it is ascertained that a specific repository is a legitimate institution that curates archaeological research collections from the collection's geographic area and that public-private ownership issues are not a problem, the next step, if possible, is to make an appointment to visit the repository. A visit is important if the archaeologist expects to be doing fieldwork or other research in the region. Establishing a working relationship with a good repository is essential to the successful completion of a field project. Repository staff can provide advice and assistance regarding curation issues throughout a project. If a repository has agreed to curate a collection, the staff typically is happy to provide such aid because in the long run, their work will be made easier if everything is in order and problems are resolved before they receive the collection.

ACQUISITION POLICIES AND PROCEDURES: WHAT TO EXPECT

Making arrangements for long-term curation of an archaeological collection need not be a frustrating or overly time-consuming process. Problems may arise, though, if the archaeologist who is making a collection does not choose an appropriate repository or is unaware of the expectations a repository is likely to have. In the past, some archaeologists assumed that all they needed to do to get a collection curated was to show up at a repository with the collection and expect that they would be greeted with open arms and profuse thanks. Such expectations typically led to disappointments and poor working relationships.

As noted in chapter 5, reputable repositories have acquisition policies and procedures that must be followed in order for materials to be accepted into their collections. Reputable repositories also have policies and procedures for using the collections they curate. These policies are not meant to thwart the researcher; rather, they are mechanisms by which repositories methodically build and care for collections. If collections policies and procedures are understood and followed, and if the repository is an appropriate one for the collection in question, curatorial transactions usually go smoothly.

ACQUISITION DECISIONS

Repository staff members consider several factors before reaching a decision on whether to acquire a collection. These include how well the collection fits the institution's scope of collections policy, the collection's ownership status, and whether the repository can adequately care for the specific collection in question (i.e., space and conservation considerations). Depending on the institution's size and organizational structure, an individual or a committee may make acquisition decisions. Decisions made by staff also may take the form of a recommendation that must be approved by a board of directors before an acquisition can proceed.

Some paperwork is necessary to initiate a curation agreement. The amount of paperwork typically varies according to the size of the institution. Larger institutions have larger and more specialized staffs and usually need more paperwork because more people are involved in acquisition decisions and collections care. In smaller institutions, only one person may be responsible for all aspects of curation. If so, the curation agreement may simply involve an exchange between the field archaeologist and the individual curator or collections manager. In any case, the agreement should always be specified in writing. A letter may suffice.

The archaeologist should be prepared to provide a written description of the collection. Because this description may be required before the collection actually is made (before fieldwork begins), the description minimally will need to include this information:

- The site or locality of origin
- Ownership status
- The collecting and recording methodologies to be used (e.g., survey, testing, full-scale excavation, number and size of excavation units, screen size, photographic records)
- The anticipated kinds and volume of materials (e.g., flotation samples, digital photos, metal objects)

Providing the repository staff with a copy of the research proposal often is the simplest way to furnish much of this information.

Depending on the institution's size and the complexity of the collection in question, acquisition decisions may take several weeks. This is yet another reason why establishing a relationship with a repository is beneficial for both the field archaeologist and the repository. Acquisition decisions often can be made more quickly if the repository knows what to expect. It may be possible to arrange curation agreements that cover more than one field project or more than one field season of the same project, which should benefit both parties by reducing paperwork.

The most common reasons a repository may turn down a collection are problems with ownership and physical considerations of the collection. Collections with unclear ownership are a bad investment for repositories. Curation fees never pay the entire cost of curation in perpetuity, and repositories understandably are reluctant to invest precious curation dollars in materials that could be claimed by other parties. As discussed in chapter 3, a collection originating from private land almost always belongs to the landowner. If the landowner has agreed to donate the collection, the archaeologist making the collection should be prepared to have the landowner sign a deed of gift that transfers ownership of the collection to the repository. A verbal agreement to donate the material to the repository is not sufficient to transfer ownership because in subsequent years, the donor's heirs may be able to claim the collection. One of the many advantages of having a curation agreement in place before beginning fieldwork is that landowners are more likely to donate materials to a specified museum than to give the materials to the field archaeologist. This situation arises because the notion of having the materials displayed or available for research often is attractive to a property owner, and it is particularly true if the landowner is unhappy about how a CRM project has affected his or her land.

Space and conservation are the main physical considerations that influence acquisition decisions. Storage space is a ubiquitous problem in repositories. It simply may not be possible for an institution to accommodate a large collection or one that includes large objects, even if the collection otherwise is an appropriate acquisition. Collections with items that need extensive conservation treatment, such as waterlogged objects, also are difficult for most repositories to accept. Few repositories have in-house conservators, and those that do typically have not had them for long. The backlog of objects needing treatment may be great, and adding to such a backlog may not be prudent. Contracted conservation treatment is expensive, and most repositories do not have much funding for conservation. Given these problems, a repository may require the archaeologist who makes the collection to include funds for conservation treatment in her or his budget.

COLLECTIONS PREPARATION AND DELIVERY

Preparing the collection for long-term curation usually is the responsibility of the archaeologist making the collection. Much of this effort is laboratory work that the archaeologist must do before analyzing the materials. Any reputable repository will have guidelines for collections preparation including how the materials are to be cleaned, labeled, cataloged, and arranged and the kinds of containers in which they are to be placed. Records—either the originals or very good copies—also must be organized and prepared for the repository. As noted in chapter 6, records that must accompany a collection include those that explain the setup and research design of a project (e.g., proposals), as well as field, laboratory, and analytical records, maps, and reports. A repository may require copies to be on acid-free, archival paper and that photographs be archivally processed. Instructions for preparation procedures and other information about vendors for specific kinds of storage supplies should be obtained from the repository staff. Should any of a repository's preparation guidelines be at odds with certain specialized analytical procedures, the archaeologist should make the repository staff aware of this problem so that accommodations can be made.

Delivering the collection to the repository also is usually the responsibility of the archaeologist making the collection. Repository staff should be contacted to schedule a mutually agreeable time. Many repositories have check-in procedures to ensure that everything is in order when a collection arrives.

SUMMARY

An archaeologist setting up a curation agreement minimally should expect to:

- provide the repository with some fairly detailed information about the collection;
- await an acquisition decision;
- pay curation fees, especially if it is a CRM project (see chapter 8);
- possibly pay for conservation treatment;
- prepare the collections according to the repository's guidelines;
- include records with the artifacts and specimens;
- deliver the collection to the repository at a mutually arranged time; and
- receive written confirmation from the repository that the curation agreement is in effect.

USING CURATED COLLECTIONS

One of the main reasons to curate archaeological collections is to make them accessible for continuing research, but this is not the only purpose for which archaeological collections are used. Repositories, especially public museums, receive requests for access to collections they curate from a myriad of individuals and other institutions. Policies and procedures for collections use are designed to identify those users who have legitimate purposes and to ensure the safety of the collections.

Unfortunately, some archaeologists believe they can show up whenever they wish to use a collection and any request for use will be granted immediately. Such expectations will only lead to a poor working relationship with the repository. Archaeologists wishing to use curated collections must be prepared to supply the repository staff with information about their intended purposes, to provide information about their facilities if they intend to borrow materials, to be somewhat flexible about the specific materials that may be borrowed, to justify requests for destructive analysis, and possibly to have their requests denied.

Another consideration of using curated collections, especially older ones, is that the researcher cannot expect to find all of the information or documentation that would result from more recent excavations. The context and time period of the fieldwork greatly influence the content of collections, including the types of specimens and data collected. For example, systematically collected botanical remains do not exist in WPA-era collections such as those from the Chickamauga Basin (see chapter 2) because archaeologists did not use flotation techniques then. Users of older collections thus must be prepared to be creative about their data needs and to structure research questions that are compatible with available data. But before collections can be used, they must be found.

FINDING COLLECTIONS

A major challenge in using curated archaeological collections often simply is identifying where the collection is curated. This problem slowly is being overcome as more repositories make their catalogs accessible though the Internet (see chapter 8). Nevertheless, it will be several decades before information about most curated archaeological collections is readily accessible electronically. Until then, word of mouth and dogged research still are standard ways to locate specific collections. Even when collections information is available online, data needed for particular research problems rarely will be available from computer databases that are designed to manage collections. Such databases generally inform users about the kinds of materials in an institution's holdings, but they rarely provide analytical data.

Collections from one archaeological site or project may be curated at different repositories. This circumstance is often unavoidable if a site or project is large and had many investigators. Knowing who has worked on a project can help a researcher find parts of some collections. Sometimes material classes sent to specialists for analysis are not returned to the repository that curates the rest of the collection. Instead, the materials become incorporated into the collections of the specialist's institution. The institution probably has no title to these materials, may not even be aware that the materials belong to another institution, and sometimes will not return the materials to the original repository if that institution requests them. Loan agreements with specialists can help avoid such problems.

Specimens also become separated from the rest of their original collections through the use of "type collections." For example, when the ceramic repository at the University of Michigan was started (see chapter 2), samples of sherds from sites all over the eastern United States were sent to Michigan. Researchers studying major older collections may well find some "missing" sherds in Ann Arbor.

Quite often, associated records are separated from collections and in too many cases are lost. Every archaeologist knows that an unprovenienced artifact is worthless for research purposes; similarly, repositories and archaeologists should view records as integral parts of archaeological collections. Often records were considered the personal property of archaeologists, or they were never sent to the institution that had the specimens. Tracing the employment history of the archaeologist and making inquiries of those institutions may find these "lost" records. The Council for the Preservation of Anthropological Records (CoPAR) is encouraging anthropologists and archaeologists to arrange for their papers and records to be placed in proper archives, either at retirement or through their will (D. Fowler et al. 1996). Another step forward is the insistence by repositories that records *must* accompany the specimens in order for them to accept an archaeological collection for curation.

Once a collection is found, the next step is to arrange to use it. This requires a previsit contact with the repository.

ARRANGING USE

Repositories usually want the collections they curate to be used, but repositories also must ensure that the use of collections is appropriate and does not endanger them. The repository also must ensure that collections will be accessible to many generations of users, not all of whom will be archaeologists. Archaeological collections may be used in three general ways: at the repository (on-site), at other institutions (loans), and for destructive analysis.

ON-SITE

Repositories prefer collections to be used in-house if practical. Some items may be available only for on-site use, such as original records and fragile artifacts. Arranging on-site use is simple. Usually a phone call or a short letter identifying the collections to be used and the general purpose (e.g., measurements, photography, etc.) is sufficient to set up an appointment with repository staff. Some institutions limit the purposes for which photographs made by nonstaff can be used, so it is important to discuss this issue if photography is an intended use. Also, the repository must approve any use that would alter an item. Discussions and approvals usually can be taken care of before a visit so that time during the visit can be used efficiently. Many institutions also have procedures intended to preserve and protect the objects when handling collections items (e.g., wearing gloves). The user should respect and follow these procedures.

LOANS

As explained in chapter 5, reputable repositories make loans of collections and individual objects to institutions, not to individuals. Depending on the size and organization of the repository, a registrar, collections manager, or curator may make loan arrangements. The individual requesting a loan usually must be an employee of the borrowing institution, although a loan can be made so that a nonemployee can use a collection at the borrowing institution. For example, a faculty member may arrange a loan from a museum to a university so that a graduate student can use the materials in the faculty member's laboratory.

Loans can be made for research, exhibition, or teaching. In general, repositories will make loans for research purposes at the request of, and for the use of, legitimate scholars at institutions that have reasonably secure laboratory facilities. Very fragile or monetarily valuable objects usually are not loaned for research purposes, nor, as mentioned earlier, are original records. Loans for exhibition can be made only to those institutions with secure exhibit facilities. For example, an institution that has guards, climate control, and locked cases would be loaned objects for exhibition, but not an institution without such facilities. Loans made for teaching purposes—materials that will be used in the classroom and handled by students—generally will be items that have little research or monetary value and are not fragile. Exceptions include loans for laboratory methods classes in which students do collections preparation work such as cataloging.

Loans are made for specified lengths of time and typically can be renewed. As explained in chapter 5, there are no permanent loans, but continuing loans can be arranged in certain circumstances. The borrower usually is expected to pay for transportation of the loaned materials. The kind of transportation depends on the kinds of materials, the distance to be traveled, and the risk and costs. Shipping companies that specialize in moving museum objects can be expensive, but transportation in a personal vehicle involves liability. The borrowing institution is responsible for damage or loss of the loaned materials.

DESTRUCTIVE ANALYSIS

Requests for destructive analysis, such as for radiocarbon assays, neutron activation analysis, or chemical tests, typically require detailed justifications and longer turnaround times for review. The reason is that destructive analysis either removes an object permanently from a collection or permanently alters the object thus changing its integrity. A repository may regard destructive analysis as akin to a deaccession. If this is the case, approval of the repository's board of directors may be required.

NAGPRA neither specifically authorizes nor prohibits destructive analysis of human remains or other items in collections that fall under its purview (25 USC 3003; 93 CFR 10.9 and 10.10). However, it does require consultation with affiliated tribes, and it instructs museums to retain unaffiliated humans remains until final regulations are promulgated or unless legally required (or recommended by the secretary of the interior) to do otherwise. Therefore, destructive analysis of any material that may relate to NAGPRA should involve tribal or legal consultation.

Sample size must be considered if the proposed analysis will remove most or all of a material category from an archaeological provenience. For example, it may not be prudent to allow destruction of all examples of one kind of seed from a site, nor may it be wise to permit destruction of all of one kind of sherd or tool from a specific provenience. Removal of these items may diminish the potential of the collection for other types of research. Where sample sizes are small, the researcher requesting permission for destructive analysis should identify several alternate samples so the repository staff can consider which alternate would have the least impact on the collection's long-term research potential.

CONCLUSION

Repositories balance the benefits of proposed uses of collections with long-term preservation needs. Different uses pose different risks to collections, risks that may affect other uses. Repositories want to ensure that by granting use for one purpose, they are not allowing the *last* use that can be made of irreplaceable materials. Archaeologists requesting use of collections should think about the impacts of their request on the long-term accessibility and availability of the materials. Requestors and repository staffs then can work together to make mutually beneficial arrangements. Uses that enhance the interpretation, cultural significance, research potential, and accessibility of collections all are in the best interests of repository staffs and archaeologists.

8

THE FUTURE OF ARCHAEOLOGICAL COLLECTIONS CURATION

In this final chapter, we focus on the future of archaeological curation in terms of three primary issues:

- Access to and accountability for collections in the computer
- Use of collections for research and public benefit
- Support for the high costs of curation over the long term

Our interest in these topics relates to several simple questions that the public and Congress might ask any archaeologist or curator:

- If we do not know what is in a collection and therefore cannot use it effectively or efficiently, why keep it at all?
- What funding mechanisms exist to cover the significant costs involved in ensuring that collections are useable, accessible, and in optimal condition?
- Is it all worth it?

ACCESS: COLLECTIONS IN THE COMPUTER AGE

We have said it before and will say it again: An archaeological collection and its documentation are the only tangible legacy of an archaeological project, whether survey or major excavation. Collections should be highly valued, preserved, and made accessible for use by researchers, educators, interpreters, and the public, especially those whose heritage is tied to the materials.

These seemingly simple goals have some important caveats. First, as noted in chapter 7, ethical and practical issues are related to balancing access and appropriate use with protecting and preserving the collections and documentation for the long term. The balance is particularly sensitive when a collection contains items that are rare or fragile, require special conservation, or have other characteristics favoring restricted access. Second, some repositories do not have adequate space for study and use or enough staff to monitor users against theft and unacceptable handling of collections. Third, some laws and repository policies control access to collections and documentation related to specific descendent groups, such as the sacred objects and human remains of Native Americans.

Effective user access to a collection and its documentation requires, at a minimum, an inventory of its contents and information on its storage location(s). Users also must know of any restrictions on access to a particular collection. It is no simple matter for a repository to assemble and organize this information for one large collection or its many different collections. Fortunately, computer technology and information management expertise can assist.

Archaeologists have used computers for decades to help manipulate and better understand data accumulated during a project (Scholtz and Chenhall 1976). Curators and managers of archaeological collections also have understood the benefits of using electronic media to catalog and inventory collections for search and retrieval purposes (Marquardt 1977b; Peebles and Galloway 1981). Budget limitations and uneven staff expertise meant that repositories entered the computer age at different times, developed their technology at different paces, had different expectations for their technology investments, and encountered different problems. To complicate matters, repository staff could not have known that the Internet would permit remote access to collections.

DATABASES

Computers allow collection managers to inventory and describe huge numbers of items and related records, to quickly retrieve basic descriptive and storage information, and to find items in storage (Chenhall and Vance 1988). These information management capabilities improve collections accountability and allow for better budgeting of future collections, including more efficient inspections, routine conservation work, and basic equipment needs.

Despite progress in the adoption and use of computerized cataloging and inventory systems at least five problems and limitations need attention: (1) changing registration systems, (2) cataloging backlogs, (3) maintenance of electronic data and the application used to access it, (4) sharing data, and (5) knowledge of and access to gray literature.

First, many older repositories retain several registration systems that evolved over decades. It can be difficult to integrate the systems into an efficient information storage and retrieval system without losing critical information, such as handwritten marginal notes on an object (Bell 1990). Nonetheless, years of effort to integrate and produce an electronic registration system for the anthropology collections of the National Museum of Natural History have proven beneficial (Krakker et al. 1999).

Second, many repositories have enormous cataloging backlogs and face significant annual increases of new archaeological collections. Automation of the process has helped considerably, but the costs are also enormous (GAO 1987; Hitchcock 1994). Trained staff must be devoted to backlog cataloging when many other tasks also have high priority.

Third, automation requires long-term maintenance of both the software applications and the data gathered. Maintaining a database of any type requires upgrading the application software as new technology develops and/or migrating all existing data to a new application as a repository develops new data management needs. The data, including digitized images, must be backed up and stored in a secure location, and periodically copied to new storage media to prevent loss (Vogt-O'Connor 1996, 1999). Use of computer technology definitely is not cost-free. Hardware and software for collections management, like the collections themselves, require baseline funding and staff expertise for long-term upkeep.

Fourth, sharing and transferring basic data about collections among repositories or between repositories and potential collection users is difficult at this time. No set of standardized data fields exists for use by all repositories with archaeological collections (Marquardt 1977b; Woosley 1992), although some agencies such as the National Park Service have one set of internal standards for all its more than three hundred park repositories. Because there is no overall set of standards, some critical data often are not collected, such as the ownership of the collections that a repository manages. It is difficult, as a result, for federal bureaus and agencies to account for and gather basic information on their collections housed in many different repositories (Ferguson and Giesen 1999). This problem would be resolved if all repositories carried an ownership field in their databases, as the Natural Science Collections Alliance successfully promoted for natural history collections. Likewise, few repositories link their accessions records to geospatial references for use in Geographic Information Systems (GIS). Doing this would aid in a wide array of research and management efforts.

Fifth, the number of archaeological reports has dramatically increased over the last twenty years. Few of the gray literature reports are available in libraries or bookstores, so they need to be systematically recorded, abstracted, and indexed. Accessibility to basic information about such reports promotes wider recognition and use of their contents and less duplication of effort (Ford 1977; Marquardt 1977b). The National Archaeological Database (NADB) Reports module is such a bibliographic database of gray literature that has been implemented at the national level (Canouts 1992). Its principal drawbacks are that it is difficult to keep up-to-date and that it lacks a standardized thesaurus of keywords for use in searching the database. Access to full reports could be achieved by digitizing each printed page of old reports and distributing via CD-ROM or the Internet. Bringing the huge numbers of gray literature reports onto desktops in this way will require considerable funding as well as careful attention to backing up the files and periodic copying to new storage media.

INTERNET

The Internet can significantly aid and transform access to and use of archaeological collections, records, and reports. The World Wide Web is a particularly useful tool for public outreach and education about archaeological collections (see Toolkit, volume 7). The Web may be accessible from a user's home, local library, office, or school. This accessibility means that a user can learn about a repository and its collections without traveling to it. For archaeologists, it can provide a critical introduction to particular collections, but it will never be a complete substitute for an on-site visit to a collection.

The Web has multimedia capabilities so that information about collections can be provided in a wide variety of formats, including text, two- and three-dimensional color photographs and images, video, and audio. Researchers can view objects from afar and make informed decisions about whether to travel to the repository for more in-depth study. For the repository, a significant advantage of remote access to specific objects is that the objects and documents, especially fragile items, are handled less and repository staff are not needed to supervise users.

With careful organization of a website, a repository can target several different audiences, such as researchers, interpreters, and the public, at the same time. Clever organization and design may offer several doorways for entry into a repository's website, one for each type of audience, including kids. A website also can be changed, updated, or enhanced with much greater ease than a book or film. New services may be added to a site when useful or removed if unsuccessful.

Table 8.1 summarizes services a repository might offer through the Web to promote access to its collections. The appendix provides a list and brief description of some websites that offer such services.

Like in-house databases, the Web can provide more information about archaeological collections than has ever been possible. However, a website is not necessarily cheap or easy to create. Large and complex sites that offer information and services to different audiences require considerable investment in hardware and software, creative thought and skills related to content organization and Web design, and funds to pull it all off. If a website is easy to navigate and users routinely find interesting materials, people were involved behind the scenes to anticipate the needs and demands of their designated audiences. Furthermore, like databases, a website must be maintained. A repository cannot expect to launch a website and sit back. The content must be reviewed periodically and the links must

Table 8.1. Possible Web-Based Services for Access to Collections and Documents

- Multiple media: text, 2D and 3D photographs and images, video, and audio
- Searchable databases, such as collections and archive inventories
- Online exhibits of one or more collections
- Object type collections for research
- Repository policies, as well as forms for access and use
- E-mail for direct communication
- Links to contextual information about the archaeological project from which a collection came, even when not provided by the repository

be checked. Finally, a repository cannot rely on a website as its only vehicle of communication because the Internet is not available to everyone. A repository should use a website as one of several strategies to inform its audiences about its collections.

USE OF CURATED COLLECTIONS

A recent study of leading U.S. anthropology departments examined the primary sources of data used for Ph.D. dissertations (Nelson and Shears 1996). The findings revealed that, in 1990, 16 percent of doctoral research used museum collections, while in 1994 and 1995, 30 percent primarily used collections. This jump suggests that fewer site excavations are being done by U.S. academics, possibly due to diminishing research funds and an increasing conservation ethos, perhaps stimulated by NAGPRA. It also suggests greater appreciation among graduate students for the research value of collections.

This trend sets up many difficult challenges. The first is getting archaeological curation introduced into graduate curriculum. Even now, with ongoing major discussions among archaeologists about curriculum reform (Bender and Smith 2000), curation is omitted as a topic to be included in graduate curriculum. Curation cannot continue to be a "back burner" issue or the responsibility or problem of others. As federal and state agencies, their partner repositories, and many other repositories work to provide good access for the use of properly managed collections, graduate faculty must understand and teach the complexities of archaeological curation, including the value and proper use of collections for research and public benefit. This necessity now is part of the ethical guidelines of the SAA (Childs 2002). Additionally, increased use requires increased collaboration between repository personnel, CRM practitioners, archaeologists in academia, and cultural groups whose heritage is represented.

THE "BIG PICTURE": CURATED COLLECTIONS AS SAMPLES OF THE ARCHAEOLOGICAL RECORD

Lipe (1974) first suggested preservation of a representative sample of regional archaeological resources as a response to the increasing loss of intact sites. Later, Dunnell (1984) and Salwen (1981) suggested that because archaeological resources are nonrenewable, preservation of a representative sample is the most ethically defensible strategy for the general management of archaeological resources. "The representation problem is . . . one of insuring that the patterns of kind, quantity, and distribution are preserved in those archaeological materials that persist beyond impact" (Dunnell 1984:71).

This approach can and should be extended to curated collections. Not only must we seek to manage collections growth, but the actual growth of collections must include samples that represent specific kinds of archaeological resources. Collections by their very nature are samples, and what is curated directly affects the character of the preserved archaeological record. Explanation of variability in the archaeological record is a main focus of archaeological research. Preservation of a representative sample of archaeological resources in curated collections can ensure persistence of this variability and allow future archaeologists to study variability and its causes. If such research is to continue, however, collections from one type of site or resource must not be curated to the exclusion of other kinds. Various kinds of sites, features, and materials must be represented in sufficient, yet practical, quantities to allow meaningful comparisons.

Defining the composition of a representative sample is fraught with difficulties but is not impossible. Artifact types must be categorized to structure the sample. Sampling plans need to consider various categorical levels and scales of analysis, such as kinds of sites, contexts (e.g., features), artifacts, and noncultural samples. Representative samples of site types have been defined successfully in some regions, yet the criteria employed to define categories for future research must transcend, to the extent possible, the restrictions of current archaeological concerns and biases (Glassow 1977:414). Dunnell (1984:71–72) notes the problem of assuring representation when blind sampling is necessary and suggests that a spatial frame of reference can provide an independent control. For example, consider that a settlement system for a certain time period in a region is not well understood. A sampling scheme that seeks to collect samples and information from sites of this time period in the represented environmental contexts would be one way to approximate a representative sample of the related archaeological resources (see Toolkit, volumes 1 and 2).

A complementary problem is determining the proportional quantities of the sampling categories. This issue can present difficulties when samples must be drawn from numerous, broadly distributed archaeological resources, as well as rare or localized occurrences (Dunnell 1984:72). For example, in the hierarchical settlement systems of complex societies, large sites with rich deposits form a small proportion of the sites in a region. Yet, given the diversity of materials and data potential of these sites, a strong argument could be made for curating collections from a larger proportion of such sites than from other types of sites. There also may be multiple reasons (i.e., scientific, humanistic, educational, and artistic) to curate the aesthetically and culturally significant artifacts that often are found at these sites. Such artifacts merit special consideration for scientific reasons because of the wide variety of information that can be derived from them (Brown 1981:181–83).

Another problem in defining samples is our inability to predict the kinds of biases that future research will discover, even in the collections resulting from today's most carefully designed data collection strategies. We cannot predict the future, but we can evaluate the past and attempt to correct existing recognizable biases in the curated data bank. Such biases severely limit the kinds of research problems that currently can be addressed with collections, and it is likely that these limitations will become more profound over time. As discussed in chapter 5, periodic review and evaluation of collections are necessary to identify and correct biases so that collections management becomes responsive to new information and changing research needs. An important component of a review and evaluation process is comparison of the curated sample with the overall composition of the archaeological record.

For example, as we saw in chapter 2, material classes now considered "standard" data sources (e.g., botanical samples recovered via flotation and chipped stone debitage collected by screening) were not collected in the 1930s. Developments in archaeological practice now make it possible and routine to recover and interpret such materials. Reviews and evaluations of curated collections could highlight sites, time periods, and geographic areas that lack such data sources in the curated sample. Consequently, collections reviews can help stimulate and justify research into these areas, especially in the context of CRM studies.

A logical way to begin developing plans for better managing collections is to determine the nature of the sample now represented in curated collections. Unfortunately, the present sample is poorly known at the regional and national levels because of inaccessibility to collections (Dunnell 1984; Salwen 1981; Sullivan 1992a, 2001) and a general lack of concern by the archaeological profession. As repositories increasingly use information management technology, and as the Internet provides new outlets for information dissemination about collections, we hope that increased accessibility will facilitate interest in and evaluation of the curated sample.

ENCOURAGING REPOSITORIES TO CURATE REPRESENTATIVE SAMPLES OF THE ARCHAEOLOGICAL RECORD

Comprehensive planning for representative data collection and curation requires well-considered input from several sources. The expertise of archaeological researchers and cultural resource managers is critical, and all plans must be in synch with repository policies. A strategic part of this coordination is to involve collections managers at the front end—that is, in making decisions about what to collect. This involvement is necessary to ensure that repositories will accept the desired range of collections and that these materials are appropriate for building data banks for continuing archaeological research.

Repository mission statements and collections management policies are essential ingredients for effective coordination of archaeological resource management activities, including shaping the curated sample into a data bank appropriate for long-term research needs. As we discussed in chapter 5, the mission statement establishes the long-term goals of a repository based on the types of collections it will curate. The mission statement considers geographic area, subject matter, and time period, as well as how the collections will be used (e.g., education, research, exhibition). Repository staff members then translate these general parameters into more specific policies for accepting and building collections (i.e., scope of collections, acquisitions, loans), as well as for other collection activities, such as research, exhibition, and deaccessioning (Malaro 1979; NPS 1990; Pearce 1996:67–68).

A good acquisition policy takes an *active* rather than *passive* approach to acquisitions (Burcaw 1997). With a passive approach, the repository merely decides what will be in its collections based on what it is offered. In contrast, active collecting means that a repository utilizes its long-term goals to make specific, conscious decisions about how its collections will grow and seeks to obtain those materials to meet those goals. For example, a repository may focus on the kinds of artifacts or specimens to be accepted, the level of documentation required, or the kind of sites or projects from which collections can or should be accepted. An active approach allows an institution to exercise considerable control over what it acquires, as well as to apply some quality control over what is curated.

Archaeologists and agency managers should become familiar with the potential for collections management policies to assist in the long-term management of archaeological resources. Many archaeologists and agencies have long regarded curation as merely a storage problem, so they may overlook the logical links among project design, collecting strategy, long-term research value, and curation (see chapters 6 and 7). The critical parameters that control the kind of collections to be made, therefore, may be set without input from appropriate repositories, which are often contacted after collections are already made. When this happens, repositories are placed in the unflattering role of critics rather than partners because repository staffs are asked to deal with a collection on whose composition or condition they had no input.

Repositories are now developing active collections policies as standard museum practice. Academic-based and CRM-based archaeologists need to become active partners in this process (Sonderman 1996; Sullivan 1992a, 2001) because collections policies may be detrimental to future research and interpretation efforts if written by persons who are unfamiliar with the long-term needs of archaeological research and heritage management. It is also appropriate that when collections involve the cultural heritage of Native American tribes, the affiliated groups are involved in the decision making concerning repository policies and long-term planning strategies.

COORDINATED REGIONAL PLANNING

Given the regional variation in the archaeological record, it is logical to develop plans for collections growth and long-term research at the regional level. State plans for managing cultural resources, which are required by federal regulation, historically have not included curated collections. The accumulating data bank of curated collections in various repositories needs to be considered explicitly in such planning efforts. Again, planning for better collections management requires coordination between the agencies and bureaus that decide which sites are to be excavated, the archaeologists who make and use the collections, the repositories that ultimately curate the materials, and the Native American tribes and other culture groups whose cultural heritage is represented by the collections.

One possible way to coordinate regional planning efforts is for statewide archaeological professional organizations to work with State Historic Preservation Offices and Tribal Historic Preservation Offices. The repositories, government entities, and the archaeologists with long-term research and collecting interests in specific regions would be brought together to conduct this planning. The repositories could solicit Native American involvement, perhaps through their Native American advisory groups. Representatives of federal programs having major influence in the regions also should be included. Input from these bureaus is necessary for coordination with their cultural resource management master plans.

The major goal of such regional task forces would be to draft complementary collections management policies among the repositories based in the region in order to shape the sample of the archaeological record that will be contained in curated collections. In some cases, interinstitutional agreements may be needed where there is overlap in the geographic areas covered by two or more museums. Such plans would neither require nor preclude designation of regional repositories (Marquardt 1977a; see chapter 3), because the plans would be regional in nature, not necessarily the repositories. Periodic review and revision of the plans and museum policies would also be necessary as collections grow and as archaeological knowledge expands and evolves. Such regional task forces could then have a positive impact on collections policies in general, especially as a wider constituency becomes aware of and involved with collections issues.

SUPPORT FOR CURATING ARCHAEOLOGICAL COLLECTIONS

Earlier in this chapter, we discussed the importance of computer hardware and software to inventory and to provide remote access to archaeological collections, records, and reports. We presented the need to build up collections data banks and to create regional task forces to develop collections plans for the future. All of this requires support, including funding, professional and experienced staff, and adequate equipment and facilities.

Several aspects of archaeological curation require significant funding support, notably:

- Inspection of existing collections and any rehabilitation, conservation, and inventory work required
- Initial processing, inventory, and long-term care of new collections
- Creation and upkeep of repository space and appropriate facilities and equipment
- Professionally trained staff to manage and accomplish these tasks

Over the years, a number of different solutions have been proposed or implemented to meet the many needs of curation. Many nonprofit repositories that curate archaeological collections, such as university museums, historical societies, and city facilities, have had to seek support through their parent institutions. They also have had to expend considerable effort to obtain grants to fund various activities, such as exhibits, conservation, collection rehabilitation, and expansion of storage facilities (e.g., Magid 1991). Fortunately, some granting agencies have provided significant aid to these institutions, such as the National Science Foundation, the National Institute of Conservation, the National Endowment for the Humanities, and the Institute of Museum and Library Services. However, federal grants are usually not available for federal collections. Designating a repository prior to receiving a federal permit per ARPA and 36 CFR Part 79 implies, though does not ensure, that long-term funding was negotiated between an agency and repository prior to starting a project. Although not backed by legislation, many concerned archaeologists and curators advocate creating a line item for curation in each research or contract project budget (Acuff 1993; Ford 1977:25; Marquardt et al. 1982; McGimsey 1977:32) or hiding the costs of initial processing in field recovery (Thompson 1977:37).

Another solution initiated in the late 1970s and regularly adopted by U.S. repositories since then has been to charge fees to collection depositors, such as a federal agency or bureau, state agency, or contracting firm, for particular services (Marquardt 1977a; Moore 1994). (See table 8.2.)

These fees have proved to be useful ways to secure financial assistance for repositories that had previously provided such services for free. However, a recent informal survey of approximately ninety repositories (Childs 1998), including at least one in every state, reveals tremendous variation in the types of fees charged, the standard units on which the fees are charged, the criteria used to assess fees,

> **Table 8.2. Curation Fees: Some Key Points**
>
> **Types of Service Fees Charged**
> - Entry or receiving (to review and process documentation on collection)
> - One-time processing and long-term curation (often referred to as "in perpetuity")
> - Processing (for cleaning, packaging, and/or cataloging to repository's standards)
> - Annual (yearly maintenance)
> - Rehabilitation or revitalization of existing collection
> - Documentation (sometimes different from object collection)
> - Oversized objects or individual objects
> - Other (e.g., NAGPRA assessment, deaccessioning, short-term storage, special conservation, inspection according to 36 CFR Part 79)
>
> **Standard Units on which Fees Are Based**
> - Box measuring about one cubic foot (12 × 15 × 10")
> - Box measuring 21 × 16 × 3"
> - Unspecified box size
> - Unspecified drawer size
> - Hours of staff time in repository or field
> - Linear inch for documentation
>
> **Principal Criteria Used to Assess Fees**
> - Conditions of curation agreement
> - Comparison of fee structures of nearby repositories
> - Evaluation of actual costs to process collections over long term (i.e., any combination of overhead, utilities, value of space, supplies, staff salaries, computer hardware and software, storage furniture, inflation)
> - Evaluation of reasonable support for curation without causing collection owners to go elsewhere
>
> **Uses of Collected Fees**
> - Personnel costs to process collections upon deposit
> - Costs of appropriate materials (e.g., shelving, boxing, computers)
> - Repository overhead
> - Meeting the standards of storage and collections care as required by 36 CFR Part 79
> - Optimizing accessibility to collections, including small exhibits and publications

and the uses of the collected fees (table 8.2). Furthermore, the actual amounts of fees charged vary considerably within and between regions (table 8.3; Childs 1998). Some of this variation relates to local differences in salaries, cost of materials, utilities, and institutional affiliation (Futato 1996). The archaeologist who is preparing a project needs to be aware of the range of fees in a region. Trying to find the repository that charges the lowest fee may not be a good strategy, especially if that repository does not provide the full range of curatorial services presented in 36 CFR Part 79.

The informal survey also yielded information on the effects of curation fees on archaeological practice (table 8.4; Childs 1998). In the

Table 8.3. Range of Curation Fees in Regions of the United States in 1998

Region	Number of States	Range of Fees (minimum–maximum)
Northeast	12, including D.C. (5 without identified repositories that charge fees)	$37.63 per box–$250 per cubic foot
Southeast	11 (3 without identified repositories that charge fees)	$68 per cubic foot–$227 per "box"
Midwest	13 (4 without identified repositories that charge fees)	$70 per cubic foot–$360 to $400 per "standard box"
Intermountain	8	$60 per cubic foot–$560 per drawer; $33.75 per person day on a project
West	7	$150 per cubic foot–$1,080 per cubic foot; $33 per hour processing

field, sampling of some artifact types and noncultural materials is increasing to minimize costs per storage unit in a repository. Before and after the field, more and more archaeologists and contracting firms are paying attention to repository policies and standards to minimize costs. These findings need to be carefully considered because they can affect the future data bank of archaeological resources, especially if

Table 8.4. Effects of Curation Fees on Archaeological Practice

Effects on Field Techniques
- Considerably more sampling and culling is done in the field, though it is unclear whether these activities are well documented.
- The policies of some agencies and bureaus are changing to not require collection of all usual artifact classes during a project.
- Objects are not picked up during surface surveys, although they are documented by mapping and photography.
- More flotation is done in the field rather than bringing soil samples to the laboratory or repository.

Effects on the Receipt of Collections by a Repository
- Collections are in good condition upon arrival, minimizing initial processing costs.
- Noncultural items are culled.
- Collection owners are taking time to learn about repository facilities, curation agreements, and long-term collections care.
- On the negative side, boxes occasionally arrive overpacked to minimize costs.

sampling is not carefully planned and documentation of the decisions made is not carefully done.

Many repositories that charge fees now place a greater burden of initial processing on the archaeologist, collection owner, or donor (Harris 1993). The repository establishes strict requirements in its collections management policy on the acceptable condition of a collection upon receipt. These prerequisites often include prior cleaning, cataloguing and inventorying using the repository's standards, specific sizes and types of bags and storage boxes, complete documentation and related records, and initial conservation work. If these requirements are not met prior to depositing a collection, then the costs are borne by the collection maker and owner, not by the caretaker repository.

The use of state, regional, or national repositories is another broad-scale solution to controlling curation costs (Bade and Lueck 1994; Marquardt 1977a; Marquardt et al. 1982; McGimsey and Davis 1977; Moore 1994; Trimble and Meyers 1991; Woosley 1992). Centralized facilities can minimize duplication of effort by sharing the costs of staff time and specialized expertise, materials, equipment, and facilities among the collection owners. Several states, such as Maryland, Maine, and Connecticut, have designated specific repositories to manage the collections from their states that are owned by state or federal agencies.

Consortia of federal agencies, state agencies, universities, tribes, and others also have been discussed to meet regional needs. Specific proposals have included building new repositories, renovating existing buildings (e.g., deactivated military warehouses and bunkers, buildings on the National Register of Historic Places), or using established, well-respected museums that only need additional support for existing staff and facilities (Chapman 1977; Ford 1977; Hester 1977; Thompson 1977). Some questions need resolution before the regional repository progresses very far (Chapman 1977; Hester 1977; Lindsay et al. 1980; McGimsey 1977):

- Who owns the collections and has the authority to loan, provide access to, and deaccession objects or collections?
- How would the various consortium members cover costs?
- Who would be in charge of and responsible for the various repository functions and services?
- Should a regional repository offer educational and public interpretation functions or just storage, conservation, and research functions?
- Who would own the equipment bought using grants?

- Will existing repositories be willing to give up their collections to a regional entity?

The costs of curating and making accessible millions of archaeological artifacts, records, and reports will not go away. The key is to control and distribute the costs among the responsible parties; increasingly this is being done. Archaeologists must have a clearer understanding of the costs involved and commit to budgeting curation and conservation activities into their projects. Happily, this essential commitment is happening more and more.

IS IT ALL WORTH IT?

We think it is all worth it! An equally compelling question is "Is it worth excavating, surveying, and managing archaeological sites if we do not properly curate the collections and associated documentation from those projects?" A well-made, well-documented, and well-curated collection such as the Chickamauga Basin collection, as noted in chapter 2, can provide fodder for numerous research projects, even if it was made over two generations ago and lacks material classes or information now routinely collected. We also have seen that it has taken the efforts of strong-minded and strong-willed archaeologists, repository staff, and heritage advocates to provide care for existing collections and to bring the curation crisis to national attention. Finally, we have seen that the crisis is far from over, but there is beginning to be some hope.

As more archaeologists learn about curation, as more repository staff members understand the multifaceted significance of the collections under their care, and as more people become aware of the irreplaceable representation of their heritage in collections, the number of advocates for quality care of collections grows and the collective voice becomes stronger. The ability to garner needed resources also will grow with this strengthened voice. To be strong and effective, all groups and individuals with interests in curated collections must respect each other's roles and views. They must attempt to work toward solutions to the crises that benefit all or that represent well-considered compromises.

Curated archaeological collections are highly significant pieces of the human past. We need to preserve that connection to peoples' heritage because it reminds us from where we came and who we are. But, we cannot and should not save everything. That which we do decide to preserve represents our legacy to future generations. The composition of that legacy deserves our careful consideration and its care our best efforts. The curation crisis is much more than a storage problem. It's about what our progeny will inherit. How will they view *us* as caretakers of their heritage, from the field to the repository, and into the future?

APPENDIX

USEFUL INTERNET SITES RELATED TO CURATING ARCHAEOLOGICAL COLLECTIONS

The following list of Internet sites is not exhaustive or definitive. It is intended to help you start searching for other helpful information pertinent to your needs and to understand the range of diversity available on the Internet.

ARCHIVES AND RECORDS MANAGEMENT

- Council for the Preservation of Anthropological Records: archaeology.la.asu.edu/copar
- National Anthropological Archives: www.nmnh.si.edu/naa
- National Archives and Records Administration: www.nara.gov
- Northeast Document Conservation Center: www.nedcc.org
- Primer on Disaster Preparedness, Maintenance and Response for Paper-based Materials, National Park Service: www.cr.nps.gov/csd/publications/primer/primintro.html

CAREER INFORMATION

- American Association of Museums Careers Information: www.aam-us.org/tis3b.htm
- Occupational Outlook Handbook for Archivists, Curators, Museum Technicians and Conservators: stats.bls.gov/oco/ocos065.htm

CONSERVATION

- Conservation Online (CoOL): palimpsest.stanford.edu
- Conserve O Grams (National Park Service): www.cr.nps.gov/csd/publications/index.htm
- International Center for the Study of the Preservation and Restoration of Cultural Property (ICCROM): www.iccrom.org
- Getty Conservation Institute: www.getty.edu/gci/conservation
- National Center for Preservation Technology and Training: www.ncptt.nps.gov
- American Institute for Conservation of Historic and Artistic Works: palimpsest.stanford.edu/aic
- South Carolina State Museum Conservation Library: www.museum.state.sc.us/Conservation/reference.htm

DEED OF GIFT FORM

- Center of Southwest Studies, Fort Lewis College: www.fortlewis.edu/acad-aff/swcenter/form-sw5.htm
- Division of Museums and History, Nevada Department of Museums: dmla.clan.lib.nv.us/docs/museums/reno/deed.htm
- Indiana University Oral History Research Center Deed of Gift: www.indiana.edu/~ohrc/deed.htm

DIGITAL ARCHIVING

- Archaeological Data Archive Project: csa.brynmawr.edu/web1/adap.html
- Arts and Humanities Data Service (U.K.), Managing Digital Collections: ahds.ac.uk/manage/manintro.html

EDUCATIONAL OPPORTUNITIES ON ARCHAEOLOGICAL CURATION AND CONSERVATION

- Managing Archeological Collections: An Online Course: www.cr.nps.gov/aad/curationcourse
- National Preservation Institute: www.npi.org
- Smithsonian Center for Materials Research and Education: www.si.edu/scmre/educate.html

USEFUL INTERNET SITES

- Cultural Resource Management Program, University of Victoria, Canada: www.uvcs.uvic.ca/crmp

FEDERAL EXPERTISE ON COLLECTIONS MANAGEMENT

- Department of the Interior Museum Property Policies and Standards: www.doi.gov/pam/mupol.html
- National Park Service Museum Management Program: www.cr.nps.gov/csd
- U.S. Army Corps of Engineers Mandatory Center of Expertise for the Curation and Management of Archeological Collections: www.mvs.usace.army.mil/engr/curation/home.htm
- Guidelines for the Field Collection of Archaeological Materials and Standard Operating Procedures for Curating Department of Defense Collections: www.denix.osd.mil/denix/Public/ES-Programs/Conservation/Legacy/Collguide/collguide.pdf

GRANTS AND FUNDING OPPORTUNITIES

- Institute of Museum and Library Services: www.imls.gov/
- National Endowment for the Humanities: www.neh.fed.us/
- National Science Foundation: www.nsf.gov
- Native American Graves Protection and Repatriation Act Grants: www.cr.nps.gov/aad/nagpra.htm

GRAY LITERATURE AND ONLINE PUBLICATIONS

- National Archeological Database—Reports Module: www.cr.nps.gov/aad/nadb.htm
- Digital Imprint (UCLA): www.sscnet.ucla.edu/ioa/labs/digital/imprint/imprint.html
- Site and Survey Reports from ArchNet: archnet.uconn.edu/sites/

HISTORIC PRESERVATION LAWS, REGULATIONS, AND STANDARDS

- Federal Laws, Regulations, and Standards: www.cr.nps.gov/linklaws.htm

- State Historic Preservation Legislation Database: ncsl.org/programs/arts/statehist.htm

INTELLECTUAL PROPERTY RIGHTS

- Copyright and Intellectual Property, Conservation Online: palimpsest.stanford.edu/bytopic/intprop
- Intellectual Property Mall, Franklin Pierce Law Center: www.ipmall.fplc.edu
- A Museum Guide to Copyright and Trademark, American Association of Museums: www.aam-us.org/museum-guide-toc.htm

MUSEUM ORGANIZATIONS

- American Association of Museums: www.aam-us.org/index.htm
- The International Council of Museums: www.icom.org

ONLINE COLLECTIONS AND CATALOGS

- Archaeological Research Institute, Arizona State University: archaeology.la.asu.edu/collections.asp
- Zooarchaeology Comparative Collection Databases, Florida Museum of Natural History: www.flmnh.ufl.edu/databases/zooarch/intro.htm
- Prehistoric Ceramics from Southern New England: archnet.uconn.edu/topical/ceramic/windsor/windsor.html
- Ohio Historical Society Online Collection Catalogs: www.ohiohistory.org/webpac-bin/wgbroker?new+-access+top

ONLINE EXHIBITS

- Illinois State Museum: www.museum.state.il.us/exhibits
- Frank H. McClurg Museum, University of Tennessee: mcclurgmuseum.utk.edu/permex/archael/archael.htm
- National Museum of Natural History, Anthropology: www.nmnh.si.edu/departments/anthro.html/anexhib.html
- Peabody Museum of Archaeology and Ethnology: www.peabody.harvard.edu/exhibitions.html

- University of Pennsylvania Museum: www.upenn.edu/museum/ Collections/ourwebexhibits.html

POLICIES (I.E., ACCESS AND USE, COLLECTIONS MANAGEMENT, DEACCESSION, FEES, LOANS)

- Alexandria Archaeology (VA) Collections Policy: ci.alexandria.va.us/oha/archaeology/ar-coll-policy.html
- American Museum of Natural History: anthro.amnh.org/anthropology/collections/access_form.htm
- Archaeological Collections Facility, Sonoma State University: www.sonoma.edu/projects/asc/acf/default.html
- Florida Museum of Natural History Collections Policy: www.flmnh.ufl.edu/admin/collect1.htm
- Kelsey Museum, University of Michigan: www.umich.edu/~kelseydb/MuseumPolicies/policies.html
- Office of Archaeological Services, University of Alabama: Museums<bama.ua.edu/~cmeyer/curation.htm
- Oklahoma State University Museum Collections Policy: www.okstate.edu/osu_policies/1-0119.html
- South Dakota State Historical Society, Archaeological Research Center: www.sdsmt.edu/wwwsarc/repos-guide/index.html
- Tehama County Museum, CA Collections Policy: www.tco.net/tehama/museum/tcmnewpolicy.htm
- Texas Archeological Research Laboratory, University of Texas at Austin: www.utexas.edu/research/tarl
- University of Alaska Museum, Archaeology Collections: zorba.uafadm.alaska.edu/museum/archeo/P&P.html

STATE CURATION GUIDELINES

- California (.pdf file):www.chris.ca.gov/PUBLICATIONS/curation_manual/curation_guide_1993.pdf
- Maryland: www2.ari.net/mdshpo/xmht-ftp.html
- North Carolina: www.arch.dcr.state.nc.us/curation.htm
- Council of Texas Archeologists, the Accreditation and Review Council: www.c-tx-arch.org/cta ARC/ARC.html

REFERENCES

Acuff, Lysbeth
    1993 Collections Management: An Ethical Responsibility. *The Grapevine* 3(3):3–4.

Alexander, Edward, David R. Brigham, John W. Durel, Ruth Helm, and Sally Gregory Kohlstedt
    1992 *Mermaids, Mummies, and Mastodons: The Emergence of the American Museum*. American Association of Museums, Washington, D.C.

Bade, Mary J., and Rhonda R. Lueck
    1994 *An Archaeological Curation-Needs Assessment for the U.S. Army Corps of Engineers, Mobile District*. U.S. Army Corps of Engineers, St. Louis District, Technical Center of Expertise in Archaeological Curation and Collections Management, Technical Report No. 3.

Bell, Jan
    1990 Getting Off to a Good Start: New Relationships between Agencies and Archaeological Repositories under the New CFR. *American Society for Conservation Archaeology Report* 17(2):7–13.

Bender, Susan, and George S. Smith (eds.)
    2000 *Teaching Archaeology in the Twenty-first Century*. Society for American Archaeology, Washington, D.C.

Browman, David L., and Douglas R. Givens
    1996 Stratigraphic Excavation: The First "New Archaeology." *American Anthropologist* 98(1):80–95.

Brown, James A.
    1981   The Potential of Systematic Collections for Archaeological Research. In *The Research Potential of Anthropological Museum Collections*, Anne-Marie Cantwell, James B. Griffin, and Nan A. Rothschild (eds.), pp. 65–77. Annals of the New York Academy of Science, Vol. 376. New York Academy of Sciences, New York.

Buck, Rebecca A., and Jean A. Gilmore (eds.)
    1998   *The New Museum Registration Methods*, 3d ed. American Association of Museums, Washington, D.C.

Burcaw, G. Ellis
    1997   *Introduction to Museum Work*. AltaMira Press/American Association for State and Local History, Walnut Creek, California.

Bush, Kent
    1996   The Fall and Rise of Fort Vancouver. *Common Ground* 1(2):46–49.

Butler, Patrick H., III
    1989   Obligations in Organizing a Museum. In *Organizing Your Museum: The Essentials*, S. K. Nichols, compiler, pp. 1–8. Technical Information Service, American Association of Museums, Washington, D.C.

Canouts, Veletta
    1992   Computerized Information Exchanges on the Local and National Levels in USA. In *Sites and Monuments. National Archaeological Records*, C. Larsen (ed.), pp. 23–47. National Museum of Denmark, Copenhagen.

Cantwell, Anne-Marie, James B. Griffin, and Nan A. Rothschild (eds.)
    1981   *The Research Potential of Anthropological Museum Collections*. Annals of the New York Academy of Sciences, Vol. 376. New York Academy of Sciences, New York.

Carnell, Clarisse, and Rebecca Buck
    1998   Acquisitions and Accessioning. In *The New Museum Registration Methods*, Rebecca A. Buck and Jean A. Gilmore (eds.), pp. 157–65. American Association of Museums, Washington, D.C.

Carnett, Carol
    1991   *Legal Background of Archeological Resources Protection*. Technical Brief No. 11, Archeological Assistance Division, National Park Service, U.S. Department of the Interior, Washington, D.C.

Carroll, Mary S. (ed.)
    2002 *Delivering Archaeological Information Electronically.* Society for American Archaeology, Washington, D.C.

Chapman, Carl
    1977 Cultural-Environmental Center Physical Facilities and Management Funding. In *Regional Centers in Archaeology: Prospects and Problems,* W. Marquardt (ed.), pp. 16–17. Missouri Archaeological Society, Columbia.

Chenhall, Robert, and David Vance
    1988 *Museum Collections and Today's Computers.* Greenwood Press, Westport, Connecticut.

Childs, S. Terry
    1995 The Curation Crisis—What's Being Done? *Federal Archeology* 7(4):11–15.
    1996 Collections and Curation into the 21st Century. *Common Ground* 1(2):25.
    1998 The Adoption and Use of Curation Fees across the United States. Paper presented at the Second Conference on Partnerships for Federally-Associated Collections, San Diego, California.
    1999 Contemplating the Future: Deaccessioning Federal Archaeological Collections. *Museum Anthropology* 23(2):38–45.
    2000 SSA Committee on Curation. Who We Are, Our Goals, and Our Issues. *SAA Archaeological Record* 1(2): 10–11, 37.
    2002 Committee on Curation Update: Implementing SAA Ethic #7, Records and Preservation. *The Archaeological Record* 2(3):6.

Christenson, Andrew L.
    1979 The Role of Museums in Cultural Resources Management. *American Antiquity* 44(1):161–63.

Conn, Steven
    1998 *Museums and Intellectual Life, 1876–1926.* University of Chicago Press, Chicago.

Cowan, Suzanne
    1998 Inventory. In *The New Museum Registration Methods,* Rebecca A. Buck and Jean A. Gilmore (eds.), pp. 117–19. American Association of Museums, Washington, D.C.

Crass, David C.
    1991 *Savannah River Archaeological Research Program—Guide to Curation Procedures.* Technical Report Series #14, Savannah River Archaeological Research Program, South Carolina Institute of Archaeology and Anthropology, University of South Carolina.

Cronyn, Janey M.
1990   *The Elements of Archaeological Conservation.* Routledge, London.

Damadio, Stephanie
1999   Linking the Past to the Future—Museum Collections and the Bureau of Land Management. *CRM* 22(4):32–34.

Department of the Interior
1993   *Museum Property Handbook, Volume I: Preservation and Protection of Museum Property.* U.S. Department of the Interior, Washington, D.C.
1997   *Policies and Standards for Managing Museum Collections.* Departmental Manual Part 411. U.S. Department of the Interior, Washington, D.C.

Dunnell, Robert
1984   The Ethics of Archaeological Significance Decisions. In *Ethics and Values in Archaeology*, Ernestene Green (ed.), pp. 62–74. Free Press, New York.

Ellis, Judith (ed.)
1993   *Keeping Archives.* D. W. Thorpe, Victoria, Australia.

Fairbanks, Charles
1970   What Do We Know Now That We Did Not Know in 1938? *Southeastern Archaeological Conference Bulletin* 13:40–45.

Farnsworth, Kenneth, and Stuart Struever
1977   Ideas on Archaeological Curation and Its Role in Regional Centers. In *Regional Centers in Archaeology: Prospects and Problems*, W. Marquardt (ed.), pp. 13–15. Missouri Archaeological Society, Columbia.

Ferguson, Bobbie, and Myra Giesen
1999   Accountability in the Management of Federally Associated Archeological Collections. *Museum Anthropology* 23(2):19–33.

Ford, Richard I.
1977   *Systematic Research Collections in Anthropology: An Irreplaceable National Resource.* Peabody Museum, Harvard University for the Council for Museum Anthropology, Cambridge, Massachusetts.
1980   A Three Part System for Storage of Archaeological Collections. *Curator* 23(1):55–62.

Fowler, Donald, and Douglas Givens
1995   The Records of Archaeology. In *Preserving the Anthropological Record*, 2d ed., Sydel Silverman and Nancy J. Parezo (eds.),

pp. 97–106. Wenner-Gren Foundation for Anthropological Research, New York.

Fowler, Donald, Nancy Parezo, and Mary Elizabeth Ruwell
1996   Wealth Concealed. *Common Ground* 1(2):31–33.

Fowler, Melvin L.
1985   A Brief History of Illinois Archaeology. *Illinois Archaeology, Bulletin No. 1, Revised*, pp. 3–11. Illinois Archaeological Survey, Springfield.

Friedman, Ed, and Brit A. Storey
1999   CRM and the Bureau of Reclamation. *CRM* 22(4):45–47.

Futato, Eugene M.
1996   A Case for Partnerships. *Common Ground* 1(2):50–53.

GAO (General Accounting Office)
1987   *Problems Protecting and Preserving Federal Archeological Resources.* RCED-88-3. General Accounting Office, Washington, D.C.

Glassow, Michael
1977   Issues in Evaluating the Significance of Archaeological Resources. *American Antiquity* 43(3):413–20.

Greene, Mary W.
1985   Report: National Science Foundation Anthropology Program; Systematic Anthropological Collections. *Council for Museum Anthropology Newsletter* 9(4):2–4.

Griset, Suzanne, and Marc Kodack
1999   *Guidelines for the Field Collection of Archaeological Materials and Standard Operating Procedures for Curation of Department of Defense Archaeological Collections.* Legacy Project No. 98-1714. Mandatory Center of Expertise for the Curation and Management of Archaeological Collections. U.S. Army Corps of Engineers, St. Louis District.

Guthe, Carl E.
1957   *So You Want a Good Museum.* American Association of Museums, Washington, D.C.
1959   *The Management of Small History Museums.* American Association for State and Local History, Nashville.

Harris, E. Jeanne
1993   The Cost of Curation Is Going Up. *The Grapevine* 3(3):4–5.

Hester, James
1977   Specialized and Generalized Models of Regional Centers. In *Regional Centers in Archaeology: Prospects and Problems,*

W. Marquardt (ed.), pp. 4–10. Missouri Archaeological Society, Columbia.

Hitchcock, Ann
    1994   Archeological and Natural Resource Collections of the National Park Service: Opportunities and Threats. *Curator* 37(2):122–28.

Hoopes, John W.
    1998   The Online Lab Manual: Reference Collections on the Web. *Society for American Archaeology Bulletin* 16(5):17–19, 39.

International Council of Museums
    1987   Code of Professional Ethics. In *ICOM Statues/Code of Professional Ethics*. ICOM, Maison de l'Unesco, Paris.

Jacobson, Claudia
    1998   Preparation. In *The New Museum Registration Methods*, Rebecca A. Buck and Jean A. Gilmore (eds.), pp. 121–25. American Association of Museums, Washington, D.C.

Jelks, Edward B.
    1989   Curation of Federal Archeological Collections in the Southwest: A Status Report. Paper presented in symposium entitled, "Using and Abusing Archeological Collections," at Preservation Challenges for the 1990s: A Conference for Public Officials, June 5–7, Washington, D.C.

Jennings, Paulla Dove
    1996   Objects of Life. *Common Ground* 1(2):39–45.

Jessup, W. C.
    1995   Pest Management. In *Storage of Natural History Collections: A Preventive Conservation Approach*, C. Rose, C. Hawks, and H. Genoways (eds.). Society for the Preservation of Natural History Collections, Iowa City.

Jonaitis, Aldona
    1992   Franz Boas, John Swanton, and the New Haida Sculpture at the American Museum of Natural History. In *Early Years of Native American Art History: Scholarship and Collecting*, Janet Catherine Berlo (ed.), pp. 22–61. University of Washington Press, Seattle.

Katz, Herbert, and Marjorie Katz
    1965   *Museums, U.S.A.: A History and Guide*. Doubleday, Garden City, New Jersey.

Kenworthy, Mary Ann, Eleanor M. King, Mary Elizabeth Ruwell, and Trudy Van Houten
    1985   *Preserving Field Records: Archival Techniques for Archaeologists and Anthropologists*. University Museum, University of Pennsylvania, Philadelphia.

King, Mary Elizabeth
    1980   Curators: Ethics and Obligations. *Curator* 23(1):10–18.

King, Thomas
    1998   *Cultural Resource Law and Practice*. AltaMira Press, Walnut Creek, California.

King, Thomas, Patricia Hickman, and Gary Berg
    1977   *Anthropology in Historic Preservation*. Academic Press, New York.

Krakker, James, David Rosenthal, and Deborah Hull-Walski
    1999   Managing a Scholarly Resource: Archaeological Collections at the National Museum of Natural History. *Museum Anthropology* 23(2):9–18.

Lee, Ronald
    1970   *The Antiquities Act of 1906*. National Park Service, Washington, D.C.

Lewis, Thomas M. N., and Madeline D. Kneberg
    1946   *Hiwassee Island: An Archaeological Account of Four Tennessee Indian Peoples*. University of Tennessee Press, Knoxville.

Lewis, Thomas M. N., and Madeline Kneberg Lewis
    1995   Appendix C: Manual of Field and Laboratory Techniques Employed by the Division of Anthropology, University of Tennessee, Knoxville, Tennessee, in Connection with the Investigation of Archaeological Sites within the TVA Dam Reservoirs. In *The Prehistory of the Chickamauga Basin in Tennessee*, 2 vols. T. M. N. Lewis and M. K. Lewis, compiled and edited by L. P. Sullivan, pp. 603–58. University of Tennessee Press, Knoxville.

Lewis, T. M. N., Madeline Kneberg Lewis, and Lynne P. Sullivan
    1995   *The Prehistory of the Chickamauga Basin in Tennessee*, 2 vols. Edited and compiled by L. P Sullivan. University of Tennessee Press, Knoxville.

Lindsay, Alexander J., Jr., and Glenna Williams-Dean
    1980   Artifacts, Documents, and Data: A New Frontier for American Archaeology. *Curator* 23(1):19–29.

Lindsay, Alexander J., Jr., Glenna Williams-Dean, and Jonathan Haas
    1979   *The Curation and Management of Archaeological Collections: A Pilot Study*. Publication PB-296. National Technical Information Service, Springfield, Virginia.
    1980   *The Curation and Management of Archaeological Collections: A Pilot Study*. Cultural Resource Management Series. U.S. Department of the Interior, Heritage Conservation and Recreation Service, Washington, D.C.

Lipe, William
  1974 A Conservation Model for American Archaeology. *The Kiva* 39:213–45.
Lipe, William, and Charles Redmond
  1996 Conference on "Renewing our National Archaeological Program." *Society for American Archaeology Bulletin* 14(4):14–17.
Lynott, Mark J., David G. Anderson, Glen H. Doran, Ricardo J. Elia, Maria Franklin, K. Anne Pybum, Joseph Schuldenrein, and Dean R. Snow
  1999 Teaching Archaeology in the 21st Century: Thoughts on Graduate Education. *Society for American Archaeology Bulletin* 17(1):21–22.
Lyon, Edwin A.
  1996 *A New Deal for Southeastern Archaeology.* University of Alabama Press, Tuscaloosa.
Magid, Barbara
  1991 *Buried in Storage: The Alexandria Archaeology Collections Management Project.* American Association for State and Local History Technical Leaflet #178. American Association for State and Local History, Nashville, Tennessee.
Malaro, Marie C.
  1979 Collections Management Policies. *Museum News* 58(2):57–61.
  1998 *A Legal Primer on Managing Museum Collections,* 2d ed. Smithsonian Institution Press, Washington, D.C.
  1994 *Museum Governance: Mission, Ethics, Policy.* Smithsonian Institution Press, Washington, D.C.
Marquardt, William H.
  1977a Epilogue. *Regional Centers in Archaeology: Prospects and Problems,* W. Marquardt (ed.). Missouri Archaeological Society, Columbia.
  1977b Prospects for Regional Computer-Assisted Archaeological Information Retrieval. *Regional Centers in Archaeology: Prospects and Problems,* W. Marquardt (ed.), pp. 11–12. Missouri Archaeological Society, Columbia.
Marquardt, William H., Anta Montet-White, and Sandra C. Scholtz
  1982 Resolving the Crisis in Archaeological Collections Curation. *American Antiquity* 47(2):409–18.
McGimsey, Charles
  1977 Discussion. In *Regional Centers in Archaeology: Prospects and Problems,* W. Marquardt (ed.), p. 32. Missouri Archaeological Society, Columbia.

McGimsey, Charles, and Hester Davis
- 1977 *The Management of Archeological Resources—The Airlie House Report.* Special Publication of the Society for American Archaeology, Washington, D.C.

McKeown, C. Timothy, Amanda Murphy, and Jennifer Schansberg
- 1998 Complying with NAGPRA. In *The New Museum Registration Methods,* Rebecca A. Buck and Jean A. Gilmore (eds.), pp. 311–19. American Association of Museums, Washington, D.C.

McManamon, Francis P.
- 1996 The Long View. *Common Ground* 1(2):2.

McManamon, Francis P., and Kathleen Browning
- 1999 Department of the Interior's Archeology Program. *CRM* 22(4):19–21.

McW.Quick, Polly (ed.)
- 1985 *Proceedings: Conference on Reburial Issues.* Society for American Archaeology and the Society of Professional Archaeologists, Washington, D.C.

Meyers, Thomas B., and Michael K. Trimble
- 1993 *Archaeological Curation-Needs Assessment for Fort Sill, Oklahoma, Fort Gordon, Georgia, Vandenberg Air Force Base, California, Camp Pendleton Marine Corps Base, California, and Naval Air Weapons Station, China Lake, California.* U.S. Army Corps of Engineers, St. Louis District, Technical Center of Expertise in Archaeological Curation and Collections Management, Technical Report No. 1.

Milner, George R.
- 1987 Contract Excavations, Collections Management, and Academic Archaeology. Paper presented at the annual meeting of the American Association for the Advancement of Science, Chicago.

Moore, Rick
- 1994 *Preserving Traces of the Past: Protecting the Colorado Plateau's Archaeological Heritage.* Grand Canyon Trust, Flagstaff, Arizona.

Morris, Martha
- 1998 Deaccessioning. In *The New Museum Registration Methods,* Rebecca A. Buck and Jean A. Gilmore (eds.), pp. 167–76. American Association of Museums, Washington, D.C.

Nason, James
- 1999 Traditional Roots, Modern Preservation. *Common Ground* Fall:20–25.

National Institute for the Conservation of Cultural Property
　　1984　*Ethnographic and Archaeological Conservation in the United States*. NICC, Washington, D.C.

National Park Service
　　1990　Chapter 2: Scope of Museum Collections, *Museum Handbook*, Part I, Museum Collections, National Park Service, Revised 1990.
　　1990　36 CFR Part 79, Curation of Federally-Owned and Administered Archeological Collections. *Federal Register* 55(177):37616–39.
　　1998　*Bureau Museum Property Management Summary Report, FY1998*. On file with the Department of the Interior Museum Management Program, Washington, D.C.
　　2000　NPS Director's Order #24: NPS Museum Collections Management. On file with the Policy Office, National Park Service. www.nps.gov/policy/DOrders/DOrder24.html.

Nelson, Margaret C., and Brenda Shears
　　1996　From the Field to the Files: Curation and the Future of Academic Archeology. *Common Ground* 1(2):35–37.

New York Archaeological Council
　　1994　*Standards for Cultural Resource Investigations and the Curation of Archaeological Collections*. New York Archaeological Council, Rochester.

New York State Association of Museums
　　1974　*The Ethics and Responsibilities of Museums with Respect to the Acquisition and Disposition of Collection Materials*. New York State Association of Museums, Troy.

Nichols, Susan K. (comp.)
　　1989　*Organizing Your Museum: The Essentials*. Technical Information Service, American Association of Museums, Washington, D.C.

Odegaard, Nancy
　　2000　Collections Conservation: Some Current Issues and Trends. *CRM* 23(5):38–41.

Parezo, Nancy J., and Ruth J. Person
　　1995　Saving the Past: Guidelines for Individuals. In *Preserving the Anthropological Record*, 2d ed., Sydel Silverman and Nancy J. Parezo (eds.), pp. 161–78. Wenner-Gren Foundation for Anthropological Research, New York.

Pearce, Susan M.
　　1996　*Archaeological Curatorship*. Smithsonian Institution Press, Washington, D.C.

Peebles, Christopher, and Patricia Galloway
  1981 Notes from Underground: Management from Excavation to Curation. *Curator* 24(4):225–51.

Puglia, Steven
  1999 Creating Permanent and Durable Information: Physical Media and Storage Information. *CRM* 22(2):25–27.

Ruwell, Mary Elizabeth
  1995 The Physical Preservation of Anthropological Records. In *Preserving the Anthropological Record*, 2d ed., Sydel Silverman and Nancy J. Parezo (eds.), pp. 97–106. Wenner-Gren Foundation for Anthropological Research, New York.

Salwen, Bert
  1981 Collecting Now for Future Research. *The Research Potential of Anthropological Museum Collections*, Anne-Marie E. Cantwell, James B. Griffin, and Nan A. Rothschild (eds.), pp. 567–74. Annals of the New York Academy of Sciences, Vol. 376. New York Academy of Sciences, New York.

Scholtz, Sandra, and R. G. Chenhall
  1976 Archaeological Data Banks in Theory and Practice. *American Antiquity* 41(1):89–96.

Scoville, Douglas
  1977 Regional Center: Opportunities for Federal-Institutional Partnership in Cultural Resources Management. In *Regional Centers in Archaeology: Prospects and Problems*, W. Marquardt (ed.), pp. 23–28. Missouri Archaeological Society, Columbia.

Sease, Catherine
  1994 *A Conservation Manual for the Field Archaeologist*, 3d ed. Archaeological Research Tools, Volume 4, Institute of Archaeology, University of California, Los Angeles.

Segal, Terry
  1998 Marking. In *The New Museum Registration Methods*, Rebecca A. Buck and Jean A. Gilmore (eds.), pp. 65–78. American Association of Museums, Washington, D.C.

Shaffer, Gary, and Elizabeth Cole
  1994 *Standards and Guidelines for Archeological Investigations in Maryland*. Maryland Historical Trust Technical Report, #2. Maryland Historical Trust, Crownsville.

Silverman, Sydel, and Nancy J. Parezo (eds.)
  1995 *Preserving the Anthropological Record*, 2d ed. Wenner-Gren Foundation for Anthropological Research, New York.

Society for American Archaeology Curation Task Force
  1993 *Urgent Preservation Needs for the Nation's Archaeological Collections, Records, and Reports.* On file at the Society for American Archaeology, Washington, D.C.

Society for Historical Archaeology
  1993 Standards and Guidelines for the Curation of Archaeological Collections. *Society for Historical Archaeology Newsletter* 26(4).

Sonderman, Robert C.
  1996 Primal Fear: Deaccessioning Collections. *Common Ground* 1(2):27–29.

Squier, Ephraim G., and E. H. Davis
  1848 *Ancient Monuments of the Mississippi Valley.* Smithsonian Contributions to Knowledge, vol. 1. Smithsonian Institution, Washington, D.C.

Sullivan, Lynne P.
  1991 Museum Records and the History of Archaeology. *Bulletin of the History of Archaeology* 1(2):4–12.
  1992a *Managing Archaeological Resources from the Museum Perspective.* Technical Brief No. 13, Archeological Assistance Division, National Park Service, Department of the Interior, Washington, D.C.
  1992b Arthur C. Parker's Contributions to New York State Archaeology. *Bulletin of the New York State Archaeological Association* 104:1–8.
  1995 Foreword. In *The Prehistory of the Chickamauga Basin in Tennessee,* vol. 1. Thomas M. N. Lewis and Madeline D. Kneberg Lewis, compiled and edited by Lynne P. Sullivan, pp. xv–xxi. University of Tennessee Press, Knoxville.
  1999 Madeline Kneberg Lewis: Leading Lady of Tennessee Archaeology. In *Grit-Tempered: Early Women in Southeastern United States Archaeology,* Nancy M. White, Lynne P. Sullivan, and Rochelle Marrinan (eds.), pp. 57–91. University Press of Florida, Gainesville.
  2001 The Curation Dilemma: A Mutual Problem for Research and Resource Management. In *Protecting the Archaeological Heritage of America,* Robert Drennan and Santiago Mora (eds.), pp. 90–98. Society for American Archaeology, Washington, D.C.

Swain, Lynn
  1998 Storage. In *The New Museum Registration Methods,* Rebecca A. Buck and Jean A. Gilmore (eds.), pp. 109–19. American Association of Museums, Washington, D.C.

Terrell, John
    1979 What Is a Curator? *Field Museum Bulletin* 50(4):16–17.

Thompson, Raymond
    1977 Discussion. In *Regional Centers in Archaeology: Prospects and Problems*, W. Marquardt (ed.), pp. 33–37. Missouri Archaeological Society, Columbia.
    2000 The Crisis in Archaeological Collections Management. *CRM* 23(5):4–6.

Trimble, Michael K., and Thomas B. Meyers
    1991 *Saving the Past from the Future: Archaeological Curation in the St. Louis District*, U.S. Army Corps of Engineers, St. Louis District.

Vogt-O'Connor, Diane
    1996 Care of Archival Digital and Magnetic Media. *Conserve O Gram* 19/20.
    1997 Caring for Photographs: General Guidelines. *Conserve O Gram* 14/4.
    1998 Care of Color Photographs. *Conserve O Gram* 14/6.
    1999 Archives—A Primer for the 21st Century. *CRM* 22(2):4–8.

Vogt-O'Connor, Diane, and Dianne van der Reyden
    1996 Storing Archival Paper-Based Materials. *Conserve O Gram* 19/15.

Warren, Winonah
    1991 A Model Cultural Center at Pojoaque Pueblo. *CRM* 14(5):3–5.

White, Nancy M., Lynne P. Sullivan, and Rochelle Marrinan (eds.)
    1999 *Grit-Tempered: Early Women in Southeastern United States Archaeology*. University Press of Florida, Gainesville.

Willey, Gordon, and Jeremy Sabloff
    1977 *A History of American Archaeology*. Freeman, San Francisco.

Wilson, Ronald
    1999 Interior's Museum Collections—Your Heritage. *CRM* 22(4):22–23.

Wilsted, Thomas, and William Nolte
    1991 *Managing Archival and Manuscript Repositories*. Archival Fundamentals Series. Society of American Archivists, Chicago.

Woosley, Anne
    1992 Future Directions: Management of the Archaeological Data Base. In *Quandaries and Quests: Visions of Archaeology's Future*, LuAnn Wandsnider (ed.), pp. 147–59. Center for Archaeological Investigations, Occasional Paper No. 20. Southern Illinois University, Carbondale.

INDEX

AAM. *See* American Association of Museums
academic archaeologist, 112
academic repository, 47–48, 52
access, collection: in academic repository, 47–48; for destructive analysis, 101–2; in government repository, 49; locating collection, 98–99; and NAGPRA compliance, 75; older collection, 98; on-site, 100; photographs, 76; problems with, 35–36, 74; 97–98; reasons for, 51; recent progress in, 43–44; through loans, 75, 100–101; use of inventory control/data management for, 71, 74. *See also* access, electronic
access, electronic: database, 44, 104–6; Internet, 106–8. *See also* access, collection
accession number, 62
accession records, 62
accessioning, 61–62
accessions register, 62
accountability, 3, 10, 21, 35–36
acquisition: definition of, 60; for long-term curation, 94–96; policy elements, 60–61

American Association of Museums (AAM), 41, 76, 124
American Indian, research on, 8
American Museum of Natural History, 7, 8, 9
Ancient Monuments of the Mississippi Valley (Squier & Davis), 8
antiquarian collection, 2
Antiquities Act of 1906, 9, 18
archaeological collection: difference from antiquarian collection, 2; importance of, 3; types of material objects/records in, 2
archaeological collections, current status of, 21; conservation ethic and, 23–24; federal legislation/policy and, 24–28, 26; historic preservation movement and, 23; Native American concerns and, 24
archaeological research: museum-based, 7–9; university-based, 9–10, 80
archaeology chronology, museology roots of, 5–6
Archeological and Historic Preservation Act (AHPA) of 1974, 24–25, 27

141

INDEX

Archeological Resources Protection Act (ARPA) of 1979, 25, 40, 60, 80, 82
archives, 50
archivist, 55, 56
Army Corps of Engineers, 34; assessment of federal collections by, 42–43; project vs. curation expenditures by, 27; reservoir salvage by, 18
ARPA. *See* Archeological Resources Protection Act (ARPA) of 1979
artifact container, 68
associated records management, 36–38
audio-/videotape storage, 70, 89, 90

Boas, Franz, 7, 8
Bureau of American Ethnology (BAE), 8
Bureau of Land Management, 49
Bureau of Land Reclamation, 18

catalog/identity number, 63–64
cataloging, 63–64, 76
Ceramic Repository for the Eastern United States, 10
Charleston Museum, 6
Chickamauga Project (case study): artifact disposition, 13–14, 28; collections/records curation, 14–15, 70; example of data available from, 16–17; extent of survey/excavation, 13; legacy of, 12–13, 16; work force for, 13, 14, 15
classification scheme: attribute-based system, 15; development of, 11; pottery typology, 11
Cole, Fay-Cooper, 10, 15
collection, as representative of archaeological record, 108–11
collection strategy: for noncultural sample, 84; in project design, 79, 82
collections manager, 54, 55, 56
condition report, 61–62
conservation, 66–67; definition of, 51; ethics of, 23–24; in field, 84–85; in laboratory, 85, 87, 88; of long-term collection, 96
conservator, 55, 56
continuing loan, 75
CoPAR. *See* Council for the Preservation of Anthropological Records
copyright, 76
Council for the Preservation of Anthropological Records (CoPAR), 44, 121
Council of Texas Archeologists, 41
cultural resource management (CRM), 20, 29, 89, 92, 110, 112
Cultural Resources—Problems Protecting and Preserving Federal Archeological Resources, 25
curation: accountability in, 3, 10, 21; crisis in, 1, 23 (*see also* curation, elements of crisis in); definition of, 2–3; impotence of, 2; lack of training in, 1–2, 108; marginalization of, 11–12, 108; as process, 1, 3; recent progress in, 42–44
curation, concerns about future of: access, 103–4; curating representational samples, 111–12; database access, 104–6; funding, 113–18; Internet access, 106–8, 107; preserving representative samples, 108–11; regional planning coordination, 112–13; using collections, 108
curation, elements of crisis in, 29–30; accountability/accessibility, 35–36; associated records

INDEX

management, 36–38; costs, 34–35; deaccessioning, 39–40; looting/illicit trading of artifacts, 39; management/care standards, 32–33; national curation initiative, 41–42; Native Americans concerns, 38–39; ownership, 29–31; personal responsibilities, 33–34; repository certification, 40–41

curation, history of: curiosity cabinet, 6; early federal programs, 11–12 (*see also* Chickamauga Basin Project); making vs. curating collections, 19–20; move from museum to university, 9–10; museum era, 5–9; New Archaeology, 19; postwar construction boom, 17–18; world's fairs, 6–7

curation, long-term: acquisition decisions for, 94–96; acquisition policy/procedure for, 94; fees for, 27, 92, 93, 95; locating appropriate repository for, 92–93; preparation/delivery of, 96–97; proposed repository standards for, 32

curation agreement, 81, 88–89, 95–96

curation costs, 114–15; controlling through facility centralization, 117; effect on archaeological practice, 115–17, 116; for long-term curation, 27, 92, 93, 95; range of in 1998, 115, 116

Curation of Federally-Owned and Administered Archeological Collections (36 CFR Part 79), 25, 42; on associated records management, 36–37; deaccessioning and, 39; on inventory/inspection of collection, 71; key components of, 26; lack of certification language in, 41; repositories and, 27, 45, 80, 81; standards and, 30, 33, 35

curator, 20, 54, 55, 56

curiosity cabinet, 8

database, for electronic access, 104–6

Davis, E. H., 8

deaccessioning, 39–40, 72–73, 83

deed of gift, 60, 95, 122

Department of the Interior, 49, 71

destructive analysis, 101–2

digital publications, 76

disposal, of object, 72, 73

document/record: storing, 37, 70, 89; submitting with collection, 88–90, 96, 97, 99

Dunnell, Robert, 108–9

electronic management, of record/report, 37–38, 89, 90

ethics, in curating, 79–80

Europe, public museum development in, 8

facility report, 76

Farnsworth, Kenneth, 33

Field Museum of Natural History, 8, 10

field school, 10, 15

fieldwork: managing records/reports from, 79, 80, 87–90; project design concerns for, 80–83, 82; sampling/conservation after, 85–87, 88; sampling/conservation during, 83–85, 84

flotation technique, 19

Frank H. McClung Museum, 15–16, 124

funding, 43, 113–18, 123

Government Accounting Office (GAO), 25

government repository, 49–50
grants/funding, 43, 113–18, 123
Guthe, Carl E., 10

Harrington, M. R., 8
Heye Foundation. *See* Museum of the American Indian
historic preservation movement: effect on collections, 23; effect on curation prejudice, 12
Historic Sites Act of 1935, 12
Historic Sites Act of 1945, 18
historical society, as repository, 48
Holmes, William Henry, 8
Hrdlička, Aleš, 8
human remains, storage of, 69. *See also* Native American Graves Protection and Repatriation Act

Institute of Museum and Library Services, 114
Interagency Archeological Salvage Program, 18
Interagency Federal Collections Working Group, 43
Internet, accessing collection via, 106–8, 107, 111
Internet site: archives/records management, 121; career information, 121; conservation, 122; curation/conservation educational opportunities, 122–23; deed of gift form, 122; digital archiving, 122; federal expertise at collections management, 123; grant/funding, 123; gray literature/online publication, 123; historical preservation laws/regulations/standards, 123–24; intellectual property, 124; museum organization, 124; online collection/catalog, 121; online exhibit, 124–25; policy, 125; state curation guidelines, 125
interpretation, definition of, 3
inventory control/data management, of collection, 71, 74

Jennings, Jesse, 13

Kneberg, Madeline, 14, 15, 16

labeling collections, 65–66
Lewis, Thomas M. N., 13, 15
Lipe, William, 108
loans, 75–76, 100–101
local historical museum, 93
looting/illicit trading of artifacts, 39

management/care, of collection: accessioning, 61–62; acquisition policy/practice, 60–61; cataloging, 63–64; cultural resource management effect on, 20, 29; deaccessioning, 72–73; inventory control, 71; labeling/conservation, 64–67; public access/use, 73–77; repository programs for, 51–52; standards for, 32–33; storage, 67–71, 68
Mandatory Center of Expertise in Archaeological Curation and Collections Management, 42
Mason, Otis, 7–8
McKern, W. C., 11, 13
Midwestern Taxonomic Method, 11
Mills, William C., 8
Moorehead, Warren K., 8
Morgan, Lewis Henry, 5–6
Moss-Bennett Act. *See* Archeological and Historic Preservation Act (AHPA) of 1974
Moundbuilder Myth, 8
museum: definition of, 3–4, 46; as repository, 46–47, 51–52; university, 8, 9, 10

Museum of the American Indian, 7
Museum Services Act, 46

NABD-Reports, 44, 123
NADB. *See* National Archaeological Database
NAGPRA. *See* Native American Graves Protection and Repatriation Act
National Archaeological Database (NADB), 44, 106
National Endowment for the Humanities, 43, 114
National Environmental Policy Act of 1969, 18
National Historic Preservation Act of 1966, 27
national initiative for curation, need for, 41–42
National Institute for Conservation, 114
National Museum of Natural History, 7, 105, 124
National Park Service (NPS), 12, 81; collection database of, 105; deaccessioning and, 39–40; repository of, 49; reservoir salvage and, 18
National Science Collections Alliance, 106
National Science Foundation, 43, 114
Native American: curation issues concerning, 24, 38–39; involvement with repository, 112, 113; museum/cultural center of, 48; ownership rights of, 31
Native American Graves Protection and Repatriation Act (NAGPRA), 25–27, 30, 31, 40, 44, 69, 71, 101, 108
Nelson, Nels C., 8
New Archaeology, 19

New Deal project, 11. *See also* Chickamauga Basin Project
New York State Museum, 9, 10
NPS. *See* National Park Service

on-site access, 100
ownership, of collection, 29–31, 71, 95

Parker, Arthur C., 8, 9
Peabody Museum of Archaeology and Ethnography, 7, 8, 47, 54, 124
Peale's Museum, 6
permanent loan, 75
photograph/negative: access to, 76; storage of, 70, 89, 90
physical space, for collection, 74, 96
pottery typology, 11
professional responsibilities, in archaeological curation, 33–34, 79–80
project design: collection strategy in, 79, 82; identifying repository in, 80–81, 91; modifications to, 83
public works project, 11. *See also* Chickamauga Basin Project
Putnam, Frederick Ward, 8–9

R. S. Peabody Museum, 47
regional repository, 43
registrar, 54, 55, 59
registration methods, 59
repository: certification of, 40–41; definition of, 3, 4, 45; identifying in project design, 80–81, 91; for records/reports, 88–89; role in project design, 80–83. *See also* curation, long-term; repository, curatorial staff for; repository, type of; repository program
repository, curatorial staff for, 53–54; typical curation-related jobs, 55

repository, type of: academic, 47–48; archives, 50; government, 49–50; historical society, 48; museum, 46–47; regional, 43; tribal museum/cultural center, 48. *See also* repository; repository, curatorial staff for; repository program

repository program: collections management, 51–52; educational program/exhibit, 52–53; research, 53. *See also* repository; repository, curatorial staff for; repository, type of

research design, 79. *See also* project design

Reservoir Salvage Act of 1960, 18

River Basin Salvage Program, 18

Rochester Museum, 9, 10

Sabloff, Jeremy, 8

Salwen, Bert, 108–9

sample, representative: encouraging repository to curate, 111–12; preserving, 108–11

sampling strategy, 109; during fieldwork, 83–84; in laboratory, 85–87

Smithson, James, 6

Smithsonian Institution, 6, 8, 18

Society for American Archaeology (SAA), 9, 34, 44, 108

Society for Historical Archaeology, 34, 44

Society of American Archivists, 56

Squier, E. G., 8

State Historic Preservation Office, 113

storage, of collection: basic standards for, 68; container for, 68–71; document/records, 37, 70, 89; environmental control, 67–68; human remains/sacred objects, 69; physical area for, 67; unacceptable container, 70

Struever, Stuart, 33

Systematic Anthropological Collections program, 43

technician, 55, 56–57

Tennessee Valley Authority (TVA), 13

Thomas, Cyrus, 8

Thomsen, J. C., 5

transfer of title, 60

Tribal Historic Preservation Office, 113

tribal museum/cultural center, as repository, 48

TVA. *See* Tennessee Valley Authority

type collection, 99

Webb, William, 13–14

Willey, Gordon, 8

Works Progress Administration (WPA), 13, 28. *See also* Chickamauga Project

world's fair, 6–7

ABOUT THE AUTHORS AND SERIES EDITORS

Lynne P. Sullivan is curator of archaeology at the Frank H. McClung Museum and research associate professor in the Department of Anthropology at the University of Tennessee. She has more than twenty years experience in making, researching, and curating archaeological collections in the southeastern, midwestern, and northeastern United States. Her main area of expertise is the late prehistory of the Southeast. Dr. Sullivan has worked for university research centers and state, local, and university museums, and she is a member of the Society for American Archaeology's Committee on Curation. She received her doctorate in anthropology from the University of Wisconsin–Milwaukee, where she also earned a certificate in museology.

Dr. Sullivan has published numerous books and articles on southeastern archaeology, the history of archaeology, and the curation of archaeological collections. These include *The Prehistory of the Chickamauga Basin in Tennessee* (University of Tennessee Press, 1995), the two-volume report of the University of Tennessee and Works Progress Administration's excavations before flooding of the Tennessee Valley Authority's Chickamauga Reservoir; and *Grit-Tempered: Early Women Archaeologists in the Southeastern United States* (University Press of Florida, 1999). Between 1987 and 1991, she directed major projects, funded by the National Science Foundation, to inventory, organize, and create a computer database for the archaeological collections at the New York State Museum. More recently, she has worked on "Archaeology of the Native Peoples of Tennessee," a major permanent exhibition at the McClung Museum. She currently is directing a project to create an online archive of photos from the WPA/TVA archaeological projects, funded by the Institute of Museum and Library Services, and serves as editor of *Southeastern Archaeology*.

S. Terry Childs is an archaeologist in the Archeology and Ethnography Program of the National Park Service in Washington, D.C. She is actively involved in federal archaeological curation efforts for objects, records, and reports (e.g., gray literature), including policy formulation. As the coordinator of several national level databases, including the National Archeological Database, she is also concerned about the proper management and curation of archaeological digital data. In her role as leader of the NPS Cultural Resources Web team, Dr. Childs has been instrumental in using the World Wide Web to highlight museum collections and issues related to their management. She just launched a self-motivated course on the Web, called "Managing Archeological Collections: An Online Course" (www.cr.nps.gov/aad/collections), which is designed to complement this book. She was recently appointed chair of the new Committee on Curation by the Society for American Archaeology.

Dr. Childs received her doctorate in anthropology from Boston University in 1986. Her primary field research interest is the Iron Age of Sub-Saharan Africa, particularly involving the anthropology of technology. She has worked in Tanzania, the Democratic Republic of the Congo, Zimbabwe, and Uganda and has published extensively on this work. Her research on collections from Africa and northeastern United States is conducted as a research collaborator with the Smithsonian Institution and a Research Associate with the R. S. Peabody Museum in Andover, Massachusetts.

Larry J. Zimmerman is the head of the Archaeology Department of the Minnesota Historical Society. He served as an adjunct professor of anthropology and visiting professor of American Indian and Native Studies at the University of Iowa from 1996 to 2002 and as chair of the American Indian and Native Studies Program from 1998 to 2001. He earned his Ph.D. in anthropology at the University of Kansas in 1976. Teaching at the University of South Dakota for twenty-two years, he left there in 1996 as Distinguished Regents Professor of Anthropology.

While in South Dakota, he developed a major CRM program and the University of South Dakota Archaeology Laboratory, where he is still a research associate. He was named the University of South Dakota Student Association Teacher of the Year in 1980, given the Burlington Northern Foundation Faculty Achievement Award for Outstanding Teaching in 1986, and granted the Burlington Northern Faculty Achievement Award for Research in 1990. He was selected by Sigma Xi, the Scientific Research Society, as a national lecturer from 1991 to 1993, and he served as executive secretary of the World Archaeological Congress from 1990 to 1994. He has published more than three hundred articles, CRM reports, and reviews and is the author, editor, or coeditor of fifteen books, including *Native North America* (with Brian Molyneaux, University of Oklahoma Press, 2000) and *Indians and Anthropologists: Vine Deloria, Jr., and the Critique of Anthropology* (with Tom Biolsi, University of Arizona Press, 1997). He has served as the editor of *Plains Anthropologist* and the *World Archaeological Bulletin* and as the associate editor of *American Antiquity*. He has done archaeology in the Great Plains of the United States and in Mexico, England, Venezuela, and Australia. He has also worked closely with a wide range of American Indian nations and groups.

William Green is director of the Logan Museum of Anthropology and adjunct professor of anthropology at Beloit College, Beloit, Wisconsin. He has been active in archaeology since 1970. Having grown up on the south side of Chicago, he attributes his interest in archaeology and anthropology to the allure of the exotic (i.e., rural) and a driving urge to learn the unwritten past, abetted by the opportunities available at the city's museums and universities. His first field work was on the Mississippi River bluffs in western Illinois. Although he also worked in Israel and England, he returned to Illinois for several years of survey and excavation. His interests in settlement patterns, ceramics, and archaeobotany developed there. He received his master's degree from the University of Wisconsin–Madison and then served as Wisconsin SHPO staff archaeologist for eight years. After obtaining his Ph.D. from the University of Wisconsin–Madison in 1987, he served as state archaeologist of Iowa from 1988 to 2001, directing statewide research and service programs including burial site protection, geographic information, publications, contract services, public outreach, and curation. His main research interests focus on the development and spread of native agriculture. He has served as editor of the *Midcontinental Journal of Archaeology* and *The Wisconsin Archeologist;* has published articles in *American Antiquity, Journal of Archaeological Research,* and other journals; and has received grants and contracts from the National Science Foundation, National Park Service, Iowa Humanities Board, and many other agencies and organizations.

ARCHAEOBIOLOGY

ARCHAEOLOGIST'S TOOLKIT

SERIES EDITORS: LARRY J. ZIMMERMAN AND WILLIAM GREEN

The Archaeologist's Toolkit is an integrated set of seven volumes designed to teach novice archaeologists and students the basics of doing archaeological fieldwork, analysis and presentation. Students are led through the process of designing a study, doing survey work, excavating, properly working with artifacts and biological remains, curating their materials, and presenting findings to various audiences. The volumes—written by experienced field archaeologists—are full of practical advice, tips, case studies, and illustrations to help the reader. All of this is done with careful attention to promoting a conservation ethic and an understanding of the legal and practical environment of contemporary American cultural resource laws and regulations. The Toolkit is an essential resource for anyone working in the field and ideal for training archaeology students in classrooms and field schools.

Volume 1: *Archaeology by Design*
By Stephen L. Black and Kevin Jolly

Volume 2: *Archaeological Survey*
By James M. Collins and Brian Leigh Molyneaux

Volume 3: *Excavation*
By David L. Carmichael and Robert Lafferty

Volume 4: *Artifacts*
By Charles R. Ewen

Volume 5: *Archaeobiology*
By Kristin D. Sobolik

Volume 6: *Curating Archaeological Collections:
From the Field to the Repository*
By Lynne P. Sullivan and S. Terry Childs

Volume 7: *Presenting the Past*
By Larry J. Zimmerman

ARCHAEOBIOLOGY

Kristin D. Sobolik

ARCHAEOLOGIST'S TOOLKIT
VOLUME 5

AltaMira
PRESS

A Division of Rowman & Littlefield Publishers, Inc.
Walnut Creek • Lanham • New York • Oxford

Acknowledgments

I would like to thank Steve Bicknell, Rick Will, Tom Green, Larry Zimmerman, Kathryn Kamp, Tom Whittaker, Vaughn Bryant, Harry Shafer, Dinah Crader, David Sanger, Tom Alex, Larry Elrich, and Heather McInnis for their help in completing this book.

ALTAMIRA PRESS
A Division of Rowman & Littlefield Publishers, Inc.
1630 North Main Street, #367
Walnut Creek, CA 94596
www.altamirapress.com

Rowman & Littlefield Publishers, Inc.
A Member of the Rowman & Littlefield Publishing Group
4501 Forbes Boulevard, Suite 200
Lanham, MD 20706

PO Box 317
Oxford
OX2 9RU, UK

Copyright © 2003 by ALTAMIRA PRESS

All rights reserved. No part of this publication may be reproduced, stored in a retrieval system, or transmitted in any form or by any means, electronic, mechanical, photocopying, recording, or otherwise, without the prior permission of the publisher.

British Library Cataloguing in Publication Information Available

Library of Congress Cataloging-in-Publication Data

Sobolik, Kristin D. (Kristin Dee)
   Archaeobiology / Kristin D. Sobolik.
     p. cm. — (Archaeologist's toolkit)
   ISBN 0-7591-0401-8 (cloth : alk. paper) — ISBN 0-7591-0023-3 (pbk. : alk. paper)
    1. Archaeology—Methodology. 2. Biology—Methodology. 3. Animal remains (Archaeology) 4. Plant remains (Archaeology) 5. Paleoethnobotany. 6. Archaeology—Field work. I. Title. II. Series.

CC75.7 .S59 2003
930.1'028—dc21   2002015058

Printed in the United States of America

∞™ The paper used in this publication meets the minimum requirements of American National Standard for Information Sciences—Permanence of Paper for Printed Library Materials, ANSI/NISO Z39.48-1992.

CONTENTS

Series Editors' Foreword — vii

1 Introduction — 1
Goals of Archaeobiology • Goals of This Book • History of Research • The Role of CRM in the Development and Future of Archaeobiology

2 Taphonomy — 17
Preservation • Cultural versus Noncultural Assemblages • Conclusion

3 Recovery Techniques — 45
Fine Screening and Flotation • Combination Method • Excavation and Sampling • Conclusion

4 Laboratory and Analytical Techniques — 75
Reference Collections • Botanical Remains • Zooarchaeology • Conclusion

5 Integration — 99

References — 119

Index — 131

About the Author and Series Editors — 137

SERIES EDITORS' FOREWORD

The Archaeologist's Toolkit is a series of books on how to plan, design, carry out, and use the results of archaeological research. The series contains seven books written by acknowledged experts in their fields. Each book is a self-contained treatment of an important element of modern archaeology. Therefore, each book can stand alone as a reference work for archaeologists in public agencies, private firms, and museums, as well as a textbook and guidebook for classrooms and field settings. The books function even better as a set, because they are integrated through cross-references and complementary subject matter.

Archaeology is a rapidly growing field, one that is no longer the exclusive province of academia. Today, archaeology is a part of daily life in both the public and private sectors. Thousands of archaeologists apply their knowledge and skills every day to understand the human past. Recent explosive growth in archaeology has heightened the need for clear and succinct guidance on professional practice. Therefore, this series supplies ready reference to the latest information on methods and techniques—the tools of the trade that serve as handy guides for longtime practitioners and essential resources for archaeologists in training.

Archaeologists help solve modern problems: They find, assess, recover, preserve, and interpret the evidence of the human past in light of public interest and in the face of multiple land use and development interests. Most of North American archaeology is devoted to cultural resource management (CRM), so the Archaeologist's Toolkit focuses on practical approaches to solving real problems in CRM and public archaeology. The books contain numerous case studies from all parts of the continent, illustrating the range and diversity of applications. The series emphasizes the importance of such realistic considerations as budgeting, scheduling, and team coordination. In addition, accountability to the public as well as to the profession is a common theme throughout the series.

Volume 1, *Archaeology by Design*, stresses the importance of research design in all phases and at all scales of archaeology. It shows how and why you should develop, apply, and refine research designs. Whether you are surveying quarter-acre cell tower sites or excavating stratified villages with millions of artifacts, your work will be more productive, efficient, and useful if you pay close and continuous attention to your research design.

Volume 2, *Archaeological Survey*, recognizes that most fieldwork in North America is devoted to survey: finding and evaluating archaeological resources. It covers prefield and field strategies to help you maximize the effectiveness and efficiency of archaeological survey. It shows how to choose appropriate strategies and methods ranging from landowner negotiations, surface reconnaissance, and shovel testing to geophysical survey, aerial photography, and report writing.

Volume 3, *Excavation*, covers the fundamentals of dirt archaeology in diverse settings, while emphasizing the importance of ethics during the controlled recovery—and destruction—of the archaeological record. This book shows how to select and apply excavation methods appropriate to specific needs and circumstances and how to maximize useful results while minimizing loss of data.

Volume 4, *Artifacts*, provides students as well as experienced archaeologists with useful guidance on preparing and analyzing artifacts. Both prehistoric- and historic-era artifacts are covered in detail. The discussion and case studies range from processing and cataloging through classification, data manipulation, and specialized analyses of a wide range of artifact forms.

Volume 5, *Archaeobiology*, covers the analysis and interpretation of biological remains from archaeological sites. The book shows how to recover, sample, analyze, and interpret the plant and animal remains most frequently excavated from archaeological sites in North America. Case studies from CRM and other archaeological research illustrate strategies for effective and meaningful use of biological data.

Volume 6, *Curating Archaeological Collections*, addresses a crucial but often ignored aspect of archaeology: proper care of the specimens and records generated in the field and the lab. This book covers strategies for effective short- and long-term collections management. Case studies illustrate the do's and don'ts that you need to know to make the best use of existing collections and to make your own work useful for others.

Volume 7, *Presenting the Past,* covers another area that has not received sufficient attention: communication of archaeology to a variety of audiences. Different tools are needed to present archaeology to other archaeologists, to sponsoring agencies, and to the interested public. This book shows how to choose the approaches and methods to take when presenting technical and nontechnical information through various means to various audiences.

Each of these books and the series as a whole are designed to be equally useful to practicing archaeologists and to archaeology students. Practicing archaeologists in CRM firms, agencies, academia, and museums will find the books useful as reference tools and as brush-up guides on current concerns and approaches. Instructors and students in field schools, lab classes, and short courses of various types will find the series valuable because of each book's practical orientation to problem solving.

As the series editors, we have enjoyed bringing these books together and working with the authors. We thank all of the authors—Steve Black, Dave Carmichael, Terry Childs, Jim Collins, Charlie Ewen, Kevin Jolly, Robert Lafferty, Brian Molyneaux, Kris Sobolik, and Lynne Sullivan—for their hard work and patience. We also offer sincere thanks to Mitch Allen of AltaMira Press and a special acknowledgment to Brian Fagan.

LARRY J. ZIMMERMAN
WILLIAM GREEN

1

INTRODUCTION

Archaeobiology is the analysis and interpretation of biological remains from archaeological sites. In the most inclusive sense, archaeobiology refers to the remains of anything that was once living: animal bone, shell, scales, plants, seeds, pollen, phytoliths, charcoal, parasites, hair, and the list could go on. In this book, I use the term *archaeobiology* in a more limited sense to refer to the analysis of animal and plant remains. I do this to separate what I discuss in this book from the analysis of human skeletal material, usually referred to as *bioarchaeology*, and other specialized biological remains, such as scales, parasites, hair, and organic residue, which are not as frequently observed in archaeological deposits. I want to discuss how to recover, sample, analyze, and interpret the most frequently excavated biological remains from archaeological sites because they are what most archaeologists encounter on a day-to-day basis.

Archaeobiology is a relatively new term. Implicit in its definition is the integration of the analysis of plant and animal remains, a synthesis that until recently was rarely attempted or achieved. Looking at the definitions of the two major disciplines incorporated within archaeobiology, zooarchaeology and paleoethnobotany, reveals their inherently cohesive nature. *Zooarchaeology* is defined as the study of animal remains from archaeological sites to understand the relationship between humans and animals at the subsistence and environmental levels (Reitz and Wing 1999). *Archaeobotany* was first defined by Ford (1979) as the collection and identification of botanical remains from archaeological sites. This term is not used as frequently as *paleoethnobotany*, which is the study of the relationships between plants and humans in prehistoric times using the collection, identification, analysis, and interpretation of plant materials recovered from archaeological sites. The difference between the two words is that *archaeobotany* refers exclusively to the technical side of such research, whereas *paleoethnobotany* refers to the scientific and interpretive arena.

GOALS OF ARCHAEOBIOLOGY

Two main goals are central to archaeobiological research. The first is to analyze the relationships between humans and plants and animals and their effect(s) on each other. Interaction between humans and the environment (including plants and animals) does not proceed in one direction only; humans influenced the environment in as many ways as the environment may have constrained or provided opportunities for humans. For example, archaeobiologists analyze various results of agricultural and animal domestication, such as effects on human health, impacts on the environment and landscape, biological changes in plants and animals, and changes in human interactions and cultural systems associated with the development of domestication.

A case study of the analysis of environmental impacts due to agricultural practices was conducted by O'Hara and colleagues (1993) in central Mexico. The authors test the hypothesis that a return to traditional, prehistoric agricultural methods would be better for the environment than plow agriculture with draft animals, as used in historic and modern times. They collected twenty sediment cores from Lake Patzcuaro in the highland regions northwest of Mexico City to analyze soil erosion through time. Their data, which include pollen and diatom studies from other cores, indicate that three periods of extensive erosional rates in the past equal and surpass the erosional degradation that is taking place today in the region due to modern agricultural techniques. Based on this analysis of humans and the environment, the authors conclude that a return to more traditional agricultural techniques would not necessarily be better for the environment of the region.

In another example, Wagner (1996) examines how an agriculturally dependent prehistoric Fort Ancient community in Ohio adapted to short- and long-term food shortages. The Fort Ancient people were "consummate maize agriculturalists" (Wagner 1996:256) who followed a seasonally based subsistence round in which communities gathered in villages with permanent houses during the spring and summer, dispersing to winter hunting camps during the fall and winter. Plant and animal remains from these sites indicate that the people focused on a few resources, such as deer, elk, black bear, corn, bean, and chenopod. In times of stress or lack of preferred resources, the Fort Ancient people relied more heavily on a wider variety of resources that were kept in storage pits. Wagner analyzed the plant remains from seasonal storage pits to conclude that during all seasons, the same types of plants were stored, both for use as food and for seed crops. Pits were also buried and concealed when villages were abandoned and families dispersed to their winter camps. Food storage of diverse plants was the main coping mechanism used to avoid short- and long-term food shortages in this Fort Ancient community.

These research directions show there are complex relationships between humans and their environment. Archaeobiological analyses can help archaeologists understand these complexities and integrate a larger database into analyses and interpretations.

The second goal of archaeobiology is to place archaeologically derived information in its anthropological context. Archaeobiology is not a list of the types of charcoal found in a hearth or the kinds of animal bone identified from a shell midden. That information is meaningless unless it is applied to the interpretation of the activities of humans at the site or the environment in which humans lived. Cultural context is imperative when analyzing archaeobiological material. Without it, the research is nothing more than a list to be placed in an appendix to a report. Like artifacts, biological remains supply clues about lifestyles and cultural patterns of past peoples. Plants and animals were not just collected at random by a group of people; they were obtained and used for specific purposes.

For example, plant and animal products were used for fuel, and often specific criteria were used in their selection, such as availability, season, inherent heat value and combustibility, and social rules based on taboo or ritual. Archaeological and ethnographic records indicate that not all potential fuels in an area were actually chosen or used in all instances by a culture. To illustrate such a case study, in conjunction with two colleagues I analyzed plant and animal remains from twenty-eight fire pits from indoor and outdoor areas of a large pueblo in the Mimbres River Valley of southwestern New Mexico (Sobolik et al. 1997). The study was done to test the hypothesis that cooking was only conducted in outdoor fire pits and that indoor fire pits were mainly for warmth. Our analysis of bone, seeds, and charcoal from the fire pits indicates that, in fact, cooking was performed in both indoor and outdoor fire pits. However, hardwood fuel sources, such as oak, walnut, ash, cottonwood, willow, and boxelder, were chosen more frequently for indoor fires. Hardwoods produce a longer-burning fire with less smoke and would have been ideal for slow cooking methods or for longer-burning fires needed for warmth. Softwood fuel sources, such as juniper, pinyon pine, ponderosa pine, and Douglas fir, were chosen more frequently for outdoor fires. Softwoods tend to burn hotter and more quickly and produce more smoke because of their pitch content. Softwoods would have been ideal for quicker cooking methods, and burning softwoods in outdoor, open areas would be optimal considering the increased amount of smoke. Therefore, Mimbres people were selectively choosing particular wood types for different burning and cooking situations.

Another example of the usefulness of archaeobiological data to determine cultural patterns of prehistoric populations is in the analysis of seasonality of site or regional occupation. For example, Sanger (1996) tested two hypotheses on prehistoric settlement patterns in the Gulf of Maine. One hypothesis stated that prehistoric people resided along the coastal regions of Maine, particularly along river mouths during summer months, and moved to interior regions during the winter. Support for this hypothesis rests in ethnographic and historic documents that state that aboriginal peoples traded with Europeans during the summer as the former occupied coastal areas. Another hypothesis stated that prehistoric peoples moved extensively around the region and resided in coastal areas of Maine during all seasons. To test these hypotheses, Sanger used extensive seasonality data obtained through the analysis of growth rings from 874 softshell clams from sixteen coastal shell midden sites. The data indicate that coastal sites were occupied during all seasons and disprove the hypothesis that prehistoric peoples occupied the coastal areas only during the summer. The study also suggests an interesting modification of the second hypothesis: the possibility that year-round seasonality on the coast is evidence for two prehistoric populations, an interior group and a coastal group. Sanger's zooarchaeological study thus lays the groundwork for future research.

GOALS OF THIS BOOK

The purpose of this book is to discuss the best and easiest ways to retrieve, identify, sample, analyze, and interpret archaeobiological remains from archaeological sites. In essence, I provide a general cookbook of tried and true recipes, containing step-by-step directions on how to treat archaeobiological material. Just like recipes, these directions can be modified according to the type of site or material that you are analyzing. I am not suggesting that there is only one way to recover small fish bones from soil samples, quantify charcoal fragments, or interpret fiber remains. Instead, I present what I believe are the best and easiest ways to do these things, point you in the direction of other literature sources, and provide case studies from diverse areas and sites in North America to illustrate specific techniques or interpretations.

I also address a problem that has long been apparent in archaeobiological work: the inferiority complex. Archaeobiologists have complained (some loud, some long) that their work is not respected and tends to be placed in the appendix sections of archaeological reports or incorporated into other chapters in tabular form only. In truth, archaeobiologists are their own worst enemies. If they treat their data as an important cultural entity, then other archaeologists will treat it so, too. Archaeobiologists should be able to incorporate their data into "the big picture" and to reconstruct what their data reveal about prehistoric life. In many instances, archaeobiological analyses tend to be conducted by different researchers, and the results are reported separately. If an integration of the results is undertaken, it is the archaeologist, usually not a biological expert, who integrates these conclusions. Because such synthesizers are often unaware of the scope and limitations of each data set, their interpretations often lack potential insight or authority (Sobolik 1994).

In this volume, I emphasize the importance of archaeobiologists being involved in excavations from the beginning. In this way, they can help determine research design, how samples are collected, and which samples will be useful to help answer the research questions being asked. The more involved archaeobiologists are, whether in planning the research design or actually excavating the site, the more informed they will be when analyzing and interpreting the data and reconstructing the prehistoric picture. When archaeobiologists are involved in helping ask and answer questions and placing their data into a cultural framework, they won't be relegated to the periphery.

That is the ideal case. In the real world, especially in cultural resource management (CRM), research questions often aren't established or refined before a site is tested or excavated (see Toolkit, volume 1). Archaeobiologists often receive boxes of bone, bags of soil for pollen analysis, or flotation samples for seed and charcoal identification without having been a part of the original planning or excavation. The archaeobiologist at this point needs to obtain information on the site and how it was excavated and any contextual information the archaeologist can provide. The archaeobiologist may need to determine his or her own research questions that may or may not end up fitting with the overall site analysis and interpretation. Archaeobiological data and interpretations need to be more highly valued at all stages of research. By providing a good, sound data set and potentially innovative analyses and interpretations, archaeobiological data should be considered an integral component of the overall research design.

Ultimately, it is best to incorporate archaeobiologists in all stages of a research and/or CRM project. However, there are a number of stages in the recovery, analysis, and interpretation of archaeobiological remains that can be conducted by archaeologists who are trained specifically in such techniques. I provide a summary table to indicate which stages of a project can be conducted by trained archaeologists and which need to be conducted by archaeobiologists (table 1.1). In

Table 1.1. Stages of a Project in which Archaeobiologists Must Be Involved*

Project Stages	Must Be Involved	Not Necessary Although Optimal
Project design		X
Analysis of taphonomy		X
Sampling design		X
Material recovery		X
Coarse screening		X
Fine screening		X
Sediment collection for flotation		X
Flotation		X
Sediment collection for pollen and phytoliths		X
Material analysis	X	
Quantification and analysis of plant remains	X	
Quantification and analysis of animal remains	X	
Pollen and phytolith processing and analysis	X	
Archaeobiological interpretation	X	

*It is always optimal, from a research project perspective, to include archaeobiologists at *all* stages of design, analysis, and interpretation to obtain the most synthetic and cohesive product.

sum, archaeobiologists should be involved in the research design process and in analysis and interpretation of plant and animal remains recovered from archaeological sites. Recovery of archaeobiological material can usually be conducted by trained archaeologists, although an understanding of potential problems and important techniques is essential. I summarize potential problems and important techniques in this book.

HISTORY OF RESEARCH

The New Archaeology of the 1960s, with its focus on cultural ecology, was significant in the development of archaeobiology. Archaeologists began to systematically save biological remains from archaeological sites as questions regarding diet, paleoenvironment, and ecology became important. With the recovery of more biological remains, due in part to methodological advances such as fine screening and flotation, came an increase in the quality of analyses and interpretation.

Although archaeobiology first became a significant focus for archaeologists as the New Archaeology was dawning, some excellent earlier studies of cultural patterns evidenced through biological remains set the stage for later systemic analyses. However, these studies tended not to be integrated with the data from other assemblages and were mainly conducted by specialists in the field. Paleoethnobotanists or botanists analyzed the plant remains from archaeological sites, and zooarchaeologists or zoologists analyzed the animal remains. These analyses usually were not integrated into the overall report, as is common with archaeobiological analyses conducted today.

PALEOETHNOBOTANY

Prior to the 1890s, the few paleoethnobotanical studies that were conducted were written mainly by botanists or people interested in natural history. This changed during the early 1900s when the emerging field of anthropology began training ethnologists to work with Native Americans on reservations and record what they supposed were the last bits of information still available about the Native American's past cultures and lifeways. When the early botanists studied North

American Indians, their purpose had been mostly utilitarian; all they wanted to record was information about plants and how those plants could be used in the present. On the other hand, early ethnologists collected several types of ethnobotanical data from the people they studied. The anthropologists' focus was on the Native Americans' point of view about the plants they used and how these plants fit into their view of the universe. Searching for a utilitarian value in the plants was not important to the early ethnologists.

A major advance in paleoethnobotanical work occurred at the Columbian Exhibition in Chicago in 1893. Part of the fair focused on the lives and artifacts of the North American Indian. Many exhibits showed different aspects of the Indians' life, including their uses of native plants. J. W. Harshberger examined dried plant materials from caves in Colorado for display at the fair. As a result of that study, he coined the term *ethnobotany* for this type of research (Harshberger 1896). After the fair, interest in ethnobotany among museums, government agencies, and universities increased. Two agencies that funded large numbers of ethnobotanical projects were the U.S. National Herbarium and the U.S. Department of Agriculture. Universities began ethnobotanical studies around the turn of the century, and the first Ph.D. in ethnobotany was awarded to David P. Barrows (1900) from the University of Chicago on the ethnobotany of the Coahuila Indians of Southern California. Barrows stressed that ethnobotanical studies must go beyond the applied or economic value of a plant and focus also on the role plants play in a group's social, religious, and folklore practices.

Melvin Gilmore's (1919) study of plant use by Plains Indian tribes was the first to note that even though some of these tribes were hunters and gatherers, their use of wild plants led to considerable modification of the environment. For example, he noted that groups often introduced plants to new regions. Through burning, Indians controlled certain weedy plants native to a particular region, and by encouraging other plants to grow, they increased the available quantity of desired plant products such as seeds or tubers.

More universities recognized the field of ethnobotany by the 1930s. In 1930, Edward Castetter established an M.A. program in ethnobotany within the Department of Biology at the University of New Mexico. Castetter and his students began to record the ethnobotany of the Indians living in the Southwest. In the late 1930s, R. E. Schultes established a program in ethnobotany at Harvard University with an emphasis on the search for new plants with medicinal merit.

Schultes and his students tended to focus mainly on the ethnobotany of Indians living in Central and South America.

In the mid-1930s, the University of Michigan created the Ethnobotanical Laboratory within the Museum of Anthropology. Gilmore and later Volney Jones headed this program, in which studies focused on plant remains from archaeological sites. In a lecture at the meeting of the American Association for the Advancement of Science in 1931, Gilmore (1932) described his research and requested that people save and send him plant remains from archaeological sites. Material from all over North America began to arrive at the lab for analysis. In most cases, he was permitted to keep the materials and send back only a report on what he had found. Gilmore believed the geographic influences and the physical environment must profoundly impact human habits and cultures. Unless the physical environment can be visualized, he believed cultural patterns could never be understood.

Gilmore's successor, Volney Jones, is often cited in archaeology and is considered the father of paleoethnobotany. Jones was head of the Ethnobotanical Laboratory for more than twenty-five years and analyzed a large number of botanical samples from sites in the eastern and midwestern United States during the 1940s and 1950s. The most significant site was Newt Kash Hollow, the first major paleoethnobotanical study conducted east of the Mississippi River (Jones 1936). In that study, Jones examined plant remains from Early Woodland deposits (later dated to ca. 700 B.P.) and reported at least eight native plants that he felt were cultivated or semicultivated. He was the first to report physical evidence of tobacco use in a prehistoric site east of the Rockies for the Early Woodland period, and he set the standard for explaining early Eastern Woodland subsistence patterns for years to come.

Other important questions then and now revolve around the origins of agriculture, which plants were domesticated, what was the character of the paleoenvironment, and how humans used their landscape. As plant remains from archaeological sites became important for testing hypotheses, other significant areas of research developed around methodological issues: What are the best techniques for recovering various plant parts from sites? How do we quantify the material that is recovered? How do we compare diverse data sets? Today, paleoethnobotanists deal with a large number of issues ranging from the technical aspects of recovery and identification to the interpretation of plant remains and their importance in broad-scale questions.

ZOOARCHAEOLOGY

Robison (1978) divided the history of zooarchaeology into three main time periods: Formative, Systematization, and Integration. The Formative period (ca. 1880–1950) saw the initial analyses of faunal material from archaeological sites. At that point, archaeologists were not systematically collecting faunal remains from sites, and the few analyses that were being conducted tended to be done by zoologists who were interested in the material for biological and environmental reconstruction rather than for archaeological purposes. Zooarchaeological research thus tended to be reported in biological publications. Archaeologists, when interested, tended to focus on one or two species, modified bone tools, or remains associated with human burials. Early studies tended to be descriptive in nature, although some studies foreshadow the types of questions and directions of study zooarchaeologists would take in the future. Such early work includes analysis of vertebrates and invertebrates from a Maine shell midden site that includes dietary hypotheses on the importance of different species based on their abundance (Loomis and Young 1912) and research on shells from an Arizona pueblo to determine trading routes (Fewkes 1896).

In the Systematization period (ca. 1950–1960) archaeologists started looking at faunal remains as a means of obtaining information on cultural behavior and adaptation, although theory and methods were starting to be designed and implemented. The most frequently cited article in zooarchaeological literature (White 1953) introduces the quantitative concept of minimum number of individuals (MNI), which is the most frequently used quantification method used on faunal material. MNI determines the minimum number of each species present at a site, and it can be analyzed according to the site as a whole or in relation to separate units, strata, or levels at a site. MNI is discussed in depth in chapter 4.

Lawrence (1957) urged analysts to change their focus from mere identification to interpretation so that meaningful and stimulating information could be obtained from faunal remains. During this period, the results of long-term, large-scale, integrative archaeological studies were being reported (e.g., Izumi and Sono 1963; Braidwood and Braidwood 1982), and the importance of faunal remains to archaeological interpretation was more widely recognized.

Zooarchaeology specialists started collecting and analyzing samples. Specialists included T. E. White, John Guilday, Paul Parmalee, and Stanley Olsen, who started to train students as zooarchaeologists. These specialists advanced zooarchaeological studies and allowed archaeologists to realize the amount of information that can be gained through the analysis of faunal material. The collections of faunal remains from excavations began to increase, and analyses started to appear in archaeological reports, although mainly as appendices. Zooarchaeology became a recognized and important field within archaeology.

All of these ideas came together during the Integration period, from the 1960s to the present, as the New Archaeology was promoted and as ecological approaches remain prevalent. Cultural ecology and environmental anthropology are the main frameworks of many analyses conducted today as zooarchaeologists integrate their research with other subfields of archaeology. Winters's (1969) analysis of faunal remains from sites of the Riverton Culture has been cited as the first significant zooarchaeological analysis of the New Archaeology era. Another important analysis was conducted by Smith (1975) on Mississipian adaptations in Missouri. He analyzed the remains from different site types in the uplands, lowlands, and swamps. He observed that the prehistoric peoples tended to have a base camp located on the ecotonal region between different microenvironmental areas. They would then exploit the different environments, depending on the season and the abundance of resources that area could provide.

The key issues addressed today by zooarchaeologists include (1) *taphonomy*, encompassing site formation processes, middle-range research, preservation and modification of site artifacts and ecofacts, and determination of cultural and noncultural site components; (2) *methodology*, entailing quantification, recovery, identification, and sampling; (3) *cultural ecology*, involving the relationship between humans and the environment, domestication of animals (which also has a strong biological component), subsistence strategies, human evolution, and human cultural lifeways; and (4) *biology*, encompassing paleoenvironmental reconstruction and ecology and morphology of various animal species.

THE ROLE OF CRM IN THE DEVELOPMENT AND FUTURE OF ARCHAEOBIOLOGY

The advent and development of CRM have taken archaeobiological analyses into the mainstream. CRM is a legal construct developed to assess and analyze cultural resources that may be impacted by economic and technological advances. Because plant and animal remains

1.1. CASE STUDY: ARCHAEOBIOLOGICAL INTEGRATION IN ANALYSIS OF CRM PHASE III EXCAVATIONS AT THE LITTLE OSSIPPEE NORTH SITE, MAINE

Archaeology at the Little Ossippee North site illustrates both a synthetic CRM Phase III project, which integrates archaeobiology at all stages of analysis (Will et al. 1996), as well as a contract-based project in which specific aspects can be used as research-focused projects (Asch Sidell 1999; Sobolik and Will 2000). Following Phase I and II study, the site was determined to be culturally significant and eligible for nomination to the National Register of Historic Places. Phase III excavations at the site were undertaken by Archaeological Research Consultants, Inc., as part of a federal dam relicensing application by Central Maine Power to mitigate anticipated erosion damage to the site over the term of the project license.

The Little Ossippee North site is situated at the confluence of the Little Ossippee and Saco Rivers in western Maine. Over one hundred square meters were excavated, representing less than 4 percent of the total site area. A large number and diversity of prehistoric remains were uncovered, including ceramics, lithics, and hearth features, as well as a diversity of archaeobiological remains, including animal bones, seeds, corn cobs, and charcoal. Geomorphological and soil analyses identified site formation processes, alluvial stratigraphy, fire history, and prehistoric flooding events. Three separate occupations were observed during excavation (figure 1.1): a Ceramic period occupation, ca. 1000 B.P. (horizon 2A); a Late Archaic period occupation, ca. 3000 B.P. (horizon 3A); and an Early Archaic period occupation, ca. 8000 B.P. (horizon 5A). Archaeobiological material was recovered from all three occupations, and analysis and interpretation of the remains were integrated into the discussion of each prehistoric occupation, not separated from site and context analysis and placed in separate chapters or in an appendix section.

During careful excavation, bone remains were recovered as a portion of large matrix samples containing many tiny fragments of bone. Bone was hand sorted from the matrix as well as recovered during coarse and fine screening. Plant remains were mainly recovered from flotation samples taken from features. Archaeobiological remains were preserved in the acidic soils because they were calcined and carbonized in addition to being covered with anaerobic alluvial deposits. These remains enhanced interpretations of the various cultural occupations at the site.

For example, the Ceramic period occupation contained many cultural features, which yielded 88.45 grams of carbonized plant remains. Flotation taken from the horizon 2A yielded an additional 174.47 grams of plant remains. Identification included an impressive list of maple, alder, birch, beech, pine, and oak wood and bark; acorn, beechnut, and hazelnut nutshells; and bunchberry, tick trefoil, strawberry, huckleberry, grass, cherry, raspberry, elderberry, blueberry, and grape seeds. The variety and amount of botanical remains recovered from the Ceramic occupation of the Little Ossippee North site is unmatched by any site in Maine. In addition, a large hearth was uncovered and 18.5 l of matrix was floated. The light fraction contained large amounts of carbonized wild plant food remains as well as two carbonized corn cob fragments. A radiocarbon date from hearth charcoal (1,010 ± 60 B.P.) indicates that these cob fragments may be the earliest evidence of domesticated plants in southern Maine. A recent AMS date on the actual corn itself returned a 570 ± 40 B.P. date, still the oldest corn in Maine (Richard Will, personal communication).

Analysis of 375 calcined turtle bone fragments from the horizon 2A in the southern portion of the Ceramic period occupation indicates that small, juvenile musk turtles were taken as food items, probably during a single, short-term procurement event. Size and age of the turtles provide another line of evidence, in conjunction with the wide variety of seeds, indicating the site was occupied during the fall. Adult turtles observed in the assemblage include the painted turtle and the wood turtle. These turtles were most likely obtained individually and in small numbers.

Analysis of the archaeobiological remains from the Little Ossippee North site was incorporated into additional research projects. The plant remains were included as a central database in Nancy Asch Sidell's (1999) analysis of the prehistoric subsistence patterns and paleoenvironment of Maine. She focused on the changing environment of Maine in the past and how subtle environmental differences would have affected human occupational patterns. She also presented evidence for the earliest agricultural practices in Maine, discussing where agriculture would have been feasible and how such practices would have affected prehistoric populations. Her analysis, in particular, illustrates that preservation of the potentially important botanical database is possible in Maine, particularly in alluvially deposited sites.

I analyzed the calcined turtle remains from the Little Ossippee North site. I was struck by the small size of the turtle bones, which led me to conduct an experimental project to assess the actual size of the turtles upon death (Sobolik and Will 2000). It is known that bone shrinks upon calcination; therefore, the turtle bones would have been larger and the turtles possibly older at death. The turtle-burning experiment involved burning modern turtle bone in an oven and in an open fire more similar to prehistoric conditions. The experiment revealed that turtle bone shrinks about 16 percent when it is burned. Therefore, 16 percent needs to be added to calcined turtle bone measurements to determine actual size of turtles at death. This addition still indicated that the turtles were small and juvenile. An analysis of turtle ecology suggested that the musk turtles were most likely captured as a group by prehistoric people when the juveniles were nestled in the muddy bottom of the slower-moving Little Ossippee River. A large quantity of small turtles would have made a more effective contribution to prehistoric diet than a few juvenile turtles caught singly over a longer period of time.

Analysis of the Little Ossippee North site illustrates CRM at its best. Archaeobiological recovery, analysis, and interpretation were essential to project goals and were incorporated at all phases of the project. Archaeobiological interpretations were included as an integral component in the discussion and write-up of cultural occupations; they were not relegated to a separate chapter or appendix with little or no integration. In addition, archaeobiological analyses and interpretations were used for other research projects that expanded from the original CRM project.

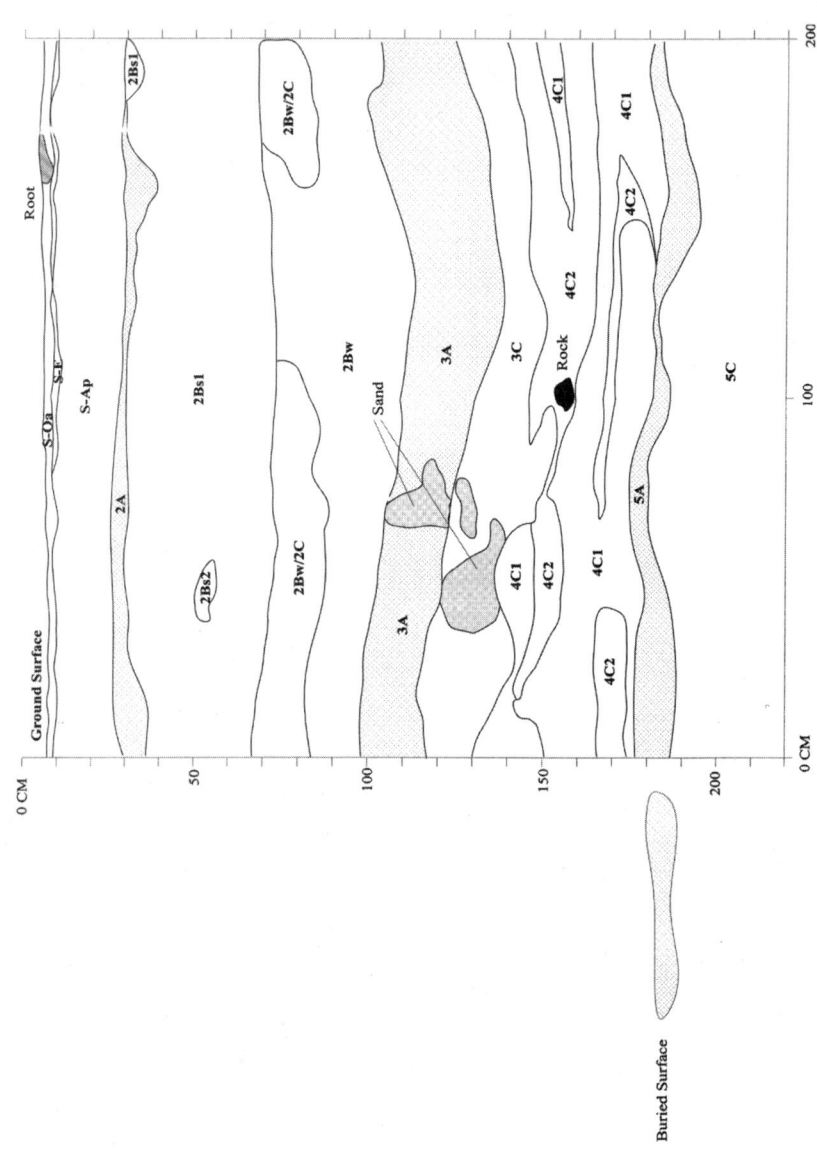

Figure 1.1. Stratigraphy of the Little Ossippee North site, Maine, illustrating three cultural horizons and time periods (2A, 3A, 5A). (From Will et al. 1996; courtesy of James Clark.)

are an integral part of cultural resources, their analysis is essential for any CRM project and report. Plant and animal remains are recovered during survey, testing, and excavation of potentially impacted archaeological sites. If plant and animal remains aren't recovered, assessments of the preservation potential of such remains are made.

Other archaeological projects, which may not be CRM driven, do not legally need to recover, analyze, and interpret archaeobiological material. Research-driven archaeological projects tend to focus on artifacts and remains that will help bolster or disprove a particular hypothesis. Artifacts and remains that are not necessary data for a particular hypothesis may not be recovered; if such artifacts and remains are recovered, they may not be analyzed. Artifacts and remains that are recovered but not necessary for hypothesis testing tend to sit in bags and boxes in laboratories or museums until they are needed as datasets for the testing of other hypotheses. In this scenario, such artifacts and remains (many of which are archaeobiological material) may never be analyzed.

Such is not the case for archaeobiological remains recovered as a part of CRM projects. As a case example, I review in sidebar 1.1 the goals, procedures, and results of a CRM project conducted by Archaeological Research Consultants, Inc., for Central Maine Power at the juncture of the Little Ossippee North and Saco Rivers. This case study not only illustrates the potential for recovery, analysis, and interpretation of archaeobiological remains as part of a CRM project but also highlights the importance of data recovered from CRM projects as tools to assess research questions and hypotheses regarding prehistoric lifeways in general. Archaeobiological analyses do not need to be limited to short discussions or appendices in CRM reports. Results from CRM-driven archaeobiological analyses may be an integral component for broader-ranging research questions that analyze prehistoric culture on a more global scale.

In addition, archaeobiological materials should always be recovered from research-driven archaeological projects. If they are not pertinent to the particular hypothesis or goals set forth by the researcher, then they should be properly curated and stored for future use. Archaeological remains are a nonrenewable resource, and excavation is entirely destructive. Therefore, recovery of archaeobiological remains from all archaeological sites is essential, not only for our present understanding of the past but also for the future's understanding of the past.

2

TAPHONOMY

Taphonomy is the study of site formation processes as they affect the preservation, inclusion, and distribution of biological components from archaeological sites. Efremov (1940), a paleontologist, defined the term *taphonomy* (*taphos*, tomb; and *logos*, law) as the study of the transition of animal remains from the biosphere to the lithosphere. Because the term was defined by a paleontologist, most taphonomic studies focus on recovery and analysis of bone remains, although taphonomy of plant remains is just as important for archaeological interpretations. Since its inception, the definition of taphonomy has been altered to fit the needs of both paleontology and archaeology. Today, the study of taphonomy has increased to include the post fossilization period (Lyman 1994), and the scope of taphonomy now includes the history of biological remains, including their collection and curation.

Because of the broad array of biological, environmental, and human agents affecting preservation, inclusion, and distribution of biological components in archaeological sites, taphonomy should be the first issue that all archaeologists think about before they even begin to recover these remains. Before walking onto a site, before screening for bones, before floating for seeds and charcoal, the archaeologist should be thinking about all the factors that could be responsible for the assemblage and could have affected the archaeological site overall. In this chapter, I discuss why it is important for the archaeologist to be aware of taphonomy at every step of investigation, what types of taphonomic factors could be influencing the assemblages, and how to account for these factors in overall analysis and interpretation.

"Taphonomic studies of modern analogs have shown the complexity of the processes that affect bones; but rather than despair, we should recognize that the processes likely to have operated at a particular archaeological site, and the likely range of variability in the patterned effects of those processes, are specifiable" (Bunn 1991:438). Every archaeologist should be able to specify which factors have influenced the assemblage he or she is analyzing and to incorporate that into site interpretations.

Archaeologists often determine taphonomy of a site or assemblage through the aid of experimental archaeology and ethnoarchaeology. Ethnoarchaeology and experimental archaeology provide methods for testing hypotheses about site formation processes and artifact assemblage preservation, movement, and origin. In experimental archaeology, scientists can conduct staged experiments as well as observe modern natural factors to examine the environmental and cultural elements that affect archaeological sites or assemblages. In this way, we can formulate ideas on how archaeological sites actually become formed and what types of impacts various factors have on the formation process.

For example, in a wood rat bone movement study, Hoffman and Hays (1987) controlled the experiment by introducing bone from six different animal taxa (dog, turtle, catfish, opossum, raccoon, and deer) into an active wood rat den in Tennessee to observe movement and modification of the bones by wood rats through time. The authors observed that six months after the bones had been introduced, 137 elements (46 percent) had been moved at least one meter; and of those moved, only 76 elements (55 percent) were recovered. The wood rat(s) had no preference for type of element, indicating that bone structure and morphology are unrelated to selection; the wood rat(s) moved bones as small as 0.3 gram (a turtle long bone) and as large as 101 grams (a deer pelvis). This research revealed that wood rats can be an important taphonomic factor in site formation and artifact movement, one that can be nonselective and can readily move cultural material from its primary context to a secondary location.

However, in a study by Bocek (1986), observation of modern rodent ecology was used to determine rodents' potential effects on archaeological sites. In that study, the experiment was not controlled; observation and interpretation of natural conditions were made. The ecology and burrowing behavior of a variety of rodents were studied to assess their possible effects on two central California archaeological sites in which rodent disturbance had been observed. In grassland environments, much of the soil surface has been disturbed by rodents, and burrowing behavior is a major soil formation factor. Rodent activity also tends to move soil and materials vertically, with smaller material accumulating near the surface and larger material moving to greater depths. The author estimates that archaeological sites have been "250% reworked by rodent activity" (p. 600) and that rodents have segregated the site into two stratigraphic horizons: an upper horizon of small materials and a lower horizon of large materials, which could easily be misinterpreted as cultural zones.

Another way in which archaeologists and archaeobiologists determine taphonomy of sites or assemblages is through the use of ethnoarchaeology. *Ethnoarchaeology* is the study of living human communities by archaeologists for the purposes of answering archaeologically derived questions. These studies are particularly useful to taphonomists because they document cultural processes involved in site formation. For example, in a study of modern Aché hunter-gatherer camps in the neotropical forests of Paraguay, Jones (1993) indicated that short-term camps exhibit a distinctive pattern that contrasts with long-term camps: The short-term camps have a fire-focused assemblage pattern in which debris is in primary context. Jones asked what types of patterns archaeologists should look for to identify short-term prehistoric camps. The author analyzed six one-night Aché camps that contained five to six fires per site for twenty-four to twenty-six people. Most of the activity at the camp took place within 1.5 meters of the fires, including butchering, cooking, and eating mainly small-sized animals. After the Aché left the camp, the main material remains were bones left in primary context; the short-term camps were not cleaned. The author suggests this pattern may be observed archaeologically when the group was foraging for immediate consumption at short-term camps and mainly used small animals. We can then infer that short-term camps of prehistoric people with a similar cultural pattern may contain artifacts in primary context, whereas long-term camps may exhibit more assemblages in secondary context.

There are limits to the use of ethnoarchaeological data because cultures of today have different behaviors and customs, and therefore different artifact and site patterns than cultures of the past. Because human behavior varies through time and across space, comparing the assemblage patterns of modern hunter-gatherers in Africa to Paleoindians in North America may be problematic. However, ethnoarchaeological studies do permit observation of cause-and-effect relationships between humans and their environment that cannot be obtained through other means.

In the rest of this chapter, I discuss factors that affect taphonomy of archaeobiological assemblages in a site. The preservation potential of each bone or seed is influenced by biological, environmental, and cultural factors (table 2.1). Of utmost importance is the determination of whether an archaeological assemblage is actually cultural (i.e., deposited or modified by humans). Biological remains can become deposited at archaeological sites in many ways that have nothing to do with humans. If you are interested in analyzing and interpreting past human behavior, it is important to ascertain which parts of your sites and assemblages are due to human behavior and which are not. After you determine which assemblages are cultural, then you need to assess whether that cultural material is in primary or secondary context. Is this cultural material in the context in which prehistoric humans placed it (primary), or has it been moved or modified by other processes (secondary) such as fluvial action, prehistoric dogs, rodents, tree throws, or pothunters? If you determine that your assemblage is

Table 2.1. Relative Preservation Potential of Archaeobiological Materials

Taphonomic Factors	Bone Organic Component	Bone Mineral Component	Shell	Plant Remains	Pollen
Biological					
Saprophytic organisms	–	0	–	–	–
Material durability	+	+	+	+	+
Carbonization/ calcination	+	+	+	+	+/–
Environmental					
Acidic soils	+	–	–	+/–	+/–
Alkaline soils	–	+	+	–	–
Mechanical destruction	–	–	–	–	–
Weathering	–	–	–	–	–
Hot, arid conditions	+	+	+	+	+
Cold, arid conditions	+	+	+	+	+
Anaerobic	+	+	+	+	+
Cultural					
Processing	–	–	–	–	–
Cultural use	+/–	+/–	+/–	+/–	+/–

+ = postive preservation factor
– = negative preservation factor
0 = neutral preservation factor

in secondary context, you need to assess whether analysis of that assemblage will be useful to your overall research agenda and/or questions. Is the time and money that will be spent on recovery and analysis of these materials going to be worthwhile in terms of information return?

PRESERVATION

There are numerous reasons why some plant and animal remains are preserved at a site and some are not. Preservation is affected by biological, environmental, and cultural factors (table 2.1). In this section, I discuss some of these factors, but this discussion is not exhaustive. It is up to each archaeologist to analyze and experiment on the types of preservational factors that may have or may be influencing their own site. Each site is different and was deposited under diverse conditions that affect whether the site becomes preserved in the archaeological record at all and whether the biological remains at that site can be recovered.

BIOLOGICAL PRESERVATION FACTORS

Many biological factors influence the preservation potential of plant and animal remains at archaeological sites (table 2.1). The most important one is the presence of *saprophytic organisms.* Saprophytic organisms are plants and animals that live on dead matter and obtain all their nutrients (nitrogen compounds, potassium, phosphates, oxidation of carbohydrates) by breaking down organic matter. Saprophytic organisms can include larger scavengers and rodents, but the term mainly refers to small organisms such as earthworms, insects, fungi, bacteria, small arthropods, and microbes. These organisms cause the decay and decomposition of most organic material, including biological materials at archaeological sites. The environment(s) in which these organisms flourish greatly influences whether biological assemblages will be preserved (see "Environmental Preservation Factors," p. 23).

Other important factors affecting whether biological remains preserve are their robusticity, durability, and density. The more durable a bone or plant part is, the longer it will survive attempted decay by saprophytic organisms and chemical decomposition. Carbonization (burning) of plants and woods makes them more resistant to destruction. During the carbonization process, chemical constituents of wood and plants are converted to elemental carbon, a durable substance that offers no nutrients for saprophytic organisms. Therefore, in many regions of the world in which archaeological plant remains are usually degraded, charcoal (of wood or plants) may be recovered. However, complete burning or incineration can destroy charcoal, turning it to ash, which is of less archaeobiological value than charcoal.

Burned animal bone, representing various stages toward calcination, also will preserve better than unburned bone under certain conditions. Burning removes protein and alters the calcium content of bone. Calcined bone is pure white, friable, and porous, whereas bone that is not quite calcined (gray to white in color) is not as fragile and is quite strong. Calcined or almost-calcined bone preserves well in areas with acidic soils where unburned bone is degraded through chemical action (see "Environmental Preservation Factors," p. 23).

Plant remains that are most frequently preserved and recovered by archaeobiologists are those containing materials that have a structural or protective role for the plant and are therefore more durable. Such constituents include cellulose, sporopollenin (the main component of pollen), silica (the main component of most phytoliths), lignin, cutin, and suberin that are found in pits, seeds, rinds, spines, woody components, resin, pollen, and phytoliths. Plant parts that do not contain these durable elements will tend not to be preserved in archaeological sites, and their potential absence should be realized.

Bone, horn, antlers, teeth, hooves, hide, and shell are the most frequently observed animal remains at archaeological sites, again due to their durability. These materials resist decay in many settings because they are made of robust structural elements such as keratin and collagen (horn and hooves), phosphatics (bone, antler, teeth), chitin (insect and crustacean exoskeletons), and/or are calcareous (shell). A shell midden site may contain tons of durable, well-preserved oyster shell, and numerous Puebloan sites in the southwest contain large amounts of carbonized corn cobs, but this does not mean that prehistoric peoples were eating nothing but oysters or nothing but corn. Preservation differentially affects biological remains; some remains will be well preserved, and others will degrade or will become totally destroyed.

Bone is the most frequently observed type of animal remains from archaeological sites. All bone, however, is not created equal. Some bone is more resistant to decay and destruction due to its density. Different bone elements and bone elements from different species vary in structural density. For example, larger animals tend to have bone with greater density than medium and small animals, so their remains tend to be preferentially recovered from archaeological sites. An exception is beaver, whose dense bone has a greater durability than most carnivores and other medium-sized mammals. Denser elements include the jaw, femur, humerus, tibia, calcaneus, and astragalus. These elements will also be recovered with increased frequency over less dense elements. Mammal bone is denser than fish and bird bone and thus is more frequently recovered (Lyman 1984).

Another important and often underrated influence on the taphonomy of archaeobiological assemblages, particularly bone, is the dog. Dogs have been "man's best friend" for at least ten thousand to twelve thousand years, and their remains are found in numerous archaeological sites around the world. Unfortunately, dogs are destructive of bone assemblages. Hudson (1993) conducted an ethnoarchaeological study on the destructive effects that dogs can have on bone remains. She analyzed bone loss due to ingestion by domestic dogs among the Aka Pygmies, a modern group of hunter-gatherers living in the tropical forests of central Africa. The author documented hunting, butchering, meat redistribution, cooking, consumption, and discard at three residential camps. After the camps were abandoned, they were excavated, revealing that 74 to 97 percent of bone elements were lost, mainly from smaller taxa, due to domestic dog consumption. Due to their density, remains of large animals survive canine assaults best, as do skulls and limb shafts. The presence and effects of domestic dogs on bone assemblages can be ascertained not only through bone loss but through the presence of gnawed bones as well.

ENVIRONMENTAL PRESERVATION FACTORS

The environment in which plant or animal remains are deposited greatly influences whether they will be preserved for a month, a year, ten years, or one thousand years. Because saprophytic organisms are the main cause of biological material destruction, preservation mainly depends on what types of depositional environments are conducive or inhibiting to these organisms (table 2.1). Carbone and Keel (1985) listed four environmental factors that influence preservation of biological assemblages: soil acidity, aeration, relative humidity, and temperature. In addition, other geological conditions may also be important for biological preservation.

Saprophytic organisms are intolerant of highly acidic soils and live almost exclusively in alkaline soils. Therefore, acidic soils will tend to preserve organic components of biological materials because they are not destroyed by microorganisms. Alkaline soils will tend to have poor preservation of organic components of biological materials due to increased saprophytic activity. This can be seen in the potential for pollen preservation in the Southwest. Because soils in the Southwest are highly alkaline, preservation of organic materials in open areas tends to be rare. For example, Bryant and colleagues (1994) analyzed 509 pollen samples from soil collected by a CRM firm along a proposed pipeline route in deposits ranging in age from one thousand to five thousand years old. The deposits all had a pH value above 6.0 with anhydrous carbonates as the most common compounds. Only 243 (48 percent) of the samples contained a significant amount of fossil pollen; the rest were almost entirely devoid of fossil pollen. The mean pollen concentration values from the pollen bearing samples was 6,545 grains/gram of soil. In addition, 90 modern surface soil samples were collected in west Texas to use as control samples for the fossil pollen record. All 90 samples contained a significant amount of pollen with an average concentration value of 21,311 grains/gram of soil. The authors state that the high-alkaline environment of the Southwest is partly the cause of poor pollen preservation in fossil soil samples, and that pollen concentration values should be determined in this region to help the analyst judge the possible amount of pollen destruction in each particular soil sample. Pollen has an outer covering (exine) made of sporopollenin, which is one of the strongest natural substances known. However, alkaline soils are conducive to fungi and bacteria that eat pollen, creating a biased array in which pollen with more sporopollenin in its exine is better preserved than pollen with less.

 Alkaline soils, however, tend to preserve mineral components better than acidic soils. Bone is made up of minerals (hydroxyapatite, calcium carbonate, trace elements) and organics (collagen, bone protein, fats, lipids) in an approximately 2:1 ratio. The organic components of bone will tend to be eaten by saprophytes in alkaline soils, leaving mineral bone components intact. Therefore, the mineral (structural) components of bone are preserved and can be recovered in alkaline conditions. Bone tends not to survive in acidic conditions because acids dissolve structural bases of minerals, leaving only organic traces of bone in the soil. Calcareous shell and antler are preserved in the same conditions as bone.

 While alkaline soils tend to destroy organic assemblages (plant remains and organic components of bone) due to saprophytic activity, such soil types tend to preserve bone and shell (mineral components) better than acidic soils. For example, soils in Maine tend to be acidic, which limits the preservation potential of the mineral component of bone. When bone is preserved at interior sites, it has been calcined or almost calcined (see "Biological Preservation Factors," p. 21). Although only small pieces of calcined or almost-calcined bone are preserved in interior sites with acidic soils, bone (mainly uncalcined) is prevalent at archaeological shell midden sites along the coast. Bone preserves in these sites because weathering and degradation of the calcareous shell matrix produces an alkaline environment conducive to preservation of mineral components of bone. Therefore, bone preservation at archaeological sites in Maine depends on site location and soil alkalinity.

Even though alkaline soils, such as in a shell midden, tend to preserve bone, bone still decays, degrades, and alters through time, as was observed in an experimental archaeology study on bone from a shell midden on the northwest coast. The mineral fraction of bones from a shell midden located in San Juan Island National Park, Washington, was measured to test the hypothesis that bone decomposes in alkaline deposits. Linse (1992) analyzed salmon vertebrae from two different deposits (facies) in the shell midden, one with high pH values (8.4–8.8) and one with pH values near neutral (7.8–7.9). Bone mineral decomposition, measured as the calcium/phosphorus (Ca/P) ratio, was assessed using inductively coupled plasma spectrometry (ICP) on the archaeological salmon bone as well as a control sample of modern salmon bone. The author observed no significant difference between the two archaeologically derived bone assemblages, but the archaeological bone mineral decomposed as compared to modern bone. She concludes, "Bones may be relatively 'safe' in a shell midden in terms of traditional morphological classification; however, the chemical integrity of bone in alkaline environments is suspect" (p. 342).

Soil alkalinity is not the only factor that influences preservation of biological assemblages. Other factors include the mechanical destruction of biological remains through seasonal freeze/thaw cycles, increased biological preservation potential in dry/arid or dry/frozen regions, and anaerobic environments which decrease saprophytic activity. For example, although soils in Maine are acidic and tend not to preserve mineral components of bone, such acidic soils, because they inhibit saprophytic organisms, should preserve other organics like botanical remains. This is not the case, however, because of the seasonal freezing/thawing cycle that mechanically destroys chemical composition. Preservation of organics in this type of environment can be achieved when they have been carbonized or burned, making them more structurally durable.

An example of excellent preservation in dry/arid regions derives from the southwestern United States. As previously discussed, alkaline soils in this region are not conducive to pollen preservation. However, the Southwest is famous for preservation of other biological remains because of high temperatures and aridity and because saprophytic organisms do not thrive in hot, dry conditions. Preservation of organics in open areas of the Southwest, however, is not as good as in enclosed areas (caves, rock-shelters, pueblos, etc.) because of increased exposure such materials have to weathering (wind, erosion, rain).

Cold temperatures also limit the decay of biological remains because saprophytic organisms do not live in such an extreme, cold environment. Biological assemblages in the Arctic can be as well preserved as in dry, hot regions; however, most archaeological sites in the Arctic are surface sites with little deposition overlying assemblages. Therefore, some types of biological remains will weather extensively and become degraded over time. For example, large animal bone remains are ubiquitous at these northern sites, whereas botanical remains are harder to recover (recovery is also due to cultural impacts; see "Cultural Preservation Factors," p. 29).

Saphrophytic organisms cannot live without oxygen, so anaerobic (lacking oxygen) environments will be conducive to biological preservation. Such environments include peat bogs, which are famous for preservation of "bog bodies" and other biological remains, and waterlogged sites, where wood is commonly preserved, such as stakes from prehistoric fish weirs. For example, numerous waterlogged pieces of wood have been recovered from the peat and silt beneath the city of Boston (Kaplan et al. 1990). Previous analyses indicate that the wood represents the remains of a number of prehistoric fish weirs used for trapping fish along the ancient shore. For this analysis, the authors examined 216 wood samples recently recovered from deposits dated to ca. four thousand to five thousand years ago. Wood taxa identified included beech, oak, alder, sassafras, hickory, maple, Canada hemlock, dogwood, birch, elm, ash, and bayberry. This assemblage indicates that prehistoric peoples collected wood from "upland and riparian habitats in fall and winter and probably constructed and repaired their weirs in the very early spring" (p. 527).

Relatively anaerobic conditions also exist under thick layers of clay or silt deposits, a good environment for preservation of biological materials. One reason carbonized botanical remains were preserved at the Little Ossippee North site in Maine (see the case study in chapter 1's sidebar) is that soon after humans made and used hearths at the site, a flooding event of the Saco and Little Ossippee Rivers capped the site with clay and silt deposits, preserving botanical remains in a relatively anaerobic environment. Depth of deposit of biological materials is an important component of preservation in such environmental conditions: The deeper material is buried, the more anaerobic the environment is, and the better the preservation potential.

We observed this principle when a nearby llama farm donated a llama skeleton to the zooarchaeology collection at the university on one condition: We dig up and collect the skeleton. The llama had been buried for three years in what the owners termed a "shallow" grave. One weekend, I took a group of graduate students to dig up the llama under the assumption that a llama buried for three years in a shallow grave in Maine's wet soils would be nothing but bones. We dug and we dug. The llama turned out to be deeply buried (1.5–2 meters) and was in very pristine condition with little decomposition of fur or muscle. We found a halter, and then we could tell why the llama was called Black Beauty. I told the owners we would be back in ten years. The deep burial was in a relatively anaerobic environment that inhibited saprophytic activity, thus slowing decay and decomposition.

Any disturbance to a relatively anaerobic environment can introduce oxygen and change the environment to one in which saprophytic organisms can thrive and cause decomposition. For example, after exposing the head portion of the llama we introduced oxygen and aerated the soil around the llama by digging up and moving the matrix. Because of this disturbance, the head portion of the llama will probably decompose at a quicker rate than the undisturbed hind portion that maintained its relatively anaerobic burial conditions. Therefore, human disturbance can influence whether soils will be more conducive to biological preservation. Similar disturbance factors include rodents, worms, and insects that dig burrows or pits, plant roots that grow into the ground, and tree throws that expose previously undisturbed, relatively anaerobic environments to the aerobic environment.

28 CHAPTER 2

Tree throws are also significant because they mix up and disturb the cultural and noncultural components of a site (figure 2.1). Most forests have long histories of disturbance, including tree throws induced by storms, floods, and other factors. However, after a tree decays, there may be little surface evidence of the disturbance. Tree throws increase the decay potential of biological materials, and they move cultural and noncultural materials around. They may mix material from deposits of different ages, making it difficult to obtain correct radiocarbon dates or even to recognize stratigraphy.

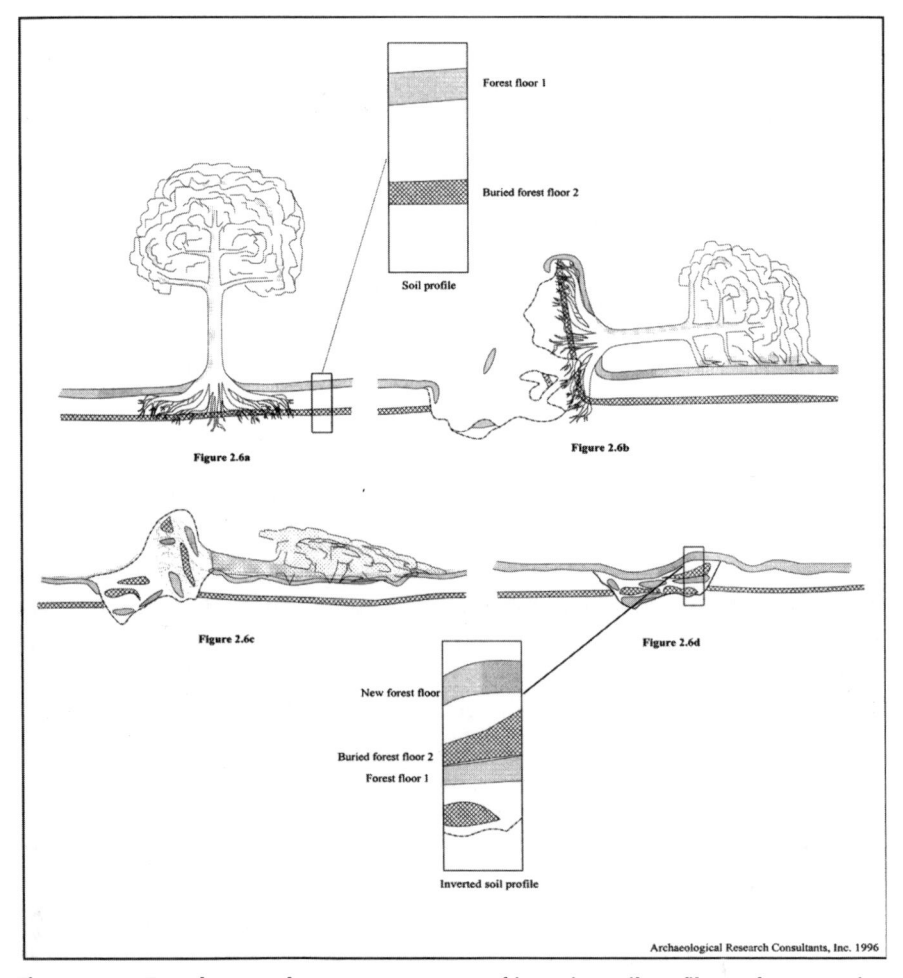

Figure 2.1. Tree throw taphonomy as a means of inverting soil profiles and transporting older materials upward in the soil column. (From Will et al. 1996; courtesy of James Clark.)

CULTURAL PRESERVATION FACTORS

Humans also affect the preservation potential of biological materials (table 2.1). Before biological remains are deposited, people can affect their robusticity and structure, decreasing or increasing their preservation potential. People burn plants (carbonization) and bones (calcination), either intentionally or unintentionally, generally increasing the probability that those remains will be preserved (see "Biological Preservation Factors," p. 21). But humans also break, macerate, pound, chop, boil, and otherwise manipulate biological materials before deposition, decreasing their chance of preservation. For example, people break bone to gain access to marrow, sometimes pounding bones into small pieces and boiling them to extract the fat.

Humans also dig pits for various purposes, exposing underlying archaeobiological materials to a more aerobic environment, potentially reducing their preservation (see "Environmental Preservation Factors," p. 23). This type of cultural transformation is seen more frequently in large, multicomponent, or stratified sites where human activity was more extensive and diverse than it is at small, single-component sites where activity tended to be centralized and less invasive.

Ultimately, the main factor determining whether biological material will become deposited in archaeological sites is whether and how humans used that material or even brought it to the site. For example, gathering meat involves disarticulation and skinning of scavenged or hunted prey, and defleshing of bone. In the case of a large animal, much of this activity may take place outside the base camp; therefore, not all bones from the animal will be brought back to the base camp and become deposited to form the potential archaeological record. This was illustrated in an ethnoarchaeological study conducted by O'Connell and Hawkes (1988) in which they reported detailed observations on the food procurement practices of the Hadza, people of southern Africa, and the subsequent creation of bone assemblages. In general, they note that vertebra, scapulae, pelves, and upper limb bones are the most likely elements to be carried back to a base camp. The number of parts taken is a product of animal size, distance carried, and number of carriers. Elements that are the least damaged by the kill and disarticulation will be transported (axial parts, scapulae, and phalanges), while those most damaged (limb and cranial elements, and upper vertebrae) tend not to be transported. In addition, they observed that small animals are killed more frequently than large animals, but small animals are underrepresented at kill sites due to total transport to base camps. When the Hadza killed larger prey, more bones were left at the kill site. Large animal butchery sites were also more complex, characterized by several clusters of debris and hearths, whereas small animal butchery sites tended to be composed of unstructured scatters of bone.

Another example of the effects humans have on preservation of biological remains and where differential parts are observed in the archaeological record is the processing of agave plants by peoples of the southern plains and southwestern deserts. Agave is a desert succulent with long, flat, sharp leaves aboveground and a compact "heart" or bulb belowground. The bulb is most nutritious just before the plant is ready to send up its reproductive stalk, so humans would dig up the plant at that time. First, the sharp leaves would be cut from the rest of the plant, and then the bulb would be dug up. The bulb would then be roasted in an earthen oven for at least forty-eight hours to make it edible (figure 2.2). People would eat agave at the earthen oven site or take it back to camp to share and eat there. Numerous chewed pieces of agave, called *quids* (figure 2.3), have been found in base camp sites. Remains of agave also will be found at procurement sites (leaves) and at and around earthen oven processing sites. In addition, people used agave leaves for basketry, sandals, paintbrushes, twining, and clothing, so fibrous remains of agave can probably be found in any site in which agave was used.

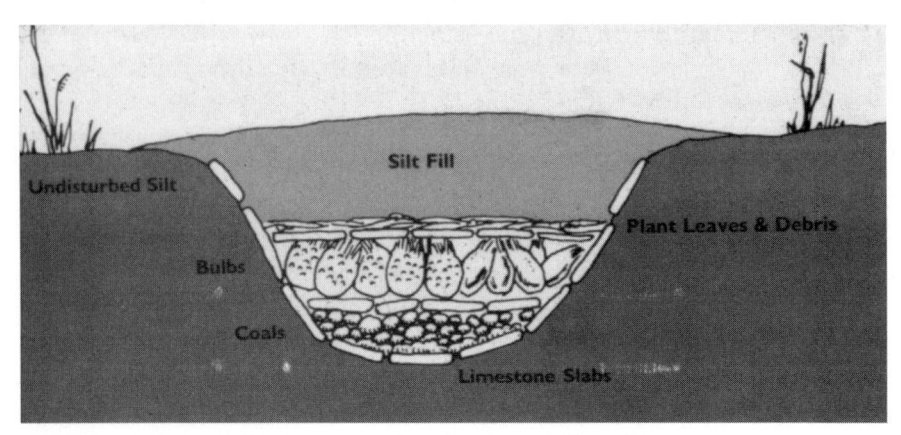

Figure 2.2. Diagram of an agave roasting pit. (From Shafer 1986.)

Figure 2.3. An agave quid. (Courtesy of Kristin D. Sobolik.)

CULTURAL VERSUS NONCULTURAL ASSEMBLAGES

When you recover biological remains, you need to assess whether the material is cultural. A *cultural context* is defined as a setting that has been physically altered or added to as a result of human activity. Evidence of human activity may include features (e.g., potholes, hearths, buildings), portions of stratigraphic layers (middens, living surfaces), or artifact concentrations. Noncultural contexts are contexts at which human activity is not indicated. Most sites have within them both cultural and noncultural contexts; that is, sites are commonly formed by a combination of cultural and noncultural processes. The archaeologist needs to determine which portion(s) of a site are cultural and which are not. The archaeobiologist can contribute greatly to this effort by determining which biological assemblages are cultural and which are not.

Most taphonomic studies addressing cultural versus noncultural deposits have focused on the accumulation and modification of faunal remains. These studies indicate that many animals, including carnivores, rodents, owls, and raptors, influence bone assemblages at archaeological sites. All of these animals collect and accumulate bones and may deposit bones into archaeological sites. Fortunately for the archaeologist, this noncultural bone generally is distinguishable from human-deposited bone. In addition, animals modify culturally used and deposited bone. This bone is still considered cultural, but it may be moved by animals out of original cultural (primary) context.

Carnivore and rodent influence on, or deposition of, a bone assemblage can be recognized by surface attributes of individual bone specimens. When chewing or gnawing, animals leave characteristic marks on bone. Microscopic examination of bone can reveal incisions, scratches, gouges, punctures, and pitting. Some of these marks are exclusively of human origin, while others are clearly of noncultural origin. Punched holes, striations, scoop marks, crunching, and splintering are examples of bone modification caused by animal activities. Canids will create shallow grooves or channels transverse to longitudinal axis on long bones because the long and thin shape of these bones prevents them from being gnawed in other directions. Punched holes, or tooth perforation marks, occur where hard bone is thin or where bone is soft, such as at the blade of the scapula or the ilium. These marks may appear as small hollows if the tooth did not fully pierce the bone surface. Striations and pitting occur on bone surfaces where canid molars scraped the surface in an attempt to reach the marrow cavity (figure 2.4). Tooth scratches tend to follow the surface of the bone, being deeper on convex surfaces and shallow on concave surfaces (figure 2.5). (In contrast, cut marks of human origin tend to be uniform in depth.) Where the bone ends have been removed, animal gnawing may produce scratches parallel or diagonal to the longitudinal axis of the bone on the shaft. Compact bone may be gnawed away to gain access to spongy bone, leaving overlapping striae and a scooped-out appearance on bone surfaces. Finally, marrow is reached by larger animals by crunching through bone, causing longitudinal splintering. Smaller canids will remove bone ends to weaken bone structure prior to crunching through the shaft.

Humans mark bone during butchering, skinning, and preparing food. Cuts are purposely placed for a desired result. For example, skinning an animal can leave cut marks on the underside of the chin and encircling the distal end of limb bones (figure 2.6). Because cultural marks are created on bone in butchering, cut marks cluster around articular surfaces or in areas of major muscle attachment. Marks will differ between species due to variation in joint strength. Bones struck with stone tools

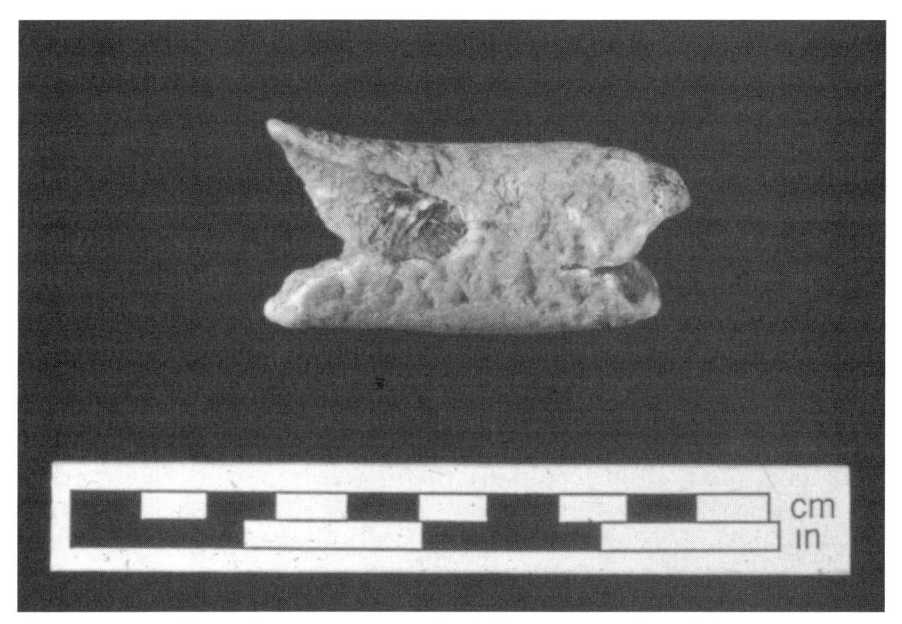

Figure 2.4. Canid gnawing, punctures, and striations on a long bone fragment. (Courtesy of Stephen Bicknell.)

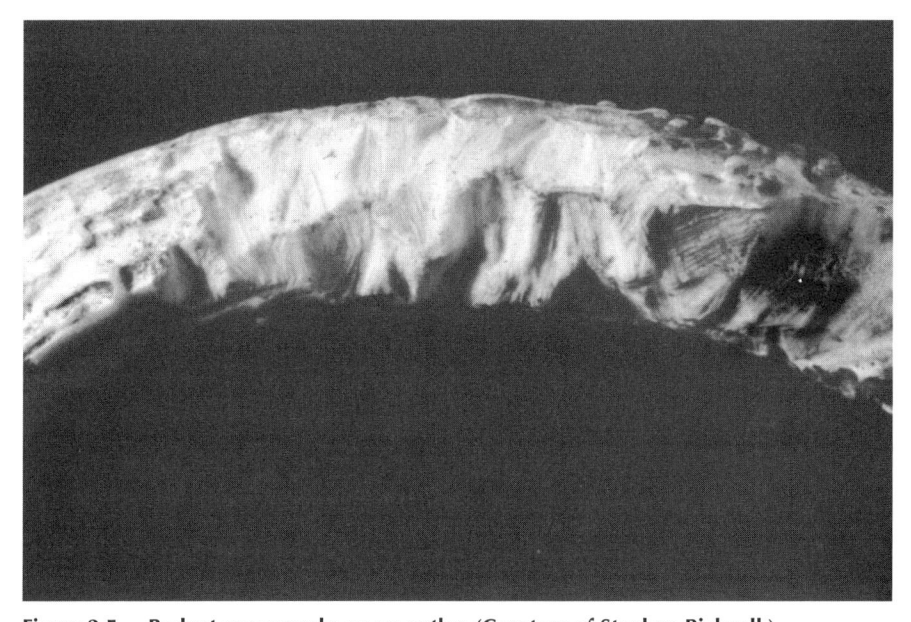

Figure 2.5. Rodent gnaw marks on an antler. (Courtesy of Stephen Bicknell.)

will exhibit crescent-shaped notches at the point of impact, and bones broken during butchering by "grooving and snapping" show a heavy incision along the broken edge. Cut marks from tools are readily distinguishable from carnivore tooth damage. Tool marks are characterized by fine striations within the furrow made by cutting action. These striations are thought to be created by irregularities on the tool's working edge. Tooth marks lack striations, but they exhibit ridges perpendicular to the direction of the mark, caused by uneven force applied by the animal to the bone. These are often called *chatter marks*.

Archaeologists and archaeobiologists also have used the presence of small animals, particularly rodents, to distinguish cultural from noncultural bone, often considering small animal remains as noncultural. However, to disregard rodents and other small animals as possible human food sources underestimates the importance of small animals to the dietary array of prehistoric peoples. Bones of a wide variety of small animals have been recovered from paleofeces, indicating that they were eaten prehistorically, so their bone remains in archaeological sites may be due to cultural factors (Sobolik 1993). For example, numerous paleofeces from archaeological sites from southwest Texas contain bone remains from small animals; 333 paleofeces have been analyzed for their macrocontents, 245 (74 percent) of which contained small

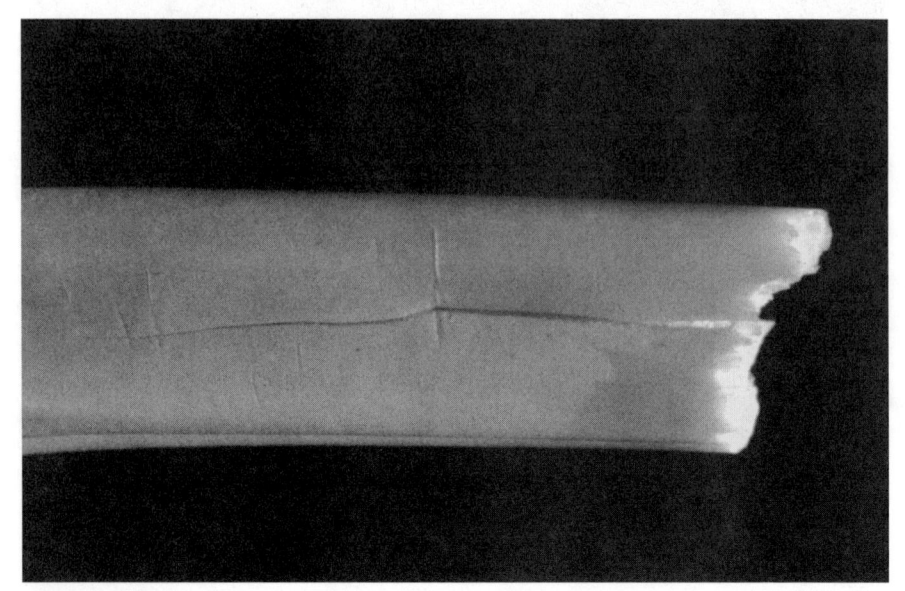

Figure 2.6. Human stone-tool cut marks on long bone. (Courtesy of Stephen Bicknell.)

bone remains, and 123 (33 percent) specifically contained rodent remains (figure 2.7).

Differentiating cultural from noncultural bone deposits also requires analysis of the potential taphonomic factors that may have influenced site depositional processes and biological assemblage preservation. This analysis must be regional- and site-specific and fairly inclusive because factors may have operated in different areas and time periods. For example, I conducted a taphonomic study of the faunal remains from a prehistoric hunter-gatherer base camp in Big Bend National Park (see the case study in sidebar 2.1). Factors that influenced faunal deposition at the site were rodent burrowing and carnivore scat deposition. Potential taphonomic factors that turned out to be not important were fluvial deposition and modification and raptor pellet deposition.

Because of the wide variety of taphonomic factors that have influenced biological assemblages and the importance of understanding taphonomic history, archaeobiologists may become fixated on data collecting at the micro- or quantitative level without seeing the big picture. Before you examine five thousand bone fragments for the presence of cut, tooth, or gnaw marks, you need to assess whether such analysis is necessary for your overall research goals.

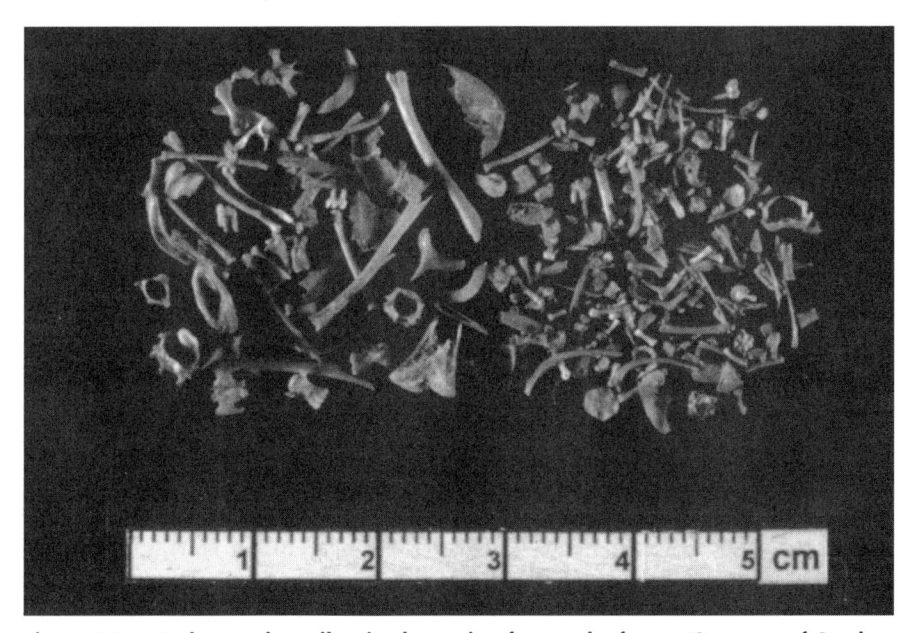

Figure 2.7. Rodent and small animal remains from paleofeces. (Courtesy of Stephen Bicknell.)

2.1. CASE STUDY: TAPHONOMIC FACTORS AFFECTING ARCHAEOLOGICAL SITES IN BIG BEND NATIONAL PARK, TEXAS

I analyzed biological factors that may have influenced faunal assemblages deposited in a Late Prehistoric rock-shelter site in Big Bend National Park (41BS921). Cultural deposition at the site represents intermittent occupation by hunter-gatherers for approximately one thousand years. The animal bone excavated from the site included a variety of fairly well-preserved mammal, reptile, and bird remains, with a majority of the sample consisting of medium and large mammal bones (table 2.2; figure 2.8). During excavation, eight rodent burrows were discerned. The contents of these burrows were analyzed separately and revealed that the frequencies of bone classes were similar to those observed in supposed cultural levels of the site (table 2.2; figure 2.9). Bones from the rodent burrows and other areas of the archaeological site also exhibited similar frequencies of burning and weather erosion. These data indicate that rodents tend to move cultural material out of primary context but incorporate little noncultural material.

In addition, twenty-seven carnivore feces (most likely from mountain lion) were collected from the surface of the site and analyzed for bone content (table 2.2). Bone from these samples included mammal, reptile, and bird and were very fragmented and acid etched. Carnivore digestive tracts have strong acids to process food. When bone passes through this digestive tract it becomes dissolved or if preserved then extremely acid-etched. Two small areas of supposed cultural deposition at the site also contain such material, indicating the presence of degraded carnivore coprolites.

I also analyzed five great-horned owl pellets and excavated three large areas of owl pellet remains from Owl Cave, located one hundred meters from the site. The pellets and excavated faunal remains revealed a wide variety of animal bones from packrats, mice, rats, rabbits, snakes, and birds (table 2.2); the bone was in pristine condition with relatively few breaks (figure 2.10). Pristine bone was not observed in supposed cultural contexts from the archaeological site, indicating that owl pellet deposition probably did not contribute to the assemblage.

This study indicates that rodent burrowing at the archaeological site influenced cultural deposition and moved remains from a primary to a secondary context. In addition, carnivore scat deposition was also important in adding noncultural bone to the faunal assemblage, whereas owl pellet deposition was not a contributing factor in taphonomy of the site. This case study also illustrates the importance of determining the taphonomy of archaeobiological materials from supposedly cultural contexts.

Figure 2.8. Bone from a prehistoric rock-shelter in Big Bend National Park, Texas. (Courtesy of Stephen Bicknell.)

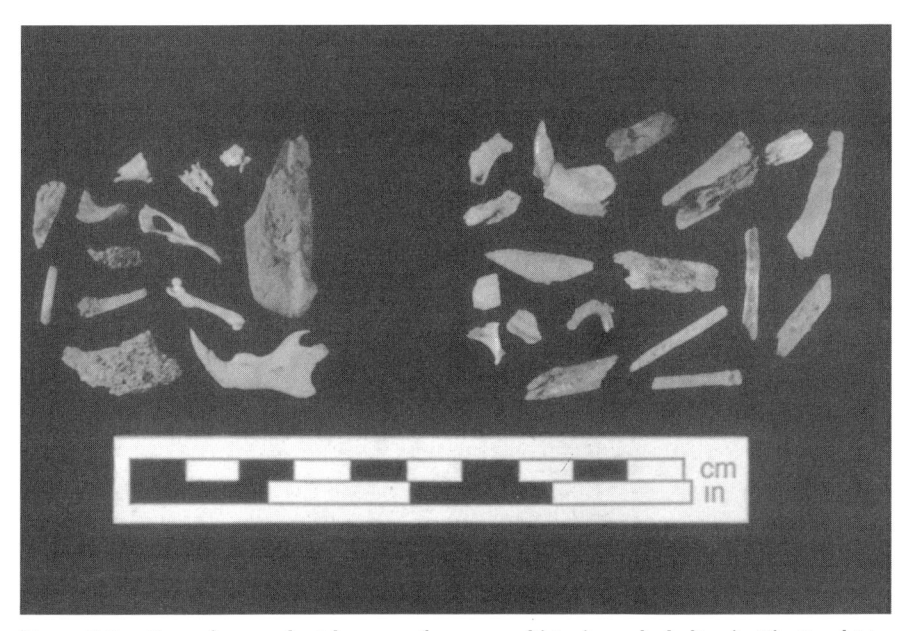

Figure 2.9. Bone from rodent burrows from a prehistoric rock-shelter in Big Bend National Park. (Courtesy of Stephen Bicknell.)

Figure 2.10. Bone from great horned owl pellets. (Courtesy of Stephen Bicknell.)

What are your research questions? Why are you doing this? What is the big picture?

Here's an example of microlevel analysis addressing a big research design. Shipman (1986) examined a large number of bones under the scanning electron microscope looking for human and nonhuman marks from early hominid sites. She observed numerous instances in which human cut marks were superimposed on scavenger tooth marks, allowing her to conclude that early hominids were actually scavengers instead of hunters, obtaining meat after other carnivores had made the kill. Shipman's painstaking analysis was for a purpose: It contributed to the big picture. Even though all of our analyses may not have such far-reaching implications as Shipman's, we need to constantly be thinking of how our data fit into project-specific research designs as well as broader research goals (see Toolkit, volume 1).

CONTEXT

Ultimately, the question of whether a biological assemblage is cultural is a question of context. Interpretation of context occurs at all stages of research, from excavation to analysis. During excavation, the direct association of biological materials with artifacts or cultural features, such as stone tools, ceramics, or stone-lined hearths, provides evidence that the biological materials may have a cultural origin. When employing artifacts to make this determination, they should be clearly of human manufacture or modification and should be found in situ within undisturbed geological deposits that demonstrate primary association of artifacts with stratigraphy.

Disturbances should be evaluated critically. Potential disturbances are numerous and can include tree throws (figure 2.1) and carnivore and other animal modification as evidenced by presence of scat, burrows, and gnawing on bone (figures 2.4 and 2.5). Rodent burrows are common intrusions into archaeological contexts and are often easy to recognize during excavation. Rodents introduce noncultural material into archaeological sites and move cultural material out of primary context. These factors and many others can displace, introduce, or remove artifacts and archaeobiological materials from their original point of deposition. As previously discussed, humans, both prehistorically and in modern times, disturb archaeological context and move cultural materials from a primary to a secondary context. They may even introduce noncultural biological material into an archaeological context in which archaeologists will consider it cultural.

For example, Miller and Smart (1984) noted that animal dung is used as fuel throughout the world and that seeds can be found in dung. When dung is burned, charred seeds become incorporated into hearth deposits and might be misinterpreted as human food refuse. The authors analyzed four modern samples of debris from a village in Malyan, Iran: one hearth, one fire pit, and two midden samples that contained the remains of cleaned hearths and fire pits. Three of the samples contained charred seeds found only in dung and not directly used by humans. These results suggest that archaeologists should consider dung as a possible vector for charred seeds when the following conditions are met: (1) the site is located in an environment where wood and food are scarce, (2) dung-producing animals are present, (3) the archaeological assemblage includes parts of burned dung or seeds that are typically eaten by dung-producing animals, and (4) the archaeological context is one of hearths or dumping areas of cleaned hearths. The authors tested their research on two archaeological sites, a prehistoric Malyan site dated to the third millennium B.P. and the Tierra Blanca site, a late prehistoric site in Texas. Seed assemblages from both sites indicate that they may have been introduced through the burning of dung as fuel.

Biological material can also be disturbed and moved into or out of a site by water and wind. For example, moving water can remove biological remains from their primary archaeological context and redeposit them elsewhere. Typical secondary contexts are channels, floodplains, lake margins, point bars, and coastal settings. Many archaeological sites have been "discovered" that, in fact, are nothing more than artifacts in secondary context. Will and Clark (1996) conducted an experimental study in which they "planted" stone artifacts in the fluctuation zone on the shores of impounded Moosehead Lake in Maine to observe artifact movement patterns due to wave action and ice movement. After recording artifact movement over a year, they observed four main results: (1) Artifacts exposed to wave and ice action in the fluctuation zone can move several meters from their primary context; (2) the direction of movement is influenced by major weather; (3) artifacts moved by ice action have less surficial damage than artifacts moved by water and may look "fresh" or recently eroded; and (4) artifacts buried at the edge of the fluctuation zone in a beach or berm can be recycled to the surface by wave and ice action. These results indicate that a cluster of artifacts in the fluctuation zone or in the beach/berm area may represent material significantly removed from its primary context rather than material eroding from an archaeological site along the shoreline. These results also helped explain why many sites along a large impounded lake turned out upon excavation to be nothing more than surficial lithic debris. Although the experiment was conducted on lithic artifacts, the same processes can affect biological materials. For example, huge amounts of bone accumulate on sand bars in rivers across North America. These bone accumulations are not archaeological sites but merely water deposited assemblages in secondary context.

Fluvial action can move cultural remains out of a site and can also deposit noncultural materials into a site. Fluvial effects on bone have been extensively studied, whereas such effects on botanical material are not as well known. Surface abrasion and rounding of bone surfaces result from transport by water. Orientation may also be a sign that bone specimens have been moved by fluvial processes; heavier ends of elongate elements point upstream. Elements with low density (see "Biological Preservation Factors," p. 21), low weight, and a high surface area to volume ratio, such as innominates, scapulae, and intact crania, are more likely to be transported long distances. Shape will also influence transportability: Long flat bones are more likely to be transported than round ones.

In a classic experimental study on the effects of fluvial action on bone, Voorhies (1969) demonstrated this using bones from medium-sized animals in flume experiments and created a chart of elements and their transportability in flowing water (table 2.3). Group I elements were immediately moved by slow moving currents. Group II elements were gradually carried away in a moderate current, and Group III elements were only moved by strong currents. If your assemblage is composed of elements representing all of the groups, then it probably wasn't modified by fluvial action. If, however, you have an assemblage consisting only of elements from one group, fluvial action should be considered as contributing to the composition of the assemblage.

CONCLUSION

Determining taphonomy of biological materials recovered from an archaeological site is the first and most important step in archaeobiological analysis and interpretation. First, ask how the botanical or faunal remains became deposited in the site, and consider all the potential factors influencing that deposition. Are the biological remains from the site cultural, or do they represent deposition through noncultural agents? If remains are determined to be noncultural, they may be useful in analyses of paleoenvironments but not for direct analysis of human activity. Even if biological materials are considered to be deposited as a direct result of cultural activity, they may be out of primary context due to postdepositional factors, such as pot hunting, other human digging, animal burrowing, or tree throws. Depending on research goals, biological materials in secondary context may or may not be useful for analysis and interpretation; even though they are cultural, it may not be worth the time, money, and effort spent on their analysis.

In all stages of an archaeological project, decisions need to be made on the effectiveness and potential of each step. It is up to the archaeologist, in conjunction with the archaeobiologist, to determine which sites are to be tested or analyzed further in Phases II and III of CRM projects, and it is up to the archaeologist to determine which sites to collect data from and which will yield the most information during their short field research season. In most cases, the archaeologist is under time and money constraints. After the archaeologist, in conjunction with the archaeobiologist, has made these decisions and the site has been excavated, it is up to the archaeobiologist to determine

Table 2.2. Bone from Taphonomic Contexts in Big Bend National Park

Context	Large Mammal	Medium Mammal	Small Mammal	Reptile	Bird	UID	Total
41BS921	593 (38%)	736 (47%)	129 (8%)	34 (2%)	9 (0.6%)	75 (5%)	1,676
Rodent burrows	7 (17%)	24 (57%)	1 (3%)	3 (7%)	0	7 (17%)	42
Owl pellets	0	2 (3%)	41 (54%)	32 (42%)	0	0	75
Owl cave floor	0	8 (2%)	177 (47%)	53 (41%)	5 (1%)	30 (8%)	273
Carnivore feces		234 (63%)		18 (5%)	10 (3%)	109 (29%)	371

UID = Unidentifiable.

Table 2.3. Transportability of Bone Elements in Flowing Water

Easily Transported, Moved by Slow Currents	Intermediate between Easily and Moderately Transported	Moderately Transported, Moved by Moderate Currents	Intermediate between Moderately Transported and Transported with Difficulty	Transported with Difficulty
Group I Elements:	Group I/II Elements:	Group II Elements:	Group II/III Element:	Group III Elements:
Ribs	Scapula	Femur	Mandibular Ramus	Skull
Vertebrae	Phalanges	Tibia		Mandible
Sacrum	Ulna	Humerus		
Sternum		Metapodia		
		Pelvis		

Source: Data from Voorhies (1969).

which biological materials from these sites are worth spending diminishing supplies of time and money on. Which biological materials will help answer research goals, which materials will help in interpretation of past human lifeways, which materials are cultural and in context and which are not? Understanding taphonomy will help you answer these questions. Archaeobiological analyses and interpretation cannot proceed without this understanding.

3

RECOVERY TECHNIQUES

Many methods exist to recover archaeobiological material from a site. Choosing appropriate methods depends on the type of site, the type of matrix and strata from which the remains will be recovered, and the kinds of research questions that are being asked. Presented here are what I consider to be the best and easiest ways to recover archaeobiological materials from most types of archaeological sites. I will also describe modifications that may be necessary depending on the type of site or matrix, emphasizing that modifications of the basic procedures for archaeobiological recovery are in the hands of project directors and archaeobiologists and depend on each individual site situation.

The most important principle in the recovery of archaeobiological material is that optimally the archaeobiologist should be involved from the very beginning (table 1.1). The archaeobiologist should participate in developing the research design and should help plan where excavations will take place and how they should proceed in reference to the recovery of archaeobiological material. During excavations, it is not necessary that the archaeobiologist be there to collect every biological sample. However, if the archaeobiologist is at the site, he or she can observe depositional conditions and recovery. In this way, the archaeobiologist will be able to assess taphonomic factors and conditions and will be able to modify the basic recovery plan if needed.

The least innovative and productive way to recover archaeobiological material is to have a field technician collect random, unspecified samples from the field and ship them to the archaeobiologist. The archaeobiologist has no indication of potential taphonomic factors; what cultural or environmental conditions may have influenced deposition and preservation; whether samples were collected from unambiguous cultural features, horizons, or zones; whether the "best" samples were collected; and how the samples relate to the entire site. If an archaeobiologist receives a bag of soil, a bag of seeds and charcoal, or a box of bones and has never been out to the site or been a part of excavation design and sample recovery, "analysis" may become nothing more than technical identification with little or no interpretation or innovation. If the archaeologist just wants a list of the types of charcoal or animal taxa present in a sample, then it is not necessary to treat the archaeobiologist as a scientist who can actually contribute to the analysis and interpretation of the site. And if the archaeobiologist wants to be treated as purely a technician rather than a scientist, he or she can keep accepting boxes of bone and bags of soil, along with a paycheck. If, however, the archaeobiologist would like to contribute to the understanding of peoples and environments and recognition of the importance of archaeobiological material in the reconstruction of past lifeways, then he or she should get involved at all stages of the recovery process.

Archaeobiology is becoming more completely integrated into archaeology. Accordingly, other volumes in this series cover certain elements relevant to archaeobiology. In particular, volume 1 discusses research design in detail, and volume 3 covers excavation methods. Therefore, I will limit my discussion of the recovery of archaeobiological material to the actual excavation of plant and animal remains.

Archaeobiological materials can be recovered in the same fashion as other archaeological material: During normal excavation and screening, some archaeobiological remains can be removed from the matrix or picked up in the screen. Some may be found during piece-plot excavations. However, because archaeobiological material often is too small to be observed during excavation and too small to be caught in normal screens, a lot of archaeobiological material is recovered during fine screening and flotation. Archaeobiological materials that are collected during normal excavation procedures should be bagged separately from those recovered during fine screening and flotation.

FINE SCREENING AND FLOTATION

Numerous studies show that using fine-mesh screens (one-eighth or one-sixteenth inch) and flotation is essential for the effective recovery of archaeobiological materials. Thomas (1969) provides baseline information on the recovery of animal bone from quarter-, eighth-, and sixteenth-inch mesh screens from prehistoric sites in the Great Basin. Bones from small mammals, weighing less than one hundred grams, were entirely recovered in the sixteenth-inch screens, but up to 54 percent of small mammal bones passed through the eighth-inch screen, and up to 93 percent were lost through the quarter-inch screen. Clearly, use of smaller-mesh screens greatly improves recovery of all animal taxa, particularly smaller taxa.

Shaffer (1992) also analyzed loss and recovery of faunal elements from quarter-inch screens. He placed modern, whole bone elements from twenty-five taxa, ranging in size from the least shrew (femur measuring 7 millimeters) to the coyote (femur measuring 176 millimeters), in a quarter-inch screen and recorded which elements passed through the screen and which were recovered. Predictably, smaller animal elements consistently passed through quarter-inch screen; rarely were elements recovered even for presence/absence determination. Some elements from larger animals consistently passed through the screen as well, such as caudal vertebrae, ribs, sternae, patellae, sesamoids, podials, metapodials, and phalanges from medium-sized animals, and sesamoids, carpals, patellae, and middle and distal phalanges of large animals. The author states that using only quarter-inch screening at an archaeological site will severely bias the sample toward larger animals and specific bone elements, thus making quantification unreliable.

The importance of flotation for the recovery of botanical remains has been recognized for many years; the study by Struever (1968) is usually cited as the first thorough North American discussion. Flotation uses water to separate lighter (less dense) material, usually organics, from heavier (more dense) material, usually inorganic matrix but also including some bone. Although flotation was used sporadically prior to the 1960s, the development of the New Archaeology, with its emphasis on ecological and environmental contexts of archaeological sites, led to increasing acceptance of flotation as the standard tool for botanical data recovery through the 1970s and its continued use today.

So, we understand that it is important to fine-screen and float sediments to recover archaeobiological materials. What are the best and easiest ways to do this? Fine screening is very simple: A known quantity of matrix is passed through a fine-mesh screen (usually eighth- or sixteenth-inch mesh), and all recognizable archaeobiological material retained in the screen is collected. Screening can be done by using a single screen (figure 3.1), inserted into coarse-screen frames (figure 3.2), or by placing the screen within or underneath the framework of an existing coarse screen (half-, quarter-, or eighth-inch), thereby screening the same matrix with coarse and fine screen. Nested screening, and dry screening in general, is easiest when dealing with dry and sandy or silty matrix. Wet or clay matrix usually is most effectively screened by water screening so that the matrix can be easily broken apart and the archaeobiological remains become more visible. Water screening involves placing a known quantity of matrix on a fine screen and using a pressurized source of water, such as a hose, to break apart the matrix and wash mud through the screen. It is important not to use too much pressure when water screening, or the archaeobiological remains will become smashed and broken and will pass through the screen. Fine screens can be purchased at any archaeological supply store or business, or they can be handmade with wood and screen purchased at the local hardware store.

Figure 3.1.   Example of a separate fine-screen system. (Courtesy of Kristin D. Sobolik.)

RECOVERY TECHNIQUES

Figure 3.2. Example of a nested fine-screen system. (Courtesy of Stephen Bicknell.)

The amount of matrix that should be fine screened and floated depends on the type of site being excavated and the project goals. It is important to obtain a representative sample of small archaeobiological remains from a site. To do this, you need to fine-screen and float matrix from features as well as from natural and cultural levels. The amount of matrix and sampling strategies for collection from five archaeological "type" sites are discussed in the section on "Excavation and Sampling," p. 53.

The advantage of fine screening and water screening is the increased recovery of small archaeobiological remains. The main disadvantage is that fragile remains, such as seeds, charcoal, and small bones, are easily broken during screening and by the impact of high-pressure water on the sample. Because of this problem, and because important material falls through even sixteenth-inch screen, flotation is usually the method of choice to recover fragile botanical remains. There are many flotation methods, and each archaeobiologist tends to prefer a specific technique. For a review of the types of flotation methods that have been employed by various projects around the world, see Pearsall (1989, 2000). I will review some of the techniques here and will then present what I consider to be the easiest technique with the fewest disadvantages.

REVIEW OF FLOTATION TECHNIQUES

The initial flotation method used by Struever (1968) was called the *immersion method* or the *apple creek method* and is useful in areas where there is ample water in which to immerse the samples. This method requires wash tubs with the bottoms removed and screens welded in their place. Partly immerse the tub in the water source (creek, river, pond, lake, etc.), then place the soil sample into the tub. Skim off all floating material with cheesecloth. Using many people and tubs simultaneously allows great volumes of material to be floated rapidly. The heavy fraction (the material that sinks to the bottom of the tub) is dumped in a bucket of zinc chloride solution (1.9 specific gravity) and separated further; some material such as bone that sinks in water, floats in zinc chloride. However, use of zinc chloride is problematic: The zinc chloride must be strained through cheesecloth after each use to prevent contamination, the process makes bones and calcium carbonate foam due to the hydrochloric acid in the zinc chloride preparation, the compound is costly, and it irritates skin and eyes.

Helbaek (1969) developed a similar technique to float samples in areas in which standing water is scarce. Soil is dumped in a bucket full of water, the soil is stirred, and the top portion of the water is poured onto a fine-mesh screen to collect the light fraction. The heavy fraction at the bottom of the bucket is either discarded or screened and dried for later examination. The process is repeated with new water for each sample. This technique, also known as the *washover method* (Greig 1989), can be used indoors or outdoors. Helbaek liked to use carbon tetrachloride rather than water because it had a higher specific gravity (1.8), which increased organic material recovery in the light fraction. Carbon tetrachloride is costly, however, and we now know that it is carcinogenic. The method works well with water, although bones tend to accumulate in the heavy fraction.

One of the most frequently used flotation methods today is the *oil drum method* or a modification of it. This method is also called the *SMAP* (Shell Mound Archaeological Project) *flotation method*, first used by Watson (1976). It involves pumping water from a nearby source into the bottom of a fifty-five-gallon drum that has a screen-bottomed bucket inset at the top. The constant addition of water agitates the drum's contents and creates a steady overflow of water. Soil is dumped into the screen-bottomed bucket; the light fraction floats, and the heavy fraction is caught at the bottom of the bucket or tub. The light fraction overflows the drum through a sluice and into a fine-mesh screen (0.425 or 0.5 millimeter), and it is then dumped onto newspaper to dry and then be sorted. This technique can process a large amount of soil even when water is limited because water in the fifty-five-gallon drum can be reused from sample to sample.

We used a similar technique at the NAN Ranch, a Mimbres Puebloan site in southwest New Mexico. A fifty-five-gallon drum was filled with water from a hose (figure 3.3). The drum had a large hole near the bottom that was closed with a screwtop. This hole was used to drain water and the heavy fraction once it had been used a number of times. Water entered the drum through a ring aeration system, creating a frothing effect that churned the soil and induced organics to rise to the top. Soil was dumped into the drum; the aeration system helped the light fraction float to the surface, where it would overflow the drum and descend a slanted spout attached to the top edge of the drum and be retained in cheesecloth set on a catchment area. The light fraction in the cheesecloth would then be hung to dry on a clothesline before being sorted and analyzed (figure 3.3). The heavy fraction would sink to the bottom of the drum, where it would be periodically drained through the large hole in the bottom side.

Figure 3.3. Example of a SMAP flotation system used at the NAN Ranch, under the direction of Harry J. Shafer. This photo reveals water source, discard area, and drying flotation samples hanging in the background. (Courtesy of Kristin D. Sobolik.)

A potential problem with this and other mechanical techniques is that because several samples are floated using the same water, cross-contamination between samples is possible—unless you can empty the fifty-five-gallon drum after every sample, which is costly and time-consuming and uses a lot of water). In addition, in an attempt to limit contamination, each sample is run or floated for longer periods of time to increase the chance of obtaining most or all archaeobiological remains from a sample, leaving less behind to potentially contaminate the next sample. This can make mechanical systems more time-consuming than others.

COMBINATION METHOD

The method that I like to use is time-effective, has minimum potential for contamination, can be run by one person, uses small amounts of water, does not use nasty chemicals, and can be adjusted to individual soil sample types. In addition, it is also a fine-screening method so that one sample can be efficiently fine screened and floated at the same time. This method does not have a particular name but originates from the first basic manual flotation techniques generated before archaeobiologists started using more elaborate procedures. This technique does not have the limitations of the more evolved methods and retains all of their positive characteristics. For descriptions of this technique and its modifications see Pearsall (1989:35–50).

It involves a plastic or metal bucket from which the bottom has been removed and a fine screen (eighth- or sixteenth-inch) attached in its place (figure 3.4). The bucket is set into a tub or directly in a sink (if flotation is being conducted in the lab), or outside in an area that will become very wet and muddy (if flotation is being conducted in the field). If flotation is conducted directly in the sink, it is best that the pipe system have a sediment trap, or the pipes will become clogged.

A hose or a faucet runs water into the bucket and in turn into the tub. A known quantity of soil is emptied slowly into the bucket. The light fraction will float to the surface where it is collected with a fishnet skimmer and dumped onto a labeled, paper-lined tray for drying. The light fraction is collected continuously until organics stop floating to the surface. Flotation is assisted as the operator

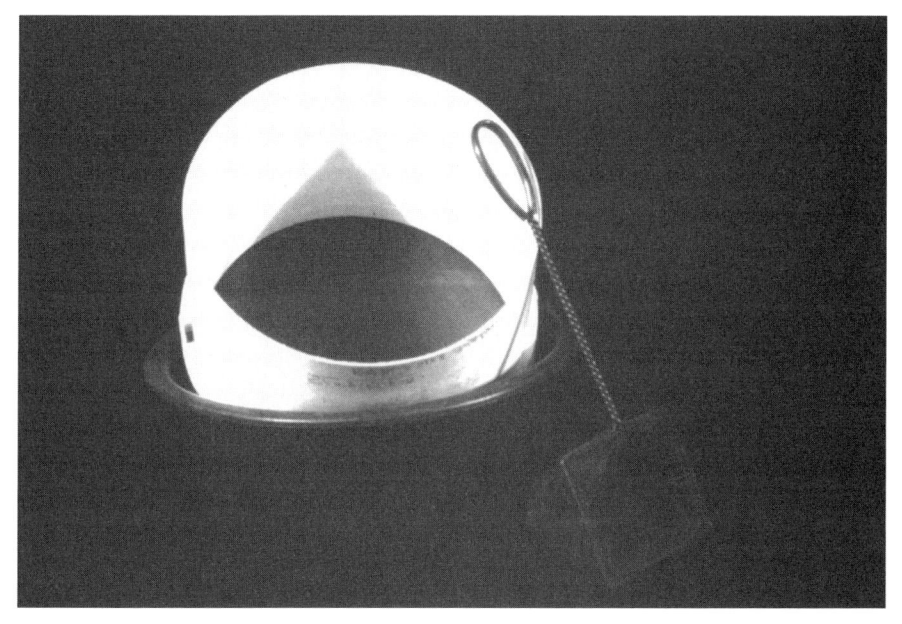

Figure 3.4. Example of combined flotation and fine-screen setup. (Courtesy of Stephen Bicknell.)

manually churns the soil to induce organics to float to the surface and soil to pass through the screen at the bottom. Each sample can be floated and manually assisted for as long as needed. After all of the light fraction has been collected, the bucket is removed from the tub, and the heavy fraction, which is automatically fine screened as well, is placed on a separate labeled tray for drying. The bucket and tub are then rinsed clean, and clean water is added to the system for the next sample, eliminating potential cross-contamination. This system can be operated by one person and can be run continuously, allowing numerous samples to be processed in a short time.

EXCAVATION AND SAMPLING

Even using the quick, easy, and efficient flotation and fine-screening method described here, fine screening and flotation are time-consuming and costly, so it is rarely feasible to run all or even a large quantity of the matrix from an archaeological site through such a system. However, in some cases, project goals are such that recovering all or most archaeobiological material preserved at a particular site is of primary interest. Such was the case at 20SA1034, a Late Woodland site in the Saginaw Valley of Michigan (Parker 1996), from which 665 flotation samples were recovered and processed from twenty-eight features, totaling over ninety thousand liters (see the case study in sidebar 3.1).

Which samples do you take for fine screening and flotation and from where do you take them? The answer to these questions depends on site taphonomy (discussed in chapter 2), research questions, and available resources. If you are interested in analyzing human diet and subsistence, you need to collect samples clearly associated with cultural areas of the site—again realizing that all portions of the site probably were not deposited or modified by humans. If you are interested in paleoenvironmental reconstruction, it is best to collect samples in areas that are not considered cultural. Paleoenvironmental sample collection therefore should take place away from the archaeological site so the information you are obtaining will have to do mainly with environment and not human selection. If you are interested in human impacts on environments and vice versa, then you should collect samples in both cultural and noncultural contexts.

Archaeobiological samples include both plant and animal remains collected from archaeological sites. As discussed previously, these samples should be collected both during normal coarse screening and with the aid of flotation and fine screening. I advocate the use of a combined fine screening and flotation system (see "Combination Method," p. 52), so in my discussions of where and how to take samples, I am referring to samples to be used for fine screening and flotation in combination. Archaeobiological samples should be collected with other artifacts during normal coarse screening and recorded and curated separately. Therefore, the following discussions refer to archaeobiological sample collection of fine-screen and flotation samples. In addition, I discuss collection of other archaeobiological samples, such as pollen, phytoliths, and paleofeces, where applicable.

Most archaeologists and archaeobiologists are interested in some aspect of human use of plants and animals. Archaeobiological samples should be collected in areas of a site that are considered cultural zones, horizons, or levels. In many cases, cultural affiliation is uncertain until artifacts and archaeobiological samples have been analyzed. In fact, archaeobiological samples can help determine whether an area of the site is cultural, whether rodents were rampant there, and

3.1. CASE STUDY: FLOTATION SAMPLING STRATEGY AT A LATE WOODLAND SITE IN MICHIGAN

The recovery of evidence of horticultural practices during the Late Woodland period in Michigan is scarce. At first, archaeologists hypothesized that the scant evidence of horticultural botanical remains was due to lack of flotation in early excavations prior to 1975. Later, archaeologists believed that horticultural plants were not an important component of Late Woodland subsistence because even when flotation samples were taken, horticultural crops were not recovered in quantity. Excavation of site 20SA1034, a Late Woodland occupation in Michigan, illustrated that when 100 percent of cultural features are floated and analyzed, significant amounts of horticultural crops and other food remains can be recovered. Of particular importance is that not all flotation samples from the same feature yielded food plant remains. If the entire feature had not been floated, wood charcoal may have been the only botanical material recovered, and evidence for horticultural subsistence practices in the Late Woodland would have remained scarce.

Excavation of site 20SA1034 was sponsored by Great Lakes Gas Transmission Limited Partnership and conducted by the Institute for Minnesota Archaeology (Dobbs et al. 1993). The project research design involved maximizing recovery of archaeological remains from all twenty-eight cultural features of the low-density site; therefore, all feature matrices were processed for flotation, yielding 665 flotation samples and a total volume of approximately 90,440 liters (Parker 1996). Botanical remains recovered through flotation were analyzed according to a timeframe priority ranking in which food plant remains were identified first and then wood charcoal was analyzed. Wood charcoal was observed in all flotation samples, and a few features contained high concentrations of food plant remains.

At least fifty different plant food taxa were identified, the highest diversity recovered from that region. The plant remains included maize from twenty-four features with a density of 2.74 fragments/ten liters. Maize kernel and cupule remains indicate several different methods of processing as well as the presence of eight-row (Northern Flint or Eastern Eight Row) and twelve-row cobs. A diversity of nutshell fragments of acorn, butternut, hickory, and walnut was identified. A very large diversity of seeds also was identified (2,591), at an average of 2.9 seeds/ten liters. Sixty-three percent of the seeds were from cultivated or domesticated species such as sunflower, tobacco, chenopod, and cucurbit. In addition, a wide variety of wild plant seeds were also identified, such as black nightshade, huckleberry, sumac, elderberry, blackberry or raspberry, bunchberry, pin cherry, plum, and panic grass.

Parker concludes that if a standard ten-liter sampling strategy for feature flotation had been implemented, large categories of botanical remains would have been missed, such as the presence of sunflower and cucurbits. Through the successful application of total flotation and recovery of features a more accurate picture of Late Woodland subsistence in Michigan was revealed. Populations occupying the Saginaw Valley of Michigan during the Late Woodland time period practiced a mixed strategy of crop production and wild plant exploitation.

whether deposition represents carnivore habitation or alluvial deposition rather than or in addition to human occupation. Because cultural affiliation may be unknown or suspect, it is best to collect more samples rather than fewer samples.

As we all know, archaeology is a destructive science, and once a site has been excavated, it is gone. Therefore, it is better to collect more samples than you might actually analyze than collect fewer samples than you really need. Curation of unused and unanalyzed archaeobiological material is the necessary cost of doing competent, progressive archaeology in which needs of the future are assessed on the same levels as needs of the present. The future of archaeology may well rest on museum shelves as archaeological sites are potted, looted, destroyed, and excavated at an alarming rate. Therefore, the costs of curating archaeobiological remains in the present are minimal compared to their potential contribution to the future.

I present a basic procedure for sample collection (figure 3.1), a procedure that should be modified to fit the needs of each researcher and the circumstances of each site. In most archaeological sites, archaeobiological samples should be collected from sequential excavation levels so that a progression of samples is obtained from a unit. Samples do not have to be collected from every level of every excavation unit, but once an excavation unit is chosen for fine-screen and flotation sample collection, then samples should be collected from every level of that unit. Archaeological statisticians believe that for data to be useful statistically, samples should only be collected in a random fashion. If archaeobiological samples are collected randomly throughout the site, they lose context comparability, a very important analytic tool. In addition, samples chosen randomly would, in effect, be collected from many noncultural contexts that may be meaningless for project goals, and many important cultural contexts would be missed. Therefore, I believe that archaeobiological flotation and fine-screened samples should be taken from nonrandom, explicitly chosen contexts that fit with project goals.

Samples should be collected from every level within a chosen excavation unit to help in cultural zone determination and to be able to develop a progression of plant and animal use through time if the deposits represent cultural zones in chronological order. I usually collect samples from the southwestern quad of each level, but any quad is acceptable as long as collection protocols are consistent. Samples should have the same volume so that concentration comparisons can be made. A two-liter sample is usually of sufficient volume to obtain

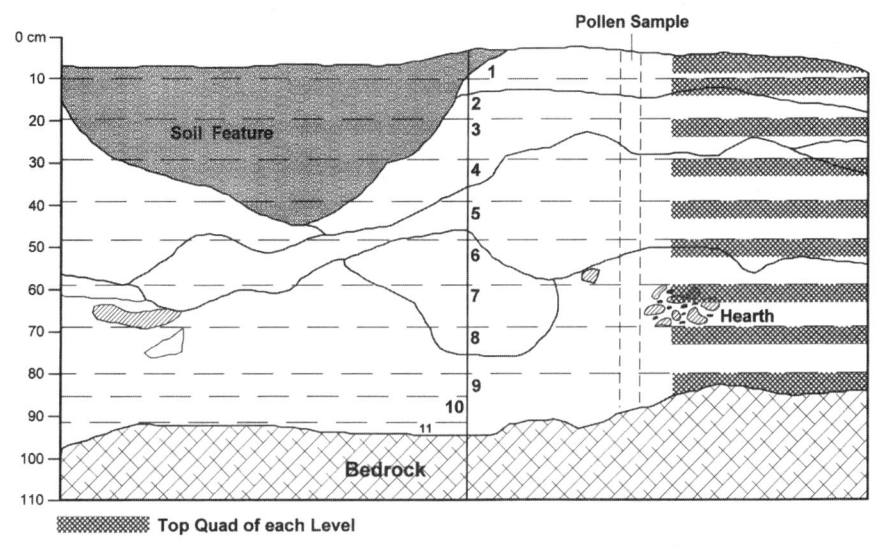

Figure 3.5. Example of basic flotation and fine-screen collection strategy in which two-liter samples are taken from the southwestern quad of each 1 × 1–meter excavation unit, as well as from sections of cultural zones or levels and cultural features. Pollen soil samples are taken from a column in the stratigraphy. (Courtesy of Stephen Bicknell.)

a representative sample from each level, but the volume collected should be increased if increased recovery is needed.

In addition to sample collection from each level, archaeobiological samples also should be taken from any feature or specific cultural context encountered, such as hearths, floors, caches, pits, and any other anomalies (table 3.1). Samples should be collected from these features even if they do not fall within units from which level samples are being collected. Most, if not all, of the matrix from small features should be collected for flotation. Large features should be sampled extensively, where possible, to take advantage of the heightened preservation potential that often exists in pits and other sealed deposits. For feature samples as well as level samples, sample volume should be carefully recorded.

The most important part of collecting samples from features is to include only matrix from the feature itself. Matrix from surrounding deposits should not be included in the sample. Therefore, excavation and sample collection surrounding and including features should proceed using natural rather than arbitrary stratigraphic control. If excavations are being conducted using arbitrary levels, excavation procedure should change to natural levels when features or cultural or natural strata are encountered, particularly if archaeobiological samples are to be taken in that area.

It is always important to record the volume collected and to catalog each sample within the project's cataloging system. Each bag is then taken to the field flotation center or back to the lab for subsequent flotation and fine screening. Bags used for collecting and recording other artifacts from the site can be used for collecting bulk samples for flotation and fine screening. In most cases, large plastic bags will be needed to hold all of the flotation and fine-screening samples in one bag. If plastic bags are used, flotation and fine screening should be undertaken relatively quickly, within a couple of days, or micro-organisms will start to flourish in the sample, potentially destroying organics. If flotation and fine screening are not going to take place relatively quickly, then soil samples should be collected in sandbags, which are slightly permeable, allowing slow drying.

Pollen and phytolith samples also should be taken throughout the site for analysis of human dietary patterns and from areas surrounding the site for paleoenvironmental analysis (table 3.1). Many researchers take pollen and phytolith samples for paleoenvironmental analysis from the archaeological site. But paleoenvironmental samples should be taken away from the site so that what is being analyzed reflects paleoenvironment and not human depositional patterns. In fact, paleoenvironmental reconstruction should proceed through the analysis of pollen and phytolith cores from lakes, bogs, or fens if such features are present. Pollen and phytoliths should be analyzed in conjunction with geomorphological analyses and with any other paleoenvironmental data that might be present.

Pollen and phytolith sample collection from archaeological sites involves collecting matrix samples from a column in an exposed profile (figure 3.5). Because the focus of sample collection in archaeological sites is understanding human activity, samples should be chosen from areas of the site that have good cultural context. In most cases, samples should not be collected randomly throughout the site. A column sampling technique from a fully excavated stratigraphic profile should be used so that potential changes through time can be delineated (see figure 3.5). In many cases, good cultural context is not be known until after excavation, so pollen column samples should be taken from a variety of areas throughout the site.

Pollen and phytolith column samples are taken after excavation has been completed in a unit and a stratigraphic profile has been exposed (Dimbleby 1985; Piperno 1988). First, you need to decide how

> **Table 3.1. Standard Archaeobiological Sampling Procedure**
>
> **Flotation and Fine-Screen Sampling**
>
> 1. Select 1 × 1–meter units for sample collection. As excavation area grows, add units from which to sample.
>
> 2. Select one quad of the unit for sample collection, such as the southwest quad. Systematically collect a two-liter sample from each level of that quad using a bucket with the two-liter volume (or whatever volume you want to use) clearly delineated.
>
> 3. Collect samples from any archaeological features encountered, such as hearths, pits, caches, floors, and other anomalies, carefully recording sample volume. Samples should include only feature matrix, not surrounding matrix.
>
> 4. Collect samples from noncultural deposits for comparison with cultural deposits, carefully recording sample volume.
>
> 5. Assign each sample a separate, unique field sack or lot number.
>
> **Pollen and Phytolith Sampling**
>
> 1. Select an exposed profile in good cultural context from a unit that has been entirely excavated.
>
> 2. Determine how many samples you are going to collect from the profile in column fashion. You can collect one or two samples from each cultural zone, from each stratigraphic level, or every five to ten centimeters along the column. Label bags according to the location of each potential sample and provide a separate field sack or log number for each.
>
> 3. Prepare for sample collection making sure all of your materials are handy. Collection needs to take place quickly and efficiently to avoid environmental pollen and phytolith contamination. Materials include a trowel, spoon or other small collecting tool, water, towel, and labeled baggies.
>
> 4. Start taking approximately one-hundred- to two-hundred-milliliter samples from the bottom of the column working to the top. The trowel and spoon should be cleaned with clean water and dried with a clean towel before each sample collection. Each sample should be taken after the profile has been scraped clean with a trowel. Take samples with a spoon and place each into a clean, labeled plastic bag. Rewash the trowel and spoon for the next sample collection.
>
> 5. Back at the laboratory the sample can be split for pollen and phytolith processing.

many samples you want to take from the profile. In most cases, you should take contiguous samples so that the samples are close together, and each will represent a range of pollen deposited from that time period. Samples are usually taken five to ten centimeters apart. Sometimes, however, samples can be taken strictly following the natural or cultural stratigraphy, particularly if the strata occur in levels thinner than ten centimeters. Mark where you want to take the samples with a measuring tape or flags beforehand.

Wind-blown pollen and phytoliths are ubiquitous in the natural environment, so contamination is an important issue. During sample collection, all tools and supplies must be cleaned and wiped to prevent airborne pollen and phytolith accumulation as well as contamination from one sample to the next. All supplies needed for sample collection should be at hand, and plastic collection bags should be prelabeled, each with its own field sack or lot number. Supplies needed include a trowel, spoon, water, cloth or towel, and prelabeled plastic bags. Sample collection should proceed from the bottom of the column up to avoid contamination from upper deposits. Use a clean and dry trowel to scrape the profile clean. At each designated sample collection spot, use a newly washed and dried spoon (or other useful collection device) to collect approximately one hundred to two hundred milliliters of newly exposed matrix and place it immediately into the appropriately labeled bag. Scrape the profile of the next sample collection spot with a newly cleaned and dried trowel, and use a newly washed and dried spoon to collect the sample. Because sampling damages the profile, this work should take place after excavation and profile mapping are complete. In addition to profile column sampling, pollen and phytoliths also can be collected from features such as pits, caches, hearths, burials, and floors.

To better understand whether pollen and phytoliths from an archaeological site are related to human activity or environmental conditions, you also should take samples off site. Such samples usually consist of modern matrix from the surface to compare with archaeological samples. For this collection method, you can either collect a number of one-hundred- to two-hundred-milliliter matrix samples, or you can use the "pinch" method and collect a number of pinches of matrix from around the site and combine them for a single modern pollen sample. Paleoenvironmental data can be obtained from pollen samples taken from a natural profile some distance from the archaeological site to use as a time/depth comparison to the archaeological samples. To do this effectively, both sample areas need to have good chronological control, usually through radiocarbon dating.

To illustrate archaeobiological sample collection techniques in different environments, I use examples from five different archaeological sites in the United States from which I have collected archaeobiological materials. Some of these examples represent actual sites, and some represent an amalgam of similar sites. The collection methods I advance are not necessarily the methods I used at each of these sites but in retrospect are those I would use if I could do it again. Again, I want to reiterate that each site is different, and the archaeobiologist or archaeologist needs to make sample collection decisions based on the conditions of each site.

The first two site types I discuss are those in which most of the material represents prehistoric cultural deposition: a Puebloan site in the Southwest and a large shell midden site on the northeastern coast. Noncultural agents may have modified parts of the site, particularly postdeposition, but most of the material in the site itself consists of cultural debris. Site type 3 is a rock-shelter in which preservation of archaeobiological materials is excellent but taphonomic factors are numerous. Site type 4 is an open, ephemeral site with shallowly buried lithic scatters and evidence of possible postdepositional modifications such as plowing, erosion, or tree throws. Site type 5 is a deeply buried site with evidence of hearths and lithic scatters, along a river or lake that has flooded or been impounded. These site types do not represent all of the environments of archaeological sites in the United States, but they do cover a wide range of environments from which archaeobiological collection procedures can be extrapolated to similar environments and sites.

PUEBLOAN TYPE SITE

Type site 1 is a large Puebloan site in the Southwest in which excavations have been conducted in both the room blocks and in the midden and plaza areas. Preservation of archaeobiological remains in the arid Southwest is excellent, and large pieces of bone and charcoal are recovered during normal coarse screening. Fine-screen and flotation samples should be taken throughout the site because of the probability of obtaining well-preserved archaeobiological remains. Collection of samples in the midden should proceed according to the basic method I use to collect most archaeobiological fine-screen and flotation samples (figure 3.5; table 3.1).

Random 1 × 1–meter units should be chosen from which to systematically collect two-liter archaeobiological samples from the southwestern quad. As the excavation grows, the number of units chosen for sample collection should grow. I recommend, on average, that for every four to five 1 × 1–meter units chosen for excavation, one unit should be chosen for archaeobiological sample collection.

This average number will change depending on a variety of factors, such as time and money constraints, environmental conditions, and research questions.

Sample collection also should be done in the room blocks themselves. Room blocks tend to contain postdepositional fill (i.e., some secondary cultural deposits and some fill representing environmental noncultural deposits). Primary cultural deposits might not be encountered until excavation has proceeded to the bottom of the room. There the collapsed roof may be found on top of the room floor. Samples from deposits near and on the room floor should be taken so that prehistoric material is recovered. However, some postdepositional room fill may include midden deposits from later occupations, so it is important to obtain samples from these deposits as well.

Throughout the midden and in the room blocks, features may be found from which additional samples need to be taken. Features may include hearths, burials, pits, caches, and anomalous matrix. In these cases, the entire feature or a large portion of the feature can be collected for fine-screening and flotation. Again, it is important to record total volume amount, or two liter samples can be collected to stay consistent with other sample volumes.

Due to the arid environment, other types of archaeobiological samples can be preserved in the Southwest, such as fiber artifacts (basketry, sandals, netting), bone artifacts and tools, and paleofeces (also termed *coprolites*). Fiber artifacts and bone artifacts are easy to recognize during excavation and screening. Paleofeces are harder to recognize if you do not know what you are looking for. Paleofeces are desiccated human feces that contain the remains of what past peoples ate (figure 3.6). They can be preserved in arid or frozen conditions. In the Southwest, paleofeces have been recovered singly from midden deposits or in large quantities from room blocks that were used as latrine areas. In a latrine situation, paleofeces may be distinguishable as separate entities, or they may be found as a large horizon. Excavation of these unique specimens should focus on the recovery of individual samples placed in separately labeled bags. Each sample represents short-time food intake by one individual, so recovery of each individual sample is preferred to the excavation of large clumps of latrine areas that would represent dietary intake of a number of individuals. We can recover DNA and hormones from paleofeces (Sutton et al. 1996; Sobolik et al. 1996; Poinar et al. 2001), providing an excellent database through which to answer research questions on diet, nutrition, and sex determination. For optimal analysis of human DNA

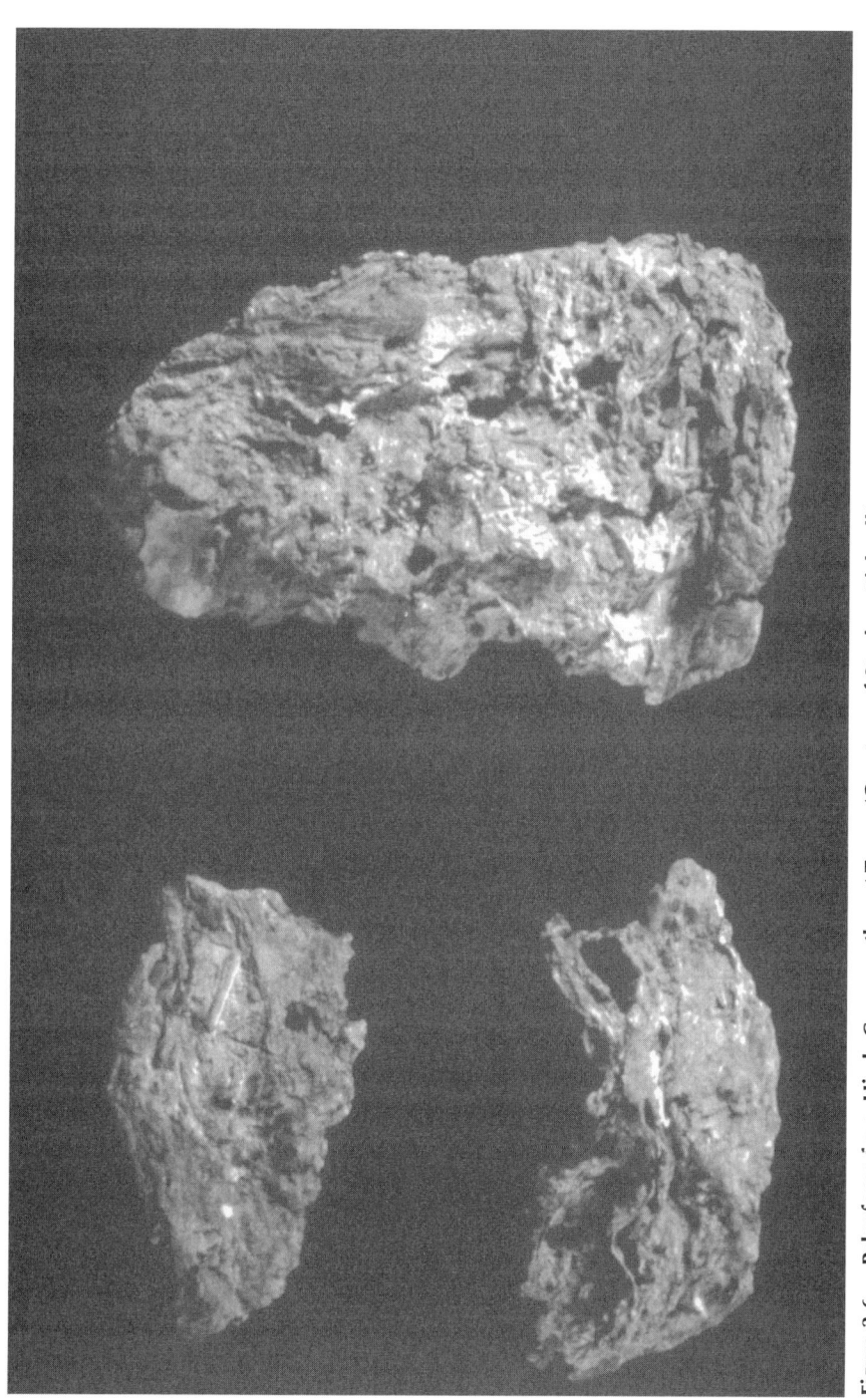

Figure 3.6. Paleofeces from Hinds Cave, southwest Texas. (Courtesy of Stephen Bicknell.)

from paleofeces, the samples must not be touched or otherwise contaminated. Therefore, paleofeces should be collected using sterile latex gloves, and each sample should be placed in a separate, clean, plastic bag to avoid contamination. The samples should not be handled, breathed on, or removed from the bag. Identification and discussion of paleofeces is covered in Fry (1985).

Pollen and phytolith collection should also proceed according to the basic guidelines already defined. Several pollen profile columns should be chosen for sample collection. In addition, other samples can be collected from a variety of features or locations encountered during excavation, such as vessels or grinding implements, to determine what uses the vessels or implements had, and from living surfaces such as room floors to determine room function and potential processing or storage areas.

For example, Bryant and Morris (1986) analyzed functions of ceramic vessels, metates, and manos recovered from Antelope House, a large prehistoric Anasazi Puebloan site located in Canyon de Chelly, through pollen analysis of matrix from the bottom of forty-two vessels and matrix adhering to one metate fragment and five manos. In addition, fourteen soil samples were analyzed for comparison so that natural, environmental pollen could be differentiated from culturally deposited pollen. The authors were able to determine potential uses of several vessels and grinding stones, indicating food preparation or storage of maize, cottonwood, beeweed, cattail, and goosefoot or amaranth. They state that vessels that had been interred in matrix surrounding hearths tended to contain culturally deposited pollen from food preparation or storage, whereas vessels found in midden deposits tended to contain natural, environmental pollen that collected in the samples through postdepositional processes.

Room function can be determined in part through analysis of pollen from matrix samples recovered from room floors. For example, Bryant and Weir (1986) analyzed numerous pollen samples collected above, on, and below floor surfaces of sixteen suspected storage rooms, eleven suspected habitation rooms, and six suspected ceremonial rooms at Antelope House. Most of the storage rooms contained large amounts of economic or culturally deposited pollen such as maize, beeweed, and cattail, indicating possible storage of these plants. Habitation rooms did not contain significant amounts of economic pollen and were thus easy to distinguish from storage rooms. Suspected ceremonial rooms usually contained significant amounts of pine, juniper, and cattail pollen, a distinctly different assemblage than pollen spectra from habitation and storage rooms. The authors state that pollen analysis should be combined with other archaeological information to assess room function.

Collecting pollen and phytolith samples from living surfaces involves a systematic procedure in which samples are taken evenly across the surface, such as every ten centimeters (figure 3.7). In the example provided, I collected pollen samples from the floor of a pithouse from Old Town Ruin, LA1113, New Mexico. As excavation proceeded through the upper fill of the pithouse, which happened to be an overlying Mimbres room block, one-hundred- to two-hundred-milliliter matrix samples were collected from the surface of the pithouse floor. To avoid modern pollen contamination, samples were collected as soon as that area of the floor was exposed, using a newly cleaned spoon or trowel. As excavation proceeded and more of the pithouse floor was exposed, sampling continued until the entire floor was exposed and sampled. Samples were labeled and identified according to a grid system in which one wall (y-axis) of the pithouse was labeled alphabetically, and another wall (x-axis) was labeled numerically (figure 3.7). Pollen sample collection proceeded normally through two hearths encountered during excavation and from which flotation and archaeomagnetic dating samples were taken.

In addition to the flotation and fine-screen and the pollen and phytolith samples collected from a variety of contexts within the Puebloan site, samples were also collected around the site to compare human versus natural deposition.

SHELL MIDDEN TYPE SITE

Type site 2 is a large shell-midden site on the coast of northeastern North America. Periodically through time, people lived at the site and collected large quantities of mollusks. The discarded mollusk shells formed a large shell midden in which other aspects of prehistoric life such as living floors, hearths, and discarded animal bones were preserved. The site contains a huge faunal assemblage in which the best-preserved fauna (mollusk shell remains) are probably overrepresented. The shells, however, produce an alkaline environment, due to the calcium carbonate content of the shell, in which bone remains are well preserved. Carbonized botanical remains are also recovered, although their preservation is not as good as bone.

Old Town Ruin LA1113, New Mexico
Pithouse Floor Pollen Sampling

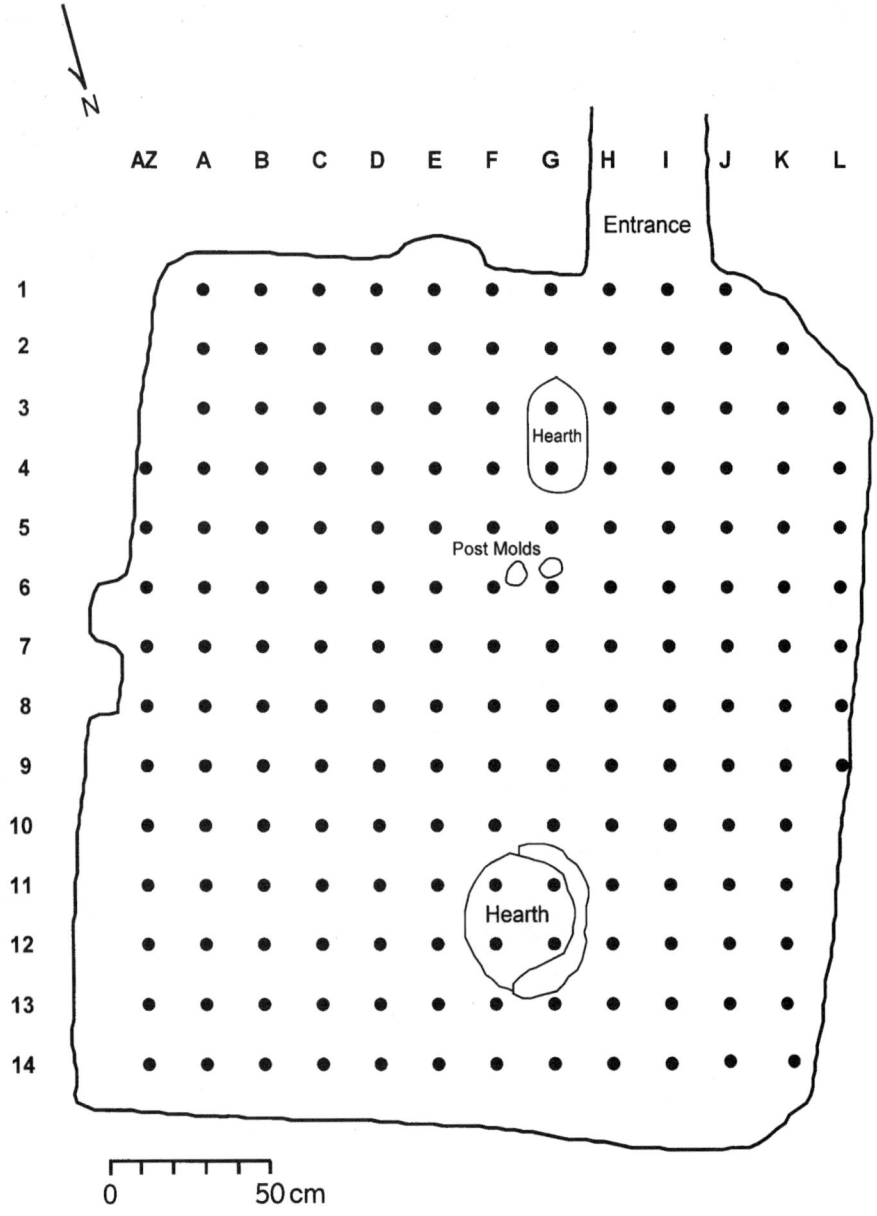

Figure 3.7. Example of pollen sampling strategy from the floor of a pithouse. Each block represents a 100- to 200-ml pollen soil samples. (Courtesy of Stephen Bicknell.)

The stratigraphy of shell middens can be confusing because cultural deposition developed as humans dumped refuse in ever-changing areas of the site (figure 3.8). People shifted living and dumping areas on and around the midden through time. Because of this, the stratigraphy of a shell midden is not uniform across the site. This makes determining cultural zones, areas, or levels difficult, particularly during excavation. Therefore, archaeobiological samples should be collected in a variety of areas in such a site and will be useful for determining cultural areas, areas that are usually determined after complete analysis of all associated artifacts.

In such a confusing site, archaeobiological samples should be collected according to the basic sampling procedure (figure 3.5; table 3.1) in which several 1 × 1–meter units are chosen for systematic two-liter matrix collection from the southwestern quad. Several units are chosen for systematic collection, and as the excavation area expands, more archaeobiological samples are collected. Most areas of a shell midden site, of course, consist of shell with some nonshell matrix. Collecting matrix samples can be difficult, so large pieces of shell should be removed so that a representative two-liter sample for flotation and fine screening can be obtained from all areas of the site chosen for sample collection. This may mean that the southwestern quad of each level (probably representing arbitrary five- or ten-centimeter levels) will be entirely included in the fine-screen and flotation sample after large pieces of whole shell have been removed.

Shell midden sites may contain hearth remains, living floor surfaces, burials, and remains of occupation or storage areas. Archaeobiological samples should be collected from all of these features with particular emphasis on samples in which midden deposits are not mixed with feature deposits. If the feature is large, more than two liters of matrix can be collected as long as the volume is recorded.

Pollen and phytolith samples can be collected from shell midden sites using the basic sampling procedure (table 3.1), although the alkaline depositional environment of a shell midden is not conducive to their preservation. Because of the presence of a large number of shells, collecting pure matrix deposits will be difficult if not impossible. Collection may need to be limited to features that contain little or no shell or areas of the site that contain less shell. Because shell midden sites mainly reflect cultural deposition and modification by humans, pollen and phytolith samples from these sites should not be used to determine paleoenvironment.

Figure 3.8. Stratigraphic profile of the Todd site, a large shell midden on the coast of Maine. (Courtesy of Stephen Bicknell and David Sanger.)

ROCK-SHELTER TYPE SITE

Cultural deposition in rock-shelters is very similar to that observed in shell midden sites. Stratigraphy and depositional patterns can be very confusing due to intense human occupation and refuse disposal and accumulation. Rock-shelters tend to exhibit many noncultural taphonomic factors that affect and modify deposition, making assessments of cultural zones, areas, or levels even more difficult than in shell middens. Because of these difficulties, it is imperative that archaeobiologists are involved in excavation and sample collection so that they can understand and help interpret taphonomic factors influencing the archaeobiological record.

Rock-shelters tend to have excellent preservation of archaeobiological materials, depending on specific environmental conditions, which warrant the time spent on clarifying taphonomic processes. Bone and botanical remains can be abundant. In some cases, such as the dry rock-shelters of the Lower Pecos Region of southwest Texas (Shafer 1986), botanical materials are so plentiful and well preserved as to be comparable to shells in a shell midden (figure 3.9). Uncarbonized perishable remains in particular can be recovered,

Figure 3.9. Archaeobiological remains recovered from the coarse screen of Hinds Cave, southwest Texas. (Courtesy of Kristin D. Sobolik.)

such as basketry, sandals, nets, digging sticks, bone tools, and paleofeces.

Archaeobiological flotation and fine-screening sample collection should proceed according to basic techniques (figure 3.5; table 3.1). Special care should be taken to assess taphonomic factors affecting the assemblage (see chapter 2). Samples should be taken in both potential cultural and noncultural areas so that differences between these two areas can be analyzed and compared. Because rodents introduce noncultural material into the deposit and move cultural material out of context, rodent burrows should be excavated and collected separately. Fine-screen and flotation samples can be collected from rodent burrows but should be considered in their context and should be used to analyze rodent activity patterns rather than cultural lifeways. Rodent activity studies are useful because they help determine site-specific and potentially regional taphonomic factors (see the case study in sidebar 2.1).

Many rock-shelter sites display microstratigraphic depositional patterns that reflect both cultural and noncultural processes. Therefore, archaeobiological samples should be collected from these cultural and natural levels, if visible, rather than from arbitrary levels that may cross-cut these strata. Samples collected from arbitrary ten-centimeter levels could reflect both natural and cultural deposition or cultural deposition from various occupations or activities.

Pollen and phytolith samples can be recovered according to basic sampling procedures (table 3.1). As with flotation and fine-screen samples, pollen and phytolith samples should be taken from a variety of contexts throughout the site, as well as from contexts away from the site for comparison of cultural and noncultural deposits. Pollen and phytolith sampling also needs to follow natural or cultural stratigraphy, if present, which may mean taking more samples than one every ten centimeters, as microstratigraphy warrants.

OPEN TYPE SITE

This type site is an open site located in plains, desert, or wooded areas with evidence of erosion and surficial modification such as plowing, tree throws, digging, or animal burrowing. The site may contain lithic scatters and evidence of hearths (e.g., fire-cracked or burned rocks or stained soils; figure 3.10). Remains of cultural activity at the site tends to consist of durable artifacts and features. Preservation of

Figure 3.10. Fire-cracked rock representing the remains of an eroded, prehistoric hearth at BIBE 57, Big Bend National Park, Texas. (Courtesy of Kristin D. Sobolik.)

organic material tends to be poor because much of the site is exposed and has been eroded or modified by cultural and noncultural agents, degrading organic material. Many sites located through CRM projects fall into this category.

Archaeobiological sampling should take place according to standard procedures (figure 3.5; table 3.1), although much of the site may consist of noncultural matrix surrounding sporadic cultural deposits represented by durable artifacts and features. Particular emphasis should be placed on sample collection in and surrounding features. Samples collected from features may include postdepositional fill but also generally contain remains of prehistoric activity. Collecting samples around features will allow you to compare samples to determine whether the feature fill is mostly cultural or postdepositional. If samples from the feature can be ascribed to cultural deposition, charcoal from these samples is an excellent source for radiocarbon dates.

At the Hulme site, an open Central Plains Tradition (CPT) homestead in central Nebraska, a large surface scatter of animal bones and artifacts was noticed during field leveling (Bozell 1991). Most of the site was excavated using quarter-inch screens in which large bones from antelope, deer, and bison were recovered. Flotation and fine-screened samples were taken from the central fireplace, and four trash-filled storage pits from which bones from a smaller and more varied fauna were recovered, such as snake, prairie dog, turtle, and fish. Analysis of the full set of remains indicates that people had a much more varied meat intake than previously hypothesized. Analysis also suggests that during the CPT, the climate became drier, so the grass forage was less preferable for bison and more preferable for antelope. Plains bison herds may have declined in number due to increasingly arid climate and human hunting patterns.

Pollen and phytolith samples can be collected from open type sites according to standard procedures (table 3.1), and, as with flotation and fine-screen samples, particular emphasis should be placed on cultural deposits such as features. Samples collected from around features and off site also are useful for comparison with cultural deposits as well as for paleoenvironmental reconstruction. Phytolith analysis is particularly useful in grassland environments because grass taxa cannot be distinguished on the basis of pollen analysis, while grass phytoliths are identifiable to various levels.

BURIED TYPE SITE

Buried sites are often hard to locate (see Toolkit, volume 2) but may contain well-preserved archaeobiological materials. Buried deposits are protected against surficial modifications that are prevalent in open sites. Buried sites also tend to have good stratigraphy, which permits defining cultural and noncultural deposits. For example, the Devil's Mouth site is a large alluvially buried terrace site located at the confluence of the Devils River and Rio Grande in Texas (figure 3.11). The site was discovered during archaeological survey for the Amistad Dam, a six-mile-long structure built for hydroelectric power, flood control, and recreational purposes. Impounded waters behind the dam created a 64,860-acre reservoir that extends seventy-three miles up the Rio Grande, fourteen miles up the Pecos River, and twenty-five miles up the Devils River (Labadie 1994). Impounded waters cover at least six hundred archaeological sites, including the Devil's Mouth site. The site is at least thirty-six feet thick, with twenty-four recognizable strata of cultural midden deposits alternating with silt, sand, and clay deposits (Johnson 1964).

Figure 3.11. Excavation of the Devil's Mouth site, an alluvially deposited site at the juncture of the Devils River and Rio Grande.

Because alluvially buried sites are located along waterways, they can be impacted (destroyed, eroded, or removed) by impoundments for energy sources or recreational facilities. Alluvially deposited sites can contain a large amount of noncultural matrix; therefore, archaeobiological sampling should focus on cultural deposits and features. Analyses of archaeobiological samples from cultural strata and from hearths, pits, and occupation floors will provide information on changing patterns of plant and animal use at the site (see the case study in sidebar 1.1). In addition, analysis of noncultural matrix samples allows comparison of paleoenvironmental conditions before and after cultural deposition. At a site like the Devil's Mouth site, two-liter archaeobiological samples should be taken for flotation, and fine screening from the southwestern quad of a number of 1 × 1–meter excavation units continuously through cultural and noncultural strata, so that changes in environment, as well as cultural patterns, can be ascertained.

Pollen and phytolith samples should be obtained through the standard procedures (table 3.1). It is important to make sure samples from cultural deposits are distinguished from those from noncultural deposits. Analysis of pollen column samples that include cultural and noncultural deposits will provide a good framework for tracing changes in environment through time as well as human impacts and plant use. Pollen analysis of matrix samples collected in column fashion from all twenty-four strata from the Devil's Mouth site revealed several climatic changes in the region over time (Bryant 1966). From 7000 to 3000 B.P., the region had a mesic environment that was probably caused by increased rainfall during spring and summer. Erosion of Rio Grande terraces occurred at the end of this mesic interval (3000–2500 B.P.), which was followed by an increasingly xeric environment up to the present.

CONCLUSION

Excavation and recovery of archaeobiological samples, including pollen and phytolith samples, can follow a standard procedure. This procedure can be modified on the basis of site type, environmental conditions, time and money constraints, and research design. It is a good idea to collect more samples than needed and even more samples than will actually get analyzed with your particular project. Although curation space is limited, excavation is destructive, and you usually don't get a second chance. In fifty years, someone may need those samples to help answer a research question, or particular samples from Phase II may be needed during Phase III. Having matrix samples sitting on a shelf for possible future analysis is preferable to having samples lost in the backdirt pile.

4
LABORATORY AND ANALYTICAL TECHNIQUES

Archaeobiological analyses require a significant amount of technical expertise that is learned through training, experience, and many, many hours of analysis. Archaeobiological analyses should only be conducted by trained archaeobiologists (table 1.1). Because these studies are time-consuming and demand a lot of experience, many archaeobiologists become specialists or experts in particular areas or on particular botanical or faunal taxa. Technical expertise is the backbone of archaeobiological analyses, and technical identification and analysis can be the most time-consuming and often tedious step. Anyone can put in the hard work necessary to become a technical expert in archaeobiology, but to become an archaeobiological scientist involves using botanical or faunal remains to answer broader-ranging questions. The most rewarding aspect of archaeobiology is not when a certain fish bone has finally been identified (although such small victories are exciting) but when the identification of an entire fish assemblage leads to new discoveries, answers previously unanswered questions, or indicates that a modification of a hypothesis is necessary.

For example, McInnis (1999) chose prehistoric subsistence practices in coastal Peru for her master's thesis topic. She helped conduct excavation at Quebrada Jaguay and chose to analyze the faunal remains. Overall, the remains were poorly preserved and consisted mainly of fish fragments and a large number of otoliths (fish ear bones). Identification of the fish bone fragments would have taken an immense amount of time and energy, from which McInnis probably would only have been able to identify one or two species or

$$Y = aX^b \text{ or } \log_{10} Y = b(\log_{10} X) + (\log_{10} a)$$
$$Y = \text{Standard Length}$$
$$X = \text{Otolith Length}$$
$$a = \text{Y-Intercept}$$
$$b = \text{Slope}$$

Figure 4.1. Allometric scaling formula. (From Reitz and Wing 1999: 175–76.)

fish elements. Instead, she focused on the otoliths and conducted an analysis of the size and type of fish taken at the site using regression formulas provided by Reitz and Wing (1999) (figure 4.1). Using the regression formulas and size of the otoliths, McInnis estimated population structure of the drum fish (Sciaenidae) at the site (figure 4.2).

The technical portions of her analysis complete, McInnis, in conjunction with the project director, used her data to address issues regarding early maritime adaptations, revealing that Paleoindian-age people in South America were adapted to marine-based environments and that Quebrada Jaguay is the earliest known site providing such evidence (Sandweiss et al. 1998). McInnis's work reveals the importance of technical analyses that ultimately feed into larger issues and broader-ranging research questions. The first steps toward being able to answer such questions, however, are technical steps.

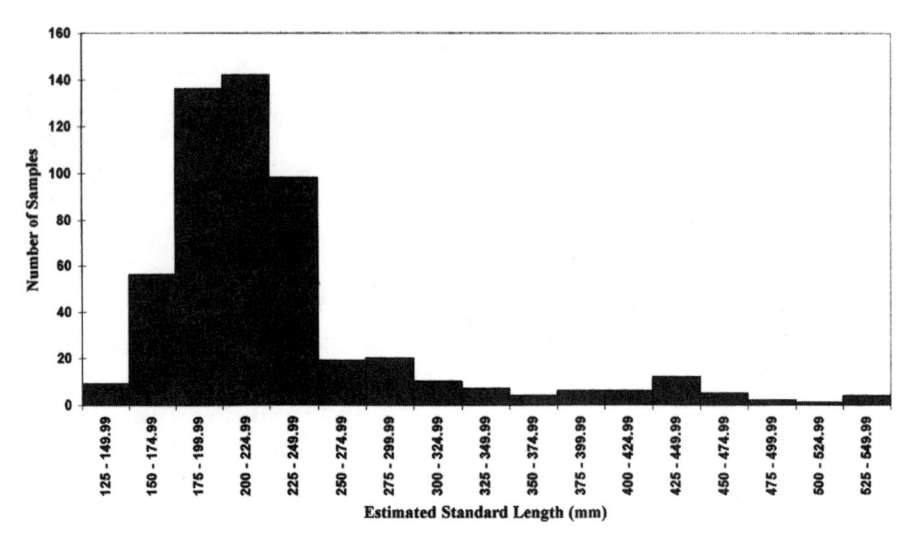

Figure 4.2. Population structure of Sciaenidae for Quebrada Jaguay. (From McInnis 1999.)

REFERENCE COLLECTIONS

The first and most important step in archaeobiological analysis is generation of an extensive and usable reference collection of modern plants, animal bones, and shells for comparative purposes. Archaeobiologists should be familiar with the plants or animals in a region and should have an extensive reference collection from that area before proper identification and analysis of archaeological samples can take place. One reason archaeobiologists should be involved in overall project design and be present during excavations is so they can collect modern reference samples from the surrounding area.

For paleoethnobotanists, developing a collection involves collecting a wide variety of whole plants during different seasons so that the life cycle of the plant is represented. Whole plants can be collected in a plant press and dried for preservation. Be sure to record where and when each plant was collected. Other reference samples should include pieces of wood (some of which can be charred for charcoal identification), nuts, seeds, berries, roots, and phytoliths from all portions of each plant. Pollen reference samples are also collected from individual plant flowers during the appropriate season. Chemical processing of modern plants needs to be undertaken to remove pollen (steps IV and V of table 4.2) and phytoliths (steps III and IV of table 4.3) for reference samples. The most important part of collecting modern plants to use as comparative reference samples is to definitively identify them. Otherwise, they are useless or potentially problematic. Be sure to consult a qualified botanist if there are any questions regarding plant identification.

For zooarchaeologists, animals can be obtained, usually with an appropriate permit, via roadkill collection, trapping, and taxidermy businesses. Animals on endangered or threatened species lists often cannot be collected, even through roadkill. A list of these animals, as well as collection permits, usually can be obtained through the state departments of natural resources or similar state agencies. Keep in mind that reference sample collecting requires defleshing animals to recover all bones or shells. This process can become an interesting exercise in ethnoarchaeology, or at least a learning experience in anatomy and morphology. Be sure to work with someone experienced in butchery and defleshing before cutting or macerating on your own. Also, record information on the animal's size, weight, sex, age (if possible), and other life characteristics.

Reference collections in museums and laboratories can supplement your collection. For example, I use the faunal collection curated at the

University of Maine Zooarchaeology Laboratory in addition to my own personal zooarchaeological reference collection. To expand my reference sample for species not available in either of those collections, I use collections at the Museum of Comparative Zoology at Harvard University. Analysts should know where such collections are located and use them if necessary.

In addition to comparative reference collections, a number of identification guides and atlases are available to help identify both plants and animals. Guides and atlases are useful for identification and should be used in conjunction with, but never in place of, a comparative reference collection. I cite identification guides in each appropriate section.

After an extensive reference collection has been accumulated or located, archaeobiological identifications can proceed using those collections in addition to the age-old process of lots and lots of time and experience. I teach a zooarchaeology class in which each student learns basic faunal identification skills, identifies a small faunal collection from an archaeological site, writes a faunal report, and presents the information to the class. Although I can lead the students in an appropriate direction or point out easy identification tips, the students are usually surprised to realize that learning identification is mainly up to them. They need to spend an enormous amount of time familiarizing themselves with animal bone and matching their archaeological bone fragments to bones from the comparative collection. It is a time-consuming and often tedious process in which there are no shortcuts. At the end of the class after their final paper is turned in, many students feel that they are now zooarchaeologists. Yet this step is only the very first stage in becoming technically competent. With further experience comes more expertise. Most archaeobiologists have probably had to reanalyze their first few research projects, correcting a number of technical errors along the way. If they haven't done this, they probably should; it is an enlightening experience. Archaeologists contracting for or hiring archaeobiologists should realize that the best analysts are usually the most experienced—and expensive—analysts.

BOTANICAL REMAINS

Botanical remains that are recovered during coarse screening should be sorted into similar categories or groups, such as seeds, charcoal, fiber, and nutshell, from which identification can proceed. This initial sort can be conducted by someone other than the archaeobiologist. All of the material should retain the particular field sack or lot number assigned in the field. Samples should not be washed or modified through brushing or removing adhering matrix unless the analyst deems this necessary.

The most tedious and time-consuming aspect of botanical analysis usually is sorting flotation samples. To facilitate sorting and quantification, samples can be graded into different size classes by screening the samples through nested geological sieves (e.g., two and one millimeter). Samples should be sorted under a magnifying lens or dissecting microscope (10×–20×). All botanical remains should be sorted into groups such as seeds, wood charcoal, fiber, leaves, and miscellaneous unidentified items. They can then be identified as to taxon and element using comparative samples and identification guides. Some useful identification guides for botanical remains include Appleyard and Wildman (1970), Catling and Grayson (1982), Core et al. (1979), Corner (1976), Dimbleby (1978), Gunn et al. (1976), Leney and Casteel (1975), Martin and Barkley (1961), Montgomery (1977), and Western (1970). See Pearsall (1989:182) for references on identification of domesticates. The samples can then be quantified to aid in analysis and interpretation. Several paleoethnobotanical recording techniques are provided in Pearsall (1989:108–22; 2000).

QUANTIFICATION TECHNIQUES

Quantification of plant remains is a significant aspect of archaeobotany. Researchers apply many different quantification methods, and few comparative papers have been published on the quantification procedures and their assessment (but see Hastorf and Popper 1988). Several of the most commonly applied procedures are discussed here.

The presence/absence or ubiquity method simply measures how many samples contain each taxon within a group of samples. Either a taxon is present in a sample, or it is absent. No matter what other quantification methods are employed, every study should use presence/absence. Because it is a universally used technique, this form of quantification permits easy comparison between samples and assemblages. Presence/absence reduces the effects of differential preservation and sampling, although the number of samples and taxa within a sample will affect results: The more taxa that are recognized in a sample, the more important a common botanical constituent will seem and the less important an infrequent botanical constituent will seem. This is also the case as the number of samples is increased.

In the *percentage weight method*, all of the botanical constituents in a sample, including both flotation and coarse-screen samples, are separated and weighed. The weights are compared directly or are reflected as a percentage weight of the total. The weight technique is often used in quantification, thus making analyses using this method easier to compare with other studies. The major drawback of this method is that it underestimates the lighter contents, such as fiber, and overestimates the heavier contents, such as wood charcoal.

Another frequently used quantification technique is the *percentage count method*, in which the botanical remains are counted and compared to the total botanical count. This method tends to overestimate botanical remains that are easily broken or contain more fragments to begin with, such as fiber particles and wood charcoal. However, this method is not time-consuming, relatively easy to conduct, and additive, so that when new samples have been quantified the percentage count simply can be added to the total (similar to the NISP [number of identified specimens] quantification method for faunal remains).

In the *percentage volume method*, all material from the sample is separated and placed into standard size containers. The number of containers each constituent fills is then compared to the total. This technique is fairly sufficient in estimating the amount of each item in a sample, although this method is very cumbersome and inexact. It is almost impossible to get irregularly shaped botanical remains, such as strands of fiber, nutshells, and wood charcoal, to fill all volumetric space of standard size containers. This method, thus, uses more "guestimation" than other quantification methods employed. It is also not widely used in archaeology, making comparisons difficult.

For the *percentage subjective method*, all of the botanical constituents in a sample are aligned according to their frequency, from most frequent to least frequent. These frequency groupings are then placed into percentage groups that provide a range of error. Each constituent is placed into these different percentage groups when the sample is being sorted and separated, making it the least time-consuming and most cost-efficient quantification technique. The percentage subjective method does not overestimate larger items or items that are broken into numerous pieces. The problems with this technique are threefold: (1) Quantities are presented as a range so data cannot be manipulated statistically; (2) this technique is not additive so that when more samples are added to the original analysis, quantification percentages cannot be added to the previous total; and (3) this technique is not widely used, making comparison between different studies often difficult.

I used the percentage subjective method in an analysis of thirty-eight paleofecal samples from Baker Cave in southwest Texas (Sobolik 1988). This quantification method was chosen because it is the easiest and introduces the least amount of bias into the analysis. The percentage subjective method does not overestimate or underestimate the importance of any botanical remains. The technique is also useful because each paleofecal sample represents an individual analysis unit; therefore, the problem of adding samples to a total was not encountered. A secondary quantification technique, the weight method, was also used for comparative purposes and to facilitate any future statistical analyses. The weights were recorded but were not analyzed because the weight method overestimates the importance of heavier botanical items and underestimates lighter botanical items. The botanical remains from each paleofecal sample were sorted, identified, and visually assessed to determine the relative percentage of each constituent as compared to the total sample (table 4.1). For example, I estimated that sample 5 contained 80 to 94 percent prickly pear fiber; 1.4 percent onion fiber, yucca fiber, sotol fiber, and prickly pear seeds each; and a trace amount of mustard seeds. The percentage subjective method provided a quick, efficient method of quantification that did not overestimate or underestimate any botanical constituent, and it could be done at the same time as the sample was sorted.

POLLEN ANALYSIS

Pollen analyses from archaeological sites and other environmental conditions can offer a diversity of information on prehistoric populations and subsistence practices that cannot be determined through the sole analysis of other archaeobiological remains. Pollen is observed and analyzed from many sample types. Such diverse uses for pollen include paleoenvironmental reconstruction, archaeological dating techniques, and paleodiet. Pollen is useful in a wide range of areas, mainly due to its prevalence in the environment and its distinctive, sturdy structure.

Pollen is a sturdy structure due to its exine (outer layer), which is composed partly of sporopollenin, a strong, resistant substance. The inner layer of pollen (intine) consists of cellulose, which is

Table 4.1. Percentage Subjective Quantification of Some Botanical Remains from Eight Paleofeces from Baker Cave, Texas

Botanical Remains	Sample Number								
	1	2	3	4	5	6	7	8
Fiber								
Onion		D	F	H	H	E	G	H
Prickly pear	B		H	G	B	H		H
Yucca					H			
Agave							H	H
Sotol			H	I	H			
Seeds								
Prickly pear					H		A	B
Juniper							H	H
Cactus						H		
Mustard					I			
Nutshell								
Acorn	H							
Walnut				G				

A = 95%–100%
B = 80%–94%
C = 65%–79%
D = 50%–64%
E = 35%–49%
F = 20%–34%
G = 5%–19%
H = 1%–4%
I = 0.1%–0.9% (trace)
Source: Modified from Sobolik (1988).

easily degraded after a short length of time, such as in archaeological deposits. When the cellulose layer of the intine is degraded, only the outer layer containing sporopollenin remains. However, this layer is often sufficient for identification because the exine contains distinct sculpturing patterns and aperture shapes (figure 4.3), allowing for pollen identification to be made to the species level in some instances.

Pollen types are divided into insect-pollinated plants (zoophilous) and wind-pollinated plants (anemophilous). Insect-pollinated plants produce relatively few pollen grains and are usually insect-specific. Insect-pollinated plants generally produce fewer than ten thousand pollen grains per anther (Faegri and Iversen 1964). These pollen types are rarely observed in the pollen record due to their low occurrence in nature and method of transport.

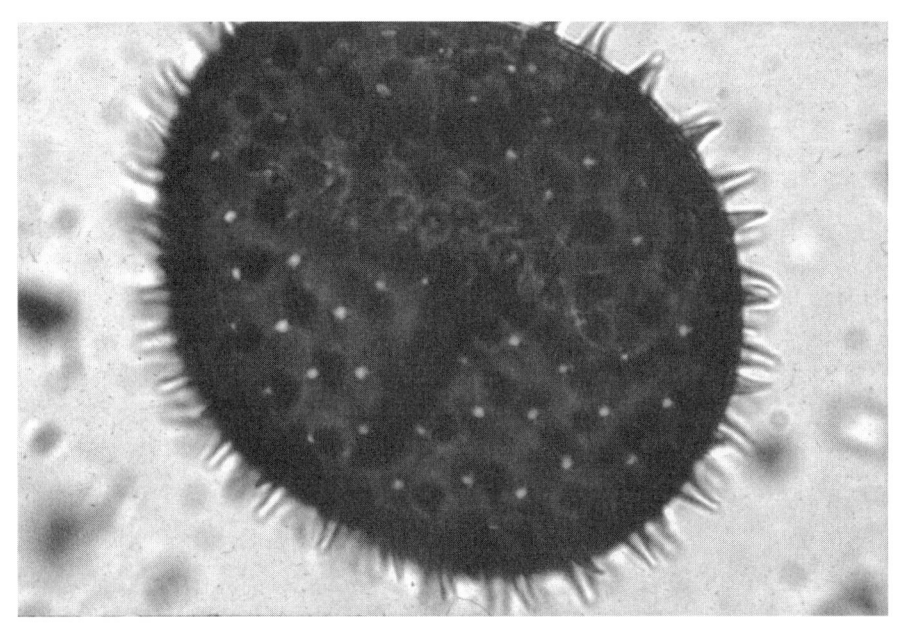

Figure 4.3. Malvaceae pollen grain showing the echinate (spiny) patterning of the exine. (Courtesy of Kristin D. Sobolik.)

Wind-pollinated plants, on the other hand, produce large amounts of pollen and are frequently found in the pollen record. Faegri and Iversen (1964) state that an average pine can produce approximately 350 million pollen grains annually, and Mack and Bryant (1974) found pine pollen percentages over 50 percent in modern deposits where the nearest pine tree was more than a hundred miles away.

A high frequency of wind-pollinated pollen types in archaeological samples most likely indicates natural environmental pollen rain rather than human subsistence activities. High frequencies of insect-pollinated pollen types, however, may indicate human use and modification of that particular plant.

Pollen analysts process samples in various ways, although the basic procedure involves removing organics, silicates, and carbonates. A standard procedure to extract pollen from soils is provided here (table 4.2). I do not recommend that untrained technicians conduct this procedure. Sample preparation involves use of highly dangerous chemicals and requires specialized equipment and safety protocols. Pollen extraction should be done only by trained technicians who realize the potential dangers at each step and take appropriate precautions to avoid damage to person, property, and pollen.

Table 4.2. Standard Pollen Extraction Procedure*

Step I: Removal of large organic or mineral particles
1. Remove thirty to fifty milliliters of soil from the sample collected. If samples come from heavily weathered areas or alluvial sediment, use one hundred milliliters of soil.
2. Screen samples through a one-millimeter mesh screen into a beaker. Discard material caught in screen.
3. Add one to two *Lycopodium* spp. spore tablets, carefully recording number of spores per tablet.

Step II: Removal of carbonates
4. Add concentrated HCl (38 percent) to remove carbonates and dissolve calcium bonding in spore tablets. Stir and allow reaction to take place. If reaction causes foam, use fine spray of ethanol to disperse.
5. Pour off and discard liquid fraction. Add a thousand milliliters of distilled water to sediment in beaker and stir. Let solution settle for two hours. Repeat this step two more times. Place remaining sample in fifty-milliliter centrifuge tubes.
6. Centrifuge the residue at 2,000 RPM for fifteen seconds. Discard liquid fraction.

Step III: Removal of silicates
7. Transfer remaining sediment into plastic beakers and add small amounts of 70 percent HF** acid until matrix sample is covered. Stir occasionally and let sit overnight.
8. Add distilled water to beaker and stir. Let solution settle for two hours. Pour off and discard liquid fraction in fume hood sink. Repeat this step at least two more times. Place remaining sample in fifty-milliliter centrifuge tubes.

Step IV: Removal of organics
9. Rinse residue in glacial acetic acid to remove water. Centrifuge and decant.
10. Prepare acetolysis mixture: nine parts acetic anhydride, one part sulfuric acid.
11. Add acetolysis mixture to samples, stir thoroughly, and place in a boiling water bath for five minutes. *Don't mix water from water bath with acetolysis mixture!* Remove, centrifuge, and decant. Repeat.
12. Wash sample in distilled water. Centrifuge and decant. Repeat.

Step V: Slide preparation
13. Place remaining residue in a small vial with glycerine for curation. Label.
14. Take a small portion of glycerine mixed residue and place on a microscope slide. Place coverslip over sample and secure with nail polish or other sealant. Identify and count pollen.

---

*Pollen extraction techniques involve the use of toxic chemicals. Extraction should never be attempted without a fully functioning fume hood and protective coat, gloves, and goggles. Processing should be done by a trained technician.

**Use of HF must be restricted to a fume hood. HF fumes are very harmful and can cause permanent damage to lungs and nose if inhaled. Contact with HF can be fatal so always wear plastic coat, plastic gloves, and plastic face mask.

As with other identifications, pollen identification must proceed using a modern comparative collection of representative pollen types from the region. Learning pollen identification is time-consuming and involves practice and experience. A standard North American identification key is Kapp (1969) and Jones et al. (1995).

Pollen data can be presented in various ways, but most data are presented as a pollen diagram. An example of a very basic pollen diagram is presented here (figure 4.4). In this diagram, cultural zones or stratigraphic levels are listed on the left axis, and associated radiocarbon dates are provided on the right axis. Most diagrams present data in stratigraphic and chronological order with the bottom of the diagram representing the deepest (and hopefully oldest) deposits and the top representing the youngest and, in many cases, modern deposits. Pollen taxa are listed along the top border with their observed percentages in each sample provided in black. Included in this particular diagram is a depiction of the AP (arboreal pollen) versus NAP (nonarboreal pollen). Because this sample represents a stratigraphic profile or column, the data are presented as a change through time, and the pollen percentages are filled in black from one sample (or pollen zone) to another.

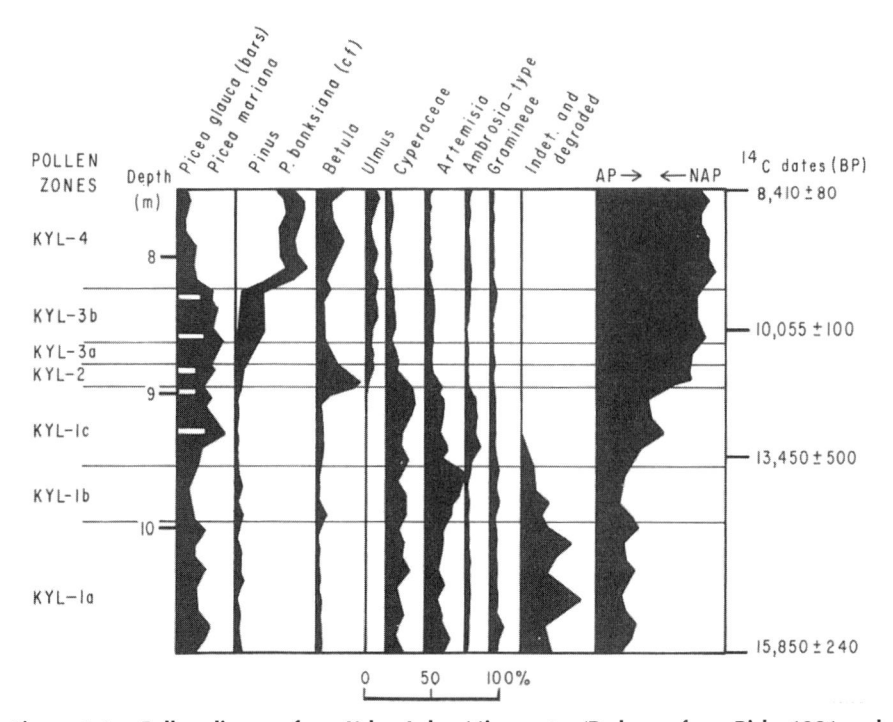

Figure 4.4. Pollen diagram from Kylen Lake, Minnesota. (Redrawn from Birks 1981 and taken from Holloway and Bryant 1985)

In another example (figure 4.5), the pollen diagram illustrates pollen identified from individual paleofecal samples rather than from a continuous profile. This type of diagram also can be used when presenting individual pollen samples, such as surface samples or samples from archaeological features, rather than stratigraphic column samples. In this diagram, the individual samples and associated radiocarbon dates are presented on the left axis. Pollen taxa are presented across the top and their percentages as observed in each sample are illustrated. In this particular case, crop pollen is separated from wild plant pollen and pollen concentration values are listed.

The amount of pollen present in a particular volume of soil can be estimated using pollen concentration values. Pollen concentration values are determined through the ratio of pollen grains to *Lycopodium* tracer spores added to each sample before processing. The number of spore grains added is multiplied by the number of prehistoric grains counted in the sample. This number is divided by the number of spore grains counted multiplied by the amount of sediment processed. Concentration values help determine the amount of prehistoric pollen present in a sample and can help assess depositional rates for soil samples and possible pollen ingestion for paleofecal samples.

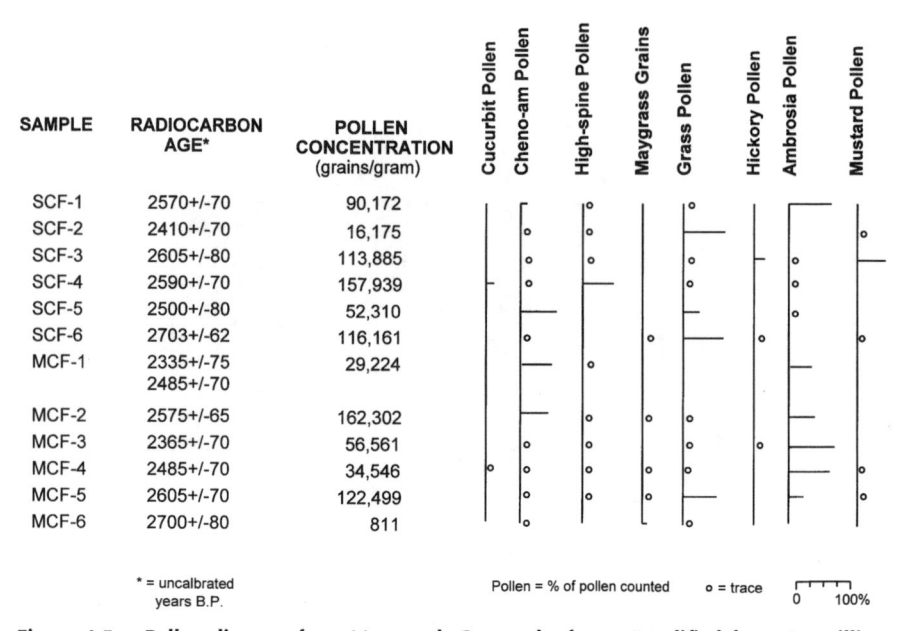

SAMPLE	RADIOCARBON AGE*	POLLEN CONCENTRATION (grains/gram)
SCF-1	2570+/-70	90,172
SCF-2	2410+/-70	16,175
SCF-3	2605+/-80	113,885
SCF-4	2590+/-70	157,939
SCF-5	2500+/-80	52,310
SCF-6	2703+/-62	116,161
MCF-1	2335+/-75 2485+/-70	29,224
MCF-2	2575+/-65	162,302
MCF-3	2365+/-70	56,561
MCF-4	2485+/-70	34,546
MCF-5	2605+/-70	122,499
MCF-6	2700+/-80	811

* = uncalibrated years B.P.

Pollen = % of pollen counted   o = trace   0–100%

Figure 4.5. Pollen diagram from Mammoth Cave paleofeces. (Modified from Gremillion and Sobolik 1996.)

PHYTOLITH ANALYSIS

Phytoliths, meaning plant stones, are "particles of hydrated silica formed in the cells of living plants that are liberated from the cells upon death and decay of the plants" (Piperno 1988:1). They are formed in plants after siliceous or calcareous substances from groundwater are absorbed and deposited in numerous cells throughout the plant (figure 4.6). Phytoliths composed of calcium oxalate also form in certain cacti and desert succulents (Jones and Bryant 1992; figure 4.7). When the plant dies and decays, these durable calcified or silicified cells remain in the soil as indicators of the plant in which they were formed. As with pollen, different plants produce diverse, and often unique, morphological phytolith types. Unlike pollen, different parts of the plant produce morphologically different phytolith types, making the use of comparative collections from all plant parts essential. However, in some areas in which pollen types are not distinctive (such as with grasses), phytolith analyses can produce excellent results. Phytoliths also can be preserved in environmental conditions in which pollen is degraded or absent, and can be used in conjunction with pollen analyses for a more complete paleoenvironmental and

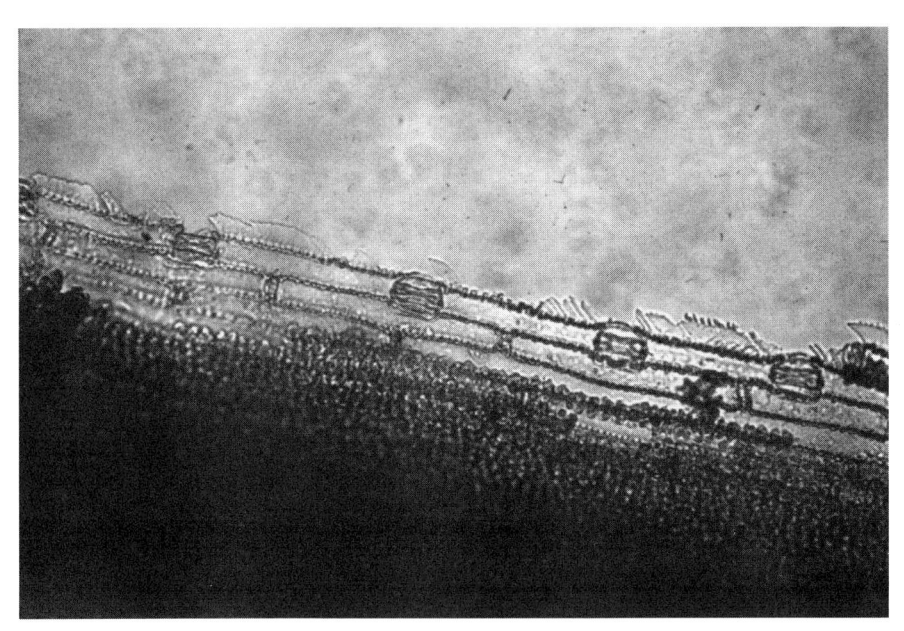

Figure 4.6. Grass fiber fragment showing dumbbell-shaped phytoliths. (Courtesy of Kristin D. Sobolik.)

Figure 4.7. Calcium-oxylate druse crystals, a form of phytoliths, from prickly pear cactus. (Courtesy of Kristin D. Sobolik.)

archaeological picture. For information on identification and interpretation of phytoliths, see Piperno (1988), Piperno and Pearsall (1993), and Rapp and Mulholland (1992).

An example of such an integrative pollen/phytolith study was conducted by Cooke and colleagues (1996) in an attempt to reconstruct the prehistoric Panamanian landscape. In this integrative analysis, vegetation history (as determined through pollen, phytoliths, and charcoal), archaeological faunal remains, and isotopic analyses of human bone were combined to provide an overall picture of changing human subsistence patterns and environmental modifications in Panama. Vegetation history was determined through the analysis of sediment cores taken from three lakes and swamps. Pollen and phytolith analyses were combined with wood charcoal research to indicate environmental changes from vegetation adapted to a cooler, dry environment during Paleoindian occupation, to extensive environmental modification when slash-and-burn horticulture became prevalent. Phytolith data also indicate that palms and arrowroots may have been cultivated in hill slope gardens and that maize started to become prevalent about 5000 to 3000 B.P. Faunal analyses and isotopic composition of human bone indicate that diet differed among groups that occupied different sites and environments.

Phytolith analysts process samples in various ways. Side-by-side tests using various processing techniques can help determine which works best under which conditions (Pearsall 1989). All phytolith processing techniques involve floating phytoliths from matrix using heavy density separation. While heavy density separation steps vary among analysts, the standard procedure used by Piperno (1988) is listed in table 4.3. Phytolith processing and analysis should only be conducted by trained archaeobiologists.

ZOOARCHAEOLOGY

Faunal remains are recovered from archaeological sites through coarse and fine screening. Bones recovered during coarse screening should be sorted as a separate category from other artifacts and placed into a separately labeled bag. All of the material should retain the particular field sack or lot number assigned in the field. It is not recommended that untrained archaeologists or field technicians presort bone into other categories, such as fish, bird, and mammal. In my experience, such presorts end up taking more time and energy and being redone later as it is determined that many of the mammal bones are in fact bird bones and vice versa. Trained archaeobiologists can easily determine which bone is mammal, bird, or fish, therefore eliminating wasted time and resources in presorts.

Samples should not be washed or modified through brushing or removing adhering matrix unless the analyst deems this necessary or in cases in which matrix is adhering to the bone, hindering identification and observation of surficial modifications. Time and energy is wasted washing every bone recovered from an archaeological excavation. Not only does this step waste time and energy, but it may cause bone to break or crumble and may add marks on the bone that can obscure prehistoric modifications such as cut marks and gnaw marks. In fact, Sutton (1994) recommends that most artifacts, including archaeobiological remains, should *not* be washed because we may be washing away important evidence such as organic and protein residues. And with our increasing awareness of nonvisual remains associated with archaeological samples, such as DNA, it becomes clear that the less we handle and modify any archaeological material, the better.

Table 4.3. Standard Phytolith Extraction Procedure*

Step I: Separation of phytoliths and removal of clay
1. Defloculate soil samples with 5 percent solution of Calgon or sodium bicarbonate.
2. Screen with fifty-three micron mesh screen. Keep the sample caught in the screen.
3. Place remaining sample (which passed through the screen) in large beakers and add water to three-quarters full. Stir vigorously. Let solution settle for one hour. Pour off and discard liquid fraction. Repeat at least five times.
4. Place remaining sample in hundred-milliliter beakers and add water. Stir, let settle for three minutes, and pour off supernatant liquid into a thousand-milliliter beaker (this separates fine and coarse silt fractions). Repeat, allow to settle for two minutes, pour off supernatant liquid into the same thousand-milliliter beaker. Repeat at least five times.

Step II: Removal of carbonates
5. Place 1 to 1.5 grams of each silt sample and the screened sand sample (three samples total) in test tubes and rinse with distilled water.
6. Add HCl (10 percent) to remove carbonates, centrifuge at 500 rpm for three minutes, and pour off liquid fraction. Repeat until no reaction is observed. Rinse with distilled water.

Step III: Removal of organics
7. Add hydrogen peroxide (3 percent) or concentrated nitric acid to remaining sample. Place in boiling water bath until reaction stops. Repeat.
8. Conduct heavy density separation with zinc bromide, specific gravity 2.3. Mix ten milliliters of heavy density solution into samples and centrifuge at 1,000 rpm for five minutes. Remove liquid (containing phytoliths) to a second centrifuge tube. Remix initial sample and repeat centrifugation. Remove liquid to second centrifuge tube. Repeat if necessary.
9. Add distilled water to liquid portion in 2.5 to 1 ratio. Centrifuge at 2,500 rpm for ten minutes, decant, and discard liquid. Repeat twice.
10. Wash in acetone.

Step IV: Slide preparation
11. Place remaining residue in a small vial with permount for curation. Label.
12. Take a small portion of permount-mixed residue and place on a microscope slide. Place coverslip over sample, and secure with nail polish or other sealant. Identify and count phytoliths.

*Phytolith extraction techniques involve the use of toxic chemicals. Extraction should never be attempted without a fully functioning fume hood and protective coat, gloves, and goggles. Processing should be done by a trained technician.
Source: Modified from Piperno (1988).

Archaeobiologists vary widely on the ways in which they initially process, sort, label, and analyze bone. I have had many chances to observe this diversity in a faunal working group of six zooarchaeologists from Maine, of which I am a member. One of the research projects the group tackled was an analysis of faunal remains from a shell midden site on the coast of Maine (Spiess et al. 1998). Each of us has our particular expertise, so we divided the bone analysis accordingly. One member analyzed the bone tools, one the fish, another the shell chondrophores for seasonality, another the small mammals, and so forth. We spent many hours discussing and arguing about the basic methods we should employ to analyze the bone. What an enlightening experience! Some of the groups wanted to take bone from a separately labeled and recognized cultural unit, sort the bone into different categories, identify taxa and surficial modifications, record these observations, and place all the unlabeled bone back into the initial bag. They indicated that it was too time-consuming and inefficient to label each bone and separate each identified faunal category into a separate bag. Some of the group wanted to take bone from an individually labeled bag; analyze each bone separately; identify taxon, portion of element, age, sex, pathology, and surficial modifications; and record that information on a separate card for each bone (figure 4.8). Each bone would be labeled accordingly and placed in a separate plastic bag with its own card, modifying the original field sack or log number to reflect what bone it was. For example, the first mammal bone from a particular field sack was separately labeled (on the bone and on the bone card) with the original field sack or log number and the addition of M.1. The third reptile bone from that same field sack would have the addition of R.3 to the original field sack or log number, again recording that number on the bone and on the bone card. They indicated that this method was best for bone curation and potential reanalysis. Only in this way could reanalysis be conducted easily and efficiently, since it would be easy to find a particular bone; you wouldn't have to search through a bag full of bone to fine the exact bone to reanalyze. Some of us employed analytical methods somewhere in between these two examples.

It is up to the archaeologist and archaeobiologist to ascertain which method is most efficient and useful for the project goals. In most CRM projects, time and money preservation are essential; therefore, it will be necessary to focus more time on bone identification, analysis, and interpretation, and less on bone labeling and intricate recording. In some research-oriented projects, time and money constraints may not be a problem—for example, in academic projects in which student labor is either cheap or free, and there are no specific deadlines in which the project needs to be completed. In such instances, it may be feasible to spend more time on bone labeling, recording, and curation.

It is not necessary to label each bone from a project, and definitely not the small fragments, as long as the bone stays in the labeled

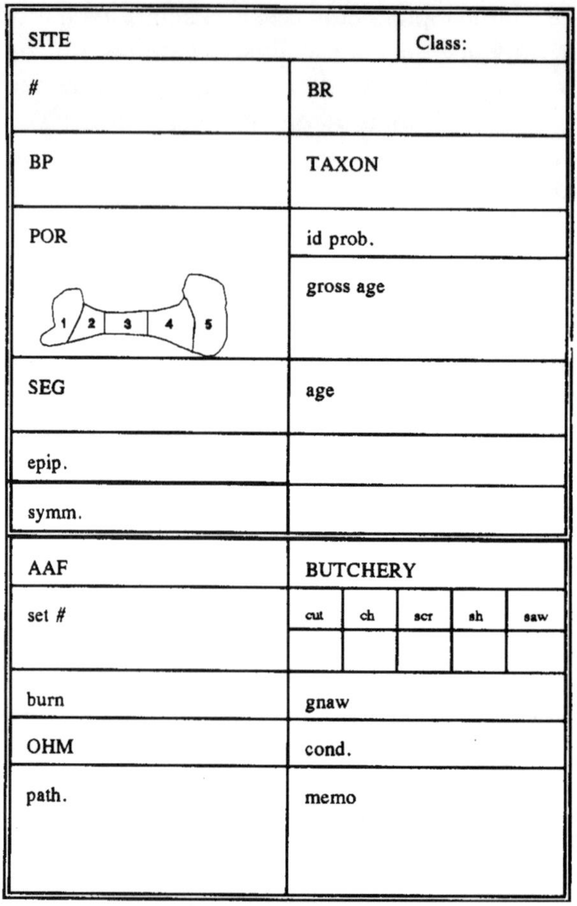

SITE		Class:
#	BR	
BP	TAXON	
POR	id prob.	
	gross age	
SEG	age	
epip.		
symm.		
AAF	BUTCHERY	
set #	cut | ch | scr | sh | saw	
burn	gnaw	
OHM	cond.	
path.	memo	

Figure 4.8. Bone coding card. (Courtesy of Dinah Crader.)

plastic bag that contains the site number and assigned field sack or log number. Some archaeobiologists feel, however, that labeling every bone from a project is essential and should be conducted at the same time that artifacts are labeled. Plastic bags are better to use than paper bags because they will preserve better through time. However, you need to be careful that provenience records written on plastic bags do not wipe off. Permanent markers are necessary for such recording.

Sorting fine-screen samples for smaller faunal remains is almost as time-consuming and potentially tedious as sorting flotation samples. If there is a lot of matrix associated with the fine-screen sample (which has been recovered with flotation samples and essentially water screened), then you may want to pass the sample through nested geological sieves to separate the sample into size grades for ease in sorting and identification. The samples recovered from fine screens are smaller and thus may be harder to identify. A good comparative collection of all size animal species is particularly important for this stage of analysis, including smaller fish, rodents, shrews, bats, reptiles, amphibians, and small birds. The species diversity recovered from a site usually increases dramatically once fine screening has been done (Reitz and Wing 1999). After the remains have been sorted, they can be identified as to particular taxon and element using comparative samples and identification guides. Some useful faunal identification guides include Cannon (1987), Casteel (1976), Claasen (1998), B. Gilbert (1990), B. Gilbert et al. (1985), Olsen (1968), and Sobolik and Steele (1996).

QUANTIFICATION

Studies of quantification in zooarchaeology have been conducted more frequently than in paleoethnobotanical studies, and papers on the subject are more prevalent (Bokonyi 1970; Casteel 1977, 1978; A. Gilbert and Singer 1982; Krantz 1968; Lyman 1979). The most frequently used techniques are ubiquity, NISP, and MNI (minimum number of individuals), all discussed later. Other quantification techniques that have been used include MNE (minimum number of elements), meat weight, and various taxonomic diversity and richness indices (Reitz and Wing 1999). For increased discussion and analysis of these techniques, see Grayson (1984) and Klein and Cruze-Uribe (1984).

The presence/absence (ubiquity) method is inherent in all faunal analyses and allows different samples to be easily compared. Presence/absence reduces the possibility of error of interpretation due to differential preservation of the sample. For example, dense bone such as beaver bone preserves well in most environments, whereas less dense bone such as small bird bone does not. If one small fragment of bird bone and 150 beaver bones are recovered from an archaeological site, the ubiquity index will not distinguish between these two samples. Both are present. The ubiquity method also does not increase the number of sample divisions. As the sample is divided into smaller groups or as sample size is increased, constituents that are more frequent will seem to be more important, whereas constituents that are less frequent will occur in fewer samples and will be considered of minimal importance. Presence/absence information has been proven useful for zoogeography and paleoenvironmental reconstruction as well as dietary purposes. I provide an example of a ubiquity table of the mammal remains from a rock-shelter site in Big Bend National Park, Texas (table 4.4).

Number of identified specimens (NISP), also common to all faunal analyses, is a basic count of the number of bones identified for each taxon category. NISP assessments can be made for an entire site or for any intrasite constituents such as cultural zones or levels, or a combination of different excavated areas; however, total NISP will always be the same for each taxon or faunal grouping used. NISP is additive: When more faunal remains are identified and quantified, the NISP is increased. One drawback to NISP is that it will tend to overestimate the frequency of taxa in an assemblage because it can increase with bone breakage thus inflating NISP. In addition, some animals contain more elements than others, such as turtles and alligators (teeth), and their NISP will therefore be higher. To illustrate NISP, I provide an example of a quantification table of the mammal remains from a rock-shelter site in Big Bend National Park, Texas (table 4.5), the same

Table 4.4. Ubiquity of Mammal Remains from a Rock-shelter in Big Bend National Park, Texas

Artiodactyla
Antilocapra (pronghorn)
Odocoileus (deer)
Carnivora
Mephitis (skunk)
Mustella (weasel)
Vulpes (fox)
Lagomorpha
Lepus (jackrabbit)
Sylvilagus (rabbit)
Rodentia
Cynomys (prairie dog)
Erethizon (porcupine)
Geomys (pocket gopher)
Neotoma (packrat)
Ondatra (muskrat)
Peromyscus (mouse)
Reithrodontomys (harvest mouse)
Sciurus (squirrel)
Sigmodon (cotton rat)
Spermophilus (ground squirrel)

remains on which I previously determined ubiquity (table 4.4).

Another frequently used quantification measure is MNI. MNI measures the minimum number of animals that are represented in a sample by determining the most abundant element of each taxon identified. MNI also may be calculated according to different sides (right and left) of the most abundant element, or by matching elements from individuals, or by sex and age differences. This type of quantification eliminates the possibility of overestimating the number of individuals, which easily can occur when one assumes that each element or fragment represents a different animal. The MNI quantification method is not biased toward animals with many bony parts (e.g., crocodiles, turtles, armadillos), bones that are highly fragmented (e.g., from bone marrow processing), or that were brought whole to the site rather than incomplete (Klein and Cruz-Uribe 1984). I provide an example of MNI from the same mammal remains I used to determine ubiquity and NISP (table 4.5).

Table 4.5. Quantification of Mammal Remains from a Rock-shelter in Big Bend National Park, Texas

Taxon	NISP	MNI
Artiodactyla		
Antilocapra (pronghorn)	1	1
Odocoileus (deer)	10	1
Carnivora		
Mephitis (skunk)	2	1
Mustella (weasel)	1	1
Vulpes (fox)	4	1
Lagomorpha		
Lepus (jackrabbit)	7	1
Sylvilagus (rabbit)	3	1
Rodentia		
Cynomys (prairie dog)	1	1
Erethizon (porcupine)	4	1
Geomys (pocket gopher)	2	1
Neotoma (packrat)	42	13
Ondatra (muskrat)	1	1
Peromyscus (mouse)	1	1
Reithrodontomys (harvest mouse)	2	1
Sciurus (squirrel)	6	1
Sigmodon (cotton rat)	6	3
Spermophilus (ground squirrel)	2	1
Unidentified mammal	222	
Total	317	31

Several problems arise from the use of MNI. One is that different aggregation techniques will produce different MNI counts (Grayson 1984). As the faunal sample is divided into smaller aggregates, such as for different analytical units or zones, the MNI for each taxon increases because the most abundant element of each taxon could be different for each aggregate. For example, using the rodent remains from the rock-shelter in Big Bend, MNI changes when cultural zones and provenience data are taken into consideration. Previous tables of rodent remains (tables 4.4 and 4.5) provided quantification of fauna from the site as a whole. When rodents are differentiated according to unit designations, rodent MNI changes (from twenty-four to thirty-two rodents), although ubiquity and NISP (sixty-seven) remain the same (table 4.6).

Another problem with the MNI method is that animals that occur in low numbers will tend to be overestimated while more commonly represented animals will be underestimated. When one bird bone is observed, the MNI for bird is one. If ten different rabbit bones were observed the MNI for rabbit would also be one even though there is a high probability that the rabbit bones are from more than one animal. Therefore, animals that occur frequently will be underestimated in the MNI tabulation.

Another problem with the MNI method is that different investigators determine the MNI differently. Some calculate the most abun-

Table 4.6. Quantification of Rodent Remains Divided by Unit from a Rock-shelter in Big Bend National Park, Texas

Rodentia Taxa	Unit S200W201 NISP	Unit S200W201 MNI	Unit S194W193 NISP	Unit S194W193 MNI
Cynomys (prairie dog)			1	1
Erethizon (porcupine)	2	1	2	1
Geomys (pocket gopher)			2	1
Neotoma (packrat)	13	8	29	9
Ondatra (muskrat)	1	1		
Peromyscus (mouse)	1	1		
Reithrodontomys (harvest mouse)	1	1	1	1
Sciurus (squirrel)	1	1	5	1
Sigmodon (cotton rat)	1	1	5	3
Spermophilus (ground squirrel)			2	1
Total	20	14	47	18

dant element, whereas others distinguish left from right elements or try to match different elements according to size, age, and sex of the animals.

CONCLUSION

There is a wide variety of ways to analyze archaeobiological remains. One way is not right and one way is not wrong; there are only better ways depending on particular circumstances, such as project goals, amount of and preservation of material recovered, potential research ideas and questions, and time and money constraints. It is up to the archaeologist in conjunction with the archaeobiologist to determine which methods will be most useful on the material at hand. It is essential to be flexible during analysis, because new techniques or ideas may be developed, or alternative avenues of study may be revealed. It is important to obtain the most information from the database, and ways in which to do that may not be understood at the beginning of a project but may be developed and changed as the archaeobiological material is recovered and analyzed.

5

INTEGRATION

No matter what types of archaeobiological analyses are conducted in a project, the most important aspect of interpretation is integration of the various assemblages and analyses (Sobolik 1994; Reitz et al. 1996). Integration of archaeobiological analyses is the means to obtaining the most complete picture of lifeways and paleoenvironments. Integrating studies of different assemblages can be difficult, given the variety of remains and the diverse ways archaeobiologists identify, analyze, and interpret them. In many of the case studies presented in this volume, one researcher or a number of researchers have attempted to integrate diverse archaeobiological analyses even recognizing that basic techniques, such as quantification, were not uniform across disciplines and/or even between analysts. Archaeobiological reconstructions, however, can rarely be effective, encompassing, and broad-based without integration.

For example, Crane and Carr (1994) analyzed faunal and botanical subsistence data from Cerros, a Late Preclassic Mayan community in northern Belize. They focused on data dating to before, during, and after a change at the site from an egalitarian village to a stratified society with an elite class. The authors tried to analyze and interpret their subsistence data (faunal remains by Carr, botanical remains by Crane) in the same fashion so changes in subsistence practices through time could be addressed in a comparable manner. They discussed the various quantification and comparative methods used by researchers analyzing botanical and bone remains and realized that there were few comparable methods. They had to use the ubiquity method, the only quantification method that would be comparable for both data sets. Their analyses indicated that during all stages analyzed at Cerros, people ate a large quantity and variety of marine resources as well as corn. Through time the consumption of deer, peccary, dog, turtles, and tree fruits increased, possibly due to the increase in consumption of rarer dietary items obtained for and by the elite. The elite also had a diet high in marine resources and corn, but they preferred rarer terrestrial and freshwater resources, thus accounting for the increase of those remains in site deposits through time. The authors conclude that "an integrated approach to the study of paleonutrition is essential because humans are broad-spectrum omnivores whose basic nutritional requirements

5.1. CASE STUDY: INTEGRATION OF ARCHAEOBIOLOGICAL ANALYSES TO ASSESS DIET, HEALTH, AND MIGRATORY PATTERNS IN THE LOWER PECOS REGION OF SOUTHWESTERN TEXAS

I conducted archaeobiological analyses of botanical, faunal, and paleofecal material recovered from two rock-shelters in the Lower Pecos Region of southwestern Texas. In addition, I synthesized results from previously analyzed botanical, faunal, paleofecal, and human skeletal analyses to present an integrative understanding of the paleonutrition of the prehistoric peoples of this region (Sobolik 1994). The archaeobiological remains (plants, bones, paleofeces) indicated that the population had access to a wide variety of dietary items that provided the necessary nutrients for good nutrition. Dietary staples, foods that were eaten almost on a daily basis and provided year-round nutrition, included agave, yucca, and sotol bulbs, prickly pear pads, a wide variety of rodents, and artiodactyls (deer and antelope). The nutritional content of these dietary staples provided essential energy (kcal) through fat, carbohydrates, and protein, as well as important fiber, vitamins, and minerals. The dietary array also included a wide variety of seeds and nuts from prickly pear, mesquite, yucca, acacia, hackberry, walnuts, acorns, and Mexican buckeye; evidence of pollen ingestion of grass, hackberry, cactus, and mustard by eating flowers or drinking teas (table 5.1); and ingestion of meat from fish, turtles, snakes, birds, rabbits, and rodents. Important vitamins, minerals, and trace elements were provided by eating these diverse dietary items. However, the human skeletal remains from this region indicate that although the people experienced few long-term pathological diseases, the population exhibited a high frequency of *enamel hypoplasias*. Enamel hypoplasias are growth arrest lines that appear as indentations in tooth enamel when the enamel stops growing, usually due to nutritional stress, but cessation of growth is short-lived and enamel is formed again. Therefore, the human skeletal remains indicate that some people in the population experienced short-term, nutritionally related stress. Since the stress probably wasn't caused by the quality of the diet, which had the necessary nutrients, it was probably caused by quantity of the diet or lack can be met by many combinations of a variety of plant and animal foods" (Crane and Carr 1994:77).

In another integrative analysis, I analyzed botanical, faunal, and paleofecal remains from the Lower Pecos Region of southwestern Texas, comparing that archaeobiological information to human skeletal analyses (see the case study in sidebar 5.1). The botanical, faunal, and paleofecal remains indicated that the diet of Lower Pecos populations was relatively stable and provided all the nutrients necessary for a healthy existence (Sobolik 1994). The human skeletal remains, however, indicated that there was dietary stress in the

---

of dietary items, possibly during the lean winter months when a wide variety of food sources were unavailable, or during weaning, a nutritionally and emotionally stressful time in a child's life.

The paleonutrition information from the Lower Pecos Region was then used to test a hypothesized seasonal round cycle for the prehistoric population (Sobolik 1996). The people occupied a territorial range, which they marked with rock art, centered on the Pecos and Devils Rivers and the Rio Grande (figure 5.1). Shafer (1986) postulated that the people followed a cyclical seasonal round based on the availability and quantity of food resources. This seasonal pattern included the desert areas and lower canyons (e.g. Frightful Cave and Conejo Shelter) during the spring to early summer, when foods such as flowers, bulbs, fruits, and plums were available. In the late summer, people moved to deeper canyon regions along the Pecos and Rio Grande (e.g., Hinds Cave) to take advantage of aquatic and upland resources. During the fall, they moved to the Devils River drainage (e.g., Baker Cave) to use acorns, walnuts, and pecans abundant in that area at that time. In the winter, when food supplies were low and available plant foods were primarily restricted to desert succulents such as yucca, agave, sotol, and prickly pear, people moved to the northern fringes (Edwards Plateau) and focused their time and energy on the acquisition of upland game.

The validity and feasibility of such a seasonal round were tested using one of the most important variables in mobility: the acquisition of resources, particularly resources related to diet. Therefore, the paleonutritional analysis of the population was essential to test the hypothesized seasonal round. The seasonal round allowed the population to maximize the food resources available in the region by following the seasonal availability and growth cycles of resources and corresponds to the nutritional information. Only through integrating many types of archaeological evidence could it be observed that populations in the Lower Pecos Region followed a seasonal mobility pattern that allowed them to maximize the dietary resources available in the environment. By maximizing seasonal resources, the population maintained a nutritionally sound and stable diet year-round.

Table 5.1. Most Frequent Dietary Botanical Items from the Lower Pecos Region

Botanical Item	Scientific Name	Portion Used
Fiber		
Agave	Agave lechuguilla	Leaf bases
Prickly pear	Opuntia sp.	Pads
Yucca	Yucca sp.	Leaf bases
Sotol	Dasylirion sp.	Leaf bases
Mesquite	Prosopis sp.	Pods
Onion	Allium drummondii	Bulbs
Seeds and Nuts		
Prickly pear	Opuntia sp.	Fruit
Mesquite	Prosopis sp.	Beans
Walnut	Juglans sp.	Nuts
Acorn	Quercus sp.	Nuts
Yucca	Yucca sp.	Fruit
Acacia	Acacia sp.	Pods and beans
Hackberry	Celtis sp.	Berries
Mexican buckeye	Ugnadia speciosa	Nuts
Pollen		
Grass	Poaceae	Achenes
Hackberry	Celtis sp.	Flowers
Sotol	Dasylirion sp.	Flowers
Agave	Agave sp.	Flowers
Yucca	Yucca sp.	Flowers
Cactus	Cactaceae	Flowers
Mustard	Brassicaceae	Flowers/achenes

From Sobolik (1994).
Note: Botanical items are presented in order from most to least frequent.

population as evidenced through enamel hypoplasias, a short-term nutritional defect. The integration of these archaeobiological data sets revealed that the population had access to necessary nutrients, but that short-term stress, most likely during weaning or during the winter months when less food was available, was a problem for this population. The paleonutritional assessment of this population was also analyzed in conjunction with a patterned seasonal round that had previously been hypothesized for them (Sobolik 1996). Therefore, in this case, the integration of archaeobiological analyses was not only conducive to nutritional and health assessments but was also used for analyses of population seasonal movement patterns. (See figure 5.3.)

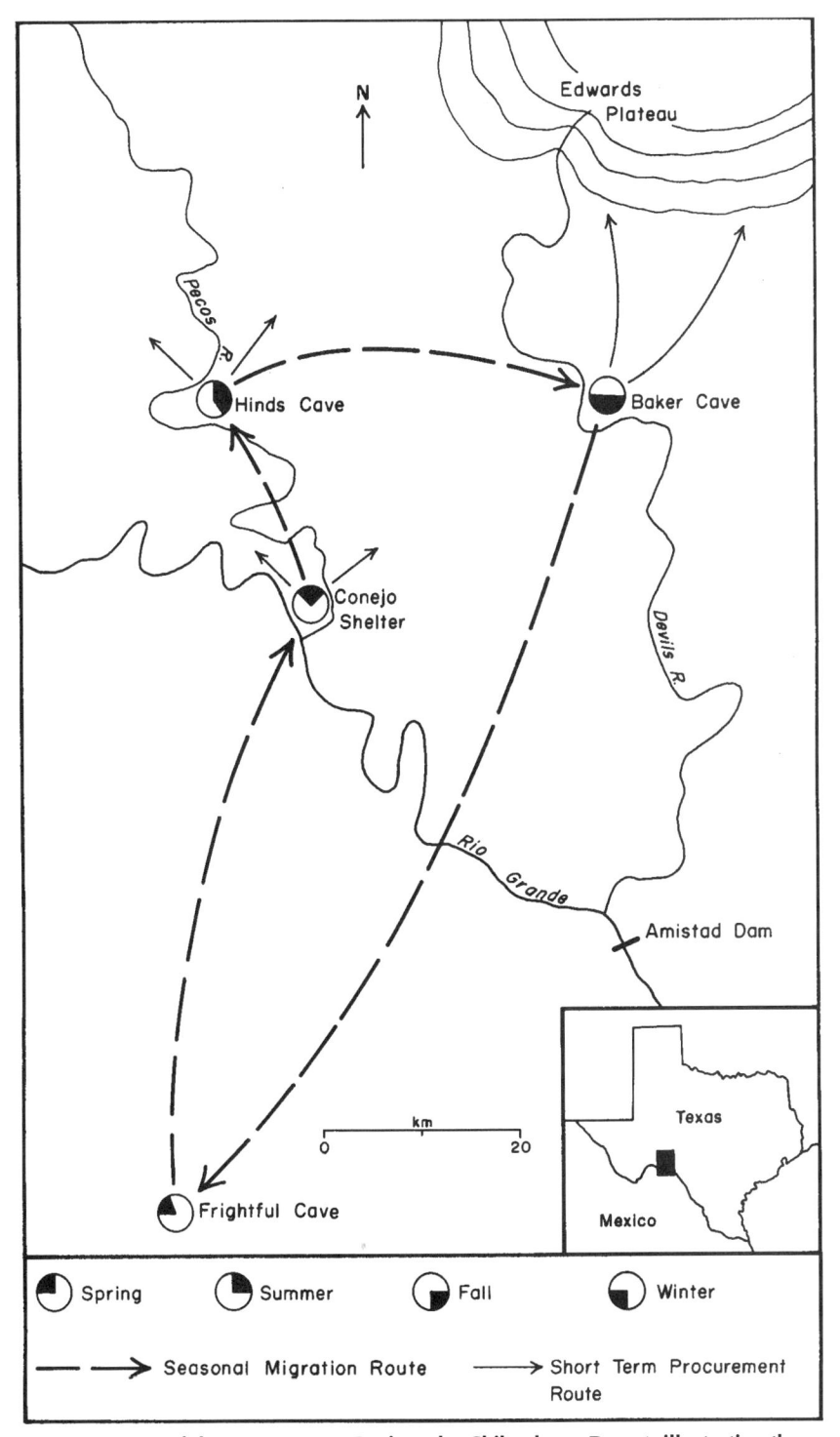

Figure 5.1. Map of the Lower Pecos Region, the Chihuahuan Desert, illustrating the seasonal round hypothesized by Shafer in 1986. (From Sobolik 1996.)

Integrative archaeobiological research does not always mean that a number of separate analyses should be combined into a comprehensive conclusion. Integration can and should involve research on other important components, such as modern species ecology, ethnography, and ethnohistory. For example, Warren (1991) demonstrates the importance of using freshwater mussels as paleoenvironmental indicators. He reviews mussel ecology and previous environmental work focusing on freshwater mussels, including habitat preferences of 133 different freshwater mussel species of the Mississippi River Basin, referencing water-body type, water depth, current velocity, and substrate composition. These preferences are given a quantitative measure that can be used in conjunction with ubiquity, MNI, and NISP and allow for statistical comparison. Warren discusses the strengths and limitations of this method and provides a number of archaeological case studies from the southeastern United States and Great Plains applying these computations and environmental interpretations.

Integrative archaeobiological work also can involve ethnographic or ethnohistoric information where feasible. For example, Snyder (1991) provides ethnohistoric, ethnographic, and archaeological evidence for the use of dogs as a dietary food item by Native Americans. Archaeological evidence of such practices was provided by faunal remains from two sites on the Great Plains. Evidence included a high frequency of cut marks on dog bones, particularly as compared to other large animals (bison and deer) that were processed at the sites. The dietary importance of dogs as food resources was assessed through a nutritional analysis of dogs as compared to other Native American foods.

In another case study involving the integration of historical information and archaeobiological analyses, Crader (1990) analyzed faunal remains found in two slave dwellings, Building "o" and the storehouse, at Monticello, Thomas Jefferson's Virginia plantation, to compare dietary practices of slaves within Monticello as compared to slaves from other habitations. Particular emphasis was placed on body part/element analysis and butchery patterns. A variety of species were identified from the assemblages, ranging from domesticated species such as pig, cattle, and sheep to wild animals such as deer, duck, and turtle. Body part representation data indicate that Building "o" dwellers obtained high-quality cuts of meat of domesticated animals (upper front limbs, upper and lower back limbs, lower back), representing a much higher frequency for quality meat than has been observed in any other analysis of slave diet (figures 5.2 and 5.3). Diet from the storehouse, however, revealed lower-quality meat part representation, what would be expected of slave quarters during this time period. The author concludes that these faunal distribution patterns may be attributed to the presence of a hierarchical system in which some slaves had higher status than others. Other hypotheses indicate that high-quality bone remains may be the result of secondary by-product reuse of leftover meat from the mansion, and that bone deposits in Building "o" merely reflect plantation garbage in general, rather than slave diet in particular.

Another important way in which to integrate archaeobiological analyses is to place them into an anthropological context. Many times archaeologists become caught up in identifying, counting, and measuring artifacts or biological remains. Their focus is on the material remains of past cultures and not on the cultures themselves.

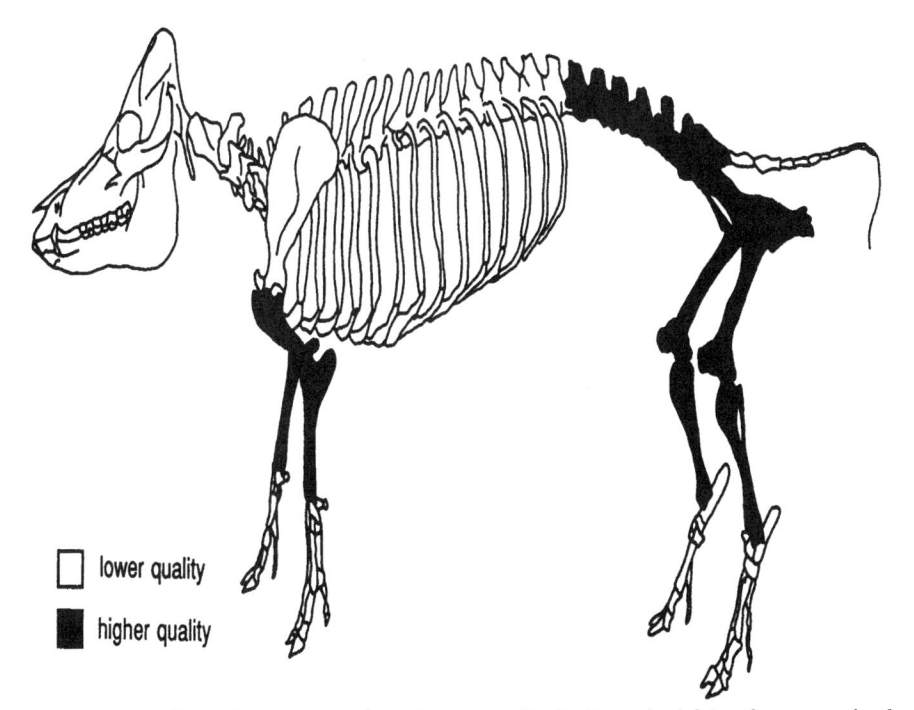

Figure 5.2. Schematic representation of meat quality in the colonial South, summarized from historical documents and archaeological information. (From Crader 1990.)

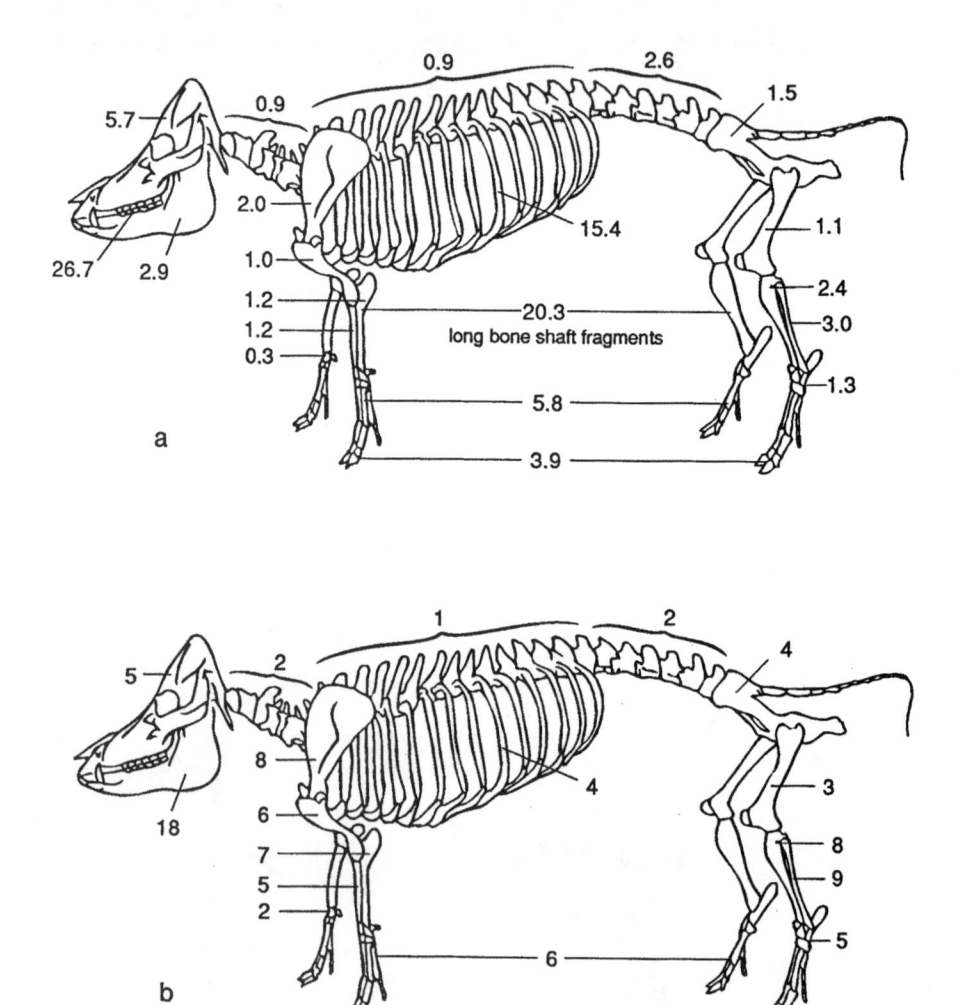

Figure 5.3. Body-part representation for pig specimens in the Building "o" assemblage: (a) number of identified specimens (N) for various body parts expressed as a percentage of the total; (b) minimum number of individuals (MNI) represented by various body parts. (From Crader 1990.)

Archaeology is a subdiscipline of anthropology, and it is important that archaeologically derived information be placed within the broader framework of cultural analysis and anthropological determination. The archaeologist and the archaeobiologist must look up from their analyses and view the bigger picture.

Kamp and Whittaker (1999) provide an excellent case study of archaeological excavation and analysis of a small Sinagua village site in Arizona (see the case study in sidebar 5.2). The authors analyze occupation, architecture, building periods, and taphonomy of the site; the types and formation of lithics, ceramics, and other artifacts; distribution and implications of botanical and faunal remains; and demography and pathology of burials. Instead of writing a typical site report detailing their analyses, identifications, and conclusions, Kamp and Whittaker provide an anthropological analysis of the people that occupied Lizard Man Village about nine hundred years ago. They compare results from their analyses to other Sinagua research, looking at stress and food shortage, subsistence strategies, social networks, and hierarchical models. Ultimately, they summarized the people's social adaptations to a marginal environment and concluded with the following:

> There were bad times when the crops failed, a child died, or your teeth hurt. There were good times: marriages, successful ceremonies, rain on the corn, a full harvest, a joke played on a friend. Your life, the lives of your ancestors, and the lives you expected for your children were no better or worse than most folk you knew, and similar to the lives of thousands in worlds beyond your ken. (Kamp and Whittaker 1999:195)

It is important to remember that people, like you and me, helped form the archaeological record, and it is their life that we are trying to reconstruct.

The focus of most CRM projects is on assessing the importance of a site before it is potentially destroyed. If a site is determined to be culturally significant, then there are two subsequent pathways. One is that the project potentially impacting the cultural site will either stop or move elsewhere. Another is that the cultural site will be excavated to retrieve the culturally significant material before it is impacted. Because the main goals of CRM projects are to assess, remove, and excavate culturally significant sites, CRM projects potentially excavate and analyze some of the most important archaeological information available today. Most archaeological excavations conducted in the United States today are CRM based. Therefore, it is

5.2. CASE STUDY: SURVIVING ADVERSITY: THE SINAGUA OF LIZARD MAN VILLAGE

The Sinagua are a prehistoric group of southwestern people who lived in present-day central Arizona at the cultural confluence of the Anasazi, Hohokam, and Mogollon. The earliest occupation of Sinagua in the area occurs at about A.D. 600 and continues until after about A.D. 1400 during which period they seem to amalgamate with modern Hopi populations. Sunset Crater volcano erupted in the area in A.D. 1064, producing a very marginal environment for human habitation. Kamp and Whittaker (1999) analyze the adaptation of the Sinagua after the eruption of Sunset Crater in this marginal environment through the analysis of Lizard Man Village, a small Puebloan and pithouse village occupied from A.D. 1065 to 1250.

Archaeobiological analyses included plant and animal remains from the site in conjunction with analysis of human burials. Plant remains, analyzed by Andrea Hunter (Hunter et al. 1999), indicate that corn, three types of beans, and squash were the dietary staples of the population, although other plant items were also important for the diet and for other cultural needs, such as medicines, dyes, and clothing (table 5.2). Faunal remains indicated that hunting supplemented farming and gathering with rabbits and artiodactyls as the primary food animals (table 5.3). The human skeletal remains from fifteen burials indicate that the population experienced nutritional stress and poor health by the presence of high childhood mortality (47 percent); the lowest mean stature of southwestern populations (Sobolik 2002); and the presence of porotic hyperostosis, cribra orbitalia, and enamel hypoplasias.

Kamp and Whittaker conclude that the Sinagua of Lizard Man Village were a small, self-sufficient village using a mixed subsistence strategy of farming corn, beans, and squash (and possibly bug seed, amaranth, and cotton) and gathering and hunting a wider variety of resources. "Food shortages were not uncommon and existence was often precarious" (1999:185), making the existence of social networks very important for survival in this marginal environment. Social organization focused on egalitarianism and a peaceful existence while spreading populations across the landscape to maximize the use and distribution of natural resources. The Sinagua lived in relatively dispersed, nonhierarchical societies in which central authority was not present.

Of particular importance in this project is the attention Kamp and Whittaker pay to placing archaeological information within an anthropological, cultural context. They discuss what the lives of the Sinagua people would have been like, how mothers and fathers would have felt at the death of their children, and the forms of interaction that were necessary to survive in such a marginal and potentially harsh environment. They also speculate about the social interactions between people on the household, local village, and regional levels. Integration of all data sets, including archaeobiological, have allowed the researchers to frame their archaeological work within the broader anthropological context.

important that archaeologists and archaeobiologists who have access to and analyze the most numerous and culturally significant archaeological material conduct such analyses in the most integrative and synthetic way possible. It is important to go beyond the most basic goals of CRM projects and to develop integrative ways to interpret and analyze prehistoric lifeways and paleoenvironments. Only through such integration will archaeologists and archaeobiologists be able to understand the broader picture of prehistoric cultures and their implications to modern people.

Fishel (1999) provides an example of integrative analysis and interpretation of remains recovered from an archaeological site that was being severely impacted, in this case by the flooding of the Little Sioux River, in addition to being periodically pothunted by local collectors and school classes (see the case study in sidebar 5.3). Fishel is the editor of this volume in which a number of archaeologists provide a synthetic report of material remains from the large site. Most important, the information gained through excavation and analysis of the site provided extensive discussion and synthesis of prehistoric and potentially historic Oneota people. This case study provides a prime example of how data obtained from impacted archaeological sites can be integrated into comprehensive analyses to provide synthetic statements and discourse on the people at an anthropological level.

Placing archaeobiological information within an anthropological context is essential to understanding the past. Integration of archaeobiological and other archaeological remains is integral to such an assessment. Ultimately, integration has to take place at the site or regional level as the archaeobiologist works with the archaeologist (they may be the same person) to synthesize all of the information obtained from site excavation into a cohesive final statement.

I hope that after reading this book you are more familiar with the techniques used in archaeobiology and have a clearer understanding of their importance in the reconstruction of lifeways and paleoenvironments. The goals of this book were to provide you with basic procedures needed to recover and identify archaeobiological remains and to help you understand the importance of taphonomy for analysis and interpretation. In addition, I hope the case studies help you understand the importance of integrative analyses in archaeobiological research.

Table 5.2. Economic Categories of Seeds from Flotation at Lizard Man Village

Category	Pithouse Occupation				Pueblo Rooms			
	Rank	Number	Identifiable Seeds%[a]	Presence %	Rank	Number	Identifiable Seeds%	Presence %
Corn (*Zea mays*)	2	459	(56.7)	86	1	160	35.3	100
Starchy-seeded, possibly cultivated (*Corispermum hyssopifolium, Amaranthus* spp., *Chenopodium* spp.[b])	1	3958	86.2	100	3	56	12.4	100
Starchy-seeded noncultivated (*Portulaca oleracea, Atriplex canescens, Sporobolus cryptandrus*)	3	115	(14.2)	86	4	52	11.5	100
Oily-seeded noncultivated (*Helianthus* cf. *petiolaris*)	4	19	(2.3)	43	7	6	1.3	40
Domesticated, dietary/technological (*Phaseolus vulgari, P. acutifolius, Gossypium hirsutum*)	6	7	(1.1)	57	2	144	31.8	20
Fruits and berries (*Juniperus monosperma, J. osteoperma, Coryphanta vivopara, Opuntia polycantha, Solanum* spp., *Arctostaphylos pringlei, Sambucus* spp.)	7	6	(.7)	14	8	5	1.1	40

Continued

Table 5.2. (continued)

	Pithouse Occupation				Pueblo Rooms			
Category	Rank	Number	Identifiable Seeds%[a]	Presence %	Rank	Number	Identifiable Seeds%	Presence %
Medicinal/dyes, low dietary (*Gutierrezia sarothrae, Polygonum amphibium, Atriplex canescens, Lepidium* spp., *Artemisia* spp.)	5	15	(1.9)	57	6	10	2.2	40
Grass and weed seeds (Poaceae, *Polygonum* spp. *Hordeum jubatum, Echinochloa crusgalli, Polygonium ramosissim*)	9	5	(.6)	29	5	18	4	80
Other	7	6	(.7)	43	9	2	.4	20
Total		4590	(100)	7[c]		453	100	5[c]

Source: From Kamp and Whittaker (1999).
[a]Percentages in parentheses calculated without *Corispermum hyssopifolium*.
[b]Includes Chenopodiaceae endosperm.
[c]Total number of discrete proveniences.

Table 5.3. Major Faunal Groups by Room at Lizard Man Village

	Lagomorphs			Artiodactyls				Large Rodents[a]				Small Rodents[b]			
Taxa	NISP	MNI	% MNI	NISP	MNI	% NISP	% MNI	NISP	MNI	% NISP	% MNI	NISP	MNI	% NISP	% MNI
Room 3	416	25	71	39	12	4	2	6	70	10	12	15	51	15	9	23
Room 4	185	10	62	26	44	5	15	13	24	4	8	10	35	10	12	26
Room 6	195	19	66	42	13	3	4	7	53	9	18	20	21	8	7	18
Room 7	82	12	47	32	5	3	3	8	50	7	29	19	25	9	14	24
Room 8	63	9	57	35	5	3	3	12	29	5	26	19	8	5	7	19
Room 11	85	8	50	24	4	2	2	6	32	4	19	12	43	14	25	42
Room 15	328	18	75	35	18	3	4	6	57	10	13	20	20	8	5	16
Room 16	194	19	80	49	6	3	3	8	22	4	9	10	14	7	6	18
Room 17	91	6	76	35	0	0	0	0	8	2	7	12	9	6	8	35
Room 19	157	13	65	37	36	3	15	9	25	7	10	20	12	8	5	23
"Room" 24	225	13	70	28	17	4	5	9	19	5	6	11	45	16	14	35
Total	2021	152	67	35	160	33	5	8	389	67	13	16	283	108	9	25

Continued

Table 5.3. (continued)

Taxa	Carnivores[c] NISP	MNI	% NISP	% MNI	Birds NISP	MNI	% NISP	% MNI	Reptiles NISP	MNI	Room Totals NISP	MNI
Room 3	4	1	1	2	37	8	6	12	0	0	590	65
Room 4	2	1	1	3	8	8	3	21	2	1	300	39
Room 6	8	2	3	4	4	4	1	9	0	0	294	45
Room 7	2	1	1	3	10	5	6	14	0	0	174	37
Room 8	1	1	1	4	4	2	4	8	1	1	111	26
Room 11	1	1	1	3	5	4	3	12	0	0	170	33
Room 15	5	3	1	6	10	8	2	16	1	1	439	51
Room 16	0	0	0	0	6	5	3	13	1	1	243	39
Room 17	8	1	7	6	4	2	3	12	0	0	120	17
Room 19	6	2	3	6	4	2	2	6	0	0	240	35
"Room" 24	5	2	2	4	7	5	2	11	3	1	321	46
Total	42	15	1	4	99	53	3	12	8	5	3002	433

Source: From Kamp and Whittaker (1999).
[a]Large rodents include prairie dogs, rock squirrels, and other squirrel-size specimens.
[b]Small rodents includes all other rodents.
[c]Carnivores include canids and mustelids.

5.3. CASE STUDY: BISON HUNTERS OF THE WESTERN PRAIRIES: ARCHAEOLOGICAL INVESTIGATIONS AT A LARGE ONEOTA SITE IN WESTERN IOWA

Excavations at the Dixon Site (13WD8) in Woodbury County, Iowa, were conducted because the large Oneota site had been extensively eroded due to expansion of the Little Sioux River during the 1993 flood, and cultural materials, including human bones, were eroding from the riverbank. In addition, the site had been and continues to be pothunted by local collectors as well as a local high school teacher who took classes out to the site to dig. Results from excavations and analysis of the cultural remains were synthesized and edited by Fishel (1999), although a number of archaeologists helped with material analysis and interpretation.

The research design included excavation and analysis of fifty pit features that were exposed and eroding from the western cutbank wall. Four research goals and questions were addressed through analysis and interpretation of the remains (table 5.4). Archaeobiological remains were useful for helping to answer all of these questions but were particularly useful to answer questions 1, 2, and 4.

The archaeobiological remains included large quantities of botanical remains from flotation, as well as faunal remains and human skeletal material. The Dausman Model A Flote-Tech Method was used for floating 972 liters of sediment from thirty-six features. The flotation samples varied from 1 liter to 53.5 liters depending on how large the features were. Botanical remains were separated into size grades of greater and lesser than two millimeters. Table 5.5 summarizes the results from flotation, including the density, weights, and counts of botanical items. Maize constituted a large part of the diet, and squash rinds were frequently observed, although other cultivated crops are not found, such as the eastern cultigens of maygrass, knotweed, and large chenopods. Wood, mainly oak and elm, was chosen from the bottomlands and along stream terraces.

Faunal remains included 3,368 bones, of which 13.7 percent were identifiable due to fragmentation of the assemblage because of breakage and processing by the Oneota (table 5.6). Bone from bison was the highest percentage of identifiable bone. A minimum of thirteen individuals were excavated revealing a preponderance of males taken as food items and for other uses, such as scapula hoes, bone tools, and furs. Deer, elk, canids, beaver, and pocket gopher were also frequently used by prehistoric Oneota. Of interest are 639 freshwater mussel shell fragments representing nine different species. Freshwater mussel was used as a food resource, for tempering ceramics, and for tool manufacture.

In response to the original four research design questions (table 5.4), some have been partially answered, and others remain to be answered. Archaeobiological remains helped realize the large extent of the Dixon site and that there may exist upwards of 9,800 additional pit features, many with human remains. Botanical processing areas were identifiable around feature 16, and more probably exist. Analysis of flotation remains indicated that Oneota diet focused on maize with little exploitation of native cultigens or wild plant foods. Hunting of bison probably took place in long-distance communal forays, indicating that bison were not frequent in the Dixon site area in the past. The Oneota of the region are better understood and it is realized that Oneota postdate Mill Creek groups in the region and that the Dixon site Oneota are one of the earliest manifestations of Oneota in the area, although it is not understood where the Oneota peoples came from. Integrative archaeological, archaeobiological, and historical analysis work together to answer research design questions.

Table 5.4. Research Design Goals and Questions for Excavation and Analysis of the Dixon Site, Western Iowa

Research Design Goals and Questions	Steps Undertaken and Archaeological Material Used
1. How did Oneota peoples first come to occupy northwest Iowa, and what were the initial settlements like?	Analysis of features and associated material remains
2. What was the relationship, if any, between Oneota groups and other prehistoric peoples who inhabited northwest Iowa?	Collect numerous radiocarbon samples from various contexts
3. What kinds of cultural remains exist at the Dixon site, and how are those remains spatially distributed? What measures can be taken to help safeguard the site against continued damage from erosion?	Magnetometer and soil resistivity surveys, topographic maps, and geomorphological assessments
4. What do the cultural remains at the Dixon site tell us about the lifeways of Oneota groups living in northwest Iowa?	Flotation samples from features, macrofaunal remains, lithics, ceramics, and human remains

Source: From Fishel (1999).

Table 5.5. Archaeobotanical Indices from Feature Flotation Samples at the Dixon Site, Northwest Iowa

Taxon	Common Name	MNI	Count	Percentage	Weight (g)	Percentage
Bison bison	bison	13	179	5.32	7,333.50	63.70
Odocoileus virginianus	white-tailed deer	1	1	.03	491.50	4.27
Odocoileus sp.	deer	6	96	2.82	802.30	6.97
Canis sp.	dog or wolf	3	64	1.90	259.70	2.26
Cervus canadensis	elk	2	10	.30	61.70	.54
Castor canadensis	beaver	1	5	.15	15.70	.14
Scalopus aquaticus	Eastern mole	1	3	.09	.90	.00
Rattus sp.	rat	1	2	.06	.70	.00
Procyon lotor	raccoon	1	2	.06	11.50	.10
Sus scrofa	pig	1	8	.24	13.10	.11
Geomys bursarius		1	2	.06	1.10	.00
Sylvilagus sp.	rabbit	1	1	.39	.90	.01
Sciurid		1	13	.39	1.50	.01
Turtle		1	3	.09	8.60	.07
Frog		1	3	.09	.20	.00
Fish (Ictalurus sp.)		6	64	1.90	16.00	.14
Tetraoninae sp.		1	2	.06	.90	.00
Anseriformes sp.		1	1	.03	.50	.00
Cf. Icterinae sp.		1	1	.03	.50	.00
Cf. Anatidae sp.		1	1	.03	.40	.00
Unidentified Aves		1	1	.03	.50	.00
Burned unidentifiable			342	1.15	192.70	1.67
Unidentifiable			2,564	76.13	2,297.20	19.95
Total			3,368		11,511.60	

Source: From Fishel (1999).

Table 5.6. Excavated Botanical Remains from the Dixon Site, Northwest Iowa

Sum raw charcoal wt (g), nonstandardized samples		125.405
Sum charcoal wt (g), standardized 10-liter samples		120.791
Average wt (g) charcoal/sample in standardized samples		1.267
Average wt (g) charcoal/feature in standardized samples		3.485
Sum flotation volumes (liters)		972.000
Average flotation volume (liters) per sample		9.818
Average flotation volume (liters) per feature		27.000
Sum raw counts, carbonized seeds, all sizes		3,027.000
Estimated sum seeds, all sizes, in standardized 10-liter samples		3,194.011
Seeds/g of charcoal in standardized samples		49.649
Seeds/g of charcoal by raw count		24.138
Sum raw counts, charcoal >2 mm		7,165.000
Sum standardized sample counts, charcoal <2 mm		6,685.088

	Count/10liters Charcoal >2mm	Frequency (%)	Feature Ubiquity	Sample Ubiquity
Nutshell	(59.967)	(0.9)	97.2	78.8
Hazelnut	14.755	0.22	11.1	19.4
Juglandaceae	4.437	0.07	75.0	57.6
Black walnut	40.775	0.61	80.6	56.6
Wood	5,202.403	77.82	100.0	100.0
Bark	69.762	1.04	58.3	47.5
Grass stems	57.887	0.87	58.3	33.3
Tubers/rhizomes	6.47	0.10	25.0	10.1
Seeds: Freq = >2mm; Ubiqu = >0.5	20.843	0.31	97.2	97.0
Squash (cucurbit) rind	10.454	0.16	83.3	76.8
Maize kernels	867.098	12.97	83.3	69.7
Maize cupules	307.943	4.61	83.3	66.7
Maize embryos	6.697	0.10	36.1	16.2
Unknowns	45.583	0.68	66.7	49.5

Source: From Fishel (1999).
[a]Frequency: % charcoal >2mm, standardized per sample to 10-liter volumes before computation. Sample ubiquity is a tabulation of the presence of a charcoal type in any sample, any fraction size, expressed as a percentage of total samples (*n*=99). Feature ubiquity is a tabulation of the presence of a charcoal type in a feature, any sample, any fraction, expressed as a percentage of total number of features (*n*=36).

REFERENCES

Appleyard, H. M., and A. B. Wildman
    1970   Fibres of Archaeological Interest: Their Examination and Identification. In *Science in Archaeology: A Survey of Progress and Research*, D. Brothwell and E. Higgs (eds.), pp. 624–33. Praeger, New York.

Asch Sidell, Nancy
    1999   Prehistoric Plant Use in Maine: Paleoindian to Contact Period. In *Current Northeast Paleoethnobotany*, John Hart (ed). New York State Museum, Albany.

Barrows, David
    1900   *The Ethno-botany of the Coahuilla Indians of Southern California*. Ph.D. dissertation, University of Chicago, Chicago.

Birks, H. J. B.
    1981   Late Wisconsin Vegetational and Climatic History at Kylen Lake, Northeastern Minnesota. *Quaternary Research* 16:322–55.

Bocek, Barbara
    1986   Rodent Ecology and Burrowing Behavior: Predicted Effects on Archaeological Site Formation. *American Antiquity* 51(3):589–603.

Bokonyi, S.
    1970   A New Method for the Determination of the Minimum Number of Individuals in Animal Bone Material. *American Journal of Archaeology* 74:291–92.

Bozell, John R.
    1991   Fauna from the Hulme Site and Comments on Central Plains Tradition Subsistence Variability. *Plains Anthropologist* 36(136):229–53.

Braidwood, L. S., and R. J. Braidwood (eds.)
    1982   *Prehistoric Village Archaeology in South-Eastern Turkey: The Eight Millennium* B.C. *Site at Cayönü: Its Chipped and Ground Stone Industries and Faunal Remains*. British Archaeological Reports International Series 138. Oxford University Press, Oxford.

Bryant, Vaughn M. Jr.
    1966   Pollen Analysis of the Devil's Mouth Site. In *A Preliminary Study of the Paleoecology of the AMISTAD Reservoir*. A Report of Research under the Auspices of the National Science Foundation (GS-667). National Science Foundation, Washington, D.C.

Bryant, Vaughn M. Jr., Richard G. Holloway, John G. Jones, and David L. Carlson
    1994   Pollen Preservation in Alkaline Soils of the American Southwest. In *Sedimentation of Organic Particles*, Alfred Traverse (ed.), pp. 47–58. Cambridge University Press, Cambridge.

Bryant, Vaughn M. Jr., and Don P. Morris
    1986   Uses of Ceramic Vessels and Grinding Implements: The Pollen Evidence. In *Archeological Investigations at Antelope House*, D. P. Morris (ed.), pp. 489–500. National Park Service, Washington, D.C.

Bryant, Vaughn M. Jr. and Glendon H. Weir
    1986   Pollen Analysis of Floor Sediment Samples: A Guide to Room Use. In *Archeological Investigations at Antelope House*, D. P. Morris (ed.), pp. 58–71. National Park Service, Washington, D.C.

Bunn, H. T.
    1991   A Taphonomic Perspective on the Archaeology of Human Origins. *Annual Review of Anthropology* 20:433–67.

Cannon, Debbi Yee
    1987   *Marine Fish Osteology: A Manual for Archaeologists*. Archaeology Press, Simon Fraser University, Burnaby, Canada.

Carbone, Victor A., and Bennie C. Keel
    1985   Preservation of Plant and Animal Remains. In *The Analysis of Prehistoric Diets*, Robert I. Gilbert Jr. and James H. Mielke (eds.), pp. 1–20. Academic Press, Orlando, Florida.

Casteel, Richard W.
    1976   *Fish Remains in Archaeology and Paleoenvironmental Studies*. Academic Press, New York.
    1977   Characterization of Faunal Assemblages and the Minimum Number of Individuals Determined from Paired Elements: Continuing Problems in Archaeology. *Journal of Archaeological Science* 4:125–34.
    1978   Faunal Assemblages and the "Weigemethode" or Weight Method. *Journal of Field Archaeology* 5:71–77.

Catling, Dorothy, and John Grayson
   1982   *Identification of Vegetable Fibres*. Chapman and Hall, London.
Claasen, Cheryl
   1998   *Shells*. Cambridge University Press, Cambridge.
Cooke, Richard G., Lynette Norr, and Dolores R. Piperno
   1996   Native Americans and the Panamanian Landscape. In *Case Studies in Environmental Archaeology*, E. J. Reitz, L. A. Newsom, and S. J. Scudder (eds.), pp. 103–26. Plenum Press, New York.
Core, H. A., W. A. Cote, and A. C. Day
   1979   *Wood Structure and Identification*. Syracuse University Press, Syracuse, New York.
Corner, Edred J. H.
   1976   *The Seeds of Dicotyledons*. Cambridge University Press, Cambridge.
Crader, Diana C.
   1990   Slave Diet at Monticello. *American Antiquity* 55(4):690–717.
Crane, Cathy J., and H. Sorayya Carr
   1994   The Integration and Quantification of Economic Data from a Late Preclassic Maya Community in Belize. In *Paleonutrition: Diet, Health, and Nutrition in Prehistory*, K. D. Sobolik (ed.), pp. 66–79, Center for Archaeological Investigations Occasional Paper No. 22. Southern Illinois University, Carbondale.
Dimbleby, Geoffrey W.
   1978   *Plants and Archaeology*. Humanities Press, Atlantic Highlands, New Jersey.
   1985   *The Palynology of Archaeological Sites*. Academic Press, New York.
Dobbs, Clark A., Craig Johnson, Kathryn Parker, and Terrance Martin
   1993   *20SA1034: A Late-Prehistoric Site on the Flint River in the Saginaw Valley, Michigan*. Reports of Investigations No. 229. Institute for Minnesota Archaeology, Minneapolis.
Efremov, J. A.
   1940   Taphonomy: A New Branch of Paleontology. *Pan-American Geologist* 74:81–93.
Faegri, Knut, and John Iversen
   1964   *Textbook of Pollen Analysis*. Hafner, New York.
Fewkes, J. W.
   1896   Pacific Coast Shell from Prehistoric Tusayan Pueblos. *American Anthropologist* 9(11):359–67.
Fishel, Richard L. (ed.)
   1999   *Bison Hunters of the Western Prairies: Archaeological Investigations at the Dixon Site (13WD8), Woodbury County, Iowa*. Report No. 21. Office of the State Archaeologist, University of Iowa, Iowa City.

Ford, Richard I.
　1979　Paleoethnobotany in American Archaeology. In *Advances in Archaeological Method and Theory, Vol. 2*, Michael B. Schiffer (ed.), pp. 285–336. Academic Press, New York.

Fry, Gary F.
　1985　Analysis of Fecal Material. In *The Analysis of Prehistoric Diets*, R. I. Gilbert Jr. and J. H. Mielke (eds.), pp. 127–54, Academic Press, Orlando, Florida.

Gilbert, A. S., and B. H. Singer
　1982　Reassessing Zooarchaeological Quantification. *World Archaeology* 14:21–40.

Gilbert, B. Miles
　1990　*Mammalian Osteology*. Missouri Archaeological Society, Columbia.

Gilbert, B. Miles, Larry D. Martin, and Howard G. Savage
　1985　*Avian Osteology*. Modern Printing, Laramie, Wyoming.

Gilmore, Melvin
　1919　*Uses of Plants by the Indians of the Missouri River Region.* Thirty-second Annual Report of the Bureau of American Ethnology. Smithsonian Institution Press, Washington, D.C. Reprinted 1977, University of Nebraska Press, Lincoln.
　1932　The Ethnobotanical Laboratory at the University of Michigan. *Occasional Contributions from the Museum of Anthropology of the University of Michigan*, No. 1. University of Michigan Press, Ann Arbor.

Grayson, Donald K.
　1984　*Quantitative Zooarchaeology: Topics in the Analysis of Archaeological Faunas*. Academic Press, New York.

Greig, James
　1989　*Archaeobotany*. Handbooks for Archaeologists, No. 4. European Science Foundation, Strasbourg.

Gremillion, Kristen J., and Kristin D. Sobolik
　1996　Dietary Variability among Prehistoric Forager-Farmers of Eastern North America. *Current Anthropology* 37(3):529–39.

Gunn, C. R., J. V. Dennis, and P. J. Paradine
　1976　*World Guide to Tropical Drift Seeds and Fruits*. New York Times Book Company, Quadrangle, New York.

Harshberger, J. W.
　1896　The Purpose of Ethnobotany. *American Antiquarian* 17(2):73–81.

Hastorf, C. A., and V. S. Popper (eds.)
　1988　*Current Paleoethnobotany: Analytical Methods and Cultural Interpretations of Archaeological Plant Remains*. University of Chicago Press, Chicago.

Helbaek, Hans
    1969   Plant Collecting, Dry-Farming, and Irrigation Agriculture in Prehistoric Deh Luran. In *Prehistory and Human Ecology of the Deh Luran Plain*, F. Hole, K. Flannery, and J. Neely (eds.), pp. 389–426, Memoirs No. 1, Museum of Anthropology, University of Michigan. University of Michigan, Ann Arbor.

Hoffman, Rob, and Christopher Hays
    1987   The Eastern Wood Rat (*Neotoma floridana*) as a Taphonomic Factor in Archaeological Sites. *Journal of Archaeological Science* 14:325–37.

Holloway, Richard G., and Vaughn M. Bryant Jr.
    1985   Late-Quaternary Pollen Records and Vegetational History of the Great Lakes Region: United States and Canada. In *Pollen Records of Late-Quaternary North American Sediments*, V. M. Bryant Jr. and R. G. Holloway (eds.), pp. 207–45. American Association of Stratigraphic Palynologists, Dallas.

Hudson, Jean
    1993   The Impacts of Domestic Dogs on Bone in Forager Camps. In *From Bones to Behavior: Ethnoarchaeological and Experimental Contributions to the Interpretation of Faunal Remains*, Jean Hudson (ed.), pp. 301–23, Center for Archaeological Investigations Occasional Paper No. 21. Southern Illinois University Press, Carbondale.

Hunter, Andrea, Kathryn Kamp, and John Whittaker
    1999   Plant Use. In *Surviving Adversity: The Sinaqua of Lizard Man Village*, Kathryn Kamp and John Whittaker (eds.), pp. 139–51. University of Utah Anthropological Papers, No. 120. University of Utah Press, Salt Lake City.

Izumi, S., and T. Sono
    1963   *Andes 2: Excavations at Kotosh, Peru, 1960*. Kadokawa, Tokyo.

Johnson, LeRoy Jr.
    1964   *The Devil's Mouth Site: A Stratified Campsite at Amistad Reservoir, Val Verde County, Texas*. Archeology Series No. 6, Department of Anthropology, University of Texas. University of Texas Press, Austin.

Jones, Gretchen D., Vaughn M. Bryant Jr., Meredith Hoag Lieux, Stanley D. Jones, and Pete D. Lingren
    1995   *Pollen of the Southeastern United States: with emphasis on melissopalynology and entomopalynology*. American Assocaition of Stratigraphic Palynologists Foundation, AASP Contribution Series No. 30.

Jones, John G., and Vaughn M. Bryant Jr.
    1992   Phytolith Taxonomy in Selected Species of Texas Cacti. In *Phytolith Systematics: Emerging Issues*, G. Rapp Jr. and S. C. Mulholland (eds.), pp. 215–38, Plenum Press, New York.

Jones, Kevin T.
 1993   The Archaeological Structure of a Short-Term Camp. In *From Bones to Behavior: Ethnoarchaeological and Experimental Contributions to the Interpretation of Faunal Remains*, Jean Hudson (ed.), pp. 101–14, Center for Archaeological Investigations Occasional Paper No. 21. Southern Illinois University Press, Carbondale.

Jones, Volney H.
 1936   Vegetable Remains of Newt Kash Hollosh Shelter. In *Rockshelters in Menefee County, Kentucky*, W. S. Webb and W. D. Funkhouser (eds.), University of Kentucky Reports in Archaeology and Anthropology 3:147–65.

Kamp, Kathryn A., and John C. Whittaker
 1999   *Surviving Adversity: The Sinaqua of Lizard Man Village*. University of Utah Anthropological Papers, No. 120. University of Utah Press, Salt Lake City.

Kaplan, Lawrence, Mary B. Smith, and Lesley Sneddon
 1990   The Boylston Street Fishweir: Revisited. *Economic Botany* 44(4):516–28.

Kapp, Ronald O.
 1969   *How to Know Pollen and Spores*. Brown, Dubuque, Iowa.

Klein, Richard G., and Kathryn Cruz-Uribe
 1984   *The Analysis of Animal Bones from Archeological Sites*. University of Chicago Press, Chicago.

Krantz, G. S.
 1968   A New Method of Counting Mammal Bones. *American Journal of Archaeology* 72:286–88.

Labadie, Joe
 1994   *Amistad National Recreation Area Cultural Resources Study*. National Park Service, U.S. Department of the Interior, Washington, D.C.

Lawrence, Barbara
 1957   Zoology. In *The Identification of Non-artifactual Archaeological Materials*, W. W. Taylor (ed.), pp. 41–42, National Academy of Science–Natural Resource Council Publication 565. National Academy of Science and Natural Resource Council, Washington, D.C.

Leney, Lawrence, and R. W. Casteel
 1975   Simplified Procedure for Examining Charcoal Specimens for Identification. *Journal of Archaeological Science* 2:153–59.

Linse, Angela R.
 1992   Is Bone Safe in a Shell Midden? In *Deciphering a Shell Midden*, Julie Stein (ed.), pp. 327–46, Academic Press, New York.

Loomis, F. B., and D. B. Young
  1912 Shell Heaps of Maine. *American Journal of Science* 34(199):17–42.
Lyman, R. Lee
  1979 Available Meta from Faunal Remains: A Consideration of Techniques. *American Antiquity* 44:536–46.
  1984 Bone Density and Differential Survivorship of Fossil Classes. *Journal of Anthropological Archaeology* 3:259–99.
  1994 *Vertebrate Taphonomy*. Cambridge University Press, New York.
Mack, Richard N., and Vaughn M. Bryant Jr.
  1974 Modern Pollen Spectra from the Columbia Basin, Washington. *Northwest Science* 48(3):183–94.
Martin, A. C., and W. D. Barkley
  1961 *Seed Identification Manual*. University of California Press, Berkeley.
McInnis, Heather
  1999 *Subsistence and Maritime Adaptations at Quebrada Jaguay, Camaná, Peru: A Faunal Analysis*. M.S. thesis, Institute for Quaternary and Climate Studies, University of Maine, Orono.
Miller, Naomi F., and Tristine Smart
  1984 Intentional Burning of Dung as Fuel: A Mechanism for the Incorporation of Charred Seeds into the Archeological Record. *Journal of Ethnobiology* 4(1):15–28.
Montgomery, F. H.
  1977 *Seeds and Fruits of Plants of Eastern Canada and Northeastern United States*. University of Toronto Press, Toronto.
O'Connell, J. F., and K. Hawkes
  1988 Hadza Hunting, Butchering, and Bone Transport and Their Archaeological Implications. *Journal of Anthropological Research* 44(2):113–61.
O'Hara, Sara L., F. Alayne Street-Perrott, and Timothy P. Burt
  1993 Accelerated Soil Erosion around a Mexican Highland Lake Caused by Prehispanic Agriculture. *Nature* 362:48–51.
Olsen, Stanley J.
  1968 *Fish, Amphibian and Reptile Remains from Archaeological Sites*. Papers of the Peabody Museum of Archaeology and Ethnology, Harvard University, Volume 55, No. 2, Peabody Museum, Cambridge, Massachusetts.
Parker, Kathryn E.
  1996 Three Corn Kernels and a Hill of Beans: The Evidence for Prehistoric Horticulture in Michigan. In *Investigating the Archaeological Record of the Great Lakes State: Essays in Honor of Elizabeth Baldwin Garland*, Margetet B. Holman, Janet G.

Brashler, and Kathryn E. Parker (eds.), pp. 307-339. Western Michigan University, Kalamazoo.

Pearsall, Deborah M.
  1989 *Paleoethnobotany: A Handbook of Procedures.* Academic Press, New York.
  2000 *Paleoethnobotany: A Handbook of Procedures*, 2nd ed. Academic Press, San Diego.

Piperno, Dolores R.
  1988 *Phytolith Analysis: An Archaeological and Geological Perspective.* Academic Press, New York.

Piperno, Dolores R., and Deborah M. Pearsall (eds.)
  1993 *Current Research in Phytolith Analysis: Applications in Archaeology and Paleoecology.* MASCA Research Papers in Science and Archaeology, Vol. 10, University Museum of Archaeology and Anthropology. University of Pennsylvania, Philadelphia.

Poinar, Hendrik N., Melanie Kuch, Kristin D. Sobolik, Ian Barnes, Artur B. Stankiewicz, Tomasz Kuder, W. Geofferey Spaulding, Vaughn M. Bryant Jr., Alan Cooper, and Svante Pääbo
  2001 A Molecular Analysis of Dietary Diversity for Three Archaic Native Americans. *Proceedings of the National Academy of Sciences* 98(8):4317–22.

Rapp, George Jr., and Susan C. Mulholland (eds.)
  1992 *Phytolith Systematics: Emerging Issues.* Plenum Press, New York.

Reitz, Elizabeth J., Lee A. Newsom, and Sylvia J. Scudder (eds.)
  1996 *Case Studies in Environmental Archaeology.* Plenum Press, New York.

Reitz, Elizabeth J., and Elizabeth Wing
  1999 *Zooarchaeology.* Cambridge University Press, New York.

Robison, N.D.
  1978 Zooarchaeology: Its History and Development. *Tennessee Anthropological Association Miscellaneous Paper* 2:1–22 (Knoxville).

Sandweiss, Daniel H., Heather McInnis, Richard L. Burger, Asuncion Cano, Bernardino Ojeda, Rolando Paredes, Maria del Carmen Sandweiss, and Michael D. Glascock
  1998 Quebrada Jaguay: Early South American Maritime Adaptations. *Science* 281:1830–32.

Sanger, David
  1996 Testing the Models: Hunter-Gatherer Use of Space in the Gulf of Maine, USA. *World Archaeology* 27(3):512–26.

Shafer, Harry J., and Jim Zintgraff
  1986 *Ancient Texans: Rock Art and Lifeways along the Lower Pecos.* Texas Monthly Press, Austin.

Shaffer, Brian
    1992 Quarter-inch Screening: Understanding Biases in Recovery of Vertebrate Faunal Remains. *American Antiquity* 57(1):129–36.
Shipman, Pat
    1986 Scavenging or Hunting in Early Hominids: Theoretical Framework and Test. *American Anthropologist* 88:27–43.
Smith, Bruce
    1975 *Middle Mississippi Exploitation of Animal Populations*. University of Michigan Museum of Anthropology, Anthropological Papers 57. University of Michigan Press, Ann Arbor.
Snyder, Lynn M.
    1991 Barking Mutton: Ethnohistoric and Ethnographic, Archaeological, and Nutritional Evidence Pertaining to the Dog as a Native American Food Resource on the Plains. In *Beamers, Bobwhites, and Blue-Points: Tributes to the Career of Paul W. Parmalee*, J. R. Purdue, W. E. Klippel, and B. W. Styles (eds.), pp. 359–78, Illinois State Museum Scientific Papers, Vol. 23. Illinois State Museum, Springfield.
Sobolik, Kristin D.
    1988 *The Prehistoric Diet and Subsistence of the Lower Pecos Region, as Reflected in Coprolites from Baker Cave, Val Verde County, Texas*. Studies in Archeology No. 7, Texas Archeological Research Laboratory. University of Texas Press, Austin.
    1993 Direct Evidence for the Importance of Small Animals to Prehistoric Diets: A Review of Coprolite Studies. *North American Archaeologist* 14(3):227–44.
    1994 In *Paleonutrition: the Diet and Health of Prehistoric Americans*. Center for Archaeological Investigations, Southern Illinois University Occasional Paper No. 22, Carbondale.
    1996 Nutritional Constraints and Mobility Patterns of Hunter-Gatherers in the Northern Chihuahuan Desert. In *Case Studies in Environmental Archaeology*, Elizabeth J. Reitz, Lee A. Newsom, and Sylvia J. Scudder (eds.), pp. 195–214. Plenum Press, New York.
    2002 Children's Health in the Prehistoric Southwest. In: *Children in the Prehistoric Puebloan Southwest*, Kathryn Kamp (ed.), pp. 125–151, The University of Utah Press, Salt Lake City.
Sobolik, Kristin D., Kristen J. Gremillion, Patricia L. Whitten, and Patty Jo Watson
    1996 Technical Note: Sex Determination of Prehistoric Human Paleofeces. *American Journal of Physical Anthropology* 101:283–90.
Sobolik, Kristin D., and D. Gentry Steele
    1996 *A Turtle Atlas to Facilitate Archaeological Identifications*. Mammoth Site of Hot Springs, South Dakota.

Sobolik, Kristin D., and Richard Will
- 2000 Calcined Turtle Bones from the Little Ossipee North Site in Southwestern Maine. *Archaeology of Eastern North America* 28:15–28.

Sobolik, Kristin D., Laurie S. Zimmerman, and Brooke Manross Guilfoyl
- 1997 Indoor versus Outdoor Firepit Usage: A Case Study from the Mimbres. *Kiva* 62(3):283–300.

Spiess, Arthur, Kristin D. Sobolik, Dinah Crader, Richard Will, and John Mosher
- 1998 Cod, Clams, and Roast Deer: Prehistoric Dining on Indiantown. In *Indiantown Island Archaeological Project*, Deborah B. Wilson (ed.), submitted to Boothbay Region Land Trust, Inc., Boothbay Harbor, Maine.

Struever, Stuart
- 1968 Flotation Techniques for the Recovery of Small-scale Archaeological Remains. *American Antiquity* 33:353–62.

Sutton, Mark Q.
- 1994 Indirect Evidence in Paleonutrition Studies. In *Paleonutrition: Diet, Health, and Nutrition in Prehistory*, K. D. Sobolik (ed.), pp. 98–114, Center for Archaeological Investigations Occasional Paper No. 22. Southern Illinois University Press, Carbondale.

Sutton, Mark Q., Minnie Malik, and Andrew Ogram
- 1996 Experiments on the Determination of Gender from Coprolites by DNA Analysis. *Journal of Archaeological Science* 23:263–67.

Thomas, D. H.
- 1969 Great Basin Hunting Patterns: A Quantitative Method for Treating Faunal Remains. *American Antiquity* 34:392–401.

Voorhies, V. R.
- 1969 Taphonomy and Population Dynamics of the Early Pliocene Vertebrate Fauna, Knox County, Nebraska. *University of Wyoming Contributions in Geology Special Papers*, 1:1–69.

Wagner, Gail E.
- 1996 Feast or Famine? Seasonal Diet at a Fort Ancient Community. In *Case Studies in Environmental Archaeology*, E. J. Reitz, L. A. Newsom, and S. J. Scudder (eds.), pp. 255–72. Plenum Press, New York.

Warren, Robert E.
- 1991 Freshwater Mussels as Paleoenvironmental Indicators: A Quantitative Approach to Assemblage Analysis. In *Beamers, Bobwhites, and Blue-Points: Tributes to the Career of Paul W. Parmalee*, J. R. Purdue, W. E. Klippel, and B. W. Styles (eds.), pp. 23–66, Illinois State Museum Scientific Papers, Vol. 23. Illinois State Museum, Springfield.

Watson, Patty Jo
    1976   In Pursuit of Prehistoric Subsistence: A Comparative Account of Some Contemporary Flotation Techniques. *Midcontinental Journal of Archaeology* 1(1):77–100.

Western, A. Cecilia
    1970   Wood and Charcoal in Archaeology. In *Science in Archaeology: A Survey of Progress and Research*, D. Brothwell and E. Higgs (eds.), pp. 178–87. Praeger, New York.

White, T. E.
    1953   A Method of Calculating the Dietary Percentage of Various Food Animals Utilized by Aboriginal Peoples. *American Antiquity* 18(4):396–98.

Will, Richard T., and James Clark
    1996   Stone Artifact Movement on Impoundment Shorelines: A Case Study from Maine. *American Antiquity* 61(3):499–519.

Will, Richard, James Clark, and Edward Moore
    1996   *Phase III Archaeological Data Recovery at the Little Ossipee North Site (7.7) Bonny Eagle Project (FERC #2529), Cumberland County, Maine.* Prepared for Central Maine Power Company, Augusta, Maine.

Winters, Howard D.
    1969   *The Riverton Culture: A Second Millennium Occupation in the Central Wabash Valley.* Illinois State Museum Report of Investigations No. 13. Illinois State Museum, Springfield.

INDEX

Aché hunter-gatherer, 19
acidic soil, 23–24
agave processing, 30
agave quid, 30, 31
Aka Pygmy, 23
alkaline soil, 24–25
allometric scaling formula, 76
anaerobic environment, 26–27
anemophilous plant, 82, 83
animal, site modification by, 18, 23, 31–32, 39, 70
animal dung study, 39
Antelope House, 64–65
apple creek recovery method, 50
Appleyard, H. M., 79
archaeobiology: definition of, 1; development/future of, 11–15; goals of, 2–4; project involvement of, 5–7
archaeobotany, definition of, 1, 2
archaeologist, project stage for involvement by, 6, 7
Asch Sidell, Nancy, 13

Baker Cave, 81, 82
Barkley, W. D., 79
Barrows, David P., 8
Big Bend National Park, 35, 36–38, 43, 71, 94, 95, 96

bioarchaeology, definition of, 1
Bocek, Barbara, 18–19
bog body, 26
bone: coding card for, 92; preservation of, 22–23, 24–25
bone assemblage: carnivore/rodent modification of, 18, 23, 31–32; differentiating cultural/noncultural, 31–38
Bryant, Vaughn M. Jr., 24, 64–65, 83
buried type site, excavation/sampling, 72–74
Burt, Timothy P., 2

calcination, 22, 25, 29
Cannon, Debbi Y., 93
carbonization, 21–22, 29
Carlson, David L., 24
carnivore/rodent, site modification by, 18, 23, 31–32, 39, 70
Carr, H. Sorayya, 99–101
Casteel, R. W., 79, 93
Castetter, Edward, 8
Catling, Dorothy, 79
Central Plains Tradition (CPT) homestead, 71–72
Cerros site, 99–101
Claasen, Cheryl, 93
Clark, James, 40

Columbian Exhibition, 8
combination recovery method, 52–53
Cooke, Richard G., 88–89
coprolites. *See* paleofeces
Core, H. A., 79
Corner, Edred J. H., 79
Cote, W. A., 79
CPT. *See* Central Plains Tradition (CPT) homestead
Crader, Diana, 104–5
Crane, Cathy J., 99–101
CRM. *See* cultural resource management (CRM)
Cruze-Uribe, Kathryn, 93
cultural context, 3, 31, 38–41
cultural ecology, 11
cultural/noncultural assemblage, determining: context role in, 38–41; through faunal remains, 31–38; in open type site, 70–71; in rock-shelter type site, 70
cultural resource management (CRM): analyzation method choice, 91; incorporation of archaeobiologist in, 5–6; and integrative analysis, 107, 109; role in development/future of archaeobiology, 11–15

Day, A. C., 79
Dennis, J. V., 79
Devil's Mouth site, 72–74
Dimbleby, Geoffrey W., 79
Dixon Site, 109, 114–18
DNA, 62, 64, 90
dog, as dietary item, 104

Efremov, J. A., 17
enamel hypoplasias, 100, 102
environmental anthropology, 11
ethnoarchaeology, 18, 19–20
Ethnobotanical Laboratory, 9
ethnobotany, 8–9

excavation/sampling, 53–54; in buried type site, 72–74; collecting sample, 56–61; flotation case study, 54, 55; in open type site, 70–72; in pueblo type site, 61–65, 66; in rock-shelter type site, 69–70; in shell midden type site, 65, 67–68
experimental archaeology, 18

Faegri, Knut, 121
fine screening recovery, 47, 48–49, 53, 58
Fishel, Richard L., 109
flotation recovery, 47, 49–52, 53, 58
fluvial action, 40–41, 42
Fort Ancient people, 2–3

Gilbert, B. Miles, 93
Gilmore, Melvin, 8, 9
Grayson, Donald K., 93
Grayson, John, 79
Guilday, John, 10–11
Gulf of Maine, 4
Gunn, C. R., 79

Hadza food procurement practice, 29–30
Harshberger, J. W., 8
Hawkes, K., 29–30
Hays, Christopher, 18
Helbaek, Hans, 50
Hoffman, Rob, 18
Holloway, Richard G., 24
horticulture practice, Late Woodland, 54, 55
Hudson, Jean, 23
Hulme site, 71–72

immersion recovery method, 50
insect-pollinated plant, 82
integration, of archaeobiological analyses: into anthropological context, 105, 107; difficulties

INDEX

with, 99; paleonutrition studies, 99–103; using ecology/ethnography/ethnohistory context, 104–5, 106
Iverson, John, 121

Jones, John G., 24
Jones, Kevin T., 19
Jones, Volney, 9

Kamp, Kathryn A., 107, 108
Kapp, Ronald O., 85
Klein, Richard G., 93

laboratory/analytical technique: importance of, 75–76; reference collection for, 77–78
laboratory/analytical technique, botanical remains, 78–79; identification guide for, 79; phytolith analysis, 87–90; pollen analysis, 81–86; quantification techniques, 79–81
laboratory/analytical technique, zooarchaeology: bone coding card, 92; identification guide for, 93; process/sort/label/analyze method, 91–92; quantification, 93–97; reference collection for, 77–78; washing/handling recommendations, 90
Lake Patzcuaro, 2
Lawrence, Barbara, 10
Leney, Lawrence, 79
Linse, Angela R., 25
Little Ossippee North site, 12–14, 27
Lizard Man Village, 107, 108–12
Lower Pecos Region, 69–70, 100–103

Mack, Richard N., 83
Martin, A. C., 79
Martin, Larry D., 93

McInnis, Heather, 75–76
Miller, Naomi F., 39
Mimbres River Valley, 3–4
minimum number of individuals (MNI), 10, 95–97, 104
MNI. *See* minimum number of individuals
Montgomery, F. H., 79
Morris, Don P., 64
Mulholland, Susan C., 88

NAN Ranch, 51
National Register of Historic Places, 12
Native American, early study of, 7–9
nested fine screening, 48, 49, 93
New Archaeology, 7, 11, 47
Newt Kash Hollow site, 9
NISP. *See* number of identified specimens
noncultural context, 31
Norr, Lynette, 88–89
number of identified specimens (NISP), 94–95, 96, 104

O'Connell, J. F., 29–30
O'Hara, Sarah, 2
oil drum recovery method, 50–52
Old Town Ruin, 65, 66
Olsen, Stanley, 10–11
Oneota people, 109, 114–18
open type site, excavation/sampling, 70–72

paleoenvironmental analysis, 58
paleoethnobotany: definition of, 1–2; history of, 7–9; reference collection for, 77
paleofeces, 34–35, 62–64, 81, 82, 86
Panama study, 88–89
Paradine, P. J., 79
Parmalee, Paul, 10–11
Pearsall, Deborah M., 49, 52, 79, 88

percentage count quantification method, 80
percentage subjective quantification method, 80–81
percentage weight quantification method, 80, 81
phytolith: analysis of, 87–90. See also pollen/phytolith
Piperno, Dolores R., 88–90
plant remains, preservation of, 22
pollen: analysis of, 81–86; preservation of, 24, 26; standard extraction procedure, 84, 89–90
pollen/phytolith: collecting/sampling, 58–60, 64–65, 66, 67, 70, 72, 73–74; Panama study, 88–89
postdepositional fill, 62
presence/absence quantification method. See ubiquity quantification method
preservation: biological preservation factors, 21–23; cultural preservation factors, 29–30; relative potential of material, 20, 21
preservation, environmental factors: oxygen/aeration lack, 26–27; soil acidity, 23–24; temperature, 25–26; tree throws, 28, 39
pueblo type site, excavation/sampling, 61–65, 66

quantification technique: minimum number of individuals, 10, 95–97, 104; number of identified specimens, 94–95, 96, 104; percentage count, 80; percentage subjective, 80–81; percentage weight, 80, 81; ubiquity, 79–80, 93–94, 96, 99, 104
Quebrada Jaguay site, 75–76

Rapp, George, 88
recovery technique: archaeobiologist involvement in, 46; combination method, 52–53; field technician as collector, 45–46; fine screening, 47–49, 53; flotation, 47, 49–52, 53. See also excavation/sampling
Reitz, Elizabeth J., 76
research, history of: paleoethnobotany, 7–9; zooarchaeology, 10–11
research-driven archaeological project, 15, 91–92
Riverton Culture, 11
Robison, N. D., 10
rock-shelter type site, excavation/sampling, 69–70
rodent ecology, 18–19

sampling. See excavation/sampling
Sanger, David, 4
saprophytic organism, 26; definition of, 21; effect of temperature/aridity on activity of, 25, 26; oxygen need of, 26, 27; soil type found in, 23–24
Savage, Howard G., 93
Schultes, R. E., 8
settlement pattern study, 4
Shaffer, Brian, 47
shell, preservation of, 22, 25
shell midden type site, excavation/sampling, 65, 67–68
Shell Mound Archaeological Project flotation recovery method, 50–52
Shipman, Pat, 38
Sinagua village site, 107, 108–12
slave dietary study, 104–5, 106
SMAP. See Shell Mound Archaeological Project flotation recovery method
Smart, Tristine, 39

Smith, Bruce, 11
Snyder, Lynn M., 104
sporopollenin, 22, 24, 81, 82
Street-Perrott, F. Alayne, 2
Struever, Stuart, 47, 50

taphonomy, 11, 41, 44; biological preservation factor, 20, 21–23; cultural preservation factor, 29–30; definition of, 17; environmental preservation factor, 23–28; ethnoarchaeology role in, 19–20; experimental archaeology role in, 18–19; importance of, 17–18. *See also* cultural/noncultural assemblage, determining
Thomas, D. H., 47
Tierra Blanca site, 39
Todd site, 65, 67–68
tree throw taphonomy, 28, 39

ubiquity quantification method, 79–80, 93–94, 96, 99, 104
U.S. Department of Agriculture, 8

U.S. National Herbarium, 8

Voorhies, V. R., 41

Wagner, Gail E., 2–3
Warren, Robert E., 104
washover recovery method, 50
water screening, 48, 49
Weir, Glendon H., 64–65
Western, A. Cecilia, 79
White, T. E., 10–11
Whittaker, John C., 107, 108
Wildman, A. B., 79
Will, Richard T., 40
wind-pollinated plant, 82, 83
Wing, Elizabeth, 76
Winters, Howard D., 11
wood, preservation of, 26
wood rat bone movement study, 18

Zooarchaeology: definition of, 1; history of, 10–11; reference collection for, 77–78. *See also* laboratory/analytical technique, zooarchaeology
zoophilous plant, 82

ABOUT THE AUTHOR AND SERIES EDITORS

Kristin D. Sobolik received her B.S. in biology from the University of Iowa (1986), M.A. in anthropology from Texas A&M University (1988), and Ph.D. (with a dissertation entitled "Paleonutrition of the Lower Pecos Region") from Texas A&M (1991). She held a postdoctoral fellowship position at Southern Illinois University, where she edited *Paleonutrition: The Diet and Health of Prehistoric Americans*, and then joined the faculty at the University of Maine in the Anthropology Department and the Institute for Quaternary and Climate Studies, where she is an associate professor and director of the Zooarchaeology Laboratory.

Kristin has conducted archaeological work in the desertic regions of North America focusing in the Lower Pecos and Big Bend regions of Texas as well as throughout the southwestern United States. In addition, she has conducted research in Maine and at Mammoth Cave, Kentucky. Her recent research has centered on analyzing paleonutrition and possible prehistoric sex differences using DNA and hormonal content from paleofeces. She is also interested in the zoogeography of extinct and extant animal species, children's health in the prehistoric Southwest, and evidence for prehistoric cannibalism. She lives with her husband, Scott, a writer and English instructor, and their four beautiful children ranging in age from 1 to 12.

Larry J. Zimmerman is the head of the Archaeology Department of the Minnesota Historical Society. He served as an adjunct professor of anthropology and visiting professor of American Indian and native studies at the University of Iowa from 1996 to 2002 and as chair of the American Indian and Native Studies Program from 1998 to 2001. He earned his Ph.D. in anthropology at the University of Kansas in 1976. Teaching at the University of South Dakota for twenty-two years, he left there in 1996 as Distinguished Regents Professor of Anthropology.

While in South Dakota, he developed a major CRM program and the University of South Dakota Archaeology Laboratory, where he is still a research associate. He was named the University of South Dakota Student Association Teacher of the Year in 1980, given the Burlington Northern Foundation Faculty Achievement Award for Outstanding Teaching in 1986, and granted the Burlington Northern Faculty Achievement Award for Research in 1990. He was selected by Sigma Xi, the Scientific Research Society, as a national lecturer from 1991 to 1993, and he served as executive secretary of the World Archaeological Congress from 1990 to 1994. He has published more than three hundred articles, CRM reports, and reviews and is the author, editor, or coeditor of fifteen books, including *Native North America* (with Brian Molyneaux, University of Oklahoma Press, 2000) and *Indians and Anthropologists: Vine Deloria, Jr., and the Critique of Anthropology* (with Tom Biolsi, University of Arizona Press, 1997). He has served as the editor of *Plains Anthropologist* and the *World Archaeological Bulletin* and as the associate editor of *American Antiquity*. He has done archaeology in the Great Plains of the United States and in Mexico, England, Venezuela, and Australia. He has also worked closely with a wide range of American Indian nations and groups.

William Green is the director of the Logan Museum of Anthropology and an adjunct professor of anthropology at Beloit College, Beloit, Wisconsin. He has been active in archaeology since 1970. Having grown up on the south side of Chicago, he attributes his interest in archaeology and anthropology to the allure of the exotic (i.e., rural) and a driving urge to learn the unwritten past, abetted by the opportunities available at the city's museums and universities. His first fieldwork was on the Mississippi River bluffs in western Illinois. Although he also worked in Israel and England, he returned to Illinois for several years of survey and excavation. His interests in settlement patterns, ceramics, and archaeobotany developed there. He received his master's degree from the University of Wisconsin at Madison and then served as Wisconsin SHPO staff archaeologist for eight years. After obtaining his Ph.D. from the University of Wisconsin at Madison in 1987, he served as state archaeologist of Iowa from 1988 to 2001, directing statewide research and service programs including burial site protection, geographic information, publications, contract services, public outreach, and curation. His main research interests focus on the development and spread of native agriculture. He has served as editor of the *Midcontinental Journal of Archaeology* and *The Wisconsin Archeologist*; has published articles in *American Antiquity*, *Journal of Archaeological Research*, and other journals; and has received grants and contracts from the National Science Foundation, National Park Service, Iowa Humanities Board, and many other agencies and organizations.

ARTIFACTS

ARCHAEOLOGIST'S TOOLKIT

SERIES EDITORS: LARRY J. ZIMMERMAN AND WILLIAM GREEN

The Archaeologist's Toolkit is an integrated set of seven volumes designed to teach novice archaeologists and students the basics of doing archaeological fieldwork, analysis, and presentation. Students are led through the process of designing a study, doing survey work, excavating, properly working with artifacts and biological remains, curating their materials, and presenting findings to various audiences. The volumes—written by experienced field archaeologists—are full of practical advice, tips, case studies, and illustrations to help the reader. All of this is done with careful attention to promoting a conservation ethic and an understanding of the legal and practical environment of contemporary American cultural resource laws and regulations. The Toolkit is an essential resource for anyone working in the field and ideal for training archaeology students in classrooms and field schools.

Volume 1: *Archaeology by Design*
By Stephen L. Black and Kevin Jolly

Volume 2: *Archaeological Survey*
By James M. Collins and Brian Leigh Molyneaux

Volume 3: *Excavation*
By David L. Carmichael and Robert Lafferty

Volume 4: *Artifacts*
By Charles R. Ewen

Volume 5: *Archaeobiology*
By Kristin D. Sobolik

Volume 6: *Curating Archaeological Collections:*
 From the Field to the Repository
By Lynne P. Sullivan and S. Terry Childs

Volume 7: *Presenting the Past*
By Larry J. Zimmerman

ARTIFACTS

Charles R. Ewen

ARCHAEOLOGIST'S TOOLKIT
VOLUME 4

AltaMira
PRESS

A Division of Rowman & Littlefield Publishers, Inc.
Walnut Creek • Lanham • New York • Oxford

AltaMira Press
A Division of Rowman & Littlefield Publishers, Inc.
1630 North Main Street, #367
Walnut Creek, CA 94596
www.altamirapress.com

Rowman & Littlefield Publishers, Inc.
A Member of the Rowman & Littlefield Publishing Group
4501 Forbes Boulevard, Suite 200
Lanham, MD 20706

PO Box 317
Oxford
OX2 9RU, UK

Copyright © 2003 by AltaMira Press

All rights reserved. No part of this publication may be reproduced, stored in a retrieval system, or transmitted in any form or by any means, electronic, mechanical, photocopying, recording, or otherwise, without the prior permission of the publisher.

British Library Cataloguing in Publication Information Available

Library of Congress Cataloging-in-Publication Data

Ewen, Charles Robin.
    Artifacts / Charles R. Ewen.
      p. cm. — (Archaeologist's toolkit ; v. 4)
    Includes bibliographical references (p.) and index.
    ISBN 0-7591-0400-X (alk. paper) — ISBN 0-7591-0022-5 (pbk. : alk. paper)
      1. Archaeology—Methodology. 2. Excavations (Archaeology) 3.
  Antiquities—Collection and preservation. 4. Antiquities—Analysis. I. Title.
II. Series.

    CC75.7 .E97 2003
    930.1'028—dc21
                                                         2002152408
Printed in the United States of America

∞™ The paper used in this publication meets the minimum requirements of American National Standard for Information Sciences—Permanence of Paper for Printed Library Materials, ANSI/NISO Z39.48-1992.

CONTENTS

Series Editors' Foreword   vii

Acknowledgments   xi

Part I Introduction   1

1 Introduction   3

2 The Material Assemblage   5
Artifacts • Ecofacts • Context • Range of Material Included • Prehistoric

3 Artifacts and Archaeological Theory   11
Culture History • Processual • Postprocessual • Cultural Resource Management • Conclusion

Part II Preparing for Analysis   17

4 Excavating Artifacts   19
Recording Context • Collecting

5 Processing   23
Artifact Check-in • Cleaning • Rough Sorting

6 Cataloging   29
Accessioning • Inventory • Labeling

7	**Further Preparation** Reconstruction and Cross-mending • Conservation • Photography	35

Part III Analysis   41

8	**Materials Analysis** Lithics • Ceramics • Metal • Glass • Organic Artifacts	43
9	**Classification** Objectives • Emic versus Etic • Classification Schemes • Quantification • Databases • Card Catalogs • Electronic Data	67
10	**Data Manipulation** Objectives • Statistical Analyses • Patterning • Dating • Spatial Analyses • Replication	89
11	**Specialized Analyses** Chronometric Dating • Faunal • Floral • Petrography • Bioarchaeology	107
12	**Curation** Artifacts • Project Records	117

Part IV Final Thoughts   119

13	**Conclusion** Future Directions • Conclusion	121

References	131
Index	141
About the Author and Series Editors	147

SERIES EDITORS' FOREWORD

The Archaeologist's Toolkit is a series of books on how to plan, design, carry out, and use the results of archaeological research. The series contains seven books written by acknowledged experts in their fields. Each book is a self-contained treatment of an important element of modern archaeology. Therefore, each book can stand alone as a reference work for archaeologists in public agencies, private firms, and museums, as well as a textbook and guidebook for classrooms and field settings. The books function even better as a set, because they are integrated through cross-references and complementary subject matter.

Archaeology is a rapidly growing field, one that is no longer the exclusive province of academia. Today, archaeology is a part of daily life in both the public and private sectors. Thousands of archaeologists apply their knowledge and skills every day to understand the human past. Recent explosive growth in archaeology has heightened the need for clear and succinct guidance on professional practice. Therefore, this series supplies ready reference to the latest information on methods and techniques—the tools of the trade that serve as handy guides for longtime practitioners and essential resources for archaeologists in training.

Archaeologists help solve modern problems: They find, assess, recover, preserve, and interpret the evidence of the human past in light of public interest and in the face of multiple land use and development interests. Most of North American archaeology is devoted to cultural resource management (CRM), so the Archaeologist's Toolkit focuses on practical approaches to solving real problems in CRM and

public archaeology. The books contain numerous case studies from all parts of the continent, illustrating the range and diversity of applications. The series emphasizes the importance of such realistic considerations as budgeting, scheduling, and team coordination. In addition, accountability to the public as well as to the profession is a common theme throughout the series.

Volume 1, *Archaeology by Design*, stresses the importance of research design in all phases and at all scales of archaeology. It shows how and why you should develop, apply, and refine research designs. Whether you are surveying quarter-acre cell tower sites or excavating stratified villages with millions of artifacts, your work will be more productive, efficient, and useful if you pay close and continuous attention to your research design.

Volume 2, *Archaeological Survey*, recognizes that most fieldwork in North America is devoted to survey: finding and evaluating archaeological resources. It covers prefield and field strategies to help you maximize the effectiveness and efficiency of archaeological survey. It shows how to choose appropriate strategies and methods ranging from landowner negotiations, surface reconnaissance, and shovel testing to geophysical survey, aerial photography, and report writing.

Volume 3, *Excavation*, covers the fundamentals of dirt archaeology in diverse settings, while emphasizing the importance of ethics during the controlled recovery—and destruction—of the archaeological record. This book shows how to select and apply excavation methods appropriate to specific needs and circumstances and how to maximize useful results while minimizing loss of data.

Volume 4, *Artifacts*, provides students as well as experienced archaeologists with useful guidance on preparing and analyzing artifacts. Both prehistoric- and historic-era artifacts are covered in detail. The discussion and case studies range from processing and cataloging through classification, data manipulation, and specialized analyses of a wide range of artifact forms.

Volume 5, *Archaeobiology*, covers the analysis and interpretation of biological remains from archaeological sites. The book shows how to recover, sample, analyze, and interpret the plant and animal remains most frequently excavated from archaeological sites in North America. Case studies from CRM and other archaeological research illustrate strategies for effective and meaningful use of biological data.

Volume 6, *Curating Archaeological Collections*, addresses a crucial but often ignored aspect of archaeology: proper care of the specimens

and records generated in the field and the lab. This book covers strategies for effective short- and long-term collections management. Case studies illustrate the do's and don'ts that you need to know to make the best use of existing collections and to make your own work useful for others.

Volume 7, *Presenting the Past*, covers another area that has not received sufficient attention: communication of archaeology to a variety of audiences. Different tools are needed to present archaeology to other archaeologists, to sponsoring agencies, and to the interested public. This book shows how to choose the approaches and methods to take when presenting technical and nontechnical information through various means to various audiences.

Each of these books and the series as a whole are designed to be equally useful to practicing archaeologists and to archaeology students. Practicing archaeologists in CRM firms, agencies, academia, and museums will find the books useful as reference tools and as brush-up guides on current concerns and approaches. Instructors and students in field schools, lab classes, and short courses of various types will find the series valuable because of each book's practical orientation to problem solving.

As the series editors, we have enjoyed bringing these books together and working with the authors. We thank all of the authors—Steve Black, Dave Carmichael, Terry Childs, Jim Collins, Charlie Ewen, Kevin Jolly, Robert Lafferty, Brian Molyneaux, Kris Sobolik, and Lynne Sullivan—for their hard work and patience. We also offer sincere thanks to Mitch Allen of AltaMira Press and a special acknowledgment to Brian Fagan.

LARRY J. ZIMMERMAN
WILLIAM GREEN

ACKNOWLEDGMENTS

When I was approached a couple of years ago and asked to produce a brief volume on artifact analysis, I was at first startled, then flattered, then dismayed. Dismayed because the series editors wanted a manuscript in nine months! "Take a sabbatical," they told me, as if this were a real option. Nevertheless, I was able to produce a rough draft, close to the deadline. Over the next couple of years, that manuscript bounced around the offices of Altamira Press until it emerged in its present form.

This book represents more of a personal philosophy toward artifact analysis than a "how-to" guide. It has been thirty years in the making, and everyone I have ever worked with or whose work I have read in the field of archaeology has had a hand in shaping it. My gratitude to so many who have helped me to a satisfying career in archaeology is far too long to list individually. However, there are several folk who deserve special mention for their particular help on this volume.

First, the students of my archaeological methods courses over the years are to be commended for providing me with insight into the teaching of artifact analysis. Many helpful comments came out of those classes. I am also indebted to my fellow ECU faculty members Randy Daniel and Jane Eastman (now at Western Carolina University) for looking over the lithic and ceramic analysis sections, respectively. Their comments helped me to better understand the finer points of these analyses. Tricia Samford and Linda Carnes-McNaughton helped to gather important references in North Carolina, and Lela Donat was helpful in obtaining information from the Arkansas Archeological Survey.

The folks at Altamira Press have also been very patient and helpful, even when I was not. Mitch Allen had the most helpful edits of the manuscript, and Grace Ebron and Lynn Weber were terrific during the production phase. I wear their Altamira Press shirt proudly.

Finally, my family (Gretchen, Kate, and Madeline) deserve mention, since several family outings had to be postponed while I worked this text. As I said, many people had a hand in producing this volume, but its shortcomings are solely my responsibility.

PART I
INTRODUCTION

This section provides the rationale for artifact analysis by defining the subject matter (the types of artifacts to be analyzed) and discussing the different perspectives concerning the interpretation of archaeological remains. We begin by defining what an artifact is and what it is not, thus setting the parameters for the remainder of this book. The importance of the concept of context, which provides meaning for the recovered artifacts, cannot be overstressed in the analysis of artifacts. The scope of the artifact assemblage is addressed, and the distinction between prehistoric and historic sites and the different approaches applied to each is noted. The significant role of archaeological theory on artifact analysis is examined, where we specifically note that it appears that the kinds of analyses performed are more a function of the questions being asked, rather than of the various technologies available. Finally, the requirements of legally mandated archaeology (cultural resource management, or CRM) and the implications for artifact analysis are discussed.

1
INTRODUCTION

Indiana Jones, in *Raiders of the Lost Ark*, makes the practice of archaeology seem relatively simple, albeit somewhat dangerous. One merely finds an old artifact or map that leads you to an amazing site, which you pillage at will. Only Nazis and fools spend long hours digging dry holes or carefully recording their finds. (Ironically, I point out in my classes that it was the Nazis who were doing the real archaeology and did the background work that allowed Jones to find the ark!) As long as the artifacts end up in a museum, it doesn't matter how you recovered them or how you arrived at your interpretations. Fortunately, most people can separate fact from fiction, although when I ask students in my introductory "Archaeology around the World" classes to name a famous archaeologist, Indiana Jones is always among the top five.

Fieldwork has always been what defines archaeology in the popular imagination. When the general public thinks about archaeology, they picture weather-beaten individuals laboring with shovels and picks in exotic locales. Even the more enlightened layperson focuses on the field aspects of archaeology. To most people, archaeologists are either working on a dig or writing a book about their exciting finds. The part between the exciting discoveries and the published report receives little, if any, attention.

The profession of archaeology is not immune to such stereotyping. If you ask most archaeologists why they got into the field, the most common answer is *not* "I enjoy endless hours of size-sorting lithic debitage." Go to the bar at any archaeological conference, and you'll likely hear anecdotes pertaining to fieldwork, not artifact analysis.

Yet, it is the analysis of what we find in the field that permits us to do our job: interpreting the past. It is the second most important thing that separates the archaeologist from the pothunter.

Curiously, relatively few books focus strictly on the analysis of the many types of artifacts recovered from archaeological sites. How can this be if this aspect of archaeology is so important? Probably, as I discovered during the preparation of this volume, because the scope of the undertaking is so daunting. There are myriad types of artifacts that archaeologists recover from excavations and countless ways of looking at them. Each question asked of the artifact may require a different analytical method to obtain the desired answer.

Answers obtained by differing methods may differ, but all may be equally valid depending on how they are used. For example, historic ceramics may be dated by applying Stanley South's mean ceramic date formula to raw sherd counts of stylistic types. By a different method of quantification and classification (i.e., calculating minimum vessel counts and a more generic typology), it is possible to approximate the status of the past owner of the same ceramic assemblage using George Miller's CC Index. Both of these methods will be discussed more fully later in this volume.

To "write the book" on artifact analysis demands that the author either be incredibly naive or possess excessive hubris. I have been accused of both. However, this book is not intended to provide the reader with a step-by-step guide to archaeological analysis or the nuts and bolts of the many arcane methods by which artifacts can be examined. Rather, it is meant to provoke the reader into thinking about the complexities of the general task of artifact analysis before embarking on that next endeavor and to provide an overview of some of the more commonly used analytical procedures and some additional source material for more detailed analyses.

2
THE MATERIAL ASSEMBLAGE

Archaeologists, like many scholars, have a bad habit of using the same term in different ways. This is especially true of archaeologists that have been educated at different institutions embracing different paradigms or working for various agencies in different states. Thus, before embarking on a discussion of the different ways to analyze artifacts, it would be prudent to clarify how the term *artifact* will be used in this book.

ARTIFACTS

I teach my students that *artifacts* are objects that are made or modified by humans. This seemingly straightforward definition can quickly become fatally vague if subjected to intense scrutiny. Is a building foundation an artifact? A trashpit? The unmodified rocks surrounding a hearth? A deer ulna found in a trashpit? Items made by *Homo erectus*? All of these items can appear on archaeological sites and aid in archaeological interpretation, but calling them all artifacts stretches the definition to the breaking point. Kenneth Feder (1996:458) offers a definition that clarifies the situation somewhat: "*artifact*—Any object manufactured by a human being or human ancestor, but usually a portable object like a stone spearpoint or a clay pot as distinguished from more complex archaeological features." This clears up some of the confusion, but what of unmodified portable objects?

ECOFACTS

Unmodified items that are found on and aid in the interpretation of an archaeological site are sometimes referred to as *ecofacts*. Actually, the only time you see this term is in texts such as this one. Still, the term is useful for those who split hairs, and it usually applies to the bulk of faunal and floral remains recovered at an archaeological site. For example, at an early nineteenth-century fur-trading post excavated in northwestern Wisconsin, I examined the faunal remains recovered from the site. Though these animal bones were not made by the inhabitants, an analysis of these remains determined the season of occupation for the site and the dietary preferences of the fur traders, and they even contributed to a determination of which cabin was occupied by the high-status company partner and which was used by the lower-status employees (Ewen 1986).

Animal bone and other objects, such as the individual rocks surrounding a hearth or pollen sealed in a trashpit, were associated with the site's inhabitants though technically not made by them. It is important to note that ecofacts are often subjected to the same analytical scrutiny by archaeologists as artifacts. Therefore, their analysis will be discussed at some length in this volume

Artifacts and ecofacts are the stuff of archaeological analysis. However, they all are meaningless if the context in which they were found was not recorded or later lost.

CONTEXT

Archaeological sites are like crime scenes of the past. Both archaeologists and detectives attempt to reconstruct past events from the physical remains of those events. It is important not only to collect the evidence but to record its precise location as well. This is why crime scenes are roped off from the public until the area has been minutely photographed and studied. In this way, archaeological sites are very much like crime scenes since the average visitor to a site is rarely permitted to jump in the pit with the excavators.

Knowing the context is what makes artifact analysis possible. Understanding where an artifact was found on a site and its relationship to other artifacts, ecofacts, and features is crucial for the archaeologist. I mentioned earlier that artifact analysis was the second most important factor separating the archaeologist from the pothunter. The concern for context is the most important factor that separates the archaeologist from the looter. It is an artifact's context that makes analysis possible by allowing you to determine its function and placement in time and space.

The importance of the archaeologist's obsessive concern with context is amply demonstrated by my excavations at the Governor Martin site in Tallahassee, Florida. Finds of sixteenth-century Spanish coins, ceramics, and beads led archaeologists to believe that they had located the site of Hernando de Soto's first winter encampment (ca. 1539–1540). However, the possibility that these materials could have been the result of the earlier (1528) expedition of Pánfilo de Narváez or salvaged and traded shipwreck material could not be dismissed. During the last month in the field for the project, the shattered jaw and several teeth from a domestic pig (*Sus scrofa*) were recovered from an undisturbed deposit in association with protohistoric Apalachee Indian pottery and a sixteenth-century Nueva Cadiz bead. The context of the pig teeth is important, because pigs are not native to the southeastern United States. Historical records indicate that de Soto, not Narváez, was the first to introduce pigs into the region. After de Soto passed through, Spaniards would not venture back into this part of Florida for another hundred years. Thus, without the sixteenth-century context established, the pig bones might have been evidence of just another southern barbecue.

As anthropologists, archaeologists are concerned with two types of context: the archaeological context and the cultural context. The cultural context is important in that, in addition to the description of human behavior and its consequences, it consists of the interplay of these actions and how they function within the larger cultural system. The archaeological context, on the other hand, *comes from* the material results of cultural behavior. It can be used to infer the lifeways characteristic of the former occupants of the site. It is from the archaeological context that we can infer the cultural context in which the artifacts were used.

RANGE OF MATERIAL INCLUDED

The range of artifacts that are analyzed and the scope of that analysis depend on several factors: the expertise of the archaeologist, the requirements of the scope of work (in the case of contracts or grants), and the funding available for the project. Often a bewildering array of material is found on an archaeological site, and there are few (if any) Renaissance individuals who are expert in them all. More often than not, an archaeologist chooses to become proficient in a certain category of artifacts, such as lithics or glass beads. Still others focus on a single time period and ethnicity, such as the artifacts of the Spanish Colonial period or Paleoindian period.

Ivor Noël Hume (1972:13–14) has stated, "The excavator whose experience is confined to English colonial sites will be out of his depth if he tries to interpret those of Spanish origin." Unfortunately, archaeologists employed in CRM-oriented archaeology usually do not have the luxury of specifying which types of sites they will excavate. Those choices are dictated by the contract and the personnel available at their company. The larger firms may be able to hire archaeologists who specialize in either historic or prehistoric archaeology, but circumstances often require these specialists to interpret types of sites outside their specialty. For this reason, it is usually a good idea to engage an appropriate specialist on a project when the variety of the artifact assemblage exceeds your ability to conduct a complete analysis.

The biggest challenge for many contract firms, in the realm of artifact analysis, is having the staff expertise to analyze the material recovered from both prehistoric and historic sites. This is because the latter requires additional research skills not always taught in the anthropology curricula to which most archaeologists are exposed. Until the latter half of this century, this was not a problem since virtually all archaeology (with a few notable exceptions) done in the United States was performed on prehistoric sites. Even as late as the 1970s, most contract archaeologists and the state historic preservation officers (SHPOs) to whom they reported paid little attention to historic sites, especially those dating to the nineteenth century and later. Though quite a bit of discussion still ensues about the significance of late historical archaeological remains (Ewen 1995; Cabak et al. 1999), today's cultural resource manager can no longer ignore them with impunity. Even twentieth-century tenant farmsteads and World War II military camps are now eligible for inclusion on the National Register of Historic Places and therefore *must* be taken into consideration by archaeological survey teams.

PREHISTORIC

Defining what is historic and what is prehistoric is not as straightforward as it may seem. In fact, this was no small matter of debate in the early days of historical archaeology. Obviously prehistoric sites are those occupied before written history. But what comprises a historic site? Even restricting this question to the Americas does not clear up all the ambiguity. If you answer, "It is a site that has historic documents associated with it," you eliminate many sites that were never documented in the historic record or whose documentation has been lost or destroyed. If you say (for the New World), "Any site dating after 1492," you are suggesting that a fifteenth-century Mississippian village in Illinois is prehistoric and that the same village a century later is historic even though the inhabitants, in both cases, never met a European and shared a similar material culture.

Is the question of whether a site is historic or prehistoric even meaningful? Of course it is. How an archaeologist classifies a site will often determine his or her approach to its analysis and interpretation. It determines the questions asked of the archaeological record and the expertise that is needed to interpret the site in question. Interpreting historic sites usually requires that the archaeologist possess the additional skills to conduct research in the documentary record or at least have a qualified historian on the research team.

Many prehistoric archaeologists feel that analyzing historic artifacts is easier to do since there are often documents describing their function, distribution, and sometimes even their price. Historical archaeologists counter that function and worth are unverifiable for prehistoric artifacts and that the analysis of historical artifacts is actually more challenging since the range of material is far greater and there is always the chance that a document will appear that refutes their initial interpretation! It is probably more prudent to say that each analysis of prehistoric and historic artifacts presents different challenges to an interpretation that must be overcome by the archaeologist in different ways.

3
ARTIFACTS AND ARCHAEOLOGICAL THEORY

Archaeology, like most disciplines, has changed its focus over time. The pace of these changes is similar to the punctuated equilibrium of evolutionary biology. There are periods during which the research questions of archaeologists follow relatively similar lines (i.e., reconstruction of past chronologies). These periods of little change are punctuated by scholarly upheavals, which some liken to paradigm shifts, where the focus of research takes a new direction. Needless to say, the reorienting of archaeological research results in an entirely new suite of questions being asked of the archaeological assemblage. It is interesting to note that it is the new research orientation that precedes the analytical technology rather than the other way around.

As the questions being asked of the artifacts change, the methods of analyses also change to accomplish these new goals. For example, a pot is recovered from a burial at a prehistoric site in the Plains. The methods of analyzing this artifact will change according to whether the archaeologist wants to know the age of the pot, whether the pot was locally made or a trade item, the function of the pot, or the gender of the individual who made it.

What factors influence archaeological interpretation? The prevailing social values at the time a site is analyzed often affect interpretation as does the training of the archaeologist. Not so long ago, most archaeologists looked to science to reveal the "truth" about the past. Today, many archaeologists see the past as a social construction with multiple, equally valid interpretations. Archaeologists needn't necessarily buck the trends in current archaeological research, but they should constantly question their validity. Being aware of the origins of research trends helps keep them in perspective and allows an assessment to be made of their potential to contribute to the long-term goals of the discipline. For this reason, it is often instructive not only to learn the different approaches utilized in archaeology but to understand something about the times in which they were conceived. The historical perspective permits the identification of subjective factors by observing how and, perhaps, why interpretations change through time.

Archaeological theory can be said to be hierarchically arranged with high, middle, and low levels (Trigger 1989:19–25). *Low theory* consists of observable, regular phenomena (e.g., typologies, seriation, dating) and does *not* refer to human behavior. *Middle theory* includes generalizations accounting for regularities between datasets or variables (e.g., population pressure and the development of agriculture). This level relates human behavior to archaeological data. *High theory* is what most students think of when they discuss archaeological theory. It can be equated with controlling models (paradigms) such as Marxism, cultural materialism, or processual archaeology. While all levels of theory have an effect on the type of analyses being performed, it is the highest level that influences all the others.

Does a discussion of archaeological theory have *any* relevance for the contract archaeologist who is surveying a right-of-way corridor for a proposed highway or a Forest Service archaeologist who is charged with managing the cultural resources of a national forest? Surprisingly, more than you might think, especially when interpreting artifacts or assessing site significance. If your goal is to reconstruct the past lifeways of a site's inhabitants, you are going to analyze the artifacts differently than if you were concerned with the processes affecting culture change or understanding how past belief systems were shaped by and reflected in a society's material culture. It is important to understand something of the history of archaeological reasoning in order to better interpret past site reports.

The history of archaeological thought can be reduced, in its most general sense, to three basic movements in which the many different approaches can be grouped: constructing culture history, processual archaeology, and postprocessual archaeology.

CULTURE HISTORY

The construction of local and regional culture histories has been, is, and will continue to be important to the field of archaeology. Essentially, this approach is characterized by descriptive work that deals with the who, what, where, when, and how of past societies and characterizes most of CRM archaeology today. It is the particularistic baseline data from which all inferences about the past are derived and tested. It is a necessary first step for all archaeology and one that many claim is still unfinished for most areas of the country. Indeed, the October 1999 issue of *North Carolina Archaeology* is devoted to "Prehistoric Pottery: Series and Sequence on the Carolina Coast" and explicitly follows "in the spirit of . . . Bill Haag's initial explorations" that established the original pottery seriation for the region (Herbert 1999:2).

PROCESSUAL

The New Archaeology was introduced in the late 1960s by archaeologists who thought that the basic regional temporal-spatial frameworks were well enough defined to start addressing questions concerning how and why cultures changed through time. This is an explicitly positivistic and nomothetic approach, which uses research design and the scientific method to analyze conditions of culture change and to generate laws governing human behavior.

Several approaches are subsumed under what has come to be known as *processual archaeology* (e.g., general systems theory, cultural ecology). *Processual* refers to the processes of culture change that take place as a result of interactions between a culture and its natural and cultural environment. This approach proved particularly well suited to many regional CRM and academic studies. The Cache River Basin Project (Schiffer and House 1975) and Deagan's (1983) study of colonial St. Augustine are good examples of processual archaeology.

Louis Binford, the leading exponent of the processual approach, concentrated on understanding culture change as an internal process, thus emphasizing the systemic aspects of culture. He differed from traditional archaeologists by focusing on the human ability to innovate while at the same time agreeing that cultures tended to remain static if undisturbed. He also agreed that human behavior was determined by forces (usually natural) of which humans were largely unaware.

Artifacts in an archaeological assemblage function in three subsystems of culture: technomic, sociotechnic, and ideotechnic. This type of approach has real repercussions when employed on artifact assemblages! Regularities in human behavior could be used to infer many aspects of prehistoric cultures that could not be directly correlated with the archaeological record. Binford denied the relevance of psychological factors and instead focused on identifying relations between technology and the environment as the key elements for determining cultural systems and, hence, human behavior (Trigger 1989:294–303). The apparent lack of concern for the impact of the individual led to the responses, which have come to be known collectively as the postprocessual school of thought.

POSTPROCESSUAL

Postprocessual archaeology subsumes theoretical approaches (largely apositivistic) that are critical of an allegedly unbiased scientific approach and that emphasize the individual and social factors in human societies. These include feminist, critical theory, and neo-Marxist approaches to archaeology. The extreme view of postprocessualism rejects the objectivity of the scientific approach and indeed whether the "truth" is really out there.

This relativistic stance claims that the interpretation of the past is dependent on the perspective of the observer or, in this case, archaeologist. Thus, multiple interpretations of the past are possible, all of which are equally valid. This is a disturbing thought for those trying to discover "what really happened." The implications for CRM-oriented archaeology are troubling, specifically with regard to the interpretation of site significance. On the other hand, the emphasis on the individual and disenfranchised groups has led to a great deal of archaeology concerning African slaves, Chinese immigrants, and other so-called people without history.

CULTURAL RESOURCE MANAGEMENT

In 1966, a bill was signed into law that marked the single most important change in archaeology as it is practiced in the United States. Section 106 of the National Historic Preservation Act states:

> The head of any Federal agency having direct or indirect jurisdiction over a proposed Federal or *federally assisted undertaking* [emphasis mine] in any State and the head of any Federal department or independent agency having authority to license any undertaking shall prior to

the approval of the expenditure of Federal funds on the undertaking or prior to the issuance of any license, as the case may be, take into account the effect of the undertaking on any district, site, building, structure, or object that is included in or eligible for inclusion in the National Register. The head of any such Federal agency shall afford the Advisory Council on Historic Preservation established under Title II of this Act a reasonable opportunity to comment with regard to such undertaking.

This legislation is so important because it means that any project that involves the federal government *in any way* might potentially require the services of an archaeologist. Section 106 catapulted archaeology from an obscure, romantic academic pursuit to a full-fledged environmental industry in less than a decade.

The National Historic Preservation Act was largely responsible for what has come to be known as cultural resource management (unless you work for the U.S. Forest Service, in which it is called heritage resource management). Cultural Resource Management (CRM) is simply archaeology performed under legal mandate and accounts for nearly 80 percent of all the archaeology done in the United States today. What does this have to do with artifact analysis? A great deal, actually, since state and federal guidelines regulating CRM archaeology often mandate what type and number of artifacts are to be collected and the types of special analyses required to satisfy an archaeological contract.

It is important for the archaeologist to know the guidelines of the agency or institution that is sponsoring or reviewing the archaeological project. For instance, artifacts collected on a project for the National Park Service must be inventoried according to their cataloging system (formerly the Automated National Catalog System, now Re:discovery). Similarly, other federal agencies and many state agencies have specified formats for cataloging and curating their data. It should be noted that there is no uniform standard across the United States. Archaeologists should contact the state archaeologist before pursuing a compliance project in their states to determine which procedures, if any, need to be followed.

This caveat does not necessarily mean that the data must be analyzed in a particular way. Rather, after the analyses are complete, the data must be left in a manner that is compatible with data from other projects sponsored by the particular agency. Ignorance of acceptable formats can result in much duplicated effort if the cataloging and reporting of the artifacts must be redone to conform to a specific agency's guidelines.

CONCLUSION

The remainder of this book will explore various methods and techniques of artifact analysis. It is not meant to be a step-by-step guide to the "right way" to analyze artifacts. There are many ways to analyze artifacts. The right way will depend on the questions being asked of the artifact assemblage. Often, archaeologists will quantify and organize artifacts in a specific way because "that's the way they learned it in grad school." The purpose of this volume is to expose the archaeologist to multiple systems of analysis and the rationale as to when they are appropriate.

The continuing theme of this volume will be the importance of relating your analytical methods to your research questions. All archaeological projects have (either explicitly or implicitly) a multistage research procedure. The stages include problem design, project implementation, data acquisition, data processing and analysis, interpretation, and publication (Fagan 1997:102–4; Hester et al. 1997:21–24). Each step refers back to the initial step: problem design. In many CRM archaeological projects, the questions to be answered are specified in the scope of work. Fieldwork, analysis, interpretation, and the production of the final goal should focus on answering those specific questions. Once they have been addressed, other questions can be considered, depending on the time and money available.

PART II
PREPARING FOR ANALYSIS

In this section, the focus is on the activities that allow the process of analysis to begin. Again, the importance of maintaining the context of the excavated artifact is paramount. Processing the artifacts is cleaning them to the extent that they can be described and identified and rough sorted into material categories. This expedites the cataloging process and identifies artifacts in need of special treatment.

Cataloging the artifact assemblage is the beginning of analysis. The goal here is twofold: (1) to inventory the collection and (2) to organize the collection to facilitate its interpretation by the archaeologist. Labeling of artifacts is a time-consuming but often necessary task for the archaeologist in the unending effort to preserve the provenience data associated with an artifact. The preservation of the artifact itself and stabilization procedures must be considered before the artifact is removed from the ground.

Finally, the use of digital imaging is discussed not only as a way of making a permanent record of the artifact that can be shared with colleagues across the country but also as a technique to enhance an artifact's attributes to aid in its analysis.

4

EXCAVATING ARTIFACTS

Getting the artifacts out of the ground is the first phase of their odyssey from excavation to curation. What happens in the field and on the way to the lab is of vital concern to the archaeologist, for it is here that specimens are most likely to be lost or commingled, which will wreak havoc with the analysis and interpretation of the site. It is crucial to the success of a project to ensure that proper care is taken of the artifacts during this preanalysis phase.

The methods employed in the excavation of artifacts is beyond the scope of this volume and, indeed, is the subject of volume 3 in The Archaeologist's Toolkit. However, the excavation of artifacts does have a bearing on the analysis phase of the project. Obviously, the recovery techniques employed will determine what there is to analyze. For example, using water flotation techniques might result in the recovery of small animal bones and seeds, which might necessitate engaging a zooarchaeologist or ethnobotanist. Conversely, improper stabilization in the field of wet, friable wooden or bone artifacts might mean that these specimens do not survive to be analyzed.

RECORDING CONTEXT

As discussed earlier, the importance of context cannot be overemphasized. The artifact loses most of its interpretive value if the context in which it is found is lost. Therefore, it is crucial to record the provenience of artifacts as they are discovered. *Provenience* (or *provenance* if you are an art historian) refers to the three-dimensional location of the artifact in space. This is usually determined in the field by referencing the artifact's horizontal location using the site's grid system (x- and y- axes) and the vertical location (z-axis) in reference to an arbitrarily established datum plane or depth from ground surface (see Tooklit, volumes 2 and 3). The provenience can be recorded either for the particular artifact (as in the case of piece plotting) or for the archaeological deposit (e.g., level within an excavation unit, feature, surface find) in which the artifact is found.

Recording of the provenience in the field is important given the inherently destructive nature of archaeological excavation. Thus, the information recorded in the field forms the basis for all that follows in the lab. There are a number of ways that archaeologists record provenience. One common method is to establish a provenience catalog, often referred to as a field specimen (FS) catalog. The field specimen number is used in the field to designate provenience and is assigned to each level or group of artifacts for which specific provenience needs to be recorded. It is assigned sequentially for each provenience and is written on the bag in which the artifacts are put as well as recorded in the FS catalog. It is absolutely crucial, to the analysis phase of the project, that the information in the provenience catalog correspond to that on the bags of collected artifacts. This may sound laughably obvious, but on a large project that may generate hundreds of FS numbers and thousands of individual artifacts, mistakes are often made, requiring hours of backtracking in the lab.

COLLECTING

What archaeologists collect and what they discard will have a bearing on the analysis of the artifact assemblage. Every archaeologist must decide at the outset of a project what they will collect and what they will leave behind in the field. This may sound like an outrageous statement to the novice since obviously everything is important and must be collected. Unfortunately, this approach is usually not practical, though sometimes the data may be recovered even if the artifacts themselves are not. An example will help clarify this seemingly paradoxical statement.

Puerto Real is a sixteenth-century Spanish Colonial outpost on the northern coast of Haiti. Archaeological excavations at this site have concentrated on the public buildings on the main plaza and on commercial and residential structures nearby (Ewen 1991; Deagan 1995).

During the excavation, thousands of pieces of glass, metal, and ceramics were found. Also uncovered were literally tons of brick and stone used in the construction of the town. To ship back all the construction debris to the archaeological laboratory at the University of Florida was not practical. Yet, the data they represented concerning the location and architecture of the site were crucial. So that all the data were not lost, the bricks and stones were counted, weighed, and, where appropriate, measured. Thus, the important information was gathered without the expense of shipping back several thousand pounds of brick and stone. Representative samples of these materials were also collected should petrographic or other specialized analyses be desired in the future.

Obviously, some data categories lend themselves more than others to the kind of sampling discussed here. The question of discarding data is a difficult one for all scientists, not just archaeologists. You must always keep in the back of your mind that new techniques may be developed or new perspectives embraced that will make seemingly useless or redundant data valuable. Yet, there is the problem that you simply can't save everything. This is also a fiscal consideration for contract firms that must pay another agency to curate the artifacts recovered on their projects. The best compromise is to save what is important now and sample the rest.

Another problem for the archaeologist is how much attention must be paid to artifacts from components other than the one in which you are interested. Agatha Christie Mallowan (1946:40) discusses this dilemma, which she encountered while helping her archaeologist husband, Max Mallowan, survey for sites in Syria. "The mounds in the immediate neighbourhood of Meyadin prove unattractive. 'Roman!' murmurs Max disgustedly. It is his last word of contempt. Stifling any lingering feeling I may have that the Romans were an interesting people, I echo his tone, and say 'Roman,' and cast down a fragment of the despised pottery." Lady Mallowan might be forgiven for such an attitude since she was a volunteer on the project, but today's archaeologist cannot. It is important to take the same care with the artifacts in which you are not immediately interested as with those that you are. Someone will be interested in those data even if you are not.

A good example of the value of other components is the recent work at the Governor Martin site in Tallahassee, Florida (Ewen and Hann 1998). The focal component was associated with the first winter encampment of Hernando de Soto. However, to get to that component, it was necessary to dig through a World War II period officer's club, an antebellum plantation, a Seminole Indian village, and a seventeenth-century Spanish mission. Below the de Soto component was an earlier Apalachee Indian village, and the lowest level showed traces of an early Archaic occupation. Although the de Soto–related artifacts were of particular interest, the other components were not ignored as they both affected and were affected by the Spanish encampment. Subsequent publications (Ewen 1989, 1996; Ewen and Hann 1998) concentrated on the contact period component, but those artifacts not related to the de Soto expedition were processed and inventoried and await future research in the curation facility at the Bureau of Archaeological Research in Tallahassee.

5
PROCESSING

The foundation of the analyses and interpretation to come is the maintenance of the context for the artifact assemblage. It has been said that archaeology differs from cultural anthropology in that we kill our informants! Since a site is destroyed by the process of excavation, it is imperative that the context be recorded for all data. This is because the site will literally be re-created on paper back in the laboratory. Old houses are often disassembled before moving and their components numbered so that they can be reassembled on a new site.

The same analogy applies to the data from an archaeological site. As computer capacities evolve, we may soon see virtual sites that can be reexcavated again and again. Indeed, this is already evident in the archaeology on the Fredricks site in North Carolina. The final report has been put on CD as *Excavating Occaneechi Town: Archaeology of an Eighteenth-Century Indian Village in North Carolina* (Davis et al. 1998). Utilizing the storage capacity of the compact disk, the user is very nearly able to re-create the dig and reanalyze the data recovered.

ARTIFACT CHECK-IN

When moving from the field to the lab, the primary goal is to ensure that all recovered artifacts and their field documentation (bag tags, field forms, notes, and photographs) are accounted for and that no discrepancies between the artifacts and their documentation exist. The easiest way to accomplish this is to arrange the artifact bags in order according to their FS number and check it against the FS catalog.

Some archaeological programs don't assign FS numbers in the field, preferring to arrange bags in a logical provenience order back in the lab (i.e., grouping all bags from the same excavation unit together) and then assigning FS numbers and producing a catalog. This practice is not recommended since bags could be lost in the field, and lab personnel would not see any discrepancy and so never realize the loss. Also, entering the catalog into a computerized database allows the investigator to sort the field specimens into any order desired based on a variety of criteria.

All of the field documentation should be duplicated and filed separately from the original documents. This provides a backup in case anything happens to the originals. It also allows the analyst to have a working copy of the field notes and maps. Scanning or manually entering these documents and images into a computerized database is becoming an increasingly popular way to archive field notes and photographs and provides a convenient way to access these data.

CLEANING

Washing artifacts is often the first exposure the student or interested layperson has to actually doing archaeology. This task is usually seen by the project director as something an untrained volunteer can do to assist the archaeologist without getting into too much trouble. While this is true up to a point, the lab assistant must beware of potential problems lest an artifact be damaged in the cleaning process. Invariably those artifacts most prone to damage are the most valuable to the interpretation of the site.

The purpose of cleaning artifacts is to remove soil or other debris that may obscure important attributes of the artifact. If all diagnostic attributes are visible or if cleaning the artifact would remove important information (e.g., cleaning a sixteenth-century copper coin may remove the faint markings on the surface that would identify it), then the artifact should be left as it was found pending consultation with a qualified conservator. Individuals involved with cleaning artifacts should always err on the side of caution and check with the lab manger or project director if they are unsure whether an artifact will withstand the cleaning procedure.

The basic tools for the artifact washer are inexpensive and readily available. They consist of a tray to empty the artifact bag into (an old cafeteria tray works well), a plastic washbasin for holding water, a variety of brushes of different sizes and stiffness (e.g., toothbrushes, nail brushes, and various paintbrushes), a drying rack, and a hand sieve. A number of other specialty brushes and picks might be suitable for removing dirt from the artifact, but they are merely variations on a theme and subject to individual preference.

The concern for context is especially important at the cleaning stage since artifacts are likely to be separated from their bags on which is written all the provenience information. One cannot be too paranoid in this situation. If the bag associated with the artifacts is lost or two bags from different proveniences are mixed, then valuable data are lost. So the first order of business is to check the provenience information on the artifact bag for completeness and to make sure that this information stays with the artifacts. Put the empty bag (or make a tag with the provenience information) into the section of the drying screen that you are using for the artifacts after they are washed.

Empty the contents of the artifact bag onto a tray so that you can see what you have before immersing the material in water. Many categories of artifact should not be put into the washbasin. Separate out plant remains such as nutshells, wood fragments, cane, even grass—do not waste time washing these obviously modern debris. Also do not wash metal objects, radiocarbon samples, bone, or any object that will degrade in water. Some prehistoric potsherds, burned clay, or daub tend to disintegrate if left too long in water. When in doubt, a gentle dry brushing is the safest course of action.

Some archaeologists prefer to do the initial soaking in warm soapy water followed by a clear water rinse. The superiority of soapy over clear water appears negligible, and one does run the risk of losing artifacts in the suds! Make sure that the artifact bag is empty before setting it aside.

Washing the artifacts is a delicate process. It is often a good idea to place glass (unless it is patinated and flaking) and historic ceramics in the water to soak. Brush them with a soft brush, making sure that all edges are clean. Do not scrub stone flakes or chipped lithic tools as you may want to conduct microscopic examination of some of these flakes and tools; vigorous scrubbing might alter their appearance. Clean these using finger pressure only. Never roughly brush prehistoric ceramics or clay objects as this may score their surfaces. Try to dry-clean them using a dry, soft brush. If this does not work, dip them quickly in water and rub with your fingertips only. Make sure the edges of all potsherds are clean so as to reveal their temper. Some prehistoric ceramics have slips applied to their surfaces that brush off easily. This is also true of overglaze decorations on historic ceramics (e.g., many Japanese porcelains or European ceramics have gilded accents that rub off easily). Just remember that the purpose is to reveal attributes important for the analyses, not to have them clean enough to eat from.

Over the years a variety of automatic washing systems have been devised, and some archaeologists swear by them, though most swear *at* them. These range from manually rotated cylinders that fit into water tanks to wire racks that fit in modern dishwashers. Most of these devices are meant for bulk cleaning of durable objects, and they suffer from lack of control over individual items and can result in a drastic increase in sherd count. A colleague had good success with his dishwasher set on delicate until the soil from repeated loads of artifacts ruined the machine's internal workings.

The washed artifacts should be placed on a drying screen to dry. A drying screen is a wooden or metal frame supporting nylon window screen. A rack usually holds several of these screens and allows for the free circulation of air to facilitate drying (see figure 5.1). The empty bag should accompany the artifacts on the screen or else a tag with the appropriate provenience information. Do not mix bags on a drying screen! After a bag of artifacts is finished, pour the water from the wash basin through a screen or hand sieve to recover all the tiny artifacts. Refill the wash basin with fresh water for each new bag.

Figure 5.1 Cleaning and drying artifacts. (Photo by C. Ewen.)

ROUGH SORTING

Rough sorting, another task that a student or lab volunteer can undertake with minimal training, can be accomplished in conjunction with the washing process. The purpose of this step is to speed the processes of cataloging and analysis by initially grouping the disparate artifacts into easily discernable, general categories. All artifacts from a common provenience can be combined into these general categories and bagged separately within the general provenience bag. These general categories might include prehistoric pottery, prehistoric lithics, unmodified raw material (e.g., rock), animal bone (faunal material), plant remains (floral material), historic ceramics, glass, metal, and a miscellaneous category. Rough sorting gives the project archaeologist a general idea of the range of materials present and facilitates the refinement of the categories into types and the removal of samples for specialized analyses.

6

CATALOGING

The goals of artifact cataloging are twofold: (1) to provide the archaeologist with a complete inventory of the artifacts present in the collection and (2) to facilitate the analytical goals of the archaeologist. These two goals are complementary, yet satisfying both is curiously difficult to achieve. This is because the goals of the registrar, who is interested in a general artifact inventory and comparability between collections, are often at odds with those of the project archaeologist, whose main interest is answering the specific questions of the project's research design. This explains why the profession has not adopted a universal artifact cataloging system.

A good example of these conflicting goals was my own experience with the Automated National Catalog System (ANCS), formerly the standard system used by the National Park Service archaeologists. The ANCS (rhymes with *angst*) consisted of rigid categories in an inflexible format. All archaeological work performed by and for the Park Service was required to use this system. I'm told that the Park Service collection managers loved it. However, many of the archaeologists contracted by the Park Service chafed under a system that allowed little flexibility in the classification of artifacts and was extremely cumbersome to use for basic analytical procedures. While I was at the Arkansas Archeological Survey, our archaeologists would first enter the artifacts into our DELOS database (a UNIX-based relational database) and then reenter the collection into ANCS after the analyses had been completed. Thus, rather than expediting the lab work, the ANCS merely added another step (which we were not allowed to charge for!). The Park Service has since switched to a new, more flexible database (discussed later).

A cataloging panacea will not be presented here. Instead, a variety of options will be presented and the merits and shortcomings of each discussed. Special emphasis will be placed on the computerization of cataloging procedures and some of the current options available to the archaeologist.

ACCESSIONING

Every institution that curates archaeological collections has (or should have) an accession catalog that numbers all the individual artifact collections in its care. An accession number, though, can mean different things to different institutions. In its broadest sense, an accession number is a way of identifying a particular collection of artifacts. This is different from a site number. A single site that has been excavated for several seasons might have several accession numbers assigned to it, each representing the artifact collection from an individual field season. Conversely, a survey project may have a single accession number, even though several sites were found during the fieldwork. So when archaeologists accession the artifacts from their latest dig, this simply refers to the process of adding the current artifact assemblage to their institution's curated collections in such a way that it can be retrieved again as a coherent group.

The assignment of an accession number is highly variable among and often within institutions. For example, a site that has been excavated for four seasons may have four different accession numbers, one for each field season. Or, especially if the field seasons were consecutive, there may be only one accession number for the combined collections. However, if that site had been tested in the 1950s and then later fully excavated in the 1980s, there may be two accession numbers, reflecting the two distinct periods of archaeological investigation at that site. Ideally, the institution should have a consistent and formal policy for accessioning collections.

There are a variety of ways to designate an accession number. East Carolina University utilizes the simplest system of all. Each collection is assigned a new number sequentially as it arrives in the laboratory. A popular variation on this simple theme is to begin the accession number with the year in which the collection was made. Thus, the third collection to enter the lab in 1998 would receive the accession number 98-3. This approach is a good way to solve the dilemma of how to split up collections from multiple field seasons. It is also vulnerable to the overhyped millennium bug problem that allegedly plagued many computer systems (i.e., how would you tell a collection made in 1902 from one acquired in 2002?).

The accession catalog (figure 6.1) is not simply a list of all the numbers that have been assigned to collections. It records this number and other pertinent information such as site name and official state num-

```
05/26/2000               ECU Phelps Archaeology Laboratory                    Page    1
                              Master Accession Catalog


Acc#  Site#   Sitename               Provenience              Recorder        Date Rec.
==========================================================================================
0001  8FR52   ST. VINCENT FERRY      ST. VINCENT ISLAND, FL   PHELPS          04-30-70
0002  8FR54   ST. VINCENT POINT      ST. VINCENT ISLAND, FL   PHELPS          05-01-70
0003  8FR60   ST. VINCENT 1          ST. VINCENT ISLAND, FL   PHELPS          04-30-70
0004  8FR61   ST. VINCENT 2          ST. VINCENT ISLAND, FL   PHELPS          04-30-70
0005  8FR63   ST. VINCENT 4          ST. VINCENT ISLAND, FL   PHELPS          04-30-70
0006  8FR64   ST. VINCENT 5          ST. VINCENT ISLAND, FL   PHELPS          04-30-70
0007  8FR65   ST. VINCENT 6          ST. VINCENT ISLAND, FL   PHELPS          04-30-70
0008  8FR66   ST. VINCENT 7          ST. VINCENT ISLAND, FL   PHELPS          04-30-70
0009  8FR67   ST. VINCENT 8          ST. VINCENT ISLAND, FL   PHELPS          04-30-70
0010  8FR68   ST. VINCENT 9          ST. VINCENT ISLAND, FL   PHELPS          04-30-70
0011  8FR69   ST. VINCENT 10         ST. VINCENT ISLAND, FL   PHELPS          05-01-70
0012  8FR70   ST. VINCENT 11         ST. VINCENT ISLAND, FL   PHELPS          05-01-70
0013  8TA35   BORKLUND MOUND         TAYLOR CO, FL            BORKLUND        05-00-70
0014  8WA10   BIRD HAMMOCK, MOUND B  WAKULLA CO, FL           HOLLIMAN        05-00-70
0015  8WA30   BIRD HAMMOCK           WAKULLA CO, FL           PHELPS ET AL.   01-00-70
0016  8WA86                          WAKULLA CO, FL    USDI   PHELPS          04-00-70
0017  8WA90                          WAKULLA CO, FL    USDI   PHELPS          04-10-70
0018          LAKE MARION            SOUTH CAROLINA           UNKNOWN         05-00-70
0019          BUNN COLLECTION        ZEBULON, WAKE CO         BUNN            05-00-69
0020          PUEBLO III             CORONADO NATIONAL MONU   SUTTON          05-00-67
0021          CERAMIC COLLECTION     TOME PUEBLO, NM
0022          CERAMIC COLLECTION     ACOMA PUEBLO, NM
0023          CERAMIC COLLECTION     OAKRIDGE, ROATAN ISLAN   SWIFT           08-16-69
0024          CERAMIC COLLECTION     "PALACIO", TLAXCALA HI   GONZALO         05-00-70
0025  8FR51   SOUTHERN DUNES
0026  8FR4
0027  8FR5
0028  8FR24
0029          J.O. SISSION HOUSE     SELMA, AL
0030          HAMMER MILL            SELMA, AL
0031  8GU11                          BLACK ISLAND, ST. JOE    STAPOR          03-26-70
0032  8SA9                           ALONG BEACH WEST OF GI
0033  8WA51
0034  8WA30   W.I. COMPOUND          FLORIDA                  BEUCE           00-00-69
0035  31BF24  WHALEN SITE            BEAUFORT CO              KERWHEL ET AL.  00-00-69
0036  31PM1   SEA VISTA              PAMLICO CO               PHELPS          09-26-70
0037  31PM1   SEA VISTA (BEACH)      PAMLICO CO               PHELPS          09-26-70
0038  31BF24  WHALEN SITE            BEAUFORT CO              PHELPS          09-26-70
0039          DOG ISLAND             FLORIDA                  STAPOR          04-14-70
0040  8FR71   PARADISE POINT         USDI; S.V.N.W.R.         STAPOR          04-14-70
0041  31NS3   THOMAS COLLECTION      ROCKY MOUNT, NC          THOMAS          11-03-70
0042  31NS3A                         ROCKY MOUNT, NC          PHELPS          11-14-70
0043  31NS3B                         ROCKY MOUNT, NC          PHELPS          11-14-70
0044  31NS3C                         ROCKY MOUNT, NC          PHELPS          11-14-70
0045  8FR72                          FRANKLIN CO, FL          STAPOR          12-12-70
```

Figure 6.1   Accession catalog. (Courtesy of the ECU Archaeology Lab.)

ber (if assigned), site location (usually by county), site excavator, the date the collection was added to the record, and any other relevant comments. Many repositories formerly had this information in an old ledger, which made finding a specific collection difficult if all the researcher knew was that the site had been dug in Greene County in the sixties—especially if there were several thousand accession records to search through. Fortunately, most repositories have entered their accession catalogs into a computerized database or spreadsheet, which allows for quick searches on a variety of categories. So, if all you knew was that Indiana Jones had left his Egyptian artifacts in your repository a long time ago, a query on the "Excavator" field in your database would quickly reveal all the accession numbers for which Dr. Jones was the excavator. It would also provide other information associated with those collections.

INVENTORY

An *artifact inventory* or *catalog* is a list of the artifacts within a particular collection. The inventory can vary in complexity depending on the use to which it will be put and the level of analysis conducted on the artifacts. A basic inventory would have the identified artifacts from a site list by provenience. That is, the contents of each FS would be listed in order for the site. Sometimes the artifact inventory is further subdivided by material (e.g., ceramics, lithics, etc.) or site component (i.e., prehistoric vs. historic), but the provenience information is always linked to the artifact.

The artifact inventory is the foundation of the analysis phase of the archaeological investigation. The subsequent manipulation of these data relies on the identification and provenience information being recorded in a logical and consistent manner. A well-constructed inventory allows the archaeologist to concentrate on those artifacts that will most affect the interpretation of the site while still having access to the entire range of data recovered from the site. For example, let's say the excavation of a late Archaic site recovers twenty-six thousand pieces of fire-cracked rock, thirty thousand bits of chert debitage, and forty-eight identifiable stone tools. The archaeologist can pull the tools from the artifact bags for further specialized analyses while shelving the rest of the collection, secure in the knowledge that most of the useable data concerning density and distribution across the site is recorded and available in the inventory.

A more elaborate inventory would include more than just the identified artifact and its associated provenience information. A complete inventory listing of an artifact could include such specific attributes as the material of which the artifact is composed; its weight, its color, and any decoration; its length and width; its age; the functional category to which it belongs; and any other information deemed pertinent by the archaeologist. The categories included in the inventory and the level of detail are determined by the research design associated with the particular project, although the basic inventory of artifact and provenience information should be recorded in all cases. The expanded archaeological catalog will be discussed at more length later in part 3 of this volume.

LABELING

In most archaeological laboratories across the country, a great deal of a lab assistant's time is spent writing tiny numbers on the surface of artifacts. This is done so that the provenience of the artifact will not be lost even if that artifact falls behind the lab table and is found months later during the annual lab clean-up day. The numbers laboriously inscribed on even tiny fragments of artifacts, at the very least, include the accession number and FS number (figure 6.2). They often include much more data.

The method and placement of this information is often rigorously prescribed by the institution. For example, the Arkansas Archeological Survey (Barnes and Cande 1994:11) requires the following information to be recorded on the artifact: the accession year and number, the FS number, the laboratory serial number (identifies subdivisions of the collection meaningful to the investigator), and the analytical serial number (identifies unique classes of artifacts such as rough sort categories, material/functional classes, or other analytical identifications). The number is applied with a drafting pen (formerly India ink on fine pen nibs was used) directly onto the surface of the artifact. If the artifact is dark colored, then white ink is used (a base coat of white correction fluid on which the number is written in black is a popular alternative at many institutions). Pieces of historic ceramics and glass are labeled on the edges, and prehistoric sherds are to be labeled on interior, undecorated surfaces. A clear coating of polyvinyl acetate, or PVA (clear nail polish is an economical alternative), is used to secure the ink on very smooth surfaces. Very small,

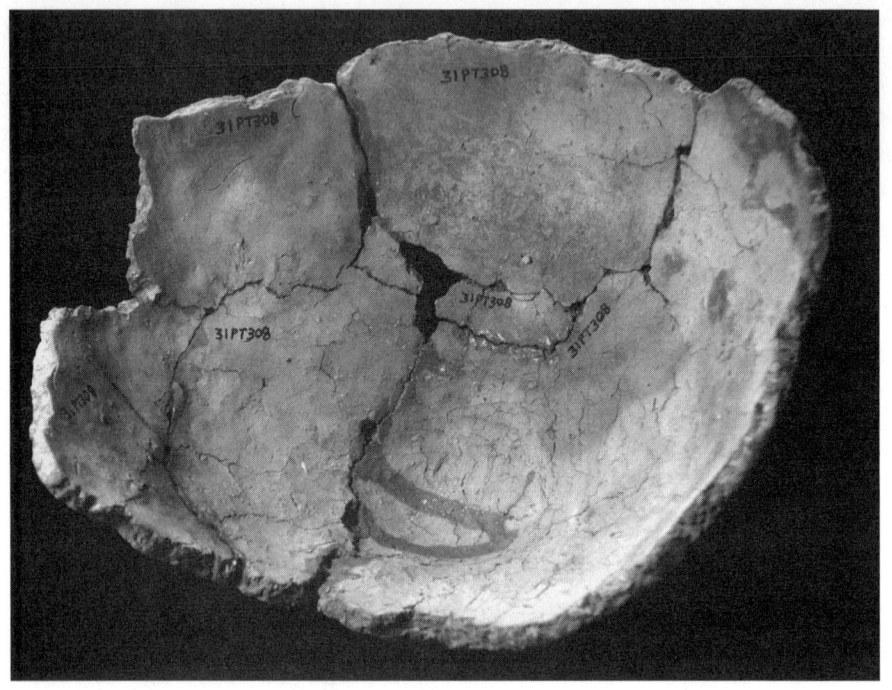

Figure 6.2 Numbered artifact. (Photo by C. Ewen.)

fragile artifacts or large numbers of small artifacts are allowed to have a tag with this information slipped in the bag with the artifact.

Not everyone agrees with this time-honored approach, and other concerns have tempered the zeal with which numbers are applied to artifacts. Though the paranoia over context is still very much in evidence, specialized analyses that retrieve information from the surface of artifacts (protein residue, AMS dating of soot, etc.) have caused some (see Sutton and Arkush 1998:28) to suggest that alternative methods of labeling be pursued. Also, the cost of this labor-intensive work is prohibitive for many small labs or private contractors.

A compromise that the author advocates is to label the bag containing the artifact(s) with an interior tag as well as writing on the bag itself. It is important to include an interior tag since the writing on the exterior of the bag tends to degrade over time. The artifact would only be physically labeled should it be separated from its bag for analytical (e.g., cross-mending) or display purposes.

7
FURTHER PREPARATION

A number of other tasks are often necessary before the analysis of the artifacts can begin. Like cleaning and cataloging, they prepare the artifact so that the important attributes are readily observable or stabilized so as to survive analysis. Some of these tasks can be performed by the nonarchaeologist with minimal training.

RECONSTRUCTION AND CROSS-MENDING

A situation where labeling of individual artifacts is important is in the reconstruction or refitting of a whole artifact from its constituent pieces. The reconstruction of ceramic vessels is a common example of this practice, though any broken artifact can be reassembled if most of the pieces are present and the lab technician possesses enough patience. The importance of labeling is apparent if the constituent pieces of an artifact come from different proveniences.

Cross-mending is an analytical tool that allows the archaeologist to determine which proveniences are linked together. If an archaeologist has differentiated three strata on a site and they all contain pieces of the same pot, then it is likely that they are contemporary (barring disturbances that drag artifacts between levels). Having sherds with the same design or other attributes is not as telling as being able to physically link pieces from the same pot.

Reassembling the artifact also presents a better indication of the original item's form. For example, a shallow bowl and brimmed plate may have similar rim shapes. However, the reassembling of the body sherds with the rim fragments gives the analyst the vessel's true shape.

Reconstructing artifacts is one of the more popular lab activities in archaeology. There is something about the puzzle solver in most people that draws them to this task. It is one of the few activities that many lab technicians will pursue on their own time! Artifact reconstruction is delicate, tedious work often involving many hours to reassemble a single, highly fragmented vessel.

One of the strictest caveats in artifact reconstruction is reversibility. The technique that is used to reassemble the pieces should be reversible and not alter the original state of the fragments. That way, should it be discovered that the pieces were misassembled or need to be returned to their original provenience, then the artifact can be disassembled without damaging the individual fragments.

The usual procedure is to put the artifact back together two pieces at a time. First, two corresponding pieces are fit together and held with an adhesive such as water-soluble white glue or an acetone-soluble glue such as Duco cement. If the refit is truly temporary, masking tape can be used. The two joined pieces are then set in a sandbox to hold them in place until the adhesive sets. This procedure is repeated until the vessel is complete or all the available pieces are joined. Missing pieces can be filled with plaster or other suitable filler or simply left open (figure 7.1).

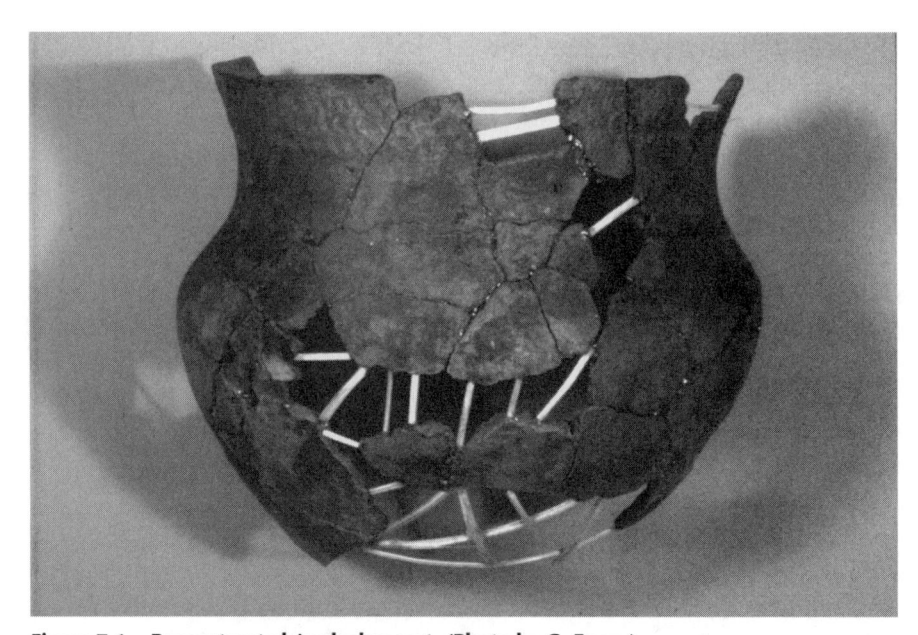

Figure 7.1 Reconstructed Apalachee pot. (Photo by C. Ewen.)

CONSERVATION

The stabilization and conservation of artifacts are important parts of the archaeological project. A complete discussion of artifact conservation is beyond the scope of this volume and is handled in more depth in another volume in this series (Toolkit, volume 6). However, a few comments here are appropriate. Artifacts made out of certain materials, such as iron, start to decay the moment they are put in a new environment (i.e., being deposited in the ground). Eventually they reach a state of equilibrium with this new environment, and corrosion slows considerably. The excavation of this artifact puts it back into another environment, which starts the decay process all over again. Plunging an artifact of this nature into a cleaning solution often further exacerbates the problem.

During the initial processing phase of a project, it is important to note which artifacts need conservation. If possible, these artifacts should be immediately sent to a conservator to stem the inexorable march of corrosion and decay. Unfortunately, that is not always possible due to funding constraints or the availability of a conservation facility and qualified personnel. In this case, steps should be taken to stabilize the artifact and the specimen flagged for conservation in the artifact catalog. The artifact should be sketched and photographed so that its original appearance is recorded in case the artifact suffers significant decay before conservation measures can be implemented.

Although it is best to have an on-site, or at least an on-staff, conservator/curator, such a position is often beyond the budget of most contract firms and small college labs. In this case, the project archaeologist will have to perform artifact first aid until the material can be sent to a specialist. Some practical advice can be founds in *A Conservation Manual for the Field Archaeologist* (Sease 1994), chapter 7 of *Field Methods in Archaeology* (Hester et al. 1997:143–58), and chapter 11 of *A Complete Manual of Field Archaeology* (Joukowsky 1980:244–75). Other useful sources include Dowman (1970), Hodges (1987), and Leigh (1978).

PHOTOGRAPHY

Photography is another useful technique in the archaeologist's toolkit and also the subject of a separate volume (Toolkit, volume 7). The proper depiction of an artifact in a report is literally worth a thousand lines of text in the final report involving some artifacts. However, there are some other good uses to which various imaging techniques can be applied during the early processing and analysis phases of an archaeological project.

CONVENTIONAL

As mentioned earlier, it is often a good idea to photograph an artifact before attempting conservation or any potentially destructive specialized analysis. The image of the artifact may be all that is left to the archaeologist to analyze if the original is somehow damaged or lost. It also saves handling and wear on the original, possibly delicate artifact (e.g., a corroded, sixteenth-century copper coin) if a photograph can be referenced rather than constantly exposing the fragile artifact to an unstable atmosphere.

Photographic techniques can also be used to enhance faint markings or decorations on an artifact, thereby aiding in its identification and interpretation. For example, the distinguishing marks on a copper coin recovered from the Governor Martin site (Ewen and Hann 1998:81) were barely legible, even when examined with a hand lens. By using oblique lighting and special film, it was possible to reveal enough detail in a photograph to identify the coin as a Spanish four-maravedi piece dating between 1505 and 1517. This helped link the site to the 1539–1540 winter encampment of Hernando de Soto.

Another type of conventional photography for analytical use is X-ray imaging. This technique often allows the archaeologist to peek beneath the corrosion to determine the nature of the artifact and assess its stability. Again, at the Governor Martin site, a rusted mass was X-rayed to reveal fourteen interconnected links, thus verifying the archaeologists' suspicions that the many separate iron rings previously found were actually bits of chain mail armor (Ewen and Hann 1998:78–79).

DIGITAL

The advent of affordable digital imaging devices and graphic manipulation software has already had a tremendous impact on the analysis of artifacts. I will try to avoid hardware- and software-specific discussions here since they will no doubt be outdated before this volume is published. The concepts involved, however, are valid, and the techniques discussed here will only be enhanced and made even more affordable as time passes.

Digital imaging simply refers to putting images (photographs, drawings, maps, etc.) into an electronic format so that they can viewed and stored on a computer. A conventional line drawing can be converted to a digital image by the use of a scanner or digitizing tablet. There are also digital cameras, which transfer images directly to a computer. These hardware devices have become increasingly affordable and are well within the budget of even the smallest archaeological laboratory. Video camcorders can also be connected to a computer equipped with a video capture board, and still images or entire movies can be saved in a digital format.

Once in the computer, digital images can be viewed and manipulated through various image-editing and -enhancing programs and computer-assisted drafting and design (CADD) software. Photographic enhancement software, such as Adobe's Photoshop, can be of great utility during the analysis phase of an investigation. Images can be enlarged, cropped, rotated, and inverted. To bring out obscured details, a photograph can be lightened or darkened, the contrast heightened, and negative images can be made positive. As with the conventional photographic techniques discussed earlier, it is often possible to extract details from an enhanced digital image that are not readily visible to the naked eye.

It is also possible to take a series of pictures of an artifact and then connect these into a single file. The image can then be rotated 360 degrees on screen to reveal all its sides. This technique is especially useful for fragile or unstable artifacts on which constant handling might damage the original specimen. This same effect can be achieved with a conventional camcorder and many repositories have made use of these technologies to record artifacts in their collections. This preserves at least some of the data should the artifact be repatriated or lost or damaged while on display.

Photographic images can also be converted to line drawings or color slides to black-and-white prints or negative images to positive. This is useful when a regular photograph has colors or shadows that obscure or overwhelm the more subtle attributes of an artifact. A good example would be the depiction of a chipped lithic artifact such as a projectile point. Archaeologists often prefer to portray these artifacts as line drawings since this method highlights fracture patterns and chipping techniques. The computer will render the surface treatment faithfully for examination, compensating for any lack of artistic talent on the part of the analyst.

Computer-assisted design programs, such as many CADD or metrical programs, have also been pressed into service by the artifact analyst.

Coupled with metrical software, it is possible to figure all the measurements of an object, including area and volume, if just one measurement is known. For example, the length of a potsherd is measured on the actual artifact, and then this dimension is indicated on an image of the artifact. The computer can now perform, with great accuracy, any other measurement desired of that artifact. More impressively, the computer can often use the measurements of the artifact fragment to extrapolate the size and shape of the original, intact specimen. For those interested in questions regarding vessel morphology, this is obviously a powerful tool.

Another use of digital imaging is the ability to store and retrieve pictures relating to all phases of an archaeological project. Thus, rather than holding multiple sheets of slides to the light or leafing through stacks of photographs to find a desired shot, the analyst can simply call up the image on a computer screen. Several computer applications allow an image to be stored along with descriptive data pertinent to that image. In essence, all the information from the field and lab photo log (photographer, date, subject, direction of the shot, provenience, and extensive commentary) can be linked to the image. By performing a query on the proper field(s), the archaeological analyst can quickly retrieve the plan view shot of a particular feature in a particular excavation unit. This image can then be printed or manipulated as necessary.

A couple of words of caution are in order regarding the use of digital imaging. First, though the cost of the hardware is coming down, the archaeologist must be aware of hidden or ancillary expenses. Using any graphics program and saving the images generated by them requires a substantial amount of computer memory, in terms of both desktop RAM and storage capacity. A single, high-density image can require well over a megabyte of space; a short video clip is usually well over a hundred megabytes. Most programs require a large amount of desktop RAM to run the graphics applications and a very large storage medium to store the images.

It would be foolish to offer advice on particular media when the technology is changing so rapidly. Not long ago a computer with four megabytes of RAM and a twenty-megabyte hard disk was sufficient for most applications. Today this doesn't even run most games! Removable storage media (e.g., tape drives, media cartridges, compact disks, and digital video disks) are evolving rapidly and offer a variety of storage solutions. It is probably prudent that these devices be purchased as external units, rather than built in, so that they can be upgraded as necessary.

PART III
ANALYSIS

Giving advice on artifact analysis to an archaeologist is a bit like telling a pitcher how to throw a baseball. How you accomplish the feat depends on what you want to do. To discuss all the intricacies of analysis of the myriad classes of artifacts is far beyond the scope of this volume. Key references devoted to the analysis of different types of artifacts are given wherever possible. Chapters 8 through 10 give a few examples of measurements and descriptions expected of artifacts in many contract reports. Chapter 11 suggests ways in which the archaeologist can organize the material assemblage from a site. Some advice and cautions are in order when using a computer to organize an archaeological database. Chapter 12 discusses how archaeologists have statistically manipulated their data to answer questions concerning pattern recognition, chronology, and spatial organization.

Part 3 wraps up with an all-too-often-ignored part of the analytical process: what to do with the artifacts after your immediate questions are answered. Others (or even you) may want to reanalyze the data at some point; how it is curated will determine how successful the reanalysis will be.

8
MATERIALS ANALYSIS

When asked to define artifact analysis, a graduate class in laboratory methods was initially nonplused. There was a great deal of confusion as to what actually constituted analysis. Some considered artifact identification and quantification to be analysis; others thought that it included the organization and interpretation of the data. One student finally opined that it was what the archaeologist did with the artifacts after they were excavated but before they were curated. In a sense, they were all correct.

David Hurst Thomas, in the third edition of his encyclopedic *Archaeology*, does not specifically define *artifact analysis*. Consulting the index, you are directed to a section concerning artifact dating. Another popular text refers to the sorting of artifacts into compositional categories and considering "the characteristics that differentiate one kind of data from others to show the ways that each can contribute to an understanding of past behavior" (Ashmore and Sharer 1996:114). Brian Fagan (1997:465), the author of many archaeological texts, simply defines *analysis* as "a stage of archaeological research that involves describing and classifying artifactual and nonartifactual data."

The actual analysis of the artifacts recovered from an archaeological project begins before the initial processing is complete. *Analysis* is defined by Webster's as a breaking up of a whole into its parts to find out their nature. In the case of archaeological analysis, an archaeologist separates the artifact assemblage into smaller categories in order to discern the nature of that assemblage. To what time period(s) does it date? What was its function(s)? Were the artifacts made locally, or did they come from some other place? How did the assemblage change through time? What does the assemblage say about the past culture's ideology? The first step toward answering these questions is the organization of the total assemblage into subsets that can shed light on these and other questions.

It would be wonderful if there were a single "right" way to organize the artifacts from an archaeological project. There would be a single text for the profession on which everyone would agree. Professors could drill their students with rote memorization until they got it right. Everyone's data would be compatible, allowing each archaeological culture to be described and fit into an "Archaeological Great Chain of Being." Unfortunately (or fortunately, depending on your perspective), this is not the case. This unrealistic view of "I love your database, and you love mine" only works if everyone is working in the same paradigm, asking the same questions and content with standardized data categories.

The last time there is very much agreement on archaeological classification is during the rough sorting stage. Here the artifact assemblage is broken down into the most basic category: material. Grouping artifacts on the basis of their composition is not only easily done by even the most untutored lab technician, but it is also a useful first step in the analytic process since many specific analytical techniques and conservation treatments depend on the artifact's composition. The basic material classes for prehistoric artifacts, discussed here, will include lithics, ceramics, metal, glass, and organic. These categories will be further subdivided chronologically into historic and prehistoric artifacts when appropriate.

LITHICS

Lithics refers to artifacts made out of stone. Sticklers for the English language would argue that archaeologists use the word improperly. It should be used as an adjective, as in "lithic artifacts," whereas most archaeologists use the term as a noun referring to the stone artifact subassemblage from their site as "lithics." You could argue the semantics if you could find someone who cared to, but modified (and unmodified) stones on a site are referred to as lithics by most archaeologists, so that is how the term is used here.

Stone artifacts are common on most archaeological sites because they are among the most durable of artifacts. The fact of their ubiquity and the fact that sometimes they are the *only* class of artifacts found on a site (e.g., lithic scatters make up the bulk of recorded sites in the offices of many state archaeologists) have caused many in the profession to claim that our view of the past is biased, and we might have termed the Stone Age the Basketry Age or Wood Age if preservation conditions on sites had been better. This may be, but since archaeologists are often limited to interpreting a site on the basis of a handful of stone flakes, a great deal of effort has been expended on the interpretation of lithic artifacts.

Though they may only represent a fraction of the total artifact assemblage that once existed at a site, lithics can tell us about many aspects of past life at those sites. The type of rock used to make the artifacts is indicative of the size of the site's catchment area, or it can be evidence of trade with other areas. How stone tools were made can reveal cultural affiliation or time period of manufacture. For example, a projectile point with a flute or channel flake removed longitudinally from its base is a good indication that it was made during the Paleoindian period (ca. 10,000 B.P.). The types of stone tools found in the artifact assemblage not only date that site but can provide the archaeologist with an idea of the site's function as well (e.g., raw material procurement, hunting, tool production, etc.). The type of analysis done on the lithic assemblage will depend not only on the questions asked but also on the kind of lithics present.

Unlike ceramics or most other artifact categories, which are formed by combining or adding material, stone artifacts are made by removing portions from the parent material. The study of this reduction process is the starting point for lithic analysis (see Carr 1994; Collins 1975; Crabtree 1972). In fact, the basic lithic category is nearly always broken down further into two categories, depending on the method of lithic reduction: ground stone or flaked stone.

It should be noted that the separation of the lithic artifacts into flaked or ground stone categories usually refers to the predominant mode of reduction on the finished artifact. A stone bowl would be classified as ground stone, even though its rough shape was pecked out, whereas an arrowhead would be classified with the chipped lithics even though the base was ground, presumably to dull the edges to permit hafting. To further muddy the classificatory waters, some types of stone, such as soapstone and catlinite, were actually carved before being polished into their final form. These are usually classified in an "Other Lithics" category or with the ground stone artifacts.

FLAKED/CHIPPED STONE

Flaked or chipped lithics are the result of the removal of stone fragments or flakes from a rock core. Most often this is accomplished by pressure or percussion being applied to the outside surface of a rock, although sometimes the heating of a rock can cause the pieces to pop off (a process known as *potlidding*). Typically, chipped lithics are made of stone that flakes in a predictable way such as flint (a type of chert), quartz, rhyolite, or a volcanic glass such as obsidian. Some types of stone are easier to work than others. The more predictable the breakage pattern, the easier the stone is to work. Obsidian and fine-grained cherts (such as from the Mill Creek formation in Tennessee) are considered easy to work and were highly sought throughout prehistory. Other kinds of raw material such as silicified sandstone or quartzite required more effort, yet they were used in the absence of more workable but less available material. It is important to know the geology of the area surrounding the site so that an assessment can be made as to where the material for the tools found at the site was from.

Humans have removed flakes from these materials by applying pressure or striking their surface either directly or indirectly. *Direct percussion* means simply striking the core or parent rock with another hard object (e.g., stone, bone, or antler). Sometimes the rock is rested on an anvil before striking, resulting in the method called *bipolar reduction*. *Indirect percussion* is accomplished by interposing another material, such as antler, to serve as a punch between the core and the hammerstone. The hammerstone strikes the punch, which indirectly transfers percussive force, thus striking off flakes. Another method of flake removal is the application of pressure. *Pressure flaking* is, as the name implies, the removal of flakes by the application of steady pressure rather than percussive force on an exposed edge. The result of either of these actions is a conchoidal fracture (shell-like flakes such as that inflicted on your windshield by a stone kicked up by the truck in front of you). Skillful placement of these conchoidal fractures removes flakes from a cobble and results in the production of a keen-edged stone tool such as a projectile point or a hide scraper that can, depending on the type of stone used, be sharper than surgical steel.

The analysis of the chipped lithic subassemblage is the subject of a great deal of diverse archaeological thought (Henry and Odell 1989; Flenniken 1984; Shott 1994). As is emphasized throughout this volume, the questions asked by the archaeologist will guide the type of analytical procedures performed. We can make some basic observations, however, that are common to many analytic approaches.

Besides noting the type of stone, it is usually important to indicate from which part of the stone the fragments derive. This means noting whether the flake has any cortex (unmodified exterior rock surface) and how much cortex is present. Primary decortication flakes have the majority of their surface consisting of cortex. Secondary decortication flakes have less cortex visible. Tertiary or thinning flakes have little or no cortex on their surface. The importance of this observation is that the proportions of the various types of flakes found at a site can indicate the stage of lithic reduction taking place. A site with a high proportion of primary decortication flakes might indicate quarrying activity with preliminary tool preparation. A lithic assemblage with a high proportion of thinning flakes would indicate tool finishing and reshaping activities.

Another qualitative observation that can be made is whether the artifact has been heat treated. It was discovered long ago, no doubt by some Paleolithic genius, that the workability of some stone types could be improved by subjecting it to intense heat (i.e., placing the cobble in a cooking fire). The reasons for this are not fully understood (see Domanski and Webb 1992; Flenniken and Garrison 1975), but the results are undeniable. Many types of stone fracture in a more predictable fashion after they have undergone heat treatment. Heat-treated stone can be recognized by a slightly pinkish color and a waxy feel to the surface.

The flake is a basic unit of analysis in lithic analysis. The landmarks on a flake consist of the striking platform, bulb of percussion, eraillure scar, and compression rings. As mentioned previously, an abundance of thinning flakes in the lithic assemblage is suggestive of tool manufacture. However, flakes can often be tools themselves. The context in which the flake is found and the presence of secondary retouching or microwear (i.e., polish) on its working edge can indicate the use of the flake as a tool.

Tools are the main concern of lithic analysis. Two general descriptive categories are bifaces and unifaces. The difference between them lies in how the working edge is achieved. *Bifacial* tools have flakes taken from both sides of the margin (the working edge), while *unifacial tools* have flakes removed from only one side of the margin. Bifacial tools include arrowheads and other projectile points, axes, cleavers, drills, and a host of other cutting or piercing tools. Unifacial tools are more restricted in their functions, serving primarily as hide-scraping and wood-shaving tools, though they were no doubt put to other uses as well. An example of a classic unifacial tool is an end scraper. Flake tools tend to be unifacial as well.

Cores are what remain of a rock or cobble after the useful flakes have been removed. Think of it as an apple from which successive bites have been taken until all the edible fruit has been consumed. However, the removal of the flakes does not necessarily end a core's use life. Often cores are converted into hammerstones as is evidenced by battering on one end. Sometimes a core will have one of its edges retouched (small flakes removed, usually by pressure flaking), suggesting that it had been pressed into service as a crude tool. This is especially likely in areas where good stone is scarce.

Analysis of chipped lithic artifacts can be as detailed as the archaeologist wants to make it. Measurements can be made of every conceivable dimension if the analyst has the time and money and the reason for doing so. However, there are some measurements that are routinely performed during analysis. The standard measurements (usually in millimeters) of lithic tools include length, width, and thickness; these should always be taken. Additional information for projectile points would include measurement of the various dimensions of the base (basal width, neck width, size and shape of notching) since the base is the most temporally and regionally diagnostic aspect of the projectile point (figure 8.1). The weight of the artifact is usually recorded (in grams) as well. Other questions concerning tool use and manufacture may require additional measurements. Consulting a lithic specialist or publications specifically relating to those kinds of questions would be necessary in those cases.

Debitage or stone waste flakes, another category of lithic artifacts, has received more attention recently (see Sullivan and Rozen 1985). It refers to the by-product of stone tool production. It includes flakes (discussed earlier), shatter (angular stone debris), and sometimes exhausted cores. A quick way to analyze your debitage from a site is to size grade the assemblage using nested screens of different size mesh (see Shott 1994 for a discussion of size grading). The size-sorted debitage can then be counted and weighed and other observations (e.g., type of stone, amount of cortex present, heat treatment) can be made for each size category. These data can be helpful in identifying activity areas and the activities within those areas at a site.

Most of the analytic techniques discussed here can be performed by the average archaeologist or lab technician. Some advanced techniques, however, will require the assistance of a lithic specialist.

MATERIALS ANALYSIS

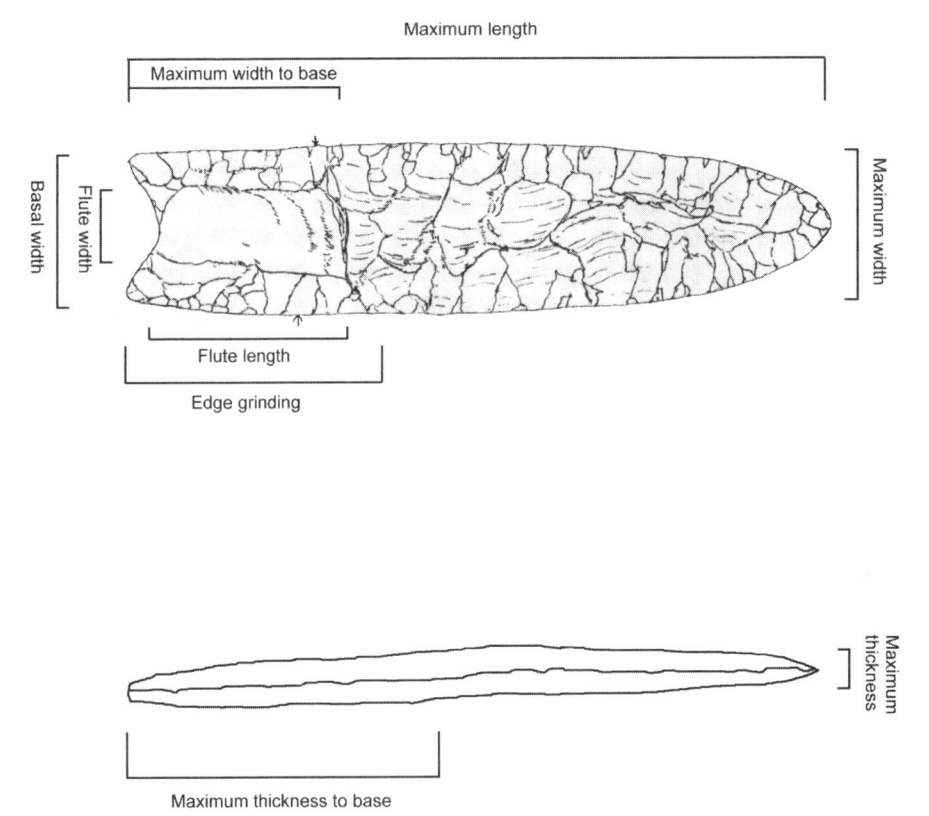

Figure 8.1 Attributes of projectile points. (Drawing by I. Randolph Daniel.)

Use-wear analysis has been conducted to help determine the function of various stone tools. Here, the edge of an archaeological specimen is examined under a microscope and compared to experimentally prepared specimens that have been used for specific tasks (e.g., wood working, hide scraping, butchering, etc.). The utility of this type of analysis has been questioned given the multipurpose nature of many stone tools as well as any postdepositional damage the artifact may have suffered. Still, the results of use-wear analysis can be useful and inspire questions that might otherwise not be considered (Vaughn 1985). For example, a site might have little in the way of faunal remains due to poor preservation in highly acidic soil. However, the presence of stone tools exhibiting wear characteristic of bone butchering might be the only indication that such activity occurred at the site.

Residue analysis is a recent development that may cause archaeologists to never wash another rock! It is possible through such techniques as PIXE (proton-induced X-ray emission) analysis or the use of SEM (scanning electron microscopy) to identify materials adhering to the working surface of stone tools. Identification of this material can conclusively identify a tool's function and may even provide information concerning the environment of use and possibly a date (if the material is organic) using AMS (accelerated mass spectrometry) dating (Loy 1990). Like many specialized analyses, it can be challenging to find a lab that performs them and they are usually fairly expensive.

GROUND STONE

Ground stone artifacts are different from flaked stone artifacts in that grinding or polishing is the method of reduction rather than flaking. Often the grinding is preceded by a percussive pecking or even flaking to rough out the initial form. Granite, basalt, greenstone, and other igneous or metamorphic rocks lend themselves most readily to this method of reduction, although some of the more durable sedimentary rocks, such as limestone, lend themselves to ground stone tool making. Often the grinding that resulted in the final form of an artifact was a result of use, such as in the case of a mano (handstone) and metate (grinding platform). These tools may have originally been pecked into a rough shape, but it was years of grinding seeds that resulted in the final shape of the artifacts. Other artifacts were purposely ground and polished to their final shape, such as a chunkey stone (figure 8.2) or discoidal (a stone disk that was rolled down a court in a Native American game). Ground stone items such as monolithic axes (hafted axes manufactured from a single piece of stone), banner stones (atlatl weights), and gorgets (ornaments worn around the throat) are often interpreted as indicators of high status due to the large amount of time and effort invested in their manufacture.

When conducting an analysis of ground stone objects, archaeologists often can separate them into two basic functional categories: ornamental/personal and utilitarian. Examples of items placed in the ornamental/personal category might include gorgets, maces/batons, beads, and figurines. The latter group would include milling stones, bowls and other vessels, banner stones, various types of weights, and axe heads. Even with this simple division there are some judgment calls to be made as utilitarian items can be highly ornamented (e.g.,

MATERIALS ANALYSIS

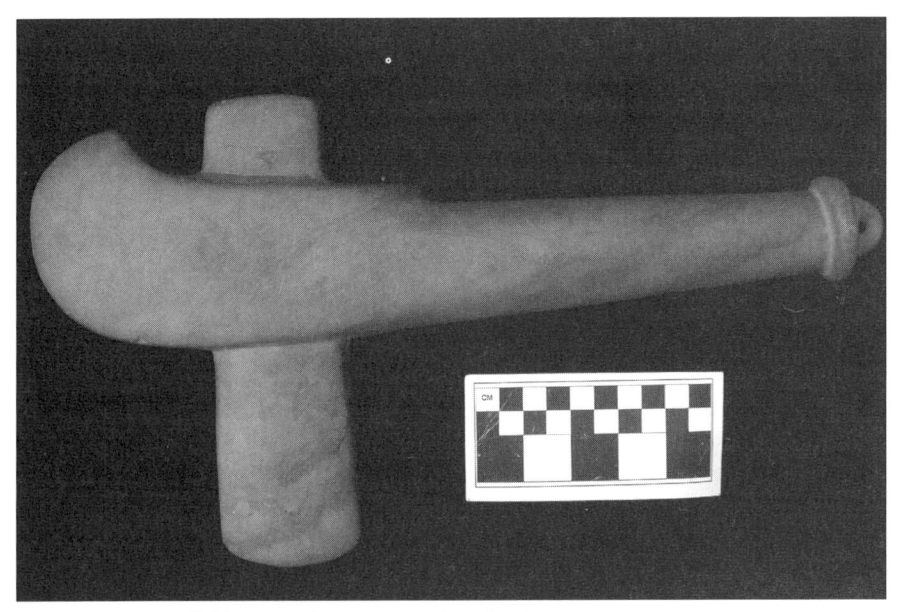

Figure 8.2 Monolithic ground stone ax. (Photo by C. Ewen.)

animal effigy plummets and banner stones), while other objects, such as gaming stones and pallets, could arguably fit into either category. This is but a glimpse of the artifact classification conundrum to be discussed later.

If one abandons a functional classification scheme, then description becomes paramount. In the case of ground stone objects, their basic form and material should be described so that someone who has never seen the item can visualize it properly. Measurements, as with flaked stone tools, should include length, width, and thickness as well as other dimensions deemed appropriate by the analyst. In the case of milling stones, depth of the concavity in the milling surface is usually measured. This is a good indication of relative duration of use when compared to other milling stones of similar material. A somewhat more difficult description concerns the method of manufacture. This involves a close examination of the surface of the stone to determine the direction of the striations (scratched lines left by the grinding process) and whether any intentional polishing (rubbing of the surface with another fine-grained substance) has taken place.

Advanced procedures such as residue analysis are also possible for ground stone tools. This is especially important for grinding stones since it should not be assumed that only corn or other types of seeds were being processed by these tools. Even if this were the case, it would be helpful to know the variety of seeds being ground. This kind of examination can be as simple as an inspection of the mill stone's surface with a hand lens for macroflora or other identifiable residue. Sophisticated techniques such as PIXE analysis are also possible with the same caveats as discussed with flake tools.

The analysis of stone tools is a basic task performed in an archaeological lab. Tools required include, at a minimum, a pan balance or electronic scale, a sliding caliper, a metric tape measure, a hand lens, and a Munsell color chart (for description of the color of the lithics). The procedures outlined earlier should be regarded as basic guidelines for the lithic analyst. A more in-depth treatment of lithic analysis can be found in chapters 4 and 5 of *Archaeological Laboratory Methods: An Introduction* (Sutton and Arkush 1998) and the various references cited previosuly.

CERAMICS

The term *ceramics* describes any item that is made of fired clay. Thus, the ceramics category can contain tea cups, chamber pots, roof tiles and the hardened exterior fragments of a wattle and daub structure that burned in antiquity. Ceramics are nearly as durable as stone and usually comprise the largest category of artifacts on sites postdating the domestication of plants, with which they are usually associated.

Because ceramics are ubiquitous on post–Archaic period sites, archaeologists try to squeeze the most data possible from them. Ceramics, for their part, are very obliging and can yield information pertaining to past foodways, the status of individuals with which they were associated, the cultural geography of a region, and they reveal trade patterns with other regions. They serve as good chronological tools and indicators of the technological prowess of their makers. Archaeologists have even used ceramics to gain insights into the belief systems and cognitive processes (e.g., Deetz 1996) of past cultures. As with every category of artifacts we will examine, the type of analysis performed on any ceramic assemblage will depend on the questions being asked.

One way of identifying the types of questions that ceramics can answer is to divide them on the basis of whether or not their primary function is food related. Food-related ceramics pertain to vessels used for storage (utilitarian wares), preparation (cooking pots and griddles), and consumption (plates, bowls, and cups). Non-food-related ceramics include those used for decoration (vases, tiles, statuary, etc.), architectural purposes (bricks, roof tiles), and personal items (clay pipes, chamber pots, dolls, etc.). This division transcends the division between historic versus prehistoric ceramics and can often help determine the primary function of a site.

Although the archaeologist's question may apply equally to both prehistoric and historic ceramics, the concern for chronology often demands that they be analyzed separately. Fortunately, they are sufficiently distinctive that even the little-trained lab technician can tell the difference. By default, in the United States, prehistoric ceramics are presumed to have been made by Native American potters (pending verification of extraterrestrial visitation or early transoceanic trade). The historic ceramics found on archaeological sites are usually imported wares from England and the other colonial powers, although, especially after the eighteenth century, there was a rising number of local potteries of which the archaeologist must be aware. Regardless of their date and place of manufacture, the ceramic analyst must be familiar with a number of characteristics.

CERAMIC ATTRIBUTES

This discussion will focus on those attributes concerned with food-related ceramics. These artifacts comprise the bulk of the literature, though other types of ceramics will be discussed where appropriate. The primary characteristics considered here include composition, manufacturing technique and firing temperature, form, surface treatment, and decoration. The classification of these attributes into types will be discussed in a later section of this book.

The importance of various attributes to consider when analyzing ceramics differs somewhat between prehistoric and historic pottery. Different aspects of the paste or matrix of the vessel is considered important to the analysis of both types of pottery.

With prehistoric ceramics, the tempering material (material added to the clay to help bind it and prevent it from cracking during firing) is an important addition to the paste. The earliest ceramics were sometimes tempered with organic material such as in the fiber-tempered Late Archaic ceramics of Georgia or Florida. Other tempering agents include sand, grit, shell, and even bits of old ceramics (referred to as *grog-temper*). The St. John's series of aboriginal ceramics in North Florida appears temperless but actually is strengthened by nearly invisible sponge spicules that are part of the marine clay matrix. The importance of noting the type of temper used is that it is often a sensitive indicator of a pot's temporal and spatial placement in an archaeological sequence. A petrographic analysis of a ceramic's paste and temper (either under microscopic magnification or PIXE analysis) can even allow the archaeologist to trace the sherd to its clay source.

Historic ceramics, on the other hand, usually have no visible tempering agent. Here, it is the hardness and the color of the paste that are the defining characteristics. The color of the paste is indicative of the clay source and the firing temperature. The hardness or degree of vitrification (how glassy the paste becomes) is also dependent on the firing temperature. Degree of vitrification is a much more important characteristic for historic ceramics and is the basis for the further subdivision of this category (figure 8.3). Coarse earthenwares, refined earthenwares, stonewares, and porcelains were each fired at progressively hotter temperatures and thus increasingly vitreous (glassy) and impermeable to liquids.

The manufacturing techniques, especially in North America, differed greatly between Native American and European ceramics. Na-

WORKSHOP FOR HISTORIC CERAMICS

PROPERTIES OF VARIOUS CERAMICS

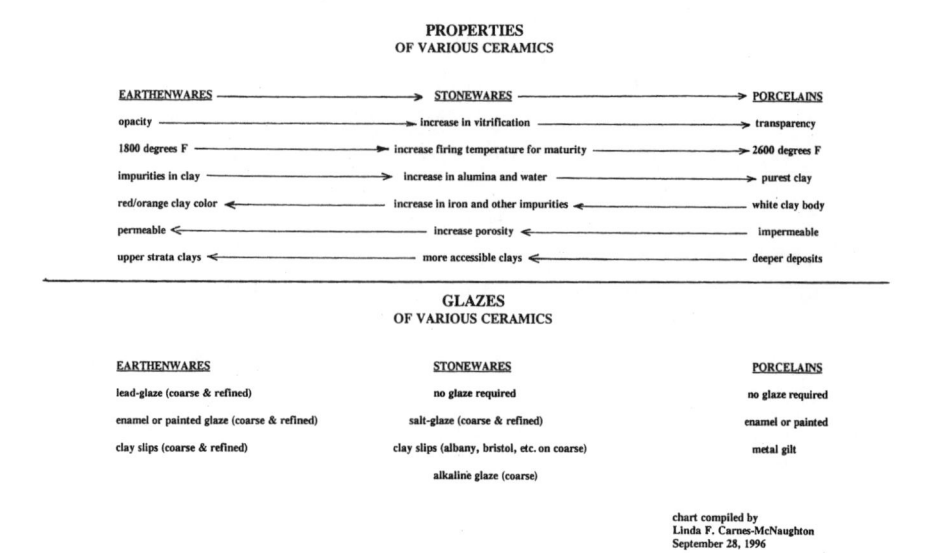

EARTHENWARES	→	STONEWARES	→	PORCELAINS
opacity	—	increase in vitrification	—	transparency
1800 degrees F	—	increase firing temperature for maturity	—	2600 degrees F
impurities in clay	—	increase in alumina and water	—	purest clay
red/orange clay color ←	—	increase in iron and other impurities ←	—	white clay body
permeable ←	—	increase porosity ←	—	impermeable
upper strata clays ←	—	more accessible clays ←	—	deeper deposits

GLAZES OF VARIOUS CERAMICS

EARTHENWARES	STONEWARES	PORCELAINS
lead-glaze (coarse & refined)	no glaze required	no glaze required
enamel or painted glaze (coarse & refined)	salt-glaze (coarse & refined)	enamel or painted
clay slips (coarse & refined)	clay slips (albany, bristol, etc. on coarse)	metal gilt
	alkaline glaze (coarse)	

chart compiled by
Linda F. Carnes-McNaughton
September 28, 1996

Figure 8.3 Physical properties of historic ceramics. (Prepared by Linda Carnes-McNaughton.)

tive ceramics, for the most part, were built using such techniques as coiling, molding, or modeling and were fired in open-air conditions (figure 8.4). European and Asian ceramics, on the other hand, were turned on potter's wheels and fired at higher temperatures in a kiln. The differing techniques of manufacture require that the ceramic analyst record different attributes for the two types of pottery.

Form is a basic category of any ceramic classification scheme. Prehistoric vessel forms can often be simply described as bowls, jars, bottles, or beakers. Historic ceramics can be described by these terms but also include plates, cups, and other terms that are more evocative of function such as serving platter or soup tureen. However, function is not always known for historic vessels, and at a basic level of classification a distinction is made between hollow wares (e.g., bowls) and flat wares (e.g., plates). Besides whole vessel form, usually the shape of different aspects of the vessel (e.g., lip, rim, neck, shoulder, body, and base) are described as well.

Surface treatment and decoration are two other important related, but separate, categories. Surface treatment is part of the manufacturing process of a pot. It includes such techniques as smoothing, burnishing, brushing, stamping, slipping (application of a thin liquid clay wash) and glazing. Historic ceramics were often glazed to make them

Figure 8.4 Open-air firing of ceramics. (Photo by C. Ewen.)

impermeable to liquids. In the United States, prehistoric ceramics were never glazed, although such surface treatments as slipping may have reduced their porosity.

Decoration applies to design elements added apart from the manufacturing process, though in some cases (such as with brushing or stamping) whether the treatment was functional or purely aesthetic is debatable. Prehistoric decoration elements were often stamped or incised on the clay before firing. Historic ceramics tended to rely on painted designs, transfer prints, or decals as their primary decoration.

The sort of measurements that should be taken of the ceramics in the assemblage can be as detailed as is necessary to answer the questions at hand. Generally, the size and volume of the vessel is approximated as closely as possible. This is done by determining the diameter of the vessel by matching the curve of a rim sherd on a chart of concentric circles (several styles of templates are commercially available). If enough of the vessel is present, then the height can be estimated. By calculating the volume of the cylinder ($\pi r^2 \times h$) a rough approximation of the vessel's volume can be achieved. Combining digital imagery with a powerful CADD package, not only can a more accurate volume estimate be made, but the computer can sometimes graphically reconstruct the entire vessel from a few sherds.

The preceding discussion merely scratches the surface of ceramic analysis. Literally dozens of books have been devoted to various aspects of ceramic analysis. However, a couple should be on the shelf of any archaeology lab. For prehistoric pottery, read Anna O. Shepard's (1980) classic *Ceramics for the Archaeologist*. First published in 1956 but reprinted and revised several times, this book discusses the basics: properties of the raw materials, techniques of manufacture, basic analytical techniques, classification, and interpretation. As comprehensive as Shepard's work is, it is certainly rivaled by Prudence Rice's (1987) encyclopedic *Pottery Analysis: A Sourcebook*. An individual with both works under his or her belt is well prepared to tackle any prehistoric pottery assemblage. Unfortunately, there are no similar treatises for their historic counterparts.

Works on historic ceramics tend to focus on a particular time period or nationality. Ivor Noël Hume (1978) makes a game try with his very useful *Guide to the Artifacts of Colonial America*, but the section of ceramics focuses primarily on English wares, and the time span is limited to the sixteenth through eighteenth centuries. George Quimby's (1980) edited volume, *Ceramics in America*, addresses many problems with ceramic analysis and interpretation, yet the larger subject matter with which the authors grapple, worldwide ceramic production, distribution, and use, is just too broad to be handled in a single volume. The analyst is encouraged to consult the literature on the time period and nationality of interest. For example, Kathleen Deagan's (1987) *Artifacts of the Spanish Colonies of Florida and the Caribbean, 1550–1800* is a good place to start for archaeologists working on Spanish sites in the southeastern United States and the Caribbean. Similar types of references exist for other regions and nationalities.

Ceramics can reveal much important information concerning an archaeological site. For example, the function of a site as well as the status of its inhabitants can be inferred from the ceramics recovered. Excavations at the sixteenth-century Spanish site of Puerto Real, Haiti, relied on ceramics to interpret the lives of their former owners. High proportions of hard-to-get imported wares delineated a neighborhood of wealthy individuals near the town plaza, whereas a prevalence of locally made wares of poorer quality on the outskirts of town defined the location of the town's lower economic class. Clandestine activity by one of the wealthy merchants was revealed by the presence of Chinese porcelain. This was only available from smugglers for most of the town's existence. Ceramics, both prehistoric and historic, can help determine a site's date of occupation, its place in the political geography of the region, whether it was engaged in long-distance trade, and even the descent system of the potters (see Longacre 1968). Some of the analytical techniques to derive these interpretations are discussed later in this volume.

METAL

Metal artifacts are most often associated with historic sites in North America. However, copper, lead (galena), silver, and gold appear on prehistoric sites as well, although their presence is relatively rare. The arrival of Europeans in the New World marks a sharp rise in the number and variety of metal artifacts on North American sites. The native inhabitants were ready customers for items made of metal, so that very early in the Colonial period metal items become fairly common on Native American sites. This sometimes makes it difficult to distinguish an historic Indian habitation from an early European homestead site on the basis of artifacts alone.

The most common metal to be introduced from Europe was iron, although artifacts made from copper and brass (an alloy of copper and tin), lead, pewter, silver, and gold were also available. Lead was molded into a variety of items, the most numerous of which were musket balls and other projectiles. The copper alloy, brass, was brought over as cooking vessels, hardware, and in sheet form, which the native peoples quickly adapted to their own uses. Copper, silver, and gold were part of the reason the Europeans came to the New World and were most often used in coinage and high-status artifacts. Various other metals and alloys such as aluminum, tin, and steel become more prevalent in later historic assemblages.

Iron has been fabricated into a bewildering array of artifacts ranging from swords to plowshares and everything in between. In fact, iron items represent so many different types of artifacts that it can be difficult to classify them. Ironically (pun intended), this popular metal is also highly corrosive and so the most common artifact to be recorded in this category is usually "unidentified iron flake or object."

Rather than basing this discussion of metal artifact analysis on the type of metal of which the artifact was made, it will probably be more useful to categorize them on the basis of the questions they can answer for the archaeologist. Among the many questions to which they might be applied, three common categories with potential for this type of artifact are as chronological indicators, as activity indicators, and as indicators of ethnicity and trade. Other possibilities include status indicators, questions relating to changes in technology, and acculturation studies.

Historic artifacts of all kinds lend themselves particularly well to questions of chronological placement, and metal artifacts are especially datable. Obviously, coins would seem a logical artifact for dating a site, yet the placing of a date on a coin is a relatively recent phenomenon. Fortunately, by using other clues (i.e., dating the reign of a monarch whose bust is featured on a coin), numismatists (e.g., Nesmith 1955) have been able to place most types of coins in a relatively narrow time frame.

Iron nails (Nelson 1963; Edwards and Wells 1994; Wells 2000) and barbed wire (Clifton 1970) have been used to roughly date sites. Nails can be chronologically divided on the basis of their method of manufacture (figure 8.5). Hand-wrought nails are the dominant type of nail until the early nineteenth century, when machine-cut nails (invented at the end of the eighteenth century) replace them in popularity. Cut nails are superceded at the end of the nineteenth century by modern wire nails. It should be noted, however, that because of recycling, distribution and other factors, these dates are truly rough approximations. However, since nails are often plentiful at many historic sites they can provide the excavator with a quick, general idea of the site's chronological placement.

Diagnostic Nail Features

Feature	Symbol	Meaning
1. Metal:	I, S	Iron, Steel
2. Mfg. Method	Hf, Ct, Dn	Hand-forged, Cut, Drawn
3. Grain (iron only)	↕, ↔	In Line, Cross
4. Point (cut only)	Rd, Ft	Round, Flat
5. Head Mfg.	Hd, Me	Hand, Machine
6. Burr (cut only)	Sf, Of	Same faces, Opposite faces
7. Pinch (cut only)	Si, Fa	Side, Face
8. Shaft Taper	4t, 2t, 0t	Taper on: 4 sides, 2 sides, no taper
9. Shaft Section	Sq, Re, Ro	Square, Rectangular, Round

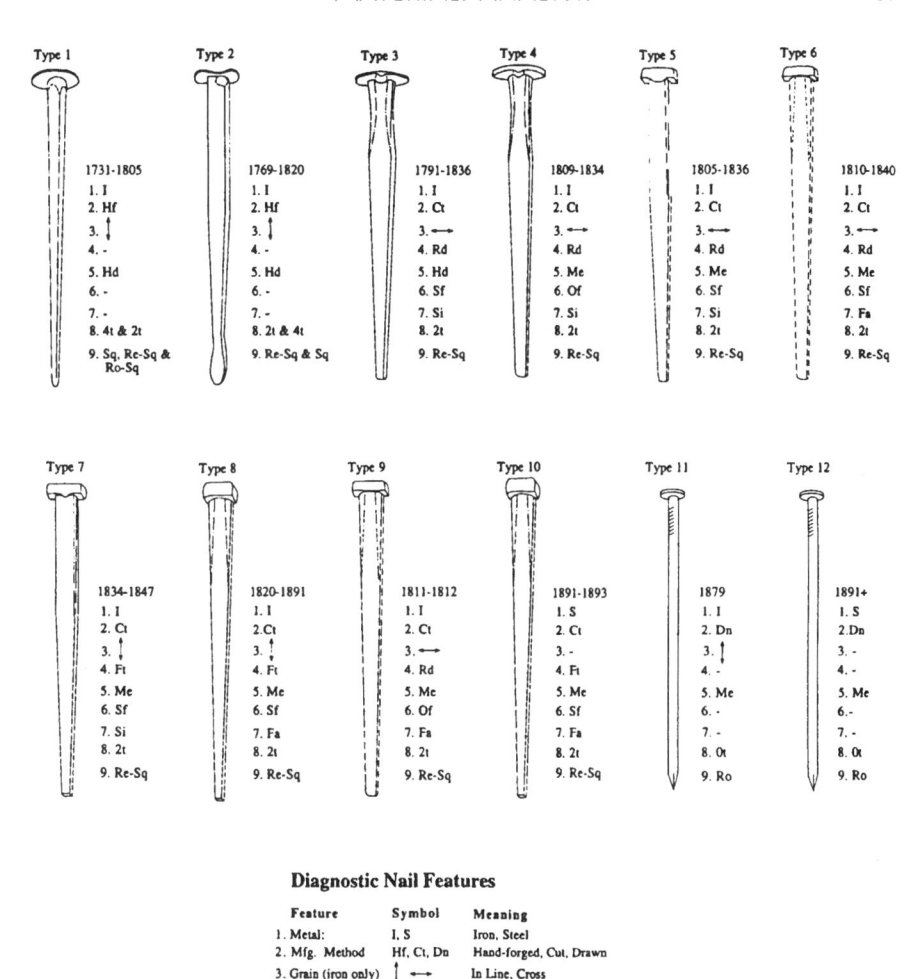

Figure 8.5 Nail chronology. (From Wells 1998. Reprinted by permission of the Society for Historical Archaeology.)

Another metal artifact that can be used to approximately date a site is the tin can, which is actually an iron container plated with tin to retard corrosion. Ubiquitous on historic sites (especially in the western United States) from the nineteenth century forward, metal containers are datable based on their method of sealing (Busch 1981; Rock 2000). These containers can reveal their contents and place of origin on the basis of their form and information that is often stamped on the can itself.

Weapons are yet another category of metal artifact whose development can be documented chronologically and which can often be traced to their place of manufacture. R. Ewart Oakeshott (1960) traces the development of weapons from the earliest atlatls and bows and arrows through use of metal in edged weapons to the development of firearms in his very readable *The Archaeology of Weapons: Arms and Armor from Prehistory to the Age of Chivalry*. Unfortunately for the North American historical archaeologist, he quits just before the discussion gets useful. It is difficult to find a single source (e.g., The Diagram Group 1980) that covers all weapons and ammunition of all nations, in the depth required by the archaeologist, from the sixteenth century to the present. Instead, the archaeologist must be more specific in the time and place of interest in order to find a suitable reference (e.g., Hamilton 1968, Brown 1980).

Questions of chronology are certainly not all that metal artifacts can answer. The function or primary activity taking place at a site is often revealed by the types of artifacts recovered. The recovery of farm implements could indicate agricultural activity; arms and ammunition might suggest a military site or evidence of hunting activity. Pins, needles, and thimbles could indicate domestic activity or even a commercial tailor's shop, depending on the quantity and context of such materials found (for an example, see South 1977:65–77).

Another important question that metal artifacts, like many other categories of historical artifacts, can answer is that of national origin and trade. With the proper references, many metal artifacts can be traced to their country and date of manufacture. Sometimes it is even possible to find the factory that made the item, the merchant that sold it, and the price he charged for it!

Where does one look to get information to identify metal artifacts or to learn about a particular class of artifact? Historical archaeologists are both blessed and cursed with a plethora of potential sources. For most labs, a selection of general secondary sources is usually the best way to start. Old catalogs, such as the Sears & Roebuck Catalog, are a good reference for many different items. It gives information concerning date (i.e., when the catalog was published), manufacturer, and price and usually has a picture of the item, thus making it a good identification tool. Collectors guides offer varying amounts of information on most kinds of artifacts and can be surprisingly specialized. If you can identify a specific type of artifact (e.g., farm implements), there is usually someone who collects it and probably at least one book on the topic (e.g., *Farm Tools through the Ages* 1973 by Michael Partridge).

These antique collectible books can be found at local and university libraries or even on-line at the Library of Congress (www.loc.gov) and are often available inexpensively through a variety of booksellers (e.g., www.amazon.com or www.barnes&noble.com). Finally, site reports and synthetic artifact guides (e.g., Hume 1978; South, Skowronek, and Johnson 1988) actually are written for the archaeologist and thus usually have the type of information in which the researcher is interested.

GLASS

Glass comprises another historic artifact category that became an instant hit with the native population of North America. Unlike most metals, glass is relatively stable and will survive with relatively little degradation (depending on its soda ash content) for long periods in most environments. Couple this with the fact that the manufacturer and contents of the bottle or jar were often molded onto the container, and one can see why historical archaeologists have devoted a great deal of study to glass artifacts.

The basic recipe for glass involves combining silica (sand), soda, and lime and heating until these ingredients fuse. This soda-lime glass is the most common type of glass, with variants depending on the proportions of the primary ingredients. Colors are the result of different impurities that were either intentionally or unintentionally added to the primary ingredients.

Although the constituent elements of glass have been known since the third millennium B.C., the techniques for working glass have evolved over time and became the jealously guarded secrets of various nations in the past. The results of this changing technology can inform the archaeologist of the origin and age of the glass artifacts that are recovered from a site (Lorrain 1968). Thus, for the analyst, it is important to observe whether the artifact has been free blown, mold blown, or machine made, as well as to note other descriptive characteristics.

Glass has been used to make a variety of artifacts including bottles, drinking vessels, window glass, and beads.

Glass beads are eagerly sought by archaeologists, especially those studying the early colonial period in North America since they were part of the trading truck brought by the first explorers (Smith and Good 1982; Karklins and Sprague 1980). Another useful form of glass artifact for the archaeologist are bottles. The evolution of the form these bottles took and their method of manufacture (blown vs. machine molded) make them useful as chronological indicators on a site (see Fike 1987). Containers of other sorts (e.g., drinking vessels and jars) are similarly useful for dating sites, and flat glass can tell us more about a building than just whether its windows were glazed.

One needn't be an expert in the history of glass to begin the basic analysis of glass artifacts in an archaeological assemblage. A basic descriptive approach is recommended as a starting point for both novice and experienced glass analysts.

Color is an important characteristic, which can deceive the unwary observer. Dark glass, which may appear to be an opaque black at first glance, often turns out to be a translucent olive-green or dark brown when back-lit. For this reason, it is usually a good idea to discern color by viewing the glass artifacts on a light table or against a light colored background. The description of color can be standardized by using a Munsell Book of Color or Pantone color wheel.

Another important descriptive attribute of glass is form. A basic division that can be made on small fragments, where the original artifact is not identifiable, is separating the fragments into flat versus curved. This roughly distinguishes window from bottle glass; however, the generic terminology is necessary so that flat, case bottle glass or curved drinking glass fragments are not misidentified. If larger, identifiable pieces are present, the analyst may want to refer to a glass nomenclature guide (e.g., Jones and Sullivan 1989 or White 1978) so that the vessel and its constituent parts can be properly identified.

The function of the glass artifact can often be determined based on its form and color. Once again, pertinent documentation is invaluable in this regard (e.g., Brauner 2000; Jones and Sullivan 1989; Hume 1978). However, some types of glass artifacts have come in for more study than others, due to their utility in dating sites or revealing the nationality of their owner/maker.

Glass beads were one of the earliest trade items in North America. Beads are made by three basic techniques: drawn cane, wire wound, and mold blown. Drawn-cane beads, made by the simplest technique, are formed by drawing a molten glass blob into a long tube or cane. The cane is then cut into smaller lengths for individual beads. Often these beads are further modified by tumbling with a polishing agent over heat. The second common manufacturing technique is the winding method. Here, a thread of molten glass is wound around a rod known as a *mandrel*. Once a suitable shape has been attained, the bead is removed and allowed to cool or is further modified by being pressed with a paddle (a technique known as *marvering*) or having additional threads of different colored glass applied. Blown glass beads are comparatively rare. Their method of manufacture is characterized by the presence of a seam running down the side from the mold into which the glass was blown (figure 8.6).

The date and place of origin for many bead types are known and have been used as chronological marker artifacts to trace the routes of explorers, such as Hernando de Soto, through southeastern United States (e.g., Ewen 1998:85). A typology for North American trade beads has been developed (Kidd and Kidd 1983) and can serve as a useful classification system in the absence of more chronologically or culturally specific typologies (e.g., Deagan 1987:156–83; Smith and Good 1982; Karklins and Sprague 1980). A cautionary note: It should never be assumed that glass beads on native American sites represent direct contact with Europeans. Once acquired by the native population, these artifacts entered into another social interaction sphere and may have been traded over hundreds of miles to an individual who never saw a colonist.

Bottles and containers are another type of glass artifact archaeologists frequently study (e.g., Switzer 1974). They (especially bottles) can be dated on the basis of manufacturing technique, form (see Hume 1978:63–68,73), and method of closure. However, one should be careful to verify the dating methods being used as many early correlations,

Figure 8.6 Blown-glass beads. (Photo by Florida Bureau of Archaeological Research.)

such as the height of the kick-up or number of layers of patination (Demmy 1967), have not stood up under close scrutiny.

I will close this brief discussion of dating glass artifacts with a mention of window glass. Flat window glass is ubiquitous on historic sites after the eighteenth century. Historically, window panes were made in three different ways (Lorrain 1968). In the first method, the glass blower made a circular disc, called a *crown*, from which rectangular window panes were cut. This technique is discernable by a concentric pattern of small bubbles in the glass. An improved technique was to blow a cylinder of glass, which was then cut down the side and the reheated glass allowed to uncurl onto a flat surface. A casting method was also used for small panes or mirrors, but the need for subsequent polishing made this an expensive alternative. Attempts have been made to use the thickness of glass window panes to date sites (Roenke 1978). However, the applicability and utility of this method is somewhat limited, making it impractical in most cases.

The care and treatment of glass artifacts is a subject often ignored by archaeologists. The conventional wisdom is that as long as the artifact is carefully packed, it should curate indefinitely. However, if the glass has lain in a wet or alkaline environment, as with metal artifacts, it will start to decay. The decay, in this case, takes the form of exfoliation in iridescent layers. This devitrification cannot be reversed and any consolidation of the artifact should be left to a professional (Sease 1994:72–73). Final curation should be in a relatively dry environment in a padded case with rigid exterior walls.

As with the other sections on particular classes of artifact analysis, I have only scratched the surface here. Several references can assist glass artifact analysts with their questions. Two essential references are *The Parks Canada Glass Glossary for the Description of Containers, Tablewares, Flat Glass, and Closures* (Jones and Sullivan 1989) and the glass section in the SHA reader *Approaches to Material Culture Research for Historical Archaeologists* (Brauner 2000). Both references, and the references they contain, should get the lab personnel off on the right foot.

ORGANIC ARTIFACTS

Organic artifacts include such items as fabric, cordage, basketry, shell ornaments, bone tools, and wooden items. They are also commonly referred to as perishable artifacts and are usually found in very dry, very wet, or very cold conditions. Many of the artifacts that fall into this category will be those that will be flagged for immediate conservation or sent to a special analyst. Unmodified organic remains or ecofacts such as animal bone or floral specimens are usually the domain of zooarchaeology and paleoethnobotany, respectively. Human remains are dealt with by specially trained bioarchaeologists. Interpretations that can be gained from these items will be discussed in chapter 11.

Of paramount importance when dealing with perishable materials are their care and handling. If you do not take care of them up front, you may have nothing to analyze later! For the most part, these artifacts should not be cleaned by lab personnel any more than to brush off loose soil. Complete cleaning and consolidation are tasks best left to the special analyst. A task that the lab personnel can and should undertake is the complete recording of the fragile artifacts prior to analysis. This would include both drawing and photographing each artifact so that a complete record of its original appearance will exist should it be damaged by shipping or handling or even during conservation.

Storage of the artifact should be in the same environment in which it was found until it can be stabilized by a conservator. For example, wooden artifacts that have been recovered from a wet environment (e.g., dugout canoes recovered from a lakebed) should be kept immersed in water until proper permanent treatment can be undertaken. Smaller perishable artifacts, such as basketry from a bluffshelter site, should be sealed in a microenvironment (e.g., a plastic bag with a desiccant package) similar to the one in which they were found. At the very least, organic artifacts should be curated in a repository with temperature and humidity control.

Other than exercising extra care in handling, the analysis of perishable artifacts is similar to their more stable counterparts. For example, shell beads are described and measured in much the same way as glass or lapidary beads, with the addition of determining the species of shell being utilized. The same is true for all organic artifacts. Determining the species and element of the plant or animal from which the artifact is made is an important aspect of the description. This information will assist in the determination of the point of origin of the artifact and possibly the environment in which it was used.

Basketry and textiles warrant special consideration because their construction utilizes a different terminology which may be unfamiliar to the average archaeologist. A good source to have on hand when analyzing this class of artifact is *Basketry Technology: A Guide to Identification and Analysis* (Adovasio 1977). Adovasio (1977:1) relates basketry to textiles except that the former is more rigid and "manually assembled or woven without a frame or loom." He asserts that "basketry may be divided into three sub-classes of weaves that are mutually exclusive and taxonomically distinct: twining, coiling and plaiting." He goes on to discuss examples of these methods and illustrates them as well. Another good, brief discussion of the analysis of textiles and basketry is found in *Archaeology Laboratory Methods* (Sutton and Arkush 1998: chapter 8). The archaeologist is also urged to consult the literature in the area in which they are working for comparative information. This is especially important when attempting to classify artifacts and fit them into an existing regional chronological sequence.

9
CLASSIFICATION

The rough sorting procedure of separating artifacts into the material categories discussed in the previous chapter represents a simple exercise in classification. An archaeological classification system or typology is simply an ordering of artifacts into classes or types according to a prescribed system on the basis of their attributes. An *attribute* is a well-defined feature of an artifact, such as color, that cannot be further subdivided. There are several kinds of attributes an archaeologist looks for in an artifact, which can be grouped into three macrocategories: stylistic, formal, and technological. An example of a technological attribute would be the kind of chert from which a projectile point was made. The design motif on a ceramic vessel is an example of a stylistic attribute, whereas the shape of the vessel itself (i.e., bowl, jar, etc.) would be a formal attribute.

In 1736, Carolus Linnaeus published *Systema Naturae,* which presented a classification system or taxonomy for all living things. While there has been some tinkering with its structure, it has generally withstood the test of time and most biologists, if not entirely happy with it, have at least come to accept its utility. As mentioned earlier, there is no archaeological equivalent to the Linnaean taxonomy, and given the diversity of opinions and interests among modern scholars, it is easy to understand why. One wonders how successful Linnaeus would be if he were to propose his scheme today!

One of the biggest obstacles to devising a classification scheme is its scope. The challenge is to make it comprehensive enough to include all the different data categories that might be of interest to the archaeologist, without having the system become so cumbersome that it actually impedes the interpretation of the artifact assemblage. Another important consideration is that the classification system have enough rigidity in its structure to ensure compatibility across datasets, while retaining the flexibility to allow the individual archaeologist to use it to answer questions peculiar to a single site or research design. This is a tall order and one that many institutions are rethinking as they move to computerize their databases.

OBJECTIVES

In designing a classification scheme, it is important always to remember the basic goals that the system is designed to achieve. This advice might seem obvious, but it is easy to lose track of the purpose of the endeavor when in the throes of subdividing artifact categories and wrestling with producing the user's manual for your laboratory's database program. So, it is important to think of the big picture of classification before, during, and after the design of your system.

The purpose of a classification system is to organize the data into manageable units—that is, to bring some semblance of order to the pile of artifacts spread out on your lab table. Once there is a logical order to the assemblage, you can begin to see patterns or correlations among different categories of your data. Which attributes occur together and which do not? The data categories include not only the composition of the artifact but also its form, function, decoration, age, and provenience, to name but a few of the fields the archaeologist must consider.

The process of classification began with the rough sorting into material categories that was done during the artifact processing stage. The next step is to subdivide these rough categories into meaningful archaeological classes. This will be the basis for relating or separating specific groups of artifacts. Once these categories are defined, then the analyst can attempt to discern relationships between the categories. Take, for example, a hypothetical artifact assemblage from a multicomponent site. Within this assemblage, you create a category for ceramics. This category can further be divided into prehistoric and historic. These categories can be even further divided, by surface treatment, into glazed and unglazed. Looking at relationships among categories reveals that only the historic ceramics are glazed. This is a simple example that nevertheless demonstrates the importance of a well-constructed database for the analysis of an artifact assemblage.

For an example of a more elaborate ceramic typology, see figure 8.3 in the previous chapter's discussion of ceramic analysis.

As stated at the beginning of this book, artifact analysis involves both the organization and description of your data. Description is the key to a good archaeological site report. A report in which an artifact assemblage has been completely misinterpreted or a key artifact misidentified can still be a valuable tool if the data have been well described. The important attributes to be described vary among the artifact classes. Some of these attributes were discussed in the previous section; others will be discussed later. In any event, it is important to be as objective as possible in your description of the artifact assemblage. The organization of your data tends to be somewhat more subjective.

EMIC VERSUS ETIC

A fundamental dispute in the construction of archaeological classification schemes revolves around the perspective taken by the archaeologist. Do the classificatory categories reflect the interests of the archaeologist (the outsider's or etic perspective), or do they reflect the categories that the people who made the artifacts had in mind (the emic or insider's perspective)? In other words, should we take the etic perspective and arrange the artifacts in an order that fits the past into a framework that is familiar to us, or should we attempt to take the emic perspective and view the artifacts as would the culture that made them? This was the essence of the Ford–Spaulding debate in the 1950s, an exchange that presaged the postprocessual paradigm by more than thirty years.

Albert Spaulding (1953) claimed that the classification of types should be a process of discovery of the combinations of attributes favored by the makers, not an arbitrary procedure of the classifier. Thus, in the true emic mode, classification should be independent for each cultural context. That is, each project should have its own classification scheme. To arrive at the emic perspective, statistical techniques would be used to discover combinations of two or more attributes meaningful to the maker. This is the essence of some types of multivariate analyses performed by archaeologists today.

James Ford (1954) responded that since culture itself is an etic construct, we would be deluding ourselves to think that reaching an emic perspective is even possible. He claimed that Spaulding's statistical analyses would reveal only the degree to which people followed a particular style at a particular point in time. The type thus "discovered" would not be useful for reconstructing a diachronic culture history. He further claimed that there were two reasons why Spaulding's techniques would not discover true emic types. First, at that time (and one could argue the same is true of today's archaeology) there simply wasn't sufficient chronological control to use the data the way in which Spaulding intended. Second, even if artifacts could be precisely placed in time, the nature of culture change is that it is not always gradual and can be affected by a variety of factors that defy prediction.

Undaunted, Spaulding (1954) replied by asking what Ford meant by "historically useful," indicating that he felt that Ford had missed the point of what he was trying to accomplish, which was not reconstructing culture history in the way that Ford understood it. He also added that any reasonably consistent and well-defined behavior pattern was historically useful. Spaulding clarified his concept of archaeological typology by claiming that there were actually three levels of types, each successive level building on the previous level and eventually revealing functional types. He countered Ford's claim that his belief that cluster analysis would automatically produce emic types was naive. He responded that his statistical tests merely revealed significant clusters of attributes. It was up to the archaeologist to interpret what these combinations meant.

Ultimately, and this is as true today as when the debate raged through the pages of the *American Anthropologist,* the debate is not really about categories but about questions. Comparing Spaulding's types with Ford's is like comparing apples to oranges—they were designed to answer different questions. Ford was interested in building a spatial-temporal framework for the lower Mississippi Valley, whereas Spaulding was interested in understanding how particular past peoples thought at a particular point in time. One could argue that Ford was promoting paleoethnology (the comparative study of past cultures), while Spaulding championed paleoethnography (the study of a particular past culture). Willey and Phillips (1958:13) in their classic *Method and Theory in American Archaeology* attempted to resolve the debate by claiming that Ford and Spaulding were both right. All archaeologically defined types were likely to possess some degree of correspondence to what a society would classify as the "right way" to make an artifact. They felt that the goal of a typology was to make the correspondence between "created" and "discovered" types as close as possible.

The challenge for the contemporary archaeologist faced with an artifact assemblage or multiple assemblages is to devise a classificatory structure that can be used to investigate both emic and etic questions. The availability of high-speed, high-capacity personal computers makes the use of complex hierarchical databases easy and practical. Statistical tests that would have taken Spaulding weeks merely to gather the pertinent data, let alone perform the calculations, can now be undertaken in a matter of minutes from existing databases.

The following section is not meant to show the reader the "right way" to classify artifacts. Rather, it is a demonstration of how other archaeologists have approached the problem of archaeological classification and the variables to be considered in this undertaking.

CLASSIFICATION SCHEMES

Depending on the questions to be asked of the data, artifacts can be classified in a number of different ways. Most classification systems are constructed on the basis of multiple attributes arranged hierarchically. There are many different basic categories to be considered for the classification of artifacts from any archaeological project.

MATERIAL/TECHNOLOGICAL

As discussed previously, the material from which an artifact is made is often the most basic means of subdividing the archaeological assemblage. It allows for a preliminary sorting of the artifacts by inexperienced personnel and provides the archaeologist with a quick assessment of the range of artifacts present. It gives a thumbnail sketch of whether the site consists mostly of lithic artifacts or whether pottery or historic materials are present. It is also useful for separating out artifacts that require stabilization (e.g., iron and perishable artifacts) as well as those artifacts that might undergo specialized analyses at another laboratory (e.g., faunal and floral remains, human skeletal remains, radiocarbon samples, etc.). Other categories will require more informed observation by a trained lab technician.

FORMAL/MORPHOLOGICAL

Another basis for separating and combining groups of artifacts is by their form or shape. In a hierarchically based system, artifacts within certain categories (such as materials) are often subdivided on the basis of their shapes. For example, in the discussion of glass artifacts in chapter 8, we saw that they could be divided into flat and curved fragments. This is a basically descriptive approach that is most often employed when the function of the artifact is not known. These descriptions are based primarily on the measurable attributes of the artifact such as height, width, length, thickness, and weight.

A good example of a subdivision based on form would be in the ceramic category. Here the form category could be subdivided on the basis of whole vessels and fragments or sherds. Sherds are usually described by their form and perceived location on the whole vessel. Most archaeologists employ such descriptive categories as rim, base, shoulder, and appendage when describing the ceramic subassemblage. These elements can be further detailed by such terms as *everted rim, conoidal base, incised shoulder,* or *effigy handle,* to name a few.

Whole vessels are rarely found outside of a burial context on most archaeological sites since if they weren't broken, they wouldn't have been thrown away! However, it is often possible to extrapolate the form of the vessel from a large sherd or series of sherds that have been reconstructed in the lab. Forms of whole vessels would include such basic descriptive categories as bowls, beakers, jars, plates, and cups or dippers. Historic vessel form descriptions may be even more elaborate, including platter, tumbler, tea cup, goblet, and tureen, to name a few. One should note, especially with historic form descriptions, that often a function is implied. This can provide the basis for another artifact categorization discussed below.

FUNCTIONAL

Functional artifact categories can be either useful or misleading. They are potentially the most power-interpretive categories that an archaeologist can devise for reconstructing past lifeways at a particular site. The function of the majority of the artifacts at a site often reveals the site's overall function as well (e.g., hunting camp, lithic extraction, domestic structure, ceremonial center, etc.). Defining the function of an object is supposed to be one of the skills that an archaeologist has over the layman.

The functions of various artifacts are derived, in the case of prehistoric artifacts, from using ethnographic analogy, replication experiments, or the direct historic approach using ethnohistoric accounts of early explorers. For example, the bannerstone or atlatl weight has no real correlate in modern Western society. However, reference to prehistoric pictographs and early accounts of tribal peoples suggests these artifacts were associated with spear-throwing devices. Replication experiments corroborate this interpretation.

Even with the corpus of ethnographic data at hand, interpretation is not always possible. Artifacts whose function had eluded the archaeologist despite their use of these investigative techniques were in the past sometimes referred to as "ceremonial objects" or items of "religious significance" to hide the classifier's ignorance. One of the unforeseen consequences of these unsupported speculations is that sacred artifacts are eligible for immediate repatriation to descendant Native American groups under the Native American Graves Protection and Repatriation Act (NAGPRA). Today most archaeologists simply describe such objects and list their functions as "unknown" rather than speculating. Other impacts of federal legislation will be discussed elsewhere in this volume.

These examples illustrate the rewards and perils of the functional categorization of artifacts. Often the function of an artifact is not known or assumed functions are either wrong or highly conjectural. For example, ceramic disks found on both historic and prehistoric sites are usually referred to as *gaming pieces*. There is usually no explanation of what the game might have been or how these disks might have functioned as part of the game other than as wagering tokens. This might be seen as harmless speculation on the part of archaeologists who just can't stand to have unidentified artifacts in their catalogs. However, these baseless speculations often get repeated again and again in the literature until they are eventually reified as "the truth" and potentially mislead all subsequent investigations.

There is yet another pitfall in the use of functional groups for artifact classification. Many, if not most, artifacts had multiple functions. Who among us has not used a pocketknife to tighten a screw or mason's trowel to excavate a feature? The multipurpose nature of various tools must have been even greater during the prehistoric period. However, there are ways, such as use wear analysis (see the "Lithics" section in chapter 8) to determine the *primary* use of many tools. Thus, some archaeologists have made productive use of this category of classification. This has been especially true in historical archaeology where many artifacts have a modern equivalent and so their function is better understood by the archaeologist.

One of the pioneers and principal exponents of functional artifact categories in historical archaeology was Stanley South. His explicit purpose for formulating these functional groups was so that he could identify "activities related to the systemic context reflected by the archaeological record" (South 1977:93). These functional categories were then used to detect patterns of regularity and variability in the site's assemblage that might be used inductively to generate statements of theory.

A modified version of South's artifact categories (table 9.1) was used to classify the artifacts recovered from the sixteenth-century Spanish site of Puerto Real, Haiti (Ewen 1991:61). The purpose of these groups was to provide a meaningful organization of the artifact assemblage in terms of human behavior as well as a basis for inter- and intrasite comparison. In this case, comparisons were made with previously excavated portions of Puerto Real as well as with sites in St. Augustine, Florida. The ceramic assemblage was subdivided into table wares and utilitarian wares so as to assess change through time as well as to use it as an index for measuring status differences within the community. The result was the successful test of a material model for Spanish Colonial adaptation to the New World (Ewen 1991:115–18).

South was cognizant of the aforementioned pitfalls of a functional classification scheme:

> Since virtually any class of artifacts can be seen to possibly serve a variety of purposes within the past cultural context, it is foolhardy to attempt to

Table 9.1. Artifact Categories at Puerto Real

Group	Artifact Category
1	Majolica
2	European Utilitarian Ceramics
3	Non-Majolica European Tablewares
4	Locally Made Ceramics
5	Kitchen
6	Structural Hardware
7	Arms and Armor
8	Clothing Related
9	Personal Items
10	Activity
11	Unidentified Metal Objects
12	Masonry
13	Furniture Hardware
14	Tools
15	Toys and Games
16	Harness and Tack
17	Religious
18	Miscellaneous
19	Unaffiliated
20	Hispanic Tablewares

arrive at a classification that has no exceptions. For this reason the artifact classifications used here are considered adequate for a wide range of historic sites. There is nothing wrong, of course, in expanding the list . . . in the face of a research design demanding such an addition. (South 1977:96)

Prehistoric archaeologists have not embraced such a classification system as enthusiastically, perhaps because, as previously noted, prehistoric artifacts are not as well understood as their historic counterparts. A notable exception is Howard Winters's (1968:184) use of functional categories for categorizing grave goods in an attempt to define a pattern that would reflect a value system for the Indian Knoll Culture (a Late Archaic manifestation in Kentucky). These artifact categories were grouped into three general classes: (1) items classifiable as general utility tools, weapons, fabricating and processing tools, and domestic implements and ornaments manufactured from locally available raw materials; (2) items generally classified as ceremonial equipment; (3) ornamental items occurring in special, segregated contexts, that are manufactured from rare materials and may serve utilitarian or ceremonial functions. Winters's categories represented values that could be inferred from their function and the raw material from which they were made. Combining simple statistical analysis with ethnohistoric data, Winters made a compelling case for the utility of this classificatory scheme for prehistoric data.

STYLISTIC

A popular approach to the classification of artifacts is on the basis of their stylistic qualities. Stylistic attributes include color, decoration, and surface finish (e.g., Munsell value 10YR 3/4, cord-marked, burnished, respectively). Prehistoric and historic pottery types are often distinguished chronologically by their stylistic attributes. Mississippian pottery is characterized by the stylistic motifs on the vessels as are the transfer-printed ceramics from the Staffordshire potteries.

The previously discussed Ford–Spaulding debate was primarily concerned with the interpretive utility of stylistic attributes. Different design elements characterize different time periods or cultural groups. In north Florida, the late prehistoric Fort Walton complex of ceramics are distinguished by incised, curvilinear designs, whereas the later Lamar-influenced Leon Jefferson series have stamped designs on the surface of the pottery. This stylistic change, interpreted as extraregional in origin, has been used as a chronological marker by archaeologists in the area.

CHRONOLOGICAL

A specialized typology can be built around these temporally sensitive or "diagnostic" artifacts. This classification would include types defined by form that are time markers. Such artifacts as Clovis points or machine-cut nails are used by archaeologists in the same manner that a paleontologist would use index fossils.

A good example of how artifacts have been used as diagnostic time markers is at the site of Puerto Real, Haiti. The site, occupied by the Spanish colonists from 1504 to 1578, was divided into two periods of occupation: early and late. It was possible to distinguish the Early Period (pre-1550) from the Late Period (post-1550) occupation primarily on the basis of the presence of such material marker artifacts as Cologne stoneware and Ming porcelain. Historical records indicate that these wares were not available in the Caribbean prior to 1550. Thus, combining these index artifacts with stratigraphic analyses permitted the deposits to be placed in one of two chronological contexts.

FOLK TAXONOMIES

Folk taxonomies are what Spaulding was after, classification systems derived by the people that made the artifacts. Ethnologists record these as a basic part of their data collecting when in the field. Unfortunately, without a time machine, archaeologists can't directly gather these data. However, archaeologists have derived methods for indirectly arriving at these categories.

The previously discussed artifact classification systems employ a primarily etic perspective. That is, the categories are determined by the archaeologist with little regard for the thoughts of the people who made the artifact. Yet, trying to gain an insider's perspective on the past is arguably one of the basic goals of archaeology. Albert Spaulding used statistical techniques in an attempt to elicit the attributes of pottery made the "right" way in the past. Historical archaeology, with the added dimension of historical records, can make a more direct inference from artifact categories.

Archaeologists in the Chesapeake region of Maryland and Virginia have developed a folk taxonomy for ceramics (Beaudry et al. 1983). The Potomac Typological System (with the clever acronym of POTS) links vessel forms with terms used in probate inventories and other colonial documents. Specifically, the authors intended to "begin to systematize the terms in the categories used to describe excavated ceramic vessels and the assemblages they comprise, in a way that will make the cultural dynamics behind them more accessible" (Beaudry et al. 1983:17)—in other words, to conduct middle-range research connecting the archaeological record with past behavior.

The Sears & Roebuck catalog is a commercial classification system with items divided by function, style, and technological criteria. Using past records like the Sears catalog of one hundred years ago, archaeologist can arrange their artifacts on the basis of cost and popularity.

These folk taxonomies are not better than the previously discussed typologies. They merely represent another perspective on the past that can help to answer a different suite of questions.

QUANTIFICATION

It has been said that if you can't quantify it, it doesn't exist. To the archaeologist, who must deal with the material remains of the past, this is especially true. So the question becomes not whether you can quantify it but how.

COUNTS

The most obvious way to quantify your artifacts is to count them. Usually the artifacts will be summarized by whichever categories they have been divided into and then those categories combined to produce a grand total of artifacts recovered from the site. For various analytical procedures, however, simple counts are not sufficient.

Ceramics is a good example of a category for which simple sherd counts are not always the best means of quantification. In some cases, sherd counts can actually be misleading! For example, two pots of identical size but having different decoration are broken. One breaks into twenty-five pieces, while the other yields fifty sherds. On the basis of this simple count of the sherds, the analyst would conclude that there were twice as many pots of the second type as of the first. This discrepancy might be obvious on the basis of sherd size when dealing with just two vessels, but when the archaeologist is confronted with hundreds or thousands of sherds, these discrepancies become less apparent. Other factors that affect sherd count include vessel size and durability of the ceramic vessel. A small stoneware bowl will break into fewer pieces than a large porcelain vase.

The archaeologist can deal with the potentially skewing affects of raw artifact counts in a couple of ways. One simple approach to correct the bias inherent in fragment size is to weigh the fragments. Thus, pots of similar size and composition would have similar weights no matter how fragmented they became. Digital scales make relatively painless the acquisition of these data, which can, with the proper equipment, be entered directly into a computerized database.

A more labor-intensive but potential valuable alternative means of quantification that addresses the issue of vessel size and durability is the estimation of the minimum number of vessels. In zooarchaeology, this statistic is referred to as the *minimum number of individuals* (MNI) and is a standard way of quantifying faunal remains. Realizing that most of a skeleton does not survive the taphonomic processes that intervene between deposition and intervention, zooarchaeologists have derived a method of determining the minimum number of animals present in the assemblage that would account for all the bones in the assemblage. This is based on the most numerous unique element (e.g., right tibia) of each species.

The ceramic equivalent of the MNI would be a minimum vessel count. This entails the examination of all sherds and separating them on the bases of distinctive morphological and stylistic attributes. This obviously involves a lot of refitting work in the lab, with the focus on the distinctive elements of the ceramics such as the rim, base, and decorative elements. A number of potential complications (e.g., how does one deal with matched sets of historic ceramics?) may arise with this method of quantification. Yet for some analytical techniques, estimated vessel counts are essential.

While discussing quantification, it is appropriate to consider the scale to be used for measurement. For most archaeologists, this is a question of the metric versus the English system. Although the United States (and Burma) steadfastly resist the metric system, most prehistoric archaeologists have adopted this system for their projects and so take their measurements in grams, liters, and centimeters. Historical archaeologists, on the other hand, often opt for the English system of feet and inches (or an engineer's scale of feet and tenths of a foot). This is because most English colonial and postcolonial sites in North America were constructed using the English system. The system of measurement should be made with an eye toward compatibility with other data sets and the availability of the proper equipment.

DATABASES

Once the classification system has been established and the method of quantification decided, it is necessary to create a means of quickly retrieving these data. It does little good to fill out reams of catalog sheets that have recorded the many variables of each artifact that you deem important if you have to laboriously scrutinize and retally each entry every time you have a question of the assemblage.

For example, say that you have recovered thirty thousand artifacts from a Late Woodland period village site in North Carolina. You are able to lump some of the artifacts into broader categories in your typology so you might have eighteen thousand entries on your artifact inventory sheets. Each sheet has twenty-five entries. If you wanted to tabulate all of the shell-tempered pottery from your assemblage, you would have to manually sort through all 720 sheets of the inventory to answer this simple question. If you wanted to know how many of these were rims, you'd have to repeat your search on the sheets that had contained shell-tempered ceramics.

Obviously this is a very labor-intensive and error-prone process (i.e., after a couple of hundred sheets, you might overlook an entry). The larger the artifact assemblage, the more time each search will take and the more likely errors will be made. Archaeologists have long sought a more efficient way to organize and retrieve their data.

CARD CATALOGS

Library card catalogs offered a workable system that was emulated by many archaeologists. The pertinent attribute data for an artifact, or groups of similar artifacts, could be placed on index cards along with provenience data. In other words, each entry from the catalog sheet was put on a separate card that could then be sorted by the archaeological analyst, such as by an attribute (e.g., composition) or provenience (e.g., unit 10, level 2).

A limitation of this system was that you could only sort on one attribute at a time, though it was fairly easy to create subsets. Returning to our Woodland village example, once you had pulled all the cards with shell-tempered pottery, you could quickly thumb through those cards to find the rim sherds (assuming you had recorded that information!). Still, searching on multiple variables was time-consuming.

A variation on the card cataloging scheme used at the University of Florida in the late 1970s involved using cards that had holes along their edges with a number printed beneath each hole. The numbers corresponded to artifact attributes (e.g., 10 = ceramic, 25 = shell tempered, 36 = rim). The holes with the numbers corresponding to the attributes possessed by the artifacts were punched open on the card. To find the artifact with attributes desired would involve running a spindle through the hole of the attribute desired. Upon raising the spindle, the cards having the artifacts with the desired attributes would drop out. This low-tech approach, while far from perfect, did speed analysis time considerably.

ELECTRONIC DATA

Computers are the most efficient way to process text, graphics, and data related to your site. Yes, they crash, develop odd glitches, and some programs are difficult to learn, but all these problems can be said to pertain to automobiles as well, and no one is still using horses for basic transportation! Every week computer technology gets faster, more powerful, easier to use, and less expensive. Today, powerful desktop units that rival the computing power of the mainframe computers of twenty years ago are well within the reach of even the most modest archaeological lab. The power of the computer is especially evident in its ability to take large quantities of disparate data and quickly sort them in a variety of ways.

It is difficult to write about specific computer solutions when that particular computer platform or software program is likely to be obsolete by the time this manuscript is published. Still, it is possible to give some general guidelines for computer applications that will apply no matter what system is in place.

Today it is possible to store large databases on a relatively inexpensive desktop computer. Many institutions are putting their entire lab's inventory on a single machine with removable storage capabilities. Powerful relational database programs allow the archaeologist to access more than one data set at a time. One file can hold the provenience data, while another contains the artifact inventory, and still another might hold images of the artifacts themselves. These data sets are linked through a common variable such as the FS number. What's more, the desired data can be quickly retrieved and manipulated with various statistical programs. Choosing the proper hardware and software for this task is an endeavor that should not be lightly undertaken.

The necessary computer hardware is the most difficult to predict. Choose a platform (e.g., Windows, Mac, Unix, etc.) that will run the currently desired programs. It is extremely difficult to predict the future of any computer company, so buy the machine that will accomplish what you want done today, not what you think you might want in the future. One caveat: Always buy the newest technology you can afford. Buying yesterday's technology may be cheaper, but it will often not run the programs you want today, let alone any future upgrades.

Storage media are difficult to predict as well. When I was at the Arkansas Archeological Survey, I ran across several boxes of keypunched computer cards. I was told they were from several past projects that were being stored in case someone wanted to reanalyze the data. Of course, that archaeologist would have had to go to the Smithsonian to find a card reader to upload the data into a computer! We ended up recycling the paper cards.

Everyone knows stories of data that are no longer retrievable because the storage media is obsolete. My master's thesis is on an eight-inch floppy disk. A colleague has his dissertation on an old Osborne computer running CPM. Computers that can read 5¼-inch floppy disks will be completely extinct by the time this book is published. The only solution is to continually transfer your data to the new industry standard as it becomes available and to *always print a hard copy of your data.* In a worst-case scenario, you can always reenter the data.

Database software is just as prone to obsolescence. The major question the lab director must confront is whether to go with a dedicated or commercial program. That is, should the database program be tailored to your particular lab, or should you buy a program that other archaeologists elsewhere are using? The answer to this question depends on how much outside support is required by the user. An organization with a great deal of in-house tech support can develop and maintain database software that fits their needs and computer hardware. The UNIX-based DELOS database used by the Arkansas Archeological Survey is a good example of a dedicated database (figure 9.1).

Organizations without extensive computer staff might want to consider purchasing a commercially available database (e.g., Access or Filemaker) or spreadsheet program (e.g., Excel or Quattro Pro) and simply develop their own data structure and program scripts for generating reports. This method presupposes a certain degree of computer literacy to design and implement the database structure, and limited tech support is possible from the commercial vendor of that program. Many large university labs and state programs have opted for this approach.

Page 11 of 16

DELOS Artifact Inventory Form

Site Number 3PU256 Site Name Ashley Recorder J. Pebworth Date Encoded

Accession Number 84-518 Project Number 00-03 Analyst J. Pebworth Date Entered 2/10/00 MW

FSN	LSN	ASN	P/H	General	Specific	Material	Weight	Count	FC	HT	DTX	FR	[illegible]	[illegible]	[illegible]	Morpho Funct	Qualifier	Location	Type/Variety	1	2	3
128	1	1	H	EUCER	RMBAS	PORCE	56 g	2								BOWL	EMBSO LAV					
128	1	2	H	EUCER	HNDL	IWARE	10.4 g	1								BOWL	BRN TRN					
128	1	3	H	EUCER	RMBDY	REWARE	13.2 g	1														
128	1	4	H	EUCER	RMBAS	PORCE	37.1 g	2								BOWL	PNK GILT					
128	1	5	H	EUCER	RMBAS	REWARE	21.2 g	1								BOWL	DECAL					
128	1	6	H	EUCER	HNDL	PORCE	13.7 g	1														
128	1	7	H	EUCER	RIM	REWARE	2.8 g	1								PLATTR	GRN SCALP SHELL					
128	1	8	H	EUCER	RMBDY	REWARE	19.3 g	1								PLATE	BLU SHELL					
128	1	9	H	EUCER	RMBDY	IWARE	26.9 g	1								BOWL	BANDED PNK GILT ✓					
128	1	10	H	EUCER	Body	SWARE	33.6 g	1								BOTTLE	BRN SIP SAL	BASE	INK BOURNE & SON			
128	1	11	H	EUCER	RMBDY	YWARE	165 g	2								MBOWL	AIK CIR AL BR ✓					
128	1	12	H	EUCER	BDBASE	SWARE	144.1 g	3									AL					
128	1	13	H	EUCER	RMBDY	EWARE	143.6 g	4								MBOWL	CIR AL					
128	1	14	H	EUCER	Body	EWARE	65.6 g	1									PNK AL					
128	1	15	H	EUCER	Body	SWARE	6.2 g	1									BR AL					
128	1	16	H	EUCER	Body	EWARE	26.2 g	1									CIR AL					
128	1	17	H	EUCER	BASE	IWARE		1									BURNED					

Figure 9.1 DELOS data sheet. (Courtesy of the Arkansas Archeological Survey.)

Another option is a generic archaeological database program that can be used as is or customized to suit a particular lab's needs. These programs (e.g., Minark, WinRelic, Re:discovery) are specifically geared to archaeologists, and the level of support is variable. Training in the use of these programs is sometimes available. The most compelling reason to go this particular route is that a professional programmer has developed this program specifically for archaeologists and has written a user interface that allows even technophobic archaeologists or untrained undergraduates to enter and retrieve data.

An example of this last approach to computerized databases is the Re:discovery database, which my lab recently acquired. I will use this merely as an example of how one lab went about computerizing their artifact collections. Working with the Re:discovery programmers, we developed a provenience database based on the excavation level sheets used by our students in the field (figure 9.2). This was linked to our artifact inventory database by the FS number field (figure 9.3). We have also linked our lab's site accession log by transferring it to a site record database (figure 9.4). Future additions being contemplated include an image database and map catalog.

A new type of database management, which will no doubt come to dominate the field, are on-line databases. The National Archaeological Database (www.cast.usark.edu/products/NADB/) is an on-line database of databases including a bibliography of CRM reports, information concerning NAGPRA, and regional GIS maps. In England, the

Figure 9.2 Re:discovery context screen. (Courtesy of the ECU Archaeology Lab.)

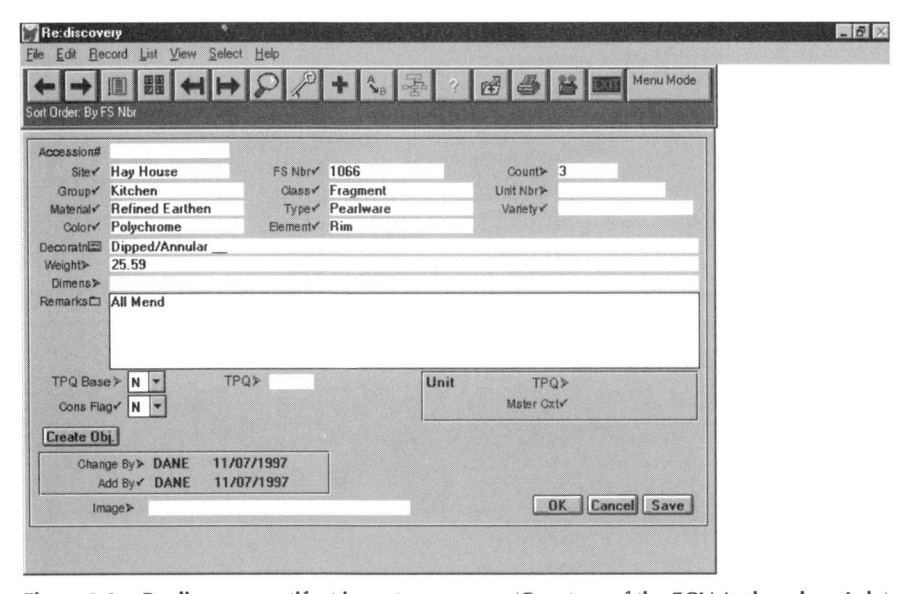

Figure 9.3 Re:discovery artifact inventory screen. (Courtesy of the ECU Archaeology Lab.)

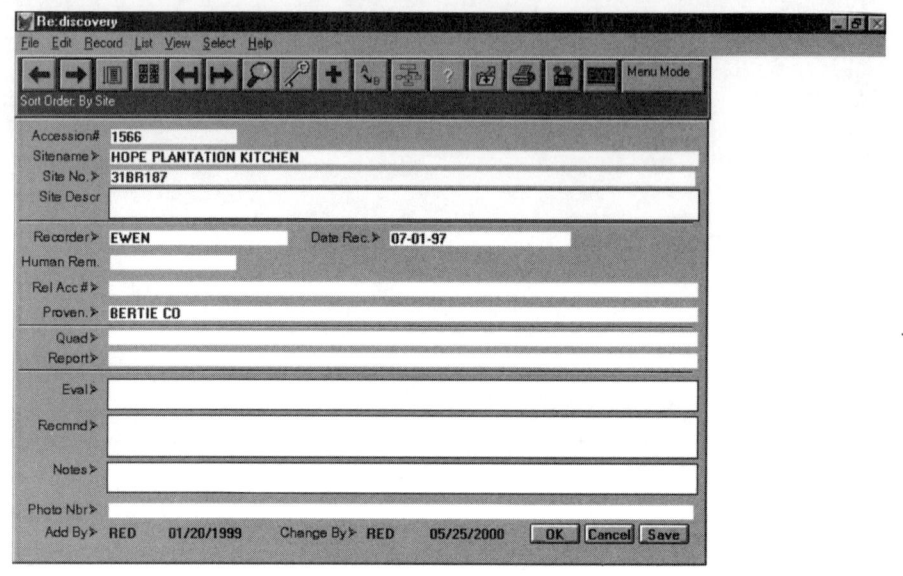

Figure 9.4 Re:discovery accession log screen. (Courtesy of the ECU Archaeology Lab.)

Archaeological Data Service (ads.ahds.ac.uk/project/general.html) will archive archaeological databases and make them available online for a modest fee. While not promoting a specific database architecture, they do advocate standard vocabulary for the artifact inventory.

The artifact inventory is the heart of the database and uses many of the different types of categories (table 9.2) discussed earlier. For instance, "Group" is a functional classification. "Class" and "Element" are morphological categories. "Color" and "Decoration" refer to stylistic attributes of the artifact, while "Material" is a technological category. The database's powerful search engine allows the archaeologist to search on several different kind of attributes simultaneously. Returning again to our Woodland Village site, it is a simple matter to construct a query that asks for all the ceramic rims that are shell tempered. You could also ask that it limit the search to feature contexts. Your answer would appear in seconds. It would also supply all the related contextual information (e.g., excavation unit, level, related contexts) and attributes (e.g., color, other decorative elements, etc.) that you specified in your request and prints it out in tabular form or imports it to your word processor for insertion into the final report.

CLASSIFICATION

Table 9.2.	Database Structure for East Carolina University Archaeology Laboratory
Group:	Functional categories following South (1977). Other categories can be added (i.e., for prehistoric assemblages or special research questions) as needed by the archaeologist.
Class:	This category is loosely based on artifact form and follows South's categories when the artifact is whole. Otherwise, it is more prudent to enter "fragment" if the artifact is not complete, especially when referring to ceramics and glass. Nails, spikes, and staples would be referred to as "fastener" in this field.
Material:	Refers to the composition of the artifact.
Type:	Similar to South's category. For ceramics, one would put such terms as *pearlware, creamware, Jackfield,* or *UID* (unidentified) *refined earthenware,* and so forth. For glass, a distinction should be made between curved and flat. For prehistoric ceramics, local type descriptions (e.g., Colington, Lamar, Weeden Island, etc.) could be used. For fasteners, the terms *spike, nail, tack,* or *staple* would be entered for this field.
Variety:	This would apply to specific clusters of attributes that distinguish a subtype. For example, Flow Blue is a particular variety of pearlware. For nails, *wrought, cut,* or *wire* would be entered.
Color:	Refers to the color of the artifact or its primary decoration. If a ceramic is plain, then nothing is entered. If it is a blue transfer-printed pearlware, then *blue* would be entered into this field. If it is a multicolored, hand-painted pearlware, then *polychrome* would be entered.
Element:	Refers to something other than a body fragment (i.e., rim, handle, base, etc.).
Decoration:	The primary decorative element characterizing the artifact. This would include such descriptions as "transfer-print," "painted," "fabric impressed," "simple stamped," and so on.
Dimensions:	This field is for measurements such as the diameter of buttons, length of projectile points, or bore diameter of pipe stems.
Remarks:	This field is for pertinent comments that don't necessarily fit in other fields. You could describe a maker's mark or note that the artifact is burned or that the type is similar to a specimen from another site.

Another advantage of computerized database systems is that you have the ability to quickly organize your entire assemblage on a variety of criteria. For example, you initially rough-sorted the artifacts on the basis of their composition. With a computerized database, you could print a summary table of your artifacts sorted by "Material." Or you could sort on the basis of provenience. This negates the advantage of waiting until you return from the field to assign catalog (FS) numbers. It wouldn't matter how they were assigned in the field since the computer could order them by excavation unit, or natural strata level, or component, or any context that made sense to the archaeologist.

One major fear that prevents many archaeologists from committing to a computerized system is the fear that their system will crash and they will lose all their data. Your system *will* crash at some point, just as the power in the lab will go out during a storm or your field vehicle will get a flat on some back road. Provisions can be made for dealing with system crashes just as they can be made to lessen the impact of any other emergency. Back up your data frequently, and have a printed hard copy as well.

Another legitimate fear that the lab manager has is that the database program selected will become obsolete. This *will* happen. It is important to look at the potential upgrade options for the program. In this case, it is probably wiser to go with a major commercial vendor with a large installed clientele rather than a small proprietary system. If a major vendor goes under, *someone* is likely to pick up the contract to service the former customers.

DATA ENTRY

The best way to make sure no data are lost is in the data entry phase of the artifact-cataloging process. It is important that the information be typed into the computer *exactly* the way the archaeologist desires. Computers do *exactly* what they are told. If you ask it find all the "shel-tempered" sherds in the database, it will find only those that were misspelled exactly in that fashion. So it is important that everyone using your database classifies the same artifact in the same way and that this information is carefully entered into the computer.

DATA ENCODING

If several archaeologists will be involved in the process of analysis, it is important that they agree on the artifact typology. This is the responsibility of the project director. She or he must ensure that everyone calls a particular piece of historic ceramic transfer-printed pearlware and not decorated refined earthenware or pearl-white ware with transfer-printed decoration or any other equally descriptive variant. A codebook available to all lab staff that describes different classes of artifacts and what the archaeologist should call them is essential.

DATA ENTRY SHEETS

Once the nomenclature is established, the data can be entered into the database structure. It is important that these data first be handwritten on separate sheets of paper with the different categories listed in columns (figure 9.5). This may seem like an extra, unnecessary step to individuals who are used to composing at the keyboard, yet there are very good reasons for doing so.

First, by sorting out the attributes on paper, the archaeologist gets a feel for how different artifacts should be classified and can check for consistency before the data are entered. Second, whereas a trained archaeologist must perform the actual artifact identification and analysis, anyone can type the data entry sheets into the computer. This is a useful task for students, who benefit by learning about artifact classification while relieving the archaeologist of the tedious time at the keyboard. Finally, and most important, these data entry sheets act as an emergency backup for your data should something go terribly wrong with your computer system.

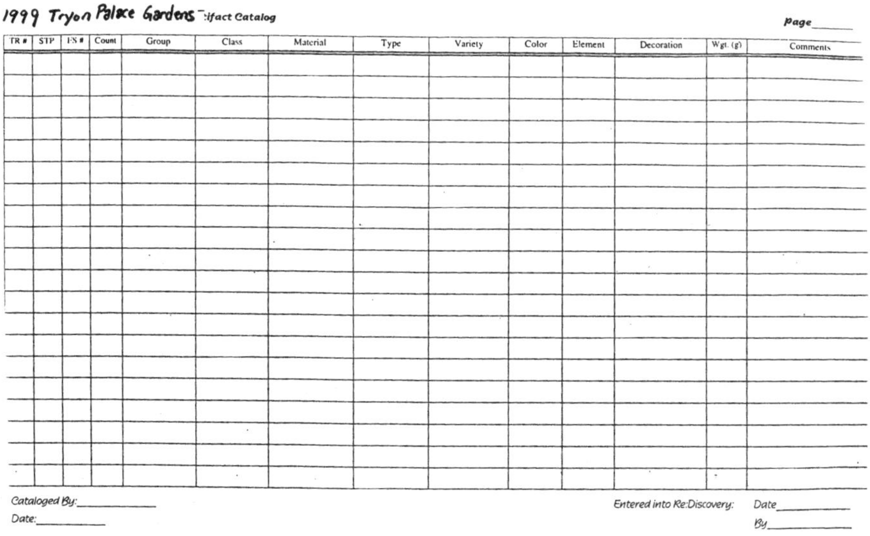
1999 Tryon Palace Gardens -ifact Catalog

page____

TR #	STP	FS #	Count	Group	Class	Material	Type	Variety	Color	Element	Decoration	Wgt. (g)	Comments

Cataloged By:__________

Date:__________

Entered into Re:Discovery: Date__________ By__________

Figure 9.5 Data entry sheet. (Courtesy of the ECU Archaeology Lab.)

PROOFING HARD COPY

A final step for data entry applies whether you are using a computer, paper catalog sheets, or index cards to organize your collection. It is important that the archaeologists check and double-check their own work and that of their assistants. This is especially true of the data entered into a computerized system. The old (in computer time) adage "garbage in, garbage out" applies here. Once the data entry specialist has completed inputting the data, the database should be printed and scanned for misspellings and inconsistencies in classification. These anomalies should be corrected and a clean hard copy produced. The archaeologist is now ready to proceed to the next phase of artifact analysis.

10
DATA MANIPULATION

OBJECTIVES

The purpose of cataloging the artifacts is essentially to prepare the assemblage for more in-depth analysis. The purpose of this in-depth analysis or data manipulation is to aid the archaeologist in interpreting the artifact collection. So, in a sense, analysis is all about asking the right questions of your data and arranging the answers in a way that allows you to achieve the interpretive objectives of your research design.

STATISTICAL ANALYSES

A complete or even cursory examination of statistical analytic procedures is clearly beyond the scope of this book. However, it is important that the archaeologist understand what statistical analyses are and how they can be used and abused on archaeological data. A good introduction to the use of statistics in archaeology is David Hurst Thomas's *Refiguring Anthropology* (1986). Other good references exist (e.g., Drennan 1996). One such manual should be on every archaeologist's bookshelf.

Statistics are numbers. They are a numeric description of the archaeological data. The basic quantification procedures discussed in earlier chapters generate statistics. So, when archaeologists count and weigh the different categories of artifacts to be listed in a summary table, they are preparing statistical tables. This probably comes as something of a shock to the mathematically impaired student who sought refuge from statistics in the social sciences! Nearly every attribute of an artifact can be quantified. Counts and weights are obviously numeric data, but other aspects can be represented statistically. Artifact shapes and areas can be measured, colors can be recorded with a Munsell chroma scale (and even more accurately with computerized color sensors), and the volume of pots can be determined.

Statistics are also a tool to infer the characteristics of a data set that are missing or not directly observable. This type of statistical analysis, called *inferential statistics,* uses the procedures that are employed by the Harris and Gallup pollsters to predict the outcome of elections with only 2 percent of the voting precincts reporting in. Since archaeologists rarely have the opportunity to excavate an entire site, they must make inferences about the entire site based on the excavated portion. Inferential statistics offer ways of assessing the degree of confidence the archaeologist can have in conclusions regarding these extrapolations

SAMPLING

Making the sample of data collected as representative as possible of the entire site is the responsibility of the field archaeologist. However, often the data collected in the field are more than can be reasonably analyzed in the lab. This is especially true of many CRM projects whose time and budget constraints dictate the level of effort that can be expended on analysis. Budget limitations also limit the size of the collection that can be analyzed by a contracted specialist, such as a zooarchaeologist. Rather than send all the faunal remains to the zooarchaeologist a representative sample is culled from the total faunal assemblage. So what constitutes a "representative sample"?

There are basically two philosophies in devising a sampling strategy: judgmental and random. Judgmental sampling relies on the past experience or special knowledge of the archaeologist in compiling the sample. For example, if you wanted to know what kinds of animals were part of the diet of the Woodland component of your site, you might select the faunal remains from undisturbed feature contexts associated with Woodland ceramics rather than the remains found scattered in a mixed sheet midden.

Random sampling, on the other hand, is preferred when little is known about the archaeological collection or the archaeologist wants to eliminate any bias in the selection of the subsample. In this case, rather than choosing the fauna recovered from proveniences from a specific part of the site, the archaeologist might want to include all the animal bones from a couple of excavation units randomly chosen from the all those excavated.

Statistical analysis picks up where classification leaves off in determining the similarities and differences between artifact categories. Which attributes are related? Which never occur together? A judicious use of scatterplots can graphically depict relationships between artifact attributes that can also be expressed mathematically (i.e., as a linear regression formula). By revealing patterns in the data (e.g., glazing is correlated with historic ceramics), statistics can be used to generate hypotheses.

Statistical tests of probability, such as the Student's t, address the question "Is the correlation between these two attributes real or could their relationship be due to chance?" In this case, our research hypothesis would be that only historic ceramics are glazed. The null hypothesis, the one we wish to disprove, would state that there is no relationship between glazing and the time period in which ceramics were manufactured. One should note that since we are dealing with probabilities, you can never prove your research hypotheses; you merely fail to reject them. You should be wary of anyone who states that something has been "statistically proven"!

Statistical association should never be confused with what actually happened. That is, statistics can point to a connection, but the archaeologist must determine what that connection means. The statistical observation that glazing occurs only on historic ceramics is merely the impetus to ask the question "Why?" Performing the statistical analyses is one of the steps leading to the interpretation of your collection.

Archaeologists should never lose sight of why they are doing the statistical test in the first place. It is often easy to forget the big picture when immersed in a sea of attributes. As part of the shift to an explicitly scientific archaeology in the 1970s, archaeologists became obsessed with finding the most arcane statistical procedures imaginable. Papers weren't published unless they contained some statistical procedure no matter how irrelevant to the questions being asked. David Hurst Thomas, one of the leading proponents of statistical applications to archaeology, was so appalled at the abuse of these procedures that he wrote "The Awful Truth about Statistics in Archaeology" (1978), in which he was especially critical of the uncritical use of multivariate statistics. Fortunately, in most cases, he found that "the analysis is needlessly complex, so alien to anything remotely archaeological, that little harm is done" (Thomas 1978:240). He advocated a return to more elementary univariate or bivariate methods before jumping into the multivariate deep end. Since then, archaeology has undergone a humanistic paradigm shift, and statistics are more judiciously used. No matter what the prevailing school of thought, it is important to remember that statistics, like computers, are only means to an end, not an end in themselves.

PATTERNING

Archaeologists are always looking for patterns in their data. As discussed earlier, this is one of the primary reasons for the classification of artifacts. It is impossible to see any patterns or omissions in the data if they are in chaos. Patterns that interest the archaeologist can take many forms. They can refer to the pattern of the designs on ceramic vessels. They can refer to the regular arrangement of postmolds on a site. Sometimes, by grouping artifacts or their attributes into statistically significant categories, patterns are revealed that can tell much about the people who made the artifacts.

Albert Spaulding used a type of statistical test called a *cluster analysis* to determine which type of stylistic attributes were linked on ceramics of the Lower Mississippi Valley. He hoped to discover the patterns the makers of those pots thought were correct and so gain some insights into the thoughts of those prehistoric potters. Stanley South looked at the proportions of artifacts in different functional categories or their distribution across a site to gain some insight into the activities of the inhabitants of the sites he excavated.

SOUTH'S PATTERNS

Stanley South, like Thomas, felt that archaeologists could get so caught up in the methodology that they would miss the point of their efforts. He argued that solutions to archaeological problems do not come directly from the observations made on the data; they come from the archaeologist. To this end, he devised a number of empirical tests to help the archaeologist recognize patterns in the archaeological data that would provide a basis for the archaeological interpretation.

South's (1977:25) goal was to

> demonstrate the patterned regularity of the archaeological record as well as the variability, under the belief that we can have no science without pattern recognition, and pattern cannot be refined without

quantification. From pattern recognition, general empirical laws can be stated, and the explanation of why these laws are operative through the hypothetico-deductive process leads to theory building through testing hypotheses with new data.

Let's explore three of these pattern recognition processes.

The Brunswick Pattern (formulated from data recovered from excavations at Brunswick Town, North Carolina) of artifact disposal seems deceptively obvious. The pattern is simply that on British colonial sites of the eighteenth century, a concentrated refuse deposit will be found at the points of entrance and exit in dwellings, shops, and military fortifications. This is useful in determining the doorways for structures where only the foundation remains.

It would be short-sighted, however, to see this pattern as merely a tool to find the door to a structure. What does this pattern say about trash disposal patterns of the people of Brunswick Town? We don't just throw our refuse out the door. What sort of refuse was being disposed of out the door? Did other ethnic groups dispose of their refuse in like fashion? Did refuse disposal habits change through time? As can be seen, identifying a pattern serves as a point of departure for other studies.

While the Brunswick Pattern relies on careful attention to the spatial distribution of the artifact assemblage, the Carolina Pattern and the Frontier Pattern utilize an organizational approach derived by classifying artifacts from a number of eighteenth-century sites into eight functional categories. It essentially is a model pertaining to percentages of artifact categories that you would find on an eighteenth-century British colonial site. With the Carolina Pattern, those artifacts classified in the Kitchen functional group would predominate over those in the Architecture functional group, whereas the Frontier Pattern would reverse the importance of these two functional groups. These patterns are examples of how the manipulation of data can reveal patterns in an archaeological assemblage.

The method of organization of the data can reveal changes in an otherwise static artifact assemblage and vice versa. A good example of this can be seen in the data recovered from the sixteenth-century town of Puerto Real on the north coast of Haiti (Ewen 1991). When viewed chronologically by descriptive types (i.e., different varieties of majolicas and utilitarian wares), the assemblage greatly changed in both kind and number of artifacts during the course of the site's occupation. However, if these same artifacts were grouped functionally rather than stylistically (table 10.1), then the assemblage could be seen to change very little. The shifts in particular artifact types that were noted pertained to trends in style and fashion *within* the functional categories. The functional categories remained relatively stable through time, indicating that the pattern of colonial adaptation of the Spaniards changed very little through time at Puerto Real.

The versatile ceramic assemblage can be manipulated in a variety of ways to investigate a variety of questions. By examining stylistic motif and mineralogical composition, a sherd's point of origin can be determined and provide information regarding trade patterns. Predominance of vessel form can provide insights on food preparation and storage prac-

Table 10.1. Early versus Late Categories at Puerto Real

Artifact Category		Early	Late
Majolica	#	1,400	7,682
	%	18.16	19.05
Hispanic Tablewares	#	139	1,513
	%	1.8	3.75
European Utilitarian Ceramics	#	1,963	10,870
	%	25.46	26.95
Non-Majolica European Tablewares	#	36	163
	%	.47	.40
Locally Made Ceramics	#	3,526	17,387
	%	45.73	43.11
Kitchen	#	232	1,037
	%	3.01	2.57
Structural Hardware	#	226	1,107
	%	2.93	2.74
Arms and Armor	#	8	15
	%	.10	.04
Clothing Related	#	116	297
	%	1.5	.74
Personal Items	#	24	169
	%	.31	.09
Activity	#	24	38
	%	.31	.42
Furniture Hardware	#	1	9
	%	.01	.02
Tools	#	2	18
	%	.03	.04
Toys and Games	#	2	9
	%	.03	.02
Harness and Tack	#	12	19
	%	.16	.05
TOTAL	#	7,711	40,333
	%	100	100

tices, which in turn can provoke questions concerning social interaction (e.g., do large vessels denote communal cooking or extended family residences?). Ceramics, as we have seen earlier, can also be used to examine patterns of status differentiation and folk classification systems.

CC INDEX

An archaeological assumption that has reached almost lawlike status is that the more exotic the ceramic, the higher the status of its owner. Thus, in prehistoric sites, individuals buried with highly decorated, imported vessels are usually interpreted as having more status than those buried with locally made plain wares. Similarly, on historic sites, structures associated with porcelain and other finely wrought imported wares are usually thought to have been occupied by individuals with higher status than those sites with locally made undecorated wares.

On historic sites of the eighteenth and nineteenth centuries, it is possible to classify the ceramic assemblage based on its relative value of the time, thanks to George Miller's (1980) CC (cream-colored ware) Index. The CC Index is an emic classification of ceramics based on decoration and price from contemporary commercial catalogs and shipping manifests, rather than the etically imposed ware types of the archaeologist. Those etic classifications, while good for dating and determining country of origin, are vulnerable to the subjective biases in classification (as anyone who has tried to distinguish a body sherd of late creamware from early pearlware can attest).

The CC Index can more easily and consistently classify pottery and answer questions about status and can be integrated with historical documents. It takes all cream-colored refined earthenwares and breaks them down into four categories: (1) undecorated, (2) minimal decoration (engine turned), (3) hand-painted, and (4) transfer print. These different categories are ranked according to their price when compared to the undecorated category, which has an index value of one. CC Indices are also scaled depending on the type of vessel form (i.e., plates, bowls, tea wares). Thus, the index reveals how much individuals were willing to spend on their tablewares. This may or may not be indicative of status but is certainly revealing of the individuals' choices and, perhaps, aspirations.

An example of how the CC Index can be used is its application to the ceramics assemblage from the Robert Hay house in New Bern, North Carolina. Hay was a carriage maker who plied his trade during the first half of the nineteenth century. Although his trade and financial records indicate that he was economically a member of the middle class, the historical record of the time suggests that he was considered in higher social status by the community. Late in life, he was forced to declare bankruptcy, allegedly because of debts incurred by his brother-in-law. The timely financial intervention of his upper-class friends saved him from complete ruin.

An analysis of the ceramic assemblage excavated at the Hay lot (Magoon 1998) revealed a much higher CC Index than expected for a working craftsman. This suggests that Robert Hay might have been living beyond his means to keep up with his society friends and that his deadbeat brother-in-law was merely the final straw that toppled his precarious fiscal situation. This example should serve notice that, as with all data manipulation, it is not the formula but how the results are interpreted in context that is important.

DATING

The concern with chronology has been one of the primary concerns of the archaeologist since the founding of the discipline. The Law of Superposition (strata on the bottom are older than the strata on top) is one of the few laws in archaeology. Chronometric techniques (discussed later) have allowed archaeologists to assign calendar dates to strata and features. These two dating strategies, the first providing a relative date, the second an absolute date, can be applied to artifacts as well.

SERIATION

One of the earliest means for determining which artifacts were older than others was the relative technique known as *seriation*. This technique arose under the Culture History paradigm when the construction of regional chronological frameworks was of paramount concern to the archaeologist. Seriation is a presumably temporal ordering of artifacts based on the assumption that certain attributes (e.g., stylistic motifs, technology, form, etc.) change at a regular rate and that the predominance of a particular attribute can be associated with a particular period in time.

For an excellent example of seriation in action, we can again return to the Lower Mississippi valley. James Ford (the protagonist in the

Ford–Spaulding debate) was interested in arranging the mostly single component sites in that region into chronological order. He recorded the percentages of the different ceramic types from a site on strips of paper. He then took these strips of paper and arranged them so that the percentages waxed and waned in popularity according to archaeological assumptions of how seriation works (figure 10.1). The assumptions guiding Ford's work in the Lower Mississippi Valley, that ceramic types evolved in a predictable manner, would be tested by historical archaeologists.

The idea of seriation seemed logical, but archaeologists lacking tight chronological control had never really tested the concept. James Deetz and Edwin Dethlefsen (1965) determined a way to test whether artifact attributes really did behave in a predictable manner by using

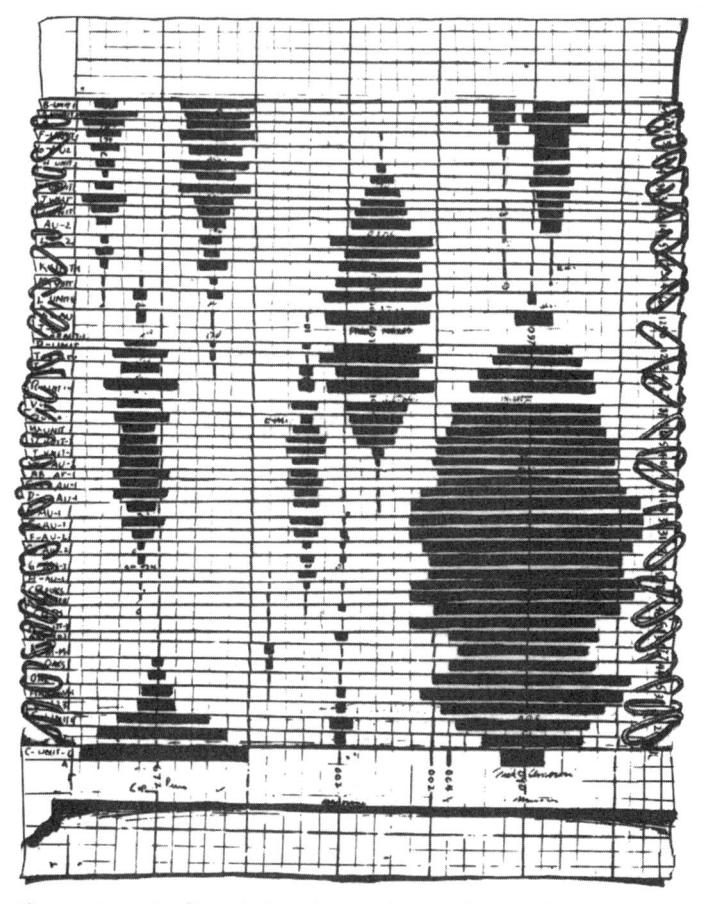

Figure 10.1 Ford's seriation chart. (After Ford 1952, figure 4.)

artifacts with stylistic attributes whose date was known: gravestones. They charted the occurrence of such different decorative motifs as death heads, cherubs, and urn and willows in several eighteenth-century cemeteries (figure 10.2). With the validity of the seriation concept established, archaeologists can note deviations from this pattern (e.g., the sudden disappearance of an attribute at its height of popularity) and investigate their causes.

DATING WITH HISTORIC ARTIFACTS

Often whole sites or individual strata are dated by using artifacts of a known age that they contain. This is similar to the paleontological dating practice of using marker fossils. Some historical artifacts and radiometrically dated prehistoric artifacts can be used in the same fashion. Historic artifacts are often used to establish a TPQ (*terminus post quem*) for an archaeological deposit.

Terminus post quem means the date *after* which the deposit was created. A simple example would be an undisturbed trash pit holding five coins dated 1920, 1921, 1924, 1925, and 1983. The TPQ for that deposit would be 1983. In other words, the trash pit could date no earlier than 1983, despite the dates on the other coins, since that is the earliest the 1983 coin could have been deposited. Archaeologists have determined TPQs for many varieties of historic ceramics, beads, and other artifacts whose date of manufacture is known.

The opposite of a TPQ is a *terminus ante quem* (TAQ), or date *before* which a deposit was created. The determination of the TAQ can be useful in establishing the end point of a certain activity. Returning to New Bern, we find recorded that a hurricane devastated the town in 1769. This event is clearly visible in the stratigraphic record as a thin layer of light-colored sand. When encountered, it establishes the TAQ of the deposits beneath it. Both TAQs and TPQs can be used in a number of ways to interpret the stratigraphy of a site.

The TPQs for various classes of artifacts have also been used statistically to determine the mean, median, and range of occupation dates for a site. This is often referred to as *formula dating*. One must keep in mind that these formulas do not yield precise dates, but simply get you in the ballpark or provoke further investigation. Your sample should at least consist of thirty specimens, and these formulas work best with collections from a relatively restricted time span.

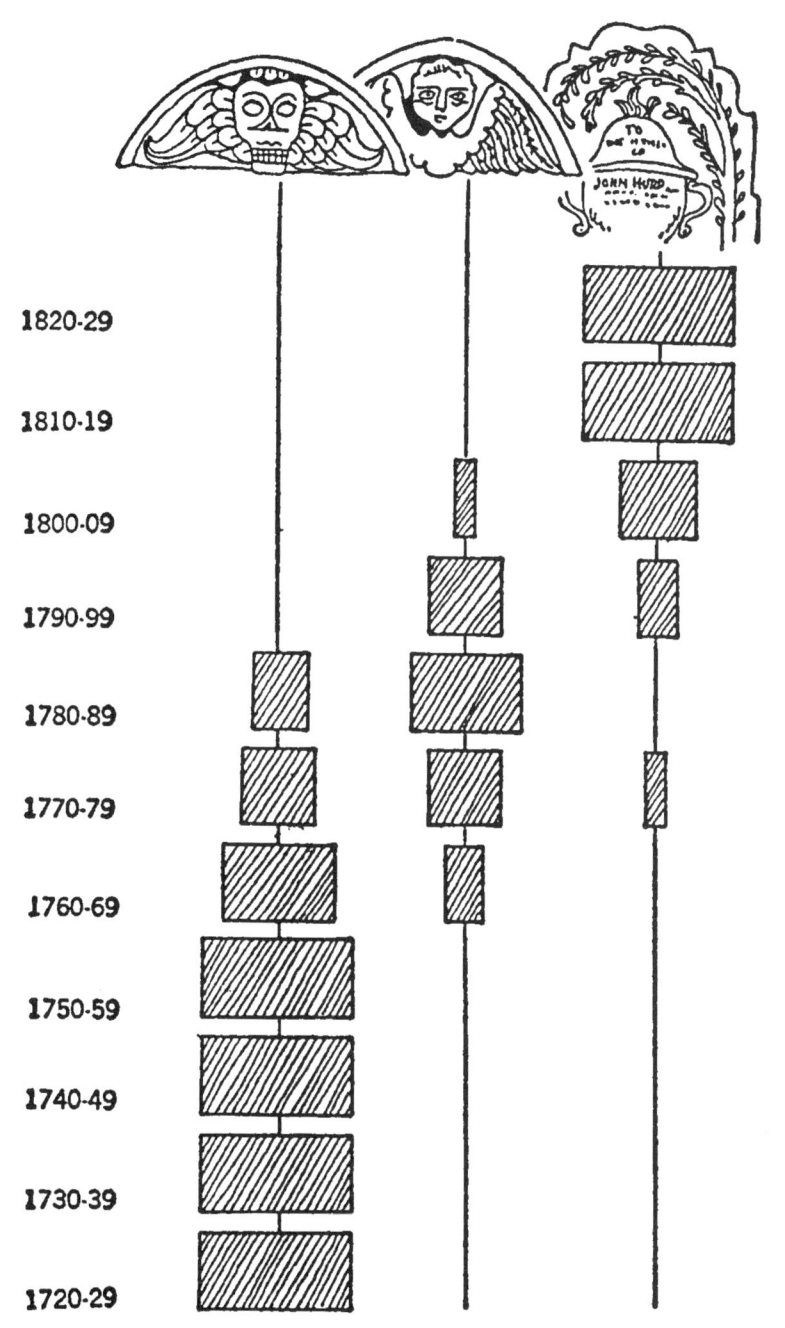

Figure 10.2 Plot of tombstone seriation, Stoneham cemetery, Massachusetts. (From Deetz 1996; used by permission.)

PIPE STEMS

Tobacco pipes have long been used to date seventeenth- through nineteenth-century historic sites. The changing form of a pipe bowl is a distinctive attribute that can often provide a TPQ and date range for that artifact's use life (see Hume 1978:302–3). However, it is the pipe stem, or the bore diameter of the pipe stem to be more precise, that can be quantified and plugged into a formula that yields a rough date for the pipe stem assemblage. There are several methods of pipe stem dating.

J. C. "Pinky" Harrington was one of the first archaeologists to recognize that the bore hole of the clay pipe stem became narrower with the passage of time. The Harrington method (1954) is more like a seriation than a strictly formulaic approach to the dating of the pipe stems. The bore diameter of each of the pipe stem fragments in a collection would be measured by inserting a standard drill bit and recording the best fit. Drill bits made in increments of sixty-fourths of an inch became the de facto standard of measurement since they were handy and inexpensive. After all the pipe stems had been measured, an average bore diameter was calculated, which was then compared to a chart (figure 10.3). This placed the main occupation of the site within an approximate twenty-five-year period and worked surprisingly well.

Louis Binford (1969) hypothesized that the bore contracted at a constant rate and so derived a straight-line regression formula to arrive at a more precise date. According to Binford the mean date of a site's occupation could be calculated by the formula $y = 1931.85 - 38.26x$, where y = desired date and x = mean bore diameter of assemblage. The

1800 1750	77%	20%	3%				
1750 1710	13%	72%	15%				
1710 1680		12%	72%	16%			
1680 1650			18%	57%	25%		
1650 1620				21%	59%	20%	
	4/64"	5/64"	6/64"	7/64"	8/64"	9/64"	

Time period vs. pipe bore diameter

Figure 10.3 Harrington's pipe stem chart. (After Harrington 1954.)

date, 1931.85, was the date when the pipe stem would theoretically disappear given a constant rate of narrowing. The practical limitations of the formula is that the bore diameter levels out toward the end of the eighteenth century and doesn't work for dates after 1780.

In the statistical quest to build a better pipe stem formula, several archaeologists questioned whether the reduction in bore diameter was as regular as had been assumed. Based on the measurements of pipe stems from assemblages of known dates (see Hanson 1971), it was determined that a curvalinear regression better expressed the technological trend.

MEAN CERAMIC DATING

Another example of formula dating is the mean ceramic date formula. Developed by Stanley South (1977), it derives the mean date of occupation for a site. This formula relies on the archaeologist knowing the date range for the manufacture of specific artifact types.

$$Y = \frac{\sum_{i=1}^{n} x_i f_i}{\sum_{i=1}^{n} f_i} \qquad \begin{aligned} x_i &= \text{median} \\ f_i &= \text{sherd count} \end{aligned}$$

Let's say you excavated a sealed privy that had twenty-two sherds of plain creamware; fifty-three sherds of white, salt-glazed stoneware; eighteen sherds of Jackfield ware; and eighty-seven sherds of slipware. By consulting a chart of manufacture dates (e.g., South 1977:210–12), you can determine the median manufacture date (MMD) for each type of ceramic. From there it is a simple matter of multiplying the MMD by the number of sherds, totaling the products of all the types, and dividing that total by the total number of sherds (see table 10.2). The

Table 10.2. Mean Ceramic Date of Hypothetical Privy Feature

Type	Mean Manufacture Date	Number of Sherds	Total
Creamware	1791	22	39,402
WSG stoneware	1758	53	93,174
Jackfield ware	1760	18	31,680
Slipware	1733	87	150,771
		180	315,027

Mean ceramic date = 315,027(total)/180(total number of sherds) = 1750.15.

resulting mean ceramic date (MCD) of 1750.15 is the interpreted midpoint of use for the privy feature.

It is also possible, using the date ranges of the various ceramics and plotting them on a chart, to estimate the span of use visually (see South 1977:214–17). Many skeptics have decried the biases (e.g., the curation effect) that would potentially skew the resulting date. Yet, the formula does seem to work and can provide a ballpark figure in the absence of associated documentation.

SPATIAL ANALYSES

Once the chronological aspect of the artifact collection has been addressed, you can turn your attention to establishing the spatial relationships. *Spatial analysis* is the study of the placement of artifacts on a site. The Brunswick refuse disposal patterns discussed earlier comprise an example of the kind of information that spatial analyses can deliver. It is while performing spatial analyses that one can appreciate the importance of context and provenience. If the exact locations of the artifacts are not known, there can be no spatial analyses.

Spatial analyses can be performed on a number of levels, from intersite to intrasite and even within an activity area. Intersite analyses are the realm of settlement pattern studies and beyond the scope of this work. Looking at the patterns in the placement of artifacts begins at the intrasite site level. Within the site, the areas of interest can be further subdivided into the determination and analyses of communities (the maximum number of people who together occupy a settlement at any one period), neighborhoods (a group of households within a well-defined area), and households (artifact patterns believed to pertain to activities in a house).

The tighter the provenience control, the more focused the research questions can be. On some sites, where artifact counts are low, it is often advisable to piece plot each find. That is, rather than merely recording the excavation unit and level in which the artifact is located, the precise grid coordinates for the find are determined—a tedious field task that often pays off when the final maps are made by revealing activity areas. An activity area is an archaeological patterning of artifacts indicating a specific activity took place (lithic reduction, food preparation, etc.). With careful plotting, it is even possible to determine where the lithic craftsman sat as he manufactured tools by noting a clear area surrounded by a scattering of chert debitage.

ARTIFACT DENSITY

The differences in density of the distribution of artifacts across the surface of a site can also reveal the location of activity areas within a site. Subsurface testing (usually performed with shovels or power auger) is routinely used in CRM surveys to find sites and define their boundaries, usually on the basis of the presence or absence of artifacts within a shovel test. The plotting of the particular types of artifact found in the individual tests can reveal specific areas of interest within a site.

The site of Puerto Real was located when a collection of sixteenth-century Spanish artifacts was brought to the attention of archaeologists at the University of Florida. One of their first activities was to dig test pits at ten-meter intervals across the area to determine the site's boundaries. The site limits were determined, but much more was accomplished with the recovered data.

The analysis of these data was undertaken with the assistance of SYMAP, an early graphic program run on the university mainframe computer. The output from this program used different symbols and shading to denote concentrations of a specified artifact category. By noting the discrete clusters of stone and brick building debris, it was possible to define the location of individual structures within the site, though no above-ground traces remained (figure 10.4). By correlating these clusters of masonry with the types of ceramics found, archaeologists could also make predictions about the status of the occupants of these structures. Some of these predictions were later borne out in more extensive excavations (Williams 1986; Ewen 1991; Deagan 1995).

One of the most visually expressive ways to examine the distribution of the density of artifacts across a site is with a contour map. Contour maps are usually associated with topographic maps, which are simply graphic representations of landforms as indicated by differences in elevation. U.S. Geological Survey quadrangle maps have topographic information on them represented by contour lines.

Contour lines are a cartographic device used to indicate gradations in value. To prepare them, the x- and y-coordinates are plotted as locational data in two-dimensional space. The z-coordinate is the third value at a particular point. Elevation isn't the only information that can be portrayed by contour lines. When examining artifact distribution, it can also represent the quantity of artifacts (or a particular type of artifact) at a particular spot on a site. The density and distribution of artifacts across a site can be displayed graphically as a contour map.

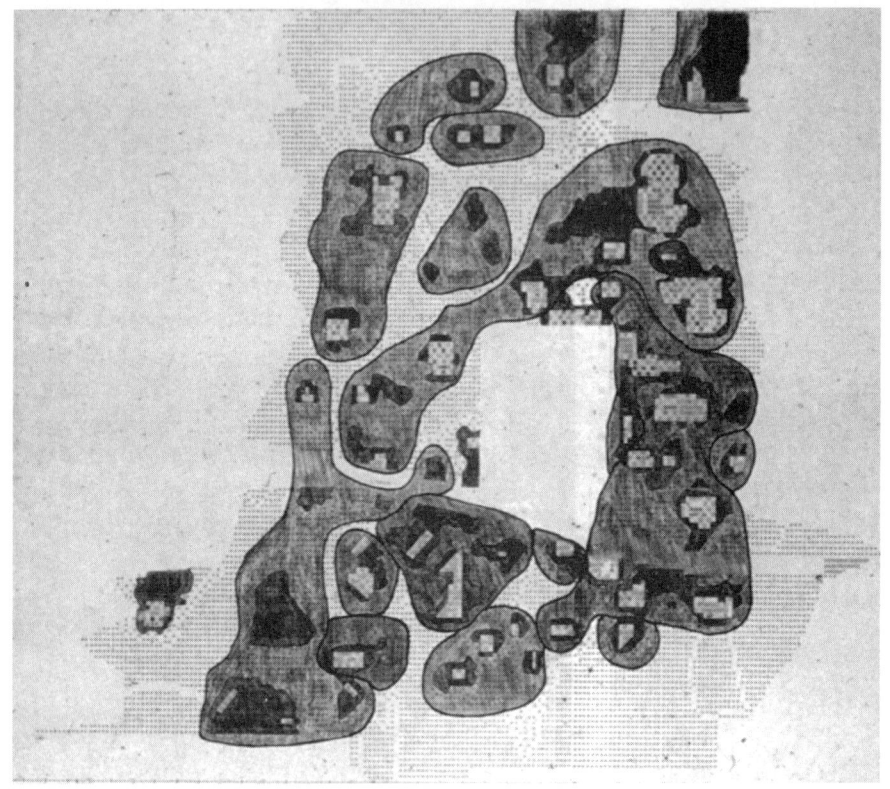

Figure 10.4 SYMAP of architectural debris at Puerto Real. (From Ewen 1991.)

The production of artifact density/distribution maps has been greatly facilitated by the increased use of personal computers. For example, Surfer (Golden Software) is an easy-to-use program that runs on desktop computers. It can quickly generate contour maps or a variety of other graphic representations (e.g., three-dimensional surface and chloropleth maps) for depicting density/distribution.

At the Governor Martin site, it was possible to produce a contour map of the density of burned clay daub with a map for the site. This was accomplished by totaling the weight of the daub found in each excavation unit. This became the z value, and the coordinates for the center of the excavation unit were used as the x and y values. The resulting map (figure 10.5) displays the distribution of daub and could be correlated with patterns of postmolds, suggesting that the structures of the Contact period Apalachee were of wattle and daub construction.

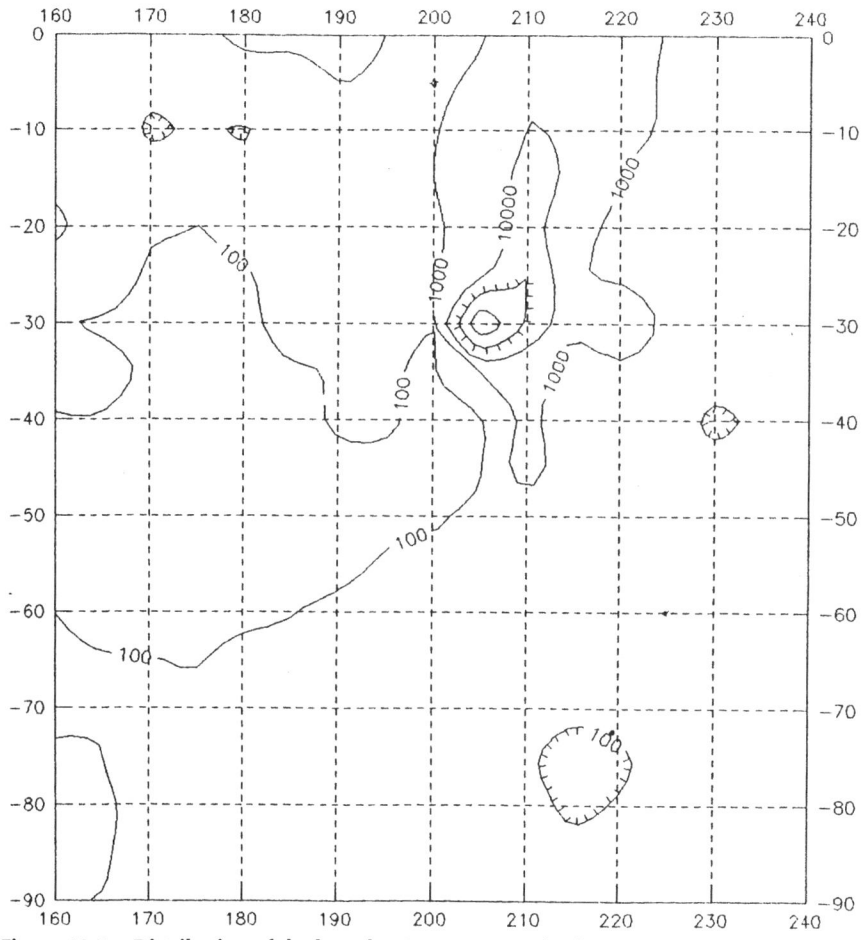

Figure 10.5 Distribution of daub at the Governor Martin site. (From Ewen 1998.)

REPLICATION

I often ask my students to question the epistemological assumptions of archaeology. How do we know what we think we know? Sharp-edged flint flakes are often called *scrapers,* but how do we know that that was their true purpose?

There are really only a few basic ways of inferring the purpose of archaeological features and artifacts. The first is the direct historical approach. Ethnographic accounts record that the earliest historic groups used an artifact in a specific way, so we infer that their prehistoric predecessors in the same area behaved in a like manner. This would constitute a sort of cultural uniformitarianism, if you will. Another way of knowing, ethnographic analogy, is bit more removed.

Ethnographic analogy posits that a prehistoric group probably behaved in a similar way to a known historic group possessing the same level of technology. This is how archaeologists can infer that Paleoindian peoples lived in egalitarian bands with few formal controls, by looking at modern groups of similar technical abilities and examining their lifestyles.

Archaeologists have analyzed modern tribal peoples in an effort to better understand the prehistoric cultures they study. This specialized branch of archaeology is called *ethnoarchaeology*. Archaeologists have actually excavated campsites of modern hunter-gatherers to see how observed behavior translates into preserved behavior. It has been possible to discover how certain tools could have been made, used, and discarded using an ethnoarchaeological approach.

A related approach to knowledge is the use of replication experiments in archaeology. This involves experimentation that attempts to recreate past conditions in an effort to gain an emic perspective on the past. It should be noted, however, that replicative experiments do not necessarily prove that an object was made or functioned in a particular way, merely that it was possible for it to have done so.

One of the questions that experimental archaeology attempts to answer is how certain artifacts were made or the effects of certain techniques on the manufacturing practice. For example, flint knapping is probably the most popular form of experimental archaeology. Individuals (e.g., Crabtree 1972) have experimented with different techniques of flake removal and shaping to gain greater insight into how lithic tools were made. Such replication experiments have demonstrated that heating chert makes it easier to work, thus providing a possible explanation for the presence of heat-treated chert on prehistoric sites.

Similarly, experiments have been conducted to determine the function of various artifacts. That is, it is one thing to say that flint hand cleavers could have been used to butcher mastodons during the Paleolithic, it is quite another to do it. Archaeologist George Frison actually butchered African elephants using only stone tools that he had made himself (Frison 1989). Did Paleoindian mastodon hunters cut up their prey in the same way? We may never know, but replication studies at least indicate what was possible and so eliminate certain blind alleys and allow the archaeologist to pursue answers in more productive areas.

11

SPECIALIZED ANALYSES

Many types of analyses require expensive equipment, special expertise, or other resources that are simply beyond the reach of most archaeological laboratories. Quite frankly, it is unrealistic for an independent archaeological contractor or small college laboratory to acquire a particle accelerator or amass a comparative collection of the mammals of North America when these resources exist elsewhere. It is a practical necessity to subcontract certain specialized analyses when they are required by your project. This section discusses some of these types of analyses and when they might be applicable.

CHRONOMETRIC DATING

One of the most commonly subcontracted specialized analyses is the chronometric dating of artifacts and ecofacts. There are several different methods for obtaining an absolute date for your sample, depending on the composition of your sample, the conditions of its deposition and recovery, and even the region of the world it is from. Each method has its own idiosyncrasies and limitations (table 11.1).

A couple of general rules of thumb apply to all methods of chronometric dating. Keep the handling of the specimen to a minimum. This lessens the chance of its contamination or alteration. Obtain as many samples as possible and submit multiple specimens for dating. If possible, use a couple of different methods. Dates obtained can vary widely, even within the same feature, so multiple specimens can help rule out an aberrant date.

Table 11.1. Chronometric Dating Methods

Method	Pros	Cons
Dendrochronology	Extremely accurate.	Samples are relatively rare on sites. Sequence length varies from region to region.
Radiocarbon	Works on organic material. Reasonably accurate when calibrated; relatively inexpensive.	Must destroy much of the artifact to date it. Large confidence interval on some materials.
Accelerated mass spectrometry (AMS)	More accurate than conventional radiocarbon. Much smaller sample required.	More expensive than conventional radiocarbon.
Oxidized carbon ration (OCR)	Works on soil humates. Can serve as cross-reference for radiocarbon.	Few labs available. Date range not known at this time.
Thermoluminescence	Works on nonorganic material such as pottery and stone	Large confidence interval. Few commercial labs.
Obsidian hydration	Works on volcanic glass	Hydration varies from region to region and through time. Wide confidence interval. Few commercial labs.

DENDROCHRONOLOGY

Dendrochronology, or tree ring dating, is the oldest and most reliable absolute dating method available to the archaeologist. It was the only absolute method available to the archaeologist prior to 1950 and is used today to calibrate the more recently developed radiometric methods (discussed later). Dendrochronology was originally developed in the southwestern United States, where wood preservation was optimal, but recent work has established tree ring sequences in other parts of the country and the world (see the Laboratory of Tree-Ring Research at the University of Arizona at www.ltrr.arizona.edu, and the University of Arkansas Tree-Ring Laboratory at www.uark.edu/misc/dendro/). In historical archaeology, dendrochronology has been applied to standing wooden architecture to validate long-cherished dates for poorly documented historic structures. Often, as in the case of the Newbold-White house, formerly thought to be the oldest house in North Carolina, the structure turns out to be decades younger than previously thought!

The dendrochronology laboratory at the University of Arkansas recently determined that there was a prolonged drought in the Lower Mississippi Valley during the sixteenth century that may have had more to do with the depopulation of the area than any diseases introduced by the de Soto expedition (Burnett and Murray 1993). They also were able to determine that a drought existed along the mid-Atlantic coast at about the same time, which may also have been a contributing factor to the demise of the Lost Colony (Stahle et al. 1998).

RADIOCARBON

The discovery of the radiocarbon dating technique by Willard Libby in 1949 revolutionized archaeology. Prior to its discovery, the only method archaeologists had for attaching a calendar date to a site was dendrochronology, and its application was severely limited. Early use of radiocarbon dating was expensive, demanded large samples, and produced confidence ranges so broad as to be almost useless. However, lacking any alternative and seeing the tremendous potential of the method, archaeologists pursued the technique, refining it until it could reliably produce precise dates from small samples at a reasonable price (see ßeta Analytic at www.win.net/~analytic/ or Waikato Lab, New Zealand, at www.radiocarbondating.com).

Radiocarbon dating is a way of measuring how long an object has been dead. So any organic material can be dated, although some kinds (i.e., wood charcoal) produce better dates than others (shell is often hard to date). The procedure measures the decay rate of radioactive C14 into stable nitrogen. Organisms intake C14 and C12 while alive. When dead, the C14 starts to break down at a known rate. By measuring the difference between C14 and C12, you can determine how long the organism has been dead (for a more in-depth discussion of this procedure see www.radiocarbon.org/info/index.html).

It was discovered that when radiocarbon samples were checked against tree ring samples, the radiocarbon samples were in error even beyond the statistical confidence interval. It was determined the carbon level in the Earth's atmosphere had not remained constant over time, a key assumption in radiocarbon dating. Rather than abandon this dating method, archaeologists used the method that detected the error, dendrochronology, to correct it.

Thus, radiocarbon dates can be calibrated with a dendrochronologically derived scale to produce calendar dates as old as 60,000 B.P. These calibration programs are freely available over the Internet (e.g. OxCal at www.rlaha.ox.ac.uk/orau/index.htm or CALIB at depts.washington.edu/qil/dloadcalib/) and are also available in tabular form through the journal *Radiocarbon*.

One of the chief drawbacks of conventional radiocarbon dating is that it requires that you destroy a fairly large portion of your sample. This is not a problem when dealing with charcoal from a hearth feature. However, when trying to date the parchment from the Dead Sea Scrolls or a bone fragment from Kennewick Man, every gram counts. Accelerator mass spectrometry dating is a method of radiocarbon dating that actually counts the C14 atoms. This technique requires much smaller samples (<0.01 grams) and produces more accurate results. It costs more money but is usually worth the extra expense, and in some cases, it is the only acceptable alternative available.

The latest twist on the radiocarbon technique is the oxidizable carbon ratio (OCR) dating. Based on the chemical analysis of charcoal within known environmental contexts, the OCR technique dates charcoal found in soil (Frink 1994:17). This technique works only on archaeological charcoal in a biochemically active, aerobic soil. This technique has been primarily used to reaffirm radiocarbon dates spanning the last eight thousand years. Its effective date range has not yet been determined, but the procedure is a promising addition to the archaeologist's dating repertoire.

THERMOLUMINESCENCE (TL)

TL dating is a widely known but not yet widely used method by North American archaeologists. This technique measures the amount of light energy released by a fired object (e.g., baked clay artifact) when it is heated rapidly so that it gives an indication of the time elapsed since the object was first fired (see Aitken 1985 for a more in-depth discussion of how this procedure works). For example, it can be used to estimate the time that has passed since a pot was made to when it was subjected to the TL process. It has the advantage of working on a nonorganic material (e.g., ceramic and even chert) that is common on most archaeological sites and can date objects as old as 8000 B.P.

Unfortunately, TL has been plagued with accuracy problems so that it has, in the past, only reliably given a ballpark estimate of an object's age. For this reason, it has chiefly been used by museums to detect modern forgeries of ancient artifacts. Refinements of this technique continue and recent work on the prehistoric ceramics of the North Carolina coastal plain (Herbert 1997) show a good correlation between TL and radiocarbon dates.

Another, more pragmatic problem for the archaeologist is finding a lab that will perform the test. There are few commercial labs (e.g., Quaternary TL Surveys at www.users.globalnet.co.uk/~qtls/ and TOSL Research laboratory at www.dal.ca/~digs/t-intro.htm), like those that conduct radiocarbon dating. Many labs that do TL dating are usually dedicated to someone else's research and will only process other samples when convenient to their schedules.

OBSIDIAN HYDRATION

Another artifact dating technique that suffers from reliability problems is obsidian hydration. The basic premise here is that a fresh break on a piece of obsidian will absorb water from the atmosphere at a constant rate. The broken edge of an obsidian artifact can be examined under a microscope and the thickness of the hydration layer measured. Thus, it should be possible to be able to calculate the time that has passed since the last flakes were removed from the artifact.

There are a couple of limitations to obsidian hydration dating. First, the hydration rate varies with the composition of the obsidian. Second, the hydration rate also varies with the climate in different parts of the world, prompting one archaeologist (Ridings 1996) to question, "Where in the world *does* obsidian hydration dating work?" Still, the procedure is straightforward and inexpensive. So, if the limitations are kept in mind, obsidian hydration can be a useful tool for the archaeologist, especially when used in conjunction with other dating techniques.

FAUNAL

Faunal analysis, or the study of animal bones from archaeological sites is one of the most commonly performed types of specialized analysis. This is covered more in depth in another volume (see Toolkit, volume 5). However, it is important for the archaeologist to understand the types of questions that faunal analysis can answer and how this can be integrated with the artifact analysis.

It is a fairly common practice for contract archaeologists to send the faunal remains recovered on their projects off to a zooarchaeologist. The expectation of many archaeologists is that they will get a species list back from the specialist and maybe a comment or two about what the inhabitants of the site were eating. Therefore, it often comes as something of a shock when the zooarchaeologist responds that they need better provenience information, including the artifacts associated with each provenience and the archaeologist's preliminary interpretations of the site.

The context in which the bone was found is incredibly important to its interpretation. Whether the faunal remains were in a sealed trash pit, a hearth, accompanying human remains, or recovered from a coprolite affects how these bones will be viewed by the zooarchaeologist. Once the archaeologist becomes aware of the wealth of information faunal analysis can provide, the need for these data becomes readily apparent.

One of the basic interpretations that zooarchaeology can provide is dietary reconstruction. This transcends merely what was being eaten and includes the diversity and equitability (the variety and proportion of species being utilized as opposed to what was available), wild versus domestic species exploitation, the variability of the diet on an inter- and intrasite basis, and on historic sites, locally raised animals versus meat imported from out of the region. Knowing the animals being exploited by the sites' inhabitants can aid in the artifact analysis by correlating food items with their corresponding food preparation artifacts.

Zooarchaeological analysis is also helpful with the environmental reconstruction. What was the environment like around the site, how did it change through time, and what species were available for exploitation? The season of occupation of the site can also be determined by the species present in the assemblage. Some animal species (migratory waterfowl, anadromous fish, etc.) are only available in certain areas at certain times. This can be correlated with the artifact assemblage to give an idea of site function. For example, a North American site with projectile points, lithic debris, and Canada goose remains, might be interpreted as a fall hunting camp.

The domestication of certain species is revealed through changes in animal morphology (i.e., physical size, horn shape, etc.). The age and sex of the animal might reveal culling activity or breeding strategy. Again, correlated with the artifact assemblage (sheep shears, harness tack) a full picture of the animal exploitation activities at a site can be derived.

A fur trading post excavated in Wisconsin (Ewen 1986) presented an interpretive problem that was solved only after both the faunal and artifact assemblages had been analyzed and correlated with each other. Three cabins within the trading post stockade were known to have housed a high-status company partner and his family, a middle-status clerk, and several low-status employees. The initial matching of cabin with occupant had been on the basis of a cursory analysis of the artifacts associated with the cabins. The faunal analysis yielded results that were not in accord with the assumed status of the cabins' occupants. This prompted a reanalysis of the artifact and architectural data which suggested a different assignation for the cabin occupants that was more in line with the faunal data.

FLORAL

Many of the same questions (e.g., dietary reconstruction, seasonality, environmental reconstruction, etc.) that are addressed during zooarchaeological analysis can be examined with floral analysis as well. Floral analyses have more subspecialties. The specialties within ethnobotany include palynology, phytolith analysis, and wood/charcoal analysis.

Palynology, or pollen analysis, while not uncommon, is not widely used in archaeology because of the difficulty involved with collecting samples and the expense involved with the analysis. It has been used to give indications of plant domestication, though the results are usually suspect without corroborating archaeological or bioarchaeological evidence. However, it can be a very sensitive environmental indicator.

Pollen has been used to reconstruct past environment for archaeological sites in the arid west, where preservation is optimal. However, in the east, where preservation is poor and human disturbance is great, pollen analysis is used for other purposes such as detecting changes in land use over time. An analysis of an area known as Scottow's dock in Boston indicated that "ethnobotanical [pollen] data in urban situations are really waste-disposal or produce-loss patterns and reflect land use more accurately than they do diet" (Kelso and Beaudry 1990:78). In this case, the pollen record from the seventeenth through nineteenth centuries corroborates the documentary record as it traced lot use from commercial periphery (shipping dock) to residential back lot to use as a row of commercial shops.

Phytoliths are siliceous particles found with the stems and leaves of plants. They have a characteristic shape for each species of plant that allows the analyst to identify their presence even after the rest of the plant has long since decayed. Though the potential for phytoliths to inform archaeologists is widely recognized, this type of analysis is not widely used owing primarily to the intensive labor required in recovering and processing a sample and also the scarcity of specialists performing this service.

However, at Harper's Ferry, West Virginia, phytolith analysis was able to provide insight into landscaping activities at a private residence over time. This study (Rovner 1994) contradicted expectations of a lack of concern with yard appearance in an industrial community. Instead, it revealed an individual's kept yard that included flower and vegetable gardens. An examination of the documentary record shows that this period coincides with the use of the site as a boarding house, where a kitchen garden would have been useful, but not recorded.

Wood/charcoal analysis can have a direct bearing on artifact analysis. This type of analysis can determine the type of wood exploited for tool handles, building, or other purposes. An interesting project that availed itself of specialized charcoal analysis was the Hardman site in southwestern Arkansas (Early 1993). The artifact assemblage indicated that this site's inhabitants were involved in salt processing from the nearby saline springs. Water from these springs was boiled off in ceramic vessels, leaving the salt behind. The charcoal analysis indicated that hardwoods were the preferred heating material. Examination of documents describing the environment at the time of early settlement indicates that the forest around the Hardman site was primarily pine, while further away hardwoods predominated. Since pines are a primary reforestation species it was possible to get a good idea of the area being exploited for firewood by the prehistoric salt processors.

PETROGRAPHY

Petrographic analyses refer to those studies involving the identification of rocks and minerals. An examination of various lithic artifacts and ceramics can reveal such information as the source of the material from which the artifact is made, technological attributes pertaining to the manufacture of the artifact, and unusual inclusions that might characterize the composition of a particular artifact. This topic is included in the specialized analyses section in that petrographic analyses are often performed by geologists who have the expertise and equipment available for such analysis.

Petrographic analysis of ceramics and lithics goes beyond the simple macroscopic that an archaeologist routinely performs. For example, an archaeologist with a hand lens can usually determine the tempering agent of a ceramic without having to send the sherd to a specialist. A petrographic analysis would involve making a thin section of the sherd and viewing it under a microscope to identify and assess the proportions of the mineral inclusions in the clay. If samples of local clay sources were available, it would be possible to match the sherd to its source.

A more sophisticated and precise approach involves the use of a particle accelerator to determine the different types and proportions of elements in a specimen. PIXE analysis requires only a very small sample for a highly accurate reading that essentially eliminates observer bias. The problem with PIXE analysis, not unlike other special analyses requiring expensive equipment and trained personnel, is trying to find a lab that will perform the service.

The utility of petrographic analysis is not restricted to prehistoric applications. For example, it was possible using petrographic analysis to determine that the olive jar sherds found at the suspected site of Hernando de Soto's winter encampment in Tallahassee, Florida (Ewen and Hann 1998:76), originated in Spain rather than from a New World pottery. This helped strengthen the association of the site with the de Soto expedition, since olive jars commonly found on the later seventeenth-century Florida mission sites were primarily manufactured at New World potteries.

Lithic analysis can derive similar benefits from petrographic analysis. An archaeologist with a little training can usually determine the type of lithic material comprising the various tools in the lithic assemblage. The same archaeologist may even be able to tell the variety of stone (e.g., Mill Creek chert). However, this will not pin down a specific locality within a geological formation. Thus, to trace a tool to a specific quarry site, advanced petrographic analysis may be necessary.

I was once confronted with an interesting artifact that required the assistance of a geologist for positive identification. A local relic collector had found a large biface that appeared to be made out of glass. Curiously, the rest of the materials he had picked up in the area were chert pieces dating to the Late Archaic period. I was skeptical of the authenticity of the artifact since glass is an easily knapped material, and relic hunters often try to dupe the local professional archaeologist with recently made artifacts. Examination of the artifact by a geologist on the faculty, however, revealed the biface to be made from a remarkably pure quartz crystal (quartz is common to eastern North Carolina).

BIOARCHAEOLOGY

The final type of specialized analysis to be discussed here has the most to tell us about the people who made the artifacts we have examined thus far. Though artifacts are indirectly involved in this type of analysis, human bone can provide complementary data concerning the past biological and social environment. Bioarchaeology is the study of human skeletal remains in order to reconstruct past lifeways. The correlation of interpretations of the artifact assemblage with bioarchaeological data can result in new insights or correct previous assumptions.

Interpretations of past social behavior that one can gain from the analysis of human remains come from several analytical realms. Social stratification and belief systems can be gleaned from a study of mortuary patterns. By examining bone modifications, both naturally and culturally inflicted, it is possible to gain insight on kinds of injuries sustained, including repetitive motion and trauma. The bones can also reveal information concerning disease and nutrition. Broadening the view to many specimens can provide demographic information for a region or time period. But how does this relate to artifact analysis?

Bioarchaeological analysis, like other types of specialized analyses, complements the data recovered from the analysis of the artifact assemblage. For example, the patterning of artifacts from a slave cemetery in New York City might offer clues regarding the retention of African cultural practices (Blakey 1998). The accompanying bioarchaeological analysis reveals the general health of the individuals, the types of work they performed, and even how well they were eating.

Bioarchaeology has much to tell us about past populations, especially with the advent of such techniques as isotope and DNA analyses. These types of analyses are answering questions concerning when domesticates such as corn became a major part of the diet and degrees of relationship between past populations. The degree of relationship of excavated human remains to present Native American populations is a dominant theme in much of archaeology today.

The passage of the Native American Graves Protection and Repatriation Act (NAGPRA) in 1990 not only regulated the excavation of human remains on public lands but also compelled those repositories using federal funds or permits (virtually all of them) to inventory items they had in their care. The items to be accounted for include human remains, artifacts found in graves, sacred artifacts, and items of cultural patrimony. Following the inventory, the descendant populations are to be contacted and offered the opportunity to reclaim the skeletons and artifacts. Needless to say, this legislation has had a great impact on the field, not the least of which is that it has forced archaeologists to pay more attention to the care and storage of their collections.

12
CURATION

One of the hidden shames of archaeology is the care (or lack thereof) accorded to artifacts after the analysis has been completed. While this subject is covered in more detail in volume 6 in this series, a few words are appropriate at this point, since the attention paid to the curation of a collection can have a direct bearing on any future analyses to be performed on the artifact collection.

Any archaeologist will testify to the importance of keeping good field notes and other records. Often the project archaeologist must rush to salvage a threatened site before conducting the analysis of the materials recovered from the last project. It is not unheard of for archaeologists at academic institutions to put off writing up the material recovered during their field schools for many seasons. In some archaeological consulting firms, the person who digs the site is not necessarily the person who writes the report.

It is good procedure to operate under the assumption that someone else will have to write up your fieldwork or that someone else will want to reanalyze your material. This means that you must ensure that all your artifacts and field and lab records are properly curated so that reanalysis is possible. There is no substitute for having dug the site yourself, but with the artifacts adequately documented and stored, the person that follows you should be able to reconstruct your fieldwork and draw valid conclusions from the data.

CHAPTER 12

ARTIFACTS

Artifacts should be cleaned and labeled as to their provenience and stored in plastic zip-top bags. Paper bags deteriorate over time and have a tendency to split open when pulled for reanalysis. The provenience information should be printed on a card and stored inside the plastic bag rather than written on the bag itself with a permanent marker. The markers are *not* permanent, and the ink flakes off over time.

The purpose of packaging is to preserve the artifacts for future research. Bags of artifacts should be stored in acid-free cardboard or plastic containers. Artifacts susceptible to moisture damage should have packets of silica gel included in the storage container. If the collection has been sorted, then there should be documentation for the basis of the sorting included in the artifact box. Don't overpack the boxes—this damages the artifacts and makes the boxes difficult to handle.

Care should also be extended to preventing vermin from destroying artifacts or the bags that hold them. Returning from Haiti with a season's worth of artifacts, we were compelled to seal them in a room and fumigate them before they would be admitted to the United States. This was a wise move on the part of the customs officials as we found that an amazing collection of insects had infiltrated the cardboard boxes during shipping as well as a small colony of bats! A colleague related that mice infiltrated the plastic box and bags in which his artifacts were packaged in order to get at the paper provenience tags. These cautionary tales apply to the accompanying archaeological project documentation as well.

PROJECT RECORDS

All project records should be assembled and stored together. The paranoid collections manager (and this is a good trait!) may want to make duplicate records and store them in a separate location. The project field records include the field notes, photographs and slides, provenience catalog, and excavation forms for units and features. The lab records will include documentation concerning the level of processing, artifact inventory sheets, and any comments concerning the conservation needs of the artifacts. Reference as to the location of these documents should be attached to the artifact boxes and vice versa. Artifacts without their documentation have lost much of their archaeological significance.

PART IV
FINAL THOUGHTS

This section reiterates the purpose of the book—to get the readers to stop and consider why they are employing different analytical techniques—and then speculates on future trends in artifact analysis. Clearly technology, especially evolving computer applications, will play an ever-increasing role in artifact analysis. The main benefit of this technology is the ability it gives the archaeologist to ask increasingly complex questions of increasingly larger data sets. This is particularly important for the contract archaeologist working under a deadline. Another benefit of this increasingly powerful (and affordable) technology is the ability to analyze and record an artifact through digital imagery. This allows the archaeologist to gain more information with less handling of fragile artifacts. Enhanced information exchange via the Internet will allow colleagues to consult and publish in a timely manner.

This book concludes on the note that you will, in all probability, not be the last one to look at the data you've unearthed. We may not be able to predict all the new questions that will be asked of the data, but we can be assured that new ones will be asked and our knowledge of the past will continue to advance.

13
CONCLUSION

This text is not intended to be a blueprint of the "right" way to conduct artifact analysis. The right way is the way that addresses the questions posed in the archaeological research design. Different questions require different types of analyses. Thus, to provide a complete "how to" book, perhaps in the Artifact Analysis for Dummies series, would require more pages than *Windows XP for Dummies* and still would not cover all contingencies. So what good is this volume?

The purpose of this volume, to reiterate the introduction, is to understand the goals of artifact analysis. It is only after those goals are defined, that the archaeologist can begin to formulate a plan to achieve them. The basic steps of artifact analysis—identification, classification, quantification, manipulation, and interpretation—have all been discussed and examples of how some archaeologists have approached these procedures were discussed.

The examples of artifact analyses provided in this text were just that, examples of analytic procedures, meant to give the reader an idea of the techniques that some archaeologists have employed. As I caution my graduate students when they go to work for another archaeologist, "If you are asked to perform a task differently than the way you were taught, don't tell your supervisors that they are doing it wrong! Learn the new way, and then, when you run your own project, you can select the way that works best for you." So it is with this text. If it gets the reader to stop and consider why they are performing different analytical procedures, then it will have served its purpose.

CHAPTER 13

FUTURE DIRECTIONS

This chapter is fraught with peril. What are the future directions in artifact analysis—same old same old, or bold new innovations? Probably both statements are true to a certain extent. Some general predictions are easy. Computers will continue to play an ever-increasing role in the archaeological laboratory. The specific predictions are harder. Which computer platform will prevail? Which computer programs will be the standards? When I received my doctorate, WordStar, Visicalc, and dBase were the standards for word processing, spreadsheets, and database applications. At this writing, those programs either have been surpassed by competitors or no longer even exist.

Predicting what questions artifact analysis will be addressing is difficult as well. Archaeology is about due for another minor paradigm shift. The postprocessual paradigm has been around for nearly twenty years, which is about the life span of a theoretical cycle. Since the artifact analysis is geared to the kinds of questions being asked, the new millennium may be ushering in new theoretical approaches as well.

MORE POWERFUL COMPUTER APPLICATIONS

There is no doubt that the increased use of more powerful computers and software applications will be the rule in various types of archaeological analysis. Database queries that took an hour or more to run ten years ago take a matter of minutes or even seconds today. However, time savings is not the big payoff from this new technology. The principal benefit is the ability to address more and more complex questions of larger and larger data sets. Questions that would not even have been considered years ago because of the time requirements can now be asked by even novice archaeologists.

The wonders of computers have been extolled throughout this text, and, while they are a great help to the archaeologist, it should always be remembered that they are only tools. The computer is only as good as the archaeologist that uses it. If the archaeologist does not understand the basic principles of statistics, then even a Cray supercomputer will not be of much assistance. Computers will not do your analysis for you, they *will* help you analyze artifacts more efficiently by performing the tedious calculations or sorting through the reams of data required by your analyses. Better computers simply allow you to concentrate on asking better questions.

LARGE DATA SETS

Probably the biggest impact of the computer for the archaeologist is the enhanced access to large data sets. As the storage capacity of desktop computers increases, so does the ability to do complex comparisons of data within large sites and even between sites. Once all the data from all the sites that a lab has excavated have been entered into a computerized database, anyone can find out every site that has yielded a particular type of pottery. It is not necessary to go through every box of artifacts or track down the individual archaeologists that worked on all those sites and then rely on their memories.

NETWORKED DATABASES

Why should your queries be restricted to your lab? Networking technology makes it possible for computers to link to one another either directly or over the Internet. The federal government and many state agencies are assembling archaeological data and making it available online. The National Archaeological Database (NADB), mentioned earlier, that is maintained for the National Park Service aims to eventually have all archaeological site data for the country in a single huge database that can be accessed over the Internet.

Two big hurdles hinder the compilation of these huge archaeological databases, and neither of them involves computer hardware. The first is one that has already been discussed and that is trying to forge some sort of compatibility between existing databases. Finding a database structure that meets every archaeologist's needs may prove as elusive as the Lost Colony.

The second problem stems from the first. Even if a common data structure can be agreed on, if existing databases cannot be made to be compatible, then we are looking at a massive data reentry project. I have no doubt that this will take place in the future. Like the massive NAGPRA inventory effort, it will slowly happen, laboratory by laboratory.

THE INTERNET

Access to databases is not the only role that the Internet will play in artifact analysis. The information highway also allows archaeologists to consult with each other. Rather than waiting for a conference to query your colleagues, you can contact them electronically, send

CHAPTER 13

them text and graphics, and have a reply the same day. At least the technology exists to do so; whether your colleagues are quick to respond is an uncontrolled variable.

DISCUSSION GROUPS

There are a number of email discussion groups (table 13.1) to which an archaeologist can subscribe. Most of these listserves have a theme

Table 13.1. Archaeology on the Internet

Discussion Groups		
ArchNet Forum	General discussion	http://archnet.org/forum/view.tcl?message_id=28
ACRA-L	CRM issues	Send email message subscribe acra-l "your name" to Listproc@listproc.nonprofit.net
ARCH-L	General discussion	Send email message subscribe arch-l "your name" to Listserve@listserve.tamu.edu
Gateways		
ArchNet		http://archnet.org/lobby.tcl
Archaeology on the Net		www.serve.com/archaeology/ring/index.html
About.com's Archaeology		http://archaeology.miningco.com/science/archaeology/index.htm
The Archaeology Channel		http://www.archaeologychannel.org
Anthropology Resources on the Internet		http:www.aaanet.org/resinet.htm
E-Journals		
Internet Archaeology		http://interarch.ac.uk/
Archaeology Magazine		http://www.archaeology.org/
Journal of Archaeological Research		http://www.wkap.nl/jrnltoc.htm/1059-0161
E-tiquity		http://e-tiquity.saa.org
Archaeological Sites		
Queen Anne's Revenge		http://www.ah.dcr.state.nc.us/qar/default.htm
Cahokia Mounds		http://medicine.wustl.edu/~mckinney/cahokia/cahokia.html
Jamestown		http://www.apva.org
Alternative Archaeology		
Quest for the Lost Civilization		http://members.aol.com/thelogo/quest
Earth/matriX		http://www.earthmatrix.com/home.html

(e.g., historical archaeology, classical archaeology, etc.) around which threaded discussions take place. If an archaeologist has a particularly hard to identify artifact or wants to know whether anyone else has ever recovered a certain type of ceramic, this information can be gathered by posting a request on the list and waiting for the responses. These lists can also provide a forum for the discussion of various theoretical approaches, analytical techniques, and the occasional flame war (i.e., an on-line heated exchange) between individuals with different opinions.

WORLD WIDE WEB

A component of the Internet is the World Wide Web. The web is especially useful for archaeologists in that it is a graphic environment. Many organizations (see table 13.1) have home pages that provide a great deal information about their activities and membership. Some organizations, such as the Society for American Archaeology, have gone so far as to discontinue printing their newsletter and instead post it electronically on their home page. Most archaeological journals either are publishing on the web in some form or have plans to do so.

Individual projects (table 13.1) have their own web pages where you can see pictures of their excavations and the artifacts that have been recovered. There are usually e-mail links to the project staff that allow the web page visitors to contact the project archaeologists directly with their questions.

Increased use of the web by archaeologists in the future seems assured. Personal webpages are becoming common and several e-zines or journals being published on the web are already happening. This has caused concern among many archaeologists about the quality of the material being posted on the web. The fear is that many people will potentially make uncritical use of unreliable data. This is a valid concern. Many class reports I receive have clearly been downloaded from the web, thus replacing the *Encyclopaedia Britannica* as the chief source of information for incoming freshmen. The only response at this point is caveat emptor: Let the user beware! Treat the web resource the same way as an unrefereed site report or a book published by a vanity press.

DIGITAL RENDERING OF ARTIFACTS

Archaeology is benefiting (as it always has) from innovations developed for other disciplines. One way that this pertains to the physical analysis of artifacts is by using computers to digitally reconstruct an artifact from its fragments.

RESIN CASTING

Archaeologists have long taken advantage of nearby hospitals to X-ray important finds that are heavily corroded or covered with incrustations of sediment. At the Governor Martin site in Tallahassee, a corroded mass of iron oxide was X-rayed to reveal a patch of chain mail armor. This allowed the archaeologists to confirm the function of the myriad single links found elsewhere on the site (Ewen 1989:38). Underwater archaeologists routinely X-ray coral-encrusted objects found on shipwreck sites since these often contain artifacts within the concretions.

New technologies have taken the X-ray technique to the next level. An artifact can be sent through a CAT scanner that produces an image of both the artifact's surface and interior. This procedure has been used to great effect in the study of Egyptian mummies since it allows the archaeologist to examine the body without unwrapping it.

Innova International, a company that primarily deals with medical clients, has combined the use of CT scanners with an advanced resin-casting technology to produce replicas of artifacts without any damage to the artifact. It is possible to isolate different parts of the artifact and reproduce each of those parts in a different-colored resin. For example, a spear point embedded in a bone can be cast in a dark resin while the bone itself is reproduced in a clear resin. Once the rendering is complete, the artifact reproduction can be examined without further handling of the original.

DIGITAL IMAGING ANALYSIS

As previously discussed, digital imaging software such as Photoshop or AutoCAD can be used to perform many metric analyses on images of an artifact so that the physical artifact need never be handled. A three-dimensional image of the artifact can be produced, and, if one dimension is known, virtually every other dimension can be calculated or extrapolated with a speed and accuracy surpassing those obtained by most lab technicians using conventional measuring devices. These images can be retained for further study should the actual artifact be removed from the laboratory.

IMPACT OF NAGPRA

The possibility that many artifacts and human remains could be repatriated has greatly impacted the way certain analyses are and will be conducted in the future. The Native American Grave Protection and Repatriation Act, besides dealing with the disposition of archaeologically recovered human remains, specifies that artifacts identified as grave goods, ceremonial items, or items of cultural patrimony be repatriated to their cultural descendants. This has introduced a new classificatory criterion for an archaeological assemblage: potential eligibility for repatriation.

Though the thought of relinquishing control of material recovered from an archaeological project disturbs some archaeologists, NAGPRA has brought many benefits. Artifact collections are analyzed more expeditiously when the threat of their return is imminent. The Native American community has become more involved with archaeologists and their analytic techniques. This in turn has encouraged an emic perspective to be considered by even the most hardened processual archaeologist. It has also fostered a greater respect and concern for archaeological remains, and so more attention has been paid to their curation and the possibilities of future analyses.

The development of some of the noninvasive analytical techniques discussed earlier was inspired or at least welcomed by those archaeologists dealing with repatriatable remains. The ability to analyze a bone or artifact without destroying a piece of it is usually important in such cases. Also important is the ability to have an enhanced graphic representation of the artifact should the artifact itself be returned. However, archaeologists must be cautioned that some Native groups are unhappy with even representations or replicas of ceremonial items being used outside of their sacred contexts.

REANALYSIS OF ARCHIVED COLLECTIONS

The final obvious trend for future artifact analyses is the analysis of previously excavated collections. In many cases, these collections have never undergone formal analysis. Many large excavation projects were conducted during the Great Depression under the auspices of the Works Progress Administration or the Civilian Conservation Corps. Funding was expended primarily during the field phase, which often employed hundreds of workers. Unfortunately, the same level of funding was not available for the analysis of the mountains of data recovered and so many of these collections were only partially analyzed or not analyzed at all.

An example of this legacy (Lyon 1996:207–8) demonstrates both the research potential of these collections and the importance of keeping good field records. Archaeologist Chester DePratter took on the task of finishing the final reports on excavations in Chatham County, Georgia, that had taken place from 1938 to 1942. His work was hindered by the fact that most of the site excavation maps were missing, and the artifacts were in some disarray after half a century of being shuttled from repository to repository. Still, the report was finished and the data were finally available in published form rather than as anecdotal memories of a vanishing generation of archaeologists.

Even collections that have been fully analyzed and published can often benefit from reanalysis. Times, and the questions being asked, change. New analytical techniques (e.g., PIXE analysis of ceramics, DNA extraction from bone) provoke new questions or the questioning of old answers. Questions concerning the gender roles of a site's inhabitants may not have seemed important a generation ago and yet are the focus of much current research.

The archaeologist can anticipate the needs of future archaeologists by publishing their work. This includes the methods and results of their analyses so that their efforts need not be duplicated. They must also ensure the availability of these data by preserving the artifact collection with sound curatorial procedures. The exact kinds of questions that will be asked of the artifact assemblage in the future are difficult to predict, except that some will be different and will require new and innovative analytical procedures.

CONCLUSION

I started this book with an Indiana Jones analogy, so it is appropriate that I end with another. At the climax of the movie *Indiana Jones and the Last Crusade,* Dr. Jones and a Nazi rival are confronted with a typological dilemma: Which of the dozens of cups in a given assemblage is the Holy Grail? The evil scientist hypothesizes that, since Christ was a person of importance to the world, the cup would be the most ornate and costly of the collection. To test this hypothesis, he fills the cup he has chosen with water and drinks from it. If his hypothesis is correct, he could expect to receive immortality. Instead, he dies a particularly gruesome death, thus disproving his hypothesis. The Guardian of the Grail turns to Jones and dead-pans, "He chose poorly." Jones, with this additional information, revises the hypothesis, makes the correct selection, and lives to make another sequel.

Archaeologists classify their artifacts, make assumptions about their makers, and test their hypotheses all the time. Fortunately, the consequences of incorrect interpretations are less dire than this example. Quite the opposite, in fact—disproving incorrect hypotheses can be as rewarding, from a knowledge-gained standpoint, as corroborating a cherished hypothesis. Either way, you have learned something about the past, which is the whole point of archaeology, after all.

REFERENCES

Adovasio, J. M.
    1977 *Basketry Technology: A Guide to Identification and Analysis.* Aldine Manuals on Archaeology. Aldine, Chicago.

Aitken, M.
    1985 *Thermoluminescence Dating.* Academic Press, New York.

Ashmore, W., and R. J. Sharer
    1996 *Discovering Our Past,* 2d ed. Mayfield, Mountain View, California.

Barnes, James and Kathleen Cande
    1994 *Laboratory Procedures.* Sponsored Research Program. Arkansas Archeological Survey, Fayetteville.

Bauman, R.
    1968 Glass and Glasswares. In *Handbook for Historical Archaeology,* J. Cotter (ed.), vol. 1, pp. 30–36. Society for Historical Archaeology, Wyncote, Pennsylvania.

Beaudry, M. C., J. Long, H. M. Miller, F. D. Neiman, and G. W. Stone
    1983 A Vessel Typology for Early Chesapeake Ceramics: The Potomac Typology System. *Historical Archaeology* 17(1):18–43.

Binford, L. R.
    1969 A New Method of Calculating Dates from Kaolin Pipe Stem Samples. *Southeastern Archaeological Conference Newsletter* 9(1):19–21.

Blakey, Michael L.
    1998 The New York African Burial Ground Project: An Examination of Enslaved Lives, a Construction of Ancestral Ties. *Transforming Anthropology* 7(1):53–58.

Brauner, D. (ed.)
    2000   *Approaches to Material Culture Research for Historical Archaeologists*, 2d ed. Society for Historical Archaeology, California, Pennsylvania.

Brown, M. L.
    1980   *Firearms in Colonial America: The Impact of History and Technology 1492–1792*. Smithsonian Institution Press, Washington, D.C.

Burnett, B. A., and K. A. Murray
    1993   Death, Drought, and de Soto: The Bioarcheology of Depopulation. In *The Expedition of Hernando de Soto West of the Mississippi, 1541–1543*, G. Young and M. P. Hoffman (eds.), pp. 227–36. University of Arkansas Press, Fayetteville.

Busch, J.
    1981   An Introduction to the Tin Can. *Historical Archaeology* 15(1):95–104.

Cabak, M., M. Groover, and M. Inkrot
    1999   Rural Modernization during the Recent Past: Farmstead Archaeology in the Aiken Plateau. *Historical Archaeology* 33(4):19–43.

Callahan, E.
    1990   The Basics of Biface Knapping in the Eastern Fluted Point Tradition: A Manual for Flintknappers and Lithic Analysts. *Archaeology of Eastern North America* 7:1–180.

Carr, P. (ed.)
    1994   *The Organization of North American Prehistoric Chipped Stone Tool Technologies*. Archaeology Series 7. International Monographs in Prehistory, Ann Arbor, Michigan.

Clifton, R. P.
    1970   *Barbs, Prongs, Points, Prickers, and Stickers: A Complete and Illustrated Catalog of Antique Barbed Wire*. University of Oklahoma Press, Norman.

Collins, M.
    1975   Lithic Technology as a Means of Processual Inference. In *Making and Using Stone Tools*, E. Swanson (ed.). Mouton, Paris.

Crabtree, D. E.
    1972   *An Introduction to Flintworking*. Occasional Papers of the Idaho Museum of Natural History 28. Idaho Museum of Natural History, Boise.

REFERENCES

Davis, R. P. S., Jr., P. Livingood, H. T. Ward, and V. Steponitis
   1998   *Excavating Occaneechi Town: Archaeology of an Eighteenth-Century Indian Village in North Carolina.* University of North Carolina Press, Chapel Hill.

Deagan, K. A.
   1983   *Spanish St. Augustine: The Archaeology of a Creole Community.* Academic Press, New York.

Deagan, K. A.
   1987   *Artifacts of the Spanish Colonies of Florida and the Caribbean 1500–1800.* Vol. 1: Ceramics, Glassware, and Beads. Smithsonian Institution Press, Washington, D.C.

Deagan, K. A. (ed.)
   1995   *Puerto Real: The Archaeology of a Sixteenth-Century Spanish Town in Hispaniola.* University Press of Florida, Gainesville.

Deetz, J.
   1996   *In Small Things Forgotten.* Anchor Books/Doubleday, New York.

Deetz, J., and E. Dethlefsen
   1965   The Doppler Effect and Archaeology: A Consideration of the Spatial Aspects of Seriation. *Southwestern Journal of Anthropology* 21(3):196–206.

Demmy, G.
   1967   A Progress Report on: Glass Dating, An Archaeologist's Evaluation of the Concept. *Historical Archaeology* 1:49–51.

Diagram Group
   1980   *Weapons: An International Encyclopedia from 5000 BC to 2000 AD.* St. Martin's Press, New York.

Domanski, M., and J. A. Webb
   1992   Effect of Heat Treatment on Siliceous Rocks Used in Prehistoric Lithic Technology. *Journal of Field Archaeology* 19:601–14.

Dowman, E.
   1970   *Conservation in Field Archaeology.* Methuen, London.

Drennan, R. D.
   1996   *Statistics for the Archaeologist: A Common Sense Approach.* Interdisciplinary Contributions to Archaeology. Plenum Press, New York.

Early, A.
   1993   *Caddoan Saltmakers in the Ouachita Valley.* Research Series 43. Arkansas Archeological Survey, Fayetteville.

Edwards, J. P., and T. Wells
- 1994 *Historic Louisiana Nails: Aids to the Dating of Old Buildings.* Fred B. Kniffen Cultural Resources Laboratory Monograph Series 2. Louisiana State University, Baton Rouge.

Ewen, C. R.
- 1986 Fur Trade Zooarchaeology: A Study of Frontier Hierarchies. *Historical Archaeology* 20(1):15–28.
- 1989 Apalachee Winter. *Archaeology* 42(3):37–41.
- 1991 *From Spaniard to Creole.* University of Alabama Press, Tuscaloosa.
- 1995 Historic Homesteads: To Dig or Not to Dig. Paper presented at the 52d Annual Southeastern Archaeological Conference, Knoxville, Tennessee.
- 1996 Continuity and Change: De Soto and the Apalachee. *Historical Archaeology* 30(2):41–53.

Ewen, C. R., and J. H. Hann
- 1998 *Hernando de Soto among the Apalachee: The Archaeology of the First Winter Encampment.* University Press of Florida, Gainesville.

Fagan, B. M.
- 1997 *In the Beginning.* 9th ed. Addison Wesley Longman, New York.

Feder, K. L.
- 1996 *Frauds, Myths, and Mysteries.* Mayfield, Mountain View, California.

Fike, R. E.
- 1987 *The Bottle Book: A Comprehensive Guide to Historic, Embossed Medicine Bottles.* Peregrine Smith Books, Salt Lake City.

Flenniken, J.
- 1984 The Past, Present and Future of Flintknapping: An Anthropological Perspective. *Annual Review of Anthropology* 13:187–203.

Flenniken, J., and E. Garrison
- 1975 Thermally Altered Novaculite and Stone Tool Manufacturing Techniques. *Journal of Field Archaeology* 2:125–32.

Fontana, B. L.
- 1978 On the Meaning of Historic Sites Archaeology. In *Historical Archaeology: A Guide to the Substantive and Theoretical Contributions,* R. L. Schuyler (ed.), pp. 23–26. Baywood, Farmingdale, New York.

Ford, J. A.
   1954   The Type Concept Revisited. *American Anthropologist* 56(1): 42–54.
   1972   *A Quantitative Method for Deriving Cultural Chronology.* Reprint of 1962 edition ed. Museum Brief 9. Museum of Anthropology, University of Missouri, Columbia.

Frink, D. S.
   1994   The Oxidizable Carbon Ratio (OCR): A Proposed Solution to Some of the Problems Encountered with Radiocarbon Data. *North American Archaeologist* 15(1):17–29.

Frison, G.
   1989   Experimental Use of Clovis Weaponry and Tools on African Elephants. *American Antiquity* 54(4):766–84.

Hamilton, T. M.
   1968   *Early Indian Trade Guns: 1625–1775.* Contributions of the Museum of the Great Plains 3. Museum of the Great Plains, Lawton, Oklahoma.

Hanson, L. H.
   1971   Kaolin Pipe Stems—Boring in on a Fallacy. *Conference on Historic Sites Archaeology Papers* 4(1):2–15.

Harrington, J. C.
   1954   Dating Stem Fragments of Seventeenth and Eighteenth Century Clay Tobacco Pipes. *Quarterly Bulletin of the Archaeological Society of Virginia* 9(1):9–13.

Henry, D. O., and G. H. Odell (eds.)
   1989   *Alternative Approaches to Lithic Analysis.* 1. American Anthropological Association, Washington D.C.

Herbert, J. M.
   1997   *Refining Prehistoric Culture Chronology in Southern Coastal North Carolina: Pottery from the Papanow and Pond Trail Sites.* Research Laboratories of Anthropology. University of North Carolina, Chapel Hill.

Herbert, J. M.
   1999   Introduction to "Prehistoric Pottery: Series and Sequence on the Carolina Coast." *North Carolina Archaeology* 48:1–2

Hester, T., H. Shafer, and F. Feder
   1997   *Field Methods in Archaeology.* 7th ed. Mayfield, Mountain View, California.

Hodges, H. W. M. (ed.)
  1987 *In Situ Archaeological Conservation*. Getty Conservation Institute, Los Angeles.

Hume, I. N.
  1972 *Historical Archaeology*. Knopf, New York.
  1978 *A Guide to the Artifacts of Colonial America*. Knopf, New York.

Jones, O. R., and C. Sullivan
  1989 *The Parks Canada Glass Glossary for the Description of Containers, Tablewares, Flat Glass, and Closures*. Studies in Archaeology, Architecture and History. Canadian Park Service, Ottawa.

Joukowsky, M.
  1980 *A Complete Manual of Field Archaeology: Tools and Techniques of Field Work for Archaeologists*. Prentice Hall, Englewood Cliffs, New Jersey.

Karklins, K., and R. Sprague
  1980 *A Bibliography of Glass Trade Beads in North America*. South Fork Press, Moscow, Idaho.

Kelso, G. K., and M. C. Beaudry
  1990 Pollen Analysis and Urban Land Use: The Environs of Scottow's Dock in 17th, 18th, and Early 19th Century Boston. *Historical Archaeology* 24(1):61–81.

Kidd, K., and M. Kidd
  1970 A Classification System for Glass Beads for the Use of Field Archaeologists. *Occasional Papers in Archaeology and History* 1:45–89.

Leigh, D.
  1978 *First Aid for Finds*. Rescue 1, Hertford, England.

Longacre, J.
  1968 Some Aspects of Prehistoric Society in East-Central Arizona. In *New Perspectives in Archaeology*, S. Binford and L. Binford (eds.), pp. 89–102. Aldine, Chicago.

Lorrain, D.
  1968 An Archaeologist's Guide to 19th c. American Glass. *Historical Archaeology* 2:35–44.

Loy, T. H.
  1990 Prehistoric Organic Residues: Recent Advances in Identification, Dating and their Antiquity. In *Archaeometry '90*, W. Wagner and M. Pernicka (eds.), pp. 645–56. Birkhauser Verlag, Basel.

Lyon, E. A.
  1996   *A New Deal for Southeastern Archaeology.* University of Alabama Press, Tuscaloosa.

Magoon, D. T.
  1998   The Ceramics of Craftsman Robert Hay and Family: An Analysis of Middle Class Consumer-Choice in Antebellum New Bern, North Carolina. M.A. thesis, East Carolina University.

Mallowan, A. C.
  1946   *Come, Tell Me How You Live.* Dodd, Mead, New York.

Miller, G. L.
  1980   Classification and Economic Scaling of Nineteenth-Century Ceramics. *Historical Archaeology* 14:1–40.

Miller, G. L., and C. Sullivan
  1984   Machine-Made Glass Containers and the End of Production for Mouth-Blown Bottles. *Historical Archaeology* 18(2):83–96.

Nelson, L. H.
  1963   Nail Chronology as an Aid to Dating Old Buildings. *History News* 19(2).

Nesmith, R.
  1955   *The Coinage of the First Mint of the Americas at Mexico City, 1536–1572.* Numismatic Notes And Monographs. American Numismatic Society, New York.

Neumann, G. C., and F. Kravic
  1977   *Collector's Illustrated Encyclopedia of the American Revolution.* Castle Books, Secaucus, New Jersey.

Newman, T. S.
  1970   A Dating Key for Post-18th c. Bottles. *Historical Archaeology* 4:70–75.

Oakeshott, R. E.
  1960   *The Archaeology of Weapons: Arms and Armor from Prehistory to the Age of Chivalry.* Dover, New York.

Partridge, M.
  1973   *Farm Tools through the Ages.* New York Graphic Society, Boston.

Quimby, G. (ed.)
  1980   *Ceramics in America.* University Press of Virginia, Charlottesville.

Rice, P. M.
  1987   *Pottery Analysis: A Sourcebook.* University of Chicago Press, Chicago.

Ridings, R.
  1996 Where in the World Does Obsidian Hydration Dating Work? *American Antiquity* 61(1):136–48.

Rock, J. T.
  2000 Cans in the Countryside. In *Approaches to Material Culture Research for Historical Archaeologists*, D. R. Brauner (ed.), pp. 275–89. Society for Historical Archaeology, California, Pennsylvania.

Roenke, K. G.
  1978 Flat Glass: Its Use as a Dating Tool for Nineteenth Century Archaeological Sites in the Pacific Northwest and Elsewhere. *Northwest Anthropological Research Notes* 12(2).

Rovner, I.
  1994 Floral History by the Back Door: A Test of Phytolith Analysis in Residential Yards at Harper's Ferry. *Historical Archaeology* 28(4):37–48.

Sanford, E.
  1975 Conservation of Artifacts: A Question of Survival. *Historical Archaeology* 9:55–64.

Schiffer, Michael B. and John H. House
  1975 *The Cache River Project: An Experiment in Contract Archaeology*. Research Series Number 8. Arkansas Archaeological Survey, Fayetteville, Arkansas.

Sease, C.
  1994 *A Conservation Manual for the Field Archaeologist*, 3d ed. Archaeological Research Tools 4. Institute of Archaeology, University of California, Los Angeles.

Shephard, A. O.
  1980 *Ceramics for the Archaeologist*. Publication No. 609. Carnegie Institute, Washington, D.C.

Shott, M. J.
  1994 Size and Form in the Analysis of Flake Debris: Review and Recent Approaches. *Journal of Archaeological Method and Theory* 1(1):69–110.

Singley, K. R.
  1981 Caring for Artifacts after Excavation—Some Advice for Archaeologists. *Historical Archaeology* 15(1):36–48.

Smith, M. T.
  1983 Chronology from Glass Beads: The Spanish Period in the Southeast, c. A.D. 1513–1670. In *Proceedings of the 1982 Trade Bead*

Conference, Charles F. Hayes III (ed.), pp. 147–58. Research Records, vol. 16. Rochester Museum and Science Center, Rochester, New York.

Smith, M. T., and M. E. Good
1982 *Early Sixteenth-Century Beads in the Spanish Colonial Trade.* Cottonlandia Museum, Greenwood, Mississippi.

South, S.
1977 *Method and Theory in Historical Archaeology.* Academic Press, New York.

South, S., R. K. Skowronek, and R. E. Johnson
1988 *Spanish Artifacts from Santa Elena.* Occasional Papers of the South Carolina Institute of Archaeology and Anthropology, Anthropological Studies 7. University of South Carolina, Columbia.

Spaulding, A. C.
1953 Statistical Techniques for the Discovery of Artifact Types. *American Antiquity* 18:305–13.
1954 Reply (to Ford). *American Antiquity* 19(4):391–93.

Stahle, D. W., M. K. Cleaveland, D. B. Blanton, M. D. Therrell, and D. A. Gay
1998 The Lost Colony and Jamestown Droughts. *Science* 280(5363):564–67.

Sullivan, A., and K. Rozen
1985 Debitage Analysis and Archaeological Interpretation. *American Antiquity* 50(4):755–79.

Sutton, M. Q., and B. S. Arkush
1998 *Archaeological Laboratory Methods: An Introduction.* Kendall/Hunt, Dubuque, Iowa.

Switzer, R. R.
1974 *The Bertrand Bottles: A Study of Nineteenth-Century Glass and Ceramic Containers.* National Park Service, Washington, D.C.

Thomas, D. H.
1978 The Awful Truth about Statistics in Archaeology. *American Antiquity* 43(2):231–44.
1986 *Refiguring Anthropology: First Principles of Probability and Statistics.* Waveland Press, Prospect Heights, Illinois.
1998 *Archaeology,* 3d ed. Harcourt Brace, New York.

Trigger, B. G.
1989 *A History of Archaeological Thought.* Cambridge University Press, London.

Vaughn, P.
    1985   *Use-Wear Analysis of Flaked Stone Tools.* University of Arizona Press, Tucson.

Wells, T.
    2000   Nail Chronology: The Use of Technologically Derived Features. In *Approaches to Material Culture Research for Historical Archaeologists*, D. R. Brauner (ed.), pp. 318–39. Society for Historical Archaeology.

White, J. R.
    1978   Bottle Nomenclature: A Glossary of Landmark Terminology for the Archaeologist. *Historical Archaeology* 12:58–67.

Willey, G. R., and P. Phillips
    1958   *Method and Theory in American Archaeology.* University of Chicago Press, Chicago.

Williams, M. W.
    1986   Sub-surface Patterning at Sixteenth-Century Spanish Puerto Real, Haiti. *Journal of Field Archaeology* 13(3):283–96.

Winters, H. D.
    1968   Value Systems and Trade Cycles of the Late Archaic in the Midwest. In *New Perspectives in Archaeology*, S. Binford and L. Binford (eds.), pp. 175–222. Aldine, Chicago.

INDEX

accelerator mass spectrometry, 110
accession catalog, 30, 31–32
accession number, 30–31, 33, 34
accessioning artifact, 30–32
Adovasio, J. M., 66
analysis, artifact: of ceramic, 52–57; discussion group role in, 123–24; of flaked/chipped lithics, 45–50; of glass, 61–64; of ground lithics, 50–52; historic vs. prehistoric, 9; literature concerning, 56–57; of lithics, 44–45; of metal, 57–61; of organic artifact, 65–66; range included in, 7–8
analysis, definition of, 43
analysis, future of: computer applications, 122; digital rendering, 125–26; Internet, 123–25; large data set, 123; NAGPRA impact on, 126–27; network database, 123–25; reanalysis of archived collection, 127–28
analytical serial number, 33
ANCS. *See* Automated National Catalog System
Approaches to Material Culture Research for Historical Archaeologists (Brauner), 64

archaeological context, 7
Archaeological Data Service, 84
Archaeological Laboratory Methods (Sutton & Arkush), 52, 66
archaeological theory, 12–14
Archaeology (Thomas), 43
The Archaeology of Weapons (Oakeshott), 60
Arkush, B. S., 52, 66
artifact: context of, 6–7, 19–20, 25; definition of, 5; ecofact, 6
artifact density, using for data manipulation, 103–5
artifact inventory/catalog, 32–33, 37
Artifacts of the Spanish Colonies of Florida and the Caribbean (Deagan), 57
attribute, 62, 67
Automated National Catalog System (ANCS), 15, 29
"The Awful Truth about Statistics in Archaeology" (Thomas), 91–92

banner stone, 50, 51, 72
basketry, 64, 65–66
Basketry Technology (Adovasio), 66
bead: glass, 61, 62–63; shell, 64, 65
bifacial tool, 47
Binford, Louis, 13, 100–101

bioarchaeology, 115–16
bipolar reduction, 46
Brauner, D., 64
Brunswick Pattern, 93, 102

Cache River Basin Project, 13
Carolina Pattern, 93
cataloging artifact: accessioning, 30–32; goals of, 17, 29, 89; inventory, 32–33; labeling, 33–34
CC (cream-colored ware) Index, 4, 95–96
ceramic: analysis of, 52–57; category of, 52–53; classifying, 68–69, 72, 76–77; definition of, 52; interpretative value of, 57; literature for analysis, 56–57; measuring, 56; open-air firing of, 55; physical properties of historic, 54; quantification methods for, 77–78
ceramic, attributes of food-related: composition, 53–54; decoration, 56; form, 55, 72; manufacturing technique, 54–55; surface treatment, 55–56
Ceramics for the Archaeologist (Shepard), 56
Ceramics in America (Quimby), 56
chronological classification scheme, 76
chronometric dating, 107–9
chunkey stone, 50, 51
classification: attributes used in, 67; card catalog for, 79–80; challenges to, 67–68; database for, 79; emic vs. etic, 69–70; Linnaean, 67; objectives of, 68–69; quantification and, 77–78. *See also* ceramic, attributes of food-related; classification scheme; electronic data, for classification; quantification

classification scheme: chronological, 76; folk taxonomy, 76–77; formal/morphological, 71–72; functional, 72–75; material/technological, 71; stylistic, 75
cleaning artifact: artifact not to be washed, 25, 37, 65; automatic system for, 26; basic tools for, 24–25; drying process, 26; organic artifact, 25, 65; purpose of, 24; retaining context during, 25; washing process, 25–26
coin chronology, 58
A Complete Manual of Field Archaeology (Joukowsky), 37
computer-assisted design program, 39–40, 56, 126
conchoidal fracture, 46
A Conservation Manual for the Field Archaeologist (Sease), 37
conservation, of artifact, 37, 65
context: of archaeological site, 6–7; recording during excavation, 19–20; retaining during cleaning, 25
conventional photography, 38
core, lithic, 48
cortex, of lithic flake, 47
counter map, 103
CRM. *See* cultural resource management
crossmending artifact, 35
crown, in window glass, 64
cultural context, 7
cultural resource management (CRM), 14–15
culture history, 12–13
curation, 117–18

data manipulation: purpose of, 89; by replication, 105–6; by spatial analysis, 102–5. *See also* dating; patterning; statistical analyses

INDEX

dating: chronometric, 107–9; historical artifact, 98; mean ceramic, 101–2; obsidian hydration, 111; pipe stem, 100–101; radiocarbon, 109–10; seriation, 96–98; thermoluminescence, 110–11
de Narváez expedition, 7
de Soto expedition, 7, 21–22, 63, 109, 115
Deagan, Kathleen, 13, 57
debitage, lithic, 48
Deetz, James, 97–98
DELOS database, 29, 81, 82
dendrochronology, 108–10
DePratter, Chester, 128
Dethlefsen, Edwin, 97–98
digital imaging, 17, 38–40, 56, 126
direct percussion, 46
discoidal, 50
discussion group, role in artifact analysis, 123–24
Dowman, E., 37

ecofact, 6
electronic data, for classification: coding data, 86–87; data entry, 86; data entry sheet (sample), 87; database software for, 81–86; hardware for, 81; proofing hard copy for, 88; strength/weakness of, 80, 84–86
ethnoarchaeology, 106
ethnobotany, 113
ethnographic analogy, 106
excavating artifacts: deciding what to collect, 20–22; recording context, 19–20
Excavating Occaneechi Town (CD), 23

faunal analysis, 111–13
Feder, F., 37

Feder, Kenneth, 5
Field Methods of Archaeology (Hester et al.), 37
field specimen (FS) catalog, 20, 23
field specimen (FS) number, 20, 23–24, 33, 34–35, 80, 86
flint knapping, 106
floral analysis, 113–14
Florida. *See* Governor Martin site
folk taxonomy classification scheme, 76–77
food-related ceramic, 52
Ford, James, 69–70, 96–97
Ford-Spaulding debate, 69–70, 75, 76
formal attribute, 67
formal/morphological classification scheme, 71–72
formula dating, 98
Fredericks site, 23
Frison, George, 106
Frontier Pattern, 93
FS catalog. *See* field specimen (FS) catalog
FS number. *See* field specimen (FS) number
functional classification scheme: for historic artifact, 73–75; for prehistoric artifact, 75; strength/weakness of, 72–73

glass: basic recipe for, 61; bead, 61, 62–63; bottle/container, 61–62, 63–64; care/treatment of, 64; color attribute of, 62; dating, 63–64; form attribute of, 62; forms of, 61–64; references for, 64; window, 64
gorget, 50
Governor Martin site, 7, 21–22, 38, 104–5, 126
grog-temper, 53
Guide to the Artifacts of Colonial America (Hume), 56

Haiti. *See* Puerto Real site
Hardman site, 114
hardware, for digital imaging, 39, 40
Harpers Ferry, 114
Harrington, J. C., 100
Hay, Robert, 96–97
heritage resource management, 15
Hester, T., 37
high archaeological theory, 12
Hodges, H. W. M., 37
human remains, 65, 126–27
Hume, Ivor Noël, 8, 56

Indian Knoll Culture, 75
indirect percussion, 46
inferential statistics, 90
Innova International, 126
Internet, 123–25
interpretation, influences on, 11–12
iron artifact, 57, 58–59

Jones, O. R., 64
Joukowsky, M., 37
judgmental sampling strategy, 90

labeling artifact, 17, 33–34, 35
laboratory serial number, 33
Law of Superposition, 96
lead artifact, 57–58
Leigh, D., 37
Libby, Willard, 109
Linnaeus, Carolus, 67
lithics: definition of, 44; reduction method for, 45, 50
lithics, flaked/chipped: core, 48; cortex of, 47; debitage, 47; heat-treated, 46, 47; measuring, 48; projectile point attributes, 49; reduction method for, 46; residue analysis of, 50; as tool, 47–48; use wear analysis of, 49, 73
lithics, ground: description of, 51; functional classification of, 50–51; measuring, 51; reduction method for, 50; residue analysis of, 51–52; tools for analysis of, 52
low archaeological theory, 12

Mallowan, Agatha Christie, 21
Mallowan, Max, 21
mandrel, in glass manufacturing, 62
mano, 50
marvering, in glass manufacturing, 62
material/technological classification scheme, 71
materials analysis. *See* analysis, artifact
medium ceramic dating (MCD), 101–2
medium manufacture date, 101
metal artifact: as activity indicator, 60; as chronological indicator, 58–60; as ethnicity/trade indicator, 60; interpretive value of, 60; reference sources for, 60–61; type of metal in, 57–58
metate, 50
Method and Theory in American Archaeology (Willey & Phillips), 70
metric vs. English system, 78
middle archaeological theory, 12
Miller, George. *See* CC (cream-colored ware) Index
minimum number of individuals (MNI), 78
minimum vessel count, 78
MMD. *See* medium manufacture date
MNI. *See* minimum number of individuals
monolithic axe, 50
Munsell Book of Color, 62
Munsell chroma scale, 90

NAGPRA. *See* Native American Graves Protection and Repatriation Act

nail chronology, 58–59
National Archaeological Database, 82, 123
National Historic Preservation Act, 14–15
National Park Service, 15, 29
National Register of Historic Places, 8
Native American, 53, 57, 61, 63
Native American Graves Protection and Repatriation Act (NAGPRA), 73, 82, 116, 123, 126–27
New Archaeology, 13–14
non-food-related ceramic, 52–53
North Carolina. *See* Fredericks site

Oakeshott, R. Ewart, 60
obsidian hydration, 111
obsidian hydration dating, 111
OCR. *See* oxidizable carbon ratio dating
online database, 82–83
organic artifact: analysis of, 65–66; care/handling of, 65; environment of, 64–65; reference source for, 66; storing, 65; types of, 64, 65
ornamental/personal ground stone object, 50
oxidizable carbon ratio dating (OCR), 110

paleoethnography, definition of, 70
paleoethnology, definition of, 70
palynology, 113
Pantone color wheel, 62
The Parks Canada Glass Glossary (Jones & Sullivan), 64
patterning: CC Index for, 95–96; effect of method used on assemblage, 93–95; South's patterns, 92–95
petrographic analysis, 114–15
Phillips, P., 70

photographing artifact, 37–38; conventional, 38; digital, 38–40, 126
phytolith analysis, 113–14
piece plotting, 20
pipe stem dating, 100–101
PIXE. *See* proton-induced X-ray emission
pollen analysis, 113
polyvinyl acetate (PVA), 33
postprocessual archeology, 14
potlidding, 46
Potomac Typological System (POTS), 76–77
Pottery Analysis (Rice), 56
preserving artifact, 17
pressure flaking, 46
processing artifact, 17; checking-in, 23–24; cleaning, 24–26, 65; rough sorting, 27, 44, 67, 68
processual, definition of, 13
processual archaeology, 13–14
project records, curation of, 118
proton-induced X-ray emission (PIXE), 50, 52, 115, 128
provenance. *See* provenience
provenience, 19–20
provenience catalog, 20
Puerto Real site, 20–21, 57, 74, 76, 93–94, 103
PVA. *See* polyvinyl acetate

quantification, 77–78
Quimby, George, 56

radiocarbon dating, 109–10
random sampling strategy, 90–91
reconstructing artifact, 35–36
Re:discovery database, 15, 82, 83–84
Refiguring Anthropology (Thomas), 89
replication, 105–6
resin casting, 126

reversibility, of artifact reconstruction, 36
Rice, Prudence, 56
Robert Hay house, 95–96
rough sorting artifacts, 27, 44, 67, 68

scanning electron microscopy (SEM), 50
scapers, 105
Sease, C., 37
Section 106, of National Historic Preservation Act, 14–15
SEM. *See* scanning electron microscopy
seriation dating, 96–98, 99
Shafer, H., 37
shatter, lithic, 48
shell ornament, 64, 65
Shepard, Anna O., 56
slipping, 55
Society for American Archaeology, 125
software, for digital imaging, 39–40, 126
South, Stanley, 4, 73–75, 101
spatial analysis, 102–5
Spaulding, Albert, 69–70, 75, 76
St. Augustine, 13, 74
stabilization, of artifact, 37
statistical analyses: attribute good for, 89–90; judicious use of, 91–92; reference source for, 89; sampling strategy for, 90–92; tests for, 91
stone artifact. *See* lithics
Student's *t*-test, 91
stylistic attribute, 67
stylistic classification scheme, 75
Sullivan, C., 64
Surfer (Golden Software), 104
Sutton, M. Q., 52, 66
SYMAP, 103, 104
Systema Naturae (Linnaeus), 67

TAQ. *See* terminus ante quem
technological attribute, 67
tempering material, for ceramic, 53–54
terminus ante quem (TAQ), 98
terminus post quem (TPQ), 98
textiles, 64, 65–66
thermoluminescence (TL) dating, 110–11
Thomas, David Hurst, 43, 89, 91–92
tin can chronology, 59
TL. *See* thermoluminescence
topographical map, 103
TPQ. *See* terminus post quem

unifacial tool, 47–48
U.S. Forest Service, 15
U.S. Geological Survey, 103
utilitarian ground stone object, 50–51

weapon, as artifact, 60
website: alternative archaeology, 124; Archaeological Data Service, 84; archaeological site, 124; bookseller, 61; dendrochronology, 108; discussion group, 124; e-journal, 124; gateway, 124; Library of Congress, 61; National Archaeological Database, 82; radiocarbon dating, 109, 110; thermoluminescence, 111
Willey, G. R., 70
Winter, Howard, 75
wood/charcoal analysis, 114
wooden artifact, 64, 65
World Wide Web, 125

X-ray imaging, 38

zooarchaeological analysis, 112

ABOUT THE AUTHOR AND SERIES EDITORS

Charles R. Ewen received his B.A. at the University of Minnesota (1978), M.A. at Florida State (1983), and Ph.D. at the University of Florida (1987). After graduation, he codirected excavations at the Governor Martin site (the site of Hernando de Soto's first winter encampment) for the Florida Bureau of Archaeological Research. The next stop was Arkansas, where he was an assistant professor and director of the Arkansas Archeological Survey's Sponsored Research Program. He joined the faculty of East Carolina University (ECU) in 1994, where he is a professor in the Anthropology Department. He also serves as director of the Phelps Archaeology Lab on the ECU campus. His research interests include archaeological method and theory, cultural resource management, and historical archaeology, specifically the Contact and Colonial periods. Like most archaeologists, however, circumstances have led him to work on nearly every kind of site, from prehistoric villages to Civil War fortifications and twentieth-century homesteads. He has established a research program in eastern North Carolina, conducting projects at Tryon Palace Historic Sites and Gardens, Hope Plantation, Historic Edenton, and the Newbold-White house.

Besides several articles and book chapters, Charlie is the author of two books about fieldwork he has conducted: *From Spaniard to Creole: The Archaeology of Cultural Formation at Puerto Real, Haiti* (University of Alabama Press, 1991) and *Hernando de Soto among the Apalachee: The Archaeology of the First Winter Encampment* (coauthor, University of Florida Press, 1998). He is currently working on a coedited volume concerning the colonies at Roanoke Island. He lives happily near the university with his wife, Gretchen, and two daughters, Kate and Madeline.

❈

Larry J. Zimmerman is the head of the Archaeology Department of the Minnesota Historical Society. He served as an adjunct professor of anthropology and visiting professor of American Indian and Native Studies at the University of Iowa from 1996 to 2002 and as chair of the American Indian and Native Studies Program from 1998 to 2001. He earned his Ph.D. in anthropology at the University of Kansas in 1976. Teaching at the University of South Dakota for twenty-two years, he left there in 1996 as Distinguished Regents Professor of Anthropology.

While in South Dakota, he developed a major CRM program and the University of South Dakota Archaeology Laboratory, where he is still a research associate. He was named the University of South Dakota Student Association Teacher of the Year in 1980, given the Burlington Northern Foundation Faculty Achievement Award for Outstanding Teaching in 1986, and granted the Burlington Northern Faculty Achievement Award for Research in 1990. He was selected by Sigma Xi, the Scientific Research Society, as a national lecturer from 1991 to 1993, and he served as executive secretary of the World Archaeological Congress from 1990 to 1994. He has published more than three hundred articles, CRM reports, and reviews and is the author, editor, or coeditor of fifteen books, including *Native North America* (with Brian Molyneaux, University of Oklahoma Press, 2000) and *Indians and Anthropologists: Vine Deloria, Jr., and the Critique of Anthropology* (with Tom Biolsi, University of Arizona Press, 1997). He has served as the editor of *Plains Anthropologist* and the *World Archaeological Bulletin* and as the associate editor of *American Antiquity*. He has done archaeology in the Great Plains of the United States and in Mexico, England, Venezuela, and Australia. He has also worked closely with a wide range of American Indian nations and groups.

William Green is Director of the Logan Museum of Anthropology and Adjunct Professor of Anthropology at Beloit College, Beloit, Wisconsin. He has been active in archaeology since 1970. Having grown up on the south side of Chicago, he attributes his interest in archaeology and anthropology to the allure of the exotic (i.e., rural) and a driving urge to learn the unwritten past, abetted by the opportunities available at the city's museums and universities. His first field work was on the Mississippi River bluffs in western Illinois. Although he also worked in Israel and England, he returned to Illinois for several years of survey and excavation. His interests in settlement patterns, ceramics, and archaeobotany developed there. He received his Master's degree from the University of Wisconsin–Madison and then served as Wisconsin SHPO staff archaeologist for eight years. After obtaining his Ph.D. from UW–Madison in 1987, he served as State Archaeologist of Iowa from 1988 to 2001, directing statewide research and service programs including burial site protection, geographic information, publications, contract services, public outreach, and curation. His main research interests focus on the development and spread of native agriculture. He has served as editor of the *Midcontinental Journal of Archaeology* and *The Wisconsin Archeologist*, has published articles in *American Antiquity*, *Journal of Archaeological Research*, and other journals, and has received grants and contracts from the National Science Foundation, National Park Service, Iowa Humanities Board, and many other agencies and organizations.

EXCAVATION

ARCHAEOLOGIST'S TOOLKIT

SERIES EDITORS: LARRY J. ZIMMERMAN AND WILLIAM GREEN

The Archaeologist's Toolkit is an integrated set of seven volumes designed to teach novice archaeologists and students the basics of doing archaeological fieldwork, analysis, and presentation. Students are led through the process of designing a study, doing survey work, excavating, properly working with artifacts and biological remains, curating their materials, and presenting findings to various audiences. The volumes—written by experienced field archaeologists—are full of practical advice, tips, case studies, and illustrations to help the reader. All of this is done with careful attention to promoting a conservation ethic and an understanding of the legal and practical environment of contemporary American cultural resource laws and regulations. The Toolkit is an essential resource for anyone working in the field and ideal for training archaeology students in classrooms and field schools.

Volume 1: *Archaeology by Design*
Stephen L. Black and Kevin Jolly

Volume 2: *Archaeological Survey*
By James M. Collins and Brian Leigh Molyneaux

Volume 3: *Excavation*
By David L. Carmichael, Robert H. Lafferty III, and Brian Leigh Molyneaux

Volume 4: *Artifacts*
By Charles R. Ewen

Volume 5: *Archaeobiology*
By Kristin D. Sobolik

**Volume 6: *Curating Archaeological Collections:
From the Field to the Repository***
By Lynne P. Sullivan and S. Terry Childs

Volume 7: *Presenting the Past*
By Larry J. Zimmerman

EXCAVATION

DAVID L. CARMICHAEL
ROBERT H. LAFFERTY III
BRIAN LEIGH MOLYNEAUX

ARCHAEOLOGIST'S TOOLKIT
VOLUME 3

ALTAMIRA
PRESS

A Division of Rowman & Littlefield Publishers, Inc.
Walnut Creek • Lanham • New York • Oxford

ALTAMIRA PRESS
A division of Rowman & Littlefield Publishers, Inc.
1630 North Main Street, #367
Walnut Creek, CA 94596
www.altamirapress.com

Rowman & Littlefield Publishers, Inc.
A wholly owned subsidiary of The Rowman & Littlefield Publishing Group, Inc.
4501 Forbes Boulevard, Suite 200
Lanham, MD 20706

PO Box 317
Oxford
OX2 9RU, UK

Copyright © 2003 by ALTAMIRA PRESS

All rights reserved. No part of this publication may be reproduced, stored in a retrieval system, or transmitted in any form or by any means, electronic, mechanical, photocopying, recording, or otherwise, without the prior permission of the publisher.

British Library Cataloguing in Publication Information Available

Library of Congress Cataloging-in-Publication Data
Carmichael, David L.
    Excavation / David L. Carmichael, Robert H. Lafferty III, and Brian Leigh Molyneaux.
        p. cm.—(Archaeologist's toolkit ; v. 3)
Includes bibliographical references and index.
    ISBN 0-7591-0399-2 (hardcover : alk. paper)—ISBN 0-7591-0019-5 (pbk. : alk. paper)
    1. Excavations (Archaeology) 2. Archaeology—Field work. I. Lafferty, Robert H. II. Molyneaux, Brian Leigh. III. Title. IV. Series.

CC76.C37 2003
930.1'028'3—dc21

2003012124

Printed in the United States of America

∞™ The paper used in this publication meets the minimum requirements of American National Standard for Information Sciences—Permanence of Paper for Printed Library Materials, ANSI/NISO Z39.48–1992.

CONTENTS

Series Editors' Foreword — vii

Acknowledgments — xi

1 Introduction — 1
Another Excavation Book? • Thematic Orientations • Organization of the Volume

2 Excavation: A Brief History — 15
The Emergence of Provenience Control • North American Developments • Direct Association and the Search for Early Humans in the New World • The Grid • Theoretical Developments • The Beginnings of CRM • The Real Indiana Jones

3 Archaeological Excavation Is Controlled Destruction — 31
Justifying the Destruction • Excavation and the Conservation Ethic • Controlling the Destruction • Site Security • Truth in Reporting • Consultation with Affected Groups

4 Site Testing, Excavation, and Characterization — 49
How Do We Know Where to Dig? • Types of Test and Excavation Units • Mechanical Excavation Techniques • Trenching as a Statistical Sampling Strategy • Mechanical Stripping • Recording Stratigraphy • Conclusion

5 Data Recovery	63
Artifacts and Things • *Collection Strategy* • *The Ways and Means* • *Conclusion*	
6 Recovery of Specialized Materials and Samples	73
Chronometric Samples • *Soil Samples* • *Human Remains* • *Looking to the Future*	
Appendix 1: Equipment, Supplies, and Sources for Specialized Analysis	91
Appendix 2: Conversion Tables and Other Useful Resources	99
References	105
Index	117
About the Authors and Series Editors	123

SERIES EDITORS' FOREWORD

The Archaeologist's Toolkit is a series of books on how to plan, design, carry out, and use the results of archaeological research. The series contains seven books written by acknowledged experts in their fields. Each book is a self-contained treatment of an important element of modern archaeology. Therefore, each book can stand alone as a reference work for archaeologists in public agencies, private firms, and museums, as well as a textbook and guidebook for classrooms and field settings. The books function even better as a set, because they are integrated through cross-references and complementary subject matter.

Archaeology is a rapidly growing field, one that is no longer the exclusive province of academia. Today, archaeology is a part of daily life in both the public and private sectors. Thousands of archaeologists apply their knowledge and skills every day to understand the human past. Recent explosive growth in archaeology has heightened the need for clear and succinct guidance on professional practice. Therefore, this series supplies ready reference to the latest information on methods and techniques—the tools of the trade that serve as handy guides for longtime practitioners and essential resources for archaeologists in training.

Archaeologists help solve modern problems: They find, assess, recover, preserve, and interpret the evidence of the human past in light of public interest and in the face of multiple land use and development interests. Most of North American archaeology is devoted to cultural resource management (CRM), so the Archaeologist's Toolkit focuses on practical approaches to solving real problems in CRM and

public archaeology. The books contain numerous case studies from all parts of the continent, illustrating the range and diversity of applications. The series emphasizes the importance of such realistic considerations as budgeting, scheduling, and team coordination. In addition, accountability to the public as well as to the profession is a common theme throughout the series.

Volume 1, *Archaeology by Design,* stresses the importance of research design in all phases and at all scales of archaeology. It shows how and why you should develop, apply, and refine research designs. Whether you are surveying quarter-acre cell tower sites or excavating stratified villages with millions of artifacts, your work will be more productive, efficient, and useful if you pay close and continuous attention to your research design.

Volume 2, *Archaeological Survey,* recognizes that most fieldwork in North America is devoted to survey: finding and evaluating archaeological resources. It covers prefield and field strategies to help you maximize the effectiveness and efficiency of archaeological survey. It shows how to choose appropriate strategies and methods ranging from landowner negotiations, surface reconnaissance, and shovel testing to geophysical survey, aerial photography, and report writing.

Volume 3, *Excavation,* covers the fundamentals of dirt archaeology in diverse settings, while emphasizing the importance of ethics during the controlled recovery—and destruction—of the archaeological record. This book shows how to select and apply excavation methods appropriate to specific needs and circumstances and how to maximize useful results while minimizing loss of data.

Volume 4, *Artifacts,* provides students as well as experienced archaeologists with useful guidance on preparing and analyzing artifacts. Both prehistoric- and historic-era artifacts are covered in detail. The discussion and case studies range from processing and cataloging through classification, data manipulation, and specialized analyses of a wide range of artifact forms.

Volume 5, *Archaeobiology,* covers the analysis and interpretation of biological remains from archaeological sites. The book shows how to recover, sample, analyze, and interpret the plant and animal remains most frequently excavated from archaeological sites in North America. Case studies from CRM and other archaeological research illustrate strategies for effective and meaningful use of biological data.

Volume 6, *Curating Archaeological Collections,* addresses a crucial but often ignored aspect of archaeology: proper care of the specimens

and records generated in the field and the lab. This book covers strategies for effective short- and long-term collections management. Case studies illustrate the do's and don'ts that you need to know to make the best use of existing collections and to make your own work useful for others.

Volume 7, *Presenting the Past,* covers another area that has not received sufficient attention: communication of archaeology to a variety of audiences. Different tools are needed to present archaeology to other archaeologists, to sponsoring agencies, and to the interested public. This book shows how to choose the approaches and methods to take when presenting technical and nontechnical information through various means to various audiences.

Each of these books and the series as a whole are designed to be equally useful to practicing archaeologists and to archaeology students. Practicing archaeologists in CRM firms, agencies, academia, and museums will find the books useful as reference tools and as brush-up guides on current concerns and approaches. Instructors and students in field schools, lab classes, and short courses of various types will find the series valuable because of each book's practical orientation to problem solving.

As the series editors, we have enjoyed bringing these books together and working with the authors. We thank all of the authors—Steve Black, Dave Carmichael, Terry Childs, Jim Collins, Charlie Ewen, Kevin Jolly, Robert Lafferty, Brian Molyneaux, Kris Sobolik, and Lynne Sullivan—for their hard work and patience. We also offer sincere thanks to Mitch Allen of AltaMira Press and a special acknowledgment to Brian Fagan.

WILLIAM GREEN
LARRY J. ZIMMERMAN

ACKNOWLEDGMENTS

David L. Carmichael

I would like to thank the colleagues and friends who have assisted me in the preparation of this volume. Researchers who have provided data and insights from their own projects include Steve Black, Peter Eidenbach, Lane Ellis, Mark Gutzman, Robert Hard, Gary Hebler, Rob Jackson, Lisa Meyer, Myles Miller, Yvonne Oakes, Gary Strobel, and Rick Wessel. Evelyn Chandler provided archival data on the archeological contracts at Eaker Air Force Base. We appreciate the patience and encouragement of the series editors, Larry Zimmerman and Bill Green. We would like to thank Scott Cutler of the El Paso Centennial Museum, Julie Hall of the U.S. Army Corps of Engineers, and Joan Oxendine of the U.S. Bureau of Land Management, California Desert District, for their assistance in securing permission to use illustrations previously published in contract reports.

Robert H. Lafferty III

I received my early training in excavation methods from J. B. Graham during three summers of salvage archaeology for the University of Tennessee. He taught me how to carry out precise and accurate excavations, to use state-of-the-art technology, and to be aggressive in using innovative field methods. Jon D. Muller showed me how to use archeological data to test hypotheses while I was a graduate assistant at the Southern Illinois University field school. I am indebted to both of them, as well as scores of fellow students, who after seven field seasons prepared me to direct excavations successfully. This volume would not have been possible without the editorial assistance of my wife, Kathleen M. Hess, who turned my parts of the manuscript into coherent English. R. W. Maringer produced some of the graphics, and Kelly Sturtevant-Murdick helped with the bibliography. I thank all of them and the hundreds of crew members who have worked on the projects I have directed.

Brian Leigh Molyneaux

I would like to acknowledge Dr. Romas Vastokas, Peterborough, Ontario, who not only presided over my first formal dig at the Trent University Field School but also let me work on weekends, because I couldn't get enough of it. And I would like to thank the many field crews I have worked with, for tolerating my relentlessly cheerful demeanor. I owe most to my family, from my mom and dad to my wife and children, for letting me spend so many summers in northern Canada, when I should have been home, digging the garden!

1
INTRODUCTION

> We cannot question seriously or effectively without reference to the past. Without the past there is no future for man, nor even a present being. All he is is the sum of his past, and all he can hope to become is written somewhere in this experience. To know the past is not necessarily to be bound by it, but some degree of knowledge can save a lot of wasted effort and set a frame for performance that gives life the meaning of shared human endeavor.
>
> —*Roderick Haig-Brown* (quoted in J. Taylor 1999)

Archaeology is unique in its ability to explore human behavior outside the limits of written records. History deals with past behavior as well, of course, but in most parts of the world, the time covered by historic documents represents a mere blink of an eye in the broad scope of the human experience. Many significant developments, such as the peopling of the New World, the development of plant and animal domestication, and the rise of complex social systems, occurred before the advent of writing. Some prehistoric phenomena have no surviving modern analogs that can be studied ethnographically. Moreover, archaeology not only allows us to study past behavior but also helps us see changes in behavior over time, or cultural evolution. We are able to measure the rates of culture change and to develop explanations to account for the observed changes.

Archaeologists most often realize this unique contribution to our understanding of the human condition through archaeological excavation. Excavating into the earth is traveling back in time, as we seek to expose soil layers containing the physical traces of the behavior of earlier human populations. Yet, archaeological excavation is much more than digging holes. It is recording and assembling a record of past behavior, giving meaning to that record, and applying the meanings in a wide variety of contexts, including academic research, education, land use planning, and public policy development. Each site and each feature we excavate is a potential window to the past. We use the collective record of such glimpses as the basis for understanding our development as a species. What we do throughout the excavation process affects our ability to explain our past and to meet the needs of society that require and justify excavation in the first place.

ANOTHER EXCAVATION BOOK?

Why do we need another book on archaeological excavation techniques? A number of how-to books have been written in recent years, and more are in preparation. Many of them are more lengthy and comprehensive than this Toolkit volume; some contain quite good discussions of the logistical and mechanical aspects of conducting a traditional archaeological dig. Most such books give short shrift, however, to a range of topics and issues that archaeologists must face every day. Archaeology has changed a great deal in the past three decades. Significant advances have been made in the technology used in the field and laboratory. The research questions of interest today are often more complex than those of the past. The reasons for selecting sites to be excavated, and the very need for excavation, are driven by the interplay of growth and development, environmental law, and conservation ethics.

During a recent revision of the anthropology curriculum, the University of Texas at El Paso consulted colleagues and potential employers in the private sector to find out what skills they felt students should obtain to be readily employable after graduation. Not surprisingly, they said that many students lack significant fieldwork training and experience. While many students have coursework in archaeological methods, and some attend summer field schools, they don't usually experience fieldwork resembling the realities of most contemporary archaeology until graduate school. Perhaps more surprising was their observation that new graduates, even from large, established academic programs, generally lacked a basic understanding of the legal requirements and ethical considerations involved in the jobs they were seeking.

Clearly, these perceived needs reflect the changing circumstances of archaeology in the United States. Since the mid-1970s, the discipline has seen a proportional decrease in the amount of grant-supported academic archaeological research and a huge increase in publicly funded studies conducted in advance of development projects. Cultural resources management (CRM) now generates most of the funds spent on archaeology and employs most of the archaeologists doing research in the United States. However, for most professional archaeologists, the field techniques, legal knowledge, and ethical skills needed were acquired on the job; they are not generally taught in most archaeology curricula or textbooks.

Most archaeology texts discuss excavation methods as they might be used on a multiyear field school project, or what Barker (1996) refers to as research archaeology. In reality, most CRM projects are conducted more quickly, and very few extend beyond a single field season in duration. Yet, textbooks typically devote little space to the demands of such studies. In two popular texts, CRM and fieldwork ethics are covered in only seven or eight pages (Hester, Shafer, and Feder 1997; Renfrew and Bahn 1991:470–77). Barker (1996) devotes an entire chapter to CRM activities but treats them in a mildly offhanded way, as rescue and salvage operations, which are not quite real research efforts.

Nevertheless, Barker (1996:143) argues convincingly that rescue archaeology (i.e., CRM) requires at least as much skill and greater rigor than traditional academic archaeology. If so, it may be the case that archaeologists have not been fully preparing students for eventual employment in the CRM setting. Students and researchers need to be familiar with excavation techniques such as subsurface testing methods, maintaining excavation units, drawing stratigraphic profiles, mapping, and dating. But they should also understand the legal requirements of CRM, as well as their implications for project schedules and the choice of research methods. It is also important to be aware of the legal and ethical issues involved in client–contractor relationships, archaeological stewardship, and interactions with the interested public.

This volume touches on all these issues, albeit lightly. The full range of fieldwork details cannot be covered in a book of this length, and the reader should understand that no text can anticipate all field contingencies. Rather, the application of field techniques is a creative process, with different projects presenting different demands and opportunities. This book includes a series of case studies from excavations that illustrate the sorts of issues the reader may encounter in archaeology, as well as a range of possibilities to consider in addressing those issues. It is a book about methods, but it is also about attitudes and approaches.

THEMATIC ORIENTATIONS

Although this volume is organized around case examples highlighting important issues in excavation, several additional overarching themes are intertwined with the discussions throughout the book. The reader is encouraged to consider these themes, discussed in the following sections, during all aspects and phases of excavation:

- CRM activities deserve our best efforts.
- Setting ethical priorities is a key aspect of modern archaeology.
- Archaeology is still part of anthropology.
- The scope of excavation activities should be warranted and appropriate.

OUR BEST EFFORTS

Years ago, a fellow graduate student, now a professional colleague, remarked that archaeologists couldn't really justify their discipline as somehow being relevant to society. It is so esoteric, so unnecessary to society at large, that we should be thankful that people tolerate archaeology, and we should just go about our business. In today's research environment, such a view is untenable. Whether researchers are comfortable with the idea or not, archaeologists have many responsibilities beyond simply conducting good research (Lynott and Wylie 1995). Archaeology is no longer simply tolerated, it is required by law and paid for primarily with public funds. Federally mandated archaeological research costs between $70 and $100 million per year (Haas 1998; Knudson, McManamon, and Myers 1995:iii, 56–60; Renfrew and Bahn 1991:475), with others giving even higher estimates, so we really do need to be accountable to the public when taxpayer dollars support our research.

Of course, this is not the only way that archaeology is relevant to society. Like other parts of culture, our knowledge of the past is an instrument used to create meaning and order. Archaeology therefore plays a social and political role in the control and use of information about the past (Thomas 1991:46–47). For example, the determination of continuity in occupation or the cultural affiliation of sites may play a role in Native American land claim cases or the disposition of human remains recovered from those sites (Ucko 1989:xiii). Site characterization and the assessment of significance have had effects on the decisions to close military bases, with enormous impacts on neighboring communities. Public interpretation of archaeological research can affect local and regional tourism markets. The involvement of local residents in research can empower communities, build pride in cultural heritage, and spread the conservation ethic (Pyburn and Wilk 1995). Archaeological studies in general, and CRM in particular, demand and deserve our best efforts because the results do matter.

However, the key laws and regulations that structure CRM fairly narrowly define the impact of most archaeological work. At the federal level, the most important of these are the National Historic Preservation Act of 1966, as amended (NHPA), the Archaeological Resources Protection Act of 1979 (ARPA), and the Native American Graves Protection and Repatriation Act of 1990 (NAGPRA). For present purposes, it is sufficient to highlight a few key points.

Section 106 of the NHPA requires the proponents of federally funded or permitted development projects to identify, evaluate, and consider the effects of their actions on historic properties (significant archaeological, historical, or traditional cultural entities). Section 110 of the act directs federal land-managing agencies to inventory and evaluate the historic properties on their landholdings to facilitate their effective management. If a proposed federal project will have an adverse effect on a significant site, a common solution is to mitigate the effect by recovering data through excavation before the site is destroyed by construction. Most archaeological excavation activities in the United States are now done to meet one or the other of these needs: site characterization or data recovery.

Most such research is conducted by contractors affiliated with private firms or university research centers. Although federal agencies, tribes, or regulatory offices employ some archaeologists, most are involved with CRM as contractors. Despite their contribution to the discipline and the labor pool, it is important to remember that contractors have no formal role and no authority in the CRM compliance process. The federal agencies' representatives are the decision makers, and they are responsible for complying with federal laws. The main procedural functions for the archaeologist are to make recommendations to the decision makers regarding the significance of sites and the appropriateness of data recovery strategies.

Significance, as slippery as the concept may be (Tainter and Lucas 1983), is something we usually determine by evaluating an archaeological site's research potential. To justify destruction of a site (see chapter 3), we need to structure data recovery efforts to make the most of that research potential, but potential in relation to what? We must bring knowledge of current regional research issues and state-of-the-art technological capabilities to our studies, as our work will help determine whether sites survive or are destroyed. We need to justify site evaluations by regional research priorities; we need to articulate data recovery plans in research designs that we link to historic contexts. To do less than our best is to abdicate the main role archaeologists have in determining how archaeological sites in this country are managed and studied.

SETTING ETHICAL PRIORITIES

Contemporary archaeologists have multiple ethical responsibilities relating to the various constituencies that are served by our work. These include the conservation ethic and a stewardship role in the public trust; the professional responsibility to meet standards of research; the need to provide cost-effective, justifiable research products for our clients; the need to provide adequate training for our students; and the responsibility to consider the effects of our research on Native Americans and members of other descendant communities. These various and sometimes conflicting demands provide ample opportunities for the development of ethical dilemmas. In fact, Fowler (1984:108) has suggested that the increase in archaeological contracting related to CRM has led to more ethical problems than any other recent development in American archaeology.

Many of these problems arise from the differences between the legitimate concerns of archaeologists, such as research goals and the conservation ethic, and project sponsors whose main objective is completion of the proposed development. Unlike that for other professionals, such as lawyers and engineers, the archaeologist's relationship with a client is not necessarily as clear as one might suppose. Many, perhaps most, project sponsors would prefer to proceed without doing any archaeological research, or at least minimizing the amount of work undertaken. But, if previous research in the region is lacking or inadequate to inform significance evaluations, a certain amount of work will be necessary to meet compliance requirements and, therefore, the client's needs. In other words, archaeologists will sometimes be in the position of informing or at least suggesting to a client what needs to be done, the reverse of what would be expected in most contracting situations. Moreover, many archaeologists would argue that their first responsibility is not to the client but to the public trust (as stewards of the archaeological record) or to indigenous peoples (because they are often the groups most affected by our research). As a result, archaeologists' recommendations can often reflect their own perceptions of the client's needs, not the client's perception of those needs.

Given the litigious nature of contemporary American law, it's not surprising that in response to a detailed proposal for consultation and data recovery on one of their projects, a client once asked, "What is our liability if we choose not to comply with the law?" Hopefully, most archaeologists will not encounter such attitudes very often, but they may be more common in business than we would like to think. A strong ethical stance is needed to advocate for the law, for the archaeological record, and to protect one's professional reputation. "Ethical responsibility is not beside the point of archaeology; it is the only thing that will keep our discipline alive" (Pyburn and Wilk 1995:76).

ARCHAEOLOGY AS ANTHROPOLOGY

Engaging nonarchaeologists in dialogs about archaeological materials and in partnerships for their management represents real change in the discipline (Zimmerman 1995:65). Consultation and cooperation with Native Americans comprise a dimension of archaeology that goes beyond the "stones and bones" training that many of us received. Just the other day, an archaeologist related a recent conversation she had with a colleague regarding their experiences consulting with Native Americans. The colleague expressed admiration for those of us who work with tribal peoples: "It's so hard; I'm an archaeologist, not an anthropologist." Of course, at least in the United States, virtually all archaeologists receive their academic degrees in anthropology, so how is it that members of our discipline can come to view themselves as archaeologists, not anthropologists?

The effectiveness of our training is fully evident in the following statements overheard at the 1993 Annual Meeting of the Society for American Archaeology (SAA) in St. Louis: "Aren't the ethics of dealing with Native Americans pretty much self evident? There sure seem to be a lot of these sessions. The ethics for responding to Native American concerns may lead to censorship and the loss of my academic freedom. I went into archaeology so I wouldn't have to deal with living people." (See sidebar 1.1 for the story of how one of the authors entered the field of archaeology.)

1.1. BECOMING AN ARCHAEOLOGIST

Although the details will differ from one individual to the next, the following scenario, an abbreviated version of David Carmichael's own career, may not be very different from how many of us were first drawn to the field and then socialized into the profession.

As a youth growing up in the suburbs of a large midwestern city, Carmichael decided early on that he would study archaeology. But, even earlier, his interest had focused on rocks. Or, more precisely, on rocks, minerals, fossils, and landforms—anything geological. He filled the crawlspace in his parents' home with rock specimens from all over the country, collected by the grocery bag full on summer family vacations. What strange caching behavior might a future archaeologist infer from this?

Then one summer, in fourth or fifth grade, while cultivating the backyard garden (the large one, before it was replaced by the garage), his hoe unearthed a chalcedony flake. Not surprising perhaps, in the rural farmlands, but pretty remarkable in the suburbs. What a revelation: Not only were there interesting rocks to be found right at home, but some of them had also been modified by humans! After spending much of the summer reading the anthropology books from the local public library, he became aware that the ancestors of today's American Indians made flaked stone tools and that flakes were their byproducts. In other words, he was drawn to the study of archaeological materials precisely because they are the products of the behavior of past Native American peoples.

How ironic that as he proceeded through his academic archaeological training, he was conditioned to view prehistoric artifacts not as the handiwork of Native Americans but as specimens, assemblages, foci, phases, branches, and material cultures. There was a hint of the dialog to come when, as an undergraduate in a western university, he became aware of the double standard in the laws relating to the treatment of human remains. It was against the law to remove bones from a marked cemetery in a ghost town, but it was all right to remove Indian remains from unmarked graves at prehistoric pueblos. Indeed, it must be important to do so, for the museum was full of them. Like the material culture items in the museum collections, the Indian remains were treated as specimens, not as the ancestors of living Native Americans.

But it was not until much later, after graduate school, that it became clear to the now-professional archaeologist how detached most of us were from the Native American essence of much of the archaeological record. As a result of his involvement in CRM research for a client working in the Northern Plains, Carmichael had the opportunity to do archaeology in a region with a large Native American population and numerous living traditions. Presumed tipi rings make up a large proportion of the archaeological features studied in that region, and he was sure that the local indigenous peoples would be able to contribute to our understanding of them. Yet, he found entire published volumes devoted to such features containing hardly a mention of ethnoarchaeology or Native American views (Ives 1986; Wilson et al. 1981). In fact, our archaeologist found a persistent, semiconscious exclusion of contemporary Native Americans from considerations of the archaeological record. For example, in Montana and Wyoming, some researchers have developed a shorthand terminology to describe archaeological sites that distinguishes *aboriginal* (Native American) sites from *historic* (Euroamerican) sites. Granted, it may be difficult, sometimes impossible, to distinguish late prehistoric tipi ring sites from historic period tipi ring sites, and this classification makes some sense on morphological grounds. However, it also embodies the idea that historic sites are Euroamerican and that modern Indian populations are not connected to the archaeology of their ancestors. Of course, this suggestion is false, but it does reflect a widely held perception among archaeologists.

In 1989, Carmichael went to the World Archaeological Congress meetings on Archaeological Ethics and the Treatment of the Dead, held in South Dakota. Not many American archaeologists attended this conference. Those who did attend often had to endure being yelled at by Native Americans; evidently, someone (actually, a lot of people) had a different kind of interest in *our* archaeological materials, and many researchers didn't yet realize it.

It is ironic that while many archaeologists practice archaeology because of an interest in prehistoric Indians, most have never worked with Native Americans, and many would rather not do so. Recent relations between archaeologists and Native Americans have often been adversarial, especially in the context of reburial and repatriation issues (Echo-Hawk and Echo-Hawk 1991; Echo-Hawk 1993). With the passage of NAGPRA, the status of Indians has begun to shift from being specimens or subjects to being partners in research. This development has been so unsettling to some researchers that it has been blamed for the demise of archaeology as a discipline (Meighan 1992a, 1992b, 1993, 1994). Is archaeology really going out of business? Probably not, but as Jonathan Haas (1993) remarked, the days of doing archaeology as we have for the last hundred years are over.

The discipline is changing, to address new levels of accountability and to accommodate collaborative approaches to understanding the past (Salazar and Barrow 2000; Magne 1997; Parker and Bevitt 1997; Swidler et al. 1997; Watkins and Parry 1997). By recognizing the valid concerns and insights Native Americans bring to such collaborations, archaeologists can gain new understandings about the past (Zimmerman 1995:66) and the living traditions that maintain them (see sidebar 6.5 for the story of a site that illustrates this point). New requirements to consult with members of traditional cultures should not be viewed as impediments to research. Isn't consultation something we should be doing with any community, traditional or not? Consultation provides opportunities to enhance and expand our knowledge of the human condition. It has been said "success is a journey, not a destination." In our discipline, one might say, "archaeology is a process of knowing, not simply digging holes." Archaeology is still part of anthropology, and there is much to be learned.

APPROPRIATE EXCAVATION

Providing our best efforts in archaeology, regardless of whether or not the project is related to CRM, does not necessarily mean doing everything that is possible at a site. Often, it will mean doing what is necessary (under the law, in relation to our research design, etc.), what is appropriate, and what is reasonable. The importance of a site is often determined by the presence or absence of intact subsurface deposits, as they relate directly to research potential (but see chapter 3). The (subsurface vs. surface) extent or boundary must be identified if one is to adequately describe a historic property for NRHP nomination.

All states are by now supposed to have in place master planning documents that summarize the existing knowledge and research priorities for each region and time period in the state (King 1977:93–94). Arkansas has one of the best such documents, *A State Plan for the Conservation of Archaeological Resources in Arkansas* (Davis 1982). One portion of the plan contains standards for fieldwork, including details of the number and sorts of test excavation units that might be

1.2. THE EAKER SITE

One of the most interesting archaeological projects Carmichael and Lafferty have worked on involves the Eaker site (3MS105), located on Eaker Air Force Base in northeastern Arkansas. The Eaker site is a large, multicomponent village covering thirty hectares (seventy-five acres) along the Pemiscot Bayou on the Mississippi River floodplain (Lafferty and Cande 1989). Artifacts recovered at the site include materials of the Woodland period Barnes and Baytown traditions, and a Middle Mississippian occupation. However, most of the site is attributable to the Late Mississippian Nodena phase, placing it among the fourteen largest Mississippian sites in the country. Subsurface deposits at the site are complex and well preserved; identified features include palisade trenches, a leveled mound, middens, burials, and an estimated four hundred to five hundred buried houses. The Eaker site is part of an important cluster of thirty-four sites near Blytheville, Arkansas (Morse and Morse 1983:289), but unlike other sites in the area, 3MS105 has been protected from looting since the 1940s by its inclusion in Eaker Air Force Base.

Most of what we know about the Eaker site is a direct result of the research undertaken at Eaker Air Force Base in the context of the proposed Peacekeeper Rail Garrison Program. Archaeological research at the site was conducted by archaeological contractors under subcontract to Tetra Tech, Inc., of San Bernardino, California, an environmental consulting firm working for the air force. The circumstances surrounding initiation of research at the site are presented here as an illustration of how contemporary CRM archaeology demands a knowledge of legal requirements for compliance, ethical concerns for our clients, and the application of state-of-the-art techniques to the research opportunities presented.

Archaeologists at Tetra Tech prepared a detailed scope of work for the initial investigation based on an understanding of project parameters, time schedules, and compliance requirements. Specifically, it was necessary to evaluate the site's significance, or eligibility for the National Register of Historic Places (NRHP). To meet this need, studies at the Eaker site were designed to identify the perimeter of the site, assess the dates and types of prehistoric use of the site, and determine the nature and depth of buried cultural deposits.

Potential subcontractors submitted bids calling for as many as 12,550 shovel tests. But, in an effort to minimize unnecessary damage to the site, and with the concurrence of the SHPO, such bids were rejected in favor of remote sensing accompanied by small-scale test excavations.

For the record, the level of effort outlined in the request for proposals (RFP) was sufficient to define and evaluate the site. Instead of making Swiss cheese of the site, the remote sensing helped identify a complex matrix of preserved subsurface features at depths of 18 to 133 centimeters (mostly below the depths reached by shovel testing) while minimizing damage to the deposits (see chapter 4). The Eaker site was successfully listed on the NRHP in 1993 and was designated a National Historic Landmark in 1996.

warranted at different kinds of sites. While the utility of such a well-conceived plan cannot be easily overstated, the tendency may be for some researchers to follow the detailed procedures like a cookbook. The guidelines themselves provide a cautionary note against this approach.

Given the geomorphology and ground cover of the region, significance evaluations would normally involve extensive subsurface testing, using techniques such as shovel tests, auger holes, coring, or test pits. The guidelines call for shovel tests or auger holes to be placed every five to thirty meters, depending on artifact densities and surface visibility. On a very large site, however, that sampling interval could result in the excavation of tens of thousands of shovel tests, with the testing constituting a major impact to site integrity. In consultation with the state historic preservation officer (SHPO), other approaches may be negotiated, such as using nondestructive remote sensing in lieu of most of the test excavation that would normally be undertaken pursuant to state guidelines (see chapter 4). The resulting scope of work might call for surface mapping, remote sensing, and some small-scale coring and test pit excavations to verify the magnetometer results. You need to be aware of situations that call for variances from guidelines (sidebar 1.2).

Preparing proposals that are beyond the scope of work is not an adaptive strategy for most archaeological contractors, as potential clients are likely to select the more responsive proposals. However, it is even more disconcerting that experienced researchers would propose a level of effort based on what was customary, not what was warranted and appropriate.

CONCLUSIONS

Archaeology is an exciting undertaking. Despite the potentially tedious individual tasks that comprise archaeological research, the excavation of material remains of the past can inform us about the human condition in ways that cannot be accomplished through any other means. Ultimately, we are studying human behavior, not features and artifacts. House floors, pottery sherds, and stone tools are the indirect indicators of those behaviors, not ends in themselves. As archaeologists, we should ask ourselves often what we are doing and why. If we remain focused on our ultimate goal of understanding human behavior, of doing the anthropology of past societies, we will be reminded that our work matters. Others care about the past, and others are affected by our research. We shouldn't propose or conduct a certain level of effort or a certain approach simply because it is customary. We should do so because it is necessary and warranted, thereby doing archaeology by design, not by default (see Toolkit, volume 1).

ORGANIZATION OF THE VOLUME

Details of the excavation activities at the Eaker site and several other interesting sites will be discussed, usually in sidebars, throughout this book. Examples come from a variety of locations and a range of site types, presenting issues, answers, and techniques that might relate to research opportunities throughout the country. Chapter 2 presents an overview of the history of archaeological excavation in the Americas and beyond. In chapter 3, we will explore the inherently destructive nature of archaeological research. Given this reality, our work can and should be justifiable in terms of research goals and ethical considerations. Also included is a consideration of ways to control the destruction by maintaining control of artifact and feature provenience within a site.

Site characterization and test excavation techniques are discussed in chapter 4. We will revisit the Eaker site to examine the results of test excavations. Traditional hand excavation techniques are compared with mechanical trenching as a basic search strategy. Chapter 5 considers data recovery. Chapter 6 discusses the treatment of special samples for dating, flotation techniques, and the treatment of human remains, including reburial. In a way, the development of these specialized techniques reflects the changes in the discipline as a whole, so at the end of chapter 6 we consider the future of excavation.

Much of what we do as field archaeologists today has a substantial history, and our techniques changed substantially as archaeology grew into a scientific discipline. What follows in chapter 2 is abbreviated history of archaeological methods, with broad geographic coverage.

2

EXCAVATION

A BRIEF HISTORY

"We have two kinds of empirical data in archaeology," W. W. Taylor lectured in his last method and theory class. "We have artifacts with their physical and chemical specification, and we have their locations in space. All else is inference."

It was the spring of 1974, and Southern Illinois University was on the forefront of the "New Archaeology" that was sweeping the nation. But whether deduction or inference was used to impute meaning to the results, Taylor had hit the empirical nail on the head. What distinguishes archaeology from pothunting is that we know where the artifacts came from. Using precise spatial information on artifact context, we can deduce or infer patterns of cultural behavior and evolutionary trajectories of technologies. These concerns govern the conduct of archaeology to the current day, and they certainly inform and even direct how we dig.

Spatial control, or provenience, of artifacts recovered in the field is critical in excavations and is one of the most powerful tools we have for investigating archaeological deposits. With good spatial control, we can tell what groups of artifacts came before another group of artifacts, and we can tell that one area of a site was used for a temple, and another was the place where canoes were made. Some kinds of analysis require spatial precision of millimeters, and other kinds require much less precision, on the order of hundreds of meters. The hallmark of modern archaeological excavation is the use of a three-dimensional X, Y, Z Cartesian grid. This grid is easily manipulated using standard mathematics, makes precise measurement easy, and can be tied into regional Geographical Information Systems (GIS) analysis.

Looters merely obtain artifacts from the ground. Archaeology does have its roots in such activities in the eighteenth and nineteenth centuries when the European empires plundered the classical antiquities of the Mediterranean, but such activities are no longer acceptable. In this chapter, we review the development of excavation and related analysis techniques as a means of defining certain basic concepts. This history focuses on the development of the techniques of spatial control, the essence of modern excavations.

THE EMERGENCE OF PROVENIENCE CONTROL

The first part of the nineteenth century saw the emergence of maps and scaled drawings that gave some provenience control, though only for large architectural elements. Austin Layard conducted the excavation of the Royal Assyrian palace at Nimrud during the 1840s (Layard 1849). His book, a best-seller, has maps of his excavations that show the location of the specific architectural remains (figure 2.1). Letters key rooms, and numbers identify architectural elements, which were enumerated in appendices. (Presumably, he also numbered the artifacts.) He also included scaled drawings of some of the architectural elements. A major find was the royal archives, containing thousands of texts in an undeciphered cuneiform script accidentally preserved by being fired when the palace burned. Three readable Egyptian hieroglyphic inscriptions were discovered as well, allowing the texts to be cross-dated to the Egyptian dynastic chronology that was rapidly being deciphered from the monuments and papyri. Suddenly, datable archaeological contexts in Mesopotamia were extended back in time, from 600 B.C. to 3000 B.C. Many of the accounts in the Bible, once thought to be mythical, were given a new reality as the capitals of many of the legendary cities were investigated over the next century. Investigators were beginning to grasp the importance of provenience as the meaning of finding an Egyptian text in the Euphrates began to sink in: These distant monarchs were corresponding with each other. This was meaningful provenience on a regional scale of hundreds of kilometers.

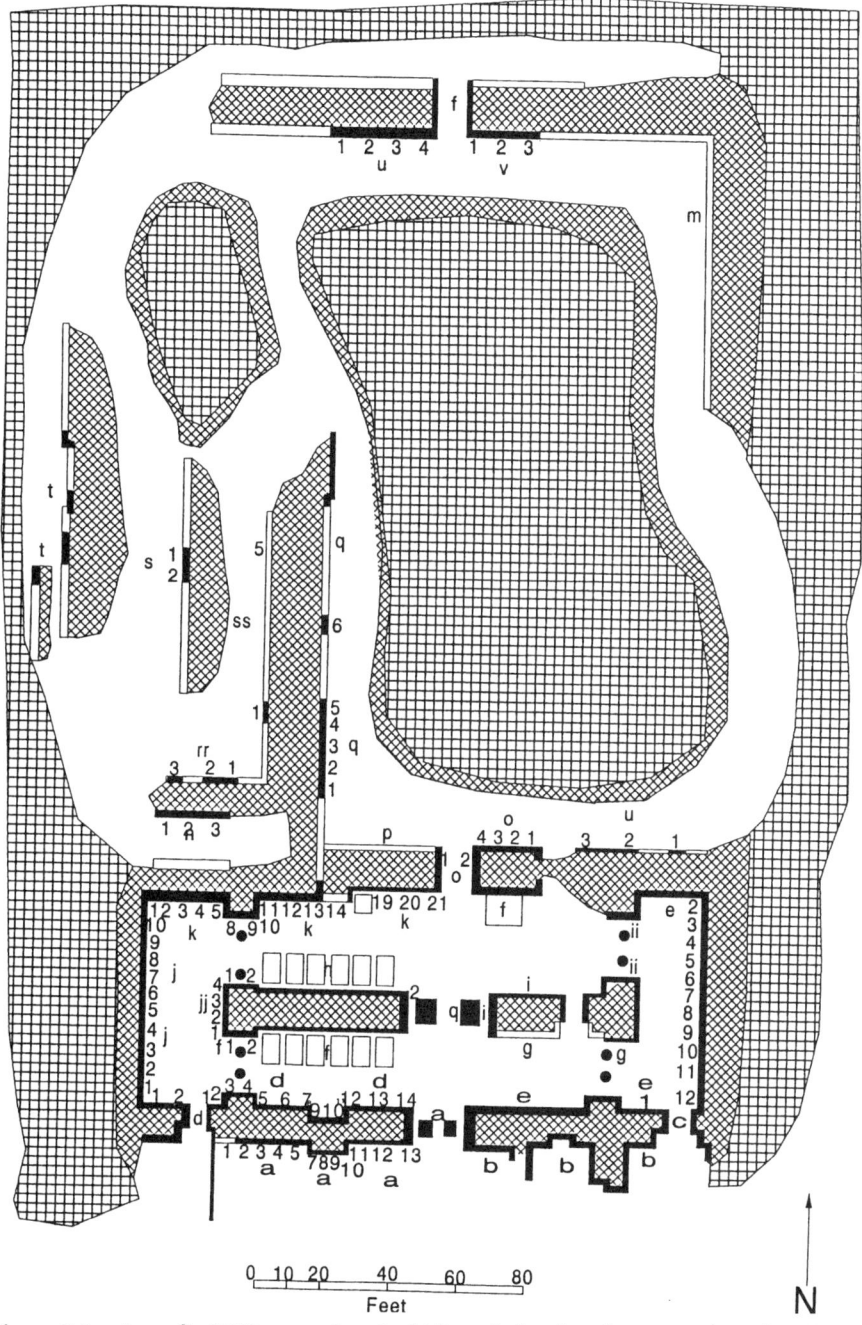

Figure 2.1. Layard's 1849 map of part of Nimrud showing the excavation plan. Room parts are lettered, and the specific architectural elements are numbered. (After Layard 1849: 35.)

Meanwhile, back in England, General Augustus Henry Pitt Rivers was developing methods of three-dimensional spatial control that would change archaeology into a science capable of dealing with prehistoric sites. He had been intrigued by the concept of progress in the 1851 exposition. The idea of progress was an early form of cultural evolutionism that saw a progression of forms leading in a unilineal fashion through stages of development that culminated in the current pinnacle of evolutionary success, Victorian Europe. This was used as a justification for the formation of empires. Pitt Rivers began collecting weapons by type to illustrate their technological development. His interest in technological development led him to excavate sites with a precision almost unparalleled today, piece-plotting even the smallest artifacts. His excavations on the Roman fortifications at Crissbury Hill Fort (1875) in southeastern England discovered a distinctive kind of pottery under the fortification wall, leading to the realization that these ceramics predated the construction of the fort. He excavated the twelve-foot-deep fortification ditch (figure 2.2), where he documented artifacts in obvious strata, then left the ditches open and reexcavated parts of them later to get data on infilling rates. About 1882, he inherited the 27,700-acre Pitt Rivers estate and began excavation in earnest. He documented this work in four lavishly illustrated quarto volumes (Pitt Rivers 1887–1898).

Pitt Rivers's procedures, including piece plotting and stratigraphic excavation, were so much more detailed than his colleagues' methods that he perceived that they thought him "eccentric" (Fagan 1985:103). However, his ideas were so current (Tylor's [1871] evolutionary theory was the academic rage) and his results so remarkable that he could not be ignored. He was tremendously influential on the young generation of archaeologists then in training, including Leonard Woolley, Flinders Petrie, Mortimer Wheeler, and Howard Carter, who would be on the forefront of developing his methods into a new archaeology. These Britons, as well as Heinrich Schliemann and others, employed stratigraphy, seriation, and cross-dating as tools to understand archaeological deposits, sites, and cultures.

Stratigraphy is perhaps the principal concept employed in archaeological excavation. Pitt Rivers borrowed the concept from geology. Stratigraphy is based on the law of superposition, which holds that layers above are more recent than those beneath; that is, the layer on

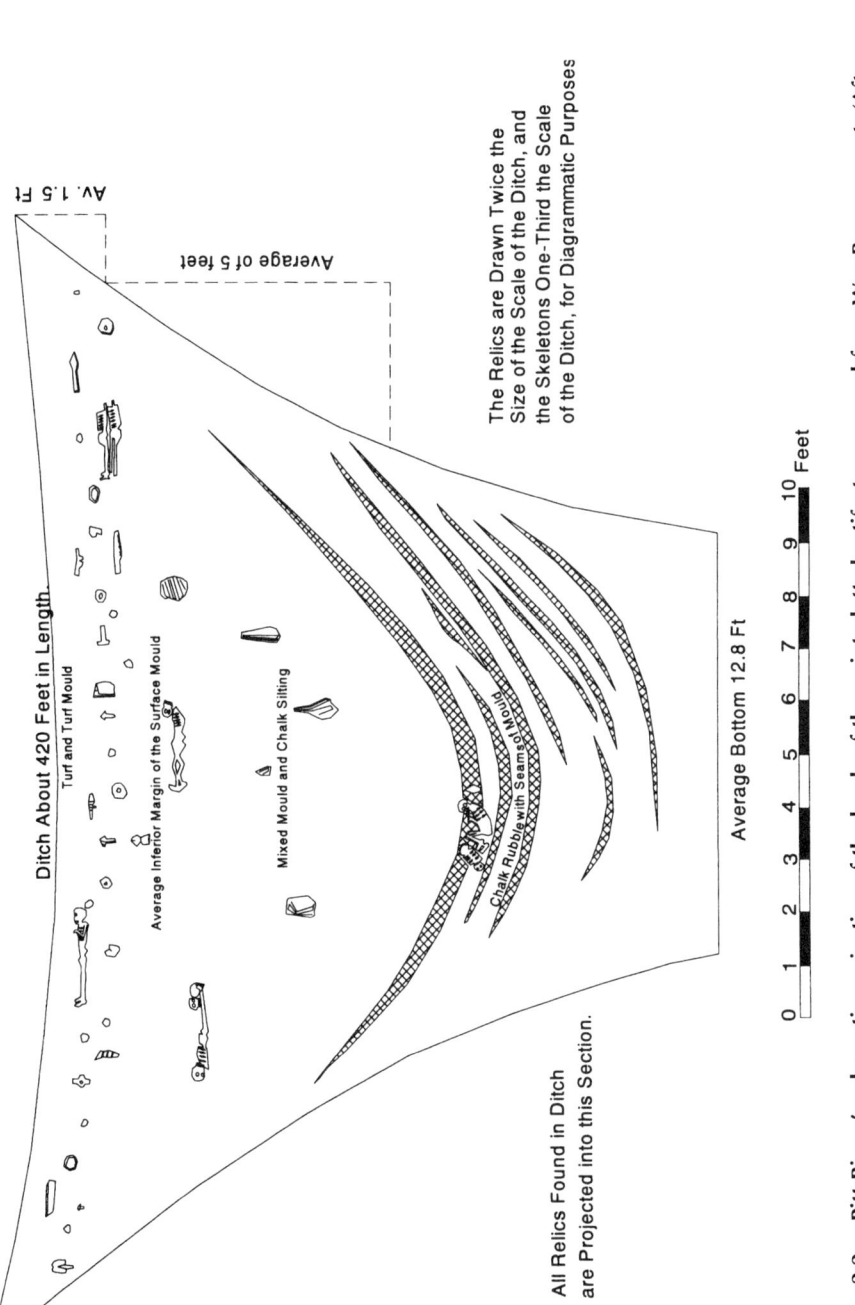

Figure 2.2. Pitt Rivers's schematic projection of the levels of the point plotted artifacts recovered from Wor Barrow moat. (After Wheeler 1954:26.)

the bottom is the oldest. In geological terms, strata usually span hundreds of kilometers, but at the scale of archaeological sites, the strata are, at most, several hundred meters wide or long. Nevertheless, the principle holds, with very few exceptions. Interpreting sequences built by stratigraphic analysis gives time depth to archaeological deposits.

Seriation refers to chronological ordering of archaeological units using changes in style or technology through time (see Toolkit, volume 4; also Lyman, O'Brien, and Dunnell 1997). Flinders Petrie (1901), working in Egypt, developed the first application when he tried to chronologically order hundreds of predynastic and early dynastic graves. Petrie's work extended the Egyptian relative chronology several thousand years into prehistory.

Petrie found two wares of Mycenaen pottery in twelfth- and eighteenth-dynasty contexts of known age, thereby dating those wares (Rapport and Wright 1964:183). This is an example of using an artifact type as a *horizon marker*. These are distinctive artifacts that have very wide spatial distributions and were produced for only a short period of time. Sites in different regions can be cross-dated on the basis of such horizon markers.

NORTH AMERICAN DEVELOPMENTS

The development of excavation methods in North America was similar to that in the Old World (Willey and Sabloff 1993). Thomas Jefferson was one of the earliest-documented excavators (Heizer 1959:218–21). He excavated a mound in 1784 to determine whether it was a mass grave from a battle, a place for reburial, or a town sepulcher. He cut a trench through the mound, described its stratigraphy, and documented the conditions of jumbled masses of bones at different levels. He observed that the burials on the top were less decayed and concluded that the mound was built through a series of periodic reburials. Jefferson's work was a precocious example of a modern approach to archaeology.

The driving impetus for subsequent excavation was to fill museums with artifacts, but there was also a specific research problem to investigate. After the American explorers and colonists crossed the Appalachian Mountains, they found large earthworks. Some conceived of a race of Moundbuilders who built the earthworks prior to the arrival of the Indians. Although there was much discussion about this "theory" and others that attributed the earthworks to lost Vikings, lost Welshmen, and the lost tribes of Israel, most serious investigators were fairly certain that Indians built the earthworks (Williams 2001).

In the 1880s, John Wesley Powell, director of the Bureau of American Ethnology (BAE), received money from Congress for surveys and excavations at mound sites of the eastern United States to test the Moundbuilder myth. Some of the BAE-affiliated excavators kept careful records and made excellent site maps. Cyrus Thomas (1894) reported his results in the twelfth annual report of the BAE. This report and the BAE would loom large in the archaeology of the following century.

Some investigators were better than others. The best excavators prepared site plans of the mounds. Some of these are the best maps we have of many of the large sites. Thomas summarized the contents of different mounds and sites and, for the most part, recorded artifact provenience only to the site level. By showing that Indians had built the burial mounds, he largely quashed the Moundbuilder myth. The resulting collections led to the first areal synthesis of eastern United States ceramics (Holmes 1903) and many other studies.

For the next several decades, Thomas's methods were the model, and data recovery was at the site level. Large museums sponsored numerous excavations, many of which recognized pit features and at least rudimentary stratigraphy (Mills 1907; Fowke 1922). Most of this work was in the Southwest, and it was there that the next major advances in the science of excavation took place.

In 1916, Nels Nelson, familiar with Pitt Rivers's work (O'Brien 1996:156), published an article in *American Anthropologist* on excavations he had carried out for the American Museum of Natural History at the Tano Ruins in New Mexico (Nelson 1916). His discoveries and methods would shake New World archaeology. Nelson cut *volumetrically controlled units* (excavation levels of the same volume or of known volume so that the quantity of artifacts can be meaningfully compared) out of a four-meter-thick profile of the midden adjacent to the Tano Pueblo. He divided the recovered sherds into different "types" based on perceived attributes and plotted their distribution by excavation level. Doing so demonstrated the presence of stratified habitation sites in North America and introduced volumetric quantification into archaeology. The former would in short order lead to spectacular discoveries of stratified deposits, but the latter would take several decades to seep into mainline archaeology. In 1919, the University of Arizona offered the first field school in archaeology (Gifford and Morris 1985). A. V. Kidder's excavations at Pecos Pueblo between 1915 and 1922 uncovered very deeply stratified deposits (figure 2.3; Kidder 1962). Kidder excavated a trench into the talus slope of the mesa, assuming that the most recent town fortification wall was on the edge of the mesa. The talus was much larger than Kidder had imagined. He had expected the work to take one season, but after two seasons, there was no end in sight. What he had thought was one ruin was at least six superimposed towns with defensive walls buried as much as seven meters deep. Kidder reserved large blocks for careful stratigraphic excavations. From the photographs, the blocks appear to be divided into 3 × 3–foot units with six- to twelve-inch cuts. The stratigraphic sequence allowed Kidder to define changes through time in ceramic styles. He confirmed and expanded Nelson's sequence at Tano, located only several miles away. It seemed clear to Kidder that the length of the pueblo period was much longer than previously suspected. How much earlier the human occupation began was about to be discovered.

During several summers, Kidder invited colleagues to Pecos to see and discuss the excavations and results. The Pecos Conference turned into an annual gathering, the first regional archaeological conference. It was no accident that Kidder was chairman of the Committee on State Archaeological Surveys of the National Research Council, which was trying to establish state, regional, and national archaeological organizations. At these conferences, archaeologists compared potsherds and results from their latest excavations, which accelerated the spread of new techniques of excavation. From that time forward, in the Southwest, excavations were increasingly conducted using volumetric control, and these techniques spread to the Southeast in the 1930s.

DIRECT ASSOCIATION AND THE SEARCH FOR EARLY HUMANS IN THE NEW WORLD

In 1926, J. D. Figgins of the Denver Museum of Natural History was seeking evidence for the antiquity of humans in the New World. At the Folsom site in New Mexico, George McJunkin discovered a spear point in direct association with bones of extinct Pleistocene bison (from the Ice Ages in which the period from forty thousand to ten

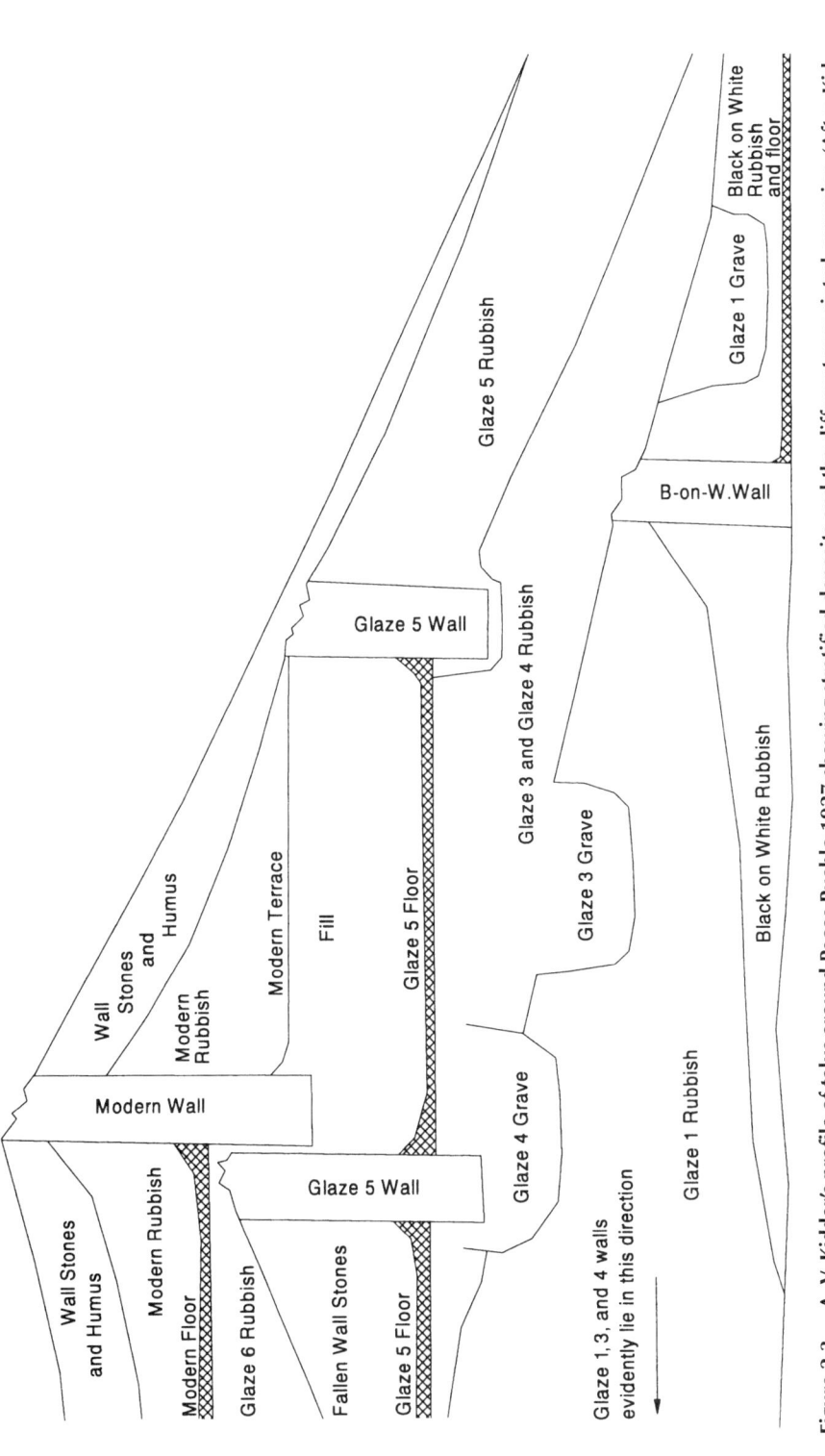

Figure 2.3. A. V. Kidder's profile of talus around Pecos Pueblo 1927 showing stratified deposits and the different associated ceramics. (After Kidder 1962:103.)

thousand years ago is the probable time of human entry to the Americas) (Figgins 1927; Wormington 1964 has an engaging account of this discovery). This was such an important association that Figgins photographed the point and associated bones and removed it as a block to his laboratory. Over the following winter, he tried to convince colleagues of the authenticity of the associations. Some accepted his evidence, but others would rather have seen the point firmly embedded in the skull before they would accept the evidence. Figgins continued working and found a point firmly embedded between two ribs. He stopped the excavations and called all interested colleagues to the site. Archaeologists from all over the country dropped what they were doing and rushed to see the find *in situ* (a find still in place in the context where it was deposited). They were convinced. Figgins's precise careful excavation techniques and precise recording techniques changed American archaeology forever. His documentation of the artifact association to the nearest millimeter was of significance.

THE GRID

At southwestern pueblo sites, it became standard practice to excavate through the prehistoric construction debris and midden deposits until room walls could be defined, at which point excavation switched to the clearing of individual rooms. In the East, sites seldom have rooms indicated on the surface, prompting use of grids for within-site control. *Grids* are units of regular size and shape laid out over a site in order to locate items conveniently in relation to each other. While the earliest use of grids was by Charles Peabody in his 1903 excavations at Jacobs Cave in Missouri (O'Brien 1996:158–69), it was not until the late 1920s that grids became routinely used in excavations. The shift to the use of grids is evident on site maps of the time, as excavations transform from round holes to a square-patterned grids. Claflin's (1931) excavations on Stalling's Island between 1921 and 1929 actually show this transition. The rectangular grid trenches are from the 1929 Harvard excavations, and the round pits and trenches were earlier.

The University of Chicago anthropology department, under the leadership of Fay-Cooper Cole (see Cole et al. 1951), played a pivotal role in the development and dissemination of the use of grids on archaeological sites. Cole started the Chicago field school in the 1920s. The first field school in the East, it was extremely influential because it trained many students who became leading archaeologists. Cole lectured on field methods at the first Southeastern Archaeology Conference, held in 1932 in Birmingham, Alabama. He advocated total recovery, even the recovery of items for which researchers had no immediate use, such as carbon and snail shells. The best way to recover everything was to grid the site into conveniently sized units, such as five-foot squares, and collect everything in a block sliced horizontally with a razor-sharp shovel. This became known as the "Chicago method" (figure 2.4). The meeting helped establish methods

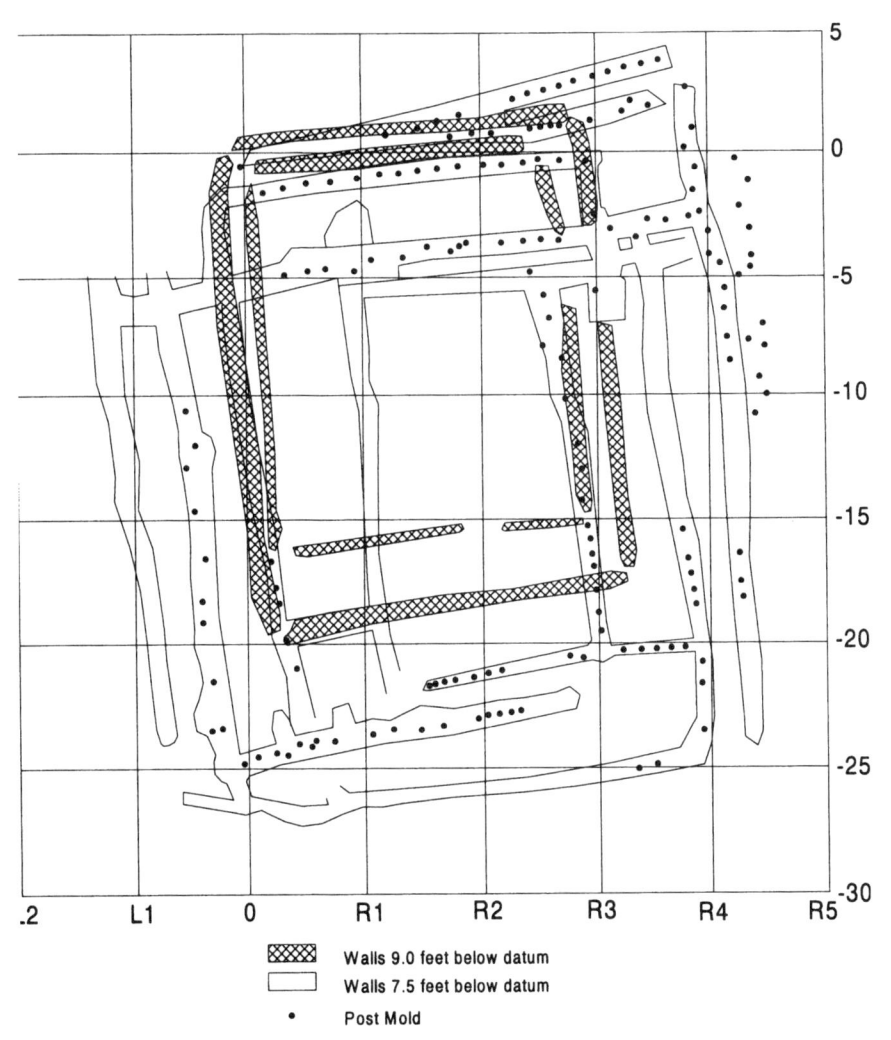

Figure 2.4. Chicago grid at the Kincaid Site of MxvlA showing the superimposed wall trenches. (After Cole et al. 1951:45.)

for the large archaeological projects of the Works Progress Administration (WPA) in the 1930s, which put thousands of people to work excavating sites throughout the Southeast. The collections resulting from these WPA projects still contain artifacts and samples that are of great importance today.

The "Chicago grid" was set up in feet and staked at five- or ten-foot intervals. A base line was run through the approximate center of the site. Parallel lateral lines were labeled L1-*n* and R1-*n* (left or right 1-*n*), and the other axis was labeled by plus or minus feet. This procedure made it possible to label each unit uniquely, such as L1-5, based on the designation of the southwest corner. Its application in the Southeast came just in time to record the large amounts of data generated when the WPA excavations began in earnest in 1934. The use of grids reached as far west as Nebraska by 1939 (see Hill and Kivett 1949).

THEORETICAL DEVELOPMENTS

In 1948, W. W. Taylor published his Harvard dissertation, *A Study of Archeology*, in which he analyzed Americanist archaeology and compared what practitioners said they were doing with what they were actually doing. He concluded that, while archaeologists said they were doing anthropology, most were in fact doing historiography. Taylor's polemic stance led him to become essentially ostracized by most of the profession, although he received thirteen National Science Foundation grants to pursue his conjunctive approach to archaeology in the dry bluff shelters of northern Mexico.

During his senior year in college, Lafferty found a copy of Taylor's *A Study of Archaeology* in a bookstore. After reading it, he asked his archaeology professor why he had never been told about this very important work, which laid out some methods that could be employed to find and define cultural patterns. Lafferty was admonished, "We do not talk about him among our people." In January 1968, he chose to attend graduate school at Southern Illinois University because Taylor was teaching there. He enrolled in Taylor's very challenging course on Old World archaeology. Taylor was terrifyingly frank in his comments in class and expected the same of everyone else. However, after one student's presentation in a methods and theory class in which the student had been very critical of a senior archaeologist, Taylor told the student, "You can say it that way, if you want to spend the next twenty years talking to yourself. . . . I know." He was also very thorough, reading every paper three times, once to get a general sense, once for English, and finally making detailed comments on content. Lafferty got excellent feedback from Taylor. He eventually was offered the chance to work with Taylor in northern Mexico, but the draft board decided he should visit Vietnam instead. After his return in 1971, he found that the world of archaeology had changed.

Lewis Binford led the new archaeology of the 1960s and 1970s. Southern Illinois University (SIU) was on the edge of this wave because Taylor, the prophet, was there and because Binford briefly worked at SIU. Carmichael also received early exposure to the New Archaeology, at the University of New Mexico (UNM), where Binford was teaching and where Taylor's book was required reading. Binford (1962) argued in an influential article entitled "Archaeology as Anthropology" that archaeologists should be doing anthropology. That is, they should be working on defining and explaining cultural processes, rather than simply generating descriptions of past cultures and sequences of descriptions of past cultures.

In his classes at UNM, Binford's critique of traditional archaeology was incisive and often polemic. He argued that our primary challenge in interpreting the past is generally *not* any inherent limitations in the archaeological record but, rather, our ability to ask useful questions of it. "The material record does not speak forth self-evident truths," he would say. Knowledge proceeds from warranted arguments, not by appeal to authority; even well-known senior archaeologists could be wrong! There are no data except in response to research questions. Archaeologists always ask such questions, either implicitly or explicitly, the latter being preferable so we can more readily evaluate our ideas about how the world works (i.e., how the material record was formed). While analogy is important to archaeological interpretation, archaeologists should not feel bound by specific historical (i.e., ethnographic) analogies because prehistory encompasses many events and processes that lack modern or historic analogs.

The New Archaeology prompted archaeologists to be concerned about sampling and sampling design. Questions about consistency of samples led to the adaptation of screening. Screening had been done sporadically since the 1930s, when the Lower Mississippi Valley Survey began screening samples to assure consistency. In the early 1970s, systematically screened samples became the standard over much of North America. Another innovation of the late 1960s and the 1970s was the use of controlled surface collections to sample plow zone deposits. These kinds of collections can tell much about the structure of a site, as discussed in volume 2 of the Archaeologist's Toolkit series.

THE BEGINNINGS OF CRM

During the 1950s and 1960s, much of the archaeological excavation being conducted in the country was salvage work in the lakes and reservoirs that were rapidly being built. This salvage work was partially modeled on the earlier WPA projects, but with smaller crews of only ten to twenty persons. The Smithsonian Institution ran the River Basin Survey and excavated many large sites in the Missouri River basin on the Plains. The National Park Service administered other salvage programs that were conducted by university departments of anthropology or museums. Usually, the sole archaeologist in the state conducted this work. This process changed radically after 1966 (see Toolkit, volume 1).

Perhaps two aspects of CRM archaeology should be highlighted here because of the effects they have on how excavations are conducted. First is the change in how sites are selected for excavation. For years it was customary to conduct excavations at the largest, most visible sites in a region to maximize the recovery of diagnostic artifacts, record the longest stratigraphic sequences, or interpret prominent sites for the public. Conversely, in CRM archaeology the study areas, and therefore the sites available for consideration, are typically determined by the extent of the proposed project. As a result, archaeologists have been forced to evaluate and excavate many smaller or less visible sites that were overlooked in the past. This has been a good thing, as these classes of sites have much to tell us about settlement patterns and the diversity of prehistoric adaptive strategies. Indeed, even in parts of the Southwest and Southeast, where excavations have been under way for more than a century, new information is gained from almost every excavation.

The other major change involves the recognition of multiple constituents and consumers of archaeological research. In addition to our colleagues, we now answer to government regulators, agency administrators, private sector clients, and the public. Clearly, the public includes a variety of ethnic groups, but a few groups have special connections to the archaeological record. The most important of these are Native Americans. Unfortunately, for much of the past two or three decades, relations between Native Americans and archaeologists have often been adversarial, due largely to disagreements over the disposition of human remains and associated artifacts. The passage of the Native American Graves Protection and Repatriation Act of 1990 (NAGPRA) substantially revised the relationship between archaeologists and Native Americans. Native American groups now have a formal role in deciding whether and how human remains will be studied and whether they will be reburied. These issues directly affect how excavations are conducted and how skeletal remains are treated in the field and laboratory.

THE REAL INDIANA JONES

Indiana Jones, realized in the personification of the swashbuckling archaeologist, seems to have captured the popular imagination, perhaps even inspiring some among us to pursue archaeology. As W. W. Taylor also said in his lectures, we should pay attention to our metaphors because they sometimes become tenets of science. Some of the footage from *Raiders of the Lost Ark* metaphorically spans the transition from using no spatial control to detailed mapping. Jones is essentially looting when he grabs the golden idol and runs, with a very large stone ball rolling after him. Yet it took precise spatial control under stressful conditions to place the Staff of Ra (which also had to be exactly the right height) in exactly the right spot on the correct day and at the correct time to find the ark's location. Then he had to translate that location into the Germans' grid of the site of Tanis to find the ark.

Archaeological excavation demands precise recovery of spatial information under demanding (or even nasty) field conditions. One must make careful observations and accurate measurements and get them down on paper while the sweat is dripping on the paper or as one's fingers are shaking with cold. Snakes, scorpions, and other vermin are real obstacles, though seldom in the quantities Indy encounters. One of us (R.H.L.) once had to hold off the looters of the past at gunpoint. Another (D.L.C.) has witnessed field crews threatened at gunpoint in Texas and Illinois because of the imagined transgressions of CRM clients.

Indiana Jones is a Hollywood creation, but the physical stress he portrays is real in archaeology, and the rewards of excavating the long-lost residue of everyday life can be every bit as exciting as finding the Ark of the Covenant. Even more exciting is seeing, understanding, or elegantly describing a cultural pattern or a previously unperceived scientific abstraction.

Modern archaeology owes a great deal to the real archaeologists who paved the way with genius and a willingness to experiment with new approaches that gave them increasingly more sophisticated spatial and temporal control in their excavations. They recognized early on that archaeology is as destructive a process as looting if done without systematic approaches to excavation. The subject of the next chapter, which gets right to the heart of the matters of stewardship and conservation of archaeological sites, is about the attitudes we bring to our work and the obligations we owe as scientists to our publics.

3
ARCHAEOLOGICAL EXCAVATION IS CONTROLLED DESTRUCTION

Excavation is inherently a destructive activity. Archaeological excavation differs from looting and other kinds of digging because of the motivation for excavation, the methods used, and the treatment of the resulting collections.

It is a professional responsibility for archaeologists to be familiar with the Indiana Jones movies, if for no other reason than to understand the stereotypes and misunderstandings nonarchaeologists often have about what we do. The movies also provide a point of departure for considering what archaeology is and how it should be done. For the most part, what passes for archaeology in the movies would be considered looting by today's standards. Images of hundreds of laborers flinging dirt in every direction may seem romantic to the average viewer, but they represent poorly controlled excavation, which would result in the loss of most of the information available at the site. Of course, the "bad guys" are merely treasure hunting, but even Indiana Jones brings back a few gold ornaments wrapped in a scarf to sell to the museum to underwrite his expedition—looting, by today's standards. Nevertheless, digging and the removal of artifacts are part of even the most cautious archaeological excavations. Archaeological excavation is controlled documented destruction.

In *The Last Crusade*, Professor Jones defines *archaeology* as the search for facts (as opposed to truth) about the past, thereby recognizing the interpretive, conditional nature of our knowledge. He further notes that most archaeology is done in the library, not the field. This view may not be far from the reality of the situation today, but library (or even laboratory) research is not the way most people, including archaeologists, would characterize the discipline.

Archaeology is the study of (mainly) past human behavior through the examination of the physical byproducts of that behavior. True, archaeology includes ethnoarchaeology and modern material culture studies, both of which involve contemporary peoples, but the discipline is defined by its access to the past. The main contribution of archaeology to the social sciences is its ability to define and analyze long-term changes in human behavior and adaptation to changing environments. Archaeology provides time depth to our understanding of the human condition beyond what can be attained from the study of historical records.

If one envisions archaeological deposits as three-dimensional archives, horizontal movement across the surface represents change in location, or the spatial dimension. Vertical movement through the archive involves the temporal dimension. To go back in time, we go down through soil layers, and we do it with archaeological excavation. Excavation is the means by which we access the past, the most basic, defining aspect of archaeology. Unfortunately, archaeological sites are finite, nonrenewable resources. Once disturbed, they cannot be replaced or restored to their original condition.

Running a bulldozer through a site to build a building or highway is destructive, but archaeological excavation is, in a sense, just as destructive. Although digging a site takes longer with shovels, trowels, and paintbrushes than with a bulldozer or backhoe, the end result is similar: The archaeological deposits are gone forever. Archaeology is one of the few scientific disciplines in which the research methods destroy the objects of our study. As a result, we have an important responsibility to justify our excavation activities and to use state-of-the-art techniques.

Because archaeological destruction proceeds more slowly and methodically than looting or construction disturbances, it is possible to make detailed records of artifacts and features before their depositional contexts are destroyed. This recording process is an absolute requirement in archaeology, distinguishing our work from the actions of others who dig in sites and to some extent justifying the consumption of our archaeological resources. However, it is not possible to record every potential observation or piece of information that could be obtained from an archaeological site. The choices of what to record and what not to record are (or should be) made in response to the questions we seek to answer; there are no data except in response to questions.

JUSTIFYING THE DESTRUCTION

Destruction of a site through archaeological excavation is justifiable if the work is done responsibly and ethically. We must make defensible choices about where to excavate, how to recover the data, and how to report our research. In addition, we must take care to protect the site during excavation, to ensure that the useful data are actually recovered and not destroyed by the elements while they are exposed.

In other words, the careful destruction of archaeological sites can be justified when it is the result of an appropriate research design. Or can it? Can contemporary archaeology be characterized as research? Some archaeologists have suggested that archaeology conducted in the service of CRM is not research at all (Dunnell 1984:67–68; Patterson 1978:134). If one accepts this view, the implication is that contemporary archaeology (which is mostly CRM) is not research oriented and, therefore, cannot be used to justify the destruction inherent in archaeological excavation. Is this position tenable?

Dunnell (1984) describes archaeological research as a process that begins with a problem or a research question to be answered. The researcher then chooses the location where excavations will be undertaken to answer the question and the methods that will be used to obtain the necessary data. It is emphasized that the archaeologist makes the *choice* of research methods and areas in response to the research problem. Dunnell argues that CRM archaeology does not follow this research model because most archaeologists do not have a choice about where they work:

> The generation of a CRM project has nothing whatsoever to do with an archaeological problem. CRM is generated by nonarchaeologists and concerns potential impact on a particular piece of real estate. The initial input is spatial. . . . The notion of relevance that guides the selection of resources in problem-oriented research is replaced by a less well-defined notion of significance. (68)

This viewpoint, intended to emphasize a dichotomy between academic and CRM archaeology, can be challenged on at least three grounds. (1) Archaeology is not the only discipline that must operate in the context of externally generated restrictions and demands. In spite of strictures on research design, scientists in other disciplines have somehow managed to conduct research. (2) Archaeology has been coping with external constraints for some time. One cannot choose to excavate a site where the landowner has not given permission for access, even if it is the "key" site necessary for answering the most important research question in the region. One cannot choose to excavate sites in a foreign country without obtaining the necessary permits from the host country. One cannot import soil samples from foreign countries for laboratory analyses without submitting to strict U.S. Customs controls on processing and disposal methods. Moreover, funding constraints have been endemic in archaeology for a long time; rarely, if ever, does a researcher have the financial resources to pursue whatever studies he might choose. Even (or perhaps especially) in the world of non-CRM archaeology, researchers have had to design their studies with funding constraints in mind. (3) Dunnell's view seems to reflect inflexibility in the use of problem-solving strategies (Adams 1986:71). Overreliance on a single, traditional approach may create unnecessary impediments to research:

> Just as we use physical tools for physical tasks, we employ conceptual tools for conceptual tasks. To familiarize yourself with a tool, you may experiment with it, test it in different situations, and evaluate its usefulness. The same method can be applied to conceptual tools. Our ability as thinkers is dependent on our range and skill with our own tools. (Adams 1986:76)

Making the best use of archaeological sites that must be consumed can be challenging, but the research opportunities provided by contemporary CRM cannot be overlooked. It is well known that the earliest investigations in many parts of the world focused on the largest and most obvious sites or those closest to major access routes. One of the main advantages of CRM work is that researchers are sometimes forced to investigate areas and sites that would have been overlooked in the past (see sidebar 3.1 and Toolkit, volume 1).

In fact, with much of today's archaeological research focusing on regional settlement pattern characteristics, small sites may even be more important than large ones, because they represent newly identified components of settlement systems (Schiffer and Gummerman 1977:242). They are also potentially more easily interpretable because their contents may be less complex than at a larger site with many overlapping occupations (Binford 1982).

It is true that project area boundaries are often oddly shaped, but even this can be used to advantage. Pipeline corridors that are several hundred miles long can be conceived of as transects to sample site

> ### 3.1. THE KEYSTONE SITE
>
> The Keystone site (41EP493), the earliest pithouse village in the western United States, dating back at least 4,500 years, was discovered through CRM (O'Laughlin 1980), as was 41EP492, an unusually late pithouse occupation dating to the "pueblo period" (Carmichael 1985). Both sites (figures 3.1 and 3.2) were discovered in an area of El Paso where construction disturbances related to the Keystone Dam flood control project were anticipated but that had not been examined during previous research in the region. The first evidence of a burned ramada ever identified in the El Paso region (figure 3.3) was discovered at 41EP751, a small site impacted by pipeline construction near the eastern edge of the U.S. Army Air Defense Center, Fort Bliss (New Mexico State University [NMSU] 1989:859–64).

characteristics in various landform settings (Carmichael 1978; NMSU 1989). Individually, a ten-acre well pad study might not contribute much information, but, over a period of years, examination of hundreds of such parcels may generate a useful regional sample for conducting a variety of synthetic and predictive analyses. Even very small-scale excavations can be creatively used to generate useful data sets. Recently, students at the University of Texas at El Paso conducted a small but fruitful research project on materials obtained from test pits excavated where orchard trees were going to be planted within an Archaic campsite on private property (see chapter 5).

Even sites that are partially disturbed or overlain with debris from modern activities may present useful research opportunities (see sidebar 3.2). Disturbance does not automatically relegate a site to an insignificant status. "The failure . . . of any project to yield research results reflects less upon the characteristics of the resources than upon the competence, background, and interests of the investigators" (King and Hickman 1973:366).

Perhaps CRM archaeology projects don't always meet contemporary research ideals, but they can and should (and the same could be said for archaeology generally). Freedom of choice does not define research; how one investigates observed patterns in the archaeological

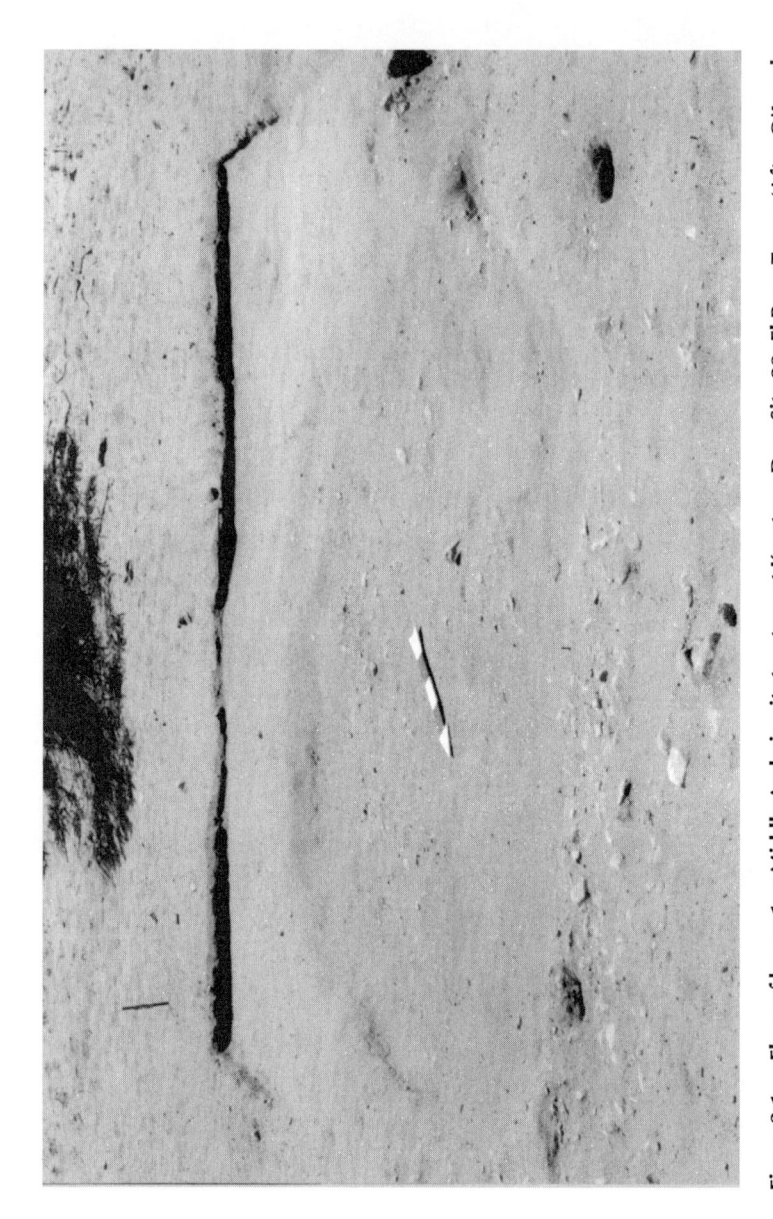

Figure 3.1. Floor of house 1, a Middle Archaic pit structure at Keystone Dam Site 33, El Paso, Texas. (After O'Laughlin 1980, courtesy U.S. Army Corps of Engineers.)

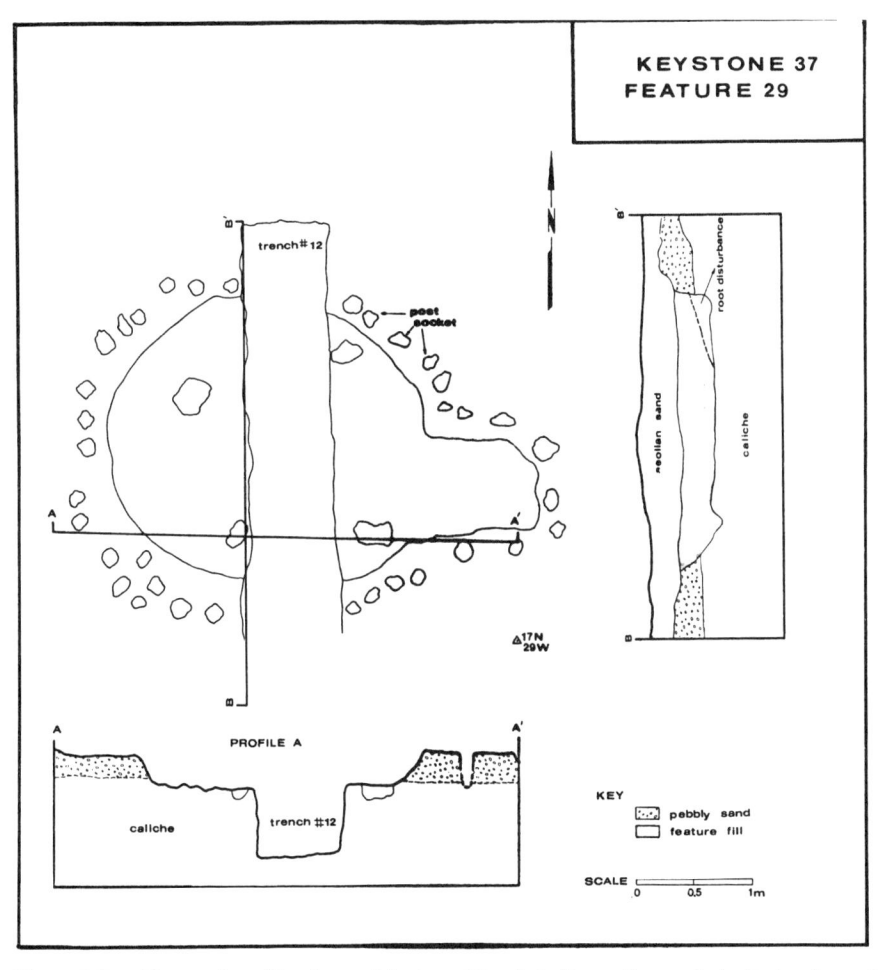

Figure 3.2. Plan and profile views of feature 29, a late Formative period pit structure at Keystone Dam Site 37, El Paso, Texas. (After Carmichael 1985, courtesy U.S. Army Corps of Engineers.)

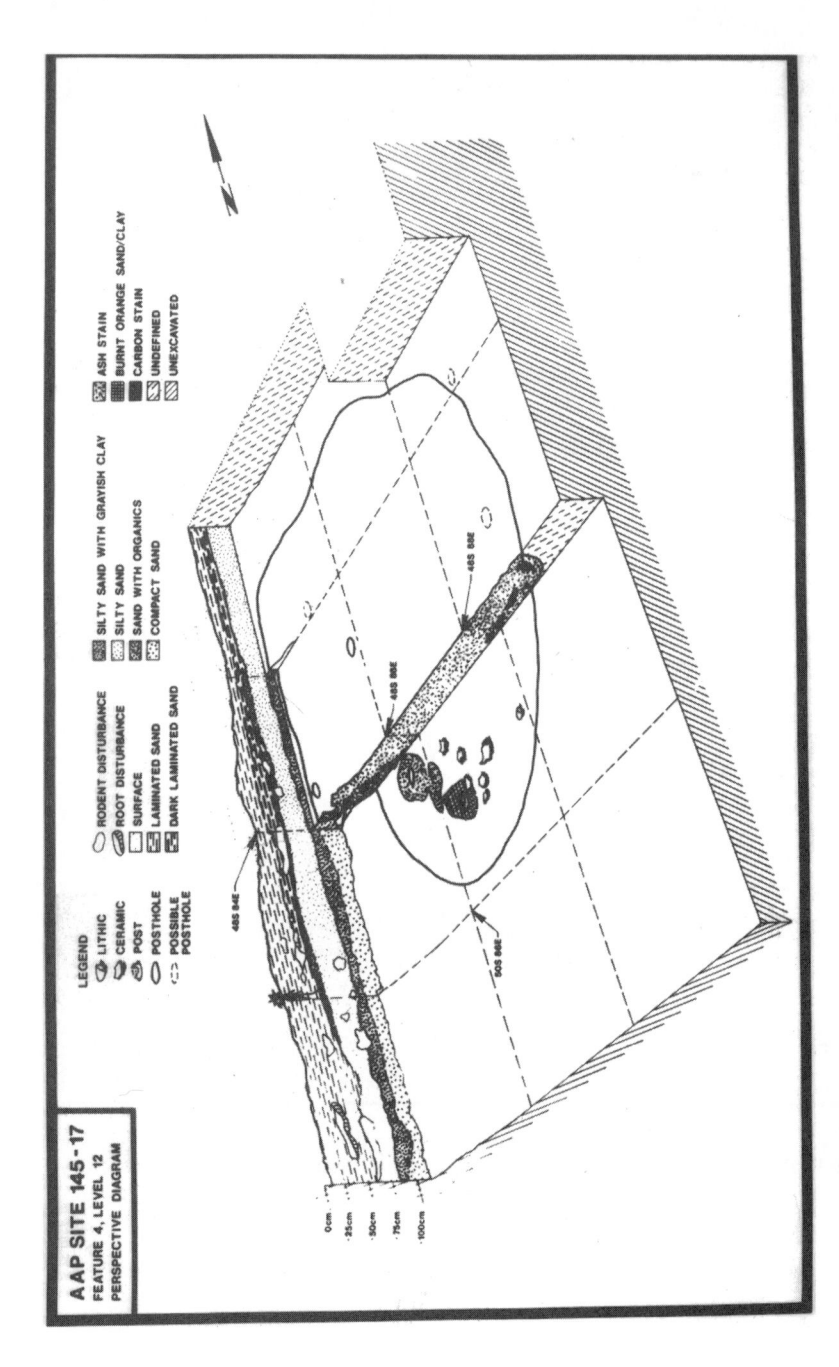

Figure 3.3. Stratigraphic block diagram of a prehistoric ramada structure (showing support posts and burned roof elements, but lacking walls), AAP Site 145–17, Fort Bliss, Texas. (After New Mexico State University 1989; courtesy U.S. Bureau of Land Management, California Desert District.)

3.2. THE FILLMORE PASS SITE (FB-1613)

FB-1613 is a complex multicomponent lithic site located near the Texas–New Mexico border on Fort Bliss. The site contains evidence of Paleoindian, Archaic, and early Formative occupations, and most of the assemblage is probably assignable to the preceramic periods. Excavations yielded the largest collection of Folsom artifacts in the region, including some two dozen Folsom points and preforms, more than sixty channel flakes, and hundreds of scrapers. Unfortunately, the deposits were shallow, and the cultural materials from different time periods overlapped. The original research design called for developing and testing techniques that might be used to spatially separate the partially overlapping assemblages from different time periods (Stiger 1986). The field methods included the standard practice of recording artifact location with point proveniencing or within 1 × 1–meter units, in order to describe any spatial patterning. It is necessary to determine whether patterns differ significantly from random or background distributions (Kellog 1987). How could this be done?

Like many places on Fort Bliss, the area around FB-1613 was once part of the impact area for an artillery firing range. As a result, shrapnel from exploded shells is scattered throughout the site. In the research design, the shrapnel distribution was defined as an independent control on the degree of disturbance to assist in the recognition and description of artifact patterning.

---

record does. Charles Darwin did not choose to visit the Galapagos Islands, nor did he make plans in advance to study variation in finches. He made good use of the opportunities presented to him. Archaeologists should be prepared to do the same.

Twenty-six years ago, Schiffer and Gummerman (1977:130) remarked, "Hopefully, it will not be necessary much longer to justify the development of explicit research designs in archaeology." Unfortunately, it is still necessary. Recently, a colleague reviewed a research design for archaeological work to be conducted in southeastern New Mexico. When the agency representative was asked what specific research questions would be addressed through hypothesis testing, the gist of the response was "Ten years ago we used to do hypothesis testing, but not anymore. It's better to just ask a general question, like whether or not there is any biface reduction technology. We'll be using quarter-inch mesh screen to look for it." Unless one specifies how biface reduction technology will be recognized and how the evidence will be recovered, it seems this project will be a hit-or-miss undertaking. Screening of this sort is likely to miss the evidence for biface reduction technology anyway, as well as whole classes of low-visibility sites such as Apache camps.

Research designs are necessary to ensure that a site's irreplaceable potential to advance knowledge is not squandered (Raab 1984:55). The scientific importance (and National Register eligibility under criterion D) of a site is determined by its research potential:

> In order to determine the significance of a site, enough testing must be done to establish the nature of the potential information that will answer research questions found in the State Plan or any other research documentation outlined in the research proposal. The fact that there are undisturbed deposits of cultural material beneath the plow zone is not in itself enough to say the site is significant. (Davis 1982:B-10)

Today, several states have plans for archaeological conservation that outline archaeological knowledge in each region of the state. The best plans (e.g., Davis 1982) are useful outlines of the research needed in a particular region, at the time they were written. While these state plans may be useful in providing "boilerplate" questions in assessing significance and structuring research questions, they are rarely up-to-date, so it is incumbent upon the archaeologist to be knowledgeable about the current state of research.

Research potential can change in relation to new research questions, new technological capabilities, or the near-threshold of data redundancy for certain resource types. To be current, one must attend conferences and read the journals. A lithic scatter lacking diagnostic projectile points might be written off as having negligible research value. But if the debitage includes many obsidian flakes, the site may have enormous potential to yield information, through chemical sourcing analysis and hydration dating, on population movements and chronology. Therefore, research designs need to be dynamic documents, not boilerplate introductory sections in a project report. It is not only acceptable but necessary to examine sites using new, creative, even experimental technologies. If we routinely avoid doing so, we will fail to define the real research potential of sites, fail to serve our client's compliance needs, fail to address archaeology's conservation ethic, and fail to advance archaeological research.

[A] little reflection will reveal that the successful research design is an outgrowth of the creative application of archaeological expertise to the problem at hand. This endeavor requires on the part of the investigator nothing less than an exhaustive knowledge of archaeological method and theory, a close familiarity with previous research on the problem, and an intimate acquaintance with available information on the archaeological resources. (Schiffer and Gummerman 1977:130)

If a single individual does not command all these areas of expertise, preparation of the research design should be a group undertaking. If you or the staff of your research facility or contracting firm lacks expertise in a particular region or analytical technique, team with or subcontract to the experts. See Toolkit, volume 1, for more details on developing research designs.

EXCAVATION AND THE CONSERVATION ETHIC

Some people believe that archaeologists want to excavate all the sites they find, but this is not generally true. Most contemporary archaeologists work within a conservation ethic, and those who don't, should. We would rather see a site preserved for future excavation, when techniques will be more advanced.

Although the boundaries of our study areas are often determined by external considerations, archaeologists do have a considerable amount of choice about where to excavate. Plog (1984:91–93) outlines a series of questions to consider before we can justify our decisions about which sites to excavate and how much to excavate. Is the site important for its potential to help answer an important research question? Excavating a site merely to train students to dig is not acceptable. The most effective investigations will be those that are directed toward answering specific questions and those that consider carefully how much excavation is required to meet the research goals. Out of economic considerations, we rarely excavate an entire site, and it is usually good practice to sample site deposits for other reasons, too, such as avoiding data redundancy, minimizing curation costs, and preserving part of the site for later researchers.

Can the research question be answered using an existing database? Museums collections are curated ostensibly so they are available to researchers in perpetuity. Often they have received minimal analysis. There may be provenience problems with museum collections, and they may have been collected using less refined techniques than would be the case today, but existing collections are underutilized (see Toolkit, volume 6). If a research question can be answered with a museum collection, unnecessary excavation can be avoided.

If you have a choice, it is better to excavate sites that are less well protected and more susceptible to looting, vandalism, or erosion damage. For this reason, it often makes sense to excavate sites on private lands, where sites have fewer legal protections. It may also be possible to obtain some kinds of data from partially disturbed sites, allowing for the conservation of better-preserved examples. While conducting research at the University of Illinois, Robert Clouse excavated portions of an upland Paleoindian site, a rare find in east-central Illinois. Although the site was in an agricultural field, it was discovered after only one pass of the plow, and Clouse was able to excavate the undisturbed ridges between the new furrows, obtaining the data he needed and leaving the unplowed portion of the site intact.

Finally, we should consider conducting excavations at sites with good potential for public interpretation. Good potential could be indicated by ease of access, proximity to other public facilities and attractions, or a location where features and the archaeological research activities can be readily observed. Such excavations are responsive to the public interest in archaeology and provide opportunities to educate visitors about the conservation ethic and proper research methods.

CONTROLLING THE DESTRUCTION

As we saw in chapter 2, excavation techniques have changed considerably in the past two centuries. It is possible to point plot every artifact, but for many research questions, such precision is not necessary. The appropriate level of precision depends on the research questions and the nature of the site. For example, at the Erbie site (Lafferty et al. 1988), the research design called for statistical sampling using 50 × 50–centimeter control columns, and then opening several 2 × 2–meter units. About a third of the way through the control column excavations, it became apparent that the site was much smaller than previously thought. In addition, a Mississippian period house, which had never been excavated in the Arkansas Ozarks, was discovered. The focus of the excavation was redirected toward the house feature, and the precision of the control was increased to one meter. If you are looking for activity areas that are two to three meters in diameter, then you must use smaller units. If you are trying to find pit features in an apparently homogeneous midden, then smaller units are called for.

Extensive documentation of the work on a site is very important. Detailed field notes are crucial. Photographs, including color slides, black-and-white prints, and digital images of the general site, taken before, during, and after excavations, are important. All excavations should be documented with detailed maps, notes, profiles, plan view drawings, detailed photographs of features, and logs to keep track of all recovered finds and samples.

It is important to take samples that you might not plan to use. Soil columns and flotation columns may prove useful for answering unanticipated questions in the future. Soil samples should be reserved from all features and burials. We recommend floating total contents of all excavated features to recover low-frequency seeds and other materials.

Be flexible but thorough. When the inevitable unanticipated discovery happens, modify your research plan to accommodate it. If you are excavating a stratified deposit and interested in the early levels, you should also properly excavate the upper levels. These deposits may hold the key to why that site was important in the earlier times. At some significant sites in the East, the upper levels were bulldozed to get at the earlier levels. The investigators later regretted doing this, but by then nothing could be done about it.

Excavation areas should be protected from the elements. Usually, plastic sheeting over the excavations will lessen damage from rain or wind and can help secure the site from the curious when excavations are not in progress.

SITE SECURITY

Site security precautions are situation-dependent. A good precaution, as well as a general safety requirement of the Occupational Safety and Health Administration (OSHA), is to put orange barrier fencing around the excavations to keep the unwary from falling into the pits. The fencing also lets the public know that they should not be inside the fence. Appropriate "Keep Out—No Trespassing" signs are also a good idea. On federal and Indian lands, a warning of ARPA violations might be appropriate. If livestock are nearby, electric fencing might prove necessary.

When conducting a large excavation, it is good to keep a low profile with the press. Eventually, the press will find out that something is going on, and we owe the public an account of our work, so plan to give a press conference or interviews at the end of the project when there is something to show. Journalists have ethics, too, and in the case of an ongoing project, they will usually understand the need to protect the site. When you do a press conference or interview, plan what your main points are going to be. See Toolkit, volume 7, for details on media relations.

Some regions entail severe threats to sites from looters and vandals. Many places have long traditions of pothunting. Pothunters have become nocturnal, especially since the passage of NAGPRA. Farmers tell of tractors dropping into four-foot-deep holes dug by looters in one night. In such situations, it is necessary to hire a security service to guard the site at night and during weekends. If the guards do not show up, one of your crew members might have to stay on the site. Provide a cell phone and lots of coffee, but tell your crew person to get out of the way quickly if she thinks she is in danger.

Vandalism is a significant problem in the western United States, especially in remote areas, on public lands, and on open lands adjacent to populated areas. One of us (D.L.C.) has experienced vandalism that seemed to be the result of frustration over new development. Apparently mistaking our grid stakes for the construction boundaries, the site's neighbors routinely removed our grid corners and datum stakes in the evenings or over the weekends. At another site, we have experienced persistent vehicular vandalism, as the local weekend warriors seemed to think that our excavation units presented a great opportunity to test their vehicles' suspensions. Despite the presence of locked fences and barriers (which were themselves vandalized and removed), trucks were purposely driven through prehistoric pithouses and storage pits.

Even the presence of military police and guards is not always effective. Lining units with plastic and backfilling each night is time-consuming but helpful. We also routinely establish a whole series of datum offsets, which are rebar stakes or steel pipes set back from several of the datum points on our grid line, hidden in the brush, or at least out of the off-road vehicle traffic area, and pounded flush with the ground surface. Then, whenever the grid is disturbed, the task of reestablishing control of the site is simplified.

TRUTH IN REPORTING

Truthfulness in reporting the facts of our research is an ethical responsibility that should be understood by all professional archaeologists. It should not be necessary to remind researchers to tell the truth. Sadly, experience suggests that veracity in reporting research results is not always the highest priority.

Some years ago, one of us (D.L.C.) conducted a well pad survey in the Gypsum Plain region, an area of gypsum karst topography in southeastern New Mexico (Carmichael 1985). The client was a mining company, and during the first day in the field, I was accompanied by one of the staff geologists. After examining perhaps half a dozen well pads, the annoyed geologist asked what I was doing—what was I looking at on the ground, and why was I taking so many notes? Why was I taking so long at each well pad? When I informed him that I had discovered prehistoric sites at half the well pads and that I was recording them according to standard archaeological field procedures, he was shocked and apologetic. He told me that the archaeological contractor his company usually hired had never found anything on any of the surveys they conducted in this area. He said, "They just drive to the well pad, get out and walk around the truck, announce that there's nothing here, and charge us $100!" Clearly, the other contractor wasn't truthfully reporting the actual field conditions, exposing his client to a possible public relations disaster, and, just as clearly, the client had a very unfortunate conception of what archaeologists do and how they do it.

This is not an isolated case. Consider the contract archaeology firm in Wyoming whose main clients are companies in the oil and gas industry. The firm printed up nice trifold brochures announcing their company motto: "We strive to prevent any client downtime!" They accomplish this goal, in part, by conducting "clearance surveys" of well pads and access roads by riding as passengers atop the earth-moving equipment and reporting that no cultural resources were observed. Oh, and did they forget to mention the eight inches of snow on the ground at the time of the survey?

A colleague working in New Mexico reported a different sort of veracity problem that occurred on a federal contract. Data recovery excavations were undertaken on a lithic and ceramic scatter that had been assigned to the late Formative prior to the start of excavation. During laboratory analysis of the artifact collections, the lithic technology specialist identified several Archaic (7500–2000 B.P.) projectile points and three channel flakes, the very distinctive debitage resulting from the manufacture of Folsom fluted points ten thousand years old. Artifact analysis clearly indicated the presence of Paleoindian (often around ten thousand years or older) and Archaic components, but the project director refused to accept this evidence. A preliminary report had already been submitted, and the project director argued that the client had already fulfilled its Section 106 responsibilities. The internal dispute over what data should be reported was eventually resolved at the level of the State Historic Preservation Office; the artifact analysis was included as an appendix, and both interpretations of the site were presented in a revised final report. Whatever the reason for submitting a report before the analysis was completed, the whole episode never would have happened had the project director simply reported the actual research findings. At least other researchers now have the benefit of the data should they choose to reevaluate the report authors' findings. Such is not always the case.

An archaeologist working near San Antonio, Texas, contacted one of us (D.L.C.) with questions about the identification of Apache sites. Test excavations in a burned rock midden believed to be late Archaic in age yielded a radiocarbon date of 140 ±60 B.P. There was no reason to suspect contamination or intrusive modern charcoal, and the researcher was understandably excited about the possibility of reporting the rare find of a protohistoric or historic Apache camp. However, the researcher's suggestion of a possible Apache affiliation was met with great resistance by most of his colleagues and his boss. The idea that the site might be attributable to the Apaches was overruled because it didn't fit the expectation that it should contain artifacts of Euro-American origin. As a result, the site was reported as being a disturbed Late Archaic site, with the late (i.e., "intrusive") radiocarbon date being the basis for inferring disturbance. Is it any wonder that Apache sites are so rare in the archaeological database, when we are so quick to assign them to the Archaic period?

Apaches were present in much of New Mexico and Texas for several hundred years; they must have left thousands of campsites behind, but very few have been documented. We are probably not recognizing them because they resemble other sorts of sites, and as long as we continue to dismiss late radiocarbon dates and other evidence because they don't fit our expectations, we will continue to be fooled by the data we create. Perhaps most distressing, if we don't at least report the "intrusive" date and fairly discuss the possibility of an Apache affiliation, it will be very difficult for researchers interested in that particular issue to locate and reevaluate the data relevant to their research interests. We are building a database for our colleagues, not just running a business.

These examples exhibit several different sorts of misrepresentation, ranging from personal biases, to procedural corner cutting, to probable fraud. Such "untruths" impede our efforts to build a reliable database and to share useful knowledge within our profession and with our sponsors and with the public. Veracity is a key component of our ethical responsibilities to our colleagues, our clients, and the public. Misrepresentation, lying, and fraud are not to be tolerated in this profession. Frauds waste time and effort, and once discovered, the perpetrators' data can no longer be trusted and are forever suspect. In archaeology, information about an artifact or its associations in the ground is ultimately based on the word of the archaeologist. As a profession, we must be totally intolerant of liars; veracity exists at the heart of the Register of Professional Archaeologists' code of conduct. Misrepresentation of research results can ruin a reputation, result in the loss of a job or career, and undermine the public's trust in the archaeological profession generally.

CONSULTATION WITH AFFECTED GROUPS

Ethical guidelines for professional archaeologists have been prepared by several professional associations, including the Society for American Archaeology, and the Register of Professional Archaeologists (see Smith and Burke 2003). All such archaeological guidelines identify our responsibility to care for the archaeological database as one of our highest ethical priorities. This is as it should be. However, we should not forget that as archaeologists we are also anthropologists, and, as such, we should be responsive to the professional ethical guidelines established for anthropologists as a whole.

Among the principles of professional ethics articulated by the American Anthropological Association, the highest priority is our responsibility to protect the interests of those people we study or those who are most affected by our research. For archaeologists working in the United States, the people affected by our research will often be Native Americans. It is an ethical responsibility, now required by law (NAGPRA), to consult with Native Americans (and others) who may be affected by our excavation projects. While working under contract to the U.S. Air Force, one of us (D.L.C.) directed consultation activities involving some forty tribal groups in fifteen states. A workable set of procedures was developed that have since been formalized as the Air Force guidelines for consultation with Native American groups in the context of land use planning. While these guidelines do not constitute a "cookbook" for consultation, they have been used in developing other agency guidelines, and they provide a useful starting point for designing culturally appropriate consultations. Stapp and Burney (2002) provide an excellent recent source on consultation with American Indian tribes.

No one can reasonably dispute the complex nature of today's archaeology. We must deal with colleagues and publics, we must work with an increasingly complex science, and we must be ethically true to ourselves, our colleagues, and those with whom we work. In chapter 4, we will see how some of these matters play out in our actual fieldwork as we identify, characterize, and test sites. We are now moving into the "down and dirty" parts of archaeology, but remembering the higher and perhaps more noble concerns of what we do remains important. Try to remember these as we move through the next chapters.

Ethical guidelines relating to the conduct of archaeology and interaction with Native Americans may be found at the following websites:

Society of American Archaeology	www.saa.org/aboutSAA/Ethics
American Anthropological Association	www.aaanet.org/committees/ethics
Register of Professional Archaeologists	www.rpanet.org
U.S. Air Force, AFI32-7065	www.epublishing.af.mil/pubfiles/af/32/afi32-7065/

4

SITE TESTING, EXCAVATION, AND CHARACTERIZATION

The harder you look, the more you find.

—Anonymous

By now, you know that excavation should recover artifacts and data to address important research questions. You know, too, that as an excavator you are answerable to several stakeholders. Yet all of our discussion in the last chapter doesn't address the nitty-gritty of actual field practice. As we move toward actual excavation, we become interested in identifying, excavating, and describing the range of features at a site and documenting the artifacts associated with them. These general goals present a dilemma, however. Features that are exposed on the ground surface, and therefore observable during surveys, are likely to be more disturbed and have less research potential than intact, buried features. But the buried features are not visible at the surface, so, as people always ask, "How do you know where to dig?"

HOW DO WE KNOW WHERE TO DIG?

Most archaeologists probably have heard at least one story about the important house or burial or storage pit that was discovered near the end of an excavation project. Sometimes this does indeed happen, but wouldn't we rather use more of our field time excavating features than looking for them? A variety of techniques, some low-tech and some high-tech, can help locate features and otherwise characterize sites. When these techniques are built into a well-conceived research strategy, the results can be remarkably informative.

Geophysical survey and remote sensing often can help you detect features and characterize the subsurface nature of a site prior to excavation. These methods are discussed in volume 2 of the Archaeologist's Toolkit series. Here, we focus on the test excavation techniques commonly used to assess sites prior to large-scale excavation. These techniques also may be used to help develop preservation plans, mitigation plans, and National Register documentation and for other purposes that require more knowledge about the site than is typically obtained during the survey phase.

Traditional testing and excavation methods use units such as shovel tests, auger holes, 1 × 1–meter units, 2 × 2–meter units, and hand-excavated trenches. For most CRM projects, it is best to begin with the quicker, less expensive techniques, such as shovel testing, to determine where to place the larger units. Shovel tests and auger holes distributed systematically across a site can help characterize basic elements of the site (e.g., artifact types, feature distribution, and soil types). Larger excavation units can give more detailed data on artifact density and range of feature types.

As Black and Jolly say in volume 1 of the Archaeologist's Toolkit series, "think before you dig," and design your test excavation strategy carefully so you meet your project goals. Figuring out a useful approach is not always easy. Some states have survey standards that require testing at regular intervals, which limits your strategic choices. Scopes of work for a CRM project may make demands for limited samples, usually a percentage of an area of potential effect (APE). Many advocate using random sampling methods to determine placement of test units, while others advocate a stratified random sample wherein a sample is drawn from each of several predetermined units, these based on such attributes as elevation, distance to water or other resources, soils, and vegetative cover. Where sites have already been located, similar approaches can be used. Grid units can be randomly selected for excavation, and they can also be stratified based on some characteristic, often density of surface materials of certain types. Where sampling is random or stratified random, the sample represents the area of survey or the site (see Orton 2000 for a thorough discussion). Many archaeologists forego this more rigorous scientific approach in favor of an intuitive approach based on their prior

SITE TESTING, EXCAVATION, AND CHARACTERIZATION

knowledge of an area. Whatever the choice, methods should be dictated by a justifiable research design.

TYPES OF TEST AND EXCAVATION UNITS

Shovel testing employs a small unit, usually a 30 × 30– to 50 × 50–centimeter square. It is good to standardize the size, so that comparative volumetric data can be computed. During excavation, take notes on the soil description, artifacts present, and depth of each excavation level. Lafferty likes to use a form that has four columns with depths shown on a scale on which excavation levels, field serial numbers, and soil descriptions are recorded. A drawing is completed for each unit, standardizing record keeping. Depths of each excavation level should be measured as excavations are in progress, and the contents should be screened to ensure consistent recovery.

Auger testing involves using a three- to fifteen-centimeter diameter auger or posthole digger that can either be hand or machine powered. Although smaller in diameter, auger holes can extend much deeper than shovel tests that seldom penetrate more than fifty to seventy centimeters. The smaller size of auger holes and problems of compaction make it essential to describe, measure, and segregate the different soil layers as they are being excavated. Each defined or auger bucketful should be screened. If you are using a machine-powered auger with screeners following behind, each stratum needs to be laid aside separately and labeled so the records coincide. A third crew member should describe the profile and record provenience data. The person making the description should fill out the provenience tag and leave it with each unit to be screened.

Large-scale units are traditionally 1 × 1 meter, 2 × 2 meters, or larger units. They provide a broader three-dimensional view of the deposits and can show the plan view and cross-sections of features and soil horizons. The orientation of the unit should always be recorded on the site map. The unit's orientation is very much dependent on the site. Generally, orientation to magnetic north makes sense, particularly if magnetometer work is anticipated. Often it makes more sense to orient the grid with the topography, such as a terrace, the opening of a bluff-edge rock shelter, or the alignment of structures visible on the surface of the site. Units can be combined end to end, to create trenches, or expanded out in several directions to provide a broad, horizontal exposure to explore activity area patterning. Some archaeologists favor these for testing where they are economically feasible because they allow the extent of features to be seen more readily than in a small unit where one can often only detect the presence or absence of artifacts.

The larger the unit, the deeper it is possible to dig but the more time it takes. For example, 1 × 1–meter units are difficult to excavate more than sixty centimeters deep with a long-handled shovel. A 2 × 2–meter unit can easily be extended to the OSHA maximum of four feet for unshored excavations, although units approaching this depth can be dangerous in unstable situations, as when sand occurs beneath fine-grained sediments. (See sidebar 4.1 for more details on excavation methods and tools.)

4.1. RESOURCES FOR EXCAVATION METHODS, TOOLS, AND SOURCES FOR SPECIALIZED TESTS

Archaeology is no longer a simple field where a shovel, trowel, screen, and a couple of tapes suffice for equipment. You have to consider a wide range of variables every time you head to the field ranging from safety requirements to specialized tests. Appendix 1 lists some of the tools you might wish to consider taking with you and sources for them. The appendix also lists places that do specialized tests from radiocarbon dating to obsidian hydration. Appendix 2 provides mathematical resources you might need in the field and lab from conversion tables for measurements to places to find or generate random number tables to use for excavation unit placement.

No single approach to excavation works for all sites, although some standard approaches need only minor modification for each site. So much depends on research design, contract scopes of work, budgets, and time. We cannot cover all bases here but would like to recommend a few current and out-of-print books, commenting briefly on each. Each provides details on excavation methods and techniques you may find useful. We have taught field schools at different times using Dancey, Hester et al., Joukowsky, and Roskams. All are clearly written, and the Hester et al., Joukowsky, and Roskams books are comprehensive.

Barker, Philip, 2nd ed., rev. and expanded.
   1983   *Techniques of Archaeological Excavation.* New York: Universe.
        Some regard this as the British standard for excavation.
Dancey, William
   1987   *Archaeological Field Methods: An Introduction.* Minneapolis: Burgess.
        This thin volume is a no-nonsense introduction to basic methods.

Drewitt, Peter L.
  1999   *Field Archaeology: An Introduction.* London: Routledge.
          A useful volume whose strength is that it presents excavation as the centerpiece of a process, starting with planning and moving through publication. The emphasis is British, but much applies to all archaeology.
Hester, Thomas R., Harry J. Shafer, and Kenneth L. Feder
  1997   *Field Methods in Archaeology*; 7th ed. Mountain View, Calif.: Mayfield.
          This hefty (nearly six hundred pages) volume is the standard field methods text in North America for good reason. It is complete, with a look at variations between Classical, Prehistoric, Historical, Nautical, and Industrial Archaeology, with part of one chapter on CRM.
Joukowsky, Martha J.
  1980. *A Complete Manual of Field Archaeology: Tools and Techniques of Field Work for Archaeologists.* Englewood Cliffs, NJ: Prentice Hall.
          Now out of print, this compendium provides a wealth of detail. The emphasis is on Classical archaeology and large-site excavation, but the techniques are well explained.
Roskams, Steve
  2001   *Excavation.* Cambridge: Cambridge University Press.
          Primarily from a British perspective with British examples, the strength of this book is that it moves from details about planning to a discussion about prospects for the future. Also good is its presentation of material on a wide range of site types.

Setting up a unit without a grid is straightforward once the orientation has been decided. Place two pins or stakes the designated distance (one or two meters) apart in the selected orientation. Pull the hypotenuse (1.4142 or 2.8284 meters) and other leg to the triangle to set the third and fourth corners (see appendix 2). String the unit, and you are ready to proceed.

There are at least four different ways to string a unit (figure 4.1). The most traditional method incorporates a square five- to ten-centimeter baulk around the stake. Gutter spikes or large nails can be angled into the corner, or the corner can be cut at a forty-five-degree angle to provide enough strength to hold the nail. In sands and poorly consolidated deposits, you can use two pins per corner, set ten centimeters or more back from the edge of the unit in line with each adjacent profile.

Excavation level thickness can be arbitrary or according to stratigraphic level. In most areas, standard arbitrary levels are no thicker

Excavation Techniques

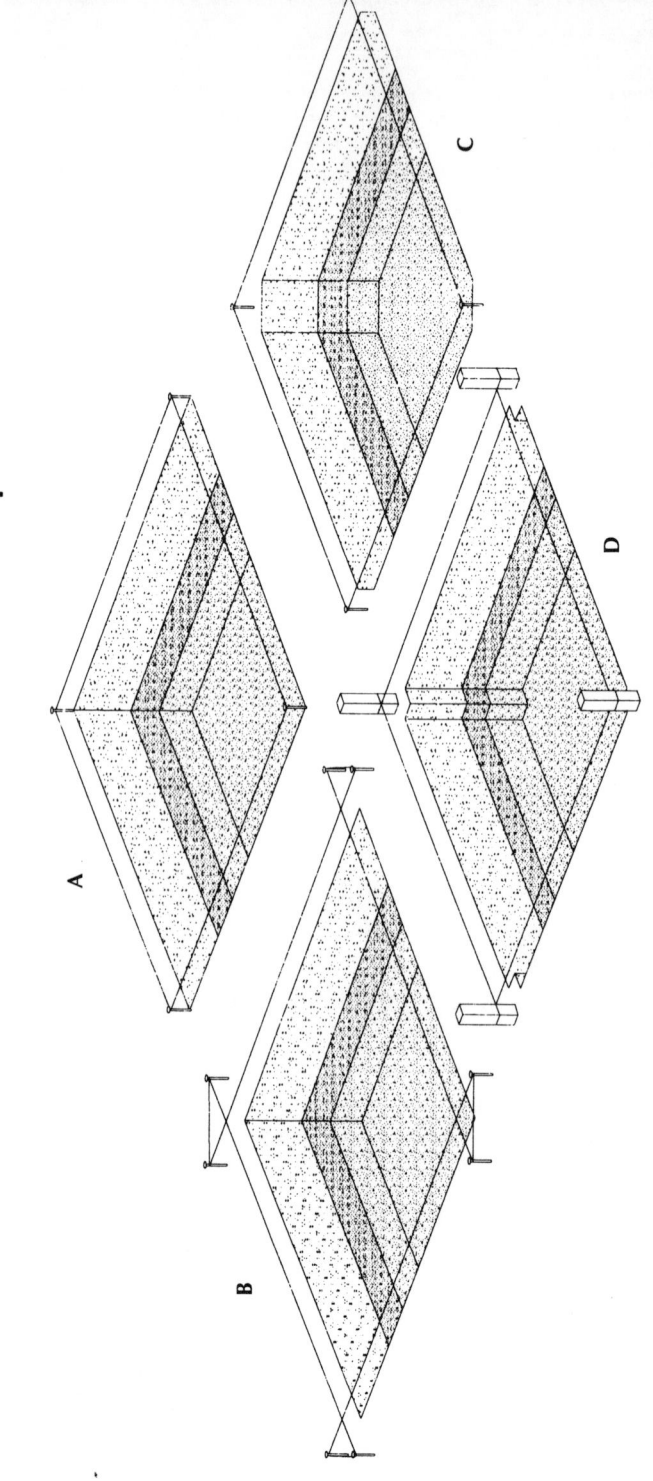

Figure 4.1. Different ways of stringing units. *A*, Gutter spikes angled into corners works under heavy soil conditions, such as clay; *B*, triangular offset is useful when the matrix is unstable such as sand; *C*, triangular baulk around stake; *D*, traditional square stake baulk.

than ten centimeters. Whatever levels are used, always note the top and bottom depths for volumetric control. If the surface is sloping and you are digging in horizontal levels, it is useful to record the depth below the surface of each corner of the unit. Under most soil conditions, the site matrix is passed through 6.25-millimeter (1/4-inch) mesh, 1/8-inch mesh, or window screen. Depending on your research design, it may be advisable to retain some of the fill from each level for water screening, to recover small artifacts (see chapter 5). If the soil is so clayey that it is difficult or impossible to screen, excavation can proceed by thin shovel scraping (or thin cutting), recovering the artifacts as they are encountered or after mapping them in situ. Indicate the method you use on your forms and in your field notes.

Generally, we use a separate level form for each level of each excavation unit. The form includes spaces for the excavators' names, date of the excavation, site number, unit number, level, depths, field serial numbers of the samples collected in the unit, excavation techniques, observations of soil color and texture, and photograph numbers. Graph paper on the back of the form is used for mapping the plan view of features, point-provenienced (piece-plotted) artifacts (artifacts that are precisely mapped and usually given their own field serial number), and disturbances observed in each level. Once the unit is completed, a drawing is made of one or more profiles. Some unit profiles are all the same, and it is a waste of time to draw more than one. Conversely, we have worked in units that had different features exposed in all four walls, and it was important to record all four profiles.

When testing a site, it is important to excavate into the underlying "sterile" deposits. A minimum of twenty centimeters will usually indicate you have reached sterile deposits, but occasionally it may be necessary to go even deeper. Without excavating below the deepest cultural deposit, you have no assurance that you have found the bottom of the site. Even if auger testing suggests that cultural deposits do not underlie a particular soil horizon, the larger volume excavated by a 1 × 1–unit can help you find deeper, low-density components that may be important. In alluvial settings, try to sink an auger or shovel probe deeper than the bottom of the test unit, just to check for any recognizable components.

When you reach what appears to be the bottom of the deposit or when you decide to bottom out your unit, trowel the profile clean, photograph it, and record the stratigraphy with a scaled drawing. The profile drawing should describe the soil levels in detail and show any features, artifacts, sample locations, inclusions such as burned clay, and postdepositional disturbances (see the later discussion of profiling techniques). When photographing a unit, include a scale, north arrow, and a menu board or chalkboard showing the site number, profile, unit number or grid location, direction, and date. The profile should be evenly lit so that harsh shadows do not detract from the stratigraphic details. Although full sun is acceptable, full shade may provide greater flexibility in the exposure setting. It is often useful to use a canopy, tarp, jacket, box, or crew member to shade the profile and to spray the profile with water from a garden sprayer to enhance color contrasts. Use a digital camera if possible, as well as good-quality outdoor color slide film. If time permits, it may be worthwhile to photograph a profile at different times of the day, when the sunlight strikes it at different angles.

Various coring tools can give you good information on site stratigraphy. Tools range in size from 1.25-centimeter-diameter hand-operated Oakfield corers (which should be standard equipment for all excavation projects) to truck-mounted 7.5-centimeter or larger core rigs. The extracted cores should be tagged and returned to the laboratory. Analyses such as sediment size, organic carbon content, flotation, pollen, radiocarbon, chemical, and many other tests can be conducted on these cores.

Cores should be sealed in polyethylene bags to prevent them from drying out. They should be labeled with their depth range and which end is up in addition to other provenience information. Where multiple cores are taken, the depth range should be measured in the core hole when each segment is pulled. This is because many coarse-grained soils compress in the core. Field notes should describe the samples and depths as they come out of the ground, and each core should be assigned a serial number. For large cores, use heavy-duty boxes to support and protect them. Professional core drillers often provide these as part of their service. (See sidebar 4.2 for more on coring.)

MECHANICAL EXCAVATION TECHNIQUES

Mechanical excavation techniques, such as backhoe trenching, have been available to archaeologists for many years, but until recently, they were generally looked down upon as a search strategy and used mainly as a last-ditch effort, so to speak. In other words, because of

> ### 4.2. CORING
>
> At the Eaker site, coring was designed to determine the depth of cultural deposits. Several cores were placed off-site to determine natural soil conditions. Three three-inch-diameter cores were excavated in segment 8, targeted to three magnetic anomalies. Core 3 was excavated into the same burned house that trench 1 intersected. This core showed a sherd below a burned area and mottling around fifty to sixty centimeters below the surface. It also showed sand to a depth of 160 centimeters below the surface where there is a wedge of silt loam, underlain by sand to a depth of two meters. In contrast, core 1, located twenty-one meters to the west in an area of low magnetic variation, was silt loam to eighty centimeters. A brown sandy loam underlay this deposit. Artifacts were present to 109 centimeters below surface in the core. The cores showed that there were cultural deposits to a depth of more than a meter in some parts of the site.

the destructive potential of mechanical trenching, the backhoe was generally brought in only after the majority of the excavation was completed, just to make sure nothing major was missed before the site was destroyed by construction. However, as archaeologists have followed this strategy over the years, many have learned that some of their most important and unexpected finds have been revealed by backhoe trenching, while other methods have been less fruitful. Furthermore, when subsurface features are discovered early in an excavation, more time can be spent describing and interpreting them, rather than just finding them.

Augers do not penetrate very deeply and do not provide a clear view of the soil strata, and any layering within the auger hole is destroyed as the soil is excavated. In addition to the time saving gained with backhoe trenching, the trenches can be excavated to a greater depth, and the long profiles obtained can reveal features that would not be observed in plan view or in small auger holes or cores. At the Meyer Pithouse Village at Fort Bliss, features were identified by augering, but their identity as pithouses was only revealed through "last-ditch" trenching (Scarborough 1986).

Excavations by the Museum of New Mexico at Luna Village (LA 45507), a pithouse village site in western New Mexico, have convinced researchers of the necessity of mechanical trenching on Mogollon sites. Yvonne Oakes reports that all five of the houses at the site were discovered in backhoe trenches, including (quite unexpectedly) one beneath a gas station and another under an abandoned historic house (Oakes and Zamora 1998). Similar results have been documented on other sites throughout the state.

At the Gobernadora site in El Paso, Myles Miller (1989) compared the rate of feature recovery from the two halves of the same site. The 2 × 2–meter tests previously excavated by Texas Department of Transportation archaeologists contained the same volume of fill as Miller's trenches, but the trenches yielded double the feature recovery.

TRENCHING AS A STATISTICAL SAMPLING STRATEGY

Trenching provides several advantages in searching for buried features (see also sidebar 4.3). It is more rapid than hand trenching, with obvious attendant cost savings. It also provides relatively easy access to deeply buried deposits. But there is another sort of advantage to trenches, related to the nature of transect sampling.

Transects are effective for identifying features because they only have to catch the edge of a feature in order to identify it. This boundary or edge effect makes transects very efficient sampling units because despite the narrow width of the trench, they sample an area about twice the width of the target feature size. Following the methods outlined by Rice and Plog (1983), backhoe trenching was used as a statistical sampling strategy during the early stages of excavation at site 41EP492 at Keystone Dam in El Paso (Carmichael 1985).

The site grid was laid out in 20 × 20–meter units, which were used as the framework for a systematic sample of transects. A trench was excavated in each 20 × 20–meter unit, alternating along the north-south or east-west centerline. The orientation of the trenches was alternated in order to reduce the chances that they might correspond to some periodicity in the distribution of features and miss them altogether. Calculations predicted that as many as twenty-eight such features might be discovered, and after additional trenching and hand

4.3. BACKHOE TESTING

At Eaker Air Force Base, we were testing sites and finding very deeply buried cultural deposits beneath clean sand. When an excavation unit was stopped because it was thought to have reached sterile sand, standard procedure was to sink a posthole test as deep as possible below the bottom of the test unit. At seven locations, we encountered dense midden deposits at depths of one to two meters below the surface. The ground-penetrating radar profile from 3MS560 showed a thirty-meter-wide depression where we found a midden buried beneath sterile sand. Using a backhoe, we cut sixty meters of trench, and we discovered that a prehistoric sand blow overlay the midden, an earthquake-related feature formed in the liquefaction zone (figure 4.2).

The trench also revealed two thirty-meter wide sand blows atop one another but separated by five centimeters of alluvial clay. In contrast to sand blows recorded on earlier projects, which were formed by the 1811–1812 New Madrid earthquake, these were highly weathered, and prehistoric artifacts occurred above and below the sand blows. By the end of the day, we had found seven more sand blows, five of which appeared to be prehistoric. Because of their association with datable archaeological materials, the sand blows at these sites have become important in dating at least two major periods of prehistoric earthquakes in the New Madrid Seismic Zone.

Figure 4.2. Backhoe trench 1 at 3MS360 showing the plow zone underlain by the light sand blow. Beneath the sand is intact midden. Flags are in situ artifacts. (Photograph by R. H. Lafferty III.)

excavations, a total of twenty-five were eventually identified. These were pithouses or huts that were very difficult to see in plan view. These were the first examples of such structures discovered in the El Paso region, but now that researchers know how to search for them in trench profiles, they are regularly encountered. Our ability to characterize the site was greatly enhanced by the judicious use of backhoe trenches. The trenches accounted for 50 percent of all the features discovered at the site, and 100 percent of the pithouses.

MECHANICAL STRIPPING

The use of mechanical stripping is an alternative to hand excavation units when it is desirable to open up a broad, horizontal exposure. Because stripping with a blade or grader is so destructive compared to shovel scraping, the approach is usually used at sites that will be largely destroyed during construction, and even then only after other efforts have been made to recover artifacts. Nevertheless, mechanical stripping can be especially useful for the removal of sterile or disturbed deposits overlying culture-bearing layers. By quickly removing overburden and by opening up large areas at a given level, stripping facilitates the recording of clusters of features.

RECORDING STRATIGRAPHY

Archaeological sites exist on and in a matrix. Sometimes the matrix is relatively simple, and sometimes it can be complex. Relatively stable surfaces with minimal deposition are at the simple end of the scale, while buildup through alluviation can lead to deeper and more complex deposits. Both the simple and complex matrices may contain evidence of soil formation and of other processes that are critical in archaeological interpretation. Therefore, it is vital to make accurate records of the soils and deposits through which you excavate, regardless of whether your site is in an upland setting, an alluvial fan, a terrace, or a floodplain.

Many natural processes cause sediments to accumulate. Chief among these in much of North America is deposition by rivers, which results in the building of terraces. Alluvial fans build up at the mouths of creeks and rivers. In the arid West, mass wasting of bedrock builds colluvial talus deposits along the bases of hills and mountains. Loess (windblown silt) deposits can be many meters thick, even in uplands. Volcanoes deposit ash over thousands of square kilometers. Earthquakes can eject sand and water over areas as large as ten hectares, as deep as three meters. All of these kinds of deposits often contain archaeological sites.

It is important to record the details of site stratigraphy whether you are working in areas with a relatively stable near-surface soil or a complex depositional history. Where soil horizons or depositional units can be discerned, the excavation should proceed with reference to these by natural strata. The thickness of each stratum (top and bottom depth of each corner) should be recorded so the artifact densities can be calculated. Stratigraphic differences can be identified in the field by a change in soil texture, color, grain size, or compactness. Some of these changes may be depositional, while others may be pedogenic (a result of soil development). When a different stratum is encountered, the unit should be cleaned off to see whether there are features intruding into the underlying stratum.

Careful documentation of intrusions and strata is essential for determining the temporal sequence of samples. Documentation of profiles should be detailed, including complete soil descriptions and point proveniences of any artifacts exposed in the profiles. If you need help with soil descriptions, learn how to identify and describe soil characteristics from a specialist or a manual such as Vogel (2002).

There are several ways of making a scaled drawing of a profile after it is cleaned and all strata of interest are scribed into the wall. One method is to string a line along the length of the profile. The string should be pulled taut and leveled with a line level, carpenter's level, or transit. Large nails, gutter spikes, or surveying pins make good anchors for the string. Draw this level line on your graph paper form. Grid locations should be shown along the top of the drawing. Place a folding engineer's rule or an extended tape parallel to the string. Make sure the tape does not sag, and don't move it until you are done drawing the profile. Using the level line, each point that you want to record can be designated by measuring a certain distance along the line and another distance perpendicular to and above or below the line. A two-person team is best for drawing profiles, one measuring and calling out the points and the other mapping and drawing them. The mapper should be positioned so she is able to see the profile as she draws it, facilitating the inclusion of details. It is most efficient to map the big picture first, starting with the top and bottom of the profile, followed by the major strata and, finally, the features, individual artifacts, and natural disturbances. Instead of making individual measurements perpendicular to your level line, you can also map a profile by stringing the excavation wall with a fifty- or hundred-centimeter grid and overlaying a portable grid with smaller units (such as ten-centimeter squares) over each section as it is mapped.

A detailed site map should be made whenever excavations are undertaken. The map must be scaled and should show the locations of all test units, shovel tests, auger holes, and trenches. Some indication of topography should be shown, preferably using contour lines. Any features such as structures, walls, and depressions should be noted. Water sources and vegetation should be indicated, and prominent trees should be mapped, identified by species, and their diameter measured at chest height. The map should also include anything else that might help future researchers relocate the site and understand what you did. Establishing an unambiguous site datum and depicting it on the map is most important. Scale and north arrow should be included, and the map should be signed and dated.

CONCLUSION

Archaeologists recognize that the landscapes of the past are often substantially different from those we see today. Geomorphological processes may work to hide the sites created by the people who lived on those landscapes, or they may work to reveal them to us. What we can learn about both the landscapes and the processes can give us some predictive ability to find sites, but mostly locating sites is a matter of technology, ingenuity, and persistence.

Still, what we do when we locate and characterize sites gets us only part way toward understanding the lives of those who created the sites and the cultural processes that influenced them. So far we have looked at dealing with sites at a macro level, but the detail of sites is often in the process of data recovery, the subject of chapter 5.

5

DATA RECOVERY

See the World in a Grain of Sand.

—William Blake

Before you start recovering anything, remember that data are not what you are after. Data are indeed the stuff of archaeology—stone, bone and other materials, spatial relationships, and so on—but as soon as you think "Clovis point" (you wish) or even "cortical flake," you are entering the realm of information, the constructs you make in the production of knowledge. Knowledge about the past is the goal, something you hope will resonate throughout the archaeological community when you publish your findings.

The first steps in archaeological knowledge production are theoretical ones. You need to understand first that artifacts and features are not data: They are just artifacts and features. Data are creations you devise based on your research design (see Toolkit, volume 1) and then extract from the artifacts and features. Artifacts and features don't speak for themselves; the archaeologist becomes their "voice." We may like to call what we do "data recovery," but it's worth remembering that it's really "data creation." As you create data, you have to define the scope of the data with which you plan to work.

While we act at a human scale, cultural data resolve at multiple scales, from global to microscopic. You may think of an artifact, for example, as something you can hold in your hand and a landscape as the cultural environment—the background for all those archaeological sites. The problem is that the thing we call a site is an artifact on the landscape, and a projectile point is a landscape, too, a surface marked by cultural impacts (literally). We are just too big to see a projectile point that way.

So, your first decision about data recovery must refer to the original research design. What are you trying to accomplish? Do you already know the nature of the culture and activity at the site, or is it still a mystery? Is this basic data recovery, or are you testing specific hypotheses? Depending on the answer, you may need to be prepared to "recover" data ranging from the spatial relationships among artifacts and the surrounding environment to pollen grains, the smallest bits of residue from toolmaking (microdebitage), or chemical residues from bone decay or foods.

If you fail to appreciate this problem of scale, you may miss data crucial to an understanding of the phenomena you hope to study. Consider a simple example: the difference between two sites along a river, one at a confluence with a creek, and the other downstream along the bank. If hunters and gatherers saw the former place as more strategic, because it provided access to two drainages rather than one, might it have attracted more people over time than the terrace downriver with only its level well-drained ground to recommend it? If you bury your nose in a site, you may not even imagine that subtle differences in the site's raw materials, artifacts, and activity areas might reveal differences of local or regional significance.

ARTIFACTS AND THINGS

Once you have sorted out your scales of analysis, you can turn to your data recovery strategy. As noted earlier, as soon as you start naming things, you are making interpretations, so now is the time and place to be concerned about classification. The point: You may want to collect "all" the data, but you will retrieve only the data you recognize from the tiny bit of data that survives.

Much of classification seems to be routine. You may distinguish material culture objects (artifacts and the residues of production), natural things used in cultural activity (e.g., hammerstones), and natural things indirectly affected by cultural activity (e.g., burnt soil or an unnaturally organic soil stratum). You may expect to find spatial relationships between and among cultural things. The problem is that you are inevitably imposing your own beliefs, implicitly and explicitly, on material things that do not advertise themselves at all. If the past speaks, it says what you want to hear!

If you describe a scatter of artifacts as a site, for example, you are assuming a cultural relationship among these things, even though many different individuals, groups, and cultures may have stopped there for many different reasons over time. Because classification is inevitable, the only solution is to think out why you are investigating this particular place and what you hope to gain from the effort.

The issue here is the extent to which you document these artifacts and relations. On an initial reconnaissance survey, for example, you may not need to describe every piece of chert or measure spatial relationships beyond the visible extents of an artifact scatter. If you are excavating, then you will.

COLLECTION STRATEGY

After thinking about the issues discussed so far, you may worry that no amount of data recovery will ever be enough. Clearly, however, there are limits to what you can do as an archaeologist. The goal is to accomplish what you set out to do; no one expects you to recover a lost world from the meager traces left to us. To resolve this dilemma in a positive way, archaeologists develop collection strategies. What you collect depends on at least two things: what you hope to learn (your research design) and how much data you need to retrieve to learn it. Have you stumbled on the site in a reconnaissance survey? Or is this part of a hypothesis-driven study dealing with some specific archaeological problem? In most excavations, you may need to sort out occupations or activity areas, so you may want to record very carefully in three dimensions exactly where everything is located and collect a representative sample.

You may consider a systematic sampling strategy if you run the risk of being overwhelmed by data—in fact, you probably already use an informal one, if you walk in linear transects across a site. More formalized sampling methods, backed by probability theory, often involve collecting only within predetermined, randomly generated grid squares. This may be particularly effective if you have a comparative problem you want to resolve—say, studying the relative distribution of bison bone types across a meat-processing site so you can speculate on butchering patterns. You may be able to arrive at a satisfactory interpretation informally, but a systematic recovery technique will help eliminate spatial bias that might deform the results.

If you make a mistake in your collection strategy, you are wasting time, perhaps money, and adding to the ever-increasing mountain of useless data. The same caution applies in the lab. Ask yourself this when you are tempted to wave a Munsell chart at a piece of debitage so you can nail down the color: "Does this type of data have any meaning in my analysis?" If the answer is no, give it a basic description and be done with it. The only reason for making collections in the long term is to provide material for someone else to study. Let him do it!

THE WAYS AND MEANS

Excavation recovers only a tiny fraction of what might have been a habitation, a production site, or other theater of cultural activity. The vagaries of preservation mean that many tools and tool parts made of perishable materials such as plant fibers, wood, leather, and bone may break down quickly under adverse conditions.

SURFACE RECORDING AND RECOVERY

An artifact exposed on the ground surface is usually not where the maker or user left it. Time is always a culprit, working with wind or water erosion, slope movement, plowing, and other natural and cultural forces. Because you tend to learn most about people, places, and events from the spatial relationships among material things, surface finds will probably not yield as much information as artifacts in situ. Sometimes, however, surface finds might be the only data you have. It is a rare archaeologist who has not found artifacts side by side that are separated by thousands of years. Such mixed assemblages are palimpsests, a word we've adapted from the printing trade to describe artifact assemblages that have become jumbled because either there was no interval of soil deposition between occupations, or the soil that did separate the occupations eroded away.

If this is the situation you face, all is not lost. Temporal and functional diagnostics such as identifiable projectile point or ceramic types, scrapers, or fire-cracked rock can help you build a picture of what happened. You may also be able to map the "horizontal stratigraphy," the spatial distribution of materials across a locality, if the artifacts may have remained more or less in the same place. This is where knowledge of the terrain and the geomorphology are invaluable. You need to be certain that a flood, sheetwash, or slope movement did not transport your artifacts from elsewhere. Finally, do not assume that all the artifacts in an assemblage derived from similar activities. Ponder this question: A toolmaker's debitage may be identical at two different places, but do these scatters offer up the same interpretation if one of the sites is a hunting blind and the other is a staging ground for ceremonial activities?

An important advantage of a surface scatter is the fact that you can map the entire exposed assemblage, if you need to, piece by piece, and therefore quickly define its extent (see Toolkit, volume 2). Careful mapping is key to any subsequent excavation. Of course, cultural material is scattered across the landscape, so what you are really doing is isolating a discrete locus of cultural activity. In a reconnaissance survey, you may find it helpful to stick a pin flag in the ground at every artifact location or, if the finds are too dense, on the outlying edges of a scatter. Pin flags are great because they allow you to see patterning in artifact distribution, but they also call attention to the site. If you are concerned that you are advertising a site location, then use discrete markers.

Once you have surveyed the entire surface, you can map in the artifacts, features, and aspects of the physical setting. If precise spatial relationships are crucial, plot every artifact and feature with a Total Station or theodolite. With very little effort you can generate a computer-aided design (CAD) image from the locational information, giving you a precise map of the area. If you only need to map the general location of artifact clusters or features, a geographic positioning system (GPS) unit is ideal. A unit with real-time differential correction will allow you to map points, lines, and polygons (enclosed shapes) with submeter precision; one without this feature will still get you within several meters, which may be okay for a reconnaissance survey.

When you have finished mapping and taking photographs, you need to invoke your collection policy and get to work. An important point: You may be the last person to see artifacts that you do not collect, especially if you are surveying in a proposed construction area. It is therefore a good idea to have a field identification procedure in place so that you can quickly record some basic data on the artifacts observed (e.g., type of object, material, function, relative size, cultural attribution). You can then pick up the objects you plan to take to the lab for further study.

There is no point in collecting for its own sake. We already know that people have left traces of daily life on the landscape for thousands of years. We do not need any more proof. Having said that, you may want to save something because it is old and uncommon, even though you know it is not going to add significantly to your study. We cannot cling to every element of the past, however, because if we did, we'd have little room to move! Never forget that the traces you find usually relate to someone else's ancestors. The descendants may not just want you to be respectful—they may not want you to interfere in any way at all, despite your legal rights or obligations. Sometimes you simply have to let things go. If you do, don't get upset when collectors move in, pick up what you left, and complain that you "missed" things.

MOVING DIRT

When you need to explore below the ground surface, you have to resolve the fundamental problem: your lack of X-ray vision! Deciding where to dig takes careful observation and planning. You may have already taken the first step, assuming that the artifacts on the surface reflect the positioning of any below. As this may not be the case, depending on the site taphonomy, the next step is to devise a technique that provides you with a good general impression of the subsurface. If your project area has cutbanks, gullies, rodent burrows, or even anthills, you can learn quite a lot without digging. If you are in an agricultural area, the county soil map will outline the soil characteristics and formation processes. Or you might be fortunate to be in an area that a geomorphologist has already analyzed. If you can, conduct a magnetometer, soil resistivity, GPR, or other geophysical survey. The information you retrieve from these preliminary efforts could lead you from aimless digging for that elusive needle in the haystack to the ground truthing of potential features. At the very least, you should have shovels and a soil corer or auger, so you can explore the subsurface in the traditional way.

The goal of subsurface data recovery is not only to retrieve artifacts but to expose the traces of cultural activity remaining on what were once ground surfaces. Visualization of these buried contexts is very difficult at the best of times. Over the tens, hundreds, or thousands of years separating components, there may have been any number of impacts that added, removed, or otherwise disturbed the deposits. It is therefore extremely advantageous if you can expose a large subsurface area quickly to help inform your subsequent excavations.

At a recent University of South Dakota (USD) Archaeology Laboratory project at Fort Riley, Kansas, for example, we faced the difficult problem of interpreting the cultural resource potential of a large open field and surrounding woods at the confluence of two creeks, an area of more than fifty acres (Molyneaux et al. 2002). To complicate matters, we knew from previous excavations that there were likely to be buried soils containing cultural material several thousand years old and numerous shallow disturbances associated with the World War I barracks and the more recent firing ranges that once occupied the site.

Rather than pick away at the surface to little effect, given the size of the site and the potential depths of cultural material, we decided to excavate a number of backhoe trenches. The goal was to expose the stratification sequence and determine whether any of these strata contained cultural material. This would inform the main testing, to be done with an eight-inch power auger that could probe more than ten meters into the ground if necessary. Given the size of the area to be explored, we did not mind that several areas approximately 4 × 2 meters each were going to be destroyed by the backhoe, perhaps disturbing cultural material. It was a risk worth taking, because it would give us a very good idea of where we should conduct our less destructive tests.

As it turned out, the backhoe trenches exposed three different buried components across the project area, ranging from a near surface component of the Central Plains tradition (surface to about fifty centimeters), to Middle to Late Woodland (one-meter deep), to Late Archaic (two-meter deep) occupations below that. We were then able to conduct auger testing much more carefully and effectively than we could have without the trenching.

You may need to use heavy equipment for other reasons as well as when you must remove overburden quickly because of time constraints (i.e., a construction crew's bulldozers are idling in the background). Mechanical stripping, using a backhoe, road patrol, or belly scraper, may expose artifact concentrations or features very quickly but at the risk of damaging portions of features and artifact scatters.

Data recovery via hand excavation is obviously a better strategy if your goals are to record material in three dimensions so you can learn more about time and space relationships. The most efficient method is shovel skimming (a.k.a. shovel scraping or shovel shining), which is removal of soil with a very sharp shovel held at an acute angle to shave off a portion of the surface. An adept shoveler can remove earth in such fine levels that very little will be missed. If you risk disturbing delicate remains or missing smaller artifacts, it is best to trowel by hand.

The choice of technique depends in part on what you expect to find and what you intend to record. Clearly, there is a difference between excavating an artifact scatter and a hearth or other intact feature. In the former case, a shovel may suffice; in the latter, very light troweling and brushing are in order.

DRY SCREENING

Whatever method you use, you always miss some artifacts, so the next step is usually screening of the soils you remove. Sifting excavated soils through a quarter- or eighth-inch screen is so simple when you are sitting in your office using your field school knowledge to create regulations for working archaeologists and so miserable when you are standing at a screen trying to break up hard lumps of earth or push damp clay through the mesh. It was a sly and cynical wag, no doubt, who gave the name "Love Slave" to one of the USD shaker screens. Screening, however, is a necessary task, because most data recovery techniques—except brushing, perhaps—miss some of the artifacts.

Carmichael and Franklin (1999) conducted a screen loss experiment at the KT site in New Mexico. Their results showed that quarter-inch screens may capture as little as 5 percent of lithic artifacts and miss many diagnostics, such as notching flakes, biface thinning flakes, initial edging flakes, and exotic raw materials. Surprised? USD archaeologists made a rather rare find in the region, a piece of obsidian from a sixty-mile highway corridor along the Floyd River in northwestern Iowa; the piece, recovered during the mechanical grade sorting of 10 percent sample bags of earth from excavation units at site 13PM407, was a piece of microdebitage of grade 5 (0.0469 inch, or 1.18 millimeters) (Molyneaux et al. 1996). The implications of the find were rather monumental in their simplicity: This chance excavation of a tiny part of a site along a large highway corridor yielded an artifact that someone chipped from a tool of material that originated far from this knoll along the Floyd River. Without screening we would never have been able to add to our knowledge of obsidian distribution and use in the region.

Basic screen design has not changed for generations. The most common form is the wooden box with a hardware cloth mesh, suspended from a tripod for long-term excavations or mounted on two legs with two handles for field surveys and short-term excavations. There are debates about which types are the most useful and easiest to use. If you work in an urban setting on large excavations and have dry earth to screen, you might consider a mechanical unit. A Gilson grade sorter with five screens will quickly reduce a pail of dirt to neatly sorted material and a tray of dust at the bottom. You might also search the websites of the industrial screening companies for innovative ideas. There are new mesh designs that work better with damp soils than the traditional square pattern; perhaps these would improve those long messy days of struggling with clay.

WATER SCREENING AND FLOTATION

Water screening is the best method for ensuring more complete recovery of cultural material, as you literally wash away the soil. It is an especially good way to retrieve small artifacts and to track artifact loss through dry screening.

A number of different water screen set-ups are in use. The most primitive method combines a high-pressure pump, a hose, and ordinary box screens. Self-contained sluice boxes and barrel systems (steel drums, fifty-five U.S. gallons or forty-five imperial gallons) pump water internally to stir up the soil. The barrel types may also double as flotation units if the light fraction can run off into a set of sieves. Do not assume that water screening makes life simple. Clay will sit in the screen unless it is physically disturbed. Pumps can be finicky. You will get wet and muddy. And you might find that there are regulations preventing the discharge of your soil back into the water.

Flotation is simpler, but don't misunderstand: If done properly it can also be very complex. You can buy or jury-rig a plastic bucket system with an inlet at the bottom and a spout at the top that draws off the waterborne fraction into a stack of sieves, or you can just fill a pail with water, dump the dirt in, stir it around, and scoop up the material with an ordinary kitchen sieve. Another approach is to pour it into a fine-mesh screen or cheesecloth. (See Toolkit, volume 5, for details on fine-screen recovery and processing methods.)

These are wonderful methods if you have carefully thought out why you are using them. If you retrieve a lot of light or heavy fraction, you may be facing long hours at a binocular microscope carefully separating the artifacts from the debris. Make sure these data will be useful. There is no point collecting seeds and other plant material unless you can differentiate between material associated with the site occupation(s) and modern material; there is equally no point in collecting microdebitage if you are only adding the numbers to your inventory. Are you going to use microdebitage to test your recovery methods? Do you think that the presence of microdebitage might indicate the precise spot that a knapper worked or reveal areas where larger pieces of debitage are absent because tool renovation (e.g., sharpening) rather than production took place there? You would miss such potentially significant data using conventional recovery methods.

CONCLUSION

Collection strategies require careful planning, derived from the key questions in your research design. Failure to do effective planning can overwhelm you with useless data or cause you to miss important data. The truth is that most of us don't think that carefully about our approaches to data recovery and come to regret it once we get to the lab. We need to be more deliberate in our approaches to use our time, money, and other resources effectively.

A final element in data collection, discussed in chapter 6, relates to a variety of specialized types of data collection, ranging from samples for dating to human remains.

6
RECOVERY OF SPECIALIZED MATERIALS AND SAMPLES

Specialized samples can and often should be taken to assist in the interpretation of sites and cultures. These samples often require specialized handling and extraction procedures. New techniques are being developed all the time, and the techniques discussed here are only some of those available. We note details of the collection procedures most relevant to interpretation of results. In the case of human remains, we focus on details that have a bearing on the eventual disposition of the remains. Specialized analytical techniques are continually undergoing refinement. Some researchers might argue it is inappropriate for archaeologists to charge our clients for the development of these advanced techniques, especially if they are still experimental and we can't yet be sure how to interpret the results. To the contrary, we believe that archaeologists must continue to try new techniques in the context of CRM excavation. Most archaeological sites excavated in the United States today are studied because they are determined eligible for the National Register of Historic Places for their information potential. A stratified lithic scatter may not inform us about the beginnings of agriculture, or ceramic production areas, but if the lithics are obsidian, the site may have considerable research potential either because it can help establish a regional hydration chronology or because it can be analyzed in the context of an existing chronology. We will never work the bugs out of new technologies if we do not apply them on real archaeological sites, and we won't be giving our clients our best efforts if we do not address the real research potential of the sites we are privileged to excavate.

CHRONOMETRIC SAMPLES

Numerous methods can be used to date deposits or artifacts. Some of these methods are easier and less expensive to use than others, but in all cases, it is good to run multiple samples. More is better, as this helps you decide which dates may be aberrant. We like to run at least ten dates from a site, and it really helps to have some of the samples ordered by stratigraphy, so the span of occupation can be dated and aberrant dates can be identified. When possible, it is also a good idea to use multiple dating techniques. Dendrochronology and radiocarbon are the oldest of these methods, and most of the other methods are interpreted by making correlations with these two.

DENDROCHRONOLOGY

Tree ring dating was developed in the early twentieth century in the Southwest, where many of the ruins contained large logs as structural support beams. A. E. Douglass (1936) observed that the widths of rings in the logs varied and that sequences of rings appeared to have patterns representing wet and dry years. He began collecting samples from living trees and archaeological sites to build a master chart from overlapping rings. By 1936, his chronology extended back to A.D. 700.

Subsequent work has developed chronologies of a millennium or more in many regions of the United States and Latin America. Dendrochronology has dated hundreds of archaeological sites and has provided the key to calibrating radiocarbon ages to calendar years. Dendrochronology also is revealing major shifts in global and regional climate patterns. Collection of old wood samples for dendrochronology is important for understanding changes in local adaptations as well as more general processes of climate–culture relationships. From this perspective, well-preserved dead trees are significant historic resources.

Dendrochronology is becoming a major method of dating log structures in the East. On historic sites and old farms, it can reveal much about landscape evolution. Projects that remove trees, such as reservoirs and highways, should have dendrochronology samples taken from the largest and most stressed trees (those whose growth is most significantly affected by annual or seasonal weather changes). Waterlogged, preserved trees in swamps should be extensively sampled.

When taking samples, it is best to consult with a specialist, as there are various ways to collect samples, depending on the allowable impact to the sampled tree or log. If trees are being cut, take the sample from about a foot above the ground. Cut the stump horizontally and then make two five- to six-centimeter-deep parallel cuts about five centimeters apart, with the center of the tree in the middle. Cut the section off, and label the provenience of the sample. Contextual data should include natural context (e.g., yard, fenceline, virgin forest, etc.), species, and map coordinates. Samples from log cabins that are being destroyed or are in an advanced state of decay may be similarly taken and also should be labeled by wall and log number.

If the tree, cabin, or pueblo will be preserved, take samples with an increment borer that removes a one-centimeter-diameter core. The living tree can heal over the wound in a year or two, but it is important to cover it with grafting tar. To get close to a true cutting date when sampling structural logs, try to take the core so that exterior rings are included. Whether or not exterior rings are included should be noted on the label. The samples may be bagged or boxed in cardboard or paper and allowed to air dry slowly.

Charred archaeological sections need to be supported, undercut, and wrapped in aluminum foil and carefully bagged in polyethylene. Samples should be carefully air dried. Waterlogged specimens need special treatment. Contact an experienced specialist whenever working on wet sites.

RADIOCARBON SAMPLES AND DATES

Radiocarbon dating is a standard method that can be applied to virtually any once-living organism (see Toolkit, volume 4, for background). Radiocarbon samples should be taken whenever suitable carbon or other organic material is observed in the excavations. These samples should be assigned unique field serial numbers (FSNs) and their locations should be plotted on plans and profiles.

Take samples with a clean trowel, forceps, or other small excavating tool. Samples should be handled as little as possible, and they should not be directly touched at all, as the oils from the excavator's hands can contaminate the sample, yielding a spurious young date. For the same reason, smoking materials should not be permitted within or near the excavation units lest they blow into an area where a radiocarbon sample is being collected. Place the sample in an aluminum foil pouch and close up the pouch. Label the sample with provenience, FSN, date, and initials and place it in a plastic bag. Note the nature of the contents, such as wood charcoal, corn kernel, leaf, or nut hull.

Standard samples should have ten grams of carbon for normal counting, which costs about $250 per sample. Accelerator mass spectrometry dates samples as small as a fraction of a gram, but analysis costs about twice as much. High precision dating of seeds from annual plants greatly increases the probability that the date is accurate for the targeted event, stratum, or feature. Scattered wood carbon is not a preferred sample type but is better than nothing. Scattered wood may derive from the interior of one or more trees or may derive from older tree falls or dead wood, which could date from hundreds of years prior to the target event, such as the burning of a building. Scattered flecks of wood charcoal often produce dates with very large standard deviations (± ranges) because they are made up of different-aged wood. On the other hand, dates on the external rings of logs or posts from structures can precisely date the erection of the building.

It is sometimes possible to identify the effects of old wood dates on archaeological sites by collecting and submitting multiple radiocarbon samples. The use of old wood may yield a pattern of radiocarbon dates in which some "unacceptably early" dates are followed by a cluster of later dates that probably estimate the actual time of occupation (Carmichael 1985:321–22). This pattern has been observed within individual features at several sites in the El Paso area, where dates for a single hearth feature span a range of seven hundred years, with a cluster of dates at the late end of the range (Hard 1983:56–57). When working in areas where the old wood problem is a potential concern, we highly recommend collecting a large number of quality samples and submitting multiple samples, even from the same feature. Your archaeobotanist should identify the species of the wood samples before they are submitted for dating. Avoiding samples from decay-resistant species and favoring short-lived species or annuals can reduce the likelihood of obtaining old wood dates. If all the collected samples are of resistant species, then at least you will be forewarned of the potential problem.

Sometimes it is advantageous to select radiocarbon samples from the analyzed flotation samples. Your archaeobotanist can select seeds or other charcoal whose direct may be important in understanding the site inhabitants' patterns of plant use. Obtain C13/C14 ratios in all dated samples to learn about their composition and to permit more accurate calibration to calendar years.

Upon completion of the fieldwork, there will likely be scores or hundreds of samples and enough money to run only a fraction of the total. Some samples will be of higher quality, and some will be from key features or stratigraphic profiles. When you can choose from among many samples, it is always good practice to run three or more dates that are in known, relative stratigraphic positions.

Dates should always be presented clearly in a table that shows the laboratory number, FSN, provenience, the lab date in radiocarbon years B.P., and the tree ring calibrated intercepts and range at the two sigma level. Use the most recent release of the standard calibration software (http://calib.org).

ARCHAEOMAGNETIC DATING

Archaeomagnetic dating relies on the fact that the Earth's magnetic poles shift over time. When iron-rich soil is fired, the iron particles will align to the then-current orientation of the Earth's magnetic field. At times during the Holocene, magnetic north has been as far south as Hawaii. Dates are determined by knowing the past orientations to magnetic north from a particular locality. The magnetic shifts are documented by correlation with radiocarbon dates. The precision of the radiocarbon dates thus limits that of the archaeomagnetic dates; however, at a particular locality, the relative dates could have a relative precision to each other of about twenty-five years.

Before taking archaeomagnetic samples, take a short training workshop with one of the method's practitioners (see www.u.arizona.edu/~slengyel). Samples are taken from very well-burned, almost brick-hard, oxidized clay. Three one-centimeter cubes are excavated in situ and oriented to magnetic north. Brass molds are placed over the samples and plaster of paris is poured into the mold and allowed to dry. Magnetic north is precisely inscribed in the plaster. Getting magnetic north correct is very important. The blocks are undercut, bagged separately in polyethylene, and labeled with provenience, date, and collector. Archaeomagnetic samples are expensive to take; however, establishing master curves for a region can result in more precise dates.

THERMOLUMINESCENCE (TL) DATING

TL dating is based on principles of quantum mechanics. When a substance is subjected to radiation, its electrons are "kicked" to higher orbits. When that material is heated the electrons return to their lower, less energetic state and give off a flash of visible light, which is measurable. To be datable, a sample must have been heated in the time period one is trying to date and then protected from further heating. Substances containing silica are good for dating purposes, particularly potsherds, fire-cracked sandstone, and heat-treated chert. The amount of radiation at the particular site also must be measured, so soil samples must be taken along with the samples to be dated, or the background radiation may be measured with a dosimeter.

TL dating has not been particularly popular because it costs about twice as much as standard radiocarbon dating, and results are not as reliable. However, TL does permit direct dating of artifacts and dating sites at which carbon is not preserved. (See sidebar 6.1 to read about how TL dating was used at one site.)

When taking TL samples, keep them cool. Bag in plastic or paper collection bags, and label the bags like other artifacts. Plot the location of the sample's FSN on appropriate plans and profiles. Don't place them in black plastic bags and leave them in the sun. As with radiocarbon dates, try to process at least three samples, preferably in stratigraphic relation to each other so it is possible to see whether they are consistent.

Use common sense in selecting samples that target the event, stratum, or feature you are trying to date. For example, with dating pottery, sherds with pristine breaks rather than eroded sherds are preferred because the eroded sherds are likely to have been kicked around the site for a long period of time before they were deposited into the archaeological record.

OXIDIZABLE CARBON RATIO (OCR) DATING SAMPLES

OCR is a recently developed and controversial method of dating pedogenesis in aerobic soils. The method, developed by Douglas Frink (1994), is based on complex relations among the parameters of soil development and the recycling of carbon through the soil to living plants. OCR has been calibrated with radiocarbon dates and has a very high statistical correlation with other dates. The method dates

> ### 6.1. TL DATING AT 3WA539
>
> At 3WA539, we realized toward the end of the project that there was virtually no carbon to date, but there was a lot of fire-cracked rock and heat-treated chert. We contacted Dr. Ralph Rowlett at the University of Missouri, who was then running TL dates at very reasonable rates. To get the background radiation counts, his preference was to set dosimeters into the site for a month. Fortunately, the remaining portion of the site had not been bulldozed, and our baulks were intact. The other alternative would have been to take a cup of soil with each sample, and we did not have soil associated with each sample of fire-cracked rock. We did have a sample of two pieces of fire-cracked rock from each provenience in which it occurred. The bulk of the rock assemblage had been counted, weighed, and discarded because most repositories do not like to maintain such collections forever. However, it is always good to keep a sample of each category discarded because one never knows what new analytical methods will be developed in the future (see Toolkit, volume 6). We were happy that we had the fire-cracked rock samples from everywhere on the site.
>
> At 3WA539, the TL dates showed that use and reuse of the large fire-cracked rock feature spanned the time from 3000 B.P. until about the year 1000. The dates suggest that at the beginning of the Mississippian period, an economic shift occurred, and this site dropped out of use. We know from other data that this was about when corn was introduced into the region.

the soil development; in the case of a pit feature or other archaeological deposit, this may or may not be the same as dating the cultural activity.

Samples should be from as thin a level as possible; one centimeter thick is ideal. Thicker samples work, but with reduced precision. The samples should be one hundred grams dry weight, about a cup. A 20 × 20 × 1–centimeter area is adequate. Soil samples should be dried out as soon as feasible to prevent mold from growing. General and specific provenience and contextual information must be supplied on the form available from the laboratory.

At the Helm site (3HS449), samples were taken in a column from each soil stratum. The results showed that the alluvial terrace began to be deposited by 6000 B.P. and that the upper thirty centimeters built up during the past thousand years. The dates were reasonable geomorphically and consistent with other data (see sidebar 6.2).

6.2. OCR APPLICATION AT THE HELM SITE

During the excavation of burials at the Helm site, the upper part of the pits was difficult to see until we actually saw the skeleton or grave goods. When the 2 × 2 unit over Burial 8 was excavated, we thought there was a pit because we saw B Horizon soils in the midden zone. It was a very small amount of soil, and once below the midden, no variation was observable across the unit floor. The floor was systematically examined with a penetrometer, with negative results. The feature was partially excavated, but no pit wall could be found. There were a few carbon flecks, but no color or hardness differences indicating a pit wall, as is the case in almost all features I (R.H.L.) had ever seen or known of. Yet, when excavation reached almost seventy centimeters below the surface, another burial was encountered. Three OCR dates were run from the soil samples that we had reserved from the flotation samples from the burials. The dates were consistent with the dates of the stratigraphic column and indicated that the different colored soils were reserved separately and were carefully refilled into the pit. This inference was consistent with findings of other investigators, which showed planned ceremonial segregation of different colored soils in Caddoan mounds.

OBSIDIAN HYDRATION

A fresh fracture in obsidian will adsorb water from the atmosphere and form a microscopic layer of hydration. During the hydration process, sodium, lithium, and magnesium are leached out of the glass while hydrogen diffuses inward. This causes a mechanical strain at the glass surface and results in a hydration rim that is measurable in thin section under a microscope. Since the formation of the hydration layer is a diffusion process, the thickness of the rim is a function of the length of time elapsed since the fracture. Assuming that an appropriate rate of hydration can be determined, the rim thickness can be converted into calendar years (Michels et al. 1983). The two primary variables that must be controlled in order to determine the hydration rate are the chemical composition of the obsidian and the temperature under which the hydration process has proceeded. Researchers are still investigating the best ways to control the key variables in the hydration process.

Source-specific hydration rates must be determined because obsidians from different sources hydrate at different rates even if they occur on the same site, under the same temperature conditions. Fortunately, physical-chemical sourcing techniques such as X-ray fluorescence (Shackley 1995) permit identification of the sources of many obsidian artifacts. Sourcing has the added benefit of revealing the distance over which obsidian was transported, either as raw material or in tool form. In addition, if obsidian artifacts at a site can be shown to belong to the same chemical source group, then variation of hydration rim thickness within the group can be used as a relative dating technique.

The real attraction of obsidian hydration is its ability to directly date artifacts and assigning a chronometric as opposed to relative date. See Miller (1996) for a detailed review of the current state of obsidian hydration dating, and contact one of the experimental labs in advance of your excavation if you wish to apply the method.

SOIL SAMPLES

Soil samples are useful for many different purposes including some of the dating methods just noted. Often they are useful for unanticipated analyses. It is a good practice to retain a cup or two of soil from flotation samples before they are floated in case the opportunity for other kinds of analyses arises. Such samples have allowed TL dating, OCR dating, pollen analysis, particle size analysis, and phytolith analysis, which otherwise could not have been preformed. Archive these samples with the collections for future use. While two-cup samples might be able to accommodate all of the following analyses, in most cases it is better to take samples for particular purposes.

Various kinds of chemical analyses can be performed on soil samples. Useful analysis can be as simple as standard pH tests or as complex as trace element analysis performed with a mass spectrometer or nuclear reactor. Required sample sizes vary from very small to a cup for most analyses. Agricultural laboratories often can analyze soil chemistry for a nominal cost. What is appropriate is dependent on research questions and preservation. It is always better to take more samples than you expect to need because new questions and different ways of testing project hypotheses always present themselves, and sometimes these samples can provide answers.

FLOTATION SAMPLES

Sampling for flotation is discussed in volume 5 of this series. Here are a few highlights to keep in mind during excavation.

Flotation samples are primarily intended to recover small, charred fragile plant and animal remains. When taking flotation samples, it is most desirable to cut out the whole sample with one or a few cuts or to chunk out the sample in pieces of manageable size. Thin-slicing the whole sample is not a good practice because it also slices the botanical remains. These remains are so fragile that rapidly changing moisture states can reduce them to unrecognizable dust. For short-term storage, store samples in polyethylene bags to help preserve their moisture stability or in sandbags for long-term storage. Lafferty, running his first project, was marveling at the great preservation of the charred walnut shell that was coming out of the flotation samples that were drying in the sun. He went to look at these half and three-quarter hulls about noon, and they were little piles of dust. Always dry samples as slowly as possible, and avoid rehydrating them once they are dry.

Label the outside of the sample bag with waterproof, rotproof tags (felt-tipped pen on flagging tape), and seal another tag inside a small plastic bag within the sample bag.

PARTICLE SIZE/GEOMORPHIC SAMPLES

Samples for geological analysis are taken like other soil samples. They can help you identify geomorphic changes such as altered river flow and location, periods of loess deposition, or prehistoric earthquakes. Have your geologist take the samples or direct the sampling program. For fluvial histories, you must have cores or profiles from several parts of the terrace system and the swamps (see sidebar 6.3 for one example). Geomorphologists like to have backhoe trenches all the way across a valley, but this is seldom possible.

POLLEN AND PHYTOLITH SAMPLES

Pollen samples from features and burials can document the use of plant materials not present in other parts of the deposits, and pollen

6.3. THE PEMISCOT BAYOU CORE

The Pemiscot Bayou core was one of over twenty cores that were taken to give the Corps of Engineers some baseline data on the likely locations of buried sites between the Mississippi River and Big Lake. The core was also selected for dating and pollen analysis to form a dated pollen chronology for the region.

The Pemiscot Bayou core was taken near the Eaker site. It took most of a day to extract this core using a commercial coring rig mounted on a 2.5-ton truck. The core driver has five-foot-long segments of driving steel. Each time a meter of core is taken, the driving segments have to be disassembled and then reassembled to proceed to the next segment. The deeper the core goes, the longer this process takes. A geomorphologist described the different strata as the core was extracted and properly labeled each core segment. It took almost all day to take the 5.8-meter-long core.

The geomorphologist subsampled the strata for particle size analysis to study the depositional environments, extracted carbon to send for radiocarbon dating, and sent twenty-three samples of the core for pollen analysis (Scott and Aasen 1987).

There were twenty strata in the Pemiscot Bayou core, which included five buried paleosols. This analysis showed change in the geological environment of Pemiscot Bayou over more than eight thousand years. About eight thousand years ago, a large natural levee was at the core site that was probably adjacent to the main course of the Mississippi River (Guccione 1987). About 6,500 years ago, the core area became a large backwater swamp as the river avulsed or migrated to the east. Around the year 1000, Pemiscot Bayou incised itself and formed a natural levee and then a channel that filled in rather recently.

---

cores from anaerobic or dry deposits can document environmental changes. Pollen and phytolith sampling is addressed in volume 5 of the Archaeologist's Toolkit series.

It is always a good idea to take pollen cores adjacent to sites that are being extensively excavated because a dated core will give you a good idea of local vegetation and landscape evolution. Pollen sequences are preserved best in undisturbed lakebed, backwater, and wetland deposits.

PROTEIN RESIDUE SAMPLES

Proteins may be preserved on stone tools for thousands of years (Loy 1983). Specialists can often determine the species of plants and animals processed by these tools. At 3WA741, we took samples and tested them for the presence of blood. While this was a rather simple test, using commercially available urine blood test "litmus" strips, it suggested that many of the ground stone artifacts usually assumed to have processed plants, such as metates and pitted cobbles, had blood protein on them (Lafferty et al. 1996). Subsequent work at another site in the Ozarks showed that rabbits were being processed with "nutting stones."

Stone tools to be sampled for protein residue should be identified in the field and not washed. They should be handled as little as possible, bagged in zip-lock bags, and labeled with appropriate provenience information. Add "DO NOT WASH" in large letters. Some investigators require excavators to use cotton curator gloves to avoid contamination.

HUMAN REMAINS

Burials exist in many archaeological sites, and you often cannot predict their presence or absence. Sometimes the presence of burials may be reasonably inferred from the context (e.g., mounds), previous studies, diagnostic burial furniture, or feature shape. In some cases, testing will have revealed the presence of human remains. In one recent case, testing with a small backhoe encountered a burial. On that basis the site was determined to be significant. Excavation of the entire site, a Mississippian village, revealed that this was the only burial on the site, and it turned out to be the remains of a nineteenth-century African American male.

Clearly, human skeletal remains have considerable research potential. Just as clearly, all peoples have the right to be involved in determining how the remains of their ancestors are treated. The Native Americans Grave Protection and Repatriation Act (NAGPRA) requires agencies to consult with tribes regarding burials affiliated with the tribe or on tribal lands. Do consultation *before* excavation to assure that the tribe's feelings are known and respected. Come to agreement on such details as excavation methods, documentation, and packaging of the remains. Some tribes require that rites of purification be carried out under certain conditions. These matters need to be settled in advance, so allot the proper budget and schedule.

Many Native Americans consider bone preservatives, once commonly used to assist in the excavation of burials, inappropriate. In addition, care should be taken to ensure that all human remains are treated with proper respect during and following excavation. If you have any doubt about what constitutes proper respect, consult with the appropriate tribal groups prior to the initiation of excavation or at least prior to the excavation of any human remains.

An experienced bioarchaeologist or paleopathologist should direct the excavation of burials. Many observations can only be made when the burials are being excavated, and the literature is extensive and specialized.

TOOLS

Excavation tools include those used in excavating features (appendix 1) as well as a specialized array for very fine work and to protect the bone from damage. Meticulous excavation is important. The field observations of bone imprints are often all that one finds from some burials. More commonly, human bone is poorly preserved. Bamboo or cane picks are good for working around bone. These can be easily shaped to different useful shapes, such as pointed or spatulated, and will not harm relatively solid bone. Dental picks are sometimes useful. Ask your dentist for his broken ones and reshape them. A set of small fine brushes is also essential. It is important to protect the excavations from the elements. Use either white or black shade cloth or tarps. Colored sunshades distort Munsell readings, and mixed colors make even consistent color observation impossible.

People react strongly when burials are exposed. In some areas of the country, the only reasonable thing to do is to hire guards so that someone is on the site twenty-four hours. If possible, keep a low profile with the press until the end of the project. (See sidebar 6.4. for the story of burials found at the Helm site.)

6.4. HELM SITE BURIALS: PROBLEMS AND POTENTIALS

The Helm site (3HS449) was a small Caddo cemetery on the Ouachita River. The Arkansas Highway and Transportation Department (AHTD) had tested it and uncovered burial-shaped rectangular pits. Preliminary excavation in the pits had revealed intact Caddo pots. Caddo burial pots are smaller than their utilitarian

6.4. (Continued)

ware and were apparently produced solely as mortuary accompaniments. The AHTD placed twenty-four-hour armed guards on the site and began negotiating a memorandum of agreement with the Caddo Tribe of Oklahoma. The Caddo wanted no remains or associations packaged in plastic nor any photographs published. If bone was still preserved, they would perform the proper cleansing. The Caddo believe that when the bone has returned to the earth, the soul is released to return to the stars, and the burials are no longer sacred.

We began the excavation on the Helm site by hand excavating 2 × 2–meter units over the areas where the testing had indicated the cemetery lay. Flotation columns were taken from each unit. When each pit could be discerned, which was not easy, special treatment of the pit fills commenced. All excavated soil was returned to the laboratory for flotation processing. Beginning with the exposed bone, sometimes with only one exposed skeletal landmark, Dr. M. C. Hill could often tell where and about how large the burials were. The bone was dust, and the Caddo were kept informed daily (a cell phone was essential). They maintained that the ancestors had gone to the stars, and no Cedar Smoke Ceremony was required. We proceeded apace with strange windstorms, guards not showing up, and intruders being convinced to leave at gunpoint. At the end of the first week, a burial was encountered with the bone intact, and the Caddo said it was time for a Cedar Smoke Ceremony.

Once the burials were exposed, they were mapped. One burial had masses of red ochre in two places. Masses of small pebbles in another burial were probably rattles. Associations were mapped and assigned their own FSN, and pollen/flotation samples were removed from beneath the skull, each shoulder, midtorso, femora, tibiae, and ankles. The excavation revealed nine individuals in eight burial pits. The bone in most of the burials was so fragmentary that all that remained were stains and teeth. Seven of the burial pits contained two to four ceramic vessels. Analysis of the vessels by Ann Early suggested that three burials dated to the seventeenth- to eighteenth-century Deceiper phase and four burials from the fifteenth- to sixteenth-century Social Hill phase. This interpretation was consistent with the superposition and intrusions of several of the burials. The eighth burial, which contained no vessels, was intruded by a Social Hill and a Deceiper burial, suggesting that it was probably from the earlier Mid-Ouachita phase.

One burial from each phase was selected for analysis of pollen and phytolith samples, including samples from inside the ceramic vessels from each burial. The pollen analysis had some surprises. Linda S. Cummings reported that burials contained cotton (*Gossypium* sp.) pollen. The Hernando de Soto chroniclers mentioned that they saw cotton cloth in this region. Could this be confirming evidence, or had the pollen percolated down one meter at a later date? A half-liter soil sample from above burial 4 contained no cotton pollen. What was the source of the cotton pollen? We're not sure.

Honeysuckle (*Lonicera* sp.) pollen was identified in the midtorso area of both burials; perhaps the individuals ingested flowers for their sugar content, or flowers were placed on the bodies. Fern spores were recovered in the shoulder areas, suggesting either a cushion of fern fronds or possibly the use of fern stems in basketry. We know of no surviving Caddo baskets, but virtually all of the ethnographic accounts speak very highly of the skill of the Caddo basket makers.

> When the last heavy flotation fraction was analyzed, the analyst asked Lafferty, "What about the nail that was in this sample?"
> "Nail? What is that doing there?"
> In a little zip-lock bag was a 21.3-millimeter-long piece of ferruginous metal. Its width tapered from 2.3 to 2.2 millimeters with a square cross-section. What was this, and how did it get there? We already knew that the site was around the right time to have been used when Hernando de Soto came through the region. Could this be the proverbial smoking gun?
> We sent it along with two pieces of iron from the Martin Farm site in Florida, a known Soto site, to Dr. S. K. Nash at the University of Pennsylvania for metallurgical and elemental analyses of the three samples using particle-induced X-ray emission spectrometry and energy dispersive spectrometry. He concluded that because of its high manganese content, the Helm site specimen had to be a product of post-1860 technology. The location of the metal at the very bottom of a burial feature remains an enigma. Such a small piece might have fallen down a deep root hole; however, virtually no historic artifacts were on the surface, and the pollen distribution suggests that the burial was intact. It is also possible that sixteenth-century European military steel was more advanced than we think. Comparative analyses from that period are lacking.
> Botanical materials were recovered and analyzed from the features, burials, and some of the overlying midden deposits. Analysis showed that the botanical remains in the grave fill were highly worn and eroded in secondary or even tertiary contexts. In composition they were similar to the midden.

REBURIAL

It often comes as a surprise to archaeologists, but there is almost no subject on which all Native Americans agree. While there may be near-universal agreement that native groups should be involved in the determination of how the remains of their ancestors will be treated, the specifics of each case can vary widely. Sometimes basic skeletal research on the remains will be acceptable, and sometimes it won't. Sometimes nondestructive analyses will be acceptable, but destructive analyses won't. Sometimes even destructive techniques (such as radiocarbon dating or bone chemistry) may be approved. Sometimes reburial will be identified as the appropriate treatment, either before or after analysis, and either with or without the accompanying grave goods recovered in the excavation. Sometimes it will be permissible to retain a sample of the remains (such as tooth enamel) for future analysis, as long as the rest of the remains are reburied. Sooner or later, you will become involved in a project where human remains are reburied. By approaching the prospect with respect and sensitivity, it can be an informative and healing process, one that may be of considerable research value (see also sidebar 6.5).

6.5. EXPERIENCES FROM THE REBURIAL ISSUE

A few years ago, one of us (D.L.C.) was called on by a Mescalero Apache holy man to assist him as an outside third party during consultations about the disposition of human remains discovered near the Mescalero reservation in New Mexico. When the question of research and reburial arose, I agreed that if he wanted the remains to be reburied, I would respect and support his wishes. He then asked what sorts of studies the archaeologists would like to do and what the possible benefits might be. I explained that standard anthropometric measurements and paleopathological observations could be made, but they would probably not teach us very much in this case. Given the state of research in the region, I suggested that it would be most useful to sacrifice some portion of the skeletal remains for radiocarbon dating and bone chemistry studies. After considering the destructive nature of these tests, the holy man agreed that they could proceed, as long as the rest of the remains were eventually reburied.

A few years earlier, we were both involved in the first reported reburial ceremony (figure 6.1) undertaken by a federal agency when the U.S. Air Force reburied the fragmentary remains of some twenty individuals recovered during archaeological testing at the Eaker site (3MS105) in northeastern Arkansas (Brown 1991). In this case, the human remains were analyzed as part of Section 106 compliance and determined to have no further research potential (Lafferty et al. 1989). Throughout the test excavations, the Air Force had consulted with the Quapaw tribe of Oklahoma, who claim descendancy from the site's occupants. Funds were provided to bring a tribal delegation from Oklahoma, including holy man Bob Whitebird and several members of the Quapaw tribal council. The Air Force halted air traffic at the base for more than an hour so the ceremony could be undertaken with hushed reverence, a simple act of respect by a federal agency that spoke volumes to people who were directly affected by our excavations.

Following the reburial ceremony, at an impromptu ethnographic interview, Bob Whitebird spoke of his people and their past. The occasion and the emotion of the moment elicited a traditional story that had never been heard by the tribal council members. So, in response to diverse interests and constituencies, a major archaeological site was preserved, human remains were studied and reburied, and archaeologists, leaders of the Quapaw tribe, and U.S. Air Force personnel consulted and cooperated in a spirit of respect. In the end, the archaeological project at Eaker Air Force Base also contributed to a fuller understanding of the Quapaw people. How satisfying for everyone concerned; how appropriate for archaeology to be anthropology.

Figure 6.1. Repatriation ceremony at Eaker Air Force Base, conducted by Quapaw holy man Bob Whitebird. (Photograph by Sergeant Christine Richards.)

CHAPTER 6

LOOKING TO THE FUTURE

As with other data recovery methods and techniques, collection of specialized samples has grown more complex as technologies have developed. For archaeologists whose careers span a few decades, the rapidly changing technologies sometimes seem amazing and a bit overwhelming because they allow us to develop data types for which we might not have dared hope at the beginning of our careers. Predictive models have increased in sensitivity, assisted by computers and GIS, allowing us to locate sites with somewhat greater reliability than our intuition and experience. Remote sensing has granted us something close to X-ray vision so that we don't always have to dig, or at least so we can place excavations more effectively. New chronometric techniques let us pry dates from materials and sample sizes we didn't imagine to be possible. Analytical techniques with environmental data have expanded our assessment of the interaction of the people of the past with their environments. It costs real dollars to collect and analyze specimens, and the increasing specialization among archaeologists has fostered fragmentation in our discipline. To be truthful, we have sometimes been incautious in our borrowing of new technologies. At the same time, whatever problems new technologies might bring, few among us would like to return to archaeology as it was practiced even a decade ago.

Changes in data recovery have also brought changes in our obligations to our publics. We need to be certain that we spend the public's money wisely as we apply these new technologies and not get carried away by our enthusiasms by thinking that these new approaches get us all that much closer to the "truth" about the past. They just provide us with more information which we still have to interpret. What we come up with can have an impact on the lives of contemporary people, and that is a responsibility of which we constantly must remain aware. All in all, however, we have a right to be excited for what will come in terms of new ways of knowing the archaeological past.

APPENDIX 1

EQUIPMENT, SUPPLIES, AND SOURCES FOR SPECIALIZED ANALYSIS

PERSONAL TOOLKITS

1 Marshalltown pointing trowel (4.5-inch)
2 3-meter tape measures graduated in millimeters
1 line level
2 pins (gutter spikes, ten- to twelve-inch spikes, or surveyor pins)
1 notebook
2–3 meters of nylon staging twine
1 compass
1 fifteen- to twenty-centimeter ruler
Leather work gloves
Pencils

These are the basic tools that everyone should have in his personal tool kit. Most are available at your local hardware store or camping goods store. Marshalltown trowels have good welds—other brands of trowels will last as little as minutes on a site. Don't waste your money on other brands. Sharpen the leading edges to razor-sharp. Metric measuring tapes are often not carried in American hardware stores. Some tapes that are labeled as metric (e.g., three-meter tape) do not necessarily have metric graduations. Check the actual tape before you buy it. Various kinds of pins, such as wire spikes, gutter spikes, and surveyors pins, are useful in pulling line levels to check the depth of excavation levels and drawing profiles. Get a fluid-filled compass. They are faster.

OTHER USEFUL TOOLS

Short-handled hoe
Ingalls pick
Ice cream scoop
Stainless steel tablespoon
Bamboo "Perino" pick
Clipboard
Whisk broom
Paintbrushes
Hand pruning shears

    A short handled hoe is useful to level floors and clean profiles. Old hoes that are worn to one- to two-inch blade heights work best and often have better steel. Sharpen the side of the blade that faces the handle. Ingalls picks are useful in cleaning foundations, masonry, and around rocks. Ice cream scoops and various-sized scoops are useful in feature excavations. The Perino pick made of a wall of a large (2.5- to 5-centimeter diameter) bamboo is useful in working around fragile specimens and bone. (It is named after its inventor, the veteran excavator Greg Perino.) Sharpen one end to about a one-centimeter-wide bevel edge and the other to a point. Whisk brooms and various-sized paintbrushes are useful for cleaning excavation units, bringing out contrasting soil textures and so forth. Hand pruning shears are necessary to cut roots in forested environments.

"COMPANY"-FURNISHED EXCAVATION EQUIPMENT

Here is a list of basic excavation equipment. This does not include the many expendable supplies, such as pencils, flagging, bags, and so forth, that are necessary. It also does not include equipment that might be required to excavate in caves, such as respirators, lighting, and hard hats, or in particularly deep units, like shoring.

Shovels
Picks/mattocks
Post hole digger
Coring equipment
Probes
Screens

EQUIPMENT, SUPPLIES, AND SOURCES FOR SPECIALIZED ANALYSIS

Scoops
Dust pans
Hundred-meter tape measures
Plumb bob
Plane/drawing table
Field desk
Cameras
Mapping/surveying equipment
Munsell soil color charts
Hammers
First aid kit
Five- to ten-gallon water jug(s)
Dust masks in dry environments
OSHA manual
Scissors
Three-hole punch

Shovels come in a variety of shaped edges that are differentially useful in different situations. Flat shovels are most useful because we dig square holes, but other shapes are useful under different circumstances. Round-tipped shovels are good when one is excavating stones, and "sharpshooters" are indispensable when excavating clays. Shovels should be sharpened daily and cleaned and oiled after a project.

Picks/mattocks are important tools when working with rocky or very dry soils.

Post hole diggers are necessary to determine whether there are archeological deposits buried deeper than the intended excavation. Always keep a set handy.

Screens are very important and commercially available. Unfortunately, most that we have seen are based on the two-legged Roberts screen. They are hard to use and really hurt the back. Lafferty likes aluminum screens that are hung on a tubular aluminum A-frame. This arrangement is portable enough for testing projects.

FORMS

Use forms to keep track of excavations data and to assure that the same type of information is recorded consistently. Forms may be created on your computer and copied. If you need thousands of copies, take them to a printer. Forms should be printed on acid-free paper. Excavation, feature, and burial forms should have green metric grid paper on the reverse. Forms generally include the following:

Field serial log or bag log
Photography log
Feature log
Excavation level form
Feature form
Burial form
Inside and outside bag tags

EXPENDABLE SUPPLIES

Expendable supplies are necessary to support excavations. It is important to get these before the fieldwork, and get them in sufficient quantities to do the job. Commonly required expendable supplies include the following:

Stakes and pins
String/twine
Flagging tape
Pencils
Disposable water cups
Waterproof marking pens (Sharpies)
Masking tape
Duct tape
Three-ring binders for completed forms
Paper clips

SOURCES OF TOOLS AND SPECIALIZED ANALYSES

The listing of a supplier or source here is not to be read as an endorsement of its services or products. Additional excellent vendors may be available.

FIELD AND SURVEYING EQUIPMENT

Ben Meadows Co.
P.O. Box 5277
Janesville, Wisconsin 53547-5277
(800) 241-6401
www.benmeadows.com

Forestry Suppliers, Inc.
P.O. Box 8397
Jackson, Mississippi 39284-8397
(800) 360-7788
www.forestry-suppliers.com

PLASTIC BAGS

Consolidated Plastics Company, Inc.
8181 Darrow Road
Twinsburg, Ohio 44087
(800) 362-1000
www.consolidatedplastics.com

Associated Bag Co.
400 W. Boden Street
Milwaukee, Wisconsin 53207
(800) 926-6100
www.associatedbag.com

Have good prices on all sizes and gauges of zip-lock bags.

ARCHIVAL SUPPLIES AND EQUIPMENT

University Products, Inc.
517 Main Street, P.O. Box 101
Holyoke, Massachusetts 01041-101
(800) 628-1912
www.universityproducts.com

RADIOCARBON DATING

Beta Analytic, Inc.
4985 SW 74 Court
Miami, Florida 33155
(305) 667-5167
www.radiocarbon.com

Beta offers standard and AMS dating with results in twenty to thirty business days. There are many other laboratories. If the dates are not calibrated, use the most recent calibration software available at http://calib.org.

ETHNOBOTANICAL ANALYSIS

A number of individuals and organizations do this kind of analysis. Most of the larger anthropology departments have one or more ethnobotanists on the staff. Most investigators have several months' backlog. It is best to find someone who is familiar with the flora in your region. One active lab:

Paleo Research Institute
2675 Youngfield Street
Golden, Colorado 80401
(303) 277-9848

DNA ANALYSIS

Paleoscience, Inc.
4989 SW 74th Court
Miami, Florida 33155
(305) 662-7760
info@paleoscience.com

ARCHEOMAGNETIC DATING

Dr. Jeff Cox
Museum of New Mexico
P.O. Box 2087
Santa Fe, New Mexico 87504
(505) 827-6343
Also see www.u.arizona.edu/~slengyel

DENDROCHRONOLOGY

Several laboratories work in different parts of the world. For a comprehensive list with interactive links, see the Ultimate Tree Ring Web Page at http://web.utk.edu/~Grissino/

THERMOLUMINESCENCE DATING

Dr. James Feathers
Department of Anthropology
Box 353100
University of Washington
Seattle, Washington 98195-3100
(206) 685-1659
jimp@u.washington.edu

OXIDIZABLE CARBON RATIO DATING

Dr. Douglas S. Frink
Archeological Consulting Team, Inc.
P.O. Box 145
Essex Junction, Vermont 05453-0145
(802) 879-2017
dsfrink@aol.com

APPENDIX 2

CONVERSION TABLES AND OTHER USEFUL RESOURCES

Basic trigonometry and geometry are necessary to lay out square units and often in site grid layout. The most important is the relationship of the hypotenuse of the right triangle. The Pythagorean theorem tells us that the square of the hypotenuse is equal to the sum of the squares of the other two sides (legs). Expressed mathematically where A and B are the legs of a right triangle and C is the hypotenuse, this is:

For a 1 × 1 unit:

$$C^2 = A^2 + B^2 = 1(1) + 1(1)$$
$$C^2 = 1 + 1$$
$$C^2 = 2$$
$$C = \text{square root of } 2$$
$$C = 1.4142135$$

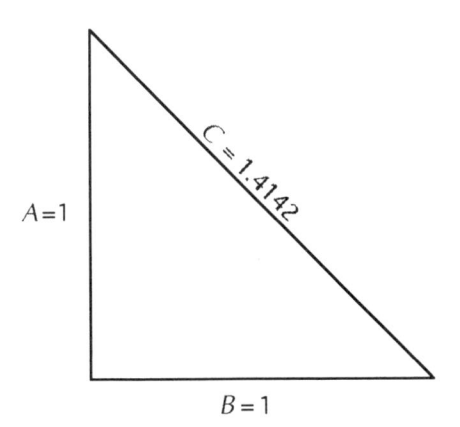

This ratio is worth memorizing as one is always using it to set up square units, extend the grid, or calculate how high a bluff is. Because we always use square units, this is a linear relationship. If one needs to pull in a 10 × 10–meter block, then pull the 10-meter leg out from the grid and pull the 14.1421-meter hypotenuse to set the new point (see tables A2.1 and A2.2).

Table A2.1. Hypotenuses of different-sized units

Unit Size	Hypotenuse
1 × 1	1.4142
2 × 2	2.8284
3 × 3	4.2426
5 × 5	7.0711
10 × 10	14.1421
100 × 100	141.4213

Table A2.2. Conversion factors

Multiply	By	To obtain
Acres	43,560	Square feet
Acres	0.4047	Hectares
Acres	4,047	Square meters
Acres	1.562×10^{-3}	Square miles
Acre feet	43,560	Cubic feet
Board feet	144 sq. in. × 1 in.	Cubic inches
Bushels	1.244	Cubic feet
Bushels	2,150	Cubic inches
Bushels	0.03524	Cubic meters
Bushels	4	Pecks
Bushels	64	Pints (dry)
Bushels	32	Quarts (dry)
Centimeters	0.3937	Inches
Centimeters	0.01	Meters
Centimeters	10	Millimeters
Chains	66	Feet
Chains	100	Links
Chains	0.0125	Miles
Cubic centimeters	6.102×10^{-2}	Cubic inches
Cubic centimeters	10^{-6}	Cubic meters
Cubic centimeters	2.642×10^{-4}	Gallons
Cubic centimeters	10^{-3}	Liters

Table A2.2. Conversion factors (*continued*)

Multiply	By	To obtain
Cubic feet	0.02832	Cubic meters
Cubic feet	0.03704	Cubic yards
Cubic feet	7.481	Gallons
Cubic feet	28.32	Liters
Cubic inches	16.39	Cubic centimeters
Cubic inches	5.787×10^{-4}	Cubic feet
Cubic inches	1.639×10^{-5}	Cubic meters
Cubic inches	2.143×10^{-5}	Cubic yards
Cubic inches	4.329×10^{-3}	Gallons
Cubic inches	1.639×10^{-2}	Liters
Cubic meters	10^6	Cubic centimeters
Cubic meters	35.31	Cubic feet
Cubic meters	1.308	Cubic yards
Cubic meters	10^3	Liters
Cubic yards	27	Cubic feet
Cubic yards	0.7646	Cubic meters
Cubic yards	202.0	Gallons
Fathoms	6	Feet
Feet	30.38	Centimeters
Feet	0.3048	Meters
Feet	0.36	Varas
Furlongs	40	Rods
Gallons	0.1337	Cubic feet
Gallons	231	Cubic inches
Gallons	3.785	Liters
Gallons	3.785×10^{-3}	Cubic meters
Gallons	8	Pints (liquid)
Gallons	4	Quarts (liquid)
Gills	0.1183	Liters
Grains (troy)	0.6480	Grams
Grams	0.03527	Ounces
Hectares	2.471	Acres
Inches	2.540	Centimeters
Kilograms	10^3	Grams
Kilograms	2.2046	Pounds
Kilograms	1.102×10^{-3}	Tons
Knots	1.152	Miles per hour
Leagues		
English	3.0	Miles
Gallic	2.2530	Kilometers
Gallic	1.4	Mile
Gallic	1.25	Roman mile
French, common	4.4481	Kilometer
French, common	2.764	Mile
French, Post	3.8977	Kilometer

Table A2.2. Conversion factors (continued)

Multiply	By	To obtain
Leagues (continued)		
French, Post	2.422	Mile
Flemish	6.2763	Kilometer
Flemish	3.9	Mile
Sixteenth-century Spanish		
League comun	5.57	Kilometers
League comun	3.45	Miles
League legal	4.19	Kilometers
League legal	2.63	Miles
Links (surveyor's)	0.01	Chains
Links (surveyor's)	7.92	Inches
Liters	10^3	Cubic centimeters
Liters	0.03531	Cubic feet
Liters	61.02	Cubic inches
Liters	10^{-3}	Cubic meters
Liters	1.308×10^{-3}	Cubic yards
Liters	0.2642	Gallons
Liters	1.057	Quarts
Meters	100	Centimeters
Meters	3.2808	Feet
Meters	39.37	Inches
Meters	10^3	Kilometers
Meters	1.0936	Yards
Miles	5280	Feet
Miles	1.6093	Kilometers
Miles	80	Chains
Miles	1900.8	Varas
Ounces	28.35	Grams
Pounds	453.6	Grams
Rods	16.5	Feet
Square feet	2.296×10^{-5}	Acres
Square feet	0.09290	Square meters
Square inches	6.425	Square centimeters
Square kilometers	0.3861	Square mile
Square meters	2.471×10^{-4}	Acres
Square meters	10.764	Square feet
Square miles	640	Acres
Square miles	2.590	Square kilometers
Square varas	7.716049	Square feet
Square varas	0.857339	Square yards
Square yards	0.8361	Square meters
Square yards	9	Square feet
Tons (metric)	10^3	Kilograms
Tons (metric)	2,205	Pounds
Tons (long)	2,240	Pounds

Table A2.2. Conversion factors (continued)

Multiply	By	To obtain
Tons (long)	2,216	Kilograms
Tons (short)	2,000	Pounds
Tons (short)	907.2	Kilograms
Varas	2.7777	Feet
Varas	33.3333	Inches
Varas	0.9259	Yards
Yards	0.9144	Meters
Yards	1.08	Varas

Excerpted, by permission, from Eidenbach (1988).

RANDOM NUMBER TABLES

Random numbers are useful for sampling of any kind and tables or generators are readily available on the web, in a range of books, and by using a software program such as Random Number Generator Pro at www.segobit.com/rng.htm (this is not an endorsement, but only an example) or even in Microsoft Excel. You can also find prepared tables online, print them, and take them to the field with you. On a web search engine, do a string search for "random number table," and you will find several as well as instructions on how to use them.

REFERENCES

Adams, James L.
    1986 *Conceptual Blockbusting: A Guide to Better Ideas.* Addison Wesley, Reading, Mass.

Barker, Philip
    1983 *Techniques of Archaeological Excavation,* 2nd. ed., rev. and expanded. Universe: New York.

Binford, L. R.
    1962 Archeology as Anthropology. *American Antiquity* 28(2):217–25.
    1982 The Archaeology of Place. *Journal of Anthropological Archaeology* 1(1):531.

Brown, Judy
    1991 A Quapaw Reburial. *CRM* 14(5):24–27. National Park Service, Washington, D.C.

Carmichael, David L.
    1978 *Archaeological Assessment of the CHICAP Pipeline Easement.* Report prepared for Union Oil of California, through Nalco Environmental Services, Chicago. On file, Illinois Department of Conservation, Springfield.
    1985 *Archeological Excavations at Two Prehistoric Campsites near Keystone Dam, El Paso, Texas.* University Occasional Papers No. 14, New Mexico State University, Las Cruces, New Mexico.

Carmichael, David L., and Terry Franklin
    1999 Archeological Screening Techniques and Their Effects on the Recovery of Lithic Artifacts. In *Archaeology of the Jornada Mogollon: Proceedings from the 10th Jornada Mogollon Conference,* compiled by Michael Stowe and Mark Slaughter, pp. 151–56. GMI Press; Geomarine, Plano, Texas.

REFERENCES

Claflin, W. H.
  1931 *The Stalling's Island Mound, Columbia County, Georgia.* Papers of the Peabody Museum of American Archaeology and Ethnology, vol. 14, no. 1. Harvard University Press, Cambridge, Mass.

Cole, F. C., R. Bell, J. Bennett, J. Caldwell, N. Emerson, R. MacNeish, K. Orr, and R. Willis
  1951 *Kincaid: A Prehistoric Illinois Metropolis.* University of Chicago Press, Chicago.

Davis, H. A. (ed.)
  1982 *A State Plan for the Conservation of Archeological Resources in Arkansas.* Arkansas Archeological Survey Research Series 21, Fayetteville (Revised 1994).

Douglass, A. E.
  1936 *Climatic Cycles and Tree-growth, Vol. III.* Carnegie Institution of Washington, Washington, D.C.

Dunnell, Robert C.
  1984 The Ethics of Archaeological Significance Decisions. In *Ethics and Values in Archaeology,* edited by Ernestine Green, pp. 62–74. Free Press, New York.

Echo-Hawk, Roger C.
  1993 Working Together—Exploring Ancient Worlds. *Society for American Archaeology Newsletter* 11(4):5–6.

Echo-Hawk, Walter R., and Roger C. Echo-Hawk
  1991 Repatriation, Reburial, and Religious Rights. In *Handbook of American Indian Religious Freedom,* edited by Christopher Vecsey, pp. 63–80. Crossroads, New York.

Eidenbach, Peter L.
  1988 *Archaeologists' Pocket Companion.* Human Systems Research, Inc. Tularosa, N.M.

Fagan, Brian M.
  1985 *The Adventure of Archaeology.* National Geographic Society, Washington, D.C.

Figgins, J. D.
  1927 The Antiquity of Man in America. *Natural History* 27(3):229–39.

Fowke, G.
  1922 *Archeological Investigations.* Smithsonian Institution Bureau of American Ethnology Bulletin 76. Government Printing Office, Washington, D.C.

Fowler, Don D.
    1984   Ethics in Contract Archaeology. In *Ethics and Values in Archaeology*, edited by Ernestine Green, pp. 108–16. Free Press, New York.

Frink, D. S.
    1994   The Oxidizable Carbon Ratio (OCR): A Proposed Solution to Some Problems Encountered with Radiocarbon Data. *North American Archaeologist* 15:17–29.

Gifford, Carol A., and Elizabeth A. Morris
    1985   Digging for Credit: Early Archaeological Field Schools in the American Southwest. *American Antiquity* 50(2):395–413.

Guccione, M. J.
    1987   Geomorphology, Sedimentation, and Chronology of Alluvial Deposits, Northern Mississippi County, Arkansas. In Robert H. Lafferty III et al., *A Cultural Resources Survey, Testing, and Geomorphic Examination of Ditches 10, 12, and 29, Mississippi County, Arkansas*. Submitted to Memphis District, COE, Contract No. DACW66-86-C-0034 by MCRA Report No. 86-5, pp. 67–99.

Haas, Daniel
    1998   *The Federal Archeology Program, Secretary of the Interior's Report to Congress, 1994–95*. Archeology and Ethnography Program, National Park Service, U.S. Department of the Interior, Washington, D.C.

Haas, Jonathan
    1993   Discussant remarks delivered as part of the symposium, Two Years After: Repatriation and its Implementation. 58th Annual Meeting, Society for American Archaeology, April 17, St. Louis.

Hard, Robert J.
    1983   Excavation in the Castner Range Archaeological District in El Paso, Texas. El Paso Centennial Museum Publications in Anthropology, No. 11. University of Texas at El Paso.

Heizer, R. F.
    1959   *The Archaeologist at Work: A Sourcebook in Archaeological Method and Interpretation*. Harper, New York.

Hester, Thomas R., Harry J. Shafer, and Kenneth L. Feder
    1997   *Field Methods in Archaeology*, 7th ed. Mayfield, Mountain View, Calif.

Hill, A. T., and M. Kivett
  1949   Woodland-Like Manifestations in Nebraska. *Nebraska History* 21(3):143–91.

Holmes, W. H.
  1903   Aboriginal Pottery of the Eastern United States. *Twentieth Annual Report of the Bureau of American Ethnology to the Secretary of the Smithsonian Institution, 1898–99*, pp. 1–200.

Ives, John W. (ed.)
  1986   *Archaeology in Alberta, 1985*. Archaeological Survey of Alberta, Occasional Paper No. 29. Edmonton, Alberta.

Kellog, D. C.
  1987   Statistical relevance and site locational date. *American Antiquity* 52(1):143–150.

Kidder, A. V.
  1962   *An Introduction to the Study of Southwestern Archaeology*. Yale University Press, New Haven, Conn.

King, Thomas F.
  1977   Resolving a Conflict of Values in American Archaeology. In *Conservation Archaeology: A Guide for Cultural Resource Management Studies*, edited by Michael B. Schiffer and George J. Gummerman, pp. 87–96. Academic Press, New York.

King, Thomas F., and P. P. Hickman
  1973   *The Southern Santa Clara Valley, California: A General Plan for Archaeology*. A. E. Treganza Anthropology Museum, San Francisco.

Knudson, Ruthann, Francis P. McManamon, and Emlen Myers
  1995   *Report on the Federal Archeology Program (1988–1990)*. U.S. Department of the Interior, National Park Service, Washington, D.C.

Lafferty, Robert H. III and Robert F. Cande
  1989   *Cultural Resources Investigations, Peacekeeper Rail Garrison Program, Eaker Air Force Base, Mississippi County, Arkansas*. Prepared for U.S. Air Force Regional Civil Engineer, AFRCE-BMS Norton Air Force Base, California by MCRA Report No. 88-5.

Lafferty, Robert H. III, A. Early, M. C. Sierzchula, M. C. Hill, G. S. Powell, N. Lopinot, L. S. Cummings, S. L. Scott, S. K. Nash, and T. K. Perttula
  2000   *Data Recovery at the Helm Site, 3HS449, Hot Springs County, Arkansas*. Submitted to Arkansas Highway and Transportation Department, Job FA3009. MCRA Report 2000-1.

REFERENCES

Lafferty, Robert H. III, Margaret J. Guccione, Linda J. Scott, D. Kate Aasen, Beverly J. Watkins, Michael C. Sierzchula, and Paul F. Baumann
    1987 *A Cultural Resources Survey, Testing, and Geomorphic Examination of Ditches 10, 12, and 29, Mississippi County, Arkansas.* Submitted to Memphis District, COE, Contract No. DACW66-86-C-0034 by MCRA Report No. 86-5.

Lafferty, Robert H. III, N. H. Lopinot, M. J. Guccione, L. G. Santeford, Michael C. Sierzchula, S. L. Scott, K. A. King, K. M. Hess, and L. S. Cummings
    1988 *Tracks in Time*, National Park Service, Southwest Region Branch of CRM, Santa Fe, N.M.

Lafferty, Robert H. III, Lawrence G. Santeford, Margaret G. Guccione, Michael C. Sierzchula, Neal H. Lopinot, Kathryn A. King, Kathleen M. Hess, and Jody O. Holmes
    1987 *The Mitchell Site: 3WA58 Archeological Investigations at a Prehistoric Open-Field Site in Washington County, Arkansas.* Submitted to McClelland Consulting Engineers and the City of Fayetteville by MCRA Report No. 87-4.

Lafferty, Robert H. III, Lawrence G. Santeford, Michael C. Sierzchula, Robert F. Cande, Kathryn A. King, Carol S. Spears, and Robert A. Taylor
    1989 *A Corridor into Ozarkian Time: National Register of Historic Places Testing of 25 Sites in the AHTD Interstate 71 Corridor, Crawford and Washington Counties, AR.* Submitted to the Arkansas Highway and Transportation Department by MCRA Report No. 89-2.

Lafferty, Robert H. III, M. C. Sierzchula, R. F. Cande, P. B. Mires, M. T. Oates, M. J. Guccione, Neal Lopinot, L. G. Santeford, H. Wagner, S. Scott, and M. Cleaveland
    1996 *Cato Springs Archeology and Geomorphology: Archeological Data Recovery at 3WA539, 3WA577, and 3WA741, U.S. Highway Relocation, Washington County, Arkansas.* Submitted to Arkansas Highway and Transportation Department, Job 4833. MCRA Report 94-6.

Lafferty, Robert H. III, M. C. Sierzchula, G. Powell, N. Lopinot, C. Spears, and L. G. Santeford
    2002 *Data Recovery at the Hillhouse site (23MI699).* Report submitted to the Memphis District, Corps of Engineers, Contract DACW66-C-0032. MCRA Report 2001-1.

Layard, A. H.
    1849 *Nineveh and Its Remains.* John Murray, London.

Loy, Thomas H.
    1989   Prehistoric blood residues: detection on tool surfaces and identification of species of origin. *Science* 220:1269–1271.

Lyman, R. Lee, Michael J. O'Brien, and Robert C. Dunnell
    1997   *The Rise and Fall of Culture History.* Plenum, New York.

Lynott, Mark J., and Alison Wylie (eds.)
    1995   *Ethics in American Archaeology: Challenges for the 1990s.* Society for American Archaeology. Allen, Lawrence, Kans.

Magne, Martin (issue editor)
    1997   Parks Canada. *CRM* 20(4):3–62.

Meighan, Clement W.
    1992a Some Scholars' Views on Reburial. *American Antiquity* 57(4):70410.
    1992b University of California Give-Away of Museum Collections? *California Scholar* (Winter 1992–1993):23–25.
    1993   The Burial of American Archaeology. *Academic Questions* 6(3):9–18.
    1994   Burying American Archaeology. *Archaeology* 47(6):64–68.

Michels, J. W., I. S. T. Tsong, and G. A. Smith
    1983   Experimentally Derived Hydration Rates in Obsidian Dating. *Archaeometry* 25(2):10–117.

Miller, Myles R. III
    1989   *Archaeological Excavations at the Gobernadora and Ojasen Sites: Dona Ana Phase Settlement in the Western Hueco Bolson, El Paso County, Texas.* Center for Anthropological Research, Report No. 673. New Mexico State University, Las Cruces.
    1996   The Chronometric and Relative Chronology Project, Section III: Obsidian Hydration Dating. Archaeological Technical Report No. 5. Prepared by the Anthropology Research Center, University of Texas at El Paso, for the Department of the Army, Directorate of the Environment, Fort Bliss, Texas.

Mills, William C.
    1907   Explorations of the Edwin Harness Mound. *Ohio Archeological and Historical Quarterly* 16(2), Columbus.

Molyneaux, Brian L.
    1983   The Study of Prehistoric Sacred Places: Evidence from Lower Manitou Narrows. Archaeology Paper No. 2. Royal Ontario Museum, Toronto.
    2002   Exploring the Landscapes of Long-Distance Exchange: Evidence from Obsidian Cliffs and Devils Tower, Wyoming. In *Geochem-*

ical Evidence for Long Distance Exchange, edited by Michael Glascock, pp. 133–51. Bergin & Garvey, Westport, Conn.

Molyneaux, Brian L., Todd Kapler, Shesh Mathur, and Nancy J. Hodgson
- 1993 A Phase II Survey of Segment I of the Highway Corridor. Prepared for RUST Environment and Infrastructure, Waterloo, Iowa.
- 1996 Highway 60 Le Mars—Minnesota Border Archaeological Resources Survey.

Molyneaux, Brian L., William Ranney, Stephanie A. Spars, and Jason Kruse
- 2002 Phase I and II Archaeological Reconnaissance and Testing on Alluvial Terraces at the Confluence of Threemile and Forsyth Creeks, Fort Riley, Kansas. Prepared for Dynamac Corporation, Fort Riley, Kan.

Morse, Dan F., and Phyllis A. Morse
- 1983 *Archaeology of the Central Mississippi Valley*. Academic Press, New York.

Nelson, N. C.
- 1916 Chronology of the Tano Ruins, New Mexico. *American Anthropologist* 18:159–80.

New Mexico State University (NMSU)
- 1989 *Cultural Resources Report for the All American Pipeline Project, Santa Barbara, California to McCamey, Texas and Additional Areas to the East along the Central Pipeline Route in Texas*. Report submitted by All American Pipeline Company to the Bureau of Land Management, California Desert District.

Oakes, Yvonne R., and Dorothy Zamora
- 1998 *Archaeology of the Mogollon Highlands: Settlement Systems and Their Adaptations*. Archaeology Note 232. Museum of New Mexico, Office of Archaeological Studies, Santa Fe.

O'Brien, Michael J.
- 1996 *Paradigms of the Past: The Story of Missouri Archaeology*. University of Missouri Press, Columbia.

O'Laughlin, Thomas C.
- 1980 *The Keystone Dam Site and Other Archaic and Formative Sites in Northwest El Paso, Texas*. Publications in Anthropology No. 8. El Paso Centennial Museum, University of Texas at El Paso.

Orton, Clive
- 2000 *Sampling in Archaeology*. Cambridge University Press, Cambridge.

Parker, Patricia L., and Emogene Bevitt
  1997 Consultation with American Indian Sovereign Nations. *Common Ground: Archeology and Ethnography in the Public Interest* 2(3/4):22–27. National Park Service, Washington, D.C.

Patterson, L. W.
  1978 Basic Considerations in Contract Archaeology. *Man in the Northeast* 15–16:132–38.

Petrie, W. M. Flinders
  1901 *Diospolis Parva.* Egyptian Research Account, London.

Pitt Rivers, General
  1887–1898 *Excavations in Cranborne Chase, 1887–98*, vol. 4. Privately printed, London.

Plog, Fred
  1984 Ethics of Excavation: Site Selection. In *Ethics and Values in Archaeology*, edited by Ernestine L Green, pp. 89–96. Free Press, New York.

Pyburn, K. Anne, and Richard R. Wilk
  1995 Responsible Archaeology Is Applied Archaeology. In *Ethics in American Archaeology: Challenges for the 1990s*, edited by Mark J. Lynott and Alison Wylie, pp. 71–76. Society for American Archaeology, Allen, Lawrence, Kans.

Raab, L. Mark
  1984 Toward an Understanding of the Ethics and Values of Research Design in Archaeology. In *Ethics and Values in Archaeology*, edited by Ernestine Green, pp. 75–88. Free Press, New York.

Rapport, S., and H. Wright (eds.)
  1964 *Archaeology.* Washington Square Press, New York.

Rathje, William, and Cullen Murphy
  2001 *Rubbish!: The Archaeology of Garbage.* University of Arizona Press, Tucson.

Renfrew, Colin, and Paul Bahn
  1991 *Archaeology: Theories, Methods and Practice.* Thames & Hudson, London.

Rice, Glen E., and Fred Plog
  1983 A Formal Method for the Use of Backhoes in Archaeological Excavations. Manuscript on file, New Mexico State University.

Salazar, Virginia, and Jake Barrow (issue editors)
   2000   Beyond Compliance: Tribes of the Southwest. *CRM* 23(9). US. Department of the Interior, National Park Service, Washington, D.C.

Scarborough, Vernon L.
   1986   Meyers Pithouse Village: A Preliminary Assessment. In *Mogollon Variability*, edited by Charlotte Benson and Steadman Upham, pp. 271–84. University Museum Occasional Papers 15. New Mexico State University, Las Cruces.

Schiffer, Michael B., and George J. Gummerman (eds.)
   1977   *Conservation Archaeology: A Guide for Cultural Resource Management Studies*. Academic Press, New York.

Scott, Linda J., and D. Kate Aasen
   1987   Interpretation of Holocene Vegetation in Northeastern Arkansas. In Robert H. Lafferty III et al., *A Cultural Resources Survey, Testing, and Geomorphic Examination of Ditches 10, 12, and 29, Mississippi County, Arkansas*. Submitted to Memphis District, COE, Contract No. DACW66-86-C-0034 by MCRA Report No. 86-5, 133–150.

Shackley, M. Steven
   1995   Sources of Archaeological Obsidian in the Greater American Southwest: An Update and Quantitative Analysis. *American Antiquity* 60(3):531–551.

Smith, Claire, and Heather Burke
   2003   In the Spirit of the Code. In *Ethical Issues in Archaeology*, edited by Larry J. Zimmerman, Karen D. Vitelli, and Julie Hollowell-Zimmer, pp. 177–97. AltaMira, Walnut Creek, Calif.

Stapp, Darby C., and Michael S. Burney
   2002   *Tribal Cultural Resource Management: The Full Circle to Stewardship*. AltaMira, Walnut Creek, Calif.

Stiger, Mark
   1986   Proposal for the Archaic Phase Project at the Fillmore. Pass Site (FB-1613), Fort Bliss, Texas. Report on files, Directorate of Environment, Fort Bliss, Texas.

Swidler, Nina, Kurt E. Dongoske, Roger Anyon, and Alan S. Downer
   1997   *Native Americans and Archaeologists: Stepping Stones to Common Ground*. AltaMira, Walnut Creek, Calif.

Tainter, Joseph A., and G. John Lucas
   1983   Epistemology of the Significance Concept. *American Antiquity* 48:707–719.

Taylor, Jeanette
    1999   *River City: A History of Campbell River and the Discovery Islands*. Harbour, Madeira Park, British Columbia, Canada.

Taylor, W. W.
    1948   [1967] *A Study of Archeology*. Southern Illinois University Press, Carbondale.

Thomas, Cyrus
    1894   [1984] *Report on the Mound Explorations of the Bureau of Ethnology*. Smithsonian Institution, Washington, D.C.

Thomas, David Hurst
    1991   *Archaeology: Down to Earth*. Harcourt Brace Jovanovich College Publishers, Fort Worth, Tex.

Tylor, E. B.
    1871   [1958] *The Origins of Culture*. Harper & Row, New York.

Ucko, Peter J.
    1989   Foreword. In *Conflict in the Archaeology of Living Traditions*, edited by Robert Layton, pp. ix–xvii. Unwin Hyman, London.

Vogel, Gregory
    2002   *A Handbook of Soil Description for Archeologists*. Technical Paper No. 11. Arkansas Archeological Survey, Fayetteville.

Watkins, Joe, and Tom Parry
    1997   Archeology's First Steps in Moccasins. *Common Ground: Archeology and Ethnography in the Public Interest* 2(3/4): 46–51. National Park Service, Washington, D.C.

Wheeler, M.
    1954   *Archaelogy from the Earth*. Clarendon, Oxford.

Willey, Gordon R., and Jeremy A. Sabloff
    1993   *A History of American Archaeology*, 3d ed. Freeman, San Francisco.

Williams, Stephen
    2001   Reviewing Some Late 19th Century Archaeological Studies: Exploding the Myth of the "Myth." Paper presented at the Mid-South Conference, Memphis.

Wilson, Michael, Kathie L. Road, and Kenneth J. Hardy
    1981   *Megaliths to Medicine Wheels: Boulder Structures in Archaeology*. Proceedings of the Eleventh Annual Chacmool Conference,

Department of Archaeology, University of Calgary, Alberta, Canada.

Wormington, H. M.
    1964   *Ancient Man in North America*, 5th ed. Denver Museum of Natural History Popular Series No. 4. Denver Museum of Natural History, Denver.

Zimmerman, Larry
    1995   Regaining Our Nerve: Ethics, Values and the Transformation of Archaeology. In *Ethics in American Archaeology: Challenges for the 1990s,* edited by Mark J. Lynott and Alison Wylie, pp. 64–67. Society for American Archaeology, Allen Press, Lawrence, Kansas.

INDEX

aboriginal vs. historic site, 9
academic archaeology, 3, 33–34
accelerator mass spectrometry, 76
accountability, 4, 10
AHTD. *See* Arkansas Highway and Transportation Department
American Anthropological Association, 47
American Museum of Natural History, 21
anthropology, archaeology as, 7–11
Apache site, 46–47, 88
archaeological excavation, as controlled destruction, 32; conservation ethic in, 41–42; justifying through research design/method, 6, 33, 39, 41, 42–43; justifying through site selection/use, 33, 34–39, 35, 36–38
archaeological excavation, use of, 2
archaeological research: academic/cultural resources management difference, 3, 33–34; constituent/consumer of, 28–29; cost of federally mandated, 4
Archaeological Resources Protection Act (ARPA), 5

archaeologist: ethical responsibility of, 6–7, 45–48; stereotype of, 29
archaeology: as anthropology, 7–11; definition of, 32; relevance of, 5
"Archaeology as Anthropology" (Binford), 27
archaeology theory, 26–28
archaeomagnetic dating, 77, 96
archival supplies/equipment, source for, 95
Arkansas Highway and Transportation Department (AHTD), 85, 86
ARPA. *See* Archaeological Resources Protection Act
auger testing, 12, 50, 51, 58, 69

backhoe trenching, 56–60, 57, 69
BAE. *See* Bureau of American Ethnology
Barker, Philip, 3
biface reduction technology, 39–40
Binford, Lewis, 27
Bureau of American Ethnology (BAE), 21
Burial. *See* human remains
Burney, Michael S., 48

117

INDEX

CAD. *See* computer-aided design
Caddo cemetery, 85–87
Carmichael, David, 70
Chicago grid, *25–26*
chronometric sampling/dating:
  archaeomagnetic, 77, 96;
  dendrochronology, 74–75, 97;
  obsidian hydration, 80–81;
  oxidizable carbon ratio, 78–80,
  97; radiocarbon, 75–77, 95–96;
  thermoluminescence, 78, 79, 97
Claflin, W. H., 24
Clouse, Robert, 42
Cole, Fay-Cooper, 24–25
Committee on State Archaeological
  Surveys of the National Research
  Council, 22
computer-aided design (CAD), 67
conservation ethic, 41–42
controlled surface collection, 28
conversion table, mathematical,
  100–103
coring, 12, 56, 57
Crissbury Hill Fort, 18, *19*
CRM. *See* cultural resources
  management
cross-dating, with horizon marker,
  20
cultural evolutionism, 18
cultural resources management
  (CRM): archaeologist as
  contractor in, 5–6; difference
  from academic archaeology, 3,
  33–34; history of, 28–29;
  law/regulation for, 5–6. *See also*
  archaeological excavation, as
  controlled destruction
Cummings, Linda S., 86

data recovery strategy, 64–65;
  choosing, 65–66; dry screening,
  70–71; subsurface
  recording/recovery, 68–70;
  surface recording/recovery,
  66–68; water screening/flotation,
  71–72. *See also* chronometric
  sampling/dating; human remains;
  soil sampling; specialized sample,
  recovering
dendrochronology, 74–75, 97
Denver Museum of Natural History,
  22
de Soto, Hernando, 86, 87
direct association, 22, 24
DNA analysis, source for, 96
documentation, importance of, 43
Douglass, A. E., 74
dry screening, 70–71
Dunnell, Robert C., 33, 34

Eaker site, 11, 57, 59, 83, 88–89
Early, Ann, 86
Erbie site, 42
ethics: of archaeologist, 6–7, 45–48;
  conservation, 41–42; and human
  remains, 8, 47–48, 84–85
ethnoarchaeology, 32
ethnobotanical analysis, source for,
  96
excavation equipment: basic, 92–93;
  miscellaneous tools, 92; personal
  toolkit, 91; supplier for, 94–95
excavation method: choosing,
  10–12; resource for, 52–53. *See
  also* subsurface testing
excavation unit, 51–56; stringing,
  53, *54*
expendable supplies, 94

FB-1613 site (Fort Bliss), 39
field/surveying equipment, source
  for, 95
Figgins, J. D., 22, 24
flotation sampling, 76, 82
Floyd River, 70
Folsom site, 22, 24
form, for excavation data, 51, 55,
  93–94

Fowler, Don D., 6
Franklin, J., 70
Frink, Douglas, 78

Geographical Information Systems (GIS), 16
geographic positioning system (GPS), 67
geomorphic/particle size sampling, 82, 83
geophysical survey, 50
Gilson grade sorter, 71
GIS. *See* Geographical Information Systems
Gobernadora site, 58
GPR survey, 68
GPS. *See* geographic positioning system
grid: Chicago, 25–26; early use of, 24; X, Y, Z Cartestian grid, 15–16
ground-penetrating radar, 59
Gummerman, J., 39
Gypsum Plain region, 45

Haas, Jonathan, 10
hand excavation. *See* shovel skimming; shovel testing
Helm site burials, 79–80
Hill, M. C., 86
historic vs. aboriginal site, 9
history, of excavation method: controlled surface collection, 28; direct association, 22, 24; grid, 24–26, *25*; provenience control, 16–20, *17*, *19*; screening, 27; seriation, 20; stratigraphy, 18, 20, 21, 22, *23*; volumetric quantification, 21–22
horizon marker, 20
human remains, 84–85; ethics in dealing with, 8, 47–48, 84–85; excavation tool for, 85; Helm site burials, 85–87; reburial of, 29, 87–89

increment borer, 75

Jacobs Cave, 24
Jefferson, Thomas, 20

Keystone site, 35, *36*, *37*, 59–60
Kidder, A. V., 22, *23*
KT site (New Mexico), 70

Lafferty, Robert H. III, 26–27, 51, 82, 87
law of superposition, 18, 20
Layard, Austin, 16, *17*
loess, definition of, 61
looting, 29, 31, 32, 44
Lower Mississippi Valley Survey, 27
Luna Village, 58

magnetic north, 51, 77
magnetometer, 8, 12, 51
map/mapping: detailed site, 62; emergence of, 16; surface, 12, 67
Martin Farm site, 87
material culture studies, 32
mathematical conversion table, *100–103*
McJunkin, George, 22, 24
mechanical excavation: auger testing, 12, 50, 51, 58, 69; backhoe trenching, 56–60, *57*, 69; stripping, 60, 69
mechanical stripping, 60, 69
Meyer Pithouse Village, 58
Miller, Myles, 58, 81
Mississippian site, 11
Mogollon site, 58
Moundbuilder myth, 20–21
museum collection, research use of, 41–42
Museum of New Mexico, 58

NAGPRA. *See* Native American Graves Protection and Repatriation Act

INDEX

Nash, S. K., 87
National Historic Preservation Act of 1966 (NHPA), 5
National Park Service, 28
National Register of Historic Places (NRHP), 10, 40, 73
National Science Foundation grant, 26
Native American:
consulting/cooperating with, 7–8, 9–10, 47–48, 84. *See also* human remains; Native American Graves Protection and Repatriation Act
Native American Graves Protection and Repatriation Act (NAGPRA), 5, 9–10, 29, 44, 47–48, 84
Nelson, Nels, 21
New Archaeology, 15, 27–28
NHPA. *See* National Historic Preservation Act of 1966
Nimrud, 16, *17*
NRHP. *See* National Register of Historic Places

Oakes, Yvonne, 58
Oakfield corer, 56
obsidian hydration dating, 80–81
Occupational Safety and Health Administration (OSHA), 43, 52
OCR. *See* oxidizable carbon ratio (OCR) dating
OSHA. *See* Occupational Safety and Health Administration
oxidizable carbon ratio (OCR) dating, 78–80, 97

particle size/geomorphic sampling, 82, 83
Peabody, Charles, 24
Pecos Conference, 22
Pecos Pueblo, 22, *23*
Pemiscot Bayou core, 83

personal toolkit, 91
Petrie, Flinders, 18, 20
photography, 55, 56, 67
phytolith/pollen sampling, 82–83
piece-plotted artifact, 55
pin flag, 67
Pitt Rivers, Augustus H., 18, *19*
plastic bag, source for, 95
Pleistocene, 22, 24
Plog, Fred, 41, 59
point-provenienced artifact, 55
pollen/phytolith sampling, 82–83
pothunting, 15, 44
Powell, John Wesley, 21
protein residue sampling, 84
provenience: emergence of, 16–20, *17*, *19*; importance of, 15
Pythagorean theorem, 99–100

Quapaw tribe, 88–*89*

radiocarbon sampling/dating, 75–77, 95–96
ramada structure, prehistoric, 35, *38*
random sampling, 50
reburial, of human remains, 29, 87–89
Register of Professional Archaeologists, 47
remote sensing, 11, 12, 50
rescue archaeology. *See* cultural resources management
research design/method, justifying controlled destruction through, 6, 33, 39, 41, 42–43
research results, misrepresenting, 45–47
Rice, Glen E., 59
River Basin Survey, 28
Rowlett, Ralph, 79

SAA. *See* Society for American Archaeology

INDEX

salvage archaeology. *See* cultural resources management
sampling: choosing strategy, 65–66; importance of, 43; random, 50; stratified random, 50; transect, 58–60. *See also* chronometric sampling/dating; soil sampling
Schiffer, Michael B., 39
screening: dry, 70–71; history of, 27; water/flotation, 71–72
Section 106, of NHPA, 5, 46
Section 110, of NHPA, 5
security, site, 43–44
seriation, 20
shovel scraping. *See* shovel skimming
shovel shining. *See* shovel skimming
shovel skimming, 69–70
shovel testing, 50, 51, 60
SHPO. *See* state historic preservation officer
site security, 43–44
site selection: change in approach to, 28; justifying controlled destruction through, 33, 34–39, 35, 36–38; methods for, 49–5
site significance, 6
Smithsonian Institution, 28
Society for American Archaeology (SAA), 8, 47
soil resistivity, 68
soil sampling, 81; flotation, 82; particle size/geomorphic, 82, 83; pollen/phytolith, 82–83; protein residue, 84. *See also* sampling
Southern Illinois University, 26–27
spatial control. *See* provenience
specialized sample, recovering: future issues in, 90. *See also* chronometric sampling/dating; human remains; soil sampling
Stalling's Island, 24
Stapp, Darby C., 48
state historic preservation officer (SHPO), 12
A State Plan for the Conservation of Archaeological Resources in Arkansas, 10–12
stratified random sampling, 50
stratigraphy: documenting, 60–62; early use of, 21, 22, 23; scaled drawing of, 55–56, 61–62; underlying basis of, 18, 20
A Study of Archaeology (Taylor), 26
subsurface testing: auger hole, 12, 50, 51, 58, 69; backhoe trenching, 56–60, 57, 69; coring, 12, 56, 57; mechanical stripping, 60, 69; shovel test, 50, 51, 60; test pit, 12. *See also* data recovery strategy; surface testing method
surface mapping, 12
surface testing method: controlled surface collection, 28; remote sensing, 11, 12, 50; surface mapping, 12, 67. *See also* data recovery strategy; subsurface testing method

Tano Pueblo, 21, 22
Taylor, W. W., 15, 26–27, 29
test pit excavation, 12
Texas Department of Transportation (TXDot), 58
theodolite, 67
theory, archaeology, 26–28
thermoluminescence (TL) dating, 78, 79, 97
Thomas, David H., 21
TL. *See* thermoluminescence (TL) dating

Total Station, 67
tree ring dating. *See*
 dendrochronology
TXDot. *See* Texas Department of
 Transportation
University of Arizona, 22
University of Chicago, 24–26
University of New Mexico, 27
University of South Dakota, 69
University of Texas at El Paso, 2–3,
 35

vandalism, 44
Vogel, Gregory, 61
volumetric quantification, 21–22

water screening/flotation,
 71–72
Works Progress Administration
 (WPA), 26, 28

X, Y, Z Cartestian grid, 15–16
X-ray fluorescence, 81

ABOUT THE AUTHORS AND SERIES EDITORS

David L. Carmichael has twenty-eight years of experience in CRM archaeology and has been conducting applied ethnographic work with Native Americans for two decades. He has been involved in more than forty projects and has directed studies in twelve states throughout the West and Midwest. He received undergraduate training at the University of New Mexico and earned M.A. and Ph.D. degrees at the University of Illinois. He worked as a CRM contractor for sixteen years, served as the first tribal archaeologist for the Hopi Tribe, and authored the U.S. Air Force guidelines for consultation with Native Americans. His recent research has included the prehistoric archaeology of southern New Mexico and west Texas, Apache land use patterns, and Native American sacred sites. He teaches archaeology and anthropology at the University of Texas at El Paso.

Robert H. Lafferty III took his Ph.D. in 1977 from Southern Illinois University and is ROPA certified. He is a co-owner of Mid-Continental Research Associates, Inc. Since 1976, he has spent more than eighty-five months in the field directing all kinds of cultural resource management projects, and he has authored or principally coauthored fifteen books and more than a hundred smaller technical reports, articles, and papers. Many of his publications have centered on the chronology and cultural adaptation of prehistoric cultures in the Lower Mississippi Valley and the Ozark Mountains. During the past decade, he has been researching the chronology of prehistoric earthquakes in the New Madrid Seismic Zone with a team of geologists for the U.S. Geological Survey.

Brian Leigh Molyneaux is an archaeologist, writer, and photographer. He is a specialist in prehistoric art and society, the human use of the landscape, and computer-aided applications in archaeology. He is director of the University of South Dakota Archaeology Laboratory, codirector of the Missouri River Institute, and a research associate of the Royal Ontario Museum, Toronto. He received his M.A. in art and archaeology from Trent University, Peterborough, Ontario, in 1977 and his Ph.D. in archaeology at the University of Southampton, England, in 1991. His extensive fieldwork includes many years of travel in northern Canada, studying Algonkian rock art, ritual and religion, and archaeological research in the northern Great Plains. He recently conducted a two-year archaeological survey at Devils Tower National Monument, Wyoming. Dr. Molyneaux has published several books: *The Presented Past* (coedited with Peter Stone, Routledge, 1994), a study of archaeology, museums, and education around the world; *The Sacred Earth* (Little, Brown, 1995), a study of spirituality related to the landscape; *Native North America* (with Larry Zimmerman, Little, Brown, 1996), a detailed survey of Native North American culture, past and present; *The Cultural Life of Images* (editor and contributor, Routledge, 1997), a study of pictures and other visual representations of the past in archaeology; a new edition of *Native North America* (Oklahoma University Press, 2000); *Sacred Earth, Sacred Stones* (with Piers Vitebsky, Laurel Glen, 2001), a compilation dealing with spirituality in landscape and architecture; and *Mythology of the Americas* (with David M. Jones, Lorenz Books, 2001), a general encyclopedia. His rock art photographs have been exhibited in the National Gallery of Canada and featured in the PBS/BBC series, *Land of the Eagle,* and he is an active contributor to web-based symposia on art, technology, environment, and culture (e.g., The Anthology of Art, www.anthology-of-art.net, School of Fine Arts, Braunschweig, Germany).

Larry J. Zimmerman is the head of the Archaeology Department of the Minnesota Historical Society. He served as an adjunct professor of anthropology and visiting professor of American Indian and Native Studies at the University of Iowa from 1996 to 2002 and as chair of the American Indian and Native Studies Program from 1998 to 2001. He earned his Ph.D. in anthropology at the University of Kansas in 1976. Teaching at the University of South Dakota for twenty-two years, he left there in 1996 as Distinguished Regents Professor of Anthropology.

While in South Dakota, he developed a major CRM program and the University of South Dakota Archaeology Laboratory, where he is still a research associate. He was named the University of South Dakota Student Association Teacher of the Year in 1980, given the Burlington Northern Foundation Faculty Achievement Award for Outstanding Teaching in 1986, and granted the Burlington Northern Faculty Achievement Award for Research in 1990. He was selected by Sigma Xi, the Scientific Research Society, as a national lecturer from 1991 to 1993, and he served as executive secretary of the World Archaeological Congress from 1990 to 1994. He has published more than three hundred articles, CRM reports, and reviews and is the author, editor, or coeditor of fifteen books, including *Native North America* (with Brian Molyneaux, University of Oklahoma Press, 2000) and *Indians and Anthropologists: Vine Deloria, Jr., and the Critique of Anthropology* (with Tom Biolsi, University of Arizona Press, 1997). He has served as the editor of *Plains Anthropologist* and the *World Archaeological Bulletin* and as the associate editor of *American Antiquity*. He has done archaeology in the Great Plains of the United States and in Mexico, England, Venezuela, and Australia. He has also worked closely with a wide range of American Indian nations and groups.

William Green is the director of the Logan Museum of Anthropology and an adjunct professor of anthropology at Beloit College, Beloit, Wisconsin. He has been active in archaeology since 1970. Having grown up on the south side of Chicago, he attributes his interest in archaeology and anthropology to the allure of the exotic (i.e., rural) and a driving urge to learn the unwritten past, abetted by the opportunities available at the city's museums and universities. His first fieldwork was on the Mississippi River bluffs in western Illinois. Although he also worked in Israel and England, he returned to Illinois for several years of survey and excavation. His interests in settlement patterns, ceramics, and archaeobotany developed there. He received his master's degree from the University of Wisconsin at Madison and then served as Wisconsin SHPO staff archaeologist for eight years. After obtaining his Ph.D. from the University of Wisconsin at Madison in 1987, he served as state archaeologist of Iowa from 1988 to 2001, directing statewide research and service programs including burial site protection, geographic information, publications, contract services, public outreach, and curation. His main research interests focus on the development and spread of native agriculture. He has served as editor of the *Midcontinental Journal of Archaeology* and *The Wisconsin Archeologist;* has published articles in *American Antiquity, Journal of Archaeological Research,* and other journals; and has received grants and contracts from the National Science Foundation, National Park Service, Iowa Humanities Board, and many other agencies and organizations.

ARCHAEOLOGICAL SURVEY

ARCHAEOLOGIST'S TOOLKIT

SERIES EDITORS: LARRY J. ZIMMERMAN AND WILLIAM GREEN

The Archaeologist's Toolkit is an integrated set of seven volumes designed to teach novice archaeologists and students the basics of doing archaeological fieldwork, analysis, and presentation. Students are led through the process of designing a study, doing survey work, excavating, properly working with artifacts and biological remains, curating their materials, and presenting findings to various audiences. The volumes—written by experienced field archaeologists—are full of practical advice, tips, case studies, and illustrations to help the reader. All of this is done with careful attention to promoting a conservation ethic and an understanding of the legal and practical environment of contemporary American cultural resource laws and regulations. The Toolkit is an essential resource for anyone working in the field and ideal for training archaeology students in classrooms and field schools.

Volume 1: *Archaeology by Design*
By Stephen L. Black and Kevin Jolly

Volume 2: *Archaeological Survey*
By James M. Collins and Brian Leigh Molyneaux

Volume 3: *Excavation*
By David L. Carmichael and Robert Lafferty

Volume 4: *Artifacts*
By Charles R. Ewen

Volume 5: *Archaeobiology*
By Kristin D. Sobolik

Volume 6: *Curating Archaeological Collections: From the Field to the Repository*
By Lynne P. Sullivan and S. Terry Childs

Volume 7: *Presenting the Past*
By Larry J. Zimmerman

ARCHAEOLOGICAL SURVEY

JAMES M. COLLINS
BRIAN LEIGH MOLYNEAUX

ARCHAEOLOGIST'S TOOLKIT
VOLUME 2

ALTAMIRA
PRESS

A Division of Rowman & Littlefield Publishers, Inc.
Walnut Creek • Lanham • New York • Oxford

ALTAMIRA PRESS
A Division of Rowman & Littlefield Publishers, Inc.
1630 North Main Street, #367
Walnut Creek, CA 94596
www.altamirapress.com

Rowman & Littlefield Publishers, Inc.
A Member of the Rowman & Littlefield Publishing Group
4501 Forbes Boulevard, Suite 200
Lanham, MD 20706

PO Box 317
Oxford
OX2 9RU, UK

Copyright © 2003 by ALTAMIRA PRESS

All rights reserved. No part of this publication may be reproduced, stored in a retrieval system, or transmitted in any form or by any means, electronic, mechanical, photocopying, recording, or otherwise, without the prior permission of the publisher.

British Library Cataloguing in Publication Information Available

Library of Congress Cataloging-in-Publication Data
Collins, James M.
  Archaeological survey / James M. Collins and Brian Leigh Molyneaux.
    p. cm. — (Archaeologist's toolkit ; v. 2)
  Includes bibliographical references and index.
  ISBN 0-7591-0398-4 (cloth : alk. paper)—ISBN 0-7591-0021-7 (pbk. : alk. paper)
  1. Archaeological surveying. 2. Archaeological surveying—North America. 3. North America—Antiquities. I. Molyneaux, Brian. II. Title. III. Series.
  CC76.3.C65 2003
  930.1—dc21
                                                2002154548

Printed in the United States of America

∞™ The paper used in this publication meets the minimum requirements of American National Standard for Information Sciences—Permanence of Paper for Printed Library Materials, ANSI/NISO Z39.48-1992.

CONTENTS

Series Editors' Foreword · vii

Acknowledgments · xi

1 The Survey Life · 1
Origins: Speculation to Survey • Theory and Practice Today

2 The Law, the Process, and the Players · 9
Why Survey? • Regulations in the United States • Regulations in Canada • Who Surveys?

3 Survey Design · 19

4 Prefield Strategies · 23
Implementing the Research Design • The Two-headed Monster • Getting Started: Map It • Background Research

5 Getting into the Field · 35
Personnel • Equipment • Local Relations • Native American, Canadian First Nations, and Ethnic Community Issues • Consulting with Other Specialists • Survey Methods • Field Evaluation to Determine Potential Eligibility for the National Register

CONTENTS

6 Indirect Exploration Techniques   77
Metal Detector • Electrical Resistivity • Electromagnetic Conductivity • Ground Penetrating Radar • Magnetometry • Magnetic Susceptibility • Conclusion

7 Recording and Mapping   99
Reading Maps and Aerial Photographs • Drafting Maps • Global Positioning Systems • Geographic Information Systems • Tomorrow's Technology

8 The Survey Report   111
Structure and Content

9 Trying to Make It Real, Compared with What?   129

References   133

Index   139

About the Authors and Series Editors   145

SERIES EDITORS' FOREWORD

The Archaeologist's Toolkit is a series of books on how to plan, design, carry out, and use the results of archaeological research. The series contains seven books written by acknowledged experts in their fields. Each book is a self-contained treatment of an important element of modern archaeology. Therefore, each book can stand alone as a reference work for archaeologists in public agencies, private firms, and museums, as well as a textbook and guidebook for classrooms and field settings. The books function even better as a set, because they are integrated through cross-references and complementary subject matter.

Archaeology is a rapidly growing field, one that is no longer the exclusive province of academia. Today, archaeology is a part of daily life in both the public and private sectors. Thousands of archaeologists apply their knowledge and skills every day to understand the human past. Recent explosive growth in archaeology has heightened the need for clear and succinct guidance on professional practice. Therefore, this series supplies ready reference to the latest information on methods and techniques—the tools of the trade that serve as handy guides for longtime practitioners and essential resources for archaeologists in training.

Archaeologists help solve modern problems: They find, assess, recover, preserve, and interpret the evidence of the human past in light of public interest and in the face of multiple land use and development interests. Most of North American archaeology is devoted to cultural resource management (CRM), so the Archaeologist's Toolkit focuses on practical approaches to solving real problems in CRM and

public archaeology. The books contain numerous case studies from all parts of the continent, illustrating the range and diversity of applications. The series emphasizes the importance of such realistic considerations as budgeting, scheduling, and team coordination. In addition, accountability to the public as well as to the profession is a common theme throughout the series.

Volume 1, *Archaeology by Design*, stresses the importance of research design in all phases and at all scales of archaeology. It shows how and why you should develop, apply, and refine research designs. Whether you are surveying quarter-acre cell tower sites or excavating stratified villages with millions of artifacts, your work will be more productive, efficient, and useful if you pay close and continuous attention to your research design.

Volume 2, *Archaeological Survey*, recognizes that most fieldwork in North America is devoted to survey: finding and evaluating archaeological resources. It covers prefield and field strategies to help you maximize the effectiveness and efficiency of archaeological survey. It shows how to choose appropriate strategies and methods ranging from landowner negotiations, surface reconnaissance, and shovel testing to geophysical survey, aerial photography, and report writing.

Volume 3, *Excavation*, covers the fundamentals of dirt archaeology in diverse settings, while emphasizing the importance of ethics during the controlled recovery—and destruction—of the archaeological record. This book shows how to select and apply excavation methods appropriate to specific needs and circumstances and how to maximize useful results while minimizing loss of data.

Volume 4, *Artifacts*, provides students as well as experienced archaeologists with useful guidance on preparing and analyzing artifacts. Both prehistoric- and historic-era artifacts are covered in detail. The discussion and case studies range from processing and cataloging through classification, data manipulation, and specialized analyses of a wide range of artifact forms.

Volume 5, *Archaeobiology*, covers the analysis and interpretation of biological remains from archaeological sites. The book shows how to recover, sample, analyze, and interpret the plant and animal remains most frequently excavated from archaeological sites in North America. Case studies from CRM and other archaeological research illustrate strategies for effective and meaningful use of biological data.

Volume 6, *Curating Archaeological Collections*, addresses a crucial but often ignored aspect of archaeology: proper care of the specimens

and records generated in the field and the lab. This book covers strategies for effective short- and long-term collections management. Case studies illustrate the do's and don'ts that you need to know to make the best use of existing collections and to make your own work useful for others.

Volume 7, *Presenting the Past,* covers another area that has not received sufficient attention: communication of archaeology to a variety of audiences. Different tools are needed to present archaeology to other archaeologists, to sponsoring agencies, and to the interested public. This book shows how to choose the approaches and methods to take when presenting technical and nontechnical information through various means to various audiences.

Each of these books and the series as a whole are designed to be equally useful to practicing archaeologists and to archaeology students. Practicing archaeologists in CRM firms, agencies, academia, and museums will find the books useful as reference tools and as brush-up guides on current concerns and approaches. Instructors and students in field schools, lab classes, and short courses of various types will find the series valuable because of each book's practical orientation to problem solving.

As the series editors, we have enjoyed bringing these books together and working with the authors. We thank all of the authors—Steve Black, Dave Carmichael, Terry Childs, Jim Collins, Charlie Ewen, Kevin Jolly, Robert Lafferty, Brian Molyneaux, Kris Sobolik, and Lynne Sullivan—for their hard work and patience. We also offer sincere thanks to Mitch Allen of AltaMira Press and a special acknowledgment to Brian Fagan.

Larry J. Zimmerman
William Green

ACKNOWLEDGMENTS

JAMES M. COLLINS

I would like to acknowledge the Foundation for Illinois Archaeology/ Northwestern University Archaeological Program (now known as the Center for American Archeology) in Kampsville, Illinois, for kindling the interest that became my career. The field training I received at Kampsville and the appreciation for good field practice instilled there have served me well for a quarter-century. In particular, I am indebted to Raymond W. Perkins for teaching me *almost* all I ever needed to know about archaeological survey.

Since coming to the University of Iowa, involvement with the Iowa Quaternary Studies Program has enhanced my understanding of the natural world and humankind's adaptations to it. Long associations with Rolfe Mandel, Art Bettis, and others have yielded many fruitful insights into Holocene landscape processes. A National Park Service workshop administered by Steve DeVore at Fort Laramie, Wyoming, concerning remote sensing and geophysical techniques for cultural resource management opened my eyes to the potential of noninvasive site survey.

During the past decade, the Office of the State Archaeologist, University of Iowa (OSA), has nurtured my research and nourished my professional development. The OSA has been at the forefront in the development of new survey methods and reporting standards in the Midwest. Many among my cohort at the OSA have generously shared knowledge, information, and skills while I worked on this book. Michele Punke is especially thanked for volunteering her graphic and editorial skills to the production of this volume. Bill Green has been extremely supportive throughout my tenure at the OSA.

Finally, and most importantly, I would like to acknowledge the support of my family, Susan, Kathleen, and Virginia, and my parents, William and Eleanor Collins.

BRIAN LEIGH MOLYNEAUX

Survey archaeology can sometimes provide wonderful silence and a deep respite from the labors of the working world. At times of repose—rare in the turmoil of a deadline-driven CRM project—I sometimes think back to people who helped me along the way: Hamilton Lang, an ornithologist and antiquarian living near Comox, on Vancouver Island, British Columbia, who let a young kid dig holes in his back yard, searching for lost treasure; Selwyn Dewdney, of London, Ontario, who took a raw undergraduate north into the Canadian Shield to look for rock art in the early 1970s; Dr. Joan Vastokas (Trent University) and Professor Peter J. Ucko (Southampton University), supervisors of my graduate theses, who hammered me into a semblance of professional form; Larry Zimmerman and Lawrence Bradley, who gave me a job and let me loose; and many fellow archaeological travelers in Canada, the United States, England, Africa, and South America. I think especially of close friends who have shared the hard work and thrill of discovery.

Most of all, in those distant, mysterious worlds archaeologists inhabit, I think of my mother, my late father, and my own family, Wendy, Freddy, Alex, and Lily. They have gotten me through narrow escapes in the field and kept me going along my chosen path.

1
THE SURVEY LIFE

There is nothing like the vast, empty expanse of a field at the beginning of a survey. Whatever else archaeology is, it begins with the land. In our search for the human past, we may walk along plowed fields that were once tundra, pleasant inland ridges that were once the shorelines of ancient glacial lakes, or grasslands that ten thousand years ago teemed with giant Ice Age bison. Our quest is for the material remnants of other lives, other times, the evidence that begins the process of bringing the past into the present. This book gives you the tools you need for your own journey in archaeology.

A big part of archaeological life is the search for stone tools, ceramics, and other artifacts, but we are not treasure hunters. We are trying to figure out the lifeways of the cultures who lived on this land before us. Who were they? Where did they come from? How long did they live here? How did they get food, make shelter, and pursue their daily lives? To learn these things, we have to see artifacts as part of a larger landscape, the natural environment, that supports all living things. This is because our real goal in studying the scattered remains of the past is to understand better the mysteries of human origins and the development of society into what we have today.

So surveyors do not walk a patch of ground aimlessly. Before we set out, we find out about its natural and cultural history, including its weather, vegetation, and animal life, and what is already known about human occupation in the area. After we get back, we take all the information we find—artifacts, photographs, field sketches, and site maps—and develop a basic picture of life there in the past. This collection of material and maps is essential information for archaeologists to plan formal excavations, and our mapping of the prehistoric landscape helps people and agencies plan to protect archaeological sites from damage or destruction by land development.

We do not operate in a social vacuum. Land and people are intertwined, so archaeologists are always on someone else's land and in someone else's community. Landowners, private and public, allow us to survey, and we have to follow their rules and regulations. We must also respect the ideas and attitudes of these communities about the past. For some, archaeology is a window into a wonderful, mysterious world; for others, it is a dangerous threat to the peace of their ancestors. Surveyors not only walk the land but also navigate through the political and social issues of our times. The most difficult relationship, however, is with those people we cannot see—those whose remains we seek. Archaeologists see and use the landscape differently than the people we study. Seeing the world through different eyes, we must be very careful not to put ourselves in their place, to imagine that our way of doing things was their way of doing them. This is the real challenge of a survey, to be able to visualize environments and ways of life that no longer exist, so that we look in the right places for the right things.

In the following pages, we discuss how today's survey life developed through changes in archaeological theory and technology over time. In subsequent chapters, we discuss how to set up and conduct a survey, from creating an initial survey strategy, doing the necessary background research, and readying equipment, to walking the land, conducting subsurface tests and geophysical surveys, to mapping and preparing for the final analysis and interpretation of results.

ORIGINS: SPECULATION TO SURVEY

Modern archaeological survey gets its vision from its own past. This history is one of changes in our perception of the past and what we think is important to know and record.

Imagine what it must have been like when the very first peoples walked into what is now the Americas. The continent had no still-warm campfires, or cast-off tools, or abandoned shelters—no human history. In time, though, human groups began to leave their marks, and other people followed them, taking the same paths and living in the same places. For instance, the town of Vermillion, South Dakota, on the edge of the Missouri River Valley, has been occupied for at least nine thousand years! After European immigrants came, subjugating Native Americans and forcing them into the more remote parts of the West, the rhythm of life changed, and what had been a familiar past was now a mystery.

The roots of survey lie in one of the greatest archaeological mysteries, the origin of the Moundbuilders. Eighteenth-century scholars, sitting in their studies in Washington and Philadelphia, London and Paris, or gathered at learned society meetings, conjured up wild theories about the origin of the Indians. Were they descendants of a second Adam, or one of the Lost Tribes of Israel, or the remnants of the fierce Scythians or Tatars of Asia? These antiquarians were even more fired up by reports from the Ohio and Mississippi Valley of massive earthworks—some burial mounds, others like temples, still others in the shapes of animals. They did not believe that the nomadic hunting and gathering people or villagers occupying these regions at the time of contact could have built them. The Great Serpent Mound in Ohio, for example, was 1,330 feet long, containing approximately 311,000 cubic feet of earth. These mounds had to be the remnants of an ancient civilization, some lost race of the same stock that had produced the most ancient cities of the Old World. It was a rather convenient fiction. Not only did it establish an Old World people as a founding race in the New World, but it justified the genocide of Indians for their destruction of this ancient civilization. Moundbuilders quickly joined giants, Phoenicians, Egyptians, Irish monks, Danes, Welsh princes, and others in popular and scholarly imaginations as original inhabitants of North America.

While these debates raged in the cities, military exploring parties, including scientists and land surveyors, were busy exploring the newly conquered territories and beginning to make systematic collections and inventories of what they found. As they started with a blank slate, the first requirement was to find out what really was there, rather than relying on hearsay. Of course, because the new nation was bent on eliminating the Indians, they were not consulted. In 1787–1788, when General Rufus Putnam established the townsite of Marietta, Ohio, he surveyed the region and found hundreds of mounds, which he mapped and briefly described. It was not until the 1840s, however, that the mounds became objects of a systematic survey. Fascinated by the Moundbuilders, Ephraim G. Squier, a newspaper editor, and Edward H. Davis, a physician, visited hundreds of earthworks, dug into over two hundred, and reported their results in the first publication of the Smithsonian Institution. Although they remained convinced that the Moundbuilders were not Indians, their great work, *Ancient Monuments of the Mississippi Valley* (1848), was a truly archaeological investigation. Rather than plundering a few mounds for their contents, they recorded as many as they could in the hopes of seeing patterns of construction and use. These remain important goals in archaeological survey today.

With the triumph of scientific evolutionism after Darwin, scientific studies based on description and classification replaced the antiquarian delight in mystery. In 1881, the Smithsonian established a Division of Mound Exploration to sort out the Moundbuilder problem and, in 1882, appointed an entomologist, Cyrus Thomas, to conduct the work. There was considerable urgency to this effort. Collectors were rampant, encouraged by a growing market in antiquities. In the spirit of Squier and Davis, Thomas surveyed over two thousand earthworks. In his *Report on the Mound Explorations of the Bureau of American Ethnology* (Thomas 1894), he showed that the mounds contained types of material culture similar to those in the more modest sites attributed to Indians. We know now that most Moundbuilders were Woodland (ca. 800 B.C.–A.D. 1000) and Mississippian (ca. A.D. 1000–1600) Indians. The Moundbuilders took their proper place in mythology. By using the evidence in the ground and understanding that people occupy regions rather than isolated places, Thomas set the standard for the archaeological surveys to come.

Archaeologists were not especially interested in chronology then, as they were in Europe. Swayed by political realities, American archaeologists generally believed that the Indians had simply remained in the Stone Age. The most urgent need was to sort through an ever-growing inventory of cultural material from all parts of the continent.

Over the next two decades, archaeologists discovered that prehistoric cultures were much more diverse than they had first thought. Cyrus Thomas divided North America into three major cultural zones: Arctic, Atlantic, and Pacific. Barely fifteen years and many artifacts later, William Henry Holmes was able to divide it into twenty-six cultural areas. While theorists continued to rearrange cultures on the North American maps, field archaeologists could not ignore the layers of occupations they found in excavations or the differences in stone tools and pottery through time. When Franz Boas argued that prehistoric cultures were really accumulations of cultural traits diffused across time and space, the continent-wide cultural maps seemed empty and pointless. Archaeologists had to study the growth and spread of individual cultural groups, and they could do this only by surveying the regions in which they developed. Alfred Kidder produced the first major regional study in 1924, a survey of nine river drainages in the Southwest. Combining geographic location and chronology, he suggested that there were four periods of cultural development, from the ancient Basket Maker people to the Pueblos of today. From these beginnings, archaeologists began to think of people, through their artifacts, as localized in time and space. The goal was no longer simply the collection of artifacts but collection in specific geographic regions.

Right after World War I, British geographer and pilot O. G. S. Crawford discovered that aerial photographs not only provided a bird's eye view of the terrain but also revealed archaeological features invisible at ground level. Infrared photography, introduced after World War II, increased this remote sensing capability. Archaeologists could now examine more territory than they could on foot and also create more accurate maps, using aerial photographs to verify spatial relationships.

Subsequent developments in survey included both the technological, enhancing the ability to detect cultural material, and the theoretical, changing or refining what we are looking for. During the late 1930s, archaeologists in North America and Europe began to realize that it was the interaction of culture and environment, not the diffusion of ideas and inventions, that gave prehistoric cultures so many common features. Julian Steward applied this new concern for function in a study of the relationship between humans and ecology in the Southwest, and the English archaeologist Grahame Clark did the same in a synthesis of European prehistory. With the development of radiocarbon dating in the late 1940s, archaeologists could date sites more confidently and develop more precise and reliable regional chronologies, giving them better understandings of settlement sequences. Drawing on these ideas, archaeologists such as Gordon Willey, working in the Virú Valley in Peru, broadened their surveys to include data on changes in subsistence, population density, and settlement patterns. Rather than working at the largest sites to extract the maximum data, the new goal of a survey was to identify networks of sites, each contributing to the function of the whole system. As environmental data were now important, archaeology required the work of the natural sciences. When Richard S. MacNeish surveyed the highlands of Mexico in the 1960s in search of the origins of agriculture, for example, his colleagues included botanists, zoologists, and geologists.

THEORY AND PRACTICE TODAY

The idea that culture functions as an adaptive system directed toward the material environment encouraged archaeologists to adopt more consciously scientific methods. In the 1950s, highly detailed excavations in Europe, such as Grahame Clark's study of the Mesolithic site at Starr Carr, England, showed it was possible to identify the seasonal activities of hunters and gatherers. This led to new interpretations of the nature of archaeological sites. In the 1960s, Lewis Binford challenged the archaeological establishment to go beyond artifact description and develop scientific theories of how and why cultures develop and change. This approach, dubbed the New Archaeology, emphasized cultural process. Sites were not accumulations of artifacts and features representing cultures but products of behavior that responded to laws governing the technical and economic aspects of cultural systems.

The assumption that such laws existed affected survey in two ways. First, it legitimized the analysis of the patterns of artifact production, use, and discard among modern hunting and gathering people as a means of understanding past behavior. Archaeological sites represented a variety of different behavioral processes, depending on the nature of the activity in various parts of the lived-in environment (e.g., temporary or longer-term habitation, and food and other resource processing). Second, it changed the nature of surveys. If laws regulated cultural behavior to maintain balance with the natural environment, then it was not necessary to gather all of the artifacts and other information available on a site to understand it. By applying a statistical sampling technique, the archaeologist could use a selection of data from a region or site to stand for the whole. This approach also demanded that archaeologists survey systematically, using a grid system, in place of the traditional intuitive method, in which archaeologists tended to survey what they considered the areas most likely to have the evidence they wanted. Although the idea of balanced systems is now outmoded, the random sampling technique persists in archaeology as a cost-effective, if often theoretically weak, means of getting what is hoped is a representative sample of the site or region.

Dissatisfaction with what some theorists complained was the programmatic and lifeless treatment of the human past began to develop in the 1970s, as part of a general attack on the objectivity of the social sciences. While these disciplines appeared rigorously scientific, they were, and are, naturally fraught with the subjectivism of their practitioners. English archaeologist Ian Hodder brought these criticisms into focus by emphasizing the role of the archaeologist as interpreter. A site was not simply a distribution of easily described and interpreted cultural material but a field of meaning, a "text" to be read, no less subject to the personal biases of the observer than any other phenomenon. The emphasis of this approach on ideas and symbols found much favor in Europe, if only because Old World prehistory covers tens of thousands of years of human culture and more than twenty thousand years in which sites yield evidence of cultural material devoted to ideology, from Paleolithic cave art to Neolithic stone monuments to complex city states. Quickly dubbed postprocessualism, this approach similarly concerned the role of societies and individuals in the formation of the archaeological record, but it rejected the assumption that natural forces were mainly responsible for controlling cultural behavior. Appealing to modern examples, as Binford had done, Hodder pointed out that material culture figures prominently in the construction of social identity and in the representation and control of power relationships. Hence, an archaeological site is a context of social activity, not simply a map of technoeconomic behavior and material formation processes. This approach has had less impact on field survey, in spite of its telling criticism. Although the processual and postprocessual approaches continue to provoke highly politicized theoretical debate, both remind survey archaeologists of the complexity of archaeological sites and the range of potential data.

Archaeology has always focused on the site as the basic unit of survey. A *site* is a place with some explicit evidence of cultural activity that an archaeologist encloses within a boundary. Things are either "inside" or "outside" sites. Drawing boundaries around activity areas makes sense in a world where all land is property, and it makes even more sense when land is managed and cultural resources must be explicitly defined and located, but it is also obvious that people occupy areas rather than places and that cultural material is actually distributed in varying densities over the entire landscape.

To rectify this problem, some archaeologists simply ignore the site concept. In its place, they record all traces of human activity in the landscape, not only the artifacts in the most intensively occupied places. While a comprehensive "nonsite" or "distributional" survey approaches the archaeological ideal, the realities of modern archaeology (cost, access to the land) make it impossible in most parts of the world. However, advances in survey equipment (electronic total stations), locational systems (Global Positioning Systems—GPS), computer mapping (Geographic Information Systems—GIS), and the speed and storage capacity of computer systems have made this type of survey more feasible. It is now possible to record the position of every artifact over a region, so that we can get an impression of human activity across landscapes rather than within sites. The very large scale and highly detailed imagery provided by remote sensing—from Landsat 7, RADARSAT, IKONOS and other satellites, the space shuttle, and high-altitude aircraft—enables vast regions to be explored in a single image and permits identification of environmental and cultural features invisible on the ground. Such technologies increase the range, density, and accuracy of information we can record and decrease the time it takes to record it.

In the end, however, the basic archaeological survey has not come much farther than when Squier and Davis tracked down their mounds. We still have to walk the land, ground-truth our expectations, and record what we find. The difference is that we have the technology to cover more ground and create maps of an accuracy unimaginable only a few years ago. Most important, we now know that the artifacts we find are only a small part of very complex human adaptations to the social and natural worlds around us.

2

THE LAW, THE PROCESS, AND THE PLAYERS

WHY SURVEY?

The vast majority of all archaeological survey undertaken in North America is mandated by the government. To be sure, surveys will always be conducted purely for research purposes, but unless you are one of the lucky few who actually land a faculty position, chances are your survey work will be conducted in accordance with federal and state or provincial regulations governing the identification and protection of cultural resources. The federal governments of the United States and Canada have laws related to heritage protection for federal lands. Federal control also extends to any project, whether private or public, local or national, that uses federal funds. All states and provinces have heritage protection laws and agencies established to manage cultural resources, as well as guidelines that control who can conduct archaeology and how they carry out their projects. Those aspiring to become cultural resource management (CRM) archaeologists should therefore familiarize themselves with the national and state or provincial laws in their region, as success in negotiating through the bureaucracy is the key to a successful archaeological project— whatever you may discover in a survey. Volume 1 in the Toolkit series addresses these issues in some detail as they apply to designing research; here, we discuss a few nuts and bolts of the CRM process as it affects archaeological survey.

CHAPTER 2

REGULATIONS IN THE UNITED STATES

The federal process is governed by several laws and regulations. Government-mandated archaeology may be conducted under many auspices, including various state laws. However, most work derives from compliance with Section 106 and Section 110 of the National Historic Preservation Act of 1966 (PL 89-665), as amended. The rules that implement this law are published as 36 CFR Part 800, "Procedures for the Protection of Historic and Cultural Properties," issued by the Advisory Council on Historic Preservation. As used by these authorities, historic properties (often referred to as cultural resources) include historic and prehistoric sites, districts, buildings, structures, or objects that are potentially eligible for inclusion in the National Register of Historic Places (NRHP). Properties eligible for the National Register must have both integrity and significance.

Exactly what sites may be considered NRHP eligible? The NRHP criteria (36 CFR Part 60.4) state:

> The quality of significance in American history, architecture, archaeology, engineering, and culture is present in districts, sites, buildings, structures and objects that possess integrity of location, design, setting, materials, workmanship, feeling, and association *and*:
> a) that are associated with events that have made a significant contribution to the broad patterns of our history; or
> b) that are associated with the lives of persons significant in our past; or
> c) that embody the distinctive characteristics of a type, period, or method of construction, or that represent the work of a master, or that possess high artistic values, or that represent a significant and distinguishable entity whose components may lack individual distinction; or
> d) that have yielded or may be likely to yield information important in prehistory or history.

Implications of these criteria as they pertain to archaeological survey are addressed more fully in chapters 3 and 5.

Section 106 of the National Historic Preservation Act mandates that federal agencies take into account the effects of their actions on properties listed in, or eligible for listing in, the NRHP. Federal undertakings that often trigger archaeology under Section 106 include construction and other land-altering projects. Section 110 requires federal agencies to locate and protect cultural resources on lands that they own.

Depending on the type and scope of the federal undertaking, provisions of the National Environmental Policy Act (NEPA) may require consideration of historic properties in the environmental impact statement process, especially in large-scale projects that will have a significant impact on the environment, such as highways and reservoirs. The Native American Graves Protection and Repatriation Act (NAGPRA) provides added protection to Native American graves and associated cultural items located on federal and tribal lands. Many individual states and municipalities have also codified protection of historic properties, complementing the federal mandates. Most cultural resource management projects, however, fall under the auspices of Section 106 of the National Historic Preservation Act.

The Section 106 process is intended to identify any historic properties that may be affected by an undertaking and to determine how to minimize adverse impacts to such properties. The insightful review of the Section 106 process by Kathryn Gourley, former Iowa SHPO archaeologist (Gourley 1995), is freely adapted here. See King (1998, 2000) for more details.

Under Section 106, federal agencies must take into account the effect of their proposed undertakings on properties listed in or eligible for the NRHP. The Advisory Council on Historic Preservation has established procedures for compliance with Section 106, published as regulations (36 CFR Part 800). Agencies must allow the Advisory Council a reasonable opportunity to comment before proceeding with the project. The federal agencies are required to do this work before the expenditure of federal funds or the issuance of any licenses or permits.

Once a federal agency has identified that it has an undertaking, the agency must define the undertaking's area of potential effect (APE). The APE must include areas directly or indirectly impacted by the action. For example, the APE for a natural gas pipeline would include not only the actual pipeline trench but also the construction right-of-way, compressor stations, meter stations, staging areas, storage yards, access roads, and other ancillary facilities.

The agency needs to consider the full range of effects that might occur. For example, a construction project might cause vibration impacts to historical archaeological sites that contain structural remains. Or undertakings along in a river valley can impact the view from historic properties on adjacent bluff tops, affecting their historical integrity, so those properties could fall within the APE of the valley undertaking.

The APE is three-dimensional, so every project must include some consideration and assessment of local geomorphology. Years of research provide convincing evidence that certain landscapes have a high potential for containing buried archaeological sites. In others, there may be no potential for cultural resources—for instance, when appropriately aged sediments have been removed by natural or cultural processes. Understanding the landscape history of a project area is vital to understanding the area's archaeological potential.

Once an agency has defined an undertaking and its APE, it is ready to begin Section 106 compliance. The regulations outline a five-step process: (1) identify and evaluate historic properties, (2) assess effects, (3) enter into consultation, (4) allow the Advisory Council to comment, and (5) proceed with the undertaking.

In step 1, the regulations require that the federal agency "make a reasonable and good faith effort to identify historic properties that may be affected by the undertaking and gather sufficient information to evaluate the eligibility of these properties for the National Register." The regulations also specify that "efforts to identify historic properties should follow the Secretary's Standards and Guidelines for Archeology and Historic Preservation" (36 CFR Part 800.4a[2]). As part of the agency's initial identification effort, the agency must seek the views of the State Historic Preservation Office (SHPO) and other interested parties.

The usual procedure is for the agency to begin with a "Phase I" survey. This survey is conducted to identify any cultural resources that exist within the proposed project area. While the majority of resources that surveys find are determined not to be significant, Phase I surveys often identify sites that are potentially eligible for the National Register. Such findings necessitate further study, often dubbed "Phase II" testing, which involves detailed research to determine the significance and integrity of the site(s) in question. If the Phase II testing results in a determination of National Register eligibility for a given property, then the project moves to step 2.

Step 2 involves determination of whether the project will have an effect on the NRHP-eligible historic properties. An effect occurs when "the undertaking may alter characteristics of the property that may qualify the property for inclusion in the National Register." If an effect is found, then the agency and the SHPO consult to determine whether the effect is adverse.

If there is an adverse effect, then the agency and the SHPO consider ways to minimize the impact of the project on the historic property. This constitutes step 3. One way of minimizing the impact is to redesign the project to avoid the historic property. If that is not possible, another way may be to excavate the site. Other alternatives might also be possible.

Once the federal agency and the SHPO have consulted, the project moves to step 4: providing the Advisory Council an opportunity to comment on the undertaking. Once the council's comments have been taken into account, the project moves to step 5 and proceeds.

The sponsoring federal agency is legally required to see that the Section 106 process is completed. If the agency does not fulfill its responsibilities, then any citizen or organization can pursue legal action to make the agency fulfill the requirements.

REGULATIONS IN CANADA

The Department of Canadian Heritage is responsible for policies and programs relating to culture, including archaeology. It administers the Historic Sites and Monuments Act (1952–1953) through the Historic Sites and Monuments Board, whose mission is to "receive and consider recommendations respecting the marking or commemoration of historic places, the establishment of historic museums and the administration, preservation and maintenance of historic places and historic museums." Among Canadian Heritage's many initiatives is the "Access to Archaeology Program," which provides financial assistance to train aboriginal people in archaeological resource protection and to promote awareness of Canada's archaeological heritage.

The provincial, territorial, and federal governments share responsibility for culture. Beginning in the 1970s, provincial governments began to establish agencies responsible for the administration of cultural heritage. Because the provinces had various forms of jurisdiction over public and private property, they were able to pass and enforce legislation extending legal protection to designated properties or sites not owned by governments or subject to federal regulation. In Nova Scotia, for example, the relevant act is the Special Places Protection Act. One of its purposes is to provide for the preservation, protection, regulation, exploration, excavation, acquisition, and study of archaeological and historical remains and paleontological sites considered important parts of the natural or human heritage of the province. Archaeologists must obtain permits to conduct archaeological work, and that work is subject to guidelines established by the province. In British Columbia, the Archaeology Branch of the Ministry of Community, Aboriginal and

Women's Services administers the Heritage Conservation Act. The act, which distinguishes between nondesignated heritage sites and designated provincial heritage sites, covers any land or water in the province that has "a heritage value to British Columbia, a community or an aboriginal people." As in Nova Scotia and other provinces, it provides legal protection for designated sites and incorporates mechanisms for protecting nondesignated sites—places that may be nominated as a result of historical, archaeological, or other heritage-related research. Typically, the Heritage Act deems it an offense to alter a heritage site; hence, the minister may order a heritage conservation inspection or heritage investigation of any property, subject to the rights of private ownership.

Significantly, First Nations communities may enter into agreements with the provincial government for the conservation and protection of heritage sites and objects, including those of spiritual, ceremonial, or other cultural value.

WHO SURVEYS?

UNDERGRADUATES

If you are an undergraduate student, you are probably wondering, "How do I get a gig surveying for these cool National Register or cultural heritage sites?" Well, perhaps surprisingly, entry-level field technician positions are relatively easy to find. You simply have to be in the right place at the right time. Persistence doesn't hurt your chances at all. Archaeological consulting groups, private or otherwise, continuously hire entry-level staff on an as-needed basis, depending on how much work they are able to capture at any given time.

Cultural resource management, sometimes referred to as *applied archaeology* or, more bluntly, *contract archaeology*, now occupies a very competitive niche within the grand scheme of national commerce and infrastructure development. Prior to 1975, "salvage archaeology," as the field was then more or less appropriately known, was the almost exclusive and usually part-time bailiwick of university anthropology departments. With the enactment in the 1970s and 1980s of federal regulations governing historic preservation, there was a general upheaval in how and by whom CRM-based archaeology was conducted. Suddenly, there was too much work for academics to perform on a part-time basis. Many universities, museums, and historical societies developed full-time CRM-based programs, and private sector firms proliferated. The legal, political, and intellectual landscape of CRM and academic archaeology changed forever. (See Toolkit, volume 1, chapter 1, and Green and Doershuk 1998 for reviews of the current state of CRM as it relates to archaeology in the United States.)

Many of today's established professionals began their careers in entry-level positions during those heady, early days of modern CRM archaeology in the 1970s. At that time, if you exhibited competence, initiative, and responsibility, you could be quickly promoted from shovel bum to crew chief and then to site director or project director, provided you could maintain an affiliation with a principal investigator who trusted your work. Those days don't exist anymore. Today, if you are a young person aspiring to direct CRM-related projects, you are almost required to possess an advanced degree (M.A. or Ph.D.)—and you cannot become a registered professional archaeologist without one.

GRADUATES

If you are a graduate student, you are probably wondering, "How do I get a gig directing a survey for these cool National Register or cultural heritage sites?" Well, a good place to start is to understand what the minimal professional qualifications are for archaeological consultants as defined by government. In the United States, for example, the Secretary of the Interior's Standards (36 CFR Part 61) read as follows:

> The minimum professional qualifications in archeology are a graduate degree in archeology, anthropology, or a closely related field, plus: (1) at least one year full-time professional experience or equivalent specialized training in archeological research, administration, or management; (2) at least four months of supervised field and analytic experience in general North American archeology; and (3) demonstrated ability to carry research to completion.
>
> In addition to these minimum qualifications, a professional in prehistoric archeology shall have at least one year of full-time professional experience at a supervisory level in the study of archeological resources of the prehistoric period. A professional in historical archeology shall have at least one year of full-time professional experience at a supervisory

level in the study of archeological resources of the historic period. In this definition, a year of full-time professional experience need not consist of a continuous year of full-time work, but may be made up of discontinuous periods of full-time or part-time work adding up to the equivalent of a year of full-time experience.

Where do you get experience? Most archaeologists start with a field school and then get work with a CRM consulting outfit. Private firms, universities, museums, and historical societies do CRM consulting in every state and province. Most SHPOs maintain lists of these consultants. Also, look in indexes such as the *AAA Guide* (issued yearly by the American Anthropological Association) or the American Cultural Resources Association's directory (www.acra-crm.org). Positions are regularly advertised at professional archaeological meetings and on archaeology- and CRM-related websites. Useful places to start are www.archaeology.about.com and www.shovelbums.org.

Call the companies and institutions doing CRM in your area. Sooner or later, you'll get a job. The only way your career will take off from there is for you to exhibit competence, initiative, and responsibility. Those characteristics, *and the ability to write well* (see Toolkit, volume 7), remain in almost universal demand. Ordinarily they will be rewarded by your employer.

CRM CONSULTING

Let's assume that you have just finished graduate school and have accepted your first more or less permanent job with a company or institution doing CRM consulting. Consultants come in a variety of forms. Ordinarily, there will be a director or company president who is mainly responsible for getting contracts from various agencies. Depending on the size of the company or institution, there may be a few or several layers of managers between the director and you, the newly hired project archaeologist.

The CRM consultant fulfills a vital role in the governmental processes outlined earlier. Agencies responsible for identifying historic properties usually assign their responsibility to the local, state, or provincial agencies that will administer the project. Those entities may (rarely) perform necessary initial surveys using their own qualified personnel. Often, they assign the responsibility to the prime engineering contractor for the project, or they may hire your firm as their archaeological or cultural resource consultant. In the cases when the agency assigns compliance responsibility to its engineering contractors, those companies may have qualified in-house personnel, or they may subcontract the work to your CRM consulting firm.

Public agencies or districts within agencies sometimes enter into long-term agreements with individual CRM consultants to perform all compliance work for that agency or district. Just as frequently, agencies will put out a request for proposals (RFP) or a request for quotations (RFQ) inviting qualified consulting groups to bid on an individual project or a package of projects. The agency may select the lowest bid in terms of dollars, but this is not always the case. Many agencies have been burned over the years by selecting strictly on the basis of cost. Contracts often are awarded to the consultant with the highest level of expertise in a given project area, the best personnel and infrastructure, or the best track record for quality work and timeliness of product delivery. The agency can suffer in terms of project scheduling if the consultant it selects cannot provide a competent and timely product that assures the agency of compliance.

The consultant provides the agency with a service (e.g., a survey of a development tract) and a report of the findings of that service. The client (agency) forwards the consultant's product (the report) to the SHPO for review, constituting the beginning of step 3 of the Section 106 process. In practice, while it is the agency and the SHPO that are the main players in the Section 106 process, the consultant as representative of the agency also works with the SHPO to ensure that the work performed will be adequate for a positive compliance review. The reason for this is simply that you do not want to provide your client with a report that the SHPO will find inadequate.

Communication between the consultant, the client, and the SHPO, provincial, or regional archaeologist throughout the course of a project generally precludes nasty surprises at the report review stage. If you provide your client with a report of investigations that is found to be inadequate when reviewed by the state or provincial agency, you have jeopardized your client's project schedule, as well as your employer's future prospects to capture more contracts from that agency. Avoid that situation. Get to know the government archaeologists and what they want to see in terms of fieldwork and reporting standards. In that way you can serve your client's and your employer's best interests, and your own. (See Toolkit, volume 1, for details on the importance of communication with reviewers and sponsors.)

3
SURVEY DESIGN

In an archaeological survey, you can never hope to find traces of all the cultural activity in a particular place. Surveyors walk in different ways and see different things. Ground visibility changes with vegetation and weather conditions. The ground may be featureless in the flat light of high noon or sharply etched in shadow as the sun rises or sets. Time is always against you. As soil accumulates or erodes with the seasons, it changes the amount of cultural material exposed on the surface. Even if you plan to walk every square meter of land, there will always be more to find. There is never a final survey.

Because the action of these variables and other subtle biases in survey technique affects the quality of the information retrieved in a survey, it is important to develop a research design and survey plan. After all, the results of your survey may determine whether a piece of undisturbed land is preserved or destroyed. The research design is your work plan for the survey, from background research to the writing of the report. To develop it you must know the area you are working in: its geological, topographical, environmental, cultural, and historical features; local and regional research trends and concerns; and previous archaeological research conducted in the project area and the surrounding region. From this background information, you can then estimate the data potential of the project area—whether it has a high, moderate, or low potential for archaeological sites. Of course, you never actually know what might lie beneath the ground surface, but by trying to estimate the potential of a land surface for archaeological sites, you must take into account the physical and cultural factors that generate human activity and consider those factors in the real world of your project area. (See Toolkit, volume 1, for details on developing a research design.)

The type of survey you do depends on whether you are conducting academic research or working under contract as part of a CRM team. Non-CRM surveys follow specific hypotheses to be tested or focus on limited types of information. You may want to conduct a comprehensive inventory, in which you attempt to identify all the cultural resources in the study area, or a problem-oriented survey, in which you search for a predefined set of cultural resources as part of research toward some specific analytical goals. For example, you may be interested only in rock art, Paleoindian environmental adaptations, or protohistoric site distributions rather than all the cultural resources of a landscape.

Problem orientation in a CRM project is a little different. The client, not the archaeologist, defines the study area. Conducting research limited in this way also requires a research design and demands creative thinking: Your interest in settlement patterns in a drainage basin is going to be difficult to accommodate when you have to work within a three-hundred-foot-wide corridor determined by construction needs. A limited survey, however, may yield some useful information if you think of your corridor as a sample of several specific environments such as floodplains, terraces, valley sides, blufftops, and uplands that may, for example, provide data on relative site density for the whole drainage basin.

Once you know what kinds of information you expect from the survey, you must decide how much area to cover. It may seem obvious that you need walk everywhere so that you do not miss any artifacts or sites. Such general reconnaissance surveys range from preliminary exploration, when you search intuitively, concentrating on likely spots, to total area surveys, covering 100 percent of the ground. The value of this approach is that you, the archaeologist, are walking in the same way as the people who walked there before, crossing similar terrain, perhaps through similar vegetation, with nothing between you and the ground—no survey theory or procedures to get in the way of what you seek. This directness of encounter and openness to the environment can help make you aware of qualities in the physical landscape that people in the past also sensed—viewshed (what you can see from a particular place), slope, exposure to sun and weather, and good places to work or camp.

But what do you do if the study area is hundreds of acres, and you have limited money, time, and personnel? Even if you can survey where you want, you may have to introduce your own constraints.

The solution is a systematic sampling survey. In this approach, you predetermine where to walk and how much ground to cover, either by establishing a grid and exploring a portion of the grid squares or by setting out transects (lines, at a set distance apart, along which surveyors will walk). Clearly, you are taking a risk here, as a significant portion of the land will go unsurveyed, so if you want to conduct this kind of survey, you will have to trust probability theory's claim that an appropriate sample will represent what you would come up with if you surveyed the entire area.

You may be somewhat suspicious of statistical claims, but this systematic method does eliminate some of the subjectivity in a survey by ensuring coverage across all areas, no matter how unpromising they look. Now it might seem like a waste of time to struggle up and down steep rocky slopes, crash though dense brush, or slog through wetlands, but biases in your technique might cause you to miss important information. For example, archaeology in the northern forests of Canada and Alaska has a strong seasonal bias: Fieldwork is possible only in the summer. Because the terrain is so rugged, with impenetrable bush and impassable muskeg bogs, surveyors tend to search along the same lakes and rivers that the native peoples inhabit. Yet, the native peoples live year-round in this environment, and their winter adaptation, settlements, hunting patterns, travel routes, and other activities may be quite different because the ground is frozen and travel overland is therefore easier. Each individual and cultural group sees the land and its potential differently. What seems uninviting to us may have been ideal in the past, when conditions and cultures were different.

When you plan a probabilistic survey, you must first divide the study area into nonarbitrary or arbitrary sample units. *Nonarbitrary units* are features with clearly defined natural boundaries, such as terrain types or ecological zones, or cultural ones, such as architectural features. *Arbitrary units* are squares (also called *quadrats*), grid intersections (fiducials or points, which become the center of a circular sampling unit), and transects. Since a simple random survey using an arbitrary method takes no account of ground-level conditions and hence cannot accommodate physical factors affecting distribution, you may want to divide the study area into natural zones and then apply the random sampling technique.

Then you have to decide on the sampling strategy. Generating random numbers to select the sample units to be surveyed is simple, but estimating the sample size is much more difficult. Ideally, the amount you survey should relate to the kind of information you expect to find. If you think you are dealing with a uniform scatter of similar sites or artifacts, you may be able to survey less to get a representative sample than you will if the density and types of sites or artifacts are variable. Unfortunately, what works well on paper may be impossible to achieve in the field because of cost. You may end up sampling much less than 10 percent of the total ground surface, with no confidence in the accuracy of this method.

A carefully planned sampling strategy, informed by background research that can help you define a set of expectations related to the nature, distribution, and density of cultural resources in the study area, might yield a good estimate of the average distribution of cultural material in the region. But you must remember, too, that there is only one Stonehenge. Not all data are susceptible to statistical analysis nor should they be. Your challenges are to determine whether the risk of a probabilistic survey is greater than the appearance it gives of scientific rigor and whether a probabilistic survey will satisfy both the agency's compliance needs and your research design.

The results of a survey are only as good as what you put into it. Finding everything you can on a plot of land may be valuable as an inventory for cultural resources management, but this information is useless unless it is part of a larger research plan.

4
PREFIELD STRATEGIES

IMPLEMENTING THE RESEARCH DESIGN

Thousands of small-scale archaeological surveys are performed each year without an explicitly defined research design. However, most surveys that are competitively bid via an RFP require a formal research design as part of the bid proposal. Depending on the size and structure of your consulting group, you may or may not be involved in the bidding process or the development of the survey research design. If your boss (normally the principal investigator) assigns the project to you (the project archaeologist), it will be your responsibility to fulfill the particulars of the research design.

Local or regionally based CRM consultants may have an advantage in developing project research designs because they may already be familiar with the project landscape and the important anthropological and archaeological questions pertinent to the region. The expertise they already possess in the region allows for quick preparation of a proposal. Trouble often arises in the compliance process when firms with no established track record in a given state, province, or region land contracts in those areas. They lack understanding of the local geography, cultural-historical sequence, relevant research problems, and idiosyncrasies of local politics and bureaucracy.

Typically, the proposal for a survey project is a relatively simple document that is geared specifically to fulfill the requirements of the scope of work (SOW) provided in the client's RFP. The responsive proposal's research design must demonstrate (1) an understanding of data categories that can be anticipated in the project area; (2) familiarity with source materials pertaining to relevant research and general theory in the social and natural sciences and the humanities that may be pertinent to the region and the project area; and (3) knowledge of previous relevant research, especially research previously conducted in the immediate project area, as well as other research on topics germane to the purpose of the survey. The survey strategy must recognize the project area's data potential.

Research problem domains should be explicitly stated. The research design must outline the survey strategy—that is, how you will approach the project in terms of research-oriented field methods. The capabilities of your consulting institution or firm must be defined, so that the client is assured you have the capability and experience to perform the work. Key personnel must be identified (hopefully, that will include you), with vitae provided. Finally, the proposal and research design must address issues of schedule and budget.

THE TWO-HEADED MONSTER

Schedule and budget—time and money—is the two-headed monster that just became your constant companion. When you enter the workaday world of CRM, concerns related to that two-headed monster are never far away. Your mission as project archaeologist and leader of the survey team, should you choose to accept it, is to do the best anthropological research possible given the constraints imposed on you by this new friend. The balance of this book is devoted to providing you the wherewithal to do just that. You have just entered the real world. Your boss just dropped a folder on your desk and said, "Sink or swim."

What are the mechanics of archaeological survey in the real world? Ordinarily, the process begins when a client or sponsoring agency forwards a set of plans to the archaeological consultant or contractor. The project plans are what just landed on your desk. Depending on the size and design stage of the given project, these plans vary greatly in their detail. Sometimes they are not plans at all but rather sketches, or simply legal descriptions, and sometimes the proposed project area is shown on aerial photographs or maps of various scales. Occasionally, the plans are quite detailed, as in the case of CADD-generated project plans that show the exact location of most natural and cultural features (e.g., streams, tree lines, transportation right-of-ways, utility lines, houses, etc.) and their relationship to the survey area.

GETTING STARTED: MAP IT

As the archaeologist in charge of the survey, your first task is to review the plans you receive from your client and determine where the project is located. This review begins by consulting available state, provincial, or county transportation maps and the appropriate index of topographic quadrangles. U.S. Geological Survey (USGS) 7.5-minute, 1:24,000 scale, National Topographic Service of Canada (NTS) 1:50,000 scale, or provincial quads (e.g., in Ontario, 1:20,000 in the north and 1:10,000 in the south) are standard.

Once the appropriate topographic quadrangle or group of quadrangles is identified from the index, the map(s) should be ordered from a convenient state, provincial, or local map repository. Visit edcwww.cr.usgs.gov/Webglis/catindex.html or http://cartes.nrcan.gc.ca/main.html to see what is available regarding map indexes, catalogs, and vendor information. Depending on the project, it may be necessary to order multiple sets. You will probably want to keep at least one clean set. These you will use as base maps when you prepare your project report. You'll also need one or more additional sets that will be taken to the field by the project archaeologist, crew chief(s), and perhaps individual members of your crew.

Prior to fieldwork, you should transfer the information from the project plans onto the field copies of your USGS, NTS, or provincial topographic maps. Double-check this work for accuracy. Lacking very detailed project plans, the topographic map will be your only reference to the landscape, your beacon in the wilderness. Develop your map-reading skills until you can read a map as well as the written word. Unless you do this, you may quite literally find yourself lost. Orienteering is now a major aspect of your work.

Large consulting firms and institutionally based archaeological consultants will often have in-house map libraries. For small survey projects, you may be able to get by with photocopies of the appropriate topographic maps for field and cartographic purposes. For larger projects, the benefit of actually having copies of the USGS, NTS, or provincial quadrangles for use in the field is almost always worth the cost of procurement. At the conclusion of your project, these maps and your narrative field records will represent your "memory" of the fieldwork. Your maps are likely the only constant during a survey project that will be on your person every time you step out of your vehicle, so cram as much information onto them as you can—site locations, landowner names, the dates you acquired permission to access property, phone numbers, local oral history, crop cover, and so forth. If your map isn't dog-eared and scribbled all over at the conclusion of fieldwork, you aren't using the resource to its full potential. Don't worry, the USGS, NTS, and the provinces make quality products that will stand up to months, even years, of rough use.

BACKGROUND RESEARCH

THE PHYSICAL SETTING

Once you have plotted your survey area on the appropriate quadrangle, regard the map carefully. Note the type(s) of terrain in the project area. Are you dealing with uplands, shorelines, floodplains, forest, range land, agricultural fields, or urban areas? The answer will help you choose your the survey strategy and techniques. Observe the location of fence lines and houses as they relate to your project area so you can determine how the land is used and who possesses it. In most cases, when you get to the field you'll need to contact these people for permission to enter their property (more on that subject later).

Next, determine the area's regional and local geologic and physiographic record. Most states and provinces have studied, mapped, and published pamphlets or books defining the geologic history and major landform regions within their borders. Ordinarily, the archives or library at your company or institution will have these materials. Review these documents to determine what bedrock geology and Quaternary landscape features might be encountered and how they might affect the character of the local archaeology. For instance, local bedrock outcrops may contain chert, a magnetlike resource for prehistoric populations. In the Northwest, tephra beds are often used as stratigraphic markers in Quaternary landscapes. Carry this knowledge into the field.

The Quaternary record governs the types of surficial deposits and often the vegetation you will encounter. For example, older, loess-mantled drift plains often supported savanna prior to European settlement in many parts of the Midwest, whereas late Pleistocene glacial drift often supported pothole terrain in which wetlands and prairie predominated. Obviously, one would expect aboriginal use and concomitant settlement patterns to have differed from one landform region to the other.

Holocene environments have been extremely dynamic over the last ten thousand years. Much of North America's Holocene environmental history is told in the climatic record. North American populations clearly responded to climatic fluctuations, just as our children and grandchildren will be forced to adapt if our colleagues are correct about global warming. Inferences about prehistoric responses to climatic and environmental change can be derived from settlement location data. Survey data are essential for such studies, but you must first understand the climatic record. Learn about regional paleoclimates from research papers by geologists and botanists.

Perhaps no portions of the Holocene landscape have had a more dynamic history than valley margins and floodplains. Colluvial aprons and alluvial fans, which develop at the base of valley walls, often bury or contain archaeological sites. By definition, the floodplain environment is constantly in flux. Episodes of stream incision and valley filling, often as a complex response to climate change, continually create and bury, and often destroy, habitable surfaces. Geomorphologists study these processes, and you should keep abreast of the latest geomorphological literature and landscape models. Review the pertinent regional geomorphological literature, as it relates to your project area, prior to fieldwork. Field trips organized by the Friends of the Pleistocene and other geology groups are great ways to obtain a hands-on understanding of earth-surface history and processes.

Holocene lithostratigraphic models can be used to predict where archaeological sites might be found on or buried within the modern landscape and the potential age of such sites. Because archaeological materials have primary context in the natural landscape, geologic processes forming the landscape determine the conditions of the archaeological record. Geomorphological information contributes essential data to determinations of significance regarding the age, preservation, and integrity of sites in colluvial and alluvial settings. Your survey strategy must account for these dynamic environments and the often-invisible (buried) surfaces they contain. If you don't understand geologic processes—and it is a sad fact that most graduate programs wouldn't have asked you to learn about them as a requirement for your M.A. in anthropology—it might be a good idea to consult with a geomorphologist familiar with your survey area before you go much further in your background preparations. Contact another archaeologist familiar with your survey area or the SHPO or provincial archaeologist to help you locate a competent consulting geomorphologist.

Once you get a handle on the general physiography of your study area, more specific information can be gleaned from the county soil survey. In the United States, the Department of Agriculture, in cooperation with state agricultural experiment stations and other federal and state agencies, has been making and publishing soil surveys since 1899. Agriculture Canada is responsible for soil surveys in Canada in cooperation with provincial departments of agriculture. These surveys furnish soil maps and interpretations needed by farmers and ranchers as well as planners and researchers. Uniform scientific and technical standards are used in soil classification, nomenclature, interpretation, and publication.

Although the information in county soil surveys is generally accurate and undeniably useful, these publications were not made with the archaeologist in mind. Their primary purpose is to assist evaluation of the land's agricultural potential. Until recently, soil surveyors did not consistently note buried soils. Recent surveys, recognizing the broader uses of the data, do a better job of identifying these hidden former surfaces, which is helpful to the archaeologist. In general, if you come across a term such as *mixed alluvium* in the definition of a floodplain soil series, be aware that buried surfaces might be present, and design your survey strategy accordingly.

Your firm may have a collection of county soil survey publications. You can also get them from county USDA offices or provincial departments of agriculture or order them directly from federal, state, or provincial publication outlets. For a list of available soil survey publications, ordering details, and other related information, explore www.statlab.iastate.edu/soils/nsdaf/. The Canadian Soil Information System (part of Agriculture Canada) provides paper and online GIS map products relating to agriculturally significant soils. It may be reached at res.agr.ca/cansis/.

Prior to fieldwork, many archaeological surveyors visit a university map library to examine aerial photographs of their project area. Since 1936, USDA has been making aerial photographs of most of the country at intervals of about every six to eight years. Canada has a similar aerial mapping program, with images dating back seventy years, and these are available from the National Air Photo Library. These photographs often reveal startling changes in land use patterns that affect the context of the archaeological sites you encounter. You can also follow patterns of natural or human-induced vegetation change. Beyond that information, careful examination of aerial photos can sometimes identify archaeological sites and individual features. Many archaeologists have found palisade lines, rows of houses, and even individual features within Mississippian sites, earthworks within Woodland sites, fortifications on military sites, and stone fences delineating historical period fields or property lines on aerial photos prior to setting foot in the field. Air photos also serve as excellent base maps for marking sites, features, and finds. They often permit more accurate plotting of survey results than topographic maps do.

Many U.S. aerial photos can be picked up at USDA county offices and can be ordered at edcwww.cr.usgs.gov/products/aerial.html in printed or digitized formats. Terraserver, a commercial enterprise, provides worldwide air photo coverage through www.terraserver.com.

Despite their value, aerial photos remain an underutilized resource in support of archaeological survey. See Deuel (1969) for a comprehensive and highly readable discussion.

Ordinarily, a background review such as that outlined here will give you enough information about the physical setting of your project area to get you into the field. You should know about the age and character of the regional landscape; potential subsistence resources available to its inhabitants through time; and the most recent diachronic models regarding regional climate, vegetation, and geomorphology. Your review of local soils data will have given you an idea of what to expect when you get to the field and how you may want to sample the project area. Aerial photographs may help you identify specific targets to investigate when you get into the field.

THE CULTURAL SETTING

Just as you must know the important characteristics of the physical landscape, you must also know your project area's cultural background. This knowledge is best obtained through systematic review of archaeological planning documents, site records, and historical archives.

The fact that you have your job more or less implies that you are aware of the important CRM and archaeological research questions pertinent to the area where you will work. This assumption is often unfounded. You may be new to the state or region, or perhaps you simply lack experience. Never fear—a statewide or regional planning document or master plan probably exists, containing the accumulated wisdom regarding cultural resources and pertinent research directions. Although always incomplete and outdated, such documents help you place your survey work within the larger CRM and cultural historical picture. You should be able to get a copy of the documents used in your area from your SHPO or provincial archaeologist. Most planning documents are divided into sections that synthesize geographic and cultural historical data. Read sections pertinent to your study area prior to doing anything else.

Archaeological site records are managed differently from state to state, province to province. They might be housed at the SHPO, the state or provincial archaeologist's office, the state Department of Natural Resources or provincial ministry of culture, a university, or a state or provincial museum. Archaeologists learn how to deal with site records simply by dealing with site records. Your boss and colleagues should be able to tell you how it's done where you will be working.

You can delegate the records search to a competent staff member, but usually it is most useful for you to conduct the site records search yourself. Many states and provinces have procedures whereby site-records personnel can conduct the search for you, usually for a fee. Some files such as Arizona's are now searchable remotely via the web. Data on many Canadian sites are available through Artefacts Canada at daryl.chin.gc.ca/Artefacts/e_MasterLayout.cgi?db=3. However you obtain the data, you must learn about any recorded sites relevant to your project area. Copy any pertinent site records and maps, and plot the locations of any nearby sites on your project maps. Ordinarily, a repository of CRM reports ("gray literature") is housed at the same agency as the site records. Examine any literature (gray or otherwise) mentioned on the site forms or that may pertain to your study area. Such works can include published articles as well as previous CRM-generated survey and excavation reports. The National Archaeological Data Base is an online guide to many of the U.S. reports (www.cr.nps.gov/aad/TOOLS/Nadb.htm).

Also review historical records pertinent to your project area. Your state or provincial historical society is an invaluable resource, as are county or other local historical societies. Make friends with the people at these places. It is their business to help you find what you are looking for, and like almost all librarians and archivists, they often possess great knowledge about relevant materials and are normally very good at their job.

Examine all available regional atlases, plat maps, and rural directories. These resources provide a wealth of information about historical land use and sometimes the locations of Indian settlements, rural homesteads, farms, ranches, schools, churches, commercial enterprises, cemeteries, and even whole townsites that may be invisible save for their archaeological signatures. Plat maps and rural directories often contain useful information about demographic and economic trends and property ownership. Cross-check information from the atlases and other documents with any published county histories. The county histories usually will contain biographical information about early settlers and leading citizens. The historical and biographical information provided in the typical old-fashioned county history warrants a degree of skepticism because biographies were normally written for subscribers, and publishers were not in business to alienate patrons. Yet these publications are not without merit and can provide important information.

Other excellent sources of historical information are military maps from the colonial and territorial periods, and, in much of the United States, the notes and plat maps derived from the General Land Office (GLO) surveys. GLO records are often available on microfilm at state historical societies. According to the system inaugurated by the Ordinance of 1785 and permanently established by Congress in 1796, the General Survey of Public Lands physically determined the boundaries of congressional townships, which were then subdivided into thirty-six sections of land one mile square. Instructions to the GLO surveyors were explicit (sidebar 4.1). Notes of the field survey were to form a full and perfect history of operations in the field.

The typical GLO surveying party consisted of a surveyor, two chain carriers, an ax man, a flag man, a hunter, and a camp keeper. These crews left a remarkable record of the American frontier that is both useful and worthy of emulation by all modern surveyors, including those of us who survey for archaeological resources. Serendipitous discovery is often the dividend accrued from a thorough review of historical period records; overlook these resources at your own peril (figure 4.1).

You never know what you might learn by reviewing historical records. It is a great advantage to be aware of the presence of potential historic-era sites before going to the field. Actually finding evidence on the ground for significant sites that you identify through background research is even more rewarding. When you find archaeological evidence of, for example, a "deadstead" (the site of a former farmstead), in many cases you will already have an idea of its occupation span, owners or occupants, and when the place was razed.

The point is simply this: Based solely on preliminary historical research conducted prior to your CRM survey project, it is possible to

4.1. ENUMERATION OF GLO RESPONSIBILITIES

Congress commissioned GLO surveyors to enter in their field notes of the survey a particular description and the exact location of the following:

1. The length and variation or variations of every line you run
2. The name and diameter of all bearing trees, with the course and distance of the same from their respective corners
3. The name of the material from which you construct mounds, with the course and distance to the pits
4. The name, diameter, and exact distance to all those trees that your lines intersect
5. At what distance you enter, and at what distance you leave every river, creek, or other "bottom," prairie, swamp, marsh, grove or windfall, with the course of the same at both points of intersection
6. The surface, whether level, rolling, broken, or hilly
7. The soil, whether first-, second-, or third-rate.
8. The several kinds of timber and undergrowth, naming the timber in the order of its prevalence
9. All rivers, creeks, and smaller streams of water, with their actual or right-angled widths, course, banks, current, and bed, at the points where your lines cross
10. A description of all bottom lands—whether wet or dry, and if subject to inundation, state to what depth.
11. All springs of water, and whether fresh, saline, or mineral, with the course and width of the stream flowing from them
12. All lakes and ponds, describing their banks and the depth and quality of their water
13. All coal banks, precipes, caves, sink-holes, quarries, and ledges, with the character and quality of the same
14. All waterfalls and mill sites
15. All towns and villages, houses, cabins, fields and sugar camps, factories, furnaces, and other improvements
16. All metalliferous minerals or ores, and all diggings therefore, with particular descriptions of both, that may come to your knowledge, whether intersected by your lines or not
17. All roads and trails, with the courses they bear
18. All offsets or calculations by which you obtain the length of such parts of your lines as cannot be measured with the chain
19. The precise course and distance of all witness corners from the true corners that they represent

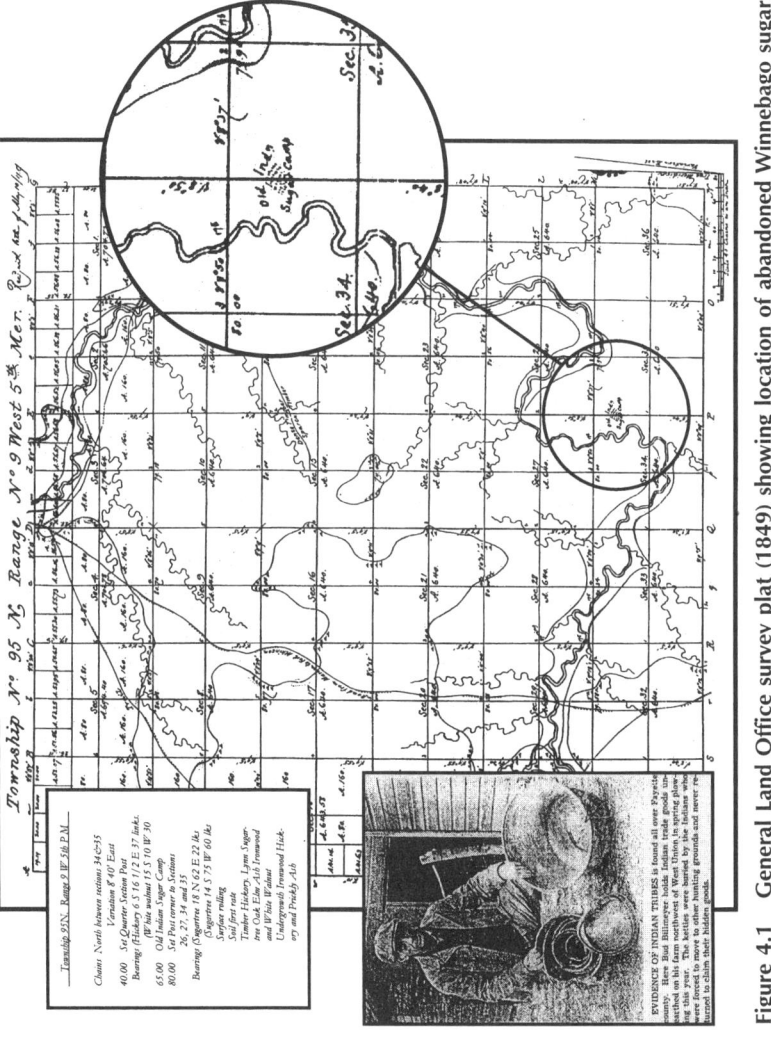

Figure 4.1 General Land Office survey plat (1849) showing location of abandoned Winnebago sugar camp near Fort Atkinson, Iowa. A sample of the surveyor's notes for the township is reproduced top left. Landowner interviews and further background research yielded the newspaper photograph reproduced bottom left. The kettles shown were found at the sugar camp site in 1949.

"discover" sites that may hold great research potential. You might also be able to predetermine that some sites are clearly not eligible for the National Register or for nomination as national or provincial heritage sites because of their age or other reasons, saving some steps in field evaluation. Do your homework before you go to the field. It is really amazing how often it pays off in a useful or effort-saving way.

5
GETTING INTO THE FIELD

Now that you are familiar with the physical, cultural, and historical background of your project area, it is time to contemplate the logistics of fieldwork. As project supervisor, you are responsible for the overall project, which includes the day-to-day, even minute-to-minute, work of every member of your survey party. Among your primary concerns will be how you manage the sometimes mundane details of transportation, equipment, property access, lodging, and meals while in the field. Your skill in managing logistical concerns and project personnel usually determines the project's fiscal success or failure. Success is measured by your ability to bring the project to a professionally competent conclusion, on time and within budget. Failure is easily quantified by missed deadlines and cost overruns.

As you begin your career, you may have little or no experience at managing anything beyond the details of your own life, but when you become the leader of a survey party, your responsibilities increase exponentially according to the number of persons on your crew. You must think not only about what tasks you will perform during the course of a project but also about the tasks every member of your crew will perform. It is generally your responsibility to keep track of their hours, get them their per diem money and paychecks, find them a place to live, provide transportation to and from the project area, and keep them productive during working hours. Depending on the initiative and character of your crew, your responsibilities may extend to deciding where they'll eat their meals, waking them up in the morning, or bailing them out of jail. Delegate responsibility whenever you reasonably can, but always remember that it's your project, and ultimately you are responsible for it.

PERSONNEL

There is only one hard and fast rule for filling crew positions: Hire intelligence and experience, in that order. From a list of experienced persons, always pick the smartest (we assume them to be physically fit if they aspire to be field archaeologists). It is easy to train an intelligent person to do archaeological survey; others are more difficult. An intelligent crew will almost always serve you better than one that is merely experienced. Experience is a generally useful secondary screen, but most really good, experienced field personnel are already employed. Unless you are very lucky and happen to catch some experienced field hands between other projects, most often you will be hiring students or people who were recently students. Select for brains and endurance rather than brawn. The benefits of this selection process are at least fourfold: You can delegate more tasks with the expectation they will be performed as you want them to be, you can influence the careers of some people who might be rising stars, it is far less likely that you will be completely sick of your crew after two weeks in the field, and you might learn something from these bright young people.

A second consideration regarding personnel is crew size. Your crew should include the absolute fewest persons necessary. Smaller crews are invariably more efficient than larger crews. If you can handle the project yourself, by all means do so, although safety considerations sometimes warrant traveling with a sidekick. Two people, you and your sidekick, can handle many projects very efficiently. Unless you've gotten yourself into a hole in terms of time, or if you have a very large or particularly labor intensive survey (e.g., shovel testing through thousands of acres of timber), we can't imagine why you would want a crew larger than five persons, including yourself. Larger groups tend toward cliquishness and other forms of interpersonal dementia; smaller groups promote camaraderie and a positive work ethic. Managing fewer people will be a blessing to you throughout your project, and the resulting efficiency will be manifest in both your bottom line and peace of mind.

The fact that you have now become a manager should not be construed as a license to be an ass. Respect every member of your crew as an individual, and don't play favorites. Let their abilities sort out the pecking order of responsibilities. Just as the individuals on GLO survey crews had specific duties, you should assign specific duties according to each individual's strengths. Yes, everyone will be performing the same tasks a lot of the time—everyone digs and screens, but some individuals will show a higher aptitude for, say, organization. Put that individual in charge of checking bags and paperwork at the end of the day and making sure that things are stored or filed appropriately. Some people will be really good at mapping, photographing, or describing soil profiles. At the end of the day, some people may simply be best suited for repairing, cleaning up, or packing away the equipment. It all has to be done, so make use of the talents individuals possess.

Make clear what you expect from each person. If you yourself are organized, you can also insist that your crew be disciplined in terms of the daily organizational details that you establish. Pretty soon, a comfortable and efficient routine is established for break-out and pack-up chores, lunch, and so on, and there won't be a lot of the goofing off and butt scratching so typical of many poorly managed crews. People will know where to find what they need when they need it, because your supplies and equipment are organized. Supplies and equipment are organized because you have delegated that responsibility to people who have the talent to make it so.

Assess your crew members' strengths and delegate assignments accordingly. You don't have to be a jerk to achieve organizational discipline, but you do have to make clear how you want things done. Establish a workable routine immediately, and stick to it or modify it as necessity dictates. Treat everyone with due respect, and usually you will have a happy, productive crew.

How do you find crew members? One of the best ways is to post a flier at the anthropology department of a nearby college or university or at professional meetings. Your best prospects will be among those people who are finishing their undergraduate degree and want to take a break from school before going on to graduate work. These people are generally smart, eager to work, adventurous, willing to learn field methods, adaptable, have the requisite interest, and are fun to be around.

You can also post employment notices online with archaeological e-mail disscussion groups such as HistArch, Arch-L, and ACRA-L or with the website www.shovelbums.org. This is kind of a grab-bag approach, so screen applicants carefully. Ask for résumés and references. There's not a thing wrong with hiring professional shovel bums—the more the better—unless they are grumpy, jaded, and too set in their ways to take direction.

More than one archaeologist has filled out a crew by walking into a local watering hole and offering patrons the alluring prospect of travel, per diem, a few steady paychecks, and the romantic notion of archaeology. You take your chances here (recall the earlier reference to posting bail), but sometimes it works out.

EQUIPMENT

Always use the best equipment at hand or that you can talk your boss into buying. If you can score a total station, a GPS unit, a laser range finder, a set of VHF five-watt two-way radios, a quality metal detector, and a power winch for your vehicle, go for it. Fluxgate Gradiomter? More power to you!

If you know how to use the instruments, technology will serve you well and usually save you time. Unfortunately, over the years we've seen an awful lot of people pretend expertise with certain instruments only to end up wasting time and effort and in a few cases actually ruining their high-end equipment and their budget.

You're an archaeologist. How do you acquire expertise with technical equipment more commonly associated with engineering and the geophysical sciences? Some field schools offer advanced instruction in the technical tools of the trade. Also, because your employer will have a vested interest in your competency, he or she may sponsor in-house workshops or send you to a course on the use of the high-end field equipment. The U.S. National Park Service sponsors workshops in remote sensing and geophysical techniques for CRM. These intensive, hands-on courses are highly recommended. Also, you may be able to find a mentor who is willing to tutor you. If you're still in school, it might be a good idea for you to go over to engineering, geology, or physics and burn some elective credits. Or, sometimes, as when you require the services of a consulting geomorphologist, you'll just want to subcontract the work.

Your company's technical expertise and the sophistication of its equipment may determine the size and nature of the contracts it can secure. And the size and nature of your project usually determine the level of sophistication required of your equipment. While it is generally true that archaeological survey is trending toward high-tech field methods, fieldwork for most projects still can be adequately completed without the frills.

Regardless of your budget, almost everything you need for fieldwork can be ordered from Forestry Suppliers, Inc. (www.ForestrySuppliers.com) or Ben Meadows Company (www.benmeadows.com). Sign on to their mailing lists the day you begin your career. Outfitters, hardware stores, farm suppliers, and lumber yards are also good sources for field equipment (sidebar 5.1).

We don't belabor every piece of field equipment you might need or find useful, but here are a few tips concerning your field notebook. Having experimented with dozens of types and styles of field books over the past quarter-century, it is our humble opinion that the "Rite in the Rain" All-Weather FIELD Spiral Notebook, Model No. 353 (www.riteintherain.com), is the best all-around notebook you can carry. You can do your own experimentation; some people prefer a book with a bound spine. Whatever field book you decide on, go ahead and doctor it up to suit your individual needs. Tape your logistics checklist onto the inside front cover for easy reference (see Toolkit, volume 1, chapter 6). Tape copies of table 1 (types and classes of soil structure) and appendix 4 (New Designations for Soil Horizons and Layers) from the *Glossary of Soil Science Terms* (Soil Science Society of America 1987:25, 42) into the back of the notebook (carry a copy of the complete glossary in your portable field file; it can be ordered from the Soil Science Society of America, 677 South Segoe Road, Madison, Wisconsin 53711; www.soils.org/). You may want to carry the Munsell charts you use most often in your notebook for quick reference. Rig up a place (e.g., a tape and paper holster) to keep a small ruler; you get the idea. Always tape your business card in the front. If you lose your notebook, heaven forbid, at least there will be a chance that someone will find it and return it to you. But *don't lose your notebook!*

As project archaeologist, organization will be your key to a simpler life. Your vehicle will be your office and command center. In it you'll keep your plans, maps, notebooks, plat books, soil survey publications, any necessary reference literature, personnel and excavation forms, photo log, copies of your project SOW and contract, computer, phone, cameras, instruments, and the myriad other necessities of working on the road. Acquire a tough portable file box, a good camera bag, a functional and durable briefcase, and something to protect your computer and other instruments. Devise a system to maintain an orderly space within this mobile office. Try to secure everything so that

5.1. BASIC EQUIPMENT FOR A CREW OF FOUR

Item	Quantity
10×12×18–inch metal toolbox	1
3-meter tapes	4
30- or 50-meter tapes	2
Trowels	4
Pipe wrenches	2
Line levels	4
Hammer	1
Mallet	1
Hatchet	1
Flathead screwdriver	1
Phillips-head screwdriver	1
Pliers	1
Channel-lock pliers	1
$7/16$-inch box/open-end wrench	1
Root-cutter	1
Bastard file	1
Plumb bob	1
Wire brush	1
Whisk broom	1
Small paintbrush	1
Dental picks	2
Screening blocks	2
Roll of string	1
Hub stakes	4
Chaining pins	20
Roll of flagging tape	1
Sharpies (indelible markers)	4+

items such as your thirty-pound file box don't become deadly projectiles in the event of an accident. Impress upon your crew that unsecured shovels and other equipment can be pretty dangerous in the event of a rollover or collision. They may scoff, but we've seen people scalped. You're on the road—accidents happen.

On small projects, your command car will double as the general field vehicle. In that case, you'll also have to store your field equipment on board. For larger, longer-duration projects, it is almost always a good idea to reserve your command car for mobile office duty, and keep the field equipment in the other field vehicle(s). Just as your

Item	Quantity
Pencils	8
Shaker screens	2–4
Shaker screen bolt and wing-nut replacements	5
Vial of wood screws	1
Vial of nails	1
Vial of survey tacks	1
Plastic vials	4
Roll of aluminum foil	1
Assorted sizes of zip-top bags	Discretionary
Assorted sizes of paper bags and storage container	Discretionary
Oakfield soil-probe	1
Tile probe	1
Scissors-type posthole digger	2–4
Seymour bucket-augers	2–4
Bucket auger extensions	3–6
Machete	1
Standard shovels	4
Short-handled, square shovels	2
Sturdy boxes for artifact storage	2
Clipboards	4
Five-gallon plastic buckets	1–4 (optional)
Munsell soil color chart book	1–2
Tobacco and/or other votive items*	small quantity

*Native groups may request that a small offering of tobacco and/or other votive material be made when archaeologists encounter mounds or cemeteries.

permanent office is normally separate from your laboratory, it is often sufficiently important to keep the details of project administration segregated from the hubbub of general field operations, equipment management, and so forth. Don't confuse this as merely a status issue—keeping administrative materials segregated from tools and constant human traffic is really a practical logistic strategy, although in practice it does afford you a certain modicum of privacy that your crew might not share. On the other hand, it also affords them ample opportunity to critique your performance—which they will do in any case.

CHAPTER 5
LOCAL RELATIONS

Doing archaeological survey requires you to place a premium on public relations. No matter how much you might wish to remain an industrious, nose-to-the-grindstone archaeologist, project public relations are now your very real concern. On small, short-duration projects, you have a chance to remain relatively anonymous out there. However, it is impossible to bring a crew of archaeologists into the field, stay for any period of time, and not be highly scrutinized by the local populace. Simply put, you will be a topic of local conversation. Try to make the best of it. Remember, if you are doing CRM archaeology, those citizens are at least indirectly paying the tab. Whether you like it or not, you have a responsibility to be an ambassador for your discipline. You'll have many opportunities.

No matter the size of your project, you'll have to make certain local contacts prior to getting started. If you'll be working on private land, you must secure permission from the landowners to survey and dig holes on their property. Nothing ruins an archaeologist's day faster than being run out of a field by an irate landowner. If you're working on public or tribal lands, there will be permitting issues, and you'll need to make contact with the appropriate agency or tribal office even after you have obtained your permit.

For larger projects, budget some time for you and perhaps a sidekick to make local contacts prior to bringing your full crew to the field. Depending on the size of the project and the number of required contacts, you may need several weeks. Don't have your full crew twiddling their thumbs in the field vehicle while you're chewing the fat with local farmers or ranchers about access to their property. On smaller projects, with smaller crews, such contacts are usually subsumed in the general course of fieldwork, and with much less legwork.

Let's assume, for the sake of discussion, that your project is a moderate- to large-size survey project—say for a forty-mile highway corridor on new alignment traversing private land. Once you've lined up your crew and equipment, how are you going to get started? Well, obviously, a lot of private landowners are going to be affected by the project. Very likely, you will have to personally negotiate access to each and every owner's property.

One way to get the ball rolling is to place an announcement about your project in local newspapers after checking with your client (sidebar 5.2). The newspapers may follow up with some kind of story for which you will be interviewed. Most local people will notice, and at least those who will be affected by the project will be expecting to hear from you. Given such forewarning, when your crew shows up, the locals might not be so quick to mistake them for the Fabulous Furry Freak Brothers (figure 5.1).

But before you bring your whole crew to the field, reconnoiter the project area yourself. This is an appropriate time for a "windshield survey." Your sidekick can drive while you juggle maps and navigate. Learn the landscape, natural and cultural—access points, roads, stream crossings, and the like. You should repeat this tour for the benefit of your crew, but you need to know the terrain before they arrive.

Your client may be able to provide you with the names and addresses of people who will be affected by the project, but more likely you'll have to locate these people yourself. Go to the county courthouse, or purchase a plat book or rural directory. Also, score a copy of the phone book(s) that covers the service area for your project area. Back at your home office, in your motel room, or at some other place where you can spread out your stuff, cross-reference your project plans, USGS, NTS or provincial maps, and the plat books or directories. The plat books and

---

5.2. NEWS RELEASE

XYZ Consulting to Conduct Archaeological Survey

XYZ Consulting will conduct an archaeological survey this summer along the proposed alternate alignments for the relocation of [insert highway identification] in [insert county name] County. The [insert client or agency name]-sponsored project will extend from [insert general project description].

The purpose of the project is to locate and evaluate all archaeological sites within each proposed alternate. The archaeologists expect to find evidence of prehistoric and historic Indian sites, as well as historic Euro-American [or -Canadian] sites. The information will be used by the [insert client or agency name] when selecting the final highway alignment.

The project is part of a phased upgrading of [insert highway identification] scheduled for completion within the next [insert general project schedule years if known]. State [or Provincial] and federal historic preservation [or heritage] laws require archaeological surveys on road projects.

Contact: [insert your name and phone number]

Figure 5.1 Survey crew, a.k.a. the Fabulous Furry Freak Brothers.

rural directories show who owns each parcel of land. The more useful maps show dwellings and their occupants. The latter may be landowners or renters, but at least you can tell who lives where. Write the pertinent information onto your field maps. Then use the telephone books to determine the addresses and phone numbers for landowners along your corridor. This list should be fairly complete save for absentee landowners. Necessary information about the absentee owners may require another trip back to the recorder's office. Regardless, you should be able to generate a working list of landowners' addresses and phone numbers fairly quickly. Once your list is made, you have several choices of how to proceed.

For relatively large projects similar to our hypothetical corridor survey, try mass mailings to affected landowners. These mailings at once announce the particulars of the project, request permission to enter the property, solicit information about local sites, and offer landowners a chance to discuss their collections with a professional archaeologist (sidebars 5.3 and 5.4). Include a self-addressed, stamped envelope. Historically, the technique has been worthwhile, generating a positive response rate of approximately 60 to 70 percent. However, almost everyone will still want to talk to you face-to-face before you start fieldwork.

5.3.
SAMPLE LETTER INTRODUCING PROJECT TO LANDOWNER

XYZ Consulting Letterhead
Month/Day/Year

Title/First Name/Last Name
Address
City, State/Province, ZIP/Post Code

Dear Title/Last Name:

XYZ Consulting has been selected by [insert client or agency name] to perform an archaeological survey of alternative alignments proposed for the [insert highway identification] highway relocation project. Our responsibility will be to locate, record, and evaluate the significance of all archaeological sites within each alternate. We expect to find archaeological evidence for prehistoric and historic Indian mound and habitation sites, as well as historic Native and Euro-American/Canadian sites. The [insert client or agency name] must consider this information when the final alignment is selected.

It is our understanding that a portion of your property lies within or near one of the possible roadway alternatives currently under consideration. This letter is to inform you that we will be in the area and to explain our survey methods. We hope that you will grant us permission to conduct our survey on your property.

We will employ a crew of [insert number] individuals to conduct the field survey. Each person on the crew has a minimum of [insert number] years of archaeological experience. If you have questions about artifacts you may have found or archaeological sites on your property, any one of us will be happy to discuss them with you. Any information you can provide to us will be greatly appreciated. We will be recognizable by our vehicles which display XYZ Consulting decals.

Our survey methods are nondestructive. The crew spaces itself at regular intervals and walks along the proposed road corridor, scanning the ground surface for evidence of past human activity. To test for the possibility of buried sites and to evaluate the condition of the sites we find, we often will dig small holes with shovels or posthole diggers. The small test holes are always filled back in immediately. Care is taken not to trample or disturb crops in any way. Gates, fences, and livestock will not be disturbed.

Our ultimate goal is to understand and conserve important resources of [insert your state's name's] past. I look forward to talking to you personally in the near future.

Sincerely,

[insert your name]
Project Archaeologist

5.4. SAMPLE LANDOWNER QUESTIONNAIRE

XYZ CONSULTING ARCHAEOLOGICAL SURVEY PERMISSION FORM
XYZ CONSULTING
XYZ CONSULTING COMPANY ADDRESS
XYZ CONSULTING PHONE AND FAX NUMBERS

Please check the appropriate line, sign, and return in the self-addressed, stamped envelope.

___ I grant permission to XYZ Consulting to conduct an archaeological survey on my property.

___ Please talk to me personally before conducting an archaeological survey on my property. The best time to contact me is_____.

___ I own an archaeological site and/or artifacts that I would like to discuss with a [insert your company's name] archaeologist.

Signature: _____
Name: _____
Address: _____
_____
_____
Telephone: _____

Comments:

XYZ Consulting requests this information for the purpose of obtaining permission to conduct an archaeological survey on your property. No persons outside our organization are routinely provided this information. Completion of this form is optional.

A second method is to pretend that you took that job with the insurance company after college instead of becoming an archaeologist. Get on the phone and start cold calling. We don't recommend this technique except when you simply can't contact people any other way. It is very difficult to explain to people over the phone what you do, what you want to do on their property, and how they might be affected by the project. Some people will confuse you with a telemarketer and hang up. Others won't know an archaeologist from a podiatrist, especially if you start blurting jargon such as "We'd like to do a pedestrian survey for cultural resources on your property."

No, you have a much better chance of explaining what you want to do if you emulate the old Fuller Brush Man and go knock on people's doors. Have your props handy—maps, plans—and be prepared to explain methodically what you propose to do and why. It is at this stage of the project that you will come face-to-face with what it means to be a salaried employee. Naturally, you'll be able to meet with some people only at night. You might be able to catch farmers only between 12:00 and 1:00 P.M. Depending on the season, if you catch them in the field, they may want to talk fast—or slow. Expect a certain percentage of people to grant you access within minutes; others will want to talk over the situation for hours. Landowner contacts for a large project can be exhausting. All of your diplomatic and social skills will be called upon. Often you'll have to negotiate a truce with a dog before you get to its owner's door. Be leery of places that seem wrong—there are a lot of meth labs in the country and a lot of twitchy people. Some people will scare you, and you'll frighten others.

There are also a lot of truly fantastic people out there, the salt-of-the-earth types. You'll literally meet all kinds. Don't be in too great a hurry to get away from these interviews; relax and enjoy the conversation. Make friends with the landowners during these contacts. You never know how they might be able to help you later on—tractors come in handy when you bury your vehicle in mud. If you've done your background research well, you'll have some idea of the history of the area, and you can use this information to prime the pump, so to speak, and get people to open up about what they know about significant historical events, local archaeological sites and collectors, and their neighbors. People may be able to produce useful artifact collections or photographs. Always ask whether they have a collection or have ever found artifacts in the neighborhood. Photograph these collections (figure 5.2), and ask to borrow historical photographs for copying whenever you can (be sure to return them promptly). You may be asked to reciprocate by presenting a talk to the person's Rotary Club or whatever. If you can work it into your schedule, do it; we owe such service (see Toolkit, volume 7).

Occasionally, you may be the first person to inform a landowner that his or her property is scheduled to be affected by the project. This can be very difficult duty, so approach it professionally and sensitively. Some people may get the idea that they can influence the outcome of your client's project to their advantage by "salting" artifacts on their property. Such people almost always tip their hand in one way or another. Sometimes they'll subtly float the idea to you during

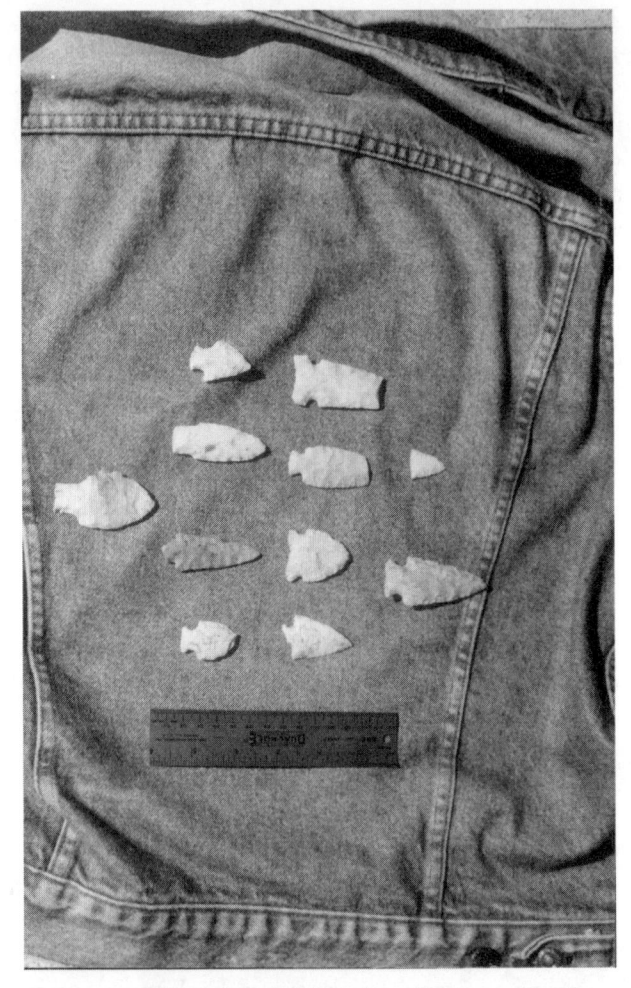

Figure 5.2 When photographing the collections of local residents, place the artifacts on a solid background, such as the back of a denim jacket, and don't forget to put a ruler, coin, or some other object into the frame for scale.

conversation in order to gauge your reaction. If they actually do try to fake a site, they might forget to wipe off the glue that had once held the artifact in its frame, or they'll tell their neighbors, who might tell you. Relatively few archaeological sites of any kind force a significant project modification, and salted sites are even less likely to be considered—spread this news as appropriate.

Keep your notebook handy and record useful information during your landowner interviews, but be prepared for anything. Once a landowner, an otherwise total stranger, told an archaeologist that her son had just committed suicide, and she wanted to talk about it. She felt that if only he had found something interesting in life, like archaeology, things might have somehow turned out differently.

Aside from landowners, it is often useful and always a welcome courtesy to contact local members of your state or provincial archaeological society, the local historical society, and archaeologists residing or working near the project area. Conversations with such interested people provide useful information about sites, landowners, recent unpublished research, and how the project is viewed locally. If you have the need or inclination, you may also solicit crew members or volunteers among these persons.

NATIVE AMERICAN, CANADIAN FIRST NATIONS, AND ETHNIC COMMUNITY ISSUES

It is especially important to find out through the contracting agency or institution sponsoring your survey whether the local or regional Native North American community has been informed or consulted about the planned archaeological work. Similarly, any recognizable ethnic community may have a stake or at least an interest in your work. Consultation is not only a courtesy and, in some areas, a legal requirement, but it opens up lines of communication that may provide valuable information about the traditional use of the land you are surveying. Many archaeologists seek to avoid direct contact, usually out of fear of the unknown or for political reasons. Indeed, some archaeologists spend their entire careers studying people of the past without ever consulting with their descendants. It seems foolhardy, however, not to consult people who have a direct, long-standing interest in the land and oral traditions that may shed light on archaeological problems.

Having said this, the process is not simple, and if you are not familiar with the community, you may find it very frustrating. If you are a member of one of these communities, you may be familiar with rules and norms. However, you may think, for example, that archaeology and its scientific knowledge of the past are essential, but such studies are not considered essential by all people. Native people have their own histories and their own priorities, and you should therefore understand their lack of interest or enthusiasm in yours. If you do consult with Native Americans or Canadian First Nations people, or ethnic community members, prepare yourself first by understanding their rules of decorum, as you should when dealing with any people of other cultures. In general, it is always best to respect your elders (and theirs), and do much more listening than talking. You may very well learn something useful.

TRADITIONAL CULTURAL PROPERTIES

Traditional Native American and Canadian First Nations cultural properties are a concern for those performing archaeological survey. In the United States, National Register Bulletin 38 (*Guidelines for Evaluating and Documenting Traditional Cultural Properties*) defines a traditional cultural property (TCP) as one eligible for inclusion in the National Register because of its association with cultural practices or beliefs of a living community that (1) is rooted in that community's history and (2) is important in maintaining the continuing cultural identity of the community. TCPs can include any type of historic properties, including cultural landscapes, which have significance in a community's or ethnic group's historically rooted beliefs, customs, and practices.

Special emphasis is often placed on American Indian communities and historic properties, but all groups may have TCPs. The identification of TCPs within a project area may require a systematic study to determine significance. A "reasonable effort" to identify TCPs should be implemented for areas that have the potential to contain such properties as determined by conducting background research on the history of the area and by identifying the cultural groups who either are living there or formerly lived there. Consultation with the identified groups should be implemented early in the planning process to determine whether the groups have any concerns about the project or activity.

CONSULTING WITH OTHER SPECIALISTS

If your project requires a consulting geomorphologist—and it does if you don't know much about soils or Holocene landscape development—have your consultant meet you in the field at the very beginning of the project. The geomorphologist should have sufficient regional knowledge and should use that knowledge to develop an appropriate investigative strategy. This approach will make the most out of the geomorphological data, providing you with information on the depositional contexts of the project area before you commit your crew to the field, and allowing you the flexibility to revise and refine the information during the course of your project.

Frequently, your geomorphologist will perform solid core sampling within the portions of your project area that hold the potential for containing deeply buried archaeological sites. This coring is typically accomplished using a Giddings or similar hydraulic drill rig (figure 5.3). You should assist the geomorphologist so that you can understand potential landscape/cultural relationships and so that you will

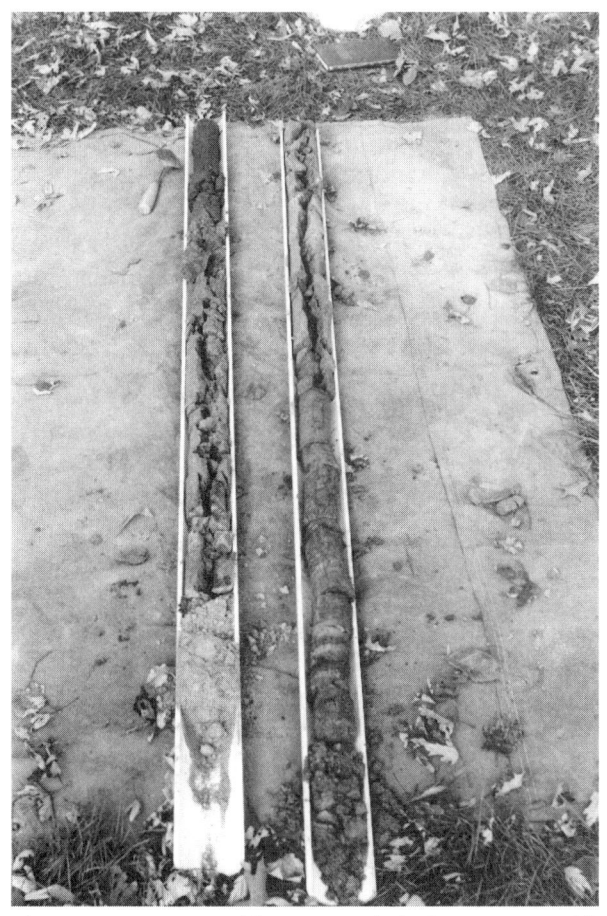

Figure 5.3 Your consulting geomorphologist may use a Giddings or similar rig to pull solid core samples from places that may contain buried archaeological sites.

be able to integrate the geomorphological data into your report. Your geomorphologist should address: the natural stratigraphy; the distribution of landscape-sediment assemblages; the environments of deposition and ages of the deposits (dated or inferred); the presence or absence of buried stable surfaces (soils), nonconformities, or gaps in the Holocene record; the nature of secondary alterations of the deposits (weathering and bioturbation); the degree of disturbance by people; and the distribution and thickness of historic age deposits across the project area.

Using this information, you won't waste time digging 60-centimeter-deep shovel tests in 120-centimeter-thick historic period alluvium. You'll know whether an upland surface is hopelessly deflated. You will understand whether and where, in your project area, you might discover the next Koster or Ozette site. These are good things to know.

On large infrastructure improvement projects, your archaeological crew may be only one of several teams of specialists (e.g., architectural historians, biologists, and zoologists) studying the cultural and environmental impacts of the project. You'll probably bump into these other crews in the field, if not at your motel or a local bar. These people will be working the same circuit that you are, so the opportunities for collaboration or mutually beneficial data sharing are great. Develop these professional relationships whenever possible. Trade information and literature. These sources will serve you well during fieldwork and when you write your report. Your relationship with these scientists also may lead to more work down the road in the form of joint ventures, contracts, or research opportunities.

SURVEY METHODS

As CRM developed, archaeologists and agencies increasingly grappled with standards and guidelines for archaeological investigations, especially surveys. In some states and provinces, guidelines have become mandates that prescribe in great detail acceptable and unacceptable field practice (e.g., requiring shovel test units every five meters in areas of low visibility, mandatory fifteen-meter intervals for survey transects). While such requirements are not universal, most states and provinces have developed some form of written guidelines for field methods. Familiarize yourself with the guidelines; be sure to get a copy from the SHPO or provincial archaeologist. Your employer also may have developed a manual of field procedures that you should read and apply.

While there is no shortage of guidelines and advice, in most places determining appropriate field methods for any given project remains, as it should, the prerogative of the project archaeologist. The reason for this is simply that no two survey projects are ever exactly alike. You have to read the landscape and determine what methods will be most effective in assessing its archaeological potential and how to go about finding sites. Rote compliance with any preordained set of generic guidelines is generally counterproductive because no set of guidelines can anticipate the specific field conditions you will encounter. Use your head, solve the problem, and do the work as efficiently as possible on a case-by-case basis. That is what you are being paid to do. (See Toolkit, volume 1, *Archaeology by Design*.)

You will have to describe and justify your field methods in your report. It is always easier to write about techniques that are conceptually systematic than it is to describe erratic, ill-conceived, or idiosyncratic methods. Be consistent in your field methods and your material recovery techniques unless you have a good reason to deviate from the norm. Avoid spur-of-the-moment decisions, such as telling one crew member to dig shovel test units in twenty-centimeter levels and another to dig in ten-centimeter levels. If you have reason to change methods, do so, but be prepared to explain why and how you did so. Always give yourself the time to make a rational decision. Bad choices will haunt you throughout the remainder of the project.

In other words, don't go off half-cocked. Survey leaders who don't have a plan waste time and effort: The crew stands around waiting for their leader to decide what they should do, or the leader, not knowing exactly what he or she wants done, impulsively rushes the crew into some unproductive task. Such "leadership" will elicit two equally unpleasant outcomes: Your crew will quickly realize you don't have a clue, and both your budget and schedule will suffer. You must have a plan for your crew each day before you even get to the project area. You can then jump out of the truck every morning and say, "OK, let's go do this thing," and your crew won't be left to scratch their butts and wonder how in the world you were ever put in charge.

In CRM-driven archaeological survey, your mission is essentially discovery and evaluation—fairly simple concepts. You use appropriate methods, governed by field contexts, to discover sites. Factors affecting selection of appropriate field method include vegetation, or lack thereof, and the geological contexts within the project area. Then you use appropriate techniques to evaluate the sites you discover in terms of their integrity and significance. Before getting into site evaluation techniques, let's outline the most common methods for site discovery.

SURFACE DISCOVERY TECHNIQUES

Almost all archaeological survey projects incorporate some form of pedestrian surface survey. Pedestrian survey is simply a walking survey of the project area. You walk along looking at the ground surface until you discover an arrowhead, a scatter of artifacts, or a foundation. But there is more to it than that. While you're looking for arrowheads, you must also, at all times, know where you are on the landscape, where your crew members are, and where your project boundaries are, all in reference to your project maps because your project area will rarely be well delineated in the field.

When is pedestrian survey an adequate technique for the discovery of cultural resources? Two main factors apply: ground cover and landscape stability and/or change.

For those who grew up on the Canadian Shield, the Pacific Northwest, or the Eastern Woodlands, it may be difficult to appreciate the fact that the ground surface is exposed to view, unobstructed by vegetation, across a large portion of North America. This was not always the case, of course. It would have been pretty difficult to see the bare ground surface in native prairie. But given postcontact of land clearance and intensive agriculture, during much of the year the land surface is all but naked in the Northeast, Midwest, Great Plains, and arid western states. There, survey seems simple: Look at the ground surface, and you are likely to see archaeological evidence from whatever groups may have previously occupied that landscape. Well, not so fast. Perhaps it isn't quite *that* simple.

The possibility of finding a site on any given surface depends completely on the age of the surface and how that surface may have been modified through time. Archaeological sites are depositional features within a dynamic natural landscape. Few landscapes, even upland landscapes, have been as stable during the past twelve thousand years as we might wish them to have been. Long considered stable or erosional throughout the Holocene, uplands are now being scrutinized more closely as areas where sites might be buried by local eolian, colluvial, and pedogenic processes (Bettis 1995; Bettis and Hajic 1995). You won't find any prehistoric sites by walking over fifty-year-old alluvial floodplain surfaces either. You must know the age and depositional environment of your survey landscape in order to determine whether pedestrian survey is a viable or adequate site-finding technique.

Circumstances when pedestrian survey is a preferred method, regardless of ground cover or landscape stability issues, include surveys geared toward the discovery of mounds, rock shelters, ruins, or petroglyph sites. Many historical period sites also exhibit ground surface features such as foundations, stone walls, cellars, or fortifications. Obviously, ground cover is not quite so material to the discovery of these types of sites, and one of the best ways to find them is to search on foot.

Look for rock shelters, caves, and rock art wherever bedrock outcrops along acute slopes, bluffs, mountains, and canyon lands (figures 5.4 and 5.5). Climbing and caving skills are extremely useful for such rugged, even dangerous, survey work. Never survey in such terrain without experience and proper tools: climbing rope, chocks, pitons, and carabiners, in case you need to tie safety ropes. If the rock is loose, wear a safety helmet. Caves and rock shelters can contain spectacular glyphs and other archaeological features, but watch for snakes and hornet's nests. Always check crevices, even those that seem too narrow to hold a site, as these were viewed by many native cultures as the homes of spirits or portals to the underworld and may contain ritual images or offerings.

Figure 5.4   Look for rock shelters wherever bedrock outcrops in bluff lands and mountains.

Figure 5.5 Check in and around rock crevices for petroglyphs and pictographs.

Don't ignore graffiti. You may be tempted to see it negatively, but it is really just a modern way of marking a special place. Names and dates left by explorers and other travelers may have historical significance. Try to look through, or past, historical period glyphs for the art of earlier peoples that may be camouflaged by the more recent graffiti. It is likely that a suitable geological medium will have been used for millennia.

An extremely effective method for finding and recording petroglyphs is working at night, using side light from a flashlight or lantern. You will be astounded at how much detail emerges in the light and shadow, so much so that you will forget the bugs and bats the light may attract.

There are two general rock art recording methods: photography and direct tracing. Use the best equipment and films for the photographs, as these photographs, not the images themselves, will be the evidence you will analyze later in the lab. If you plan to trace the images, you must be careful not to damage the rock surface. It may be fragile, and if you contaminate it with foreign substances, such as tape or chalk, you may affect the chemistry of the paint or surface, either destroying valuable information or hastening its destruction. A safe recording method is to press Mylar against the rock and trace the images with a fine-tipped black pen. (For details, see Whitley 2001.)

You may treat cave floors as you do other surfaces in a survey, but cave deposits may be much more complex to test and interpret because of their complicated stratigraphy and formation processes. Inspect any deposits in front of the shelter and confine your initial investigations inside the shelter to small samples, as the entire deposit may be relatively small. A shelter is more than a protected floor, so you will have to record its basic interior shape—a tricky job for archaeologists used to planar surfaces, but one accomplished in the same way, with measurements in all three dimensions.

If you find offerings in caves, crevices, or shelters, it is best to leave them untouched and, perhaps, undescribed, except in very general terms. Such offerings may be sacred to the people who left them, and you must respect their privacy. In the interests of protection, however, you should note their location as you would any other cultural resource.

When performing a pedestrian survey, be systematic about covering the ground. Work along contours in bluff lands and mountains. In more open landscapes, most people work in linear transects spaced close enough together so that it is unlikely that a significant site would be undetected between any two transects. Generally, pedestrian survey transect intervals vary from five to twenty-five meters, rarely wider. This is a somewhat subjective call, and you are the one who has to make it and, if necessary, defend it. What kinds of sites are you looking for? How large or small are they likely to be? These questions, and a large literature on survey and sampling methods, will help you design your survey transect plans (see, e.g., Zeidler 1995).

Your crew will take their lead from you. Take a simple example: a square, forty-acre borrow area. Start in one corner, spacing your crew at the determined appropriate interval, and walk parallel transects to the opposite end of the borrow area. There the person farthest from the starting point (call this person the anchor) holds his or her ground, while the other members of the crew walk past the anchor and establish the appropriate spacing again on the anchor's opposite side. Once everyone else is situated at the appropriate transect interval, the anchor walks past the rest, to one interval beyond the last person in line. The crew then walks parallel transects back to the starting baseline. This is repeated in both directions until the entire borrow area is covered. Unless people have convenient crop rows or another gridlike pattern to follow, they inevitably wander off track while walking their transects. Although you may encourage a certain amount of meandering along the transects, you must maintain the general integrity of the transect intervals by monitoring the progress of the entire party. It helps if you place yourself on a transect near the middle of the pattern and ask your crew to occasionally monitor their distance from you. This tip seems very simple, but you will be surprised how often people, head to the ground, wander astray.

It is even easier to wander off course in more irregular survey areas, such as highway or transmission line corridors that may themselves occasionally be somewhat serpentine. However, the principle is the same. Using basic orienteering skills and the landmarks shown on your topographic quadrangle map, you will navigate an appropriate course along the project corridor, and your crew will maintain appropriate parallel transect intervals relative to your lead transect. You must adequately survey your own transect and monitor the transect intervals of the rest of your crew, while performing map navigation along an unfamiliar and unmarked landscape. It isn't hard, but you have to be able to read a map. When it comes to herding your crew around, it doesn't hurt if your gene pool includes a bit of border collie.

Upon finding an artifact or feature along a transect, the lucky crew member generally hollers, "Flake," or "Point," or "Unexploded ordnance," or whatever is applicable to the situation, thereby alerting you and the rest of the crew that the game is afoot. At this point it is a good idea to convene the crew at the site of the discovery and perhaps mark the spot with a wire flag. Deploy your crew at narrow (two- to five- meter) transect intervals, and perform a methodical walkover of the area near the find. Crew members might verbally announce the discovery of other artifacts along their transects, or have them use wire flags to mark each artifact location. This method tends to use up flags pretty quickly, but it is helpful if you plan to piece-plot each artifact or photograph the scatter.

A variation on this technique is to have your crew walk along their transects until they are well beyond the site scatter, turn back, and flag the first artifact they encounter upon reentering the site. This results in delineation of the perimeter of the artifact scatter or site upon completion of your walkover. Once the perimeter is defined, you can then accurately measure the site by pacing, taping, or using a range finder. Draw a sketch of the site and any prominent nearby landmarks (distance to water, fences, etc.) in your field notebook, and write a narrative description of what you did and what you found. Using available landmarks and topography, plot the site location on your air photo and topographic quadrangle map. Obtain several GPS readings to aid in mapping and revisiting the site. Take some photographs, if there is something worthwhile to show. Collect any diagnostic artifacts you encounter; perhaps make a collection of representative material classes. Sometimes all surface artifacts will be collected during a survey, but remember, they'll all have to be curated. Make sure the benefits of making a total surface collection are justified by your research design and budget. Collection procedures may be spelled out in your research design. It is always a good idea to be consistent from site to site because it will allow you to make legitimate intersite comparisons during analysis.

At this point, you may have the choice of evaluating the site immediately or continuing with the survey, to return later to evaluate the site. Because site evaluation entails an array of equipment and a level of effort beyond what is usually necessary for pedestrian survey, we usually opt for the latter. Another reason for continuing with the rest of the survey immediately is that surface conditions may change radically during the course of a long survey project. Hopefully, you will have had the luxury of starting your survey in the spring of the year when surface conditions are optimal—but crops grow. Where crops don't grow, rank vegetation often does. You should complete as much of your pedestrian surface survey as possible when the surveying conditions are good. Ground cover is usually a less critical factor in site evaluation. However, if you are incorporating intensive surface collection techniques in your survey or evaluation design (e.g., surface collection in 2×2–meter units), you may want to go ahead and finish your surface collection immediately upon identifying site limits. (Site evaluation techniques are discussed in detail later in this chapter.)

Cultivated agricultural fields, badlands, and sagebrush flats aren't the only types of terrain that afford good surface visibility for pedestrian survey. Always, *always* inspect any stream cutbank you encounter. Don't just inspect the exposures; scrutinize them in the finest detail. Get right in and wade the stream. If a site is present, you may find artifacts sticking out of the cutbank or exposed in slumped deposits at the base of the exposure. Take a shovel, scrape a clean profile, and examine it (figure 5.6). On high or unstable sections, sometimes it is best to scrape the cutbank down in stepped sections, each a meter or so in height. Considerable digging may be involved, requiring the effort of your whole crew.

Work safely, though. Be sure the cutbank exposure is stable and won't slump while you're working on it. Dirt trickling down the face of a cutbank is one sign of instability. Dirt is heavy; you don't want to find yourself at the base of a cutbank in the middle of nowhere with four hundred pounds of it on your legs.

Identify discrete landform sediment assemblages and buried surfaces. Archaeological sites often are associated with buried soils that are invisible on the current floodplain surface. Sites in buried contexts often hold high research potential because they may be relatively undisturbed. Whether you find a site or not, draw and describe the cutbank soil profile. Minimally this will prove to your reviewers that you at least made the effort to identify buried soils and any associated sites. The more soils you describe, the more adept you will become at the task.

Mapping and describing soils are critical skills you must develop. We highly recommended that you squeeze in a soil genesis class while you're still in school. If you don't, you'll have to pick up an understanding of soils the way most archaeologists do: by paying attention when you have a geologist in the field, and through experience, lots of experience. A good place for the novice to start accumulating soils wisdom is Schoeneberger et al. (1998).

Figure 5.6 The archaeologist examines alluvial stratigraphy along a cutbank. Trowel marks spot where ceramics were found. Note dark soil band between shovel handles indicating a developed soil related to a former buried surface.

Shorelines, whether along a sea coast, lake, river, or man-made impoundment, generally provide good visibility for pedestrian, canoe, or boat surveys. Waves tend to erode beach lines, leaving lag deposits of deflated artifacts together with other flotsam and jetsam along the shore. Follow these lag deposits inland, beyond the destructive influence of the waves, and you may find relatively undisturbed portions of the sites that you first discover along the strand line.

Lakes and rivers were and are major transportation routes for most of the year. Pay close attention to the nature of the shoreline, looking for sheltered coves, portages, good fishing spots, and other natural resource areas that may have attracted occupation nearby, or features such as fish dams and weirs (figure 5.7). In areas such as the vast Canadian Shield, Algonkian peoples regarded the massive granite outcrops along lakes and rivers as the homes of spirits. Many of these striking formations have rock paintings and offerings, and there is much left to explore.

You may assume that the best water craft for a survey is a motor boat, because of its speed. But speed and noise reduce your ability to read the landscape. If you travel as the original inhabitants did, by canoe, you will begin to see as they saw, and this may lead you to a deeper understanding of how they used shorelines.

Perhaps it should go without saying, but don't forget that water can be dangerous. Two archaeologists once conducted a shoreline survey along the upper Mississippi River. They had just pushed off from where they had parked their john boat along the shore, below a U.S. Army Corps of Engineers lock and dam, when the locks unexpectedly opened and a giant whirlpool, big enough to swallow

Figure 5.7 Many archaeological sites are directly related to rivers and streams, such as mill sites, dams, and weirs. A fish weir near Amana, Iowa, is pictured here.

their boat, appeared about twenty feet away. One sat nervously, watching the vortex with dread, as his buddy fiddled to get the outboard started. Luckily, they escaped, but more than one archaeologist has drowned while doing survey, so be sure to take all necessary precautions for water safety.

Wherever you survey, don't ignore the smaller patches of bare ground afforded by cow paths, rodent holes, soil slumps, around tree trunks, and so forth. Many an arrowhead has been found in the middle of a cow path, and you never know when you'll find a trade bead or some other artifact on the piles of dirt around a rodent burrow, crawfish chimney, or anthill. It does happen!

SUBSURFACE DISCOVERY TECHNIQUES

When your project area affords limited or no surface visibility, you must make a good-faith effort to discover sites that may be present on the obscured surface. If an area has been plowed in the past, it may be worthwhile to have a fallow field plowed prior to your survey. Obviously, this method is contingent on landowner cooperation, but in situations where you are dealing with large tracts of functionally condemned land, such as in the case of reservoir construction, replowing your survey sample units may make very good sense.

By far, the most common survey method used in ground-obscured conditions is some form of shovel testing. This technique is suitable for situations where archaeological sites are expected at or near the surface, although its effectiveness has been intensively scrutinized and hotly debated (e.g., Kintigh 1988; Krakker et al. 1983; Lightfoot 1986, 1989; McManamon 1984; Nance and Ball 1986, 1989; Shott 1985, 1989; Zeidler 1995). Essentially, everyone realizes that shovel test sampling has severe limitations as a site-finding technique. In particular, it has been demonstrated that shovel test sampling is biased against the discovery of small sites or sites with few artifacts. Nevertheless, with few exceptions (discussed later), practical and economical alternatives for everyday CRM survey work have not been forthcoming, and shovel testing remains the technique most often used to find near-surface sites in low surface visibility situations.

Although shovel-testing techniques vary widely within and among regions, most archaeologists recognize a distinction between lower effort types of shovel testing, often identified as *shovel probes*, and higher effort, generally larger shovel test units. In general experience, shovel probes are expedient excavations of one, two, or three shovels full of dirt that are then hand sorted with a trowel. The higher effort units are more like miniexcavation units, generally between thirty and fifty centimeters square, and normally excavated in arbitrary levels deep enough to penetrate the B soil horizon. Fill from these more formal shovel tests is usually screened through quarter-inch hardware cloth using shaker-type screens. A given project's research design and specific field conditions normally determine the level of effort employed.

Any subsurface sampling strategy is most effective when done systematically. You must determine unit intervals and transect spacing appropriate to the sometimes-conflicting considerations of site discovery probability, scheduling requirements, and budget. There is no question that excavation of larger units, on tighter grid intervals, with fill screened through quarter-inch hardware cloth, will result in discovery of more sites, smaller sites, and sites of lower artifact density than a less intensive survey will find. Remember, though, your mission as a CRM survey archaeologist is, as the U.S. regulations state, to "make a reasonable and good faith effort to identify historic properties that may be affected by the undertaking and gather sufficient information to evaluate the eligibility of these properties for the National Register" (36 CFR Part 800.4a[2]).

Consult with your SHPO or provincial archaeologist to determine whether any guidelines exist relating to the level of effort required for shovel test surveying in your state. Unit and transect intervals of between ten and twenty meters seem to be most common. It has been demonstrated that use of a staggered or offset grid, rather than a square grid (figure 5.8) is most efficient for site finding (Krakker et al. 1983; Shott 1985; Wobst 1983). Be organized. You can assign letter designations to transects and number designations to units along those transects (e.g., transect A, shovel test unit 1; or transect C, shovel test unit 5). These designations can be referred to as unit A1 or C5. Sketch your project area in your field notebook and show the location of all transects and units, along with whatever designation you assign to each. Be sure to make note of any positive test unit on your sketch. You may also want to plot your transects and shovel test units directly on your project maps or aerial photos.

When you find artifacts (sites) by shovel testing, the logistics of artifact bag control are critical. Be sure that everyone on your crew is labeling their bags in the same way throughout the project. Use a Sharpie or some other indelible marker to label your bags, and on

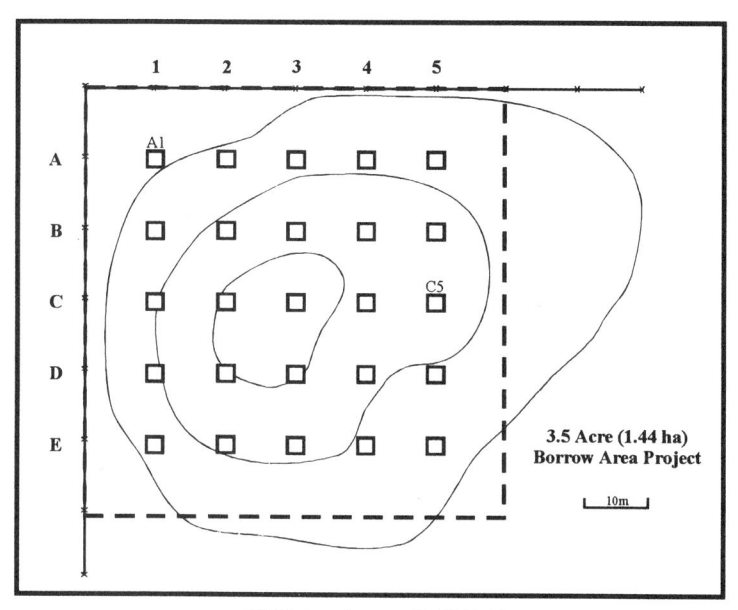

10 Meter Square Grid Pattern

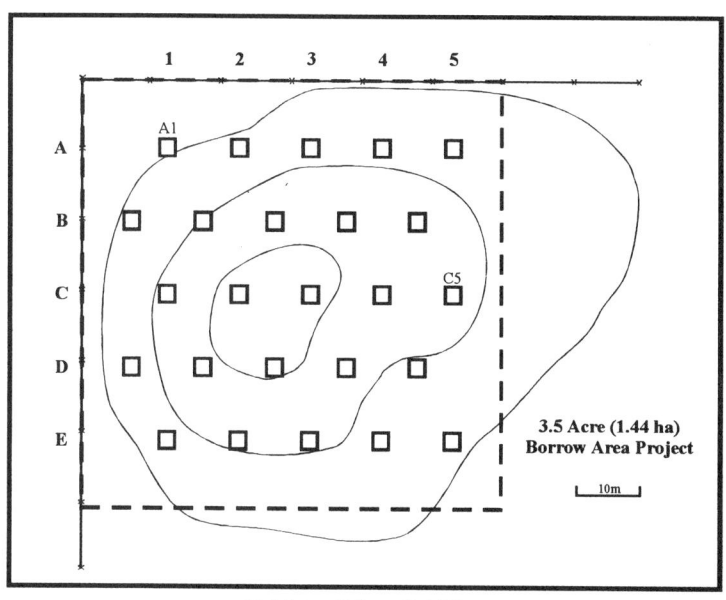

10 Meter Offset Grid Pattern

Figure 5.8 Schematic diagram showing square versus offset shovel test grid patterns.

every bag write all pertinent provenience information so that it can be cross-referenced with the data entered into your field notebook. Prevent hassles later in the lab by including the following data on every bag:

project number;
project designation;
county or township name;
survey parcel designation (landowner's name or legal description of parcel) or field site number (official site numbers usually are obtained from the appropriate state or provincial archaeologist's office immediately upon completion of fieldwork);
date;
excavator's name or initials;
transect or unit designation (e.g., shovel test unit D15);
level or depth (e.g., twenty to thirty centimeters); and
contents of the bag (e.g., all artifacts, projectile point, etc.).

Multiple bags (e.g., levels) from a single unit should be enclosed together within a unit bag, appropriately labeled. Periodically during the day, or minimally at the end of each day, all bags from that day's work should be logged on a bag list and stored appropriately for transport to the laboratory. If practical, store all the unit bags from a single site in one large bag, appropriately labeled. The extra effort will pay off in fewer headaches during analysis.

Normally, if you find artifacts during shovel test sampling, you should tighten the transect and unit intervals around the positive unit to try to define the site's horizontal limits. Carefully map the locus and record the circumstances regarding the change in grid interval. One way to allay confusion is to identify any such satellite excavations by a lowercase letter designation (e.g., shovel test unit G2a, G2b, etc.), and map them on an independent grid relative to the original positive unit, in this case G2 (figure 5.9). Try to put in enough units around the original positive unit to be confident that you have identified a sterile, nonsite buffer around all the positive units. This can be quite difficult, but often sites are confined by landscape characteristics (e.g., terrace scarps, bluff edges) that can be read intuitively. Experience in the field helps in this regard.

Shovel test survey through dense vegetation, such as second-growth forest, is among the most physically and technically demanding of archaeological survey techniques. Transect intervals have to be

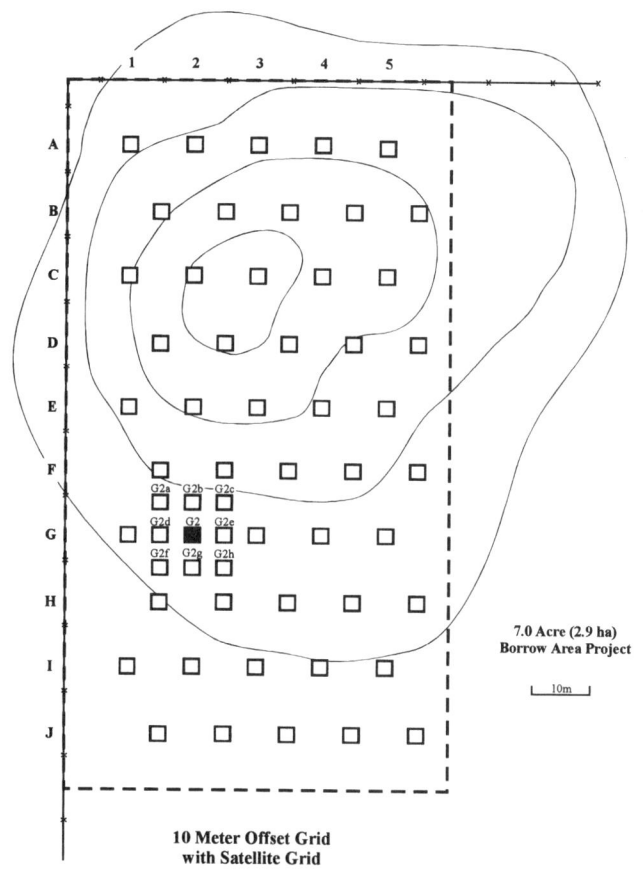

Figure 5.9 Schematic diagram showing typical offset grid with satellite grid around positive unit.

close enough for crew members to remain in visual contact, and great concentration is required to maintain grid integrity. It is advisable for every member of the crew to carry a compass, using it to stay on the appropriate directional tangent you wish to follow.

Shovel test profiles can be relatively easily drawn and described. The excavator should record soils data and any other pertinent data. Crew members can record this information in their own field journals, but it is easier if you develop a simple form that can be attached to a clipboard, with single or multiple generic, scaled, blank unit profiles on which the excavator can simply record the date and provenience data and sketch pertinent depths of soil horizons, soil colors, presence or absence of artifacts, and so forth (figure 5.10). The forms can be collected and filed at the end of each transect or at the end of the day, and they are easily reconciled with data you record in your field book.

Other near-surface testing methods popular in various regions include posthole testing and bucket auger testing (Abbott and Neidig 1993). The bucket auger is also a very useful tool for deep subsurface testing (see our discussion later). Both methods follow essentially the same logistical procedures (grids, screening, etc.) outlined for shovel test survey. Both methods are also useful for quickly evaluating sites you discover during your survey.

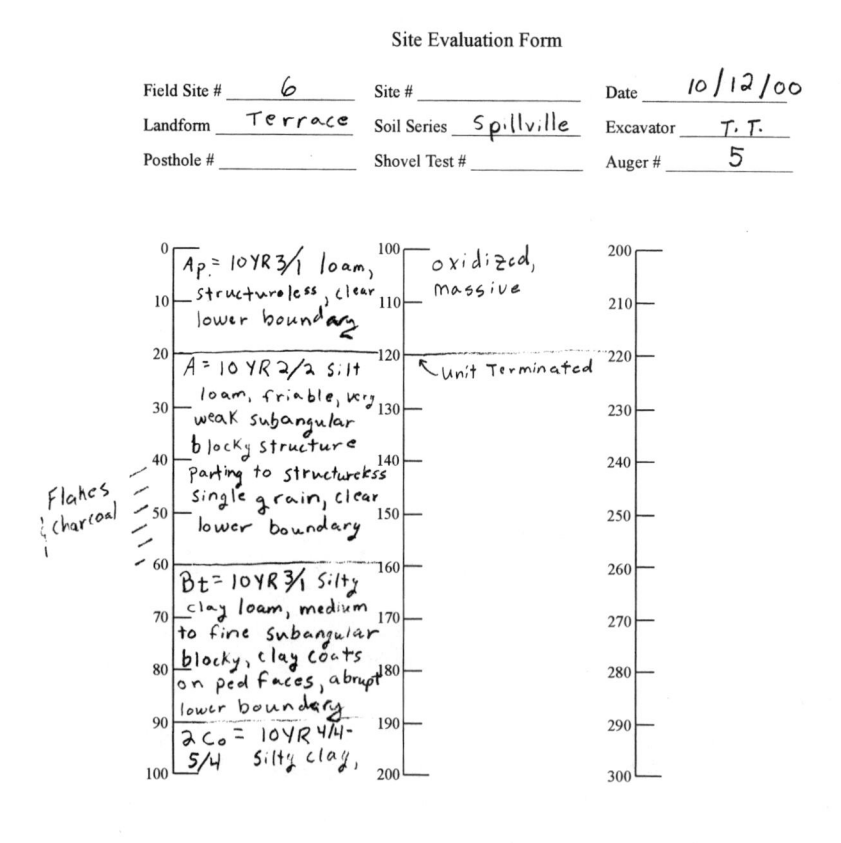

Site Evaluation Form

Field Site # __6__ Site # _____ Date __10/12/00__
Landform __Terrace__ Soil Series __Spillville__ Excavator __T. T.__
Posthole # _____ Shovel Test # _____ Auger # __5__

Column 1 (0–100 cm):
- Ap = 10YR 3/1 loam, structureless, clear lower boundary
- A = 10YR 2/2 silt loam, friable, very weak subangular blocky structure, parting to structureless single grain, clear lower boundary
- Bt = 10YR 3/1 silty clay loam, medium to fine subangular blocky, clay coats on ped faces, abrupt lower boundary
- 2Co = 10YR 4/4-5/4 silty clay

Notes at left: Flakes & charcoal

Column 2 (100–200 cm):
- oxidized, massive
- Unit Terminated

Column 3 (200–300 cm): [blank]

Remarks: Unit exhibits Late Holocene sediments (0–90 cm) inset over truncated Early-Mid Holocene sediments (90–120 cm). Flakes and charcoal recovered from the 40–60 cm level near the interface of A/Bt

Figure 5.10   Example of multifunctional subsurface evaluation form.

Posthole testing employs a standard scissors-type posthole digger. These tools have a cylinder-shaped bit that recovers a plug of earth fifteen centimeters in diameter and twenty centimeters in depth. Marking the side and handle of the tool with a ten- or twenty-centimeter scale using an indelible marker helps with vertical control. Excavators can comfortably excavate to a depth of about one meter in standard ten- or twenty-centimeter levels using a posthole digger. Posthole diggers are particularly well suited for quickly determining the degree of plow disturbance on sites and for locating near-surface buried soil horizons. Their limited volume per level makes them less useful than shovel testing for artifact recovery, and the smaller-diameter hole precludes detailed examination of side-wall profiles. However, it has been argued that posthole testing provides better systematic control than shovel probing (Abbott and Neidig 1993:42), and it is common experience that when the tool is in the right hands (there is a knack to it), the technique is more time- and cost-efficient than shovel testing.

It is important to match your subsurface investigation methods to the requirements of the landscape. Not long ago, archaeologists routinely limited their site discovery techniques to pedestrian surface survey or shallow shovel or posthole testing of their project areas. Through blissful ignorance, cynical disinterest, or simply a lack of adequate methods for subsurface testing, they paid little regard to the age or depositional environment of the surface being traversed. In most places, those days are over.

With the growing sophistication of geoarchaeological landscape settlement models (e.g., Bettis 1992; Bettis and Hajic 1995), archaeologists now realize that survey must be conducted in three dimensions. In response to the need to test certain landforms more deeply, many archaeologists now routinely use bucket augers as a primary subsurface testing tool.

The bucket auger is a manual digging tool that rotates horizontally as it penetrates the ground and collects excavated soil in a bucketlike bit (an excellent model that can be ordered through your local lumber company is made by the Seymour Manufacturing Company, Seymour, Indiana). The cylinder-shaped working end is twenty centimeters in diameter and roughly twenty centimeters in depth. The design of the tool facilitates recovery of controlled twenty-centimeter plugs of soil that can be extracted and screened in normal archaeological fashion.

The tool is particularly effective for sampling alluvial deposits and documenting the presence of buried archaeological remains. The handle and shaft of the bucket auger is 110 centimeters long and can accommodate a number of extensions of similar length. Alluvial fan deposits have been probed to depths exceeding six meters using a bucket auger with multiple extensions. Bucket augering is a particularly effective technique when your project traverses alluvial bottomland, as in the vicinity of most bridge replacements. For long traverses across bottomlands, simply excavate bucket auger units at appropriate intervals and to appropriate depths along one or more transects that parallel the traverse of your right-of-way. Transects such as these provide geological and cultural data that can be easily mapped for your report. The technique is useful for surveys of highway corridors, pipelines, levees, and many other types of projects.

In some parts of the country, bucket augering is considered the premier method for the discovery of buried sites in alluvial settings. The technique has a proven track record and can be used in any situation where a local landscape sediment assemblage requires a deep-testing survey strategy. Buried archaeological sites are often associated with buried, formerly stable surfaces in alluvial settings. Bucket auger testing can detect these buried surfaces and can document the presence of archaeological components that may be associated with them. Large projects with extensive deep-testing needs may benefit from use of a gasoline-powered drill rig such as a Giddings coring system. Sediment cores ten to twenty centimeters in diameter can be extracted, and deposits can be described, sampled, and collected quickly. You can also use a gas-powered screw auger for rapid subsurface testing.

Under careful, controlled, and safe conditions, backhoe trenching is a useful method of site survey, although some landowners will not allow it. Usually with your consulting geomorphologist, you can use a backhoe to explore terraces, floodplains, and alluvial fan deposits in order to identify and investigate depositional history and stratigraphic relationships. When buried soils are encountered, so too might be cultural materials or features associated with those soils. Such materials and features can be easily mapped along with the stratigraphic profile.

Backhoe trenches are inherently dangerous. Archaeologists in the United States must be aware of, and comply with, Occupational Safety and Health Administration (OSHA) regulations on excavation safety to ensure the safety of themselves and their crew members, and also to avoid fines and lawsuits if someone is injured. The supervisory archaeologist, or someone else on site, must be qualified as a "competent person" according to OSHA standards. See Mickle (1995) for a concise review of regulations concerning excavation safety and a pamphlet issued by the Trench Shoring and Shielding Association (1994) for a discussion of shoring techniques. More detailed information and sample plans for excavation safety is available on the websites of OSHA (www.osha.gov/) and the National Institute for Occupational Safety and Health (www.cdc.gov/niosh/). In Canada, the Canadian Centre for Occupational Health and Safety serves as a clearinghouse for information on relevant federal and provincial legislation and procedures (www.ccohs.ca/). It will be useful for you to review several safety plans before initiating your survey, especially if your work will take you to remote areas. Also, several archaeological excavation safety plans designed to meet OSHA standards have been published (e.g., Merry and Hedden 1995).

Learn about safety requirements before you commit yourself to a field program that includes backhoe trenching. In many cases, protective shoring systems will be required. Working in a backhoe trench is among the most deadly of occupational situations. Try to avoid becoming a statistic.

As noted earlier, cutbanks are valuable in survey. They offer greater exposure of sediment than can be obtained by any subsurface sampling technique. Additionally, stream cutbanks often afford continuous, sinuous, even serpentine exposures along an entire reach of bottomland. Used in conjunction with other sampling methods, examining such lengthy exposures frequently offers the best view of bottomland stratigraphy and is one of the best techniques for determining the presence or absence of buried surfaces and archaeological deposits, surpassing the usefulness of most other techniques in such settings. The horizontal and vertical extent of any site identified on the basis of cultural material observed within or along a cutbank should be defined by additional subsurface investigations.

Survey crews should carry one-inch-diameter Oakfield soil probes in their equipment inventory (Oakfield Apparatus Co., Oakfield, Wisconsin). These tools are useful for quickly assessing near-surface soil stratigraphy. In fact, the Oakfield soil probe is the essential tool of professional soil surveyors. Using Oakfield sediment probes can be an effective survey technique, particularly in mound or cemetery contexts. (In many jurisdictions, you must receive official authorization before initiating investigations into mounds and cemeteries; laws and procedures vary.) In known or suspected mound, cemetery, or site areas, close interval Oakfield sampling can document disturbances, which present as unnatural variations or unconformities when compared to the normal local soil profile. Atypical or disturbed profiles are characteristic of mounds, grave shafts, and other cultural features. Oakfield core sampling is among the fastest and least intrusive ways of documenting the presence of such features. But again, you must understand soils and be able to describe them in order to get the most out of Oakfield sampling as a supplemental survey technique. Also, bear in mind that soil probing is not an artifact discovery technique despite its value in soil-stratigraphic documentation.

Finally, a simple tile probe (a metal rod soldered to a handle) is also a very useful tool for subsurface investigation. This probe is most useful for following and mapping foundations, walkways, and other features that lie just below the surface at many historical period sites.

FIELD EVALUATION TO DETERMINE POTENTIAL ELIGIBILITY FOR THE NATIONAL REGISTER

Documenting the presence of archaeological sites is only half the equation of CRM-driven archaeological survey. Sites also must be evaluated to determine their potential significance based on NRHP criteria, or, in Canada, criteria for nomination to the register of National Historic Sites and Monuments or the lists of provincial heritage sites. In both countries, site identification, analysis, and mitigation are conducted in stages related to the nature and extent of information recovered.

In the United States, it is important to note the distinction between the identification of *potential* and *actual* NRHP significance. Ordinarily, at the reconnaissance survey (Phase I) level of investigation, documentation of potential significance is all that is required to recommend moving on to a more formal intensive survey or testing program (Phase II). The purpose of Phase II testing is to determine whether a site actually meets NRHP eligibility criteria. In some cases it may be obvious, based solely on Phase I evaluation, that a given site will meet NRHP eligibility criteria. In those cases, a recommendation for Phase II site testing would still be warranted to acquire additional information about the site to develop a mitigation strategy.

Properties eligible for the NRHP (as defined in chapter 2) must have *integrity* and *significance*. Minimally, Phase I survey must determine the horizontal and vertical extent of the site and its level of physical integrity. Ideally, you will also determine the site's age, cultural affiliation, and significance.

On sites that appear as a surface scatter of artifacts under conditions of good surface visibility, horizontal site boundaries can be relatively easily determined by walkover, as discussed earlier. Occasionally, a survey research design will call for systematic, intensive surface collection as an element of the site evaluation plan. Such programs assume excellent surface visibility and surficial expression of archaeological materials. A common procedure for systematic intensive surface collection is to establish a grid using lath or flags and tapes across the site area. Unit size within the grid typically varies from 2×2 meters to 10×10 meters, and crew members normally pick up and bag every artifact within each unit. Provenience data are recorded on each unit bag. Expedience sometimes warrants sampling only every other or every fourth grid unit or employing a more elaborate sampling scheme. On sites with low density of surface material, it is often more efficient to flag and piece plot every artifact, mapping their locations with a transit or total station. Maps of the surface collected materials often indicate features or activity areas: Concentrations of fire-cracked rock may suggest the location of hearths, or concentrations of daub may indicate the presence of houses. Such are the building blocks leading to a recommendation that a site is potentially significant.

For sites that afford less than ideal surface visibility or for sites that include buried components, horizontal limits must be determined by subsurface testing. Systematic subsurface sampling must be performed in nearly all cases to determine the depth of the site and the physical integrity of deposits. You must dig enough holes to determine the boundaries of the site and assess its degree of integrity, but you don't want to make Swiss cheese out of a potentially significant site that may be the subject of further study or protection.

Identifying physical integrity is a prerequisite to a determination of significance. Subsurface testing will allow you to determine whether a site is deflated, as is often the case in erosional environments, or confined to a disturbed surface layer such as a plow zone. In either case, a determination of ineligibility for the NRHP usually would be warranted.

Your subsurface testing program should identify evidence for in situ cultural remains if present at the site. On historical period sites, such remains include foundations, cellars, cisterns, privy pits, and the like. On prehistoric sites, in situ cultural remains include pit features, house basins, and a myriad of other feature types. The subsurface testing program should be comprehensive enough to identify buried surfaces if present and any associated archaeological components. This work is particularly important in depositional environments such as floodplains or alluvial/colluvial fans. Archaeological materials encountered on buried surfaces hold considerable potential for integrity and significance. With few exceptions, it is generally true that to be considered potentially eligible for the NRHP, a site must yield evidence of in situ, undisturbed features or other cultural materials. When Phase I evaluation procedures document the presence of such features or materials, consideration of the site as potentially eligible for the NRHP would be warranted at that stage.

So, just as when conducting your reconnaissance survey, your intensive survey must employ subsurface testing methods that are appropriate to the landscape (figures 5.11 and 5.12). Appropriate methods might be systematic, relatively close-interval shovel testing for near-surface archaeological sites, or bucket auger testing for more deeply buried sites. Your evaluation program should be systematic; units should be placed at relatively close intervals and unit levels should be screened.

Figure 5.11 The archaeologist defines an Archaic period feature within a shovel test unit during site evaluation. This feature provides excellent evidence of the site's physical integrity, a prerequisite to a determination of National Register of Historic Places eligibility.

Figure 5.12 Sometimes site evaluation requires excavation of larger, deeper test units.

One of the most difficult questions concerning site evaluation is the level of testing that is sufficient at the Phase I level. The answer depends on whether your survey is intended to evaluate the significance of each site or merely to locate the sites that are present and leave the evaluation to a later stage. Many CRM surveys are designed to identify all sites *and* to distinguish between those that are NRHP-eligible—or potentially so—and those that are clearly not eligible. Generally speaking, if you know the horizontal and vertical extent of the site, have documented the presence of in situ features or materials, and have enough information about the site to develop a testing strategy (Phase II), that is enough. If you're fairly sure of the site boundaries and believe that the site retains integrity, you can recommend further evaluation or (if consistent with your research design) conduct it yourself. If you have clearly demonstrated that the site lacks physical integrity, that is usually enough to recommend no further work. Ordinarily, you will know what your recommendation will be for any given site before you leave the field. If you aren't sure what your recommendation for the site will be, chances are you haven't done enough.

6
INDIRECT EXPLORATION TECHNIQUES

Circumstances may preclude direct subsurface exploration and testing. For example, excavations in known or suspected cemeteries, mounds, or ossuaries may be prohibited in some regions. Fortunately, many types of sites and landscapes can be productively surveyed and evaluated in minimally invasive ways. Archaeologists find geophysical methods of site prospecting to be viable complements to pedestrian and subsurface survey. This chapter presents a general introduction to some of the most useful techniques for noninvasive, indirect subsurface exploration.

Archaeological prospecting by indirect methods of subsurface exploration has been a long-featured element in European archaeological research, and North American archaeologists also are using remote sensing and geophysical techniques more frequently. As technologies have matured, these methods have become increasingly cost-effective for archaeological survey. It is certain that they will be used more, not less, in the coming decades.

Remote sensing using aerial photography has a long history in archaeology and is useful in many CRM contexts (e.g., Avery and Lyons 1981; Camilli and Cordell 1983; Limp 1989; Lyons and Hitchcock 1977; Maxwell 1983; Scollar et al. 1990; Weber and Yool 1999). However, the focus in this chapter is on geophysical methods—those that assist archaeological survey through on-the-ground detection of subsurface anomalies.

The science of geophysics as applied to archaeology can be described in one word: *contrast.* The reason geophysical methods work well in archaeology is that cultural influences disrupt magnetic and electrical fields, creating measurable contrasts (figure 6.1). There are limitations on all geophysical techniques, and lack of sufficient contrast is a fundamental limitation. In some cases, instrument sensitivity may not be adequate. Background noise is a real problem in some areas. Noise that might disrupt the ability of instruments to measure contrasts may come from the instrument itself; the operator (e.g., metal buttons on clothes); geological bodies or layers; or nearby features such as overhead power lines, railroads, and microwave transmitters. Yet, geophysical and remote sensing techniques can be used on a variety of scales to identify individual features within a site or to examine phenomena that occur on continental scales. In the context of archaeological surveys, these techniques can be used for both discovery and evaluation. A common scenario encountered by the survey archaeologist is the historically documented but long-forgotten cemetery that exhibits no visible surface expression. Certain geophysical techniques are useful in locating grave shafts associated with such cemeteries. As an evaluative technique, geophysical methods can help document the physical integrity of sites you discover during your survey.

Applying geophysical techniques in support of archaeological survey always requires site-specific planning. At what resolution will you expect to find features? Will you target individual artifacts, hearths, grave shafts, or stockades? How do you pick a method? Is the

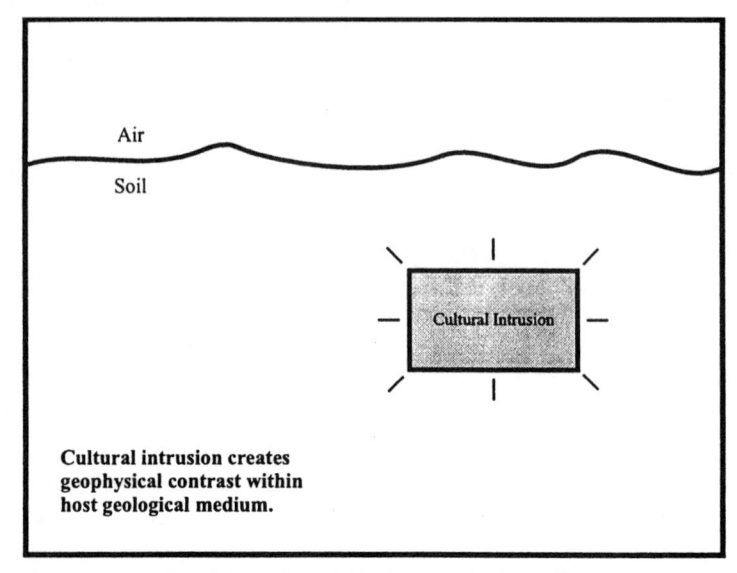

Figure 6.1   Cultural intrusion within host geologic medium.

cost/benefit ratio warranted based on your research design and scope of work?

Although more archaeologists are developing their own expertise, you will probably hire or contract with a consulting geophysicist to perform or direct the fieldwork and to provide you with a formal report of the investigation. Your geophysical contractor will consult with you about the central questions just noted, and together you will come up with an appropriate plan of action. It remains your responsibility to make sure your consultant knows exactly what you want to learn. Before even contacting a geophysical contractor, you must have a good idea of the local geology, geomorphology, soils, topography, depth to water table, weather, site access, potential maneuverability of equipment on site, what the target features will be, the contrast you expect and what signals there will be, the presence of noise-producing features such as power lines or railroads, knowledge of previous geophysical work in the area, and how much time and money you can expend on geophysical methods. Given that information, you and your consultant will be able to develop a workable, site-specific survey strategy.

Always do site-specific survey by design. How do you develop a survey design for geophysical archaeology? Determine what questions you want to answer and what data are available. Think about of what your targets are and what size they are apt to be. Use appropriate equipment for the anticipated target. For instance, you don't want to use electromagnetic equipment to find a nail, although you could.

Once you determine your anticipated target size, think in half-scale terms; that is, use a line and station grid that is half the size of the smallest feature you are targeting. Conveniently for archaeologists, the standard grid for geophysical methods is 1×1 meter. Make sure your geophysical data are collected in sufficient quantity and quality to be representative of the material and features present. If you have the time and money, a combination of methods may provide complementary data because of their differential ability to locate various types of targets.

Does a geophysicist have to collect the data, or can an archaeologist do it? Experience with the instrumentation is key. Some archaeologists possess limited experience with the instruments, which gives a feel for the equipment, but you will probably want to hire a geophysical consultant who really understands what's going on. Would you want a geophysicist to run an archaeological excavation?

As when working with any specialist, scheduling is critical. If possible, get in touch with your contractor a month or several months before you expect him or her in the field. Weather can be a real concern, so schedule contingency dates.

Costs of geophysical consulting are site-specific and contingent on transportation to and from the site, shipping costs for instruments, and the consultant's time. Similar to an old rule of thumb for archaeology, one full day in the field usually requires at least two days interpreting the data. Costs for your consultant can run up to $100 or more per hour when dealing with large geophysical consulting firms, but independent contractors usually run a fraction of that. You may want to negotiate an explicit agreement about billing for downtime due to equipment failure. Make sure you are paying for, and dealing with, experienced people who are familiar with archaeological problems and are comfortable working with archaeologists.

For comprehensive reviews of geophysical methods and equipment as they are applied to archaeological problems, see Clark (1990) and Bevan (1998). Heimmer and De Vore (1995) have authored a very useful complementary volume, and Weymouth (1986) and Wynn (1986) provide significant reviews of archaeological applications of geophysical methods. Techniques and equipment in common use by North American archaeologists are briefly described here. These techniques can be divided into active and passive methods. Active methods induce a measurable reaction from the host geological medium, and utilize a variety of electrical and electromagnetic instruments such as metal detectors, resistivity meters, conductivity meters, and ground-penetrating radar. The most commonly used passive method (i.e., one that measures existing geophysical conditions within a host medium—the soil), is magnetometry.

METAL DETECTOR

The metal detector is a generally underrated survey tool for archaeologists. It has had to overcome a bad reputation, derived from its use by treasure hunters to plunder sites for artifacts. But the metal detector is a very good tool for defining the boundaries of historic sites as well as activity areas within them (e.g., Connor and Scott 1998).

The sole purpose of the metal detector is to find objects. Metal detectors normally consist of a single or double coil, forming part of a tuned oscillator circuit. The presence of metal artifacts upsets this circuit, producing an out-of-balance signal (Clark 1990:121). Anomalous signals are registered by either audio (beep) or digital readers.

The equipment detects the resistivity of both ferrous and nonferrous metals, and the better instruments can be set to discriminate certain metals, such as brass, silver, or aluminum.

Metal detectors can be used as random or systematic survey tools. As always, it's best to work systematically on a grid. Transect intervals at three to five meters, resulting in about a 35 percent sample, are most common, but make your own site-specific decisions. Metal detectors are usually used in combination with a total station instrument. The standard procedure is to use a flag to mark each target, then to excavate as necessary to expose and recover the target. The total station is used to piece-plot the identified artifact. The total station data can then be downloaded into a CADD or GIS program to produce a site map. If you don't have access to a total station instrument, you can do the same thing a little more slowly using a transit and stadia rod.

The use of metal detectors has several advantages. Detectors are fairly inexpensive as geophysical instruments go, running from about $250 to $2,000. It is also a fairly fast survey technique. The 780-acre Little Bighorn battlefield was surveyed with eight metal detectors in ten days. On most projects you will be able to recruit volunteers among local hobbyists who are willing to help with their own equipment. The underlying physics and practical methods are easily comprehended by the average archaeologist, so specialized consultants are not needed.

The major limitation of the equipment is depth. Relatively unsophisticated equipment will detect objects only to an average depth of twenty to thirty centimeters. However, that depth is usually sufficient for preliminary work, as material on most historic sites is located within twenty centimeters of the surface.

Analysis of metal artifact distribution patterns can lead to sophisticated anthropological interpretations. A classic example of the archeological survey potential of metal detectors is the work of Scott et al. (1989) at Little Bighorn Battlefield National Monument. At many sites, though, metal artifacts are either so rare or so abundant that other survey techniques are more appropriate.

ELECTRICAL RESISTIVITY

This method measures the electrical resistance of the material properties of cultural intrusions vis-à-vis a surrounding matrix. These properties are measured by introducing an electrical current through current electrodes (C1, C2) into the subsurface medium and measuring reductions in current density near features and measuring the increasing potential gradient that is sampled by potential electrodes (P1, P2). The distortion of voltage resulting when a resistant object is encountered is measured by a voltmeter.

Results depend on the electrode configuration. Most archaeological applications are based on the principles of the Wenner configuration (Clark 1990; Wenner 1916) where two current electrodes (C1, C2) introduce current into the ground setting up a potential gradient, and two potential electrodes (P1 and P2) placed between the current electrodes measure the difference between the expected and actual potential. Clark (1990:37–48) reviews a variety of resistivity arrays, most of which are variations on the Wenner configuration.

The resistivity equipment most commonly used by North American archaeologists is a twin-electrode configuration, the RM-15, which was developed specifically for archaeological applications by Geoscan Research. The RM-15 includes two electrodes (C1, P1) attached to a voltmeter mounted on a single portable housing. A remote electrode (C2, P2) placed at a distance of thirty meters acts as a control (figure 6.2). An onboard data logger automatically records readings at each station along a grid.

Figure 6.2   RM-15 resistivity meter and twin electrode array.

With the RM-15, electrons enter the ground through the current electrodes, causing voltage to course through the soil in an electron scatter that exhibits radial geometry similar to the inside of an onion. These concentric rings are voltage contours that develop in a perpendicular plane relative to the introduced electron scatter (figure 6.3, left). When material or a feature with high resistivity is encountered, the feature distorts the introduced current and the distortion is read as the variance from 90 degrees, or the distortion from perpendicular between voltage contours and the introduced electron scatter. The surface contour is similarly distorted. These distortions (figure 6.3, right) are read according to the formula: Resistivity equals Volt measure over Injected Current (R=Vm/I). The variance is usually read as parts per hundred, read in ohms.

The depth of your investigation is governed by the distance between the twin electrodes. A horizontal probe (electrode) distance of one meter will provide readings to a depth of one meter. If you wish to investigate deeper features, you must move the electrodes farther apart. But if you go much deeper, you will encounter the limitations of the equipment. For deeper surveys, consider an alternative geophysical method such as electromagnetic conductivity.

Resistivity survey has many advantages. It is especially useful for identifying near-surface anomalies with high resistivity, such as

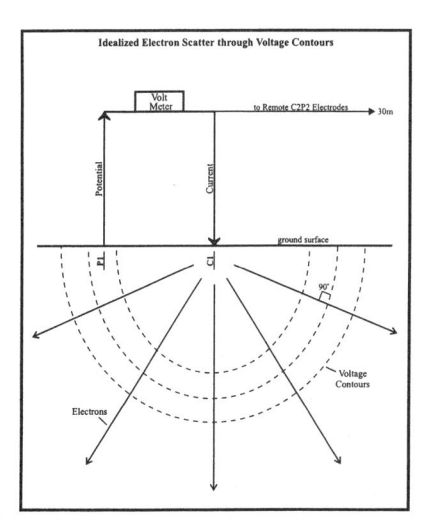

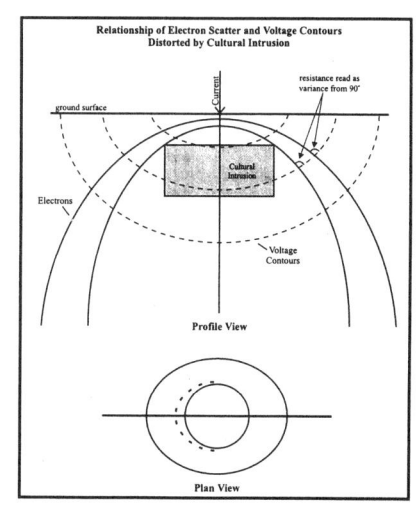

Figure 6.3 Schematic diagram showing idealized electron scatter through voltage contours, and relationship of electron scatter and voltage contours as they might be distorted by the presence of cultural intrusions.

fire-cracked rock features, hearths, stone walls, and filled excavations such as pit features or houses. The technique requires a straightforward systematic grid that maintains spatial integrity and is easily incorporated into existing archaeological baselines. The method is speedy and easily understood. With minimal basic training and adequate supervision, most archaeological crew members are capable of using the equipment and recovering field data. Under proper conditions, data can be recovered from a 20×20–meter grid in about thirty minutes.

Equipment built specifically for archaeological applications, such as the RM-15 with its onboard current source, voltmeter, and data bank, has made resistivity survey quicker, more cost-effective, and more straightforward in terms of analysis. Stored data can be downloaded at the end of the day, and the data can be converted into graphic formats such as profiles and contour maps using relatively inexpensive software. Your geophysical contractor usually prepares these graphics for you as part of his or her report of investigations. The data can then be used to guide management decisions or future site investigations. For these reasons, resistivity survey is becoming a relatively common component of CRM work.

ELECTROMAGNETIC CONDUCTIVITY

Conductivity is the opposite of resistivity: When electrical resistance is high, conductivity is low, and vice versa. Electromagnetic conductivity provides a measure of porosity. The electromagnetic (EM) array includes a transmitting antenna and a receiving antenna that measures introduced amps over resistance. That is, whereas Resistance equals Volts over Introduced Amps, Conductivity equals Introduced Amps over Resistance ($C = I/R$). Conductivity meters measure the conductivity of an electrical current through the subsurface medium, recording the measurement in units of millisiemens per meter (mS/m) (Bevan 1983:51).

The most common EM induction instruments used for archaeological work in North America are the EM-31 and EM-38 conductivity meters made by Geonics Limited of Mississauga, Ontario. Unlike the resistivity array, EM instruments do not have to come in contact with the ground. They are carried a few centimeters above the ground along established grid lines, and measurements of conductivity are recorded by an onboard data logger at either timed or stationed intervals. The EM-31 (figure 6.4) has a boom length of about four meters (between transmitting and receiving antennas) allowing investigation of subsurface anomalies that are at least one to two meters wide and within six meters of the surface (Bevan 1983:50). The EM-38 is more compact, with a boom length of about 1 m, and is handier for investigation of anomalies situated within a meter or so of the surface (figure 6.5).

Both instruments work on the same geophysical principle (figure 6.6). Simply described (following Bevan 1983:51), antenna (coils of wire) are situated at both ends of the instrument's boom. Current in the transmitting antenna creates a fluctuating (sinusoidal) magnetic field. The magnetic field penetrates into the ground causing electrical

Figure 6.4   Electromagnetic survey using the Geonics EM-31 conductivity meter.

Figure 6.5 Electromagnetic survey using the Geonics EM-38 conductivity meter.

currents to flow through underground conductors. The more porous the conductor, the higher the rate of current. In general, the ground itself is a good conductor. When a resistant or magnetic feature is encountered, it creates eddy currents that distort the transmitted field, generating a second magnetic field. The receiving coil (antenna) at the other end of the boom senses the second magnetic field. The magnetic field at the receiving antenna represents a weighted average of the conductivity of the ground near the instrument. Circuitry attached to the antenna distinguishes between buried conductors and buried magnetic objects. "This distinction is possible because the received signal from magnetic objects is proportional to the amplitude of the transmitted magnetic field, while the signal from conductors is proportional to the rate of change of the transmitted field" (Bevan 1983:52). As further explained by Dalan (1989:44), "The equipment divides the signal measured by the receiver into two components differing in phase by 90 degrees; one component is in phase with the signal generated in the receiving coil by the transmitter in absence of the secondary field, and the other is out of phase (the quadrature-part of the signal) by 90 degrees." EM conductivity meters can therefore be used in two ways (Clark 1990:34). When operated at high frequencies (e.g., forty kilohertz), they measure conductivity by examining the quadrature component that will respond to the conduction of moisture. This is useful for identifying sump features such as filled ditches, pit features, or pit houses. However, when operated at low frequencies (e.g., four kilohertz), they examine the in-phase component that responds to the magnetic susceptibility of ferrous objects, like a glorified metal detector.

Electromagnetic conductivity surveying is well suited to archaeological exploration in many of the same situations where resistivity survey is viable. Advantages of the EM technique include relatively high resolution of subsurface features and speed of data recovery. Because the instruments require no direct contact with the ground and no external wiring for remote electrodes, EM surveying is sometimes more practical than resistivity, particularly in brushy areas. EM surveying also requires less time and personnel than are generally required by resistivity surveying. Disadvantages are that the equipment is considerably more expensive and more susceptible to outside sources of interference, such as metal pipes, fences, and incidental

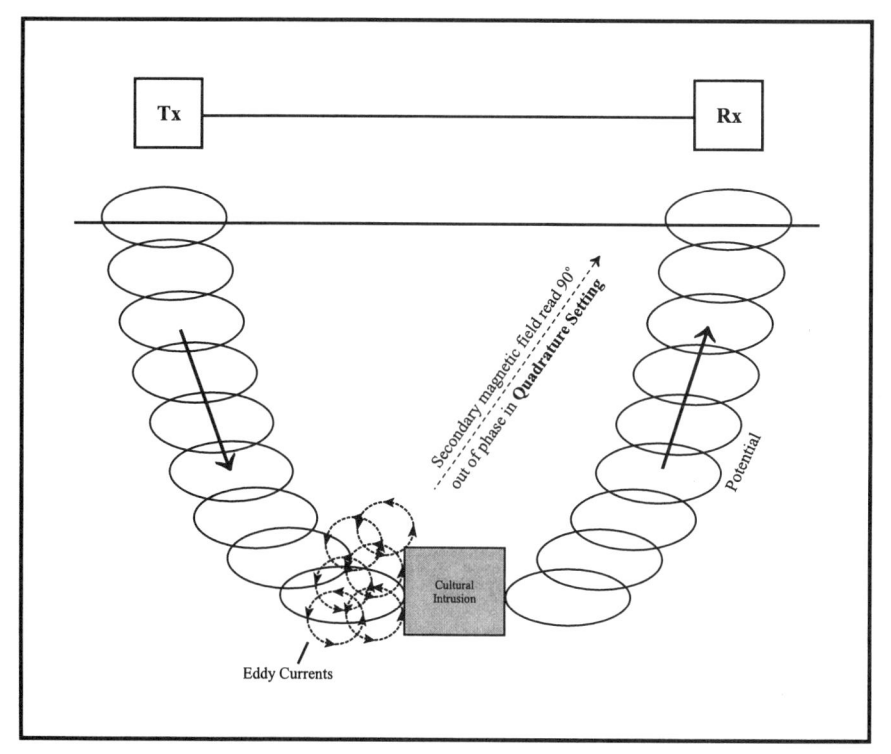

Figure 6.6 Schematic diagram illustrating the principles of electromagnetic conductivity surveying.

metal on the person conducting the survey (zippers, jewelry, nose rings, etc.).

Like modern resistivity equipment, EM conductivity instruments now include onboard data loggers, the contents of which can be downloaded at the end of the day. Software packages can then convert the data into conductivity profiles, or you can use various plotting programs to produce contour maps of sensed subsurface anomalies. Dalan's electromagnetic work at the Cahokia site provides excellent examples of successful applications of the technique (Dalan 1989, 1991; Holley et al. 1993).

GROUND PENETRATING RADAR

Although everyone is generally familiar with the concept of radar, Ground Penetrating Radar (GPR) is a rather complicated technology. Under ideal conditions, GPR can be a very useful technique for the archaeologist because it can detect a wide range of buried features and items (Conyers and Goodman 1997). Unfortunately, ideal conditions are rarely encountered in the real world. Of all the geophysical techniques, GPR data may be the most difficult to interpret. Ironically, GPR is often the first geophysical method many archaeologists think of using, but GPR really isn't justified in many situations. It is also a comparatively expensive method.

GPR equipment consists of a low-power antenna mounted on a chassis that can resemble a vacuum cleaner. Pulses of electromagnetic energy are transmitted into the ground at a frequency within the range of one to one thousand megahertz. A receiving antenna, also mounted on the chassis, measures reflections from subsurface interfaces and from objects that produce electrical or magnetic contrasts. Depth of objects is determined by the measurement, in nanoseconds, of the time it takes for the reflection of the transmitted wave to arrive at the receiving antenna (figure 6.7). The chassis is pulled across the surface of the site along a known baseline or grid traverse. As this is done, a continuous, real-time record of the subsurface readings is produced. A graphic plotter is attached to the antenna chassis by an electronic umbilical cord, generating an almost instantaneously gratifying scaled profile of the subsurface (figure 6.8). Under the best of circumstances, any archaeologist can see features in the GPR profiles. More frequently, however, the profiles can be virtually indecipherable to all but experienced geophysicists.

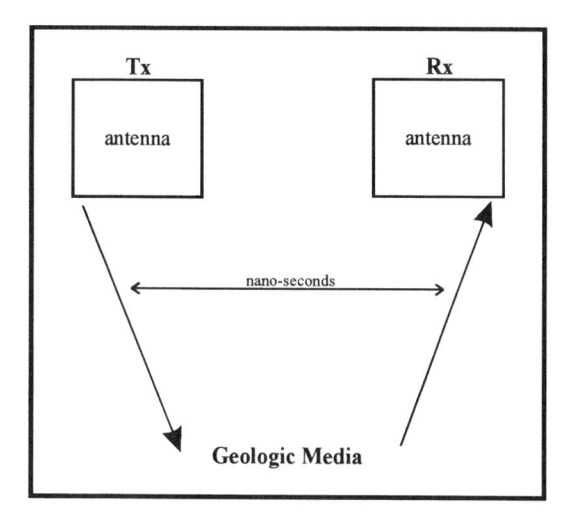

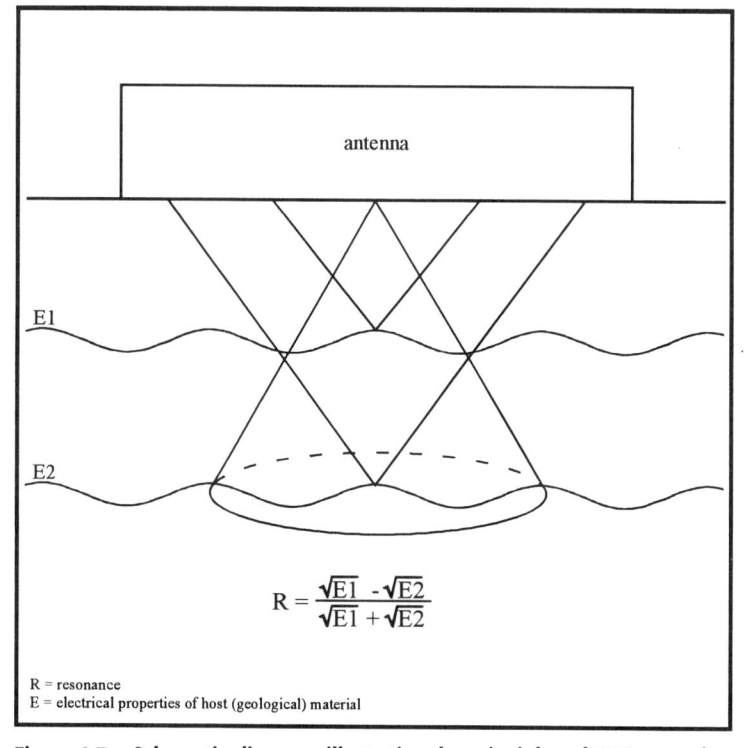

Figure 6.7 Schematic diagrams illustrating the principles of GPR surveying.

$$R = \frac{\sqrt{E1} - \sqrt{E2}}{\sqrt{E1} + \sqrt{E2}}$$

R = resonance
E = electrical properties of host (geological) material

Figure 6.8 Subsurface profile emerging from graphic plotter during GPR survey.

Anticipated target depth, the electrical properties of the target(s), and the three-dimensional geometry and electrical properties of the host geological material are considerations that will determine what size antennas are required. Generally, radar will work well when conductivity is below thirty millisiemens per meter. Low-frequency antennas will penetrate deeper but yield poorer resolution. The amount of local ground moisture will affect results. Wet conditions will produce a more focused scan, which is sometimes good, but very wet or saturated conditions can render the equipment almost totally ineffective. A host matrix of predominantly clay or saline soils is also a serious handicap for GPR.

GPR works best when applied to the problem of finding subsurface voids such as pipes and, in archaeology, features such as mounded tombs or historical period cemeteries. Foundations and other below-ground disturbances also can be detected. GPR is good at documenting the presence of grave shafts and especially coffins, because coffins represent a classic subsurface void (Bevan 1991). Only the best examples of GPR applications to archaeology find their way into the literature (e.g., National Center for Preservation Technology and Training 1998). The failures are rarely published. Use your geophysical dollar wisely.

MAGNETOMETRY

In North America, magnetic survey probably has the longest and strongest tradition of all the geophysical methods used in support of archaeology. Magnetic survey is effective because there is a magnetic field everywhere. The earth's magnetic field has a magnitude and direction that in the United States and southern Canada is approximately 70 degrees vertical and toward north (this is why compasses function the way they do). Cultural conditions or intrusions alter, distort, and modulate local magnetic fields by chemical, thermal, and mechanical processes. In prehistoric sites, it is the magnetic fields of the soils that are of interest. Distortions in these magnetic fields are measured at the surface with a magnetometer.

The most commonly used instrument for measuring magnetic fields on archaeological sites for many years was the proton, or total field, magnetometer. The magnetometer measures the Earth's total magnetic field in units called nanoteslas (nT). The Earth's magnetic field can range from seventy thousand nanoteslas in the polar regions to twenty-five thousand nanoteslas near the equator (Heimmer and De Vore 1995). In the Central Plains of North America, the Earth's total field varies from fifty-five thousand to sixty thousand nanoteslas (Weymouth and Nickel 1977:106). Against the background magnetic field, archaeological features may produce anomalies in a range from less than one to several hundred nanoteslas. An anomaly must exhibit at least 0.1 nanotesla of contrast to be detected by the most sensitive of magnetometers.

The background or total magnetic field can change through time, even during the course of a day. Magnetic distortion within a given project area can change by tens of nanoteslas throughout the day. This phenomenon is known as *diurnal variation*. Magnetic storms, which produce much higher variations, occasionally occur, rendering magnetic survey fruitless at those times.

Proton magnetometers are normally used on an established grid, with magnetic readings recovered at intervals of about one meter. No contact with the ground is required, and data can be automatically logged at each data point. At every data station, it takes about five seconds to read the immediate magnetic field. So essentially, it is a halt-and-go procedure. Random (diurnal) variations, intrusions (e.g., a passing train), and other "noise" must be controlled and removed from the data in order for magnetic surveys to produce desired results. Therefore, when doing a magnetic survey, you actually need two magnetometers. The second magnetometer is placed off-grid, normally within visual distance of the grid magnetometer, and acts as a control to measure the local magnetic noise. The noise is subtracted from the survey data, and, ideally, the remaining variations from the total magnetic field represent the cultural features for which you are looking (figure 6.9).

Another type of magnetometer, the fluxgate gradiometer, now is commonly used for magnetic survey in archaeology. The fluxgate gradiometer enjoys a major advantage over the proton magnetometer in speed of data recovery. With the fluxgate gradiometer, data sensing is accomplished in about one-thousandth of a second, so operation is virtually continuous (Clark 1990:69). The investigator can simply walk along established grid lines carrying the instrument. Instrument sensitivity is comparable to that of the proton magnetometer, and data are automatically logged as the surveyor walks (figure 6.10). Gradiometers measure the immediate magnetic field using two sensors mounted in a vertical mode. This array minimizes strong gradient influences and solar or diurnal effects, providing

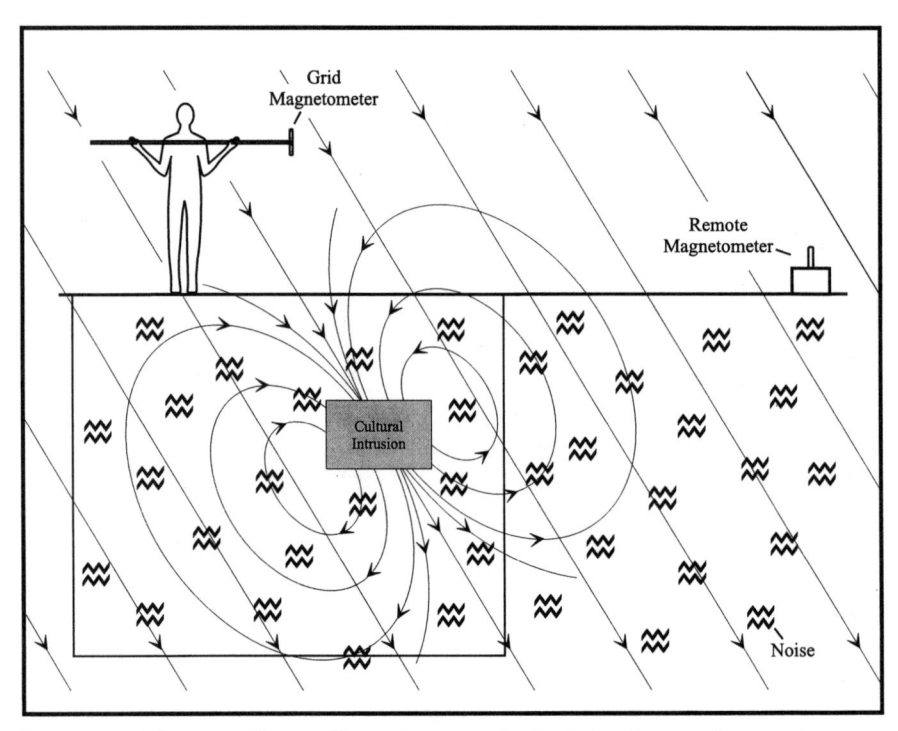

Figure 6.9 Schematic diagram illustrating general principles of magnetic surveying.

Figure 6.10 Magnetic survey using a fluxgate gradiometer. The archaeologist walks along an established site grid while the instrument automatically records data.

greater resolution of features and clarification of anomalies with greater precision (Heimmer and De Vore 1995).

A minor disadvantage of the fluxgate gradiometer is that it is highly directional. It measures only the component of the Earth's magnetic field parallel to its axis or along its length. The instrument must be held in a consistent direction as the operator traverses the grid. The proton magnetometer, by contrast, measures the total field without regard to direction (Clark 1990:69).

Geoscan Research has developed a line of fluxgate gradiometers that share characteristics of high sensitivity, compactness, and relative simplicity. These instruments include built-in data loggers and digital displays of grid, line, and measuring points.

As in many other geophysical exploration methods, data are stored in the field for later downloading. At the end of the day, the data can be dumped into a preprocessing program (you can use ASCII text files to edit data) and then transferred to a mapping program. During preprocessing, you can select for or against trends by using data filters. You may wish to use a high-pass filter for extreme contrasts or a low-pass filter for smoothing. You will end up with a plotted graph, gray-scale map, or contour map of your site showing magnetic anomalies that hopefully represent cultural features.

Weymouth's (1986; Weymouth and Huggins 1985; Weymouth and Nickel 1977) work on Plains Village sites provides a compelling advertisement for the use of magnetic surveying in support of archaeology.

MAGNETIC SUSCEPTIBILITY

Unlike most of the other geophysical sensing techniques, magnetic susceptibility does not sense the presence of subsurface objects or features. Instead, it measures magnetic conditions near the surface of the ground. Organic matter and bugs, worms, and other burrowing creatures create measurable magnetic conditions in the topsoil. These introduced magnetic conditions decrease with depth and are generally confined to the A soil horizon. Magnetic activity in the topsoil is conveniently increased by cultural processes, such as fires, that add organic material to the A horizon. Magnetic susceptibility therefore is useful for determining the presence and limits of surface anomalies that might be representative of midden or habitation areas within sites, rather than a method of identifying the location of individual features.

Clark (1990:99-105) reviews the complex marriage of chemistry and physics on which magnetic susceptibility is based. Suffice it to say here that in this type of survey, you are interested in measuring magnetic concentrations and the mineralogical signature of the topsoil. The standard instrumentation, currently a unit called the Bartington MS2, consists of a small electronics box, carried by the field surveyor, that is connected by cable to a canelike apparatus with a sensor attached to the bottom (figure 6.11). Power is supplied to an oscillator circuit at the sensor that generates an alternating magnetic field in the sensor coil. A probe at the instrument's distal end is inserted into the ground surface, and soil within the influence of the magnetic field (a few centimeters) will change the frequency of the oscillating current (Clark 1990:102). The magnetic susceptibility bridge coil measures alternating current frequency as it changes according to the magnetic susceptibility of the soil. The instrument is capable of data point or continuous data logging. Data processing and presentation are similar to that of the other geophysical methods discussed in this section.

Figure 6.11 Photograph showing a Bartington MS2 magnetic susceptibility bridge coil.

CONCLUSION

Geophysical prospecting techniques are used on a small but growing percentage of archaeological survey projects. There will come a time in the career of most archaeological surveyors when one or more of these methods will be appropriate and useful. If you have reason to believe you might encounter a cemetery or other site containing human remains, a geophysical survey program should be considered as part of your research design. In most instances, geophysical methods will be most useful during the site evaluation phase of your survey project.

Select your geophysical method carefully on the basis of anticipated targets. Consider what limitations your project area will place on types of equipment you can use. Budget realistically to account for your chosen method's rate of data recovery, effort needed in interpretation, and cost. Refer to Sidebar 6.1 for a quick summary of the comparative strengths and weaknesses of each geophysical method.

6.1. RELATIVE STRENGTHS AND WEAKNESSES OF GEOPHYSICAL EQUIPMENT

	Potential Targets	Rate of Data Recovery
Metal detector	Metal objects, historical period sites, battlefields	Fast
Resistivity	Near surface features, rock features, hearths, pits, houses, mounds	Moderate
Electromagnetic conductivity	Sump features, pits, houses, trenches, metal objects	Fast
Ground Penetrating Radar	Voids, grave shafts, tombs, coffins, foundations, cellars, cisterns	Moderate
Magnetometry	Subsurface anomalies, pits, houses, trenches, foundations, wells	Moderate
Magnetic susceptibility	Middens, hearth areas, habitation zones	Moderate

With a little work, it is easy to understand and use geophysical methods in support of archaeological survey. The process is straightforward, from field acquisition of data through download, processing, and visual representation (figure 6.12). Many archaeologists are becoming trained in these techniques. Skill in interpretation remains the key, and in most cases it is very important to interpret your results in consultation with a geophysical specialist. Colleagues can recommend geophysical contractors with successful archaeological track records in your region.

INDIRECT EXPLORATION TECHNIQUES 97

Interpretive Difficulty	Relative Cost	Limitations
Low	Low	Shallow depth
Low → moderate	Low → moderate	Thick brush trees, only moderate depth
Moderate	Moderate	Environmental interferences
High	High	Wet matrix clay, saline soils
Moderate → high	Moderate	Magnetic storms, diurnal variation, random intrusions
High	Low → moderate	Depth, feature definition

Geophysical Data Processing Flow Chart

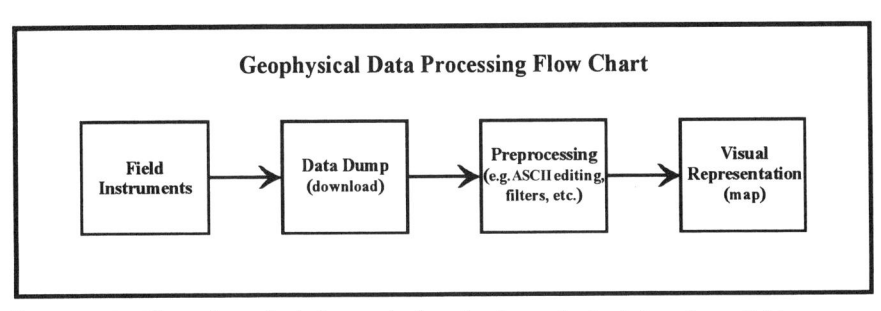

Field Instruments → Data Dump (download) → Preprocessing (e.g. ASCII editing, filters, etc.) → Visual Representation (map)

Figure 6.12   Flow chart depicting typical path of geophysical data from field recovery through graphical representation.

7
RECORDING AND MAPPING

READING MAPS AND AERIAL PHOTOGRAPHS

A map gives us an image of the world that we can never get from the ground. It reduces a vast and complex environment to a set of symbols in a manageable paper space, something we can fold up and throw into our field pack. Some maps of the terrain, compiled from satellite images or aerial photographs, can even take the place of a ground-level survey if you are looking for large features such as field systems, trails, or architectural remains. A map can be limiting, too. CRM archaeologists know a world of engineering drawings, of narrow survey corridors that bear little or no relation to the terrain, where the project area is an absolute boundary. We all carry a map in our heads, a vague but important aspect of our thinking about what we know as meaningful space. This is important because the people whose sites we seek had their own ideas of the landscape and their own ways of remembering and recording. We have to be careful not to impose our understandings of what might be a good place for a site on people with an entirely different way of life. Reading and mapping the landscape is therefore a practical and intellectual task. Maps and aerial photographs—and, increasingly, satellite images—are essential tools in every aspect of a survey, for preliminary research and planning, for field navigation, and for recording your results.

While you are planning the survey, you need to learn all you can about the land you'll be walking through. Understanding the terrain, vegetation, animal life, drainage features, soils, and bedrock geology of a place over the last ten thousand years helps you visualize how native and immigrant cultures may have used it. With this background knowledge, you can manage survey resources more effectively, as you can identify areas with greater site potential. You may even be able to identify sites before you set foot in the field!

Aerial photographs provide the most detailed view of the land surface. As noted in chapter 6, aerial remote sensing has a long and distinguished history in archaeology. Air photos using false-color infrared (IR) film may reveal the presence of buried features and sites. IR films detect variation in reflected heat (infrared radiation) and are especially effective in detecting shallow subsurface features that retard or enhance the growth of vegetation above them. Healthy vegetation is a strong reflector of infrared, appearing bright red on a photograph. Damaged or stunted vegetation will stand out, too, often as yellowish hues.

Photogrammetric mapping based on air photos also is an important archaeological survey tool (Creamer et al. 1997; Eddy et al. 1996; Fowler 1997). Even if you will not use them to make maps, air photos are useful tools to familiarize yourself with new landscapes. Chapter 4 discusses the importance of air photos in helping you understand your study area's physical setting. At first, air photos may appear to depict a flat and rather unfamiliar world. Where photographic images partly overlap, however, you can take two consecutive pictures (a stereopair) and view them with a binocular magnifier that will give you a three-dimensional image. With practice, you will be able to see what the actual ground surface looks like along with the general land use and features you want to know about in a survey, such as rock outcrops, eroded areas, buildings, and other structures. The 3D images often enhance the subtle changes in terrain produced by buried foundations, abandoned roads, and other cultural features not visible at ground level.

Satellite imagery captures whole regions in a single image ideal for use in large-scale regional studies of topography and ecology and for the detection of trail networks, settlements, and other large archaeological features. Multispectral images produced by the Landsat 7 satellite provide views of virtually all of North America at a resolution of three meters. The Canadian satellite RADARSAT-1, carrying a synthetic aperture radar (SAR) sensor that can penetrate clouds, darkness, smoke, and fog, produces stereographic images at resolutions as high as eight meters. RADARSAT-2, to be launched in 2003, will increase the resolution to a maximum of three meters. The U.S. company Space Imaging sells color images of North America that have a resolution of one meter.

Chapter 4 notes the importance of topographic, soils, geological, and plat maps. Use these maps to obtain further background on your study area and to help you make your own maps. The essential map is the topographic quadrangle. It shows basic natural and cultural features, including drainages and physical relief, tree cover, transportation features, and buildings. An area with closely spaced contours suggests a rugged landscape with steep slopes and deep gullies; farther apart, contour lines indicate a more gentle terrain. Note any conspicuous landmarks in your survey area. You may need them once you are limited to your ground-level view.

Soil maps give a close-up picture of the land's history because soils developed in specific environmental situations. Each soil type is a product of original matrix (e.g., glacial till, alluvial sediment, bedrock), vegetation, slope, and relation to drainage (upland, slope, river bottom). Geological maps may lead you to potential lithic sources, and county plat maps and other property maps preserve local settlement history. As humans are much the same in terms of their basic material needs whatever the millennium, don't be surprised if the ideal prehistoric site locality was, or is, the site of more recent occupation.

Once you begin fieldwork, your navigation skills come into play. It looks easy on a map. On the ground, however, your view is limited, and you may have a problem figuring out exactly where you are. Surveyors in much of North America thoughtfully marked the land in one-square-mile units, as you can tell from the checkerboard pattern you see from an airplane. Yet one square mile of featureless terrain can look like the next square mile, and there are dusty, unmarked roads from the Arctic to the tip of South America. If you are in forested or mountainous terrain, the problem may be that you cannot see anything but trees. Of course, you have a compass with you, and perhaps a GPS unit, so you may not get truly lost, but to know where you are, you need to be able to translate the scale of features on a map or air photo to the same features in the real world.

Here is the problem: If someone wished to mark your exact location on a map and he gave you the space of a real-world meter to occupy, it would be represented as one twenty-fourth of a millimeter, one of the standard 1:24,000-scale topographic maps published by the U.S. Geological Survey (USGS). This is the problem of scale. So what looks like a small, easily visible gully on a map may in fact be a broad valley quite unlike what you imagine. It takes practice to change your way of seeing. In difficult terrain, follow the map carefully, as one landform may look like the next. Your aerial photographs can help you find clearings or other distinctive features that can keep you on track. The terrain looks awfully easy to traverse when it is simply rows of contours on paper. Because USGS 1:24,000 maps are drafted to be accurate to within a twelve-meter range, don't rely on the map to locate anything or any place with greater precision.

Once you have found a site, you need to locate it on the topographic map. This can be difficult, unless you are next to features already marked on the map—once again, because of the problem of scale. Imagine you have found a small lithic scatter, about ten meters square. On a 1:24,000 map, this site would occupy a space of less than one-half millimeter square. Put a dot on the map to mark the site. An average HB pencil dot is a millimeter, and that equals about twenty-four meters, or almost eighty feet. But now that basic GPS units (discussed later) are priced for virtually anyone's budget, all you need to find yourself and your site is to take a reading, and you will have a location accurate to about ten or twenty meters (one pencil dot). With GPS error correction through real time or postprocessing, you can achieve accuracy of less than a meter. This is great for site mapping, but you can tolerate a less accurate unit for locating sites on the topographic map, as you have that twenty-four-meter pencil dot to work with.

Even with electronic equipment, you still need to read the landscape and know where you are on the map, as you may not be able to get a reading because of the terrain or positioning of the satellites. It is also good practice to locate the sites on aerial photographs. With air photos and quads in hand, revisiting the sites should be no problem.

DRAFTING MAPS

Once you have identified a site, it is time to draft your own map. What kind of survey are you doing? If it is a basic inventory, a quick sketch may do if you plan to return. You will need this map anyway for the site form that you submit for every site you find. A sketch map should include enough detail to convey a basic impression of the site in its setting, so you can find it again. You might be tempted to make a highly detailed map, using a transit or other precise surveying instrument, but you must remember the purpose of your survey. If the site will be revisited anyway, your main goal is to get the location right. This is relatively easy if you have a GPS unit, as you will be able to navigate back to very close to your original reading. If you don't, you will need to map to landmarks such as geological survey benchmarks, buildings, or permanent natural features. If you shovel tested or otherwise disturbed the site, you will also have to include these data (see chapter 5). It is especially important to record additional detail at this stage if the site is small and does not appear to be archaeologically significant because this may be the only time an archaeologist ever sees it. Your record will be the only record.

How you map the site depends on your survey goals. Most archaeologists learned at field school to lay out grid systems, oriented north, with squares anywhere from one to twenty meters a side. This type of arrangement makes sense in a survey if you are trying to study intrasite variations in artifact density or if you have a research design built around a systematic sampling technique. It may be less practical if you are simply trying to determine site boundaries and collect a small sample of material as a prelude to additional work. Remember that the main goal of survey mapping is to locate the site and map its general features and your subsurface testing accurately and to scale. You could place the squares randomly and superimpose a grid later—as long as you have the squares oriented consistently. Alternatively, why not simply establish a datum and record the bearing and distance (and, perhaps, the elevation) from that datum to the objects you want to map (whether natural or cultural features, or test pits), ending up with a pattern of rays? You can easily reconstruct the map back in camp or in the lab.

Surveying instruments range from simple compasses and tapes to electronic theodolites (total stations) and GPS units. The fastest and most precise field mapping today relies on the advanced electronic capabilities of GPS units and GIS computer programs. For a more than three-hundred-mile survey project along the Cheyenne River valley in South Dakota and Wyoming, the University of South Dakota (USD) Archaeology Laboratory used a GPS unit with real-time differential correction, accurate to less than a meter; a total station that gave azimuth, horizontal distance, and relative elevation readings at the touch of a button; and laptop computers with a GIS incorporating electronic versions of topographic maps and aerial photographs. Computer software automatically transformed the field data into electronic maps, and the field crew transmitted these (via e-mail or ftp) to the lab at USD, where they were added to a master GIS that produced all the paper maps for the project report.

Does this high-tech equipment make other methods obsolete? Not at all. You may not need submeter accuracy, especially if you are simply conducting a reconnaissance survey. One traditional solution is the Brunton pocket transit, combining a compass, sights, a level bubble, and a sighting mirror. If it is mounted on a tripod, it will help you make a suitably accurate field map. If you have the luxury of time, you may carry a plane table and alidade (basically a telescope mounted on a protractor sitting on graph paper on a flat board and tripod) and produce your finished map on site. The value of this nearly obsolete though real-time method is that you avoid recording errors, a tragedy if you only discover them when you are reconstructing a map from your data in the lab. A good compromise for work requiring more speed and less accuracy is a sighting compass. You look through a sight at your target, and the azimuth appears superimposed in the view. If you think the angle or degree of slope may be important, bring along a clinometer. The same goes for altitude. You can now buy a small, electronic altimeter with an LCD display that gives you the altitude, the current barometric pressure, a visual readout of the barometer over the previous twelve hours, symbols predicting the weather over the next twelve hours, the temperature, the date and time, a timer, and an alarm to wake you up in the morning! Perhaps the most effective way to create a simple map by hand in the field is to locate the site on an aerial photograph, as you can then easily translate this to a topographic map or GIS in the lab.

Sometimes you may adjust your mapping system for historic sites, if you are dealing with standing structures or foundations. It makes sense to record these in the English system, as that is the system used to build them. Jumping from metric to English systems is something you need to be comfortable with anyway, as most of your topographic maps still have contour intervals in feet. For prehistoric sites, it depends entirely on what you, or future archaeologists, plan to do with the information you collect.

Greater precision does not necessarily produce better understanding of the phenomena you are studying. Archaeology will always be a subtle intertwining of science and imagination. You need to produce a map that will bring life to both your analysis and your interpretation.

GLOBAL POSITIONING SYSTEMS

The essence of Global Positioning Systems (GPS) is triangulation, just like ground-level surveying, but instead using one terrestrial position and three or more satellite (or SV—space vehicle) positions. There are twenty-four SVs orbiting the Earth to provide these positions. Each satellite has an atomic clock that broadcasts a continuous signal of the time. The handheld GPS unit has an antenna (either built-in or external) that receives the time signal and compares it to its own internal time clock. The difference between the two is then calculated as distance, through the old Velocity × Time = Distance formula. Following the principles of triangulation, if you get directional data from at least three positions (ideally, four or more), you can compute the position of the GPS unit, rendered by the software into the coordinate system of your choice. Archaeologists tend to work with Universal Transverse Mercator (UTM) coordinates, because you can easily compute and mark your locations on a topographic map.

Giving precise locations is fantastic enough, but the GPS can do even more. As it is able to take positions approximately every second, the unit does not have to be stationary. You can use it to record linear features and areas, and, most important, you can instruct the unit to give you readings in relation to a predetermined point. This means you can record a site in the midst of a featureless plain and then navigate back to it easily in the future.

That's basically it—a matter of pushing a few buttons. If you are going to make effective use of the GPS as a mapping tool, however, you must know two important things about how the system works so you don't get into trouble in the field.

First, the system relies on access to satellite signals. You can't take readings any time you want, because the level of accuracy depends on the positions of the satellites relative to each other and to you, holding the GPS. Sometimes there are not enough satellites in view—usually because something blocks part of the sky. You have to have a direct, unobstructed path to at least four satellites.

The second point is accuracy of the signal. Sometimes, the available satellites are simply in bad positions—they may be too close to each other in relation to your antenna's ground position—so the accuracy of the readings will be low. This measure of position accuracy is rendered numerically in your GPS unit as a DOP (dilution of precision). Of several DOPs, PDOP (position dilution of precision) provides the most general estimation of potential accuracy. When the PDOP is too high (e.g., a value over 6), you'll just have to wait until the SVs move into better positions. Fortunately, as the orbits of all the SVs are known, you can (with appropriate software) obtain an hourly readout of the measurement conditions for any day in the future—even many months ahead. This way you can plan your observation times strategically.

When the time is right, however, you have another problem to deal with: If you take a single position with your GPS, all but the most expensive single-frequency instrument may give you an accuracy of no better than five to ten meters. You can improve accuracy by taking multiple readings at this position and averaging them, but you still may not achieve accuracies below several meters. The problem is errors in satellite signal transmission. These may be related to the satellites (clock timing and orbit errors), receiver noise, or signal interference caused by conditions in the ionosphere, storms and other atmospheric disruptions, or multipath effects caused by buildings or other obstacles that can influence the direct transmission of radio waves.

The solution: If you take your reading from a known point on the Earth's surface (e.g., a benchmark), you can calculate the error for that reading—simply the difference between where you know you are and where the instrument says you are. This is practical only if you have two GPS instruments reading simultaneously, one you are using to map (the rover) and another stationary in a known location (the base station). With two sets of readings taken at the same time, in roughly the same geographic area (up to three hundred miles, actually), you can calculate the degree of error for every second of observation time and correct your instrument accordingly, a process called *differential correction*. This is how it works: When you begin to record a set of positions, the GPS unit creates a file and names it with a set of numbers that indicate the year, date, and hour of the file creation. The data for each position also include the time, to the second. While you are busy taking readings, the base station is also creating files and recording positions every second. When you have obtained your readings and downloaded both the rover and base station files into the computer, a postprocessing program automatically matches the second-by-second readings, calculates the error, and makes the correction.

If you don't have the luxury of two suitable GPS units, you can still correct your readings by obtaining base station data from a separate source. Various governmental agencies and commercial outfits maintain their own base stations. They offer (free or for a subscription price) base station files to public users, either by ftp or, increasingly, from the World Wide Web. As long as the base station is within three hundred miles of your project, you should be able to obtain files with sufficient accuracy for differential correction.

The best GPS units have a built-in receiver that is capable of detecting the signals of base stations that broadcast the error correction continuously. If you are in range of a base station beacon or subscribe to the radio frequency of a commercial differential correction satellite (e.g., Landstar, Omnistar), you can obtain real-time differential correction. Using such instruments, you know where you are to a meter, right on the spot. This is clearly a great advantage when you are using the GPS to navigate back to a site.

Once you have the planning out of the way, you can get down to the mapping. All but the cheapest GPS units allow you to record data as points, lines, or polygons (areas). Recording a position is rather easy. Once you initiate a reading, the GPS will record positions (normally, one per second) until you stop it. With averaging, you can refine your accuracy by taking more readings, but unless you have time to burn, one or two minutes at a very important location, such as a datum, is enough. Lines and polygons are different, as the unit records positions while you are moving. Hence, all positions are individual and they cannot be averaged. If you have an expensive unit, this is not much of a problem, as the accuracy with differential correction will be less than a meter. If you don't, it may be better to take a series of single points along a line or around a boundary. Each of these can be averaged, and you can connect the dots when you make the map.

As the spatial accuracy of GPS is coarse compared to transits and other surveying instruments, choose carefully what spatial features to map via GPS, so that the information required suits the accuracy of the unit. GPS is ideal for plotting site locations and for work on sites where features are more than a couple of meters apart and absolute precision is not important. You may think that this eliminates much archaeology, but think about what precision means. If you are doing a basic inventory survey and shovel-test a site, and your readings are off by two to five meters, will this dramatically misrepresent your work effort or your interpretation of the site? It may, but only if your research design uses the precise location of these tests for some analytical purpose. If this is so, limit the GPS to locating your datum, and use surveying instruments to map your site.

Agencies see the world in terms of fixed boundaries, so you are always required to determine archaeological site boundaries. A GPS with the capability to record polygons is ideal, as the boundaries of most archaeological sites are fuzzy enough that the error is meaningless. One cautionary note: If the sky is blocked on part of your site, resulting in a loss of signal, the GPS automatically joins the first and last points you take. This can cause a great problem if the points are widely separated, as you will end up with a straight line between them. In such circumstances, using the line function is better, as it is open-ended. In both cases, always remember to mark the position that you started from. It is easy to forget to do so after you have trudged around in a big irregular circle.

If a GPS allows you to locate a feature to within a meter and return to this point at any time in the future, guided by its navigational program, what is the point of a much less accurate paper map? Well, you can make notes much more easily on a sketch map—important for discussion of the site environment—and a paper map can reside safely in your backpack, unaffected by battery failure or file corruption.

GEOGRAPHIC INFORMATION SYSTEMS

Another major innovation in mapping is the suite of computer-based Geographic Information Systems (GIS). A GIS is, essentially, a tool for displaying spatial data and fostering the analysis of those and related data.

The advantage of a GIS is obvious. Rather than being stuck with a pile of paper maps, each with a separate theme—history, topography, vegetation, elevation, physical relief, geology, hydrology, and so on—you can incorporate all the information in individual databases and define them as layers in the GIS. You might have a basic topographic map as your foundation, and layers of relief, drainage features, and the other desired map information above them.

The most powerful aspect of a GIS is its analytical capability. You can apply all the operations of a database, such as selecting records with a certain condition (e.g., in a field containing a count of artifacts in a test unit, selecting all the test units with more than five artifacts), and these will be distinguished by color from records lacking the conditions. Because the information is tied to geographic position, you can do more complex operations related to the spatial relationships of features. For example, you can select all the sites within a certain distance from water and above a certain elevation, in your project area or drainage. With such capabilities, the GIS is an immensely powerful visualization tool.

The most relevant aspect of GIS for survey mapping is that GPS systems will convert their locational data to the formats of the major GIS software, so that you can seamlessly incorporate the results of your own GPS survey directly into a GIS. In 1997 and 1998, for example, a Trimble Pro-XR GPS unit was an essential part of the USD Archaeology Laboratory Phase I survey of Devils Tower National Monument, Wyoming. During the survey, we took locations on every site datum, test unit, and linear feature and on a number of site boundaries. Back in the lab, we set up a GIS with a digital elevation model (DEM) of the USGS Devils Tower, Wyoming, quadrangle as background, contoured it at ten-meter intervals, and quickly translated all the differentially corrected Trimble files into the GIS format. In a matter of minutes, the entire season's work appeared in its precise geographic location within the National Monument boundaries. To this map we added a digital orthophoto quarter quadrangle (DOQQ), which is an electronic version of an aerial photograph, and a digital raster graphic (DRG), which reproduces a standard USGS 7.5' quadrangle. Finally, using the DEM, we constructed a three-dimensional model of the tower environment. With locational uncertainty gone, now all we had to worry about was what all these hard-won data really meant!

TOMORROW'S TECHNOLOGY

It is hard to look more than a year into the future of mapping technologies. Remote sensing satellites continue to be launched, each more sophisticated and sensitive than its predecessors, so we gain access to an increasingly diverse range of imaging products. In October 2001, DigitalGlobe, a commercial earth information company, launched Quickbird, a satellite that provides the highest-resolution imagery of any current space vehicle: 61 centimeters in panchromatic (grayscale) and 2.44 meters in multispectral (color and near-infrared) forms. Significantly, this information is freely available commercially—too costly at this point, perhaps, for most individual archaeologists, but certainly within the means of larger institutions and CRM firms.

Five months later, the European Space Agency launched a gigantic satellite, ENVISAT (ENVIronmental SATellite), designed to retrieve atmospheric and terrestrial data and to enable monitoring of changes in these systems. Archaeologists should benefit as this information helps earth sciences gain better understanding of environmental and climatic changes.

GPS units continue to get more precise, smaller, and cheaper. Many have detailed topographic base maps built in, making paper maps virtually obsolete in the field (as long as you have lots of batteries and a backup system!). No archaeologist can claim today that the price of a good GPS survey unit is beyond his or her means.

Also, GIS is expanding exponentially. While it has its roots in geography, GIS is becoming indispensable in such widely divergent fields as crime mapping and land use studies. Working archaeologists in the future will ignore GIS at their peril!

Of course, no matter how beautiful and seductive all this technology is, it cannot free us from the essential dilemma: We are still stuck between the earth and sky, trying to make connections between places and cultural landscapes that have long since disappeared. So, while it is always good to dream about the future, do it with your feet planted firmly on the ground!

8

THE SURVEY REPORT

You have had a great field project, found lots of neat sites, and enjoyed your crew, and all concerned wouldn't mind doing it again. But for you, the job isn't even half finished. In from the field, you may find yourself directing a laboratory team as large as your field crew. They might be the same people. There are artifacts to be washed and cataloged, analyses to be done, tables to be made, graphics to be drawn or created, and curation standards to be met (topics covered in Toolkit, volume 4, *Artifacts*, and volume 6, *Curating Archaeological Collections*, and therefore not considered in detail here). Finally, though, you must find a way to translate all the data you've assembled, from the beginning of your background research through your field, laboratory, and analysis programs, into a coherent statement of the effects your client's project will have on cultural resources.

The culmination of every field survey project is the report of investigations. This is the product you will supply to your client. Your client will pass it on to the SHPO or provincial archaeologist for review. In the United States, it may be passed on to the Advisory Council on Historic Preservation if the SHPO and your client disagree about management decisions based on your data or recommendations. It may even be entered as evidence in a court case. Therefore, you must present the results of your survey in the clearest, most concise way possible. No review and compliance officer likes to wade through pages and pages of fluff. There is no reason to wax eloquent in your background research section about any subject unless there is a compelling reason to bring up that topic for contextual, analytical, or other reasons. Cite existing literature, if necessary, to guide you and your readers to important background material. But focus on the most effective possible presentation of *your data*. That way, the overworked reviewers can make real-life judgments in the most expedient manner possible as to whether the project's legal mandates related to cultural resources have been fulfilled.

This is not meant to imply that you should not explore to the fullest any serious anthropological or archaeological problem to which your data are germane. Indeed, you would be negligent if you did not. Your client and the review agency need this information to make appropriate management decisions. Just don't fluff up your report. Your client and the reviewer will recognize unnecessary report padding for what it is—an expensive irritation.

The quality of reports that reviewers receive is uneven. Strive not to be among those who submit obtuse, even incomprehensible reports. Such reports are rarely accepted without further clarification, which invariably requires a time-eating, three-way dialogue. Such discussions often lead to delays and bad feelings. On the other hand, if you do the job correctly the first time, by presenting your data clearly and competently, the reviewer can quickly check off on the project or enter into timely consultation with your client regarding further work. Your report will have armed both your client and the reviewer with a clear understanding of all relevant archaeological facts regarding the project or site. Competence demonstrated in this way will make your client happy and willing to send you more work.

Volume 7 in the Toolkit, *Presenting the Past*, focuses on communicating archaeology to broad audiences of colleagues and the public. The present chapter therefore stresses preparation of the survey report to document an entity's compliance with cultural resources requirements—a necessary springboard for any further use of CRM-generated data and interpretations.

STRUCTURE AND CONTENT

Be systematic and organized when you prepare the project report. While certain fundamental content is needed for the document to move smoothly through the bureaucracy, there remains plenty of room for personal innovation in your analytical style, data presentation, and interpretive synthesis (given an interesting and productive project area).

The level of effort required for report preparation varies widely depending on the scope and results of the survey. No template can adequately express the range of variability encompassed by all survey projects, and sections of your report may be rearranged to better fit your data set, so what follows is a working formula. Many scopes of work and contracts specify the report's format, and most states and provinces have issued report guidelines, too. Discuss report organization and content with your sponsor and reviewer if you're receiving mixed signals form these sources.

TITLE PAGE

The title of your report should explicitly state the type of investigation and identify the project and its location. Identify the author by name and title. Most CRM enterprises have a numbered sequence to identify each of their reports (e.g., *XYZ Consulting Report #798*). As this will be one of the easiest ways to identify the report for future reference, the report's number and date of release should be provided on the title page. Also identify the sponsoring agency (your client) and, if applicable, the contract, job, or permit number of the project. The principal investigator, if other than the author (normally your boss), should be identified on the title page, along with your company's name and address. Some states and provinces may require a governmental review and compliance number on the title page of the document.

If you have a graphic in the report that quickly identifies the project, it may be appropriate to put that figure on the cover of your report. This is entirely optional and is rarely if ever done for small or negative surveys. However, if you have located a significant site or if the project is regional in scope and you feel your report may be frequently cited, a cover illustration is useful to future users of the document because it makes the report easier to locate on a shelf full of otherwise achromatic examples of "gray literature." If your cartographic skills are good, it doesn't hurt to occasionally show them off on the cover of your report.

FRONT MATTER

Substantial survey reports should include front matter including a table of contents that identifies page numbers for report headings, subheadings, individual site discussions, references, and appendices; a list of figures with page numbers identified for each; and a similar list of tables. Short documents (generally less than fifteen pages or so) reporting small or negative surveys may not require these pages.

ABSTRACT

The abstract is a very important component of the report and should always be included. Often, agencies will require a management summary at the beginning of the report, and that is essentially the same as an abstract. The idea of the abstract is to boil down all of the report's essential information, conclusions, and recommendations from the pages of your report into a brief synopsis. Reduce the content of each major section of the report into a sentence or two. Taken together, these sentences comprise your abstract.

Your client may never read beyond the abstract before shipping your report to the SHPO or provincial archaeologist. The abstract will be the first thing the reviewer will turn to in the report, so it should encapsulate the substantive issues covered in the project. Make the abstract a straightforward and credible summary of your report and the project.

The abstract should identify the undertaking, the type of project, and the purpose of the study. You should summarize the area surveyed in acres and hectars (2.47 acres = 1 hectare) and briefly summarize field procedures. Summarize any historic properties identified by the survey, their significance, and your recommendations for the sites. If you are recommending project clearance from an archaeological perspective, so state; if further work is recommended for any historic property, state that, too.

INTRODUCTION

The introduction should identify the contractor (your company) and the project sponsor (your client). It should specify the purpose of the study, the date(s) on which it was accomplished, and principal personnel (you, if you were in charge of the fieldwork). Your client may ask you to add a statement that the opinions and recommendations are yours (or your company's) and not necessarily theirs. Also state where materials and field records generated by the project are curated.

THE SURVEY REPORT

The introduction must include a detailed description of the undertaking—a clearly worded summary of the development project or other reason for the survey. Include a concise legal description of the project area (section or quarter section, township, range, and county), total project length (for corridor surveys), or total area (acres/hectares) surveyed. Provide a map, referenced in the text, that shows the general location of the project. Normally, the location map is drafted at a scale of 1:100,000 or smaller. Your second, larger scale map details the project location and area surveyed in relation to the surrounding topography. Site locations are also typically included on these larger scale maps. The USGS 7.5' quadrangle, printed at the scale of 1:24,000, is used almost everywhere in the United States as a base map. Most reviewers require that you depict your project area on a 7.5' USGS, NTS, or provincial quadrangle (figure 8.1).

Figure 8.1 Typical large scale map (1:24,000) showing location of project.

Describe the project area and any ancillary construction in the area of potential effect, such as borrow areas or stream channel modifications. The reviewers will want to know how you defined the limits of the project, so indicate whether you worked from detailed project plans, aerial photographs, or a verbal description of the project area in relation to existing landmarks. Some reviewers and agencies require UTM coordinates for the project area. As noted earlier, UTM readings can be obtained from your GPS; also, UTM grid lines are shown on the USGS 7.5', NTS, and provincial quadrangles.

PROJECT AREA DESCRIPTION, A.K.A. ENVIRONMENTAL CONTEXT

Since all culture groups must adjust to the physical environment they occupy, and archaeological deposits inhere in the natural landscape, environmental processes that form the landscape determine the conditions of the archaeological record and influence our perception of the human past. Therefore, a discussion of the environmental conditions relevant to both the past and present of your project area is an essential component of your report.

In this report section, reiterate what you learned about the project area's physical setting during your background research and what you determined about the physiography of the project area during fieldwork. Under a subheading such as "Regional Context," provide a brief discussion of regional landforms. The regional description includes a concise summary of bedrock geology, Quaternary landforms, topography, drainage patterns, and surficial deposits. This review can be followed by a general discussion of lithologic, stratigraphic, and age-specific relationships of Holocene Age valley fills in the area.

With this background established, you can move on to a more detailed, project area–specific consideration of physiographic conditions. You can use a subheading such as "Local Context" to distinguish this relatively detailed discussion from the more general, regional data provided earlier.

SMALL PROJECTS

For small projects in uncomplicated landscapes, the project area discussion might be a fairly simple description of local topography and terrain. If applicable, you can discuss hydrologic factors; for example, does water sit in closed depressions on the landscape, or is it drained by dendritic stream patterns?

Project area soils should be described and compared to typical profiles mapped in the county soil survey. What is the parent material for local soils? Till? Loess? Residuum? Did local soils develop under prairie, savanna, forest, or something else? You should attempt to explain any discrepancy between your field data and that published in the soil survey. Mapped boundaries of soil series are not always accurate at the site-specific level, and soil profiles might have been badly eroded since the survey was published.

Discuss recent and modern cultural and natural factors that bear on the physical state of the project area. The area may have been disturbed by processes such as earth moving, stream channeling, construction, or agriculture. Soil horizons of the upper solum (i.e., the A and B horizons) are often missing from upland settings, especially in agricultural areas where plowing has exacerbated erosional processes. In such circumstances, you may find artifacts representing the location of site, but they often appear only as lag deposits, in secondary rather than primary context. Such information will influence your recommendations about site significance.

Finally, you should discuss current project area conditions. Is the area plowed, in crops, sagebrush, fallow grasses, forested, or in some other kind of cover? What is the surface visibility, described as a percentage? Because ground cover influences your investigative strategy (see chapter 5), reviewers need details on ground conditions in order to determine the adequacy of your survey methods for all parts of the project area.

LARGE PROJECTS

Large projects that encompass several physiographic zones generally require more extensive discussions of environmental contexts. The more complex the physiography of a project area, the more complex may be the human adaptations to the landscape. Large survey projects can contribute important new data about any region. For these data to be presented most usefully, a thorough discussion of the physical context of the project area is necessary. Our advice is to view research and writing about the physical environment not as a chore but as a great opportunity to learn the physiography of the area. By understanding the environmental contexts of your project area, you will acquire information necessary for interpreting human adaptations to those contexts—subjects about which large, regional surveys can provide a great deal of essential data. Do this research for your own edification and so you can serve your client and the discipline more competently. Your report will be better for it.

Start with a discussion of the basic geology of the project area. Subsections under this heading should cover the bedrock and the Quaternary geomorphological processes responsible for the current landscape. You will have source material for these topics from having done thorough background research on the physical setting of the project area before you started fieldwork (see chapter 4).

Does the local bedrock contain culturally useful minerals? Are there chert-bearing outcrops where you might expect a concentration of short-term resource procurement sites? How might the Quaternary record affect human settlement patterns? Would inhabitants of the area have enjoyed life in a settlement on the edge of an oak grove, or would they have had to manage the challenges and opportunities of a wetland setting? How has the Holocene geomorphological record affected the visibility and preservation potential of sites in valley settings? Have relatively stable floodplain surfaces been destroyed by lateral stream migration, or are they merely invisible because they are buried by historic period alluvium? These and other important questions must be answered in this section of the report.

If you contracted with a consulting geomorphologist, this section is the place to incorporate his or her information. Incorporate the specialists' expertise as usefully as possible into the fabric of the survey and the report. Most consulting geomorphologists will provide you with a report of their investigations. You can attach the report as an appendix to your archaeological reports, but often when this is done, there is no attempt to integrate the geomorphological data with the archaeological interpretations. The archaeologists and the geomorphologists were simply talking past each other. One wonders, in such cases, why bother? The disciplines should work hand in glove. To get the best out of both, the geomorphologist's data must be fully integrated into the "Environmental Context" section of your report.

Use high-quality illustrations liberally to help clarify important points. Such graphics can enhance your narrative presentation immeasurably (figure 8.2). Be as creative with visuals as you like, as long as they augment your discussion. Poor graphics are both pitiable and the mark of inexperience. Don't try to cram too much information onto any one illustration. It is better to make two maps that clearly support the ideas you want to express than to give the reviewer a

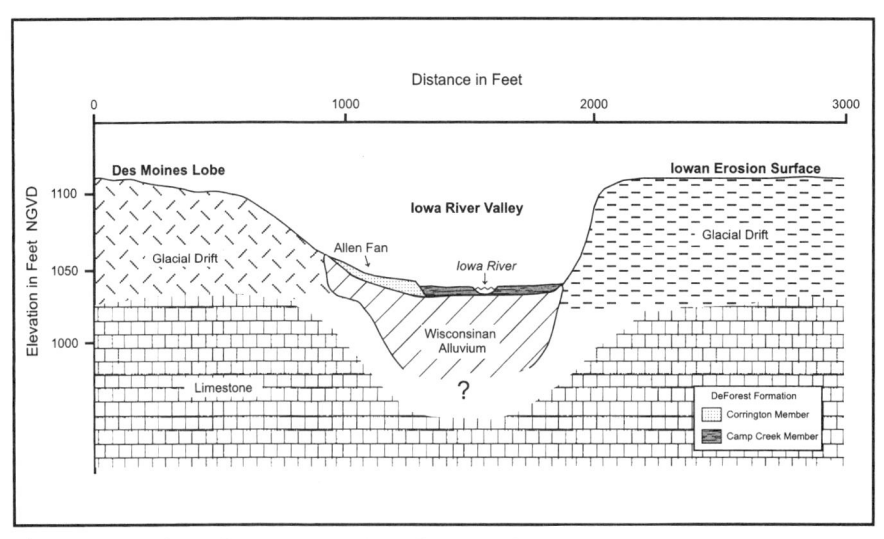

Figure 8.2 Schematic representation showing the relationship of Quaternary and Holocene deposits along an Iowa River valley transect.

headache by forcing him or her to distinguish among several data categories squeezed onto a single map.

After laying the environmental groundwork in the geology section, you can move to subsections on hydrology, soils, climate, vegetation, fauna, and the modern environment. The sections on climate, vegetation, and fauna are particularly important. Changes in climate and vegetation are frequently synchronous or nearly so. Culture groups that occupied the region at different times may have experienced and adjusted to starkly different climatic and biotic patterns. It can be useful to map these patterns diachronically because they provide excellent clues to long-term changes in settlement patterns (e.g., Collins 1991; Warren 1982). A solid discussion of environmental contexts is not fluff; as noted earlier, survey data are extremely useful for settlement pattern studies and in identifying changes in human responses to environmental variability.

In summary, the "Project Area Description" or "Environmental Context" section of your report will fix the physical background against which you will apply your data. The larger the scale of your project, generally the greater the detail you should present in this section. You should learn and present these data so you can make informed judgments and interpretations of your survey data, not because it is a bureaucratic requirement or a rote exercise. Pay attention to what you are doing. This is the essence of research, even in a CRM environment.

PREVIOUS ARCHAEOLOGICAL INVESTIGATIONS

A formal heading for this topic may not be required for most small archaeological survey projects. Generally speaking, and there are exceptions, the smaller the project, the less likely the need to discuss a large body of previous literature in detail.

For the large survey project, the opposite is the case. Just as you should provide a detailed environmental review for a large or complex survey area, a review of previous archaeological work in the region will set the stage for your new data. Your project should supplement, expand, or modify existing knowledge about the region, but it can't do this unless you know what has already been learned.

If you did a good job in your background research on the cultural setting (chapter 4), this is the place to reiterate what you found in the literature and archives. Depending on your project area, pertinent literature may go back one hundred years or more. A review of nearby sites that are recorded in the official state or provincial site file may be warranted. Who recorded the sites? Are temporal or cultural parameters for these sites secure, approximate, or negligible? Check the National Archaeological Data Base and state and provincial bibliographies for lists of relevant CRM reports. Read those reports and cite them here. Also review any other unpublished manuscripts such as theses and dissertations, as well as all relevant published archaeological literature and unpublished records and archives you consulted. Finally, discuss any local properties listed on the National Register of Historic Places, Historic Sites and Monuments register, or lists of state or provincial heritage sites. Often it is useful to supplement your discussion of the literature and previous archaeological investigations with a summary table.

RESEARCH DESIGN

Small projects are routinely conducted without the benefit of a formal, explicitly defined research design. Therefore, a separate heading to address the topic would not necessarily be found in all project reports. Still, your research design is an essential element of your project (see Toolkit, volume 1), so be sure it is presented in the report's introduction, at least.

For large survey projects, a formal section outlining the research design should be included. In this section you reiterate the research design that governed the project. State what questions or problem domains the project was to address and your assumptions about the data. If you used any kind of predictive models to guide a sampling strategy, identify them. Describe how your sampling strategy intended to obtain data pertinent to the identified problem domains. If your research design was modified during the course of the project, discuss that here.

METHODS

All reports must include a "Methods" section. For many small survey reports, it is convenient to subsume "Previous Archaeological Investigations" and "Research Design" sections within the "Methods" section. Regardless of whether the project is large or small, your "Methods" section must explicitly describe what you did to accomplish the survey. Make clear each method used during every step of the project. You will realize how important it was to have used systematic methods throughout the project, as it is always easier to write about systematic methods than those that are not.

Review the historical documents and other resources you checked, whom you consulted for background information (e.g., your landowner interviews, or discussions at the local historical society), photo collections or map libraries consulted, and any other pertinent methods used in conducting background research. If you sent mailings to landowners, discuss the particulars here.

Describe all of the field methods you used and the circumstances under which you used them. Discuss your pedestrian survey techniques, shovel tests, posthole tests, bucket augers, backhoe trenches, and how you investigated cutbank exposures. How did you define site limits? Report your artifact collection strategy. Did you recover all artifacts or a sample? Why? Define the strategy you used for site evaluation. Did the strategy differ for prehistoric and historic period sites? If so, how?

Provide details on your laboratory and analytical methods. Did you size-grade your material categories? Weigh them? Why? Discuss the classification and technological parameters used for analysis of prehistoric and historic artifacts. Did you attempt to identify the sources of the lithic material you found? How did you accomplish this? On what did you base your ceramic classification? Published sources? Comparison with extant collections? Spare no details on your methods and techniques—these are essential data for follow-up studies locally and in the general region.

Define the categories you use to organize your data on artifacts, features, and sites. Cultural categories may derive from temporal or developmental frameworks standard in your region. Functional categories can include a wide range of classifications based on any of numerous systems (see Toolkit, volume 4). The fact that you will normally be dealing with data from surface collections and limited subsurface testing often renders your interpretations tentative, but this should not deter you from describing the taxonomies you use.

Finally, state the criteria used for determining the eligibility or ineligibility of sites for the National Register of Historic Places or Historic Sites and Monuments register. Criteria applied to particular sites are discussed on a case-by-case basis in the "Results" section of the report.

RESULTS

This is where you finally get to write about what you actually found during the course of the survey. Begin with a discussion of any previously known sites in the project area. Did your examination of GLO records or other historical documents identify cultural properties in or near the project area? If so, discuss these sites. Provide illustrations keyed to the text that indicate the spatial relationship of the project area to previously recorded sites and to properties identified from historical sources. Discuss any pertinent information gleaned from local residents or from other interviews. Essentially, start this section by covering any significant information you gleaned about the project area while doing your "cultural setting" background research (chapter 4).

After that, simply present the field results—what you found on and in the ground. If the survey found no historic properties, this should be clearly stated, and you can skip to the "Summary and Recommendations" section of your report.

Under separate headings, discuss each site recorded by the survey. In the United States, headings should correspond to the site's designation as provided to you by the curator of your state's official site file. A trinomial designation used in most states identifies the state's alphabetical rank (e.g., 13 = Iowa), the county abbreviation (WA = Warren), and the order in which that site was recorded in that particular county (532). If there is reason to associate a name with the site, you may include the name in the heading of the site discussion (e.g., 13WA532, the Townsite of Wheeling). In Canada, you must number all sites according to the Borden System, a grid system based on geographic coordinates. The country is divided into coordinate blocks identified by two uppercase alpha characters, forming a grid sequence that runs from south to north and from east to west; following each uppercase alpha is a lowercase alpha that designates a subdivision based on latitude and longitude. Sites within each lowercase block are assigned sequential numbers by the Canadian Museum of Civilization or the responsible provincial ministry (e.g., AdHk-1, the Hind site).

In the first paragraph of each site discussion, identify the legal location of the site in reference to section, township, and range wherever that system is used (e.g., S½, SW¼, SW¼, SE¼ of Section 32, T77N-R21W; or NW¼, and in the NW¼, NE¼, and N½, SW¼, NW¼, NE¼ of Section 5, T76N-R21W). Provide a figure reference to a map showing the site in relation to the project area on the appropriate USGS 7.5', NTS, or provincial quadrangle. Provide specific UTM zone, easting, and northing coordinates for the site (e.g., Zone 15, E474940, N4585240). You may provide UTM coordinates for the site's center or perimeter, as your scope of work, sponsor, or reviewer may require. Also provide the site's elevation in relation to mean sea level.

Next, specify the basis on which you identified the site. For example, did you observe a surface scatter of prehistoric or historic artifacts? Was the site designated on the basis of positive shovel tests? Perhaps a local resident told you about the site. Or did you determine the presence of a deeply buried site through excavation with bucket augers? Describe the site's specific physical setting, and the condition and ground cover encountered at the time of the survey. Identify the procedure by which you determined site limits. Clearly state the spatial relationship between the site and the project right-of-way or boundaries. Is the site entirely within the project boundaries, partially within, or outside the right-of-way?

State whether or how you evaluated the site. Describe your collection strategy, and refer to a data table that summarizes the materials you collected. Did you perform an intensive, systematic surface collection? Artifact distribution maps generated from any such collections should be included and referred to in the text. Identify your subsurface sampling methods and show the location of your units on a detailed map of the site. If the site-specific geomorphological situation is complex, include plan view and cross-sectional figures showing the geological context of the archaeological materials.

Discuss the soils at the site and whether they are typical of the mapped soil series. Sites in upland settings often exhibit eroded profiles that reflect disturbance to the physical integrity of the site and its assemblages. This is often the critical factor in terms of significance determinations. If you can demonstrate that the soil profile is relatively intact and that the artifacts or features are relatively undisturbed on the surface or within the A and B soil horizons, the site retains a degree of physical integrity and a higher level of significance. If you can establish that the site retains physical integrity, it likely warrants serious consideration for potential National Register or national or provincial heritage site eligibility. For this reason, it is especially important to document any features you were able to define at the site. The presence of intact features is a clear indication that the site retains some degree of physical integrity.

Ordinarily, the site's cultural affiliation will be determined through analysis of its diagnostic artifacts or features. Be sure to cite the sources on which you base your interpretations. Significance is not necessarily precluded for sites that you are unable to assign to a particular cultural or temporal period, but it is more difficult to make the case when you cannot do so. Occasionally, cultural or temporal affiliation can be inferred from the geomorphological context of the archaeological material. For instance, if you know that the alluvial sediments containing the site are early- to mid-Holocene in age, then the artifacts themselves are also very likely to be that old. Historical records may be more reliable than features or artifacts for determining the age and significance of many historical period sites. Make the best case you can based on all lines of available evidence.

Ultimately, every archaeological evaluation in the United States should include the author's opinion of the site's eligibility or ineligibility for listing on the National Register of Historic Places. This recommendation must be made based on the significance, potential significance, or lack of significance according to National Register criteria (chapter 2). Criterion D, most frequently applied to archaeological sites, is that the site has "yielded, or may be likely to yield, information important in prehistory or history." Certain properties also may be eligible under criteria A through C. For sites that you are recommending as potentially eligible, be sure you refer to the site's significance in relation to the relevant State Plan or Historic Properties Plan (see chapter 4). Generally, these planning documents will discuss pertinent cultural contexts within which the site's significance can be considered.

Many of the sites you encounter during your survey career will not be eligible for national, state, or provincial recognition. The most common reason is that the site retains neither physical integrity nor significance. Wherever this is the case, so state, and justify your conclusion. If you have done a competent job of establishing that a site lacks secure physical context, or that the site is not likely to yield significant information or otherwise fails to meet eligibility criteria, then this should be a fairly easy thing to do.

Your individual site discussion should conclude with a paragraph summarizing the site. Reiterate the physical relationship of the site to your client's project. State concisely the temporal and cultural affiliation of the site. Include a definitive assessment of pertinent factors such as physical integrity and site significance. State whether the site does or does not have the potential for inclusion on the National Register of Historic Sites or the Historic Sites and Monuments register. Finally, recommend further appropriate action (e.g., additional testing) or, alternatively, no further work.

Repeat this process for every site discovered during your survey. For many sites, the information you provide will constitute the only record that they ever existed. Give each site the benefit of your best effort.

ARCHAEOLOGICAL SYNTHESIS

Typically, a synthesis section does not appear in reports for small survey projects. However, large survey projects can add to, modify, or change the perception of what is known about a state or a region. For this reason, a section of your report devoted to bringing all of your data together can be extremely useful to readers and researchers. An archaeological synthesis can be personally gratifying to write, as well as edifying to future readers of your work. In this section, you should bring the data together into an interpretive synthesis that addresses the problem domains set forth in your research design. Topics such as culture history or site location modeling can be approached in an integrated manner that will provide a better understanding of the archaeology of the project area than was possible before you conducted your survey.

Existing state plans or regional historic properties contexts may provide baseline data on what was known or suspected about the project area's culture history prior to your investigation. This information can serve as a point of departure for the synthetic interpretation of your data. Contrasts between the findings of your project and the baseline data can both illustrate and validate the contributions of the CRM process. Along with the identification and protection of significant resources, CRM aims to gain useful knowledge for society. So enjoy writing your synthesis section—this is real archaeology. A job well done will enhance your reputation as both a solid archaeologist and a reliable CRM practitioner.

SUMMARY AND RECOMMENDATIONS

Your report must conclude with a summary of the survey and your clearly stated management recommendations. Briefly recap the nature of your client's project and the objectives of the survey. Restate the total area surveyed in acres and hectars.

If no sites were identified and you are confident of your survey coverage, no further work should be recommended. If you found sites but none are considered eligible or potentially eligible for the National Register of Historic Places, Historic Sites and Monuments register, or provincial list of heritage sites, you should state this and recommend no further work for those sites. Still, summarize basic data including the size of each site and whether the entire site, or only a portion, lies within the area of potential effect. If you could not determine eligibility at the Phase I level, recommend additional survey, testing, archival research, geomorphological studies, or other types of investigation as appropriate, explaining why in each case. Outline the types and scopes of these additional studies.

Include a short summary of each site, with recommendations. If you have carefully constructed the summary paragraphs in each individual site discussion within your report, you can simply cut and paste these paragraphs into your "Summary and Recommendations" section. All of the pertinent management recommendations should be included (e.g., no further work, avoid site if possible, Phase II testing, etc.).

A table summarizing survey results will be useful to you, your client, and the reviewers. This table should list the sites recorded by the survey, along with information on landform position, site size, cultural affiliation, range of occupation, site function, NRHP (or HSM) potential, and recommendations.

ACKNOWLEDGMENTS

Depending on the size and complexity of your project, you may be indebted to a great number of people who helped you. It is proper to acknowledge those individuals in an acknowledgments section. This section can appear at the beginning or end of the report, but here, between the summary and the references, is often a convenient placement. Give credit to whom it is due.

REFERENCES CITED

The "References Cited" section should list all reference materials cited in the body of the report. Most archaeological and CRM reports follow the *American Antiquity* style guide for references (www.saa.org/Publications/Styleguide/styframe.html). It is very important that your references are complete and cited correctly. Nothing shouts "beginner" more clearly than an incomplete or improperly structured references section. If the "References Cited" section does not contain references that were actually cited in the text, it is impossible for the reader to examine the works you had deemed important enough to mention. A shoddy references section leads the reader to wonder whether you actually read the cited literature or merely regurgitated information and ideas out of someone else's report. Do your own research; extend the discipline. You'll be better served in your career by doing so. The better command you maintain of the literature, the more successful you will be.

FIGURES

Figures can either be collated into the body of the report (usually a good idea for large reports) or appended in order after the references section. Figures for most small survey reports are gathered at the back of the report.

TABLES

Tables also can be collated into the body of the report or appended in order after the figures. As is the case with figures, tables in most small-project reports are typically printed at the back of the report.

APPENDICES

Attach any other necessary materials at the end of your document in an appendix or series of appendices. Appended material may include the project scope of work, relevant correspondence, copies of permits, review comments on the draft report, any other pertinent forms or documents, and reports from subcontractors.

9
TRYING TO MAKE IT REAL, COMPARED WITH WHAT?

A classic jazz tune, recorded most famously by Les McCann and Eddie Harris on their *Swiss Movement* LP (Atlantic Recording Corporation, 1969), asked the question that is the title of this chapter. It seems an apt question with which to close this volume. Just what realities do we address by conducting an archaeological survey?

Archaeology is constantly changing. We use technologies today that were unthinkable even a decade ago. We also deal with social and political issues that have had a deep effect on archaeological method and theory, issues that today are perhaps as profound as the debunking of the Moundbuilder myth was a century ago.

What will the future hold? Will the North American public continue to support the ideals of historic preservation? Constantly increasing demands for more intensive levels of survey, and the development of new and better, but also more costly, techniques, offers the promise of technically more efficient site finding. Simultaneously, there is resistance from many agencies, and the public at large, concerning the costs of this technical efficiency. County officials, for instance, occupy a pragmatic world and deal with hard issues of infrastructure maintenance and generally parsimonious budgets. They are accountable to a demanding yet tax-intolerant public. It can be difficult to explain to a county engineer why the costs of archaeological surveys for their bridge replacement projects continue to increase, consuming ever larger percentages of their project budgets, while the return to the public may not be apparent. Can the archaeologist's typical response—a shrug of the shoulders and a vague reference to higher agency demands—carry water for CRM in the twenty-first century? Doubtful.

The discipline will have to do a better job of explaining the social benefits of archaeology to its patrons—the public. CRM archaeologists, including those of us who survey, can support such a movement in many ways. Once a project report is written, it doesn't take much effort to edit the manuscript, remove the professional jargon, and distill its interesting conclusions into a public-oriented, nontechnical publication or pamphlet. Some may argue that the consultant is not paid to do this. Really? If you are not providing a tangible product for your constituency, the interested public, exactly for whom do you presume to be working? All CRM archaeologists also should feel an obligation to make public presentations, especially when invited to do so by archaeological and historical societies and local project area groups. If you are unwilling to do so, it just seems very difficult to justify your work. (See Toolkit, volume 7.)

When you do media interviews, make an effort to clearly explain the meaningful archaeological or anthropological context of your project. Don't assume that the reporter will understand: Help him or her understand what you are doing and why. It is up to you to ensure that when Jane Q. Public reads the piece, she says, "Gee, that's interesting!" Otherwise, the alternative might be "What a waste of money!"

You also have obligations to your peers to present the results of your work at professional meetings and through publication outlets. If your work has resulted in something interesting, get it out there. Few professionals working outside your region will ever read the gray literature you produce. Most regional and interregional networking and information exchange occurs at conferences. Attend them. The Internet also is helpful for data exchange, and Internet publishing will no doubt increase in the future.

Throughout this book an ideal relationship has been assumed between you, the archaeological contractor, and the government archaeologist. This ideal assumes that the agency archaeologist is a partner in making a better archaeology. Many archaeologists have experienced this ideal. However, for some consulting archaeologists, the relationships with governmental review and compliance officers have been more adversarial. Keep cool, stay professional, and you'll find that nearly everybody wants to promote high-quality work despite the compromises that are often necessary.

Whatever methods we use in archaeological survey, we are often, figuratively, trying to find a needle in a haystack. Imagine a scatter of lithic artifacts, a handful of flakes, that we find in an open patch of ground on a grassy knoll above a river. Do these artifacts represent a moment's knapping along a hunting trail used millennia ago? A single day's occupation? Or are they a tiny bit of a large, multicomponent site that is spread over a large portion of the bluff?

Consider this example: Suppose our project area encompasses 2,500 square meters—a little over a half acre or a quarter hectare. Shovels and augers at the ready, we set up a systematic testing strategy and start to move dirt. Each of our eight-inch-diameter bucket auger units explores a little over four hundred square meters of surface, or about 4 percent of a square meter. Our shovel tests, at 50 × 50 centimeters, expose 25 percent of a square meter. Since we suspect that aeolian deposits cover the bluff, and we can investigate more easily to depth with the bucket auger than we can with the shovel, we decide to combine both methods of subsurface investigation. To test 4 percent of the project area would require 2,500 bucket-auger units. A 25 percent sample obtained by shovel testing would provide a fair evaluation of our site but would virtually destroy it (and take a very long time to complete). Our compromise is to sample every ten meters, instead of every one meter, thereby excavating about thirty-six shovel test units. Our shovel tests sample nine square meters, less than 0.4 percent of the site. With the bucket auger, we'll sample a little over one meter of surface area on the site, or about 0.04 percent of any near-surface or buried components that may exist. Therefore, even "intensive" survey often opens only tiny windows into the archaeological record.

Keep your perspective. No matter how systematically we plan our methods or how scientific they appear, we don't always know what we're getting into when we survey and evaluate sites.

Will new technologies change the way we evaluate sites? Remote sensing gives us a representation of the subsurface without digging. Geophysical techniques work best when distinctive contrasts are present—buried metal, an abrupt change in soil moisture, an interruption in the pattern of magnetic fields—so they are effective for documenting disruptions produced by historic structures; trenches; and the more obvious prehistoric pits, hearths, or houses, particularly when they occur within a relatively uniform geological matrix. If the technology continues to develop at the present rate, however, it might not be long before our ability to detect subsurface features will significantly curtail the need to excavate for site assessments.

At the same time, we are now able to view and analyze three-dimensional models using computers. More than a decade ago, researchers such as Dr. Paul Reilly at IBM in England began to construct experimental archaeological sites in which all artifacts were positioned within a transparent, three-dimensional matrix. If, in the future, we combine remote sensing and three-dimensional imaging technologies, similar to the way GPS and GIS are now combined, we may find that archaeological survey will be as much a computer visualization skill as it is a matter of physical labor.

Such technological progress may have ancillary benefits. By limiting physical testing, especially when investigating sensitive sites (e.g., cemeteries, mounds), we may assuage some of the concerns held by descendant communities about our work. We may never agree on how the past should be represented, and for what purposes, but we may be able to come to terms on how it might be investigated.

With all the promise of higher technology, however, field archaeologists do what they do because of their fascination for discovery and the love of the land. These emotional forces may lurk behind the protective masks provided by jargon and professionalism, but they are always present. We will always survey, no matter what government policies are in effect, or what issues are relevant to special interest groups, or what the new technologies have to offer. As long as there remains among people an inherent interest in who they are and where they have been, archaeologists will continue to seek the past's silent remnants, which remain hidden, buried beneath our feet.

REFERENCES

Abbott, Larry R., and Craig A. Neidig
    1993 Archaeological Postholing: A Proposed Subsurface Survey and Site-Testing Method. *Illinois Archaeology* 5:38–45.

American Anthropological Association
    Issued yearly *The AAA Guide*. American Anthropological Association. Arlington, Virginia.

Avery, Thomas E., and Thomas R. Lyons
    1981 Remote Sensing: Aerial and Terrestrial Photography for Archeologists. Cultural Resources Management Division, National Park Service, U.S. Department of the Interior, Washington, D.C.

Baker, Craig, and George J. Gumerman
    1981 *Remote Sensing: Archeological Applications of Remote Sensing in the North Central Lowlands*. Cultural Resources Management Division, National Park Service, U.S. Department of the Interior, Washington, D.C.

Bettis, E. Arthur III (ed.)
    1992 Soil Morphologic Properties and Weathering Zone Characteristics as Age Indicators in Holocene Alluvium in the Upper Midwest. In *Soils in Archaeology*, edited by Vance T. Holliday, pp. 119–44. Smithsonian Institution Press, Washington, D.C.
    1995 *Archaeological Geology of the Archaic Period in North America*. Special Paper 297. Geological Society of America, Boulder, Colorado.

Bettis, E. Arthur III, and Edwin R. Hajic
    1995 Landscape Development and the Location of Evidence of Archaic Cultures in the Upper Midwest. In *Archaeological Geology of the Archaic Period in North America*, edited by E. Arthur Bettis III, pp. 87–113. Special Paper 297. Geological Society of America, Boulder, Colorado.

Bevan, Bruce
- 1983 Electromagnetics for Mapping Buried Earth Features. *Journal of Field Archaeology* 10:47–54.
- 1991 The Search for Graves. *Geophysics* 56:1310-19.
- 1998 *Geophysical Exploration for Archaeology: An Introduction to Geophysical Exploration.* Special Report No. 1. U.S. Department of the Interior, National Park Service, Midwest Archaeological Center, Lincoln, Nebraska.

Camilli, Eileen L., and Linda S. Cordell.
- 1983 *Remote Sensing: Applications to Cultural Resources in Southwestern North America.* Cultural Resources Management Division, National Park Service, U.S. Department of the Interior, Washington, D.C.

Clark, Anthony
- 1990 *Seeing Beneath the Soil: Prospecting Methods in Archaeology.* Batsford, London.

Collins, James M.
- 1991 *The Iowa River Greenbelt: An Archaeological Landscape.* Office of the State Archaeologist, University of Iowa, Iowa City.

Connor, Melissa, and Douglas D. Scott
- 1998 Metal Detector Use in Archaeology: An Introduction. *Historical Archaeology* 32(4):76–85.

Conyers, Lawrence B., and Dean Goodman
- 1997 *Ground-Penetrating Radar: An Introduction for Archaeologists.* AltaMira Press, Walnut Creek, California.

Creamer, Winifred, Jonathan Haas, and Thomas Mann
- 1997 Applying Photogrammetric Mapping: A Case Study from Northern New Mexico. *American Antiquity* 62:285–99.

Dalan, Rinita, A.
- 1989 *Geophysical Investigations of the Prehistoric Cahokia Palisade Sequence.* Illinois Cultural Resources Study No. 8. Illinois Historic Preservation Agency, Springfield.
- 1991 Defining Archaeological Features with Electromagnetic Surveys at the Cahokia Mounds State Historic Site. *Geophysics* 56:1280–87.

Deuel, Leo
- 1969 *Flights into Yesterday: The Story of Aerial Archaeology.* St. Martin's Press, New York.

Eddy, Frank W., Dale R. Lightfoot, Eden A. Welker, Layne L. Wright, and Dolores C. Torres
- 1996 Air Photographic Mapping of San Marcos Pueblo. *Journal of Field Archaeology* 23:1–13.

Fowler, Melvin L.
- 1997 *The Cahokia Atlas Revised: A Historical Atlas of Cahokia Archaeology.* Studies in Archaeology, Volume 2. Illinois Trans-

portation Archaeological Research Program, University of Illinois, Urbana.

Gourley, Kathryn
1995 The Section 106 Process. Unpublished manuscript in possession of the author. Office of the State Archaeologist, University of Iowa, Iowa City.

Green, William, and John F. Doershuk
1998 Cultural Resource Management and American Archaeology. *Journal of Archaeological Research* 6:121–67.

Heimmer, Don H., and Steven L. De Vore
1995 *Near-Surface, High Resolution Geophysical Methods for Cultural Resource Management and Archaeological Investigations.* Interagency Archaeological Services, National Park Service, Denver, Colorado.

Holley, George R., Rinita A. Dalan, and Philip A. Smith
1993 Investigations in the Cahokia Site Grand Plaza. *American Antiquity* 58:306–19.

King, Thomas F.
1998 *Cultural Resource Laws and Practice: An Introductory Guide.* AltaMira Press, Walnut Creek, California.
2000 *Federal Planning and Historic Places: The Section 106 Process.* AltaMira Press, Walnut Creek, California.

King, Thomas F., Patricia Parker Hickman, and Gary Berg
1977 *Anthropology in Historic Preservation.* Academic Press, New York.

Kintigh, Keith W.
1988 The Effectiveness of Subsurface Shovel Testing: A Simulation Approach. *American Antiquity* 53:686–707.

Krakker, James J., Michael J. Shott, and Paul D. Welch
1983 Design and Evaluation of Shovel-Test Sampling in Regional Archaeological Survey. *Journal of Field Archaeology* 10:469–80.

Lightfoot, Kent G.
1986 Regional Surveys in the Eastern United States: The Strengths and Weaknesses of Implementing Subsurface Testing Programs. *American Antiquity* 51:484–504.
1989 A Defense of Shovel-Test Sampling: A Reply to Shott. *American Antiquity* 54:413–16.

Limp, W. Frederick
1989 *The Use of Multispectral Digital Imagery in Archeological Investigations.* Research Series No. 34. Arkansas Archeological Survey, Fayetteville.

Lyons, Thomas R., and Robert K. Hitchcock (eds.)
1977 *Aerial Remote Sensing Techniques in Archeology.* Chaco Center, National Park Service, U.S. Department of the Interior, Albuquerque, New Mexico.

McManamon, Francis P.
  1984   Discovering Sites Unseen. In *Advances in Archaeological Method and Theory*, Volume 7, edited by M. B. Schiffer, pp. 223–92. Academic Press, New York.
Maxwell, Gordon S. (ed.)
  1983   *The Impact of Aerial Reconnaissance on Archaeology*. Research Report No. 49. Council for British Archaeology, London.
Merry, Carl A., and John G. Hedden
  1995   Excavation Safety Plan for the Iowa Office of the State Archaeologist. *Journal of the Iowa Archeological Society* 42:5–7.
Mickle, Jack L.
  1995   Occupational Safety and Health Administration Regulations on Excavation Safety. *Journal of the Iowa Archeological Society* 42:1–4.
Nance, Jack D., and Bruce F. Ball
  1986   No Surprises? The Reliability and Validity of Test Pit Sampling. *American Antiquity* 51:457–83.
  1989   A Shot in the Dark: Shott's Comments on Nance and Ball. *American Antiquity* 54:405–12.
National Center for Preservation Technology and Training
  1998   Ground Penetrating Radar: New Developments in Data and Image Processing Techniques. *NCPTT Notes* Number 26. National Center for Preservation Technology and Training, United States Department of the Interior, National Park Service, Washington, D.C.
Schoeneberger, P. J., D. A. Wysocki, E. C. Benham, and W. S. Broderson
  1998   *Field Book for Describing and Sampling Soils*. National Soil Survey Center, Natural Resources Conservation Service, United States Department of Agriculture, Lincoln, Nebraska. (Online at www.statlab.iastate.edu/soils/nssc/field_gd/field_gd.htm)
Scollar, Irwin, Allain Tabbagh, Albert Hesse, and Irmela Herzog
  1990   *Archaeological Prospecting and Remote Sensing*. Cambridge University Press, New York.
Scott, Douglas D., Richard A. Fox Jr., Melissa A. Conner, and Dick Harmon
  1989   *Archaeological Perspectives on the Battle of Little Bighorn*. University of Oklahoma Press, Norman.
Shott, Michael J.
  1985   Shovel-Test Sampling as a Site Discovery Technique: A Case Study from Michigan. *Journal of Field Archaeology* 12:458–69.
  1989   Shovel-Test Sampling in Archaeological Survey: Comments on Nance and Ball, and Lightfoot. *American Antiquity* 54:396–404.
Soil Science Society of America
  1987   *Glossary of Soil Science Terms*. Soil Science Society of America, Madison, Wisconsin.

Squier, Ephraim G., and E. H. Davis
  1848 *Ancient Monuments of the Mississippi Valley.* Smithsonian Contributions to Knowledge, Volume 1. Smithsonian Institution, Washington, D.C.

Thomas, Cyrus
  1894 *Report of the Mound Explorations of the Bureau of Ethnology.* Twelfth Annual Report, Bureau of American Ethnology, Smithsonian Institution, Washington, D.C.

Trench Shoring and Shielding Association
  1994 *Excavation Safety Guide.* Trench Shoring and Shielding Association, Tarrytown, New York.

Warren, Robert E.
  1982 Prehistoric Settlement Patterns. In *The Cannon Reservoir Human Ecology Project: An Archaeological Study of Cultural Adaptations in the Southern Prairie Peninsula,* edited by Michael J. O'Brien, Robert E. Warren, and Dennis E. Lewarch, pp. 337–68. Academic Press, New York.

Weber, Scott A., and Stephen R. Yool
  1999 Detection of Subsurface Archaeological Architecture by Computer Assisted Airphoto Interpretation. *Geoarchaeology* 14:481–93.

Wenner, F.
  1916 A Method of Measuring Earth Resistivity. *Bulletin of the United States Bureau of Standards* 12:469–78.

Weymouth, John W.
  1986 Geophysical Methods of Archaeological Site Surveying. In *Advances in Archaeological Method and Theory,* Volume 9, edited by Michael B. Schiffer, pp. 311–91. Academic Press, New York.

Weymouth, John W., and R. Huggins
  1985 Geophysical Surveying of Archaeological Sites. In *Archaeological Geology,* edited by George Rapp and J. A. Gifford, pp. 191–235. Yale University Press, New Haven, Connecticut.

Weymouth, John W., and Robert Nickel
  1977 A Magnetometer Survey of the Knife River Indian Villages. *Plains Anthropologist* 22(78), part 2:104–18.

Whitley, David S.
  2001 *Handbook of Rock Art Research.* AltaMira Press, Walnut Creek, California.

Wobst, H. Martin
  1983 We Can't See the Forest for the Trees: Sampling and the Shapes of Archaeological Distributions. In *Archaeological Hammers and Theories,* edited by James A. Moore and Arthur S. Keene, pp. 32–80. Academic Press, New York.

Wynn, Jeffrey C.
   1986   Review of Geophysical Methods Used in Archaeology. *Geoarchaeology* 1:245–57.

Zeidler, James A.
   1995   *Archaeological Inventory Standards and Cost-estimation Guidelines for the Department of Defense.* USACERL Special Report 96/40. U.S. Army Corps of Engineers, Construction Engineering Research Laboratories, Champaign, Illinois. (Online at www.denix.osd.mil/denix/Public/ES-Programs/Conservation/Legacy/AISS/usacerl1.html)

INDEX

AAA Guide, 16
"Access to Archaeology Program," 13
Advisory Council on Historical Preservation, 11, 12, 13, 111
aerial photograph, 5, 28–29, 77, 100, 102, 104
Agriculture Canada, 28
alidade, 104
altimeter, electronic, 104
American Anthropological Association, 16
American Antiquity style guide, 127
American Cultural Resources Association, 16
Ancient Monuments of the Mississippi (Squier & Davis), 4
APE. *See* area of potential effect
applied archaeology. *See* cultural resource management
arbitrary unit, 21
archaeologist, obligations of, 130. *See also* fieldwork, public relations in
area of potential effect (APE), 11–12
Artefacts Canada, 30

backhoe trenching, 70–71
Barrington MS2, 94
Benham, E. C., 60
Beven, Bruce, 80
Binford, Lewis, 6, 7
Boas, Franz, 4
Borden System, 123
Broderson, W. S., 60
Brunton pocket transit, 104
bucket auger, 68, 69–70, 74

Canadian Centre for Occupational Safety and Health, 71
Canadian First Nations people, 14, 49–50, 62
Canadian Museum of Civilization, 123
Canadian Soil Information System, 28
cave, surveying, 56–57
cemetery, surveying, 77, 78, 90
Clark, Anthony, 80, 82, 94
Clark, Grahame, 5, 6
clinometer, 104
computer, as imaging technology, 131–32
computer mapping. *See* Geographic Information Systems
Conner, Melissa A., 81
consultant: cultural resource management, 16–17; geomorphologic, 50–52;

139

geophysical, 79–80; minimal qualifications for, 15–16
contact archaeology. *See* cultural resource management
county history, 31
county soil survey, 27–28
Crawford, O. G. S., 5
CRM. *See* cultural resource management
cultural area, in North America, 4–5
cultural resource management (CRM): consultant for, 15–17; contract for, 17; educational requirement for, 15; entry-level position for, 14; growth in, 14–15; locating reports, 30; regulations for, in Canada, 13–14; regulations for, in United States, 10–13
cutbank, surveying, 60, 61, 70–71

Dalan, Rinita A., 86, 88
Davis, Edward H., 3–4, 8
De Vore, Steven L., 80
deadstead, 31
DEM. *See* digital evaluation model
Department of Canadian Heritage, 13
design, survey: background research for, 19–20; sampling strategy for, 21–22; type of survey for, 20–21
Devils Tower National Monument, 108–9
differential correction, 106
differential correction satellite, 107
digital evaluation model (DEM), 109
digital orthophoto quarter quadrangle (DOQQ), 109
digital raster graphic (DRG), 109
DigitalGlobe, 109
dilution of precision (DOP), 105
distributional survey, 7–8
diurnal variation, 91
Division of Mound Exploration, 4
DOP. *See* dilution of precision

DOQQ. *See* digital orthophoto quarter quadrangle
DRG. *See* digital raster graphic

electrical resistivity: meter, 82–83, 84; strength/weakness of, 84–85, 96–97; voltage distortion in, 81–82, 83
electromagnetic conductivity: difference from resistivity, 84; meter, 84–85, 86; principles of, 87; strength/weakness of, 87–88, 96–97
email discussion groups, archaeology, 37
ENVISAT (ENVIronmental SATelitte), 109
European Space Agency, 109

field crew, 36–38
field notebook, 39
fieldwork, 35–36; basic equipment list, 40–41; consultant for, 50–52; to determine National Register eligibility, 72–75; equipment for, 7–8, 38–41; Native American/Canadian First Nations/ethnic issues in, 49–50; personnel for, 36–38
fieldwork, prefield strategy: cultural setting research, 29–31, 33; implementing research design, 23–24; mapping, 25–26; physical setting research, 26–29; scheduling/budgeting, 24
fieldwork, public relations in: contacting local archaeologist, 49; contacting local society, 49; gaining permissions, 42, 44; introductory letter (sample), 45; landowner questionnaire (sample), 46; making initial contacts, 42–44, 46–47; news release (sample), 43; photographing collection, 47, 48; salted site awareness, 47–48

fieldwork, survey method in: determining appropriate, 53; guidelines for, 52. *See also* pedestrian survey; shovel testing; subsurface survey
First Nations people. *See* Canadian First Nations people
fluxgate gradiometer, 92–93
Forestry Suppliers Inc., 39
Fox, Richard A., Jr., 81
Friends of the Pleistocene, 27

General Land Office (GLO), 31, 32–33
General Survey of Public Lands, 31
Geographic Information Systems (GIS), 8, 28, 103, 108–9, 110
geomorphologist, 27, 50–52, 118
Geonics Limited, 84
Geonics meter, 84–85, 86
geophysical sensing technique: consultant for, 79–80; cost of, 80; electrical resistivity, 81–84; electromagnetic conductivity, 84–88; ground penetrating radar, 88–90; magnetic susceptibility, 94–95; magnetometry, 91–94; metal detector, 80–81, 88–90, 96–97; process in, 96, 98; site-specific planning in, 78–79; standard grid for, 79; strength/weakness of, 77–78, 131; strength/weakness of equipment for, 96–97
Geoscan Research, 93
Giddings coring system, 51–52, 70
GLO. *See* General Land Office
Global Positioning System (GPS), 8, 102–3, 104–8, 109
Gourley, Kathryn, 11
GPR. *See* ground penetrating radar
GPS. *See* Global Positioning System
gray literature, 30
Great Serpent Mound, 3

ground penetrating radar (GPR): graphic plotter for, 88, 90; principles of, 89; strength/weakness of, 90, 96–97
Guidelines for Evaluating and Documenting Traditional Cultural Properties, 50

Heimmer, Don H., 80
Heritage Conservation Act, 14
Historic Sites and Monuments Act, 13
Historic Sites and Monuments register, 120, 125, 126
historical records, 30–31
Hodder, Ian, 7
Holmes, William Henry, 4
Holocene environment, 26–27, 50, 52, 54, 118, 119, 124
hydraulic drill rig, 51–52

Indian, American. *See* Native American
indirect subsurface survey. *See* geophysical sensing technique; remote sensing
infrared film, 100
infrared photography, 5

Kidder, Alfred, 5
King, Thomas F., 11

Listserv, archaeological. *See* email discussion groups, archaeological
locational system. *See* Global Positioning System

MacNeish, Richard S., 5
magnetic susceptibility, 94–95, 96–97
magnetometry, 91–94; principles of, 92; strength/weakness of, 93, 96–97; using fluxgate gradiometer in, 92–93; using proton magnetometer in, 91–92

INDEX

map: soil, 28, 101; topographic, 25, 26, 101–2, 109, 115; uses of, 99–100
mapping/recording: aerial photograph, 5, 28–29, 77, 100, 102, 104; drafting survey map, 102–4; future of, 109–10; Geographic Information Systems, 8, 28, 103, 108–9, 110; Global Positioning System, 8, 102–3, 104–8, 109; photogrammetric mapping, 100; as prefield strategy, 25–26; satellite imagery, 8, 100; strength/weakness of, 99–100
metal detector, 80–81, 96–97
Moundbuilder, origins of, 3–4

NAGPRA. See Native American Graves Protection and Repatriation Act
nanoteslas (nT), 91
National Air Photo Library, 28
National Archaeological Data Base, 30, 120
National Environmental Policy Act (NEPA), 11
National Historic Preservation Act (1966), 10, 11, 12–13, 17
National Historic Sites and Monuments, 72
National Institute for Occupational Safety and Health, 71
National Park Service, 38
National Register Bulletin 38, 50
National Register of Historic Places (NRHP), 10, 12, 33, 72–75, 120, 124–25, 126
National Topographic Service of Canada (NTS), 25, 26
Native American, 3–4, 11, 49–50
Native American Graves Protection and Repatriation Act (NAGPRA), 11

NEPA. See National Environmental Policy Act
New Archaeology, 6
nonarbitrary unit, 21
nonsite survey, 7–8
NRHP. See National Register of Historic Places
NTS. See National Topographic Service of Canada

Oakfield soil probe, 71–72
Occupational Safety and Health Administration (OSHA), 70–71

PDOP. See position dilution of precision
pedestrian survey: basic method for, 57–58; of cave/crevice/shelter, 55–57; to determine National Register eligibility, 73; flagging artifact, 59; inspecting stream cutbank, 60, 61; mapping/describing soils, 60; as preferred method, 55; recording rock art, 57; shoreline observation, 61–63; viability of, 54–55
personnel, 36–38
petrograph/pictograph, 56–57
photogrammetric mapping, 100
planning documents, 29–30
plat book, 43–44
plat map, 31, 33
position dilution of precision (PDOP), 105
posthole testing, 69
postprocessualist archaeology, 7
probabilistic survey, 21, 22
"Procedures for the Protection of Historic and Cultural Properties," 10
project area description: for large project, 117–19; for small project, 116–17

project plan, variations in, 24
proton magnetometer, 91–92
provincial archaeologist, 17, 27, 30, 52, 64, 111
public relations, 130. *See also* fieldwork, public relations in
Putnam, Rufus, 3

quadrat. *See* arbitrary unit
Quaternary environment, 26, 118, 119
Quickbird satellite, 109

RADARSAT, 8, 100
radiocarbon dating, 5
random sampling, 6
recording/mapping. *See* map; mapping/recording
regulation, government: in Canada, 13–14; in United States, 10–13, 15–16
Reilly, Paul, 131–32
remote sensing: by aerial photography, 5, 28–29, 77, 100, 102, 104; by satellite, 8, 100, 107, 109
report, survey: abstract, 114; acknowledgments, 127; appendices, 128; area description of project, 116–17; clarity/conciseness of, 111–12; environmental context of project in, 117–19; figures, 127; front matter, 113–14; illustrations, 118–19; introduction, 114–16; literature review, 120; map, 115; methods used, 121–22; references cited, 127; and research design, 120–21; results, 122–25; structure/content of, 112–13; summary/recommendations, 126; synthesis of, 125–26; tables, 127; title page, 113; uses of, 111

Report on the Mound Explorations of the Bureau of American Ethnology (Thomas), 4
request for proposals (RFP), 17, 23
request for quotations (RFQ), 17
research design, implementing, 23–24
RFP. *See* request for proposals
RFQ. *See* request for quotations
Rite in the Rain notebook, 39
RM-15 resistivity meter, 82–83, 84
rural directory, 31, 44

salted site, 47–48
salvage archaeology. *See* cultural resource management
SAR. *See* synthetic aperture radar (SAR) sensor
satellite, 8, 100, 107, 109
Schoeneberger, P. J., 60
scope of work (SOW), 23
Scott, Douglas D., 81
Section 106, of National Historic Preservation Act, 10, 11, 12–13, 17
Section 110, of National Historic Preservation Act, 10
shovel probe, 63, 64
shovel testing: to determine National Register eligibility, 74; identifying satellite excavation in, 66, 67; labeling/storing bags in, 64, 66; lower vs. higher effort, 63–64; offset grid vs. square grid in, 64, 65; requirements for, 52; site evaluation form for, 67–68
SHPO. *See* State Historic Preservation Office
sighting compass, 104
site, definition of, 7
site records, 30
Smithsonian Institution, 3, 4
soil map, 28, 101
Soil Science Society of America, 39

soil survey, 27–28
solid core sampling, 51–52
SOW. *See* scope of work
Space Imaging, 100
space vehicle, 104–5
Special Places Protection Act, 13
Squier, Ephraim G., 3–4, 8
State Historic Preservation Office (SHPO), 12–13, 16, 17, 27, 30, 52, 64, 111
statistical sampling technique, 6
Steward, Julian, 5
subsurface survey: backhoe trenching, 70–71; bucket auger, 68, 69–70, 74; of cutbank, 70–71; to determine National Register eligibility, 73–74; Oakfield core sampling, 71–72; offset grid with satellite grid, 67; posthole testing, 69; safety in, 70–71; site evaluation form (sample), 68; square vs. offset shovel test grid patterns, 65; tile probe for, 72. *See also* geophysical sensing technique; remote sensing; shovel testing
survey, archaeological: current theory/practice in, 6–8; future of, 129–32; non-CRM, 20; origins of, 3–5; proposal for, 23–24; reasons to perform, 9; responsibility of leader for, 35–36
surveyor: CRM consultant as, 16–17; graduate as, 15–16; undergraduate as, 14–15
SV. *See* space vehicle
synthetic aperture radar (SAR) sensor, 100
systemic sampling survey, 21

TCP. *See* traditional cultural property
Terraserver, 29
36 CFR Part 60.4, 10
36 CFR Part 61, 15–16
36 CFR Part 800, 10, 11, 12, 64
Thomas, Cyrus, 4
tile probe, for surveying, 72
topographic map, 25–26, 101–2, 109, 115
total stations, 7–8, 73, 103
traditional cultural property (TCP), 50
Trench Shoring and Shielding Association, 71

Universal Transverse Mercator (UTM), 105, 116, 123
U.S. Department of Agriculture (USDA), 28, 29
U.S. Geological Survey (USGS) map, 25, 26, 101–2, 109, 115
UTM. *See* Universal Transverse Mercator

viewshed, 20
Virú Valley (Peru), 5

website: aerial photograph, 29; *American Antiquity* style guide, 127; American Cultural Resources Association, 16; Artefacts Canada, 30; CRM consulting, 16; CRM report, 30; employment notices, 37; Forestry Suppliers, Inc., 39; National Archaeological Data Base, 30; occupational safety, 71; Rite in the Rain notebook, 39; soil map, 28; Soil Science Society of America, 39; Terraserver, 29; topographic map, 25
Wenner electrode configuration, 82
Weymouth, John W., 80, 94
Willey, Gordon, 5
windshield survey, 43
Wynn, Jeffrey C., 80
Wysocki, D. A., 60

ABOUT THE AUTHORS AND SERIES EDITORS

James M. Collins is an archaeologist on the staff of the Office of the State Archaeologist, University of Iowa. He has been engaged in Midwest and Plains archaeology since 1974. In addition to hundreds of archaeological survey and excavation reports, recent publications include articles in *American Antiquity, Illinois Archaeology, Plains Anthropologist, The Minnesota Archaeologist, Midcontinental Journal of Archaeology,* and the *Journal of the Iowa Archeological Society;* chapters in *Mississippian Communities and Households* (University of Alabama Press, 1995) and *Cahokia: Domination and Ideology in the Mississippian World* (University of Nebraska Press, 1997); and four monographs: *The Archaeology of the Cahokia Mounds ICT-II: Site Structure* (Illinois Historic Preservation Agency), *The Iowa River Greenbelt: An Archaeological Landscape, The Archaeology of the Dolomite Ridge Site,* and *Prehistoric Archaeology of the Marriott Site* (Office of the State Archaeologist, University of Iowa). Among his interests are the prehistoric and protohistoric archaeology of the Midwest and Plains; the archaeology of settlements and settlement systems; the social, economic, and political parameters that define territories; the psychology and structure of complex societies; geoarchaeology; geophysical prospecting; and cultural and natural resource management. Most recently, his attention has focused on the Woodland tradition of the Upper Mississippi River Valley, ongoing research in the Iowa River Greenbelt, and the iconography and distribution of late prehistoric shell masks on the Prairie Peninsula and Plains.

Brian Leigh Molyneaux is an archaeologist, writer, and photographer. He is a specialist in prehistoric art and society, the human use of the landscape, and computer-aided applications in archaeology. He is director of the University of South Dakota Archaeology Laboratory, codirector of the Missouri River Institute, and a research associate of the Royal Ontario Museum, Toronto. He received his M.A. in art and archaeology from Trent University, Peterborough, Ontario, in 1977 and his Ph.D. in archaeology at the University of Southampton, England, in 1991. His extensive fieldwork includes many years of travel in northern Canada, studying Algonkian rock art, ritual, and religion, and archaeological research in the northern Great Plains. He recently conducted a two-year archaeological survey at Devils Tower National Monument, Wyoming. Dr. Molyneaux has published several books: *The Presented Past* (coedited with Peter Stone, Routledge, 1994), a study of archaeology, museums, and education around the world; *The Sacred Earth* (Little, Brown, 1995), a study of spirituality related to the landscape; *Native North America* (with Larry Zimmerman, Little, Brown, 1996), a detailed survey of Native North American culture, past and present; *The Cultural Life of Images* (editor and contributor, Routledge, 1997), a study of pictures and other visual representations of the past in archaeology; a new edition of *Native North America* (Oklahoma University Press, 2000); *Sacred Earth, Sacred Stones* (with Piers Vitebsky, Laurel Glen, 2001), a compilation dealing with spirituality in landscape and architecture; and *Mythology of the Americas* (with David M. Jones, Lorenz Books, 2001), a general encyclopedia. His rock art photographs have been exhibited in the National Gallery of Canada and featured in the PBS/BBC series *Land of the Eagle,* and he is an active contributor to web-based symposia on art, technology, environment, and culture (e.g., The Anthology of Art, www.anthology-of-art.net, School of Fine Arts, Braunschweig, Germany).

Larry J. Zimmerman is the head of the Archaeology Department of the Minnesota Historical Society. He served as an adjunct professor of anthropology and visiting professor of American Indian and Native Studies at the University of Iowa from 1996 to 2002 and as chair of the American Indian and Native Studies Program from 1998 to 2001. He earned his Ph.D. in anthropology at the University of Kansas in 1976. After teaching at the University of South Dakota for twenty-two years, he left there in 1996 as Distinguished Regents Professor of Anthropology.

While in South Dakota, he developed a major CRM program and the University of South Dakota Archaeology Laboratory, where he is still a research associate. He was named the University of South Dakota Student Association Teacher of the Year in 1980, given the Burlington Northern Foundation Faculty Achievement Award for Outstanding Teaching in 1986, and granted the Burlington Northern Faculty Achievement Award for Research in 1990. He was selected by Sigma Xi, the Scientific Research Society, to be a national lecturer from 1991 to 1993, and he served as executive secretary of the World Archaeological Congress from 1990 to 1994. He has published more than three hundred articles, CRM reports and reviews, and is the author, editor, or coeditor of fifteen books, including *Native North America* (with Brian Molyneaux, University of Oklahoma Press, 2000) and *Indians and Anthropologists: Vine Deloria, Jr., and the Critique of Anthropology* (with Tom Biolsi, University of Arizona Press, 1997). He has served as the editor of *Plains Anthropologist* and the *World Archaeological Bulletin* and as the associate editor of *American Antiquity*. He has done archaeology in the Great Plains of the United States and in Mexico, England, Venezuela, and Australia. He has also worked closely with a wide range of American Indian nations and groups.

William Green is the director of the Logan Museum of Anthropology and an adjunct professor of anthropology at Beloit College, Beloit, Wisconsin. He has been active in archaeology since 1970. Having grown up on the south side of Chicago, he attributes his interest in archaeology and anthropology to the allure of the exotic (i.e., rural) and a driving urge to learn the unwritten past, abetted by the opportunities available at the city's museums and universities. His first fieldwork was on the Mississippi River bluffs in western Illinois. Although he also worked in Israel and England, he returned to Illinois for several years of survey and excavation. His interests in settlement patterns, ceramics, and archaeobotany developed there. He received his master's degree from the University of Wisconsin at Madison and then served as Wisconsin SHPO staff archaeologist for eight years. After obtaining his Ph.D. from the University of Wisconsin at Madison in 1987, he served as state archaeologist of Iowa from 1988 to 2001, directing statewide research and service programs including burial site protection, geographic information, publications, contract services, public outreach, and curation. His main research interests focus on the development and spread of native agriculture. He has served as editor of the *Midcontinental Journal of Archaeology* and

The Wisconsin Archeologist; has published articles in *American Antiquity, Journal of Archaeological Research,* and other journals; and has received grants and contracts from the National Science Foundation, National Park Service, Iowa Humanities Board, and many other agencies and organizations.

ARCHAEOLOGY BY DESIGN

ARCHAEOLOGIST'S TOOLKIT

SERIES EDITORS: LARRY J. ZIMMERMAN AND WILLIAM GREEN

The Archaeologist's Toolkit is an integrated set of seven volumes designed to teach novice archaeologists and students the basics of doing archaeological fieldwork, analysis, and presentation. Students are led through the process of designing a study, doing survey work, excavating, properly working with artifacts and biological remains, curating their materials, and presenting findings to various audiences. The volumes—written by experienced field archaeologists—are full of practical advice, tips, case studies, and illustrations to help the reader. All of this is done with careful attention to promoting a conservation ethic and an understanding of the legal and practical environment of contemporary American cultural resource laws and regulations. The Toolkit is an essential resource for anyone working in the field and ideal for training archaeology students in classrooms and field schools.

Volume 1: *Archaeology by Design*
By Stephen L. Black and Kevin Jolly

Volume 2: *Archaeological Survey*
By James M. Collins and Brian Leigh Molyneaux

Volume 3: *Excavation*
By David L. Carmichael and Robert Lafferty

Volume 4: *Artifacts*
By Charles R. Ewen

Volume 5: *Archaeobiology*
By Kristin D. Sobolik

Volume 6: *Curating Archaeological Collections:*
From the Field to the Repository
By Lynne P. Sullivan and S. Terry Childs

Volume 7: *Presenting the Past*
By Larry J. Zimmerman

ARCHAEOLOGY BY DESIGN

STEPHEN L. BLACK
KEVIN JOLLY

ARCHAEOLOGIST'S TOOLKIT
VOLUME 1

ALTAMIRA
PRESS

A Division of Rowman & Littlefield Publishers, Inc.
Walnut Creek • Lanham • New York • Oxford

ALTAMIRA PRESS
A Division of Rowman & Littlefield Publishers, Inc.
1630 North Main Street, #367
Walnut Creek, CA 94596
www.altamirapress.com

Rowman & Littlefield Publishers, Inc.
4501 Forbes Boulevard, Suite 200
Lanham, MD 20706

PO Box 317
Oxford
OX2 9RU, UK

Copyright © 2003 by ALTAMIRA PRESS

All rights reserved. No part of this publication may be reproduced, stored in a retrieval system, or transmitted in any form or by any means, electronic, mechanical, photocopying, recording, or otherwise, without the prior permission of the publisher.

British Library Cataloguing in Publication Information Available

Library of Congress Cataloging-in-Publication Data
Black, Stephen L.
 Archaeology by design / Stephen L. Black and Kevin Jolly.
  p. cm. — (Archaeologist's toolkit ; v. 1)
Includes bibliographical references and index.
 ISBN 0-7591-0397-6 (alk. paper) — ISBN 0-7591-0020-9 (pbk. : alk. paper)
 1. Archaeology—Research. 2. Archaeology—Methodology. I. Jolly, Kevin. II. Title. III. Series

CC83 .B58 2002
930.1'028—dc21

2002014277

Printed in the United States of America

∞™ The paper used in this publication meets the minimum requirements of American National Standard for Information Sciences—Permanence of Paper for Printed Library Materials, ANSI/NISO Z39.48-1992.

CONTENTS

Series Editors' Foreword   vii

1 Designing Archaeological Research with an Attitude   1
The Process of Designing Archaeological Research • Professional Archaeological Research • Why Design Archaeological Research? • The Cost of Archaeology • Your Obligation to the Public • Archaeology by Default • Designing Research Is *Archaeology*

2 Building Your Professional Toolkit   19
Anthropological Archaeology • Current Archaeological Method and Theory • Multidisciplinary Studies • Keeping Current • Developing Regional Expertise • Navigating the Political Playing Field • Why the Professional Archaeologist's Toolkit Is Heavy

3 A Quick Look at the Research Process   35
The "Pure" Research Process in Academic Archaeology • The "Directed" Research Process in CRM Archaeology • The Parts of CRM Research • Conclusion

4 The Scope of Work   53
The Project Area • The Work • Work Standards • Time • Money • Logistics • Conclusion

v

5	**Research Questions**	67
	Typical Approaches to Archaeological Research • Creating Research Questions • Sources of Questions	
6	**Developing Your Research Strategy**	83
	Think First, Dig Later—Archaeological Strategy • Making Tough Choices—Operational Strategy • Conclusion	
7	**The Written Research Design**	99
	Whom Are You Writing For? • Projects Big and Small • Critical Components of Your Written Research Design • Additional Considerations • Conclusion	
8	**Putting Your Research Design to Work**	113
	Implementing Your Research Design • In from the Cold • Evaluating Your Research Design • Now It's Up to You	

Appendix A. Groups You Should Join	127
Appendix B. Journals You Should Read	131
Appendix C. The ABCs of CRM	133
Appendix D. The Logistics Checklist	135
References	143
Index	149
About the Authors and Series Editors	155

SERIES EDITORS' FOREWORD

The Archaeologist's Toolkit is a series of books on how to plan, design, carry out, and use the results of archaeological research. The series contains seven books written by acknowledged experts in their fields. Each book is a self-contained treatment of an important element of modern archaeology. Therefore, each book can stand alone as a reference work for archaeologists in public agencies, private firms, and museums, as well as a textbook and guidebook for classrooms and field settings. The books function even better as a set, because they are integrated through cross-references and complementary subject matter.

Archaeology is a rapidly growing field, one that is no longer the exclusive province of academia. Today, archaeology is a part of daily life in both the public and private sectors. Thousands of archaeologists apply their knowledge and skills every day to understand the human past. Recent explosive growth in archaeology has heightened the need for clear and succinct guidance on professional practice. Therefore, this series supplies ready reference to the latest information on methods and techniques—the tools of the trade that serve as handy guides for longtime practitioners and essential resources for archaeologists in training.

Archaeologists help solve modern problems: They find, assess, recover, preserve, and interpret the evidence of the human past in light of public interest and in the face of multiple land use and development interests. Most of North American archaeology is devoted to cultural resource management (CRM), so the Archaeologist's Toolkit focuses on practical approaches to solving real problems in CRM and

public archaeology. The books contain numerous case studies from all parts of the continent, illustrating the range and diversity of applications. The series emphasizes the importance of such realistic considerations as budgeting, scheduling, and team coordination. In addition, accountability to the public as well as to the profession is a common theme throughout the series.

Volume 1, *Archaeology by Design*, stresses the importance of research design in all phases and at all scales of archaeology. It shows how and why you should develop, apply, and refine research designs. Whether you are surveying quarter-acre cell tower sites or excavating stratified villages with millions of artifacts, your work will be more productive, efficient, and useful if you pay close and continuous attention to your research design.

Volume 2, *Archaeological Survey*, recognizes that most fieldwork in North America is devoted to survey: finding and evaluating archaeological resources. It covers prefield and field strategies to help you maximize the effectiveness and efficiency of archaeological survey. It shows how to choose appropriate strategies and methods ranging from landowner negotiations, surface reconnaissance, and shovel testing to geophysical survey, aerial photography, and report writing.

Volume 3, *Excavation*, covers the fundamentals of dirt archaeology in diverse settings, while emphasizing the importance of ethics during the controlled recovery—and destruction—of the archaeological record. This book shows how to select and apply excavation methods appropriate to specific needs and circumstances and how to maximize useful results while minimizing loss of data.

Volume 4, *Artifacts*, provides students as well as experienced archaeologists with useful guidance on preparing and analyzing artifacts. Both prehistoric and historic-era artifacts are covered in detail. The discussion and case studies range from processing and cataloging through classification, data manipulation, and specialized analyses of a wide range of artifact forms.

Volume 5, *Archaeobiology*, covers the analysis and interpretation of biological remains from archaeological sites. The book shows how to recover, sample, analyze, and interpret the plant and animal remains most frequently excavated from archaeological sites in North America. Case studies from CRM and other archaeological research illustrate strategies for effective and meaningful use of biological data.

Volume 6, *Curating Archaeological Collections*, addresses a crucial but often ignored aspect of archaeology: proper care of the specimens

and records generated in the field and the lab. This book covers strategies for effective short- and long-term collections management. Case studies illustrate the do's and don'ts that you need to know in order to make the best use of existing collections and to make your own work useful for others.

Volume 7, *Presenting the Past*, covers another area that has not received sufficient attention: communication of archaeology to a variety of audiences. Different tools are needed to present archaeology to other archaeologists, to sponsoring agencies, and to the interested public. This book shows how to choose the approaches and methods to take when presenting technical and nontechnical information through various means to various audiences.

Each of these books and the series as a whole are designed to be equally useful to practicing archaeologists and to archaeology students. Practicing archaeologists in CRM firms, agencies, academia, and museums will find the books useful as reference tools and as brush-up guides on current concerns and approaches. Instructors and students in field schools, lab classes, and short courses of various types will find the series valuable because of each book's practical orientation to problem solving.

As the series editors, we have enjoyed bringing these books together and working with the authors. We thank all of the authors—Steve Black, Dave Carmichael, Terry Childs, Jim Collins, Charlie Ewen, Kevin Jolly, Robert Lafferty, Brian Molyneaux, Kris Sobolik, and Lynne Sullivan—for their hard work and patience. We also offer sincere thanks to Mitch Allen of AltaMira Press and a special acknowledgment to Brian Fagan.

LARRY J. ZIMMERMAN
WILLIAM GREEN

1

DESIGNING ARCHAEOLOGICAL RESEARCH WITH AN ATTITUDE

Archaeology is expensive, time-consuming, and, as Kent Flannery put it, "the most fun you can have with your clothes on." Envisioning and articulating a research strategy is a creative exercise that forces you to consider carefully all the contexts of an archaeological project: the players, the laws, the physical environment, the cultural setting, the logistical constraints, and the state of our shared archaeological knowledge. A well-designed research project allows you to target, acquire, and analyze the data needed to address meaningful, interesting research problems while coping with the often competing interests of sponsors, regulators, peers, and the public. The task of designing your research is one of the most essential, intellectually stimulating, and engaging aspects of archaeology. And you can do it with your clothes on.

This volume explains the process of designing archaeological research. It begins with an overview of the archaeological research process, emphasizing the value of carefully built research design in all projects. The book then identifies the conceptual tools needed to do archaeology effectively, and it focuses on ways to plan for and implement successful projects.

While the focus here is on research mandated by federal and state laws, or cultural resource management (CRM) archaeology, much of the approach we advocate applies equally to academic and grant-funded archaeological research. The steps presented here aren't the last word in designing archaeological research, but they reflect the best of what we've seen work and the worst of what we've seen fail. What we've noticed more than anything else is the importance of attitude. The key difference between good, interesting research and boring, worthless projects can be traced right back to attitude.

Every time we archaeologists go to the field, we are given an opportunity to learn something new about past humans and their landscapes. And to gain those hard-won insights, we often destroy or make way for the destruction of the very remains, deposits, and patterns that provide those insights. We need to take the attitude that each research opportunity is precious, that our shared archaeological heritage is built of thousands of small things, that every time we attack the archaeological landscape, the archaeological record will gain more than the landscape has lost. We just need to make the most of each and every research opportunity.

The way that we do that is by stating our assumptions, our plans, and our intentions before we hit the field. We lay out our approach to a particular project or research question and then explain what we're going to do, how we're going to do it, and what we think we will find. Designing archaeological research allows our peers, sponsors, and regulators to become engaged in our research, rather than to simply be consumers of the bare facts and arcane interpretations we generate at the final bell. More important, though, when we design our research, we engage ourselves. We give ourselves the ability to make the most of the research opportunity, for our own satisfaction and for posterity as part of the ever-growing archaeological record.

It's that attitude—making the most of opportunity—that is archaeology by design.

THE PROCESS OF DESIGNING ARCHAEOLOGICAL RESEARCH

Designing archaeological research is a process. The first product generated from that process is the research design. Research designs don't all look alike. Sometimes a research design may be part of your response to a request for proposals (RFP), or it could be part of a permit application, grant proposal, internal memo, or even a term paper. Like any other tool, what it looks like isn't important. Marshalltown and Goldblatt trowels both do the job equally well in skilled hands. What is important is that the cutting edge is sharp, the handle strong, and, most critically, that it is guided by a keen observer. Research designs can take many different forms and can be written (or remain unwritten) for quite different audiences. But to accomplish effective archaeological research, your research design should entail five critical things:

The Five Essential Elements of a Research Design

- A research context
- Explicit research questions
- Definitions of the data you plan to collect
- A plan to present your work and results
- Accommodation to the real world

Your research design should reflect a larger set of goals—a research context for your work. In CRM archaeology, many projects will seem so small that you might think they don't need a research design. But multiply those small projects by a hundred or a thousand a year, and you're looking at a sizable chunk of archaeology. That's why you always need to identify a research context for your work. It might be a favored theoretical framework, a regional research design developed by your state historic preservation officer (SHPO), a personal research agenda, or an institutional research focus. Whatever it is, likely a combination of several of these, this larger research context is what allows you to fit your specific research design (and your work) into the larger body of archaeological knowledge and contribute to the field. No matter what kind of project it is, large or small, your research design can help ensure that your work contributes to our understanding of the archaeological record.

Your research design should spell out a set of explicit research questions. These can be framed as questions, problems, or hypotheses—different forms of the same thing—a device for giving your research direction. The focus of your questions can be broad or narrow but should match the scope of the project. While a ten-mile pipeline survey is not likely to produce data bearing on the origins of plant domestication, it might help evaluate a regional model of Early Archaic site distribution (or, more likely, produce the data needed to convince a utility company to consider an alternative route). By making your research questions explicit and appropriately focused, you can create a research environment where your planned data collection methods and analytical strategies are firmly and rationally linked to larger goals (scientific goals, management goals, regulatory goals, preservation goals, etc.). The kicker here (and the fun) is to create research questions that are interesting and meaningful within a larger research context.

Your research design should define the data that will be collected to address the research questions. Tied inexorably to the data itself are the methods used to collect it. This portion of the design is really a methodological overview and a guide to how you hope the work will be done. Once in the field, as all experienced archaeologists know, reality takes precedence. Changes in soil conditions, access arrangements, weather, unexpected deposits, and equipment problems all conspire to thwart the most finely laid plans. It may sound backward, but this is where explicit plans are the most critical. If you've carefully defined your methods and the data you hope to collect, then when changes need to occur, your research design will serve as a guide for making systematic changes in the methods (and data), so that the research focus of your project remains intact.

Your research design should provide an after-the-field plan for analyzing and reporting your work. An often overlooked component of an effective research design is that it should spell out an intended analytical approach. The worst time to begin thinking about how you're going to analyze the materials from an archaeological project is when the fieldwork is complete. Once you've hit the lab, there is too much to do and never enough time to do it. And if there is no plan in place, you'll likely fall back on standard descriptive "analyses" that produce reams of paper but precious little data that actually speak to an identified research problem. By defining the analytical approach up front (and modifying it as the fieldwork progresses), you can leave the field ready to begin the analytical chores at hand. An added plus is that by planning for analysis and focusing analyses in purposeful directions, you can often free up time on the back end to explore the new and different questions that inevitably arise through the course of your work and to pursue innovative ways of presenting the work.

Finally, your research design should accommodate the real world. Your research doesn't exist in a scientific vacuum. It happens or fails to happen in a swirling world of business, politics, media, government, personal interaction, tight schedules, juggled budgets, and capriciously enforced bureaucratic policies. The CRM arena is everyday life itself, a world changing so fast that most of us feel like we can't keep up. A preoccupation with the human past may be a soothing antidote to the modern world, but you can't sustain your research unless you pay keen attention to the real world. And reality impacts even the archaeologists who enjoy the sheltered academic world. Like it or not, the real world is a critical part of a successful research design.

With the attitude that every project is a research opportunity, these five core elements build a solid base for an effective research design. But don't lose sight of the real end product of the process of designing archaeological research and the ultimate reason you create a research design in the first place: the research.

PROFESSIONAL ARCHAEOLOGICAL RESEARCH

The real world of archaeological research for most of us employed as archaeologists in North America today lies in the burgeoning field of cultural resource management (CRM). CRM archaeologists in both the public and private sectors spend over half a billion dollars a year complying with and enforcing the various federal, state, and local laws in place to protect and manage archaeological resources. Academic-based and grant-funded archaeology comprise a much smaller piece of the research pie, both in the number of practitioners involved and in the pool of funding available. CRM archaeology and academic research are two different worlds, and the disparities in funding, in research strategies (or the lack thereof), and in perceived status have led to well-developed animosities between practitioners in the two worlds. The animosities can be very real, and they stem from very real differences in the research worlds within which the two groups operate. To design effective research in either world, you must understand the research world in which you operate and work within its peculiar constraints.

ACADEMIC RESEARCH

In the academy, designing research is often initiated in one of three ways. In archaeology-as-science, things get started when the professorial protagonist (and her graduate students) pose some scientifically scintillating research problem tied to an often dense theoretical discourse. They submit a formal research design to the National Science Foundation and, if funded, sally forth (often in some Third World country) to test their ideas in a rigorous fashion. In archaeology-as-culture-history, the research begins with a promising place, a region, or a time period about which not much is known. The professorial protagonist (and his graduate students) may prepare a formal research design and seek government funding or, more likely, develop a short problem statement and a flier for a university field school. Either way, he takes his graduate student assistants and undergraduate student workforce into the field (usually not too far from home), secure in the assertion that their hard work will surely fill in worthwhile gaps in the ever-accumulating record of archaeological knowledge. In classical archaeology, and in the methodologically similar archaeology often done in the centers of high civilization in the New World, things first unfold when the professorial protagonists (and their graduate students) set out on a foreign expedition to uncover some unexplored tell, ancient metropolis, or unplundered tomb. Here the strategy is discovery on a grand scale, though the work may be couched in scientific terms. A prospectus is prepared, grants are sought from the National Geographic Society, and donations are solicited from wealthy philanthropists. Funding in hand, they head out (usually to some Third World country) year after year to dig big holes and find something spectacular enough to fund the next season.

In academic research, you (or the big cheese who heads the project) have the opportunity to choose where to work and what problems to work on. You may need permission or permits from national, state, or local governments, and your chances for tenure, another permit, or another grant may depend on your publication record, but you rarely have a legally binding obligation to publish your results or even to finish your research project (although you do have an ethical obligation!). As a professional archaeologist with graduate school in your rear-view mirror, you will be able to take academic approaches to archaeological research only if you manage to land a tenure-track teaching job or have independent means.

CULTURAL RESOURCE MANAGEMENT RESEARCH

In the CRM world, research design starts in government offices, business suites, corporate headquarters, and law offices with a decision that will impact or destroy archaeological deposits. Usually the driving force is progress—new highways, buildings, drainage improvements, pipelines, and a hundred other reasons to move dirt and lay steel and concrete. These agency heads, CEOs, lawyers, and politicians rarely care or think about the human past: they want the future, and they want it now. Archaeology is not even an afterthought until someone down the bureaucratic food chain realizes that their project won't move forward, and their large budgets will languish unspent until those pesky archaeological sites have been dealt with. Then the archaeology happens. In a hurry.

The research design starts with the area of potential effect (APE), drawn by an engineer or an architect whose interest is the compressibility of the soil, not the depth of the A horizon or the boundaries of an ancient campsite. Depending on the size of the construction budget—and in proportion the size of the budget for archaeological research—a government archaeologist may get involved to focus the work more finely. An advertisement in the *Commerce Business Daily* may announce the project, and if you're interested, you request a copy of the RFP. This document can call for anything from a general bid to "survey four hundred acres for cultural resources" to a massively detailed data mitigation plan (such as a large-scale excavation to salvage archaeological information in advance of an impending construction project). To get the money to do the work and win the opportunity to learn more about the archaeological record of the APE, you must successfully meet the requirements of the RFP. No ifs and no buts, not much wiggle room, and no extra points for a brilliant theoretical framework. That's the way it is in the CRM research world.

Assuming you meet the RFP's technical specification, you must also figure out how to do the work for less than what your competitors charge. Once you've successfully won a project, there are binding contracts with nonperformance clauses, line-item budgets, and deliverables. If all this sounds frighteningly restrictive, keep in mind that as a CRM archaeologist, you will be dealing with budgets and resources that most academic archaeologists can only dream about. For most budding professional archaeologists today, these things will form the reality of your research world, and they will profoundly shape and color the way that you design your research.

A VERY SHORT HISTORY OF CRM ARCHAEOLOGY

Beginning with the Depression-era federal make-work programs in the 1930s and continuing after World War II with the Smithsonian River Basin surveys, large-scale government-funded archaeological research was done and managed by academic archaeologists and their best students. This was the era of "salvage" archaeology, rescuing or salvaging what could be learned and collected before government-funded reservoir projects flooded many of the nation's rivers. The money was modest, and the archaeology of the day was basic culture history—big digs at big sites. Most of the money was spent on fieldwork, resulting in massive and minimally studied artifact collections and comparatively few published accounts. There was little competition among the universities because the money was doled out to the relatively few established programs in parts of the country where reservoirs were being built. There was a close working relationship between the academic archaeologists interested in North America and the small number of government archaeologists at the Smithsonian and the National Park Service.

As the salvage era drew to a close in the 1960s, the consensus approach of culture history came under attack by young academic hotheads like Lewis Binford. For the next decade or more, the ivory tower swayed in turmoil as the "New Archaeology" blossomed on the scene. Amid lots of incomprehensible rhetoric, "archaeology with a capital S," as Flannery put it—Science—took hold. With the manifesto that archaeologists needed to follow an explicitly scientific approach and seek to learn about cultural processes instead of culture history, the need for purposeful research design came to the fore. It wasn't enough to seek out interesting places to dig, expose architecture, amass artifacts, and establish sequences. The archaeological record was far more complex and difficult to understand than had previously been acknowledged. Dead cultures could not be reconstructed, even on paper. Understanding cultural processes and the complicated patterns of human evolution required a new paradigm—a different model of how archaeology should operate. And academic archaeologists began to explore, argue about, and champion all sorts of scientific and pseudoscientific investigative pathways.

While this theoretical and methodological fray was taking place, "contract" archaeology was born with the passage of the Reservoir Salvage Act of 1960, the National Historic Preservation Act (NHPA) of 1966, and the National Environmental Protection Act of 1969. The NHPA was particularly influential because it said that when the federal government is involved in any way in land development and land use, cultural resources must be taken into consideration. Implementing regulations and amendments to this act and the passage of other federal and state laws extended and institutionalized the new field of cultural resource management (see sidebar 1.1). Over the next several decades, funding for archaeology grew by leaps and bounds. Government money coupled with overall economic and population growth also led to the establishment of departments of anthropology and the creation of new contract programs at many smaller state universities and colleges. Few of the university-based contracting programs were integrated with the academic departments. Most were stand-alone (or weakly linked) programs that depended solely on contracts for continued existence.

In the late 1960s and early 1970s, the university-based contract programs did almost all of the research projects that the swelling

1.1. WHAT THE HECK IS A CULTURAL RESOURCE?

Cultural resources can be all kinds of things, but the ones that matter in CRM are the ones that are recognized by the National Register of Historic Places (NRHP) and spelled out in section 106 of the National Historic Preservation Act (NHPA). For these purposes, cultural resources are things and places. The cultural resources that section 106 seeks to preserve are the physical manifestations of cultural activities—stuff that can be managed and preserved. Archaeology is only a part, and actually the smaller part, of the national historic preservation effort. A much larger part of what are considered cultural resources under state and federal laws consists of the built environment. (See King 1998 for a comprehensive treatment of cultural resource laws and practice.)

The National Register places cultural resources into five classifications: buildings, structures, objects, sites, and districts. *Buildings* are buildings—houses, barns, and train stations, for example. If there's only part of a building or only a foundation, it's usually considered a ruin and is classified as a site, not a building. Most historic archaeological sites are classified as sites rather than buildings. *Structures* are things like bridges, dams, and grain elevators—basically anything built that can't be classified as a building. *Objects* are things that are built but that are smaller than structures or are decorative in nature, such as boundary markers, monuments, or fountains. *Sites* are places on the landscape that are either associated with some significant event or person or contain information important to history or prehistory. Sites can be natural features such as a spring or a hill as well as cultural features such as a battlefield, a place where important ceremonies took place, or an archaeological site. *Districts* are areas with a concentration of buildings, structures, objects, or sites that are thematically linked in some way. The individual features in a district don't have to be particularly significant in and of themselves because a district is viewed as more than the sum of its parts.

Section 106 of the NHPA deals with all of these cultural resources. While archaeologists are concerned mostly with archaeological sites, a whole raft of managers, historians, architects, and even cultural anthropologists work with other pieces of the cultural resource pie. Historians and architects focus primarily on documentation and on preserving and sometimes restoring cultural resources as they appeared during the time that they were historically significant. In contrast, archaeologists are often interested in preserving the *information* represented by the patterns of materials at a site. In many parts of the CRM world, practitioners have developed interests or expertise in several areas and may wear several cultural resource hats.

ranks of government archaeologists couldn't do themselves. As more programs sprang up, academic contractors began competing head-to-head with one another for projects. But many academic-trained archaeologists, both those in government who were issuing the contracts and those who were contractors, proved to be poor administrators and poor judges of the real costs of doing solid research. There was little quality control, few established rules, and not much accountability. For a while it didn't matter because the pace was so fast—always a new contract to spend from while the old one was being finished. This vicious spiral eventually caught up with many programs, and university officials began pulling the plug. By the mid-1970s, it was obvious that the universities couldn't meet the increasing demand for work, and private contracting firms were springing up all across the country.

By the mid-1970s, CRM archaeology had already developed a bad reputation from the harried work pace, cutthroat bidding, atheoretical approaches, excessive government regulation, and shoddy reporting. It is equally important to recognize that it wasn't all thorns—lots of good archaeology was being done in the contract world of the day. Some contract programs were well administered and effectively exploited the many opportunities to do meaningful archaeological research. Excellent scientific research designs were drawn up and executed. Many young professional archaeologists learned firsthand how to read the dirt—they investigated more archaeological sites in a decade than most academic archaeologists did in a lifetime. They also wrote some fine reports and, to a lesser extent, scholarly articles that presented solid data and keen insight into the archaeological record. Dedicated government archaeologists fought within their agencies to put genuine concern for cultural resources in the mainstream of decision making. A conservation ethic—the idea that sites should be protected for the future whenever feasible—was widely embraced. The seeds of both the worst and best of today's CRM archaeology were firmly planted.

By the mid-1980s, much of the archaeological work across the country was being done by private consulting firms. These for-profit businesses ranged from one-person firms to large environmental and engineering consulting firms in which archaeology was just one of the services offered. The growth of private consulting firms was paralleled by increasing numbers of federal government archaeologists in the big, land-holding agencies (like the Forest Service and the Bureau of Land Management [BLM]), big land development agencies (the U.S.

Army Corps of Engineers), and many smaller agencies. It wasn't just the federal government that hired archaeologists; state governments and even some local ones did, too. Government archaeologists did regulatory work, they did archaeological contracting, they did public service, and some of them even did research. Archaeology had irrevocably escaped the academic world.

WHY CRM ARCHAEOLOGY HAS A BAD REPUTATION

It's the truth. Academic archaeologists have looked askance at CRM or "contract" archaeology for years. Even within the ranks of CRM archaeology, there is a certain pervasive and perverse self-loathing that diminishes and minimizes the work of thousands of professional archaeologists as not quite "real" archaeology. But this bad reputation is only partly deserved, and it is completely within the power of the current archaeological generation to refute.

A primary reason for its bad reputation is that CRM archaeology generally happens in a hurry. Slice it how you will, haste really does make waste when it comes to learning about the human past. Academic archaeologists take their sweet time because they have to: Good research ideas take time to develop and a heck of a lot of concerted effort to investigate. In the academic world, scholarship really does matter, and it is consistently rewarded. It's a lot harder to come up with thought-provoking questions and carefully honed scholarly publications when progress means looking over your shoulder and considering a lower bid. The speed with which CRM archaeology proceeds has led to the worst aspect of CRM: archaeology by default. Too much archaeological research that is rushed to the field, through the lab, and out the door is ill-considered, boilerplate technical garbage that adds little or nothing to the archaeological record.

If you want to design and carry out effective archaeological research in the world of CRM archaeology, you'd better know what you are getting into. CRM archaeology is young, and many of the problems and pitfalls can be traced to its youth. Its history is still unfolding, its future uncertain. It also has incredible potential to do interesting and effective archaeological research—but only if you learn the system and position yourself to take advantage of the opportunities that will come your way.

Most professional archaeologists today and in the foreseeable future will make their living in the world of CRM archaeology. Whether you're a regulator, an agency archaeologist, or a private contractor, your world will revolve around compliance-driven research. As the twenty-first century begins, the world of CRM archaeology is continuing to mature and become institutionalized. The formation of a trade association, the American Cultural Resources Association (ACRA) in 1995, and the 1998 establishment of the Register of Professional Archaeologists (ROPA) are signs that the field may be coming together to deal with the problems and potentials in modern CRM archaeology. You have a chance to help by being part of that maturing process.

But you can't just trot out your academic training, puff out your chest, and claim the high road of science. You have to be pragmatic and smart and figure out how to balance science with government and business. It is not an easy path to follow; in fact, there really isn't much of a path at all. It's up to us practicing professional archaeologists, CRM and academic alike, to change CRM's bad reputation. The good news is that the surest way to do this is to make our work more interesting, more challenging, more useful, and more fun, by designing our research with an attitude.

WHY DESIGN ARCHAEOLOGICAL RESEARCH?

Designing archaeological research—creating a concrete plan—is where every research project should start and end. If it's done well, it focuses your thoughts and energies so you can make the most of your time and resources. It provides a context for making difficult field decisions. It directs the analysis into areas that are (or should be) interesting and productive. It provides a structure for reporting your work and results. A good research design can help keep archaeological research interesting, productive, and fun. More than that, though, as a researcher involved in the destruction of the archaeological record, as a steward of (mostly) public money, and as a member of a professional community, you have extra obligations that should color the attitude you bring to the work.

Modern archaeology is expensive, and you are responsible for delivering value for the money that you will spend. For CRM archaeologists, the American public supports the preservation laws that provide the opportunity to explore the archaeological record, and you owe them much more than a slap and a tickle when it comes to presenting your research. When methods are unconsidered, data are collected by default—and the archaeology, the analysis, and the conclusions are short-changed. Finally, if designing research is approached as a hurdle, a technical hoop, a box to be checked, then the whole point of doing archaeology is missed. It's thinking, and planning, and analyzing, and understanding that is archaeology—real archaeology that does a service to the archaeological record, the public, and the discipline.

THE COST OF ARCHAEOLOGY

Modern archaeology costs a lot because it is labor-intensive. The federal government alone spends more than a quarter-billion dollars each year on CRM archaeology, and state and private sponsors may spend twice that or more. Most CRM labor is provided by workers with college degrees. Lots of them have acquired highly specialized knowledge in graduate school and through years of on-the-job training. While archaeologists are not paid very well by the standards of many other professions, most of us with graduate degrees and experience get paid $30,000 to $60,000 per year. Archaeology often requires a large labor force. The work is usually far enough from home so that you have to pay to house and feed these folks. Even if the average "field technician" gets paid poorly, when you start multiplying the number of weeks times the number of crew members and then add in the salaried supervisors, travel, housing, food, equipment, and supplies, we're talking serious money.

After the fieldwork ends, the analysis begins. It can be labor-intensive as well, particularly for excavation projects. The longer you dig, the more you haul back to the lab for processing and analysis. In the heyday of culture history archaeology, the analytical emphasis was on describing, classifying, and illustrating the artifacts, usually only the complete tools and whole vessels. This trend carried over into CRM archaeology. Over time archaeologists started looking at and analyzing (or attempting to) all the artifacts they recovered—whole, broken, tools, flakes, sherds, everything. There is, however, only so much you can learn from artifacts alone. Simply processing those collections takes time and money. Since the 1960s, archaeologists have spent more and more time studying bones, plant remains, soils, rocks, snails, residues, and a dozen other types of evidence. Radiocarbon dating has become routine and often costs over $500 for a single sample. A lot has been learned, but many of these materials require specialized study and hard-science analytical techniques. Like everything, analytical and curation costs have gone up and up.

Modern archaeology takes lots of skilled people, lots of specialized analyses, and lots of time—and money. But the expenditure of big piles of money does not guarantee that the research will be worthwhile. In fact, a substantial portion of the money being spent on archaeology today is misdirected to pro forma archaeology done by technicians who rely on default ways of doing things because the boss didn't invest the time to develop a thoughtful research strategy.

YOUR OBLIGATION TO THE PUBLIC

As a CRM archaeologist, you have a weighty obligation to the public. They are the ones whose interest and support for the human past and historic preservation caused the laws to be passed that brought CRM archaeology into existence. They are also the ones who pay for almost all CRM archaeology. In a very real way, your future as a professional archaeologist depends mightily on the continued support of the American public. If we archaeologists screw up our opportunities much longer, there is a very real chance that the laws and regulations will be weakened as the public loses its interest and faith in what we do.

Sounds heavy, doesn't it? It is, and, as a group, CRM archaeologists are just starting to own up to our obligation to the American public. It's finally dawning on us that the golden goose is starting to get pissed.

First, we have to consistently strive to spend our research dollars wisely. This means we have to think carefully about the work we propose and make sure that the research we accomplish is worthwhile. We have to cultivate the attitude that we should get the most we can from each opportunity—to provide value for the money, to justify the faith the public puts in us as archaeologists, and to help understand and preserve the past. When the public sees a project wasting money, they are quick and ruthless. At the same time, most folks are very reasonable if there is a good reason for the money to be spent, a plan in place, and evidence that the project is moving in a positive direction. You're spending the public's money, and unless you'd like to join the $600 hammer and the $1,000 toilet seat in the annals of government waste, you must have a well-conceived plan for performing your research.

Second, we have to give back to the public a much greater share of what we are learning. Our dull technical prose won't cut it. Neither will our endless preoccupation with hair splitting. The public wants to see the big picture and enough of the details to make it interesting, understandable, and believable. In plain English. On the TV. On the Web. In popular books. We have to make our research relevant and interesting to the American public. Not all of it, especially not all the split hairs and abstract discussions, but enough of it so people will know why understanding and preserving the human past is important. If you want your project to succeed in the increasingly public arena of CRM archaeology, and if you want the public to continue to support our profession, you'd better have a plan for sharing some of what you learn with the interested public. It's part of your research design.

ARCHAEOLOGY BY DEFAULT

Many decisions in archaeology are made simply out of habit, following established traditions for reasons that may or may not make sense today. It's worth remembering that CRM archaeology is, to a significant degree, a continuation of the culture history approach to archaeology. So is academic archaeology. Many of the field traditions that characterize American archaeology today, academic and CRM alike, were originally developed to address the chronological and artifact-oriented questions that characterized American archaeology in the mid–twentieth century.

The classic example is excavation by uniform and arbitrary metric units like the one-meter square and the ten-centimeter unit level. Such rote digging became widespread in this country to permit minimally skilled field workers to maintain basic provenience while digging deep or big holes. It replaced the wholesale strip mining for artifacts and unsystematic digging that characterized archaeology in the early twentieth century and represented an important improvement. It persists today in this country to an extent unknown in virtually all other areas of the world. We personally have dug hundreds of square holes, sometimes to good effect, sometimes not, but almost always out of habit, convenience, and tradition.

The square hole method works pretty well when you are digging test pits in unknown territory to answer basic exploratory questions. It sometimes works well in block excavations, particularly those where the stratigraphy is uncomplicated. But all experienced archaeologists realize that arbitrary units don't work very well when the stratigraphy is complex and subtle. The standard units invariably result in the mixing of things from different strata that should be kept separate. European archaeologists have developed much more sophisticated, if sometimes difficult-to-execute, systems of stratigraphic excavation and recording. Yet the one-meter square is still the first thing that most beginning archaeologists in this country learn to recognize as the basic excavation unit. Square hole archaeology is a tradition and a very widespread default method in American archaeology. The implications of the method—constrained data sets, limited spatial resolution, and arbitrary artifact associations—are seldom mentioned in either research design or analysis.

While the square hole and many other field traditions are common to both academic and CRM archaeology, default methods are particularly rampant in all aspects of CRM work: field, lab, and reporting. Unthinking acceptance of the status quo often results in rote, mechanical work in which the participants are merely going through the motions of the process rather than concentrating on making and conveying the critical observations.

The fast-paced, competitive world of CRM archaeology provides many reasons to have standard methods. If we had to reinvent the one-meter square or a laboratory data entry form on every project, we wouldn't get very much dirt moved or data processed. In each archaeological region there are standard approaches to typical problems. For instance, archaeologists in the arid West often rely on surface survey and collection to locate and begin to evaluate sites, while those in the wooded East often use shovel tests for these purposes because sites there are buried or shrouded in dense vegetation. It's not the approach, standard or innovative, that matters but how you apply that approach to the particular circumstances for the research you're doing.

Standard methods should be explicitly justified just like anything else: by weighing the costs against the benefits. If there are compelling reasons for using them, fine. But if there is a potentially better, less costly, more informative way to learn what we need to and intelligently sample the archaeological universe in front of us, then let's try it. When we approach problems by falling back on the intellectually lazy use of archaeology by default, we shortchange the archaeology and ourselves.

DESIGNING RESEARCH IS ARCHAEOLOGY

Archaeology is more than digging holes, counting potsherds, classifying arrowheads, and obtaining radiocarbon dates. We have to recognize the complexity of the human past in light of the fact that most of it is unknowable. The archaeological record is static, dead as a doornail, and decaying and being destroyed continually. We can, however, learn real verifiable, replicable things about the static archaeological record. Using a scientific approach, we can formulate and test ideas for explaining the cultural processes and causal links that account for the patterns we find in archaeological deposits. But there are also widely accepted ways of doing archaeological research that aren't based in science. Archaeology survives and even thrives as a hobby, culture history, a technical service, and a business. These models work fine if that's all archaeology is meant to be. But for the true professional archaeologist, the fun, the excitement, the joy of archaeology is in learning new things and discovering patterns we didn't know existed—coming to human terms with tiny bits of an almost unknowable past. We design archaeological research to develop a strategy for learning those new things.

The attitude you bring to designing your research should acknowledge these things and use them to focus your work. The design you create is just a tool. If you need to clean a profile for a photograph, the first thing you'll likely do is grab a file and sharpen the edge of your trowel. A research design is like that file: You use it to hone your ideas, your methods, your strategy, so when you bear down on that face (or site), you're not gouging valuable (and costly) hunks from the archaeological record. A research design is a tool you use to organize and present your ideas, lay out your plan of attack, and connect the results that will be produced by your chosen field techniques to the problems you want to address.

2
BUILDING YOUR PROFESSIONAL TOOLKIT

To design effective research—research that will contribute to our knowledge of the archaeological record, research that allows you to reach achievable goals, and research that will be engaging, fun, and interesting—you need more than attitude: you need tools. You need tools that provide a good grounding in archaeology in general and the regional archaeology where you plan to work in specific. You need tools to understand the range of multidisciplinary options available for examining and analyzing archaeological deposits. You needs the tools to place your work in the context of the regional and fieldwide understanding (or misunderstanding, as the case may be) of the archaeological record. You need tools to manage the bureaucratic and political environment in which your research lives and dies. And you need the tools to manage the day-to-day tactical and long-term strategic business planning that ensures the crew is paid and the truck has gas.

Before anything starts, you need to have your tools sharp and ready to work, and designing archaeological research is no exception. Good, interesting, effective archaeological research flows directly from the time you spend investing in these tools. It's all too easy to spot a project that's been done by folks who view archaeology as little more than hunting for arrowheads and slapping together a table of artifacts and a sketchy site map. To design and execute effective research and make a contribution to the field, you have to first load and then lug your professional toolkit. Here's a quick overview of the tools you'll need.

ANTHROPOLOGICAL ARCHAEOLOGY

The very first thing to load into your toolkit is a good, up-to-date understanding of the goals, history, theory, and methods of modern archaeology and its mother discipline, anthropology. While many archaeologists trained in the Old World see archaeology as part of history and classics, most American archaeologists are trained in anthropology. Archaeologists with a solid foundation in anthropology are more capable of understanding and explaining the material things we dig up than are those who have only technical training in archaeology.

Included in this anthropological foundation should be a basic overview of physical anthropology. Humans represent just one peculiar branch of millions of years of primate evolution, and we share many behaviors and physical characteristics with our fellow apes (members of the superfamily Hominoidea). We need to understand behaviors as controlled by biology and have a good grasp of how the human body responds to challenges like disease, pregnancy, poor nutrition, and extreme climates. Knowledge of primate behavior and human physiology will help you ask interesting questions about the human past. You will also gain insights into why leading experts—like male silverbacks defending their territory—often act like such thuggish jerks.

By studying sociocultural anthropology you learn that biology doesn't explain it all. Humans are social critters who develop and pass along complex behaviors that override and go beyond our genetically conditioned habits. Culture is learned and socially transmitted human behavior. You need a good working knowledge of how human societies, large and small, work—how people are organized or not, how they make a living, how they interact with others, how their belief systems are (or are not) related to the material things that will become the archaeological record.

Along with a good introduction to sociocultural and biological anthropology, you will need to know more about the many aspects of the subdiscipline of archaeology. An appreciation of the history of archaeology is essential because it gives you a strong sense of how young the discipline is relative to most fields of inquiry and because you'll learn how rapidly archaeology has changed over the past century. A strong grounding in field methods is obviously essential. We are endlessly amazed (depressed, really) at how many so-called archaeologists can barely wield a trowel, recognize a soil change, or use a transit. But technical skills are all too mechanical if you are clueless about theory. Sure, too much of what passes for theory in archaeology is abstract word juggling and half-baked borrowing from the hard sciences, but you really do need a theoretical perspective—an explanatory framework within which you will operate—and you need to know how to articulate that perspective.

Beyond method and theory is analysis—the process of assembling and assimilating the field data and figuring out how to wring useful information from the sherds, bones, and stones that we dig up. All archaeologists should have at least a basic introduction to the gamut of analytical techniques that may prove essential.

A solid high school education followed by a four-year college degree in anthropology should give you most of what you will need to get started and get your archaeological feet wet. Once you have some real field experience and an inkling of what it will take to stay gainfully employed in archaeology, you will realize that most career paths involve graduate training. Increasingly, the more responsible and better paying positions in archaeology are held by those with at least a master's degree. In one sense, an advanced degree is just a piece of paper, a jumped hoop. But more importantly, graduate school is a chance to learn the things you didn't learn in college. To make it really count, you need to know what you want to learn before you get there so you can figure out how to structure your experience to your advantage.

CURRENT ARCHAEOLOGICAL METHOD AND THEORY

Method and theory are often trotted out like a cute pair of twins that get "oohed" and "ahhed" over but then get wheeled back into the house when the real work starts. To do effective and interesting research, your methods must be explicit, well defined, and replicable. You employ your methods within a research context defined by one or many theoretical perspectives. Theory is much more than the abstract ideas jawed over in a graduate seminar: It is the foundation for productive archaeological research. Your theoretical approach helps you define the data you expect to recover and how you will utilize the data in your analysis. Your methods determine the data you do recover. This is worth saying again: Your methods determine your data. You must have an explicit and clear understanding of the implications of the methods you choose on the data you hope to collect. All effective work starts with a solid foundation in method and theory.

Things change, constantly. The way we look at sites, the way we analyze archaeological deposits, and the questions we ask change through time. They all change because we've got new people, new theories, and new methods coming down the pike every day. You don't want to chase every fad that gets pumped into the mix, but you do want your research to be fresh and interesting. The only way to avoid stale, overworked (boring) approaches and to steer clear of the current "pet rock" approach is to maintain an active interest in the field as a whole. It's important, although often difficult, to maintain intellectual energy when doing CRM work. One way you can keep the mental buzz going is by paying attention to new ways of doing things. Are there new dating techniques that might apply to questions you'd like to ask about your site? Are there new ways of collecting and analyzing archaeological information that you might make use of? New ideas and ways of doing things feed the process of creating new, interesting, and exciting research.

MULTIDISCIPLINARY STUDIES

Even an archaeological generalist should have an appreciation for and a basic understanding of other archaeologically related disciplines: biology, botany, paleontology, geology, geography, mathematics, chemistry, and physics as well as history, literature, art, and sociology. Don't squawk yet—you don't need to dedicate your life to these things, but you should be familiar with them and how they impact the work you do in archaeology. For instance, archaeometric dating techniques often come from methods pioneered in paleontology, high-energy physics, or even lunar exploration. You'll be a more effective designer of archaeological research if you have an appreciation for the many different fields that archaeology can involve. Learn to take advantage of opportunities to learn from and work with folks in other related fields.

Enough? Heck no. Newcomers are often surprised to learn that much of the daily life of an archaeologist revolves around things that seem to have absolutely nothing to do with archaeology. Things like managing a budget, hiring and firing people, and running an effective meeting are generally taught in business school. Things like writing and producing an interesting report mean training in writing and a basic understanding of the mechanics of publishing. Archaeologists need many skills beyond the ones that you will get with a degree in anthropology. A smart and far-thinking archaeologist-in-the-making (or one trying to catch up) needs to focus on the whole pie rather than just a single slice. Most of us can't do everything and shouldn't try. But all of us can benefit from understanding how all the pieces and parts fit together.

If you're unfamiliar with managing a budget or hiring people, pick up one of the thousands of how-to books on the market, and give yourself a quick course. The business details of managing an archaeological project are just as critical as the research plan. If you botch a project because of poor management, the damage to the archaeology is just as real as if you'd bulldozed the site. When you can manage your resources effectively, you'll produce more and better work and make a greater, more positive impact on the body of archaeological knowledge.

KEEPING CURRENT

Keeping your knowledge current is just as important as a having a good background. New ideas, theories, and methods appear on the scene almost daily. It's important to keep up with these changes, because staying engaged in the discipline—talking, writing, and e-mailing with folks about real archaeology—keeps you energized and makes archaeology more stimulating. Keeping current simply means staying engaged in the field—engaged with archaeology—with the people and the literature. You should join national, state, and local archaeological societies and participate as actively as you can (see appendix A, "Groups You Should Join"). Meetings are where you meet colleagues, hear what others are learning about, check out the newest books, and otherwise "network" (Latin for rubbing elbows while drinking beer and stretching the truth). One of the best professional favors you can do for yourself is to give professional and popular talks to archaeological groups and community groups. It is the best way to focus your ideas, hone your communication skills, and engage yourself—and others—with what fascinates you about archaeology.

The journals and newsletters that archaeological societies produce are your first line of reference (see appendix B, "Journals You Should Read"). There is no better place to stay current on the thinking, research contexts, and new insights in a region than the local and regional society journals. The importance of keeping up with the literature can't be overemphasized. If an archaeologist loses touch with the field, it is all too common that his work becomes stale, his reports and papers become stale, and he himself becomes stale and bored—"There's just nothing new out there." Just a few hours a month skimming the journals and reading a few articles of interest can keep your interest and imagination piqued.

DEVELOPING REGIONAL EXPERTISE

To do truly effective research, you will need to develop an intimate familiarity with the region where you work—the landscape, its history, its environment, and its material culture. Regional experts make their name because they are interested in and knowledgeable about the total package of things that influence the archaeological record: environment, climate, geology, history, and politics. This kind of wide-ranging appreciation for a region takes sustained interest and a long-term investment. We have seen more than one archaeologist who has a deservedly top-notch reputation in their research region but who has an equally well-deserved reputation for sloth and incompetence outside that region. Understanding your research region is a critical tool in your professional toolkit.

First and foremost, you need to have a clear understanding of the physical landscape and the dynamic natural and cultural processes that form and define the landscape. While the principles that govern such forces are pretty much uniform worldwide (remember Sir Charles Lyell), the landscape of any region on Earth has distinctive characteristics and patterns that you will need to understand (see sidebar 2.1). That way you won't waste time sinking fifty-centimeter-deep shovel tests across terraces where the archaeological deposits are likely buried by ten meters of Holocene sediment.

Natural landscapes are more than just rocks, soils, and streams. Archaeologists also need to develop an appreciation for the unique regional mix of plants and animals that form the ecological systems in which past peoples lived (see sidebar 2.2). You need to adopt an ecological perspective if you are to make much progress in understanding the archaeology of your region, particularly when dealing with prehistoric remains. This means you need to learn about how climatic changes have affected the landscape and the distribution of its plants and animals.

You also need a firm grasp of the culture history of the region. Who were the peoples who lived and settled in your region (see sidebar

> **2.1. GEOARCHAEOLOGY**
>
> *Geoarchaeology* is the application of geological science to archaeology. Archaeologists in some parts of the world have long worked closely with geologists. Today, many archaeologists are being trained in both fields. This is a natural development because few geologists are trained in and truly interested in the relatively short-lived phenomena and recent time frames that archaeologists typically deal with. Well-trained geoarchaeologists understand archaeological questions and situations and are able to apply appropriate geological methods. Today, geoarchaeologists have their own scholarly journal (*Geoarchaeology*) and hold academic positions in anthropology and geological science departments as well as private and government CRM jobs. Geoarchaeologists are often integral members of archaeological research teams. Their knowledge of the physical landscape often provides crucial contextual information for archaeologists.
>
> The growth of geoarchaeology has caused archaeologists to become more aware of the depositional circumstances in which archaeological deposits are found and more knowledgeable about how such deposits are transformed by erosion, soil formation, and other processes. For instance, it is now being recognized that many of the artifact-rich sites and deposits once simplistically regarded as "occupations" are *palimpsests*—accumulations from many occupations that formed on stable land surfaces and that are often blended in the active biological zones known as soils. Such contexts can yield important information but only when studied for what they are. Through geoarchaeological work, we are now learning how to find and target context-rich, though often artifact-poor, deposits that have greater potential to address many behavioral and ecological questions.

2.3)? What is the current perception of the sequence and timing of events and basic patterns in the archaeological record? What are the important gaps in knowledge? These are the historical contexts within which your analysis and conclusions must operate.

Last, but certainly not least, you need a good handle on the material culture of the region. Not every archaeologist needs to be a Zen master at pottery typology, but you must be able to recognize and evaluate the artifacts, features, and other materials you'll be encountering. You need this knowledge so you can grasp the significance of what you find in the field and use that information to make informed decisions.

The place to start becoming a regional expert is with the current regional experts and what they write. Their reports contain references

2.2. ECOLOGY AND ARCHAEOLOGY

Ecology is the study of organisms and their environmental relationships. Ecological approaches to archaeology and anthropology, often termed *human* or *cultural ecology*, are robust scientific frameworks. Ecological approaches are particularly useful when archaeologists deal with the vast stretch of preindustrial human existence. Hunter-gatherers and other technologically "primitive" peoples had a complex relationship with the natural world, much more complex and interesting than the simple, romantic notion of the "noble savage." Cultural-ecological studies around the world show that this idyllic view of ancient people at one with nature is a modern fantasy. Past societies often failed or were drastically altered by their own ecologically unsound practices. Similarly, natural disasters such as volcanic eruptions and prolonged droughts have profoundly altered human history, recent and ancient.

For the archaeologist, an ecological approach is often attractive because, unlike most ancient cultural systems, many natural ecosystems have survived, albeit in ever-smaller patches. Ecological studies of plants and animals, as well as climatic and other kinds of environmental studies, provide concrete, quantifiable data—baselines against which past human patterns can be plotted. Studies of paleoecology and paleoclimate often complement archaeological studies because the relevant data sets are intermingled and inform one another. For instance, archaeological deposits often yield preserved plant and animal remains that reveal natural biotic distribution patterns as well as patterns of human exploitation (see Toolkit, volume 4). Evolutionary ecologists have developed conceptual approaches such as optimal foraging theory that archaeologists borrow and adapt with varying degrees of success.

to the key sources with which you should familiarize yourself. Many of these will be nonarchaeological sources such as historical accounts, soil surveys, ecological studies, geological guidebooks, and natural history accounts. Once again, the local and regional archaeological society publications are often the best sources for the most current work in a region. Another resource is regional research designs produced by the SHPO. These often have good overviews of work on a regional level, as well as landscape and material culture information.

Most important, spend your time in your research region wisely—keep your eyes open for opportunities to learn from knowledgeable people who live and work there. If your archaeological project em-

> ### 2.3. ETHNOLOGY TO ETHNOARCHAEOLOGY
>
> In the heyday of culture history archaeology, most sociocultural anthropologists were primarily concerned with *ethnography,* the field study of living cultures, and *ethnology,* the comparative study of such cultures. Today sociocultural anthropology, like archaeology, is a much more diversified field, and relatively few anthropologists are interested in the kinds of questions and lines of evidence that interest most archaeologists. Yet ethnology and ethnohistory (the study of living or once-living cultures through firsthand accounts in historical documents) remain critical sources of interpretive models and tools for archaeologists seeking to understand cultural landscapes Archaeologists still make much use of the ethnographic and ethnohistoric data compiled by old-style cultural anthropologists as well as a few modern ones.
>
> In recent decades, archaeologists have begun to conduct their own ethnographic fieldwork to focus on the dynamic links between material and behavioral aspects of living cultures. The growing subfield of ethnoarchaeology helps us explain the static archaeological record by providing "middle-range" studies of many different aspects of human behavior. While most ethnoarchaeological studies have been done by academic archaeologists, their results are equally relevant for CRM archaeologists. By looking at how living peoples generate garbage, make pottery, and survive blizzards, for example, archaeologists gain interpretive clues that help them make better inferences about the archaeological record. Such studies often show how the hot spots we call archaeological sites are but small pieces of the larger landscapes within and upon which humans live their lives.

ploys a specialist on plants, animals, or geology, make sure you are present when this expert visits. Ask questions and pay attention to what they find noteworthy about your research area. Some of your most exciting archaeological moments will come when visiting specialists share observations that help you fit together important pieces of your research puzzle. You will find that most people enjoy sharing what they know with others in informal settings, particularly if you ask good questions. And don't overlook the rancher, the landowner, and the old-timer. They may not have the scientific lingo and they may harbor folk beliefs, but most people have an innate interest in their surroundings. Experience does yield wisdom. So build rapport, show interest, and be prepared to learn from unexpected opportunities. As you become a regional expert, you'll find yourself giving back as much as you've gotten.

NAVIGATING THE POLITICAL PLAYING FIELD

The ability to navigate the political playing field for your project is a critical component of your toolkit. The world's best-planned and most significant research project cannot survive a loss of funding or a pulled permit. Academic research and CRM research navigate different waters here, but the results of political missteps are much the same. The tools you need are a good understanding of the players and their roles and how they affect your work.

The politics of the academic world are as legendary as the egos. To accomplish successful research, the academic archaeologist must placate or avoid alienating fellow professors in the department as well as various deans and university administrators, each a self-styled master of a small empire. Since much of the labor will be provided by students, the academic must also win a student following, competing with fellow professors for the best graduate students or the right to run the summer field school. Beyond the intrigue and bureaucracy of the university, academic researchers must win the support of many other players. Granting agencies and foundations operate by very different rules. To win a grant from the National Science Foundation (NSF), for instance, one has to comply with page after page of tightly worded requirements and correctly anticipate the political and theoretical biases of anonymous peer reviewers. Many private foundations have much simpler but no less precarious grant approval processes that require salesmanship of a very different sort. Cultivating foundations and wealthy individual benefactors often requires high-brow personal schmoozing and the willingness to emphasize glory and goodies over science. And even with the sabbatical, the student workforce, and the funding in hand, the academic archaeologist must still woo foreign bureaucrats, suspicious colleagues, landowners, and local officials before the first test pit can be dug. Many an academic research project has ended in ruin long before it really began because of ill-played politics on one of many different levels.

CRM archaeologists must navigate an equally strange sea. Most of the work you'll do is generated by federal laws (NHPA, NEPA, etc.; see appendix C, "The ABCs of CRM"), often administered and regulated by state agencies. You'll be working for sponsors obligated to do archaeology under these laws as well as state and occasionally local laws (see sidebar 2.4). Throw in federal and local agencies, time pressure from construction schedules, and political pressures from all levels of government, and you've got a very crowded playing field. All of these factors form the framework within which you have to do your work. Spending time learning the lay of the regulatory and political landscape can save you time, money, and heartache later. Bureaucracies are nothing if not bureaucratic, and the *i*'s you don't dot and the *t*'s you don't cross can come back and bite you on the butt. If you are going to play the game, you must learn the rules.

The first step is a basic grasp of the National Historic Preservation Act (NHPA), especially section 106 of that law and its regulations (see sidebar 2.5). Other federal laws also govern the course of sponsored archaeological research. There is a full alphabet soup of federal CRM laws, rules, and jargon (NEPA, TCP, NAGPRA, etc.) that you will encounter and need to recognize. These laws and regulations, and the way they are implemented by various federal agencies, determine when, where, why, and how archaeological research will be done and funded. This is the guts of most CRM research, and the better your understanding of the process, the better you'll be able to steer your project safely home.

One of the best ways to get up-to-date on these laws and regulations is to sign up for intensive multiday CRM training seminars and workshops. These programs are sponsored by the Advisory Council on Historic Preservation (ACHP), and various federal agencies, as well as private enterprises. The University of Nevada at Reno has developed a good reputation for offering such courses. Look for the training sessions taught by well-respected experts, especially those with experience in the agencies you'll be working with. These people can give you the no-holds-barred advice that you will need. All good training sessions, private and government sponsored, will provide you with copies of the current laws and regulations and take you through practical exercises.

In addition, reams of printed and electronic literature are put out by the federal government. The Advisory Council maintains a website with all the current federal laws, short tutorials on the section 106 process, and recent news and information on how these laws are applied (www.ACHP.gov). The National Park Service also maintains a set of excellent resources for CRM archaeologists (www.nps.gov).

Many federal regulations are administered at the state level by the SHPO and specifically by the state historic preservation officer (SHPO, also) or the deputy SHPO and his archaeology staff. Each state has developed rules or procedures to administer the section 106 process, as well as to enforce state preservation laws. It's a critical part of your toolkit to know who the players are in your state SHPO

> ## 2.4. KNOW YOUR SPONSORS
>
> Your sponsors are the public agencies or private organizations that pay the bills for most CRM research. It's important to recognize that different sponsors have different expectations for your work. The differences in how sponsors view your work will directly affect how you handle the inevitable changes, problems, and potentials you encounter.
>
> Public sponsors, in general, have a better understanding of the CRM laws and the process and pitfalls of CRM research than those in the private sector. Many federal and state agencies have professional staffs to coordinate this kind of research. This means that the RFPs and scopes of work you'll see from these agencies are usually fairly thorough. The staff archaeologists can usually be counted on as in-house advocates for the archaeological resource. If you encounter unexpected opportunities (say a well-preserved buried component where none was expected), these folks can often go to bat at their agencies to extend the scope of work and hopefully the budget to take advantage of these research surprises. On the downside, you may also work with sponsors who set out RFPs and scopes of work so anal and detailed that there is no room for you to apply your experience and imagination. In general, governmental sponsors are concerned with meeting the requirements of the law first, with time and money considerations following closely behind.
>
> There are two kinds of private sponsors. The largest group consists of developers, mining companies, and construction firms that find themselves under a regulatory or contractual obligation. A smaller group consists of organizations that contract for archaeology because they have an interest in archaeology, feel that it's the right thing to do, or see the valuable public relations aspect of the work.

and how the state rules are implemented to manage the process. Many SHPOs offer seminars on the section 106 process and their state and local regulations, which is an excellent way to get up to speed on the factors that will shape your work. All SHPOs also publish guides to the state regulations and processes, and current copies should be part of your library.

One of CRM archaeology's favorite sports is SHPO bashing. Many SHPOs complain (rightly) that some archaeologists are clueless when it comes to the laws and regulations they are paid to implement. Many archaeologists complain (rightly) that the SHPO has lost touch with what's going on in the field. Most of this bashing can be attributed to miscommunication. Archaeologists need to understand that the SHPO's job is to administer and manage the federal regulatory process within the context of state government and local politics. The SHPO sees more of the archaeology done in a state than anyone else—good, bad, and ugly . . . very ugly. A little communication can go a

> Sponsors who are simply fulfilling their regulatory obligations see the archaeology as a hurdle to be overcome. The budget is important, but in the context of a development project, the thing that matters most is the schedule. They usually don't have a great deal of interest in the work itself, other than getting it done on time and getting clearance to proceed. This sounds like a bad situation, but in fact it can be the basis for a very good working relationship as long as you recognize the sponsor's needs. For example, in a privately sponsored cemetery relocation, we found that a section of the ninety-year-old cemetery wasn't recorded on the plat map. The SHPO required that we locate all the burials in that area before construction could proceed. The sponsor supplied heavy equipment and operators, three-wheeled all-terrain vehicles, a professional surveyor, and laborers to help move the project forward. By working together, the sponsor was able to start his project on time, we were able to locate six additional burials, and the SHPO was satisfied.
>
> Private sponsors who are funding the work even though it isn't required are a different case. They're often interested in what you find, and they want to feel like they've done a good thing. The budgets may be small for the work they expect, and there may also be tight schedules to maintain. The challenge for you is to work closely with the sponsors and educate them on what archaeological research really is—that it's the information that's important rather than the artifacts. Then the onus is on you to do the work in a timely manner, knowing that you probably can't expect many project extensions or budget increases. It's important to produce a well-written, readable, interesting report. These sponsors want to feel that they've made a contribution.

long way to smoothing what can be a bumpy process. Get to know your SHPO as a colleague rather than a bureaucrat to open those lines of communication. If your SHPO doesn't make regular presentations to the archaeological community, ask him or her to, and give the archaeological community at large the opportunity to interact with the SHPO.

Archaeologists working on Indian lands may need to operate under tribal laws and regulations that may take precedence over state and federal laws. Many landholding tribes now have their own Tribal Historic Preservation Office/officer (THPO), who performs the same function as the SHPO at the state level and is responsible for administering the section 106 process and applying tribal laws and regulations.

Cities, municipalities, and some counties have preservation ordinances that may control or affect research done within their boundaries. Your SHPO or THPO is generally up-to-date on local ordinances

2.5. SECTION 106 AND YOU

The National Historic Preservation Act (NHPA) of 1966 forms the cornerstone of the federal historic preservation system. Section 106 of the NHPA requires all federal agencies to take into account the effects of their actions ("undertakings") on "historic properties." The law applies to any undertakings in which federal money is spent or federal permits or licenses are issued.

The NHPA also created an independent federal agency, the Advisory Council on Historic Preservation (ACHP, usually just called the Council), to oversee the review process. The rules for the section 106 review and compliance process were developed by the Council and are published in the U.S. Code of Federal Regulations (36 CFR 800).

Under section 106, one of the first things a federal agency is supposed to do when it's going to do, fund, or permit land alteration is to see whether the proposed work will cause changes in the character or use of "historic properties" that are listed on, eligible for, or potentially eligible for the National Register of Historic Places (NRHP), including structures, objects, and districts, as well as archaeological sites. If the agency thinks that the project may affect these resources, it (or its partner state agencies or the entity that needs the permit) is required to consult with the SHPO in the relevant state or states and determine what if any kind of work needs to be done. This determination may lead to requests for proposals (RFPs) and contracts to perform surveys, testing, or mitigation. If this work finds sites that weren't previously identified or listed on the NRHP, these new sites (or new data about known sites) are evaluated

and regulations, so that is always the first place to look (good communication is paying off already!). It is increasingly common in larger cities for there to be a local preservation office that administers local preservation ordinances. Boston, for example, has its own city archaeologist and its own rules under which city-sponsored archaeology takes place.

On the archaeological side, being plugged into the grapevines that monitor federal and state politics is important to stay abreast of pending changes. At the national level, the American Cultural Resources Association's e-mail list (ACRA-L) and *The Archaeological Record* of the Society for American Archaeology (SAA) are sources where you can find current information. At the state level, check with your SHPO and make sure you receive any newsletters or publications it regularly publishes. As we've said before, join your state and regional societies. Most states also have professional organizations of CRM archaeologists; check with your SHPO to find out how to join yours.

by the agency in consultation with the SHPO. This is where determinations of eligibility and potential eligibility for the NRHP are generated.

Once a property is determined eligible, the agency works with the SHPO to determine the effect of the proposed undertaking on the property. If they determine that there will be an adverse effect, the agency again consults with the SHPO to find ways to mitigate the damage—reduce its effects. These consultations may include other "interested parties" that may be affected by the work: local governments, tribes, property owners, and preservation groups. The goal is to develop a memorandum of agreement (MOA) between the SHPO and the agency that details the work to be done to mitigate the damages to the historic properties. Federal agencies that do routine actions often strike a longer term deal with the SHPO (a programmatic agreement) that spells out and simplifies the review and compliance process. In cases where the feds and the SHPO can't agree, the Advisory Council weighs in, but ultimately the federal agency makes the call.

Archaeological contractors involved in the section 106 process are called on to perform reconnaissance, survey, or testing to determine the existence and evaluate the significance of properties within an area of potential effect (see Toolkit, volume 2). Also, you may be involved with data collection efforts designed to mitigate the effects of development on eligible properties (see Toolkit, volume 3). In either case, while the archaeological consultant is an unspoken player in the section 106 process, legal responsibility and authority for decision making falls to the federal agency and the SHPO. It requires your hard work and expertise, but they call the shots.

Every CRM archaeologist needs to know the ins and outs of section 106.

Staying current with federal, state, and local laws and regulations is always a problem. One of the best ways is to maintain an active professional network. Remain involved in professional organizations, and maintain close working relationships with your archaeological colleagues, including those who work for the sponsoring and regulating agencies. Make it part of your job to get to know the folks you'll be dealing with in the SHPO office. That way it will be easy and comfortable for you to call them when you have a question about laws, regulations, or processes. They would much rather spend a few minutes with you on the phone than have to untangle a mess six months down the road. They can also clue you in to changes in the laws and generally point you in the right direction for more information. To keep up with local practices and the political scene, there is nothing better than talking with other archaeologists. It is at meetings, brown-bag talks, and after-hours bull sessions that you'll learn more about what's happening in and out of archaeology. This advice comes directly from cultural anthropology: Be a participating observer of your profession.

As simple as this sounds, far too many CRM archaeologists have little or nothing to do with archaeology beyond what they get paid to do. It shows. Understanding the interplay between the laws and the people who administer and regulate them is critical to the successful completion of your research. The tools to deal with the political and bureaucratic landscape may not seem important when you're in the field, but when the bureaucratic crap hits the political fan, you'd better have the shovel to dig yourself and your project out. That shovel is a good understanding of the laws and the players.

WHY THE PROFESSIONAL ARCHAEOLOGIST'S TOOLKIT IS HEAVY

Sound like a lot? It is! No one should ever think that being a competent professional archaeologist is easy. The old saying "Know a man by his tools" is absolutely true. Your colleagues, regulators, and sponsors will judge you by your intellectual and practical tools and how you wield them. If you keep them sharp and well used, it will show in the quality of your work, in the way your work is received, and in the satisfaction of your sponsors. Becoming a good researcher of any stripe takes lots of training, experience, and the attitude that education is an ongoing and never-ending process.

As a professional archaeologist, realize that the tools you bring to your work shape and color the work you do and how it impacts the entire archaeological record. Just as important as your tools is how you wield them: the attitude that you bring to each project, each opportunity to explore another fragment of the archaeological record. By doing the best job you can of designing your research, you'll help make the most of each of those opportunities.

3
A QUICK LOOK AT THE RESEARCH PROCESS

In the broadest sense, archaeology is archaeology no matter where or by whom it is conducted. But there is no denying that North American archaeology is a two-track field: academic archaeology and CRM archaeology. There are fundamental and seemingly irreconcilable differences in the process of research in the academic and CRM worlds. In academic research, the researcher's interest (or something less noble like tenure, fame, or a convenient field school locality) is usually the genesis for a research project. In contrast, the driving force in CRM research is a sponsor's interest in completing some land-transforming construction project and a government mandate that any inconveniently situated cultural resources must be located, inventoried, evaluated, and possibly removed or excavated.

Three key differences between the research process envisioned and taught by most academic archaeologists and the one encountered in CRM archaeology are (1) the CRM process begins with a solicitation—"How much will you charge to come here and do this for me?"—instead of an initial desire or problem on the part of the researcher; (2) the CRM process has built-in, government-mandated funding mechanisms, while funding for academic research is almost always a major limiting factor; and (3) the CRM process involves legally binding contractual obligations instead of ethically binding academic promises. These fundamental differences color almost all aspects of the research process.

CHAPTER 3
THE "PURE" RESEARCH PROCESS IN ACADEMIC ARCHAEOLOGY

While academic archaeologists are sometimes viewed as high-minded scientists with a pure interest in advancing the state of knowledge, the reality is often very different. The ideal is to be able to independently develop and follow one's personal research agenda wherever it takes you. And some archaeologists have the talent, drive, good fortune, and institutional support to do just that. But most academic archaeologists take much more pragmatic routes and, like CRM archaeologists, must take advantage of research opportunities as they arise.

For pragmatic reasons, many academic archaeologists develop an interest in the archaeology of the region in which their college or university is located. It is much cheaper and logistically easier to carry out underfunded archaeological research near home than far away. And most academic research projects, near or distant, are done with budgets that pale by comparison to those in the CRM world.

Other academic archaeologists become specialists in a topical subfield such as geoarchaeology, zooarchaeology, or archaeobotany. For most such specialists, the research process begins when they are called in as consultants to an ongoing research program. Today, many of these ongoing research programs are CRM projects. In fact, most academic archaeologists are significantly involved with CRM research at some point in their careers and many count on consulting work to supplement academic salaries and provide research funds.

Because of this diversity, it's hard to outline the "typical" academic research process. Most academically based research projects in archaeology do begin with a desire, however pure, to learn more about a topic or area. Many academics build upon research interests to which their own professors introduced them in graduate school. The ability to follow through, year after year, on a research topic or in a region is one of the greatest advantages that academic archaeologists have. CRM archaeologists have to go where the work is, while academic archaeologists can go wherever they want as long as they can solve the myriad logistical, political, and budgetary problems along the way. So let's take a real-life example involving an academically based research project that one of us (Black) took part in and look at key elements of the research process.

In the late 1970s, Thomas R. Hester from the University of Texas at San Antonio and Harry J. Shafer from Texas A&M University began a research project at a little-known Preclassic and Classic Maya center called Colha in northern Belize, a Central American country bordered by Mexico and Guatemala. Both professors were stone tool experts, lithicists, drawn to Colha because of the site's impressive evidence of large-scale stone tool manufacture. After visiting the site as part of a small conference of archaeologists interested in Maya lithics, Shafer and Hester decided to launch what ultimately evolved into a multidecade research program that is still winding down. Initially, they envisioned only a few seasons of work, but even this was a big undertaking in an underdeveloped country some 1,300 miles from home.

The first step was recognizing, articulating, and deciding to take on a research problem: Were the large mounds of chipping debris at Colha evidence of industrial-scale specialized craftsmanship? If so, when did this phenomenon arise, and what role did it play in the economy of the ancient Maya? Problem in hand, Shafer and Hester next had to figure out how they would get the money and labor to undertake the initial explorations and excavations. Luckily, Hester was acquainted with an Italian benefactor who agreed to provide a modest amount of start-up funds. The funding was stretched by attaining university support, mainly time off for Shafer and Hester, and equipment, and by involving graduate students through field schools as well as volunteers, many of them CRM archaeologists in search of an exotic field experience. Most staff members were nonsalaried and were paid only transportation costs and all the rice, beans, and canned meat product they could eat. While figuring out the funding and logistics, the lead archaeologists also had to win the permission of the archaeological authorities in Belize. Each year they negotiated a research permit that spelled out what they could and could not do and what they would do for the country, such as hire so many local workers and donate equipment. They also had to gain the cooperation of the landowners even though archaeological sites in Belize were in principle controlled by the country.

After the first few years, a serviceable field camp had been established and enough had been learned about the site to realize that much more work was needed to thoroughly address the starting problems and begin to face the many new problems that had cropped up. With private funding waning, Hester and Shafer concentrated on applying for grants from the National Endowment for the Humanities (NEH) and other granting agencies and private foundations such as the National Geographic Society and Earthwatch. Some years they succeeded in cobbling together adequate funds; some years they did not, and the field season had to be postponed a year. Each year their research designs grew more sophisticated, the research agendas more complicated, and field crews larger. While Hester administered the project, he shared the field directing duties with Shafer and several other academic archaeologists. Specialists and graduate students from their own universities and other academic institutions joined in the project, each adding his or her own research interests. Black, for instance, was a graduate student and became interested in carrying out stratigraphic excavations in the site's main plaza, an open space surrounded by small temples and elite residences.

As the work evolved, the research design was represented by many different documents, each justifying and outlining a research plan for different audiences—funding agencies, the Belizean government, graduate advisers, and field school participants. Some of the plans were carried out, others died for want of funding or because the ideas were too grandiose or otherwise flawed. Fieldwork at Colha continued intermittently into the early 1990s as new problems, new funding, new graduate students, and new field schools came along. Several graduate student veterans of the early years at Colha came back as young academic archaeologists in pursuit of their own goals. Functionally, the Colha Project was really a series of overlapping smaller research projects united by a network of personal relations linked to the lead investigators, Hester and Shafer. The results of the investigations have been presented at dozens of research conferences, academic meetings, and popular venues including a traveling museum exhibit. Numerous interim reports, journal articles, book chapters, theses, and dissertations have presented many aspects of the work, addressing the original research problems and many others that cropped up along the way. More than twenty years after the work was initiated, a final report is still in progress.

The research process in academic archaeology depends mightily on circumstance and a host of other factors that have much more to do with politics, personalities, logisitics, marketing, and the accidents of history than with the pure ideal of advancing knowledge. Although there really isn't a "normal" sequence of academic research, here are some common elements, even if they rarely happen in a neat sequence:

Problem/Opportunity → *Plan* →
Funding → *Permit* → *Work* → *Dissemination*

At Colha, after the initial opportunity and problem were defined, the other elements of the research process were repeated over and over as the research program evolved. While not all academically based research projects span twenty years, some are even longer, and most see many years between initiation and completion. While CRM projects tend to be of fixed scope and finite duration, academic projects are typically more organic, evolving, and never-ending. Whereas CRM projects almost always result in a final report, academic projects are disseminated in more diverse ways and may never culminate in the ideal outcome, a scholarly monograph.

If you are planning on becoming an academic archaeologist or if you find yourself involved in a long-term research project as a graduate student, we have two pieces of advice. The first is to carve yourself out an interesting niche that allows you to follow or develop your own interests as opposed to those of your mentor or project director. The second is to pursue finite goals and outcomes that you can reasonably expect to achieve in the time you have available. Many an academic archaeologist has gone to his or her grave leaving behind a stack of unfinished research projects and ugly holes in the archaeological record.

THE "DIRECTED" RESEARCH PROCESS IN CRM ARCHAEOLOGY

For better and worse, the CRM world is much more cut and dried, although it, too, is very diverse. Cut to the bone, the typical contracted research process in the CRM archaeological world proceeds in five parts. It begins when an archaeological firm or organization receives a *request* from a sponsoring entity or client: "Tell us what it will take for you to solve our archaeological permitting problem." The archaeological firm replies with a *proposal:* "For so-many bucks, we will do blah-blah archaeological research work toward gaining your archaeological permit clearance." The proposal is accepted by a signed, legally binding document—a *contract*—and the archaeological firm does the *work* (field, lab, etc.) as per the contract. The process ends when the *report* is delivered to and accepted by the governmental entity that issues the permit that the client or sponsor needs. Although there are many permutations, smaller pieces, and complications, the CRM research process usually follows these five sequential parts:

Request → *Proposal* → *Contract* → *Work* → *Report*

The real world of CRM research is just as twisted and complex as the academic world, and by way of example we'll take a look at the

Wurzbach Project. In the early 1990s, the Texas Department of Transportation (TxDOT) and the city of San Antonio had decided to build a much-needed crosstown expressway—the Wurzbach Parkway—through an existing urban environment. The only practical route lay along and across the upper reaches of a major drainage system where there were stretches of undeveloped land, including what was left of an area known as the Walker Ranch, the last major holdout to urban development in north-central San Antonio. Several decades earlier, the area had been designated as the Walker Ranch Historic District and placed on the National Register of Historic Places because it contained numerous prehistoric archaeological sites as well as several historic sites, including what was left of a Spanish Colonial ranch associated with Mission San Antonio de Valero, better known today as the Alamo.

Despite the Historic District designation, the Walker Ranch area and its archaeological sites had suffered all sorts of disfiguring impacts as it was parceled up into subdivisions, paved over, and otherwise incorporated within the city. As urbanization encroached, some archaeological investigations did take place, including surveys, testing projects, and one substantial excavation. The building of a small flood control dam in the early 1980s had been preceded by the excavation of a prehistoric campsite named after the Panther Springs Creek, which it overlooked. As it happens, Black had directed that excavation and had made many of the mistakes in designing archaeological research that we are urging you to avoid. Fourteen years later, in 1993, he found himself back in the same spot because the route of the planned Wurzbach Parkway went right through Walker Ranch and what was left of the Panther Springs Creek site. Even though a last-minute shift in the dam location had spared the site from being gouged out by a footing trench, trespassing looters had churned up the site in search of chipped stone tools. But the looters had all but overlooked another, less conspicuous prehistoric campsite just across the creek, the Higgins site.

TxDOT asked Black to assemble a small research team (including Jolly) to investigate the Higgins site and several others and to evaluate their potential for continued listing on the National Register. This is section 106 talk for "Do we have to excavate the site before we blow it away?" To make a long story shorter, our team answered yes, and TxDOT entered into negotiation with the SHPO, Texas Historical Commission (THC), and the Texas Archaeological Research Laboratory (TARL) of the University of Texas at Austin where Black worked.

Because TxDOT and UT Austin were both state agencies, the request for proposals (RFPs) was not a formal document and not subject to competitive bidding. Instead, TARL presented a contract proposal that included a scope of work, a timetable, and a budget. The actual research design was developed as part of the contract and included within a technical report summarizing the results testing at the Higgins site and several others. After some negotiation, the SHPO agreed that the proposed work would constitute an acceptable mitigation program for the damage the Wurzbach Parkway would cause to the archaeological information represented at the Higgins site. Because the land was owned by the city of San Antonio, a political subdivision of the state, a Texas Antiquities Permit was also required and was issued by the THC. Shortly thereafter, the archaeological excavations that came to be known as the Higgins Experiment were under way.

By this point, several critical differences between the CRM research process and that typical of academic-based research are apparent. The process started with a looming construction project, not a burning research question. The sponsor and funding source (TxDOT on behalf of the Federal Highway Administration) sought out the archaeological research team, not the other way around. Although the researchers wrote the research plan, it was formally approved and negotiated between the sponsor (TxDOT) and the SHPO (THC), who could have altered any part of the plan without the researcher's approval. The scope of work spelled out that the work would be confined to the actual right of way of the proposed road, meaning that the research team was strictly limited to a linear research universe. Furthermore, the research timetable was relatively short (three years from start to finish) and set to accommodate the sponsor, not the researcher. Still, as CRM projects go, the researchers had a pretty free hand and a generous budget, far larger than most multiyear NSF archaeological grants.

With contract and permit in hand, we were required to start work within ninety days, and all of that time was filled with preparations. A crew was hired, and crew housing and field lab space (a portable building designated the Wurz-shack) had to be acquired and renovated. Field vehicles were scrounged and rented, computers purchased, and software written and revised for data management. TxDOT provided on-site power, a field office, and twenty-four-hour security to protect the site from both the neighborhood juvenile delinquents and the homeless vagrants who supported their drug habits by selling looted artifacts. There was also heavy equipment at our disposal: cherry pickers for high overhead photos, a Gradall excavator for pulling off overburden, and front-end loaders and large trucks for moving dirt. The logistical support that a large agency like TxDOT can provide is a hidden asset to many CRM projects.

The pace of the work was determined by both the budget and the construction schedule. If we decided that we hadn't really got it quite right the first time, there would be no option to come back in another year or two to take a different tack. After the end of the project, construction of the Parkway—the reason that the budget, the Gradall, and the crew were there in the first place—would destroy what we had left of the site. Like many CRM projects, it was a one-shot deal. We had one opportunity to use the generous resources we had been given to make a contribution to the archaeological record. Not just a little pressure.

The fieldwork was completed in about ninety days, and lab work and analysis continued intermittently for another three years. The Higgins Experiment itself was part of a larger research project covering work at several other sites, the final reports of which were issued in a series of volumes from 1995 through 1998. Part of the work was used by one of the consulting specialists in his dissertation work, but no other theses or dissertations came out of the project. The results were presented at local, state, and national archaeological meetings but largely remain known only to a small CRM community.

The real, fundamental difference between academic research and CRM funded research is control. In both worlds you have budgets, permits, and bureaucracies to limit and confound your well-crafted ideas. In the academic world, though, the researcher exerts more control over the research process. The researcher decides where and, generally, when to work. The researcher develops the questions of interest, the scope of work, and the research priorities. The academic researcher also decides how and when the research results will be disseminated. In the CRM world, many or most of these decisions are out of the researcher's hands. The study area is prescribed by the sponsor, and the schedule is at the mercy of funding and the construction schedule. Research questions may be designed by the researcher, but the SHPO and the sponsoring agency have the final word on what is significant and what the research priorities will be. The sponsoring and regulating authorities also have a large hand in determining how and when the results will be reported, usually within a few short years or even months. It's this difference in control that is at the heart of the distinction between academic and CRM research.

The rest of this volume will focus specifically on designing research for CRM archaeology. The reality is that the vast majority of funded archaeological research in the United States is CRM archaeology. Most archaeologists will work in CRM archaeology all or part of their professional careers. To be effective, you must understand the details of how the research process really works. One steadfast rule is that from academic to CRM, from agency to agency, sponsor to sponsor, state to state, project to project, and year to year, the pieces and parts of the research process are rarely exactly the same. We will focus here, and in the rest of this volume, on the CRM side of the coin, with some quick characterizations of the most common aspects of the process that you will encounter as you design CRM research. We will follow the simplified five-part research process from start to finish, adding mention of some of the permutations you need to know about.

THE PARTS OF CRM RESEARCH

Your call to action is a request from some potential sponsor or client that has archaeological obligations. This request can begin with a simple phone call, a faxed letter of inquiry, or a formal document often called a *request for proposals* (RFP). When the sponsor is encountering the need to do archaeology for the first time, the request is usually a call or letter from someone who knows little more than what the form letter from the governmental permitting or regulating agency states: "We need a cultural resource inventory that will fulfill our obligations under section X of. . . ." The archaeologically unsophisticated sponsors do not care about the quality of the archaeological research; they are looking for a cheap, fast solution to a bureaucratic requirement. They are the most likely to go with the lowest bid and often suffer the consequences because they don't know any better. If an engineering or environmental consulting firm is contacting you on behalf of a sponsor, then your contact will be more archaeologically sophisticated and will probably have a good idea of what they are asking you to do. The engineer won't care much about the research, either, but they will want to work with competent archaeological firms that can do acceptable work in a timely fashion. For this reason, they may go sole-source (i.e., no competitive bid) and choose a reputable archaeological firm they know they can work with. If a state or federal agency is sponsoring the work, it will probably have an archaeologist or archaeologically knowledgeable contracting officer issue an RFP.

Such documents will spell out in some detail what they want and expect from you. Depending on the agency, they may pay more attention to the bottom line or to the qualifications of the firm and the responsiveness of the proposal.

The most critical part of any request, formal or informal, is the *scope of work*. As we explain in detail later in the book, the scope of work lays out what is to be done, why, when, where, and under what conditions. Naturally, you cannot design effective archaeological research in the CRM world unless you have a clear understanding of what your sponsor wants from you and of the physical, legal, and logistical research universe you will be working within. Unless you have a comprehensive RFP in hand, it will often be up to you to take what the sponsor says they want and craft a scope of work as part of your proposal.

The *proposal* is your written response to the formal or informal RFP. Proposals consist of different parts depending on the particular circumstances, but usually they contain a revised scope of work, a budget, and a schedule. The *revised scope of work* or work plan is where you say how you will address the stated scope or, in cases where the sponsor doesn't know enough to provide you with an adequate scope, where you tell the sponsor what the work will involve (i.e., you provide the scope). The *budget* tells the sponsor how much you will charge. Sometimes your budget is a fixed price for a set amount of work, and sometimes it is a cost-plus budget that says you will bill the sponsor on the basis of your actual/direct costs plus your overhead and profit margin (usually a multiplier such as 150 percent of the direct costs). The *schedule* lays out the time frame—when you will "deliver" (do) what. Formal proposals submitted to government agencies often must contain other elements that are intended to help the sponsor evaluate your proposal. For instance, you may need to provide *qualifications* such as the vitae or résumés of the key individuals who will do the work and examples of comparable work your firm has done.

Naturally, the pieces of your proposal are directly linked to the design of your research. The scope, the budget, and the schedule tightly constrain what you can and cannot do during your research. So can the qualifications and abilities of your research team. You also need to realize that in many contractual situations, your proposal may serve as the "research design." As we will keep stressing, putting the proposal together is only part of the ongoing process of designing effective research. For now, however, just keep in mind that while academic archaeologists often envision (and produce) research designs as stand-alone, formal documents (often prepared in the process of applying for competitive grants), in the CRM world the proposal often constitutes the only written research design. The smaller and more routine the work, the more likely that this is the case.

In some CRM projects the contract may call for you to prepare a formal *research design* as part of the initial stages of your work. Usually this requirement is done to satisfy the federal or state agencies that are mandating or regulating the work. Formal research designs are likely to be required in larger projects, particularly multiphase projects where many sites are involved or that involve mitigation (excavation or data recovery). In such circumstances, the research design builds on and refines the general approach outlined in your proposal. Often the research design must meet the approval of archaeological bureaucrats in state or federal agencies or both. This step can create contractual difficulties when the regulating or permitting archaeological bureaucrats demand changes that have financial consequences beyond what your budget and schedule allow (i.e., they want you to do more work than you have planned to do). If you have correctly anticipated the potential impact of the third party (the archaeological bureaucrat whose name does not appear in your two-party contract, but who plays a critical role in the outcome of your research), you have will have covered your firm's butt by inserting clauses that allow for contract modifications. (If not, you will next time!) A lot more needs to be said about the various elements that are typically involved in a formal research design, but we will save that for later.

A defining piece of the CRM research process is the *contract*. This is the signed document that legally binds the sponsor and archaeological firm or organization doing the work. In small, uncomplicated situations, the contract may be a one-page letter that says the two parties who sign on the dotted lines at the bottom of the page (i.e., the sponsor and the archaeological contractor) agree to the terms spelled out in the attached proposal. In other projects, the contract may be a lengthy document that subsumes the scope of work and the proposal (and its parts) and that spells out the contractual agreement. State and federal agencies often attach all sorts of standard clauses that cover everything from intellectual property rights to dispute resolution, safety requirements, nondiscrimination, subcontracting, and so on. In all circumstances it behooves you, the archaeological contractor (or you, the project director charged with completing the contract), to read the fine print and make sure you know what you are legally obligated to do.

You may think that with the contract signed and the research design approved, the process of designing your research project is over. This is untrue in most situations. The core of the research process is reexamining ideas and strategies as new information becomes available. As you work, you'll encounter unexpected situations, new kinds of information, and other things you didn't imagine you would encounter when you designed your research. In a perfect world, with unlimited time and money, you might be able to jot down a quick note about what you've seen and plow right ahead with your original plan, with the expectation that some time in the future you'll revisit the anomaly and attack it with its own special research plan. The reality is, though, that the work you're doing may be the last work any archaeologist gets to perform on that patch of dirt, so there is an ethical obligation to deal with unexpected circumstances as you find them. The kicker here—and one of the things that makes fieldwork fun—is that you usually can't (and shouldn't) just abandon your original plan to jump ship to some new plan for success. To do the best work you can, you need to deal intelligently with the unexpected, not by default. This means taking some time to revise or add to your research plan to handle new wrinkles. Most important of all, it means documenting—writing down—exactly what the changes are and why they're being made. If you don't document your changes, you could (rightly) be accused of misrepresenting the work that was done to fit more closely with your final conclusions. You'll be revisiting and probably revising your research plan all along the way.

THE RESEARCH PROJECT

Archaeological research projects in the CRM world come in all sizes and many different types. The stereotypical CRM project (or series of projects involving one project area) is often seen as consisting of three sequential phases, sometimes termed Phases I, II, and III. The first phase is the *survey* or *cultural resource inventory*, in which the contractor inspects the project area on foot, sinks shovel tests, and maybe some backhoe trenches. The aim is to make a reasonable effort to identify factors present in the project area (or some quantified sample of the total area) and to determine which of these are obviously insignificant and which need further work.

The next phase is *testing* or *site evaluation*, in which those identified sites that have not yet been written off (dismissed as unimportant) are looked at in more detail. The aim of testing is to determine which, if any, of the potentially significant sites are, in fact, likely to contain information that is judged to be significant by criteria such as those listed in section 106 of the National Historic Preservation Act (see sidebar 3.1).

The final phase is *mitigation* or *data recovery*, in which the sites that have survived the winnowing process—those judged to be the most significant and that will be adversely impacted by the planned actions of the sponsoring entity—are partially or (rarely) wholly excavated. The aim of data recovery is to offset or mitigate the impending impact by extracting enough data from the site so that something useful can be learned before it's too late.

Many CRM projects go through the sequential phases of archaeological research more or less as outlined. But the Phase I–III "system" is an artifact of archaeological convenience, not an immutable rule. It appears in no federal regulation. At the dawn of the twenty-first century, there are increasing numbers of projects that do not neatly fit into any of these three phases. Some agencies define "classes" instead of phases, while others stick to the descriptive labels (survey, testing, etc.). And most now recognize all sorts of permutations.

The first step might be a *literature survey* or *background research* phase in which existing records are combed for previously investigated sites, archival documentation, and research on closely comparable areas and topics. After this phase and before a full-scale survey is done, there may be a *reconnaissance* project intended to identify high- and low-probability areas so that later *intensive survey* work can be selectively directed. A fairly recent trend in projects involving large tracts of land is a *geoarchaeological* or *geomorphological evaluation*. This work is done by geologists, soil scientists, or archaeologists trained in geology. The aim is not to find archaeological sites, per se, but to understand the evolution and depositional history of the landscape in question so that the best places to find well-preserved sites or deposits of a particular age can be determined. Such evaluations often identify large areas of the landscape where little potential exists for intact, buried archaeological sites.

Similarly, testing may be broken out into several different phases of fieldwork. *Intensive testing*, for instance, may be intended to "test" a site out of existence or otherwise avoid mitigation by demonstrating that not enough remains intact to warrant a full-scale mitigation project. Or intensive testing may be done in cases where the initial testing was intended to evaluate the site but failed to obtain adequate information to plan an effective data recovery program.

3.1. WHY IS "SIGNIFICANCE" SIGNIFICANT?

You won't do CRM work for long before you confront the notion of significance. This isn't ordinary garden variety significance that we're talking about. The term has special meaning in CRM. For an archaeological site to be significant—worthy of preservation or data recovery (excavation) under section 106—it must be deemed eligible or potentially eligible for inclusion on the National Register of Historic Places. This determination is made by the federal agency in consultation with the SHPO, using your recommendations, following National Register guidelines. It's this formal significance that is significant to you.

Determining a site's significance involves two parts. First, the reviewer or the archaeologist identifies one or more "historical contexts" within which a site's significance is to be judged. Historical contexts are rationales for why the site and the information it contains are significant, but there is no commonly agreed-on definition of a historical context. It can be a broad historical theme (e.g., the late nineteenth-century western frontier), a general research problem (how did drought impact Archaic life on the Plains?), or a very narrow knowledge gap (timing of the adoption of maize agriculture in the Caddo area). Sometimes historical contexts are presented in regional or statewide research designs. They can be the research questions you've developed for your project, or they could be hypotheses developed by an academic archaeologist interested in human ecology. However they are conceived, historical contexts are used to place the site or district in perspective, to provide the reviewers with the yardstick they need to assess the site.

When you present recommendations to the SHPO or another agency for sites you think are eligible for the National Register, you should also provide suggested

---

Mitigation or data recovery projects also may be divided into several phases. Perhaps not enough money is available in a given funding year to excavate a large enough sample of a site, or the planned mitigation proves inadequate because of new information that comes to light. There are also *alternative mitigation* projects that may not involve data recovery at all. For instance, sometimes it is desirable to set aside a property (often one that lies outside the impact area) as an archaeological preserve in return for accepting the loss of a site within the impact area. Or maybe the effects of construction can be mitigated by covering the site in concrete or fill so it is preserved for the future. Another alternative mitigation might be preparation of a regional or topical synthesis where instead of digging up yet another example of type X site, the research team pulls together what has already been learned.

The design of your project will depend on the specific types of research that you will be expected to do. This is, of course, obvious to any experienced archaeologist, but it may not be so obvious to someone contem-

> contexts for the evaluation. If there are applicable standard contexts in a regional or state research plan, use those, but you don't have to stop there. A historical context is the research design process distilled to its essence. A site can be judged under multiple contexts. This is a great opportunity to develop your research questions and build an argument for how information from a given site can illuminate these questions and deepen our understanding of the archaeological record.
>
> Once the contexts are identified, a site is judged significant (or not) within each context with regard to one of four National Register eligibility criteria: (A) events, (B) people, (C) distinctive architectural or artistic character, or (D) information potential. Since most National Register and section 106 actions involve the built world, this accounts for criteria A through C. The vast majority of archaeological sites and districts are judged for significance under criterion D. The information potential is judged within the contexts you provide.
>
> For a site to be significant, it must meet two requirements. First, it must contain (or likely contain) information that contributes to our understanding of human history or prehistory. Second, the information must be considered important relative to one or more historical contexts. In other words, it addresses gaps in current knowledge, informs theories that challenge or go beyond current thought, or focuses on priority areas identified by state or federal management plans.
>
> Site significance is where your research design becomes a true management tool. The research design is a map to developing effective arguments for or against determining whether a site is significant and thus eligible for inclusion in the National Register of Historic Places. This significance determination is the critical bureaucratic step at which certain sites are consigned to oblivion and others are tagged for preservation or data recovery.

plating or beginning a career in CRM archaeology. You can't just lock in on a limited set of ways of doing archaeological research or on one set of fits-all terms. There are many different paths to doing effective archaeological research even within the confines of government mandates. We expect that the plethora of different kinds of research projects will continue to expand. Keep in mind that because archaeology is only one aspect of the wider CRM field—containing architects, historians, architectural historians, and sociocultural anthropologists, all players whose research domains may overlap with yours—interdisciplinary approaches will open new research directions.

THE RESEARCH REPORT

No matter what kind of archaeological project you undertake, your research results must be reported in some formal document.

In the vast majority of CRM projects, your report will be a technical one written by and mainly for an archaeological audience. Although archaeologists are finally beginning to break out of this rut and report their results in more creative ways to wider audiences, the technical research report is still an expected product of most projects. Your sponsor may not care, but the regulatory forces will not be satisfied with anything less for the good reason that the reporting of archaeological research is a fundamental ethical responsibility (see sidebar 3.2).

3.2. THE SIX COMMANDMENTS

Archaeological deposits are fragile, and when we excavate we destroy those deposits forever. Because we destroy what we study and because the public pays for most of our work, CRM archaeologists must follow a well-defined ethical path. Most professional archaeological societies, associations, and councils have their own codes of ethics or standards of conduct. Boiled down, most are covered by these six commandments:

Thou Shalt Not Work with Your Head Up Your Ass

You have a professional responsibility to undertake only work for which you're qualified and to do a competent job of any research project you take on. You should stretch your skills, but make sure that you have qualified people in positions of responsibility to make the critical decisions. Before you accept a project, you need to be sure that you have the people, the equipment, and the wherewithal to take it through to completion.

Thou Shalt Mind the Public's Interests

Government-mandated archaeology depends heavily on the support and interest of the public. You have an obligation to spend the public's money and its trust well and to do your best to learn something worth knowing. The public includes many groups, but you have to pay particular attention to the groups whose ancestors created the archaeological record you investigate. This means you must consult with others and do your best to reach compromises that balance the public's compelling and competing interests with your own research interests.

Thou Shalt Write Your Damn Field Notes

Your most important job in the field is to make and record accurate observations of the things you find. Once a site or portion of a site is excavated, all

Research reports come in various sizes and types. For small tasks, especially those where little or nothing of archaeological significance is found, a *letter report* may be all that is needed. These are fine if they are accessible to other researchers, but often they are filed away, and so one more set of potentially useful observations is lost for most practical purposes. Larger, multiphase research projects often issue *preliminary* or *interim reports*, which are intended to be followed by *final reports* (but too often aren't for all sorts of reasons, some of them legitimate). Sometimes projects issue reports for each phase of the

---

that is left are the bare tidbits of artifacts, some samples, and your observations of what was there. You should strive to make accurate and complete observations and measurements. After you leave, your recorded observations *are* the site.

Thou Shalt Not Sell Out the Archaeology

Don't do anything to encourage or abet those who buy and sell artifacts and loot sites. Don't offer an opinion on what an artifact might be worth (and you will be asked). Don't get involved with treasure hunting schemes and other activities that destroy the archaeological record without proper documentation. Avoid the dark side.

Thou Shalt Not Screw the Archaeology (or Your Client)

If you take a job you need to strive to complete it honestly, on time and within budget. In the real world, things don't always work out. If you suspect that you will have trouble completing the scope of work for your project, sit down with the client or sponsor and the SHPO and figure out what to do. If you're having a conflict with the client over payment, don't hold the archaeology hostage. At a minimum, do your best to properly close up the project: fill in exposed units, organize the artifacts and notes, assemble the film and photo logs, and don't obstruct any follow-up studies.

Thou Shalt Share What You Learn with Your Colleagues

Your observations and analyses gain value when they become part of the body of archaeological knowledge. The information you add to the archaeological record helps other archaeologists understand the region and evaluate sites. Your reports should be timely, and your data—notes, samples, databases, artifacts—should be available to other researchers.

work, and sometimes they combine two or more phases of work in one report. Truly massive research projects like reservoir surveys or major excavations are often reported in multivolume *report series* that might culminate with a *summary* or *synthesis* volume. Increasingly, there are also contractual requirements for *journal articles, education curricula,* and *popular accounts* in addition to the technical reports.

Before your eyes glaze over completely, consider this: *Writing your report is one of the most important aspects of designing archaeological research!* Yes, that's right—the process of designing research culminates when you complete and assess your research. Because scientific research is a cyclical and, hopefully, progressive process, it is essential that you evaluate your research design as the end of each cycle draws nigh. Otherwise, you and your colleagues are doomed to repeat your mistakes and unproductive strategies. By critically assessing both the positive and negative aspects of your research strategy, you will fulfill one of your obligations as a scientist and set the stage for the next round of research.

You might want to pause and think about the archaeological reports you have read and perhaps helped write. Is the research strategy effectively evaluated? Does the report convey the impression that the researchers made any mistakes? If not, do you think this was really the case? Finally, do you think the reader would learn anything useful from an honest and thorough evaluation?

CONCLUSION

This quick overview of the research process should point you to the places where you may need to beef up your foundation, fill in some gaps, and prepare yourself for the job at hand. Effective and successful archaeological research is a careful blend of intellect and experience. As with anything else, designing archaeological research is about what has come before—the methodological, theoretical, and practical perspectives you bring to bear on the specific research problem. Your job as a professional is to make sure that you have the knowledge and experience you need to succeed. You'll be judged both by the quality of your research and by your success at clearing bureaucratic and logistical hurdles gracefully and efficiently.

4
THE SCOPE OF WORK

The scope of work (the scope) defines the universe within which your research design, field work, lab work, analysis, and write-up will operate. It defines the limits of what you can do and sets the baseline for what must be accomplished. It outlines the time frame in which the work should be accomplished, the type of work to be done, and the products that will be delivered when the work is complete. The scope of work defines or helps you decide when you work, how many people you hire, how long you'll work, and how much work you'll be able to do. In the CRM world, the scope of work allows you to prepare a budget and response to an RFP that the sponsor will use to choose a contractor. The scope of work is where your research starts.

There is a diverse range of projects, from tiny one-day surveys to multiyear, multisite mitigation projects with budgets that some small countries might envy. The archaeological research carried out by these diverse projects differs in scope, result, and presentation. But all archaeological projects have or should have a common goal: to learn something worth knowing from the archaeological record.

Designing effective archaeological research—large, small, academic, or CRM—is a process. In the CRM world, it generally involves three essential and more or less sequential tasks. You first define and document the scope of work—when and where the work will be accomplished, why it is being done, and its logistical and legal constraints. Second, you develop and document a set of research questions and objectives—what you plan to learn from the work. Third, you figure out your methodology and research strategy— how you will collect and analyze the data you need to address your research questions. It's only after this third step that you actually have a real research design—a complete plan of action.

The same three elements are equally critical in academic research. In the academic world, the process of designing research usually begins by posing research questions because it's the researcher's initiative or an opportunity that drives the process, not that of an external sponsor. The research strategy and methodology often come next. The scope of work may not be worked out until many months (or even years) later once you know how much money and labor you can muster.

The scope of work is the place to map out the things that will define and limit your research universe. The scope determines where on the landscape you'll work and the basic parameters of what you will do (i.e., survey, testing, mitigation, etc.). Vague scopes of work may do little else. Comprehensive scopes may specify when you'll work, how long you'll work, and how much work you will do. Scopes written by an archaeological bureaucrat with very specific expectations may go as far as spelling out how many samples you will take, what methodology you will follow, and what research problems you are to address. More typically, the scope of work will set minimum expectations and ask you to follow a set of guidelines or standards developed by your SHPO or other archaeological organization, such as your state's archaeological council.

In CRM research, the scope may be prepared by your sponsor. It might be vague and amorphous and might even ask for the impossible—survey and test sites on 1,200 acres of prime archaeological ground within the next two weeks. When a scope of work is prepared by the sponsor, it could be the work of an engineer or manager who simply doesn't understand what's involved in the CRM aspect of her project. Her interest may be with the time line and the budget, not the archaeology. Or the scope of work might not even physically exist. It might be simply a phone call or meeting where the sponsor lays out what needs to be done to fulfill its requirements under the law and to keep its project on time and on budget.

Regardless of what the sponsor provides as a scope of work, your first task is to thoroughly understand and document the parameters—the scope—of the work to be done. If a written scope is provided, you will likely need to talk with the sponsor to clarify anything that is unclear and make sure that you and the sponsor are on the same page regarding the work to be accomplished and the time frame involved. In a competitive bid situation, agency or state bidding rules may not allow you and the sponsor to meet and clarify issues in the scope of work. In that case, you're the one who must make the decision, but it is doubly critical that you explicitly define *in writing* the assumptions you're making about the scope of work. If you're awarded the contract, the first item of business should be a memo to the sponsor outlining your understanding of the scope of work and listing any issues that should be resolved and documented in writing.

If a written scope of work does not exist, it is your job as an archaeological researcher to create one. It should be written down, as a memo, a letter, a part of a formal contract—something you can share with the sponsor, regulators, or grantors so everyone is literally reading from the same page. If a sponsor gives you the scope of work over the phone or in a meeting, create a memo reiterating what you talked about and send it to him. If there is simply no scope of work, create one with what information you have using your best archaeological judgment. Without a real written document, you, your research design, and the success of your entire project are at the mercy of notoriously fallible human memory. Get it down on paper: dated, signed, and delivered.

If you're new to creating scopes of work or even if you've been doing it for years, the first place to start is with your SHPO office. The folks there see more scopes of work in a year than most of us will see in a lifetime. Because they often have the opportunity and responsibility to follow these projects from start to finish, they have seen what works and what doesn't. They're as busy as the rest of us, so don't expect a half-day seminar on what a good scope of work looks like, but a quick phone call might shake loose a couple of examples of the kinds of scopes of work that they're seeing from your area. This will be a much smoother process if you have invested the time to develop a good working relationship with your SHPO. The SHPO office is an underutilized research resource in most states, and you should take advantage of their expertise. A comprehensive scope of work is to everybody's advantage.

To develop a good scope of work, you need to focus on six primary areas: the project area, the work, work standards, time frame, budget, and logistics. First, critically read and understand what material you already have in hand. If it's in your head, write it down. Then read it through and mark sections that are unclear or for which you need more information. If there are any missing sections, note them, and sketch out the information you need to complete them. For example, if the work involves excavation of sites that can be reached only by boat, will the sponsor provide transportation to and from the site area? Will you need permits to perform water screening in a public waterway? Does the schedule allow for lost time over holidays or bad weather? Where you have questions or need more information, talk with the sponsor if possible and document the answers (or lack thereof) so that you and your sponsor have a clear understanding of the project area, the work to be done, the work standards to be applied, the time frame for completing the project, the project budget, and any special logistical problems or needs.

THE PROJECT AREA

This is the physical place where the work will take place—not the approximate location, but the exact boundaries of the property where the work is to be done. In the section 106 process, one has to deal with the area of potential effect (APE), which is supposed to include all areas that will be affected directly and indirectly by the proposed project. Some agencies take a much narrower view than others. Usually the project area (and APE) is depicted on one or more maps that outline the project boundaries. If some work has already been conducted in the area, the maps may show identified site locations, or it may be up to you to determine whether previously recorded sites exist within the project area. This seems pretty obvious, but if your idea of the project area isn't exactly the same as your sponsor's, you could be doing unnecessary—possibly illegal—work or leave work undone that needs to be completed. From a research perspective, the most basic research questions focus on the landscape: If it is unclear where you'll do your work, your research will be unfocused. From the contractual perspective, poorly defined project areas are red flags that alert the experienced archaeologist that he may be in for some very ugly and possibly costly surprises. Pin it down.

Keep in mind that the neatly bounded areas marked on your project map may be difficult or impossible to accurately locate on the ground. How will the project boundaries be marked? Who is responsible for marking them? Will the sponsor provide a surveyor to mark boundaries and establish a permanent datum, or will this be your responsibility? If you're not familiar with the area, it's a good idea to get input on the actual conditions on the ground. The thirty-year-old U.S. Geological Survey quad map or five-year-old U.S. Department of Agriculture air photo may show a nice clear terrace slope for your survey. If there is actually a hundred-acre bramble patch there, you will be in for a very thorny surprise. These kinds of easily overlooked details can make a big difference in your budget and schedule.

THE WORK

Knowing exactly what work is expected is a critical component of the proposed scope. What needs to be done archaeologically? Where along the CRM continuum does the project fall? Is it a survey project, testing, data collection, or some combination thereof? Are there other cultural resources to be dealt with in addition to the archaeological ones? Will you be expected to carry out nonarchaeological tasks such as archival research, oral history, or architectural documentation? With what federal and state laws and regulations will the work need to comply? What specifically will you be contractually obligated to do? You need to understand exactly what is expected and work to clarify anything that is unclear.

You also need to make sure the scope of work is understood and accepted by all of the critical players involved. The sponsor may have no real idea what it is asking you to do and what the regulators require, and it may expect you to perform miracles for less than $10,000. The regulators may not know what you have contracted to do. It could fall to you to educate the sponsor and inform the regulators (see sidebar 4.1). It's in your interest to make sure everybody is on the same page by writing or rewriting the scope of work and getting it into the hands of those who need to know.

Watch out for overly rigid specifications that assume a perfect world. On a fairly large-scale testing project in the Southwest, the scope of work prepared by a federal agency required excavation of a particular volume of soil. Unfortunately, the project area consisted of deflated terraces and hill slopes where the soil depth reached a maximum of five to ten centimeters, and there simply wasn't that much soil to be excavated from the sites tested. With a few weeks left to complete the project, the archaeological contractor made the decision to trench through portions of a site simply to increase the soil volume moved to meet the scope of work. Nothing archaeologically was learned from the trenching, and while the site was deflated, that portion of it was destroyed because they had not taken the time to understand the character of the project area or to question the work that was required in the scope.

4.1. AVOIDING THE VICIOUS REGULATORY CYCLE

An all too familiar and very unhappy pattern can characterize CRM projects: The sponsor does not know what the regulators expect. The regulators don't know what the sponsor is having its archaeological contractors do. The unethical contractor plays both ends to fatten his middle. He accepts a contract from the sponsor to do what he knows the regulators will consider inadequate work. He does the work, milks every possible dime from the sponsor, files his report, and waits for the regulators to raise objections. The regulators review the report and call for more work. The sponsor is forced to shell out more dollars to the contractor who pretends to be outraged by the evil and unfathomable SHPO or federal or state agency. The sponsor calls the SHPO and screams bloody murder. The SHPO tries to explain the requirements and points out that the contractor should have been aware of these (in a nice way because most SHPOs are too politically attuned to call a crook a crook). The sponsor is confused but finds it easier to blame the governmental regulator than the kiss-ass contractor, who swears loyalty to the client. In extreme cases, the sponsor calls his legislators and demands that they lean on the regulatory agency. In a recent case we witnessed, this vicious cycle resulted in the SHPO's office having to finish the archaeological work out of state-appropriated funds.

Guess who loses in all this? Everybody except the unethical contractor who moves on to the next unsuspecting client. The archaeological record gets the shaft. So does the archaeological profession. We all suffer for the sins of our unethical colleagues. The sponsoring and regulatory worlds grow distrustful of archaeological contractors. If enough powerful people become involved, preservation laws and regulations can be weakened.

You can avoid this vicious cycle by being an ethical archaeologist. Educate your sponsor right from the get-go and let her know what the regulators expect. You've been hired for your expertise—demonstrate it. Head off conflicts and problems by talking with your sponsors and regulators on a regular basis. If you see a problem on the horizon, get on the phone, send a fax, shoot an e-mail, or wave your semaphore flags. The single best way to head off conflict and contention is to communicate effectively. If conflict does loom, do your best to be up-front and honest with all the parties and get problems resolved early on.

You can also help by becoming a registered professional archaeologist (RPA) and living by the register's Code of Conduct, a set of standards that states, among other things, that an RPA will not "engage in conduct involving dishonesty, fraud, deceit or misrepresentation about archaeological matters." The American Cultural Resources Association (ACRA) also has a strong Code of Ethics, and you can expect CRM firms that are members of ACRA to be reputable places to work.

If the scope calls for excavation of so-many cubic meters, are there provisions for modifying the scope if something especially time-consuming (e.g., unexpected burials) is encountered? How about allowances for bad weather? These and other factors can prevent your team from being able to complete the scope of work. You can protect your interests by asking the sponsor to modify the scope of work or by spelling out contingencies in your proposal and outlining how such difficulties will be resolved.

WORK STANDARDS

To what standards will your work have to measure up? A comprehensive scope will probably specify that your work (and the qualifications of your key team members) must past muster with relevant federal and state requirements or professional standards. You may also be expected to follow the established or mandated practices in the region. These can take the form of published standards or guidelines, or they may be less formal, but no less real, unwritten expectations on the part of the regulators. If your sponsor does not understand these standards and expectations, you should carefully explain them because they often affect schedules and budgets. Pointing these out up front may help the sponsor realize that your more costly bid is a better deal than the cheaper (and competing) one that does not purport to meet standards.

TIME

In most cases, CRM contracts are designed to run for a specific length of time. The project time span determines to a large extent the size of the budget you'll have and how much work you'll be able to do. Time is highly correlated with money. The kinds of methods you can use in the field are controlled to a large extent by the time–money relationship. You may intend to water-screen every bucket of dirt through 1/8-inch mesh with deionized water, but if you've got only six weeks to test a dozen sites, that's not bloody likely.

In many areas of the world, weather plays a big factor in determining when fieldwork can be done and how long it will take. Make sure the scope of work will not have you trying to excavate in frozen earth (or, if so, that you have budgeted and planned accordingly). Surface surveying is largely a waste of time when the ground is covered in snow, mud, or seasonally thick vegetation. Your sponsor may not realize this and may need to be educated. In the CRM world, fieldwork rarely seems to take place during the optimal times of the year despite the best laid plans. Play it smart and plan for this by adding contractual clauses that modify the scope of work (and/or the budget) if predictable (or unpredictable) climatic or seasonal conditions occur that adversely affect your work.

MONEY

Few formal scopes of work spell out the amount of money that will be spent—it is your job to propose to do the specified work for a price you set. But often the sponsor has a budgeted or otherwise restricted amount of money to spend on archaeology. You may be told something like "I sure hope you can do this work for under $20,000; that's all we have." Or you may become aware of financial limits through less direct means. However it happens, the cost of doing a given project is directly related to the scope of work. The more you are asked to do, the more demanding the logistical conditions are, the more rigid the work standards—the more it will cost to do the work. Money, of course, also comes into play in competitive bidding.

Much CRM work today is contracted through competitive bidding. Ideally, the bids will come from ethical and competent companies that are all bidding on the same scope of work. The price differences should reflect differences in experience, efficiency, labor-unit costs, overhead, innovation, and concern for profit. But in the real world, the playing field is often uneven, and competing firms use a variety of strategies to win bids. Some may calculate what they think the sponsor is willing to pay and bid no more, even if it is not enough to do the work properly. Others may calculate what their competitors will bid and set their own price lower. Still others may be inexperienced contractors who tender unrealistically low bids based on little more than wild guesses. All of these tactics sometimes win contracts, but often the archaeological record suffers as a result of significantly underbidding the cost of doing a good job.

For the ethical bidder the only reasonable approach is to do your best to calculate the true costs of successfully addressing the scope of work (see sidebar 4.2). There is no adequate substitute for experience. Experienced archaeological contractors who carefully manage their projects can usually predict about what it will actually cost to do a set amount of work under known conditions. They also know what contingencies to allow for and what fatal flaws to be on the lookout for. Such individuals will read a scope of work very carefully and may not elect to bid on poorly or unrealistically conceived projects.

LOGISTICS

Scopes of work may spell out (or fail to spell out) other logistical provisions that will directly affect your ability to do the work in a timely and cost-effective manner. A prime example is access: Who owns the land, and who controls access? While this may not be an issue with many CRM projects, it can be a real problem when private landowners are involved or when more than one party controls access to the property. Make sure the scope of work spells out who will provide access and what conditions are involved. Usually it is the responsibility of the sponsor to secure access, but not always. We have experienced costly delays when access issues have cropped up at the last moment—the crew is twiddling thumbs and drawing per diem, rented vehicles sit idle, and contractual obligations loom. Cover yourself on this one; make sure you have contractual provisions that protect you in the event that access problems arise.

Another logistical consideration is whether heavy machinery or other specialized pieces of equipment are needed. The scope of work and/or the contract should spell out who provides what. If the sponsor says it will provide all the needed equipment, you should make sure that this is written down, again so that you're covered.

Images of heavy machinery and covering yourself conjure up another thing you need to consider: Occupational Safety and Health Administration (OSHA) regulations, especially those regarding trench safety (see sidebar 4.3). Traditionally, archaeologists have been lax, but the CRM world is becoming increasingly aware that OSHA safety regulations apply to archaeological work. This is particularly true in trenching and deep excavations. Planning safe excavations is always important, even though meeting OSHA requirements may go far beyond what you consider safe. You need to educate yourself about these requirements and the logistical consequences of following OSHA standards. Such considerations may well alter your research design and significantly impact your budget.

4.2. HOW MUCH IS TOO LITTLE?

Underbidding a project is bad—on several fronts. The most obvious problem is that you'll run out of money before the work is finished. You may end up with boxes of materials in your garage that you've obligated yourself to analyze. You may have an unfinished report dangling over your head and dropping into your lap every time you think you've got a little slack. Aside from the personal angst of leaving a job unfinished, you could also jeopardize your ability to get new contracts or the required permits for new projects.

We archaeologists have a stewardship responsibility to the archaeological deposits that we disturb. When we destroy a deposit via backhoe trenches, excavation units, test pits, shovel tests, surface collecting, or any other destructive technique, we have an absolute obligation to systematically record all relevant observations, curate the collected material properly, and analyze and report on the materials recovered. The only way to fulfill your obligation to yourself, the regulations, and the archaeological record is to complete your work.

Here are a few tips on how to avoid underestimating the money required to fulfill a scope of work:

1. Unless you know the general project area from firsthand experience, try to take a quick trip there and see the lay of the land for yourself. Failing that, do some homework and read up on work that others have done nearby. Consult with people who are experienced in the area. This effort may be more than rewarded by preventing you from badly misjudging the logistics of working in an unfamiliar area.

CONCLUSION

Once you've worked through your scope and satisfied yourself that all the bases have been covered, you've started the process of designing effective archaeological research. The purpose of the scope of work is to make sure that everyone involved—you, your sponsor, the regulators, interested parties, everyone—has the same understanding of what is to be done, where, and when. It may be that you share your revised scope of work with the sponsor as part of the bidding process and submit it as part of your proposal. Or in more informal situations, it may be enough to fax over a copy and discuss the revisions and additions over the

2. If you don't know what it will cost to do a given set of work, find out. Call on someone who is experienced, and ask for help. Hire an experienced hand to help you. Beg if you have to, but get help before it is too late. Once your proposal has been accepted and you have signed a contract, you are stuck with your bid.
3. If you have to guess, guess high. This may lessen your chances of getting the work, but if you land it, you'll have a better chance of breaking even or making a profit.
4. Word your proposal and contract carefully to acknowledge and allow for as many contingencies as possible. If possible, give yourself some wiggle room, so you can make adjustments when reality rears its ugly head. One way to do this is to prioritize your research plan and distinguish between "must do" and "intend to do (if all goes well)" items. Make sure the contract obligates you only to complete the "must do" list.
5. If you do fail to accurately estimate the cost of doing everything the scope of work calls for and you find yourself in a rapidly deteriorating circumstance (the money is running out, and there is lots of work still to do), the best policy is honesty. Go to your sponsor and explain what happened as soon as you possibly can—you screwed up, or, less judgmentally, you were dealt an unexpected hand by nature or the archaeological record. Try to renegotiate the contract or scope. Your sponsor may or may not be willing to give you a break, but cutting needed corners, lying, and other forms of deceit are sure paths to archaeological perdition. If all you care about is making a buck, why not pick a more profitable line of work?

phone. Expect some negotiation as part of this process. Remember that time is more precious than money to many sponsors. If your project is part of the environmental review process for a major development or construction project, the schedules for contracted work can extend over half a decade or more. Be prepared to be flexible where you can and hard-nosed where the research is concerned.

4.3. OH, MY GOSH, IT'S OSHA!

OSHA, the Occupational Safety and Health Administration (www.osha.gov), is concerned with your and your crew's safety, even if you aren't. Plus, OSHA is equipped with the clout to make you become concerned, to make you take action to protect your workers, and to shut your project down dead. But OSHA is nothing to be scared of if you do your homework and follow the rules as closely as you can.

You'll most likely encounter OSHA regulations and their strict enforcement when you're working on a project for a federal agency or a large state agency that gets most of its money from the feds, like our various state departments of transportation. While OSHA has enforcement responsibility for any business with more than one employee, the odds are against ever encountering an OSHA inspector in the wild. It's much more likely that you'll come up against an agency or corporate safety officer. Their job is to make sure that the folks working for them—including contractors—obey OSHA regulations to head off possible lawsuits.

While there are literally tens of thousands of pages of OSHA safety regulations, most archaeologists encounter problems only when we start to trench (i.e., excavate deeply) or use heavy equipment. Before heading to the field, you need to check on the possible location of any utility lines or pipes. This may not be a big deal in the wild yonder, but it only takes one nick in a high-voltage cable to ruin your backhoe operator's whole day. Many states and utility companies have a central hot line for utility line location. The companies will flag any lines in your project area for free within a few days of your call. Call before you dig! When trenching, you need to ensure that there are no surface hazards in the way of the trench. Another common-sense rule is that no employees can be under the backhoe bucket. Nor should anyone be near the area where the excavated material is dumped (unlike one of our favorite crew chiefs who liked to have us hold the wheelbarrow while he tried to dump a backhoe bucketful of dirt into it). Make sure anyone in the trench is wearing an approved hardhat and that all equipment and vehicles remain more than two feet from the edge of the excavation so stuff doesn't fall on the heads of the folks in the trench.

If you do plan to trench, and you expect that your trenches will be deeper than five feet (1.5 meters), you'll be responsible for developing a plan to slope or bench the sides of the trench or to install shoring to protect folks from cave-in. Trench collapses aren't accidents—they are manifestations of physics and soil mechanics. The exact formula for benching or sloping or shoring depends on the soil type and projected depth of the trench. The regulations require that these calculations be made by a registered engineer or other competent person. Your best bet in these situations, and really any time you work for a large agency, is to contact the agency's safety officer and get his input on the planned work. Don't guess: Ask the experts or get the OSHA training yourself. If you plan to do much trenching more than 1.5 meters deep, be sure to ask your sponsor about meeting OSHA regulations when you are preparing your proposal. Renting shoring systems or benching an excavation can radically change the schedule and budget for what you might assume would be a simple stratigraphic trench or excavation block.

Most OSHA regulations are common sense and are there for your safety. The simplest and best way to handle site safety is like everything else: Plan for it. Take an OSHA short course. If your sponsor or company has a safety officer who works with contractors, contact her, get her input, and take it to heart. For the slight inconvenience it might be to wear that hardhat on a hot day, it's a lot hotter with a metal plate in your head.

5
RESEARCH QUESTIONS

The research questions you pose and attempt to tackle will form the core of your research plan. If you are able to frame good research questions, you'll be much more likely to learn something worth knowing. Conversely, if your questions are vague or purely technical in nature ("Are there sites here?"), the odds are slim that you will actually be doing much research. You really can't do any kind of research without asking good questions.

Research questions, problems, and hypotheses are essentially the same thing. You may prefer to structure your research in terms of problems to be solved or addressed or hypotheses to be tested. Or you might start with a grand, overarching problem or hypothesis and break this down into narrower questions or sets of expectations and test implications. It really doesn't matter what you call them. What matters is that you develop and articulate clear, structured ideas about what you are trying to learn, so that your research strategy will be focused and aimed at acquiring specific sets of information that will inform your research questions, problems, or hypotheses.

No matter what you call them, research questions are simply what you plan to learn about the archaeological record. As all archaeologists know, you may not learn what you want to from your archaeological work. You may learn other things, or you may disprove something you set out to prove. If your mind is fully engaged, you will always end up with more new questions than definitive answers. But by planning to learn something worthwhile, you're focusing your work, focusing your approach on something concrete. Such strategic planning is the essence of all science—no less so for archaeology done to satisfy government requirements. If archaeologists are to continue to suckle the CRM teat in the twenty-first century, we sure better be learning things worth knowing. So a research question is really the core of a plan for learning.

In many cases in CRM, particularly those involving small, routine projects, the scope of work in your proposal may be as far as your formal research design goes. Even so, you need to elaborate the work portion of your scope and develop some explicit research questions. Otherwise, you are not really doing archaeological research—you are just performing a technical service that has comparatively little chance of making a meaningful contribution beyond the most basic site records. It's the difference between going through the motions and doing archaeology with your mind fully engaged. Often, the choice is yours.

Over and over we've observed colleagues who just "shovel it out and write it up," who often lack basic research skills, and who make little or no effort to keep current (which is easy to tell since they can't hold up their end of a research conversation with a two-by-four). They may be perfectly nice people, but they wouldn't know an interesting research question if they stepped on it. And not surprisingly, most of their "research" questions are seldom more profound than "How much can I get away with charging my client to dig these holes?" Conversely, our colleagues who are well prepared, involved in their research, and in tune with what's current are blazing productive trails into the archaeological record. These are the folks who everyone wants to talk to at the annual meeting, who everybody wants on their research team, and who rarely have trouble getting their research designs and draft reports accepted. These are the real archaeologists, the ones who embrace CRM as an opportunity for doing research.

Most research questions are posed to address holes in current archaeological knowledge. CRM research often begins with very basic and mundane holes: Are there any sites in this previously unexamined area? How old are they? What time periods and archaeological cultures do they represent? What are the depositional circumstances and preservation conditions? Has the site's recent history left enough traces of its ancient history intact to warrant more archaeological investigation? As work progresses from survey and preliminary testing, holes begin to emerge that are potentially more profound. Does this Paleoindian site preserve the data needed to shed light on the mobility patterns of Clovis peoples? Can this ruined nineteenth-century textile mill tell us something worth knowing about the Industrial Revolution in New England?

Your research questions will be framed within many contexts. Your approach to doing archaeology, the type of project, the nature of the project area, and even the folks who are on your crew will determine the kinds of questions you can ask and the kinds of results you can expect. To start developing your research questions, you need to understand these contexts and how they interrelate. At the base of everything you do is the approach you take to archaeological research.

TYPICAL APPROACHES TO ARCHAEOLOGICAL RESEARCH

While archaeologists often conceive of themselves as eclectic scientists, most tend to follow a narrow path from one research project to the next. Here we contrast four typical approaches and consider their impact on designing research. Keep in mind that ideas and attitudes are rarely exclusive to one approach or another. The first two approaches, scientific archaeology and culture history archaeology, are common to both academic and CRM archaeology. The second two, compliance and management archaeology and archaeology-as-business, are unique to CRM, although some academic archaeologists participate in these approaches as well.

SCIENTIFIC ARCHAEOLOGY

Much verbiage has been expended debating what archaeology as a science is and how it should proceed. As we alluded to in the introduction, we like the simple dictum: "Science is a strategy for learning." Pressed for a label, most academic archaeologists in North America call themselves processualists or cultural ecologists. Processualists consider themselves to be scientists engaged in research aimed at understanding the cultural processes by which human societies evolve and adapt. Cultural ecologists also consider themselves to be scientists but focus on the dynamic interaction between human societies and their environments. While archaeologists, postprocessualists also employ other scientific strategies (and may prefer to call themselves evolutionists or various other kinds of -ists), these have in common the goal of developing bodies of knowledge based on systematic observation, experimentation (or repeated sampling), and evaluation. A well-considered research design is fundamental to all scientific approaches because it lays out one's research problems and links these to method and theory. Yet scientific archaeological research designs can be densely written documents in which buzzwords, rhetoric, grand ideas, and abstract concepts take precedence over pragmatics. Done well, a scientific approach highlights explicit research questions or hypotheses to be tested and evaluated through a formal and achievable process. As should be obvious by now, we think a scientific approach is critical for the success of all substantive archaeological research projects, including those in the CRM world.

CULTURE HISTORY ARCHAEOLOGY

Archaeologists of almost any stripe share a common need to place archaeological finds in time and space. Prior to the 1960s, most American archaeologists thought of themselves as (pre)historians bent on reconstructing past cultures and past events. While the culture history approach is no longer the discipline's central focus, it still provides the basic chronological framework and nomenclature that most explanations about the human past require. Although these days few academic archaeologists younger than age fifty champion culture history as a productive strategy for learning, most archaeologists still follow field traditions that were developed to address basic cultural-historical problems.

The best culture historians have an enviable command of the ethnographic and archaeological record and often take a detectivistic and functional research approach aimed at reconstructing the human history of particular areas and tracing broader patterns of human behavior. "Typical" culture historians emphasize categorization, description, and chronology following the premise that once enough details are amassed, clear patterns and truths will emerge. This enduring sort of blind empiricism and typological wallow is not a viable part of any real science, and it was never the culture historian's ultimate goal. Today many CRM research designs continue to focus on traditional cultural-historical goals such as refining chronology and studying little-known cultural periods and regions. When linked to more robust scientific frameworks, such goals can be fruitful, but as ends in themselves, they often contribute little to an improved understanding of the archaeological record.

COMPLIANCE AND MANAGEMENT

In archaeology, like any arena in which government is involved, the process of following, implementing, and monitoring regulations has taken on a life of its own. Curse them if you will, but bureaucracies have their own logic, a logic embodied by the term *cultural resource management*. Instead of learning about the human past, the prime directive is to comply with the legal and bureaucratic mandates and manage culture resources, of which archaeological sites are only one category. Government archaeologists often prepare the scopes of work and requests for proposals that contracting archaeologists respond to. The language of section 106 has become the lingua franca of CRM research.

Excellent archaeological research can be accomplished much more easily within the framework of compliance and management when you learn to how to couch your research strategy in terms of the bureaucratic process. The downside, of course, is that satisfying the process often becomes the sole objective rather than the means by which worthwhile research is accomplished. Without linkage to scientific objectives, compliance and management research designs are often incredibly boring and nonproductive exercises in government excess.

ARCHAEOLOGY-AS-BUSINESS

Over the past twenty-five years, the epicenter of archaeological research has shifted from the university to private enterprise. The positive side is that for-profit firms have brought more efficient business practices to an inefficient discipline. Private companies can respond quickly to the changing contexts, client needs, and deadlines of government-mandated archaeology. The best consulting firms employ well-trained archaeologists and specialists in other fields who are committed to achieving meaningful and cost-effective research within a viable business environment. The downside is epitomized by low-bid "archaeologists" whose main goal and chief skill are exploiting the process for their own profit. Such unethical individuals and firms survive largely because the discipline has not yet figured out how to effectively regulate itself as a profession. In designing archaeological research, a businesslike approach forces you to look closely at the true costs of a research plan. When business and science are appropriately linked, excellent research can result. When the linkage is weak or absent, archaeology-as-business utterly fails as a worthwhile research endeavor.

LANDSCAPE: A CROSS-CUTTING APPROACH

All approaches to archaeological research share a common need to understand the physical, ecological, and cultural landscapes within which we find the patterns we study. Understanding these contexts, and the processes that shape them, is at the core of all archaeological research, no matter its theoretical, political, or bureaucratic focus. Without these contexts, field research is a mindlessly routine mechanical process.

The landscape provides the real-world context for the things we study. A landscape approach is not an alternative to scientific archaeology (indeed, it's critical to cultural ecology) or to one of the other typical approaches we outlined but is a necessary and integral element of most productive archaeological research projects. This is especially true of CRM projects that seek to evaluate and make life-and-death judgments about small, arbitrary segments of the surviving in-the-ground archaeological record. Few intelligent management and research decisions are possible unless they are informed by a keen appreciation of the wider landscapes to which all cultural resources belong. You literally have to know the lay of the land and its culture history, depositional environments, and paleoecology to make an informed judgment about the information potential of a newly tested or discovered site.

Landscapes are defined and understood through studying the distributions and interactions of plants and animals; the geological processes and formations on and within which plant and animal communities exist; the climatic regimes that govern geological and biological processes; and the land use patterns of modern, historic, and prehistoric peoples. Developing an understanding of these interrelated landscape contexts is a key part of any archaeological study. With a clear understanding of the local landscape, an archaeological team is able to devise intelligent and productive research strategies. The key here is *team*.

Successful landscape studies are usually accomplished by archaeological research teams that include experts in diverse fields such as geology, soils, geography, botany, and zoology, among others. Some are archaeologists who have developed specialties, while others are specialists from other fields. The composition of your team will vary greatly, of course, depending on the research circumstances. For instance, when undertaking research in unknown or poorly known territory, many archaeological projects today employ geologists, soil scientists, or geoarchaeologists whose task it is to study the physical landscape before them and figure out how it formed and changed during the time spans in which humans may have used it. Archaeologists are often able to draw on existing studies made by botanists, paleontologists, hydrologists, and many other kinds of scientists whose research interests conveniently intersect with those of archaeology.

In taking a landscape approach, archaeologists are often able to focus research effort on the landscape contexts themselves. Obviously in an archaeological research project the emphasis is on understanding or exploring the archaeology of the study area. But by recognizing and targeting important depositional and environmental data, we can place our archaeological data in more meaningful contexts. Such an approach helps ensure that every project, no matter how small or seemingly insignificant, can make a contribution to our knowledge of the archaeological record. If we focus only on archaeological aspects of our research, then a survey that finds no sites could be (and often is) considered a waste of time. The same survey, though, is an opportunity to document a small part of the landscape and add to our overall understanding of the region. Such information informs future research.

The contexts of archaeological site deposits are too complex and too varied for us to depend on the random interest of other professionals for the data we need. A landscape approach allows archaeologists to take the lead in developing the rich contexts we need to usefully study archaeological deposits. By broadening our reach, we can ensure that our work is useful and interesting.

DEVELOPING YOUR OWN APPROACH

Quite likely your own approach to archaeological research will be only partially under your control, unless you own a consulting company or have marched far enough up the career ladder so that you are calling the shots. Even then, the contexts in which you are doing research will determine many aspects of your approach. It is your choice either to blindly follow the pack to do archaeology by default or to approach your archaeology actively and creatively.

Developing your own approach is something you will work on most of your career. Early on you'll be told what to do as you experience the shock of trying to reconcile the archaeology you learned in college with the archaeology you encounter in the trenches and offices of the CRM world. Your idealistic mind asks, What the heck is scientific about digging shovel tests day in, day out? And what does coming in under budget so the company can make a profit have to do with doing research? Lots of would-be archaeologists never make it beyond this point and quit in disgust or disillusion. Others abandon their academic ideals and learn to walk the CRM walk without thinking much about the connection between what they do and the scientific goals of the archaeological discipline. But many of us carry on the struggle as our careers unfold. We see things that work and things that don't, opportunities that pay off and those that are squandered, and we vow to do our best when our turn comes to make decisions.

It's up to you to develop your own approach and do your best to find ways of achieving meaningful archaeological research within the contexts and confines of the CRM world. Your own approach will evolve from its idealistic and naïve beginnings as you mature and gain experience. If your commitment to the discipline of archaeology stays strong, you will figure out how to reconcile theory with practice and craft an approach that works for you and your situation. You will see that a pure scientific approach is difficult, if not impossible, to follow in CRM but that scientific goals can be pursued and reconciled with compliance, management, and business goals. You will also see how hard yet essential it is to understand the past landscape contexts within which lived the people who created the archaeological remains you study.

The approach you take to archaeological research will probably suggest to you several relevant research questions for just about any project. If you're interested in culture history, filling gaps in a regional chronology might be a likely question. If compliance needs loom large, you're probably asking basic questions like site location and condition. If you take a scientific landscape approach, you may be more interested in the evolutionary relationships between climate and social complexity in your region. Whatever your approach, whatever the scale of your project, you'll need to develop a guiding set of research questions.

CREATING RESEARCH QUESTIONS

If you've already got a research question (or several) in mind, you're ahead of the game. If you're unsure where to get started, what would work for your project, or your questions are hopelessly vague, take heart, because you're not alone. Most archaeological practitioners have been in that very same spot, poised to go forth and do research without a clear question in mind.

Fortunately, there are numerous places to look for ideas, direction, and even fully formed research questions. Keep in mind that a research question doesn't usually have an expiration date. One of the most successful ways we push forward our knowledge of the archaeological record is by examining, reexamining, and breaking apart common questions that have not been satisfactorily answered. Already-asked questions are all around you, waiting to be gleaned for your project.

An obvious place to start in most circumstances is in published regional research designs and research syntheses. If you are willing to dig, you can also find great questions posed in reports on previous research efforts, often hidden amid the dull technical writing that distinguishes far too much CRM reporting. Your firm or institution may have a particular research focus that lends itself to your current work. Although we don't like to admit it, some (too many) research questions are the result of the academic fad of the moment. But for most of us, the most interesting, fun, and useful questions are the ones that you pose anew for yourself based on your foundation, regional knowledge, and personal research interests. These are the questions you'll care the most about and the ones that define the work that will give you the most satisfaction.

SOURCES OF QUESTIONS

The following aren't the only sources for research questions, but if you're at a loss as to where to begin, start with these leads.

- *Regional research designs.* One generally positive consequence of the sustained growth of CRM research is the existence of documents that synthesize what is known about the archaeology of a given region or topic and spell out research questions. These documents are variously termed *syntheses, overviews, regional research designs,* or *state* or *regional plans.* In many states, the SHPO has produced or commissioned such studies as part of its federal mandate. Other studies have been produced by or for federal agencies such as the Corps of Engineers (COE) and the Bureau of Land Management (BLM) or state agencies such as your Department of Transportation. Still other syntheses at the state and regional levels have been written by academic archaeologists on their own dime for their own reasons. In active regions, there may be many such sources of background knowledge and research questions. To find out what (if anything) exists in the area where you'll be working, the place to start is with the SHPO's office. The staff there will be aware of most of the syntheses and regional plans written for your state.

In addition to the SHPO, check also with any land-holding federal agencies that sponsor work in your area, such as the U.S. Army Corps of Engineers (COE), Bureau of Land Management (BLM), Forest Service, or National Park Service (NPS). They may have regional research designs which cross-cut state boundaries that the SHPO doesn't deal with regularly.

Lastly, don't forget your peer network. A quick phone call or e-mail to a few of your cohorts can lead you to obscure, newly published, or soon-to-be published syntheses or research designs that might fit the bill. Be forewarned that such documents can become outdated rather quickly in active research areas. You also need to find out whether the SHPO uses the syntheses and regional plans that might be in place. If the SHPO doesn't accept a regional research plan, using it as the foundation of your research effort could complicate your life.

- *Institutional focus.* Many archaeological organizations develop their own research focus. This can be a region, a time period, an archaeological culture, or a particular analytical or theoretical approach (see sidebar 5.1). If your institution has one or several such specialties, these should help you develop good research questions by building on what has already been done (e.g., filling in gaps or testing existing hypotheses). Often such institutional foci help strengthen your research proposals because the track record is already there. It is easier to sell your research ideas when your team includes experienced analysts who have a proven ability and the needed research tools such as comparative collections and specialized equipment. Don't panic if your organization has no research focus, but do keep in mind that experience and established expertise count a great deal.
- *Personal research interests.* Yes, you'd better have your own research interests. If you wake up one day after years of archaeological employment and find yourself without any particular research interests, it's past time for you to start considering an-

> ### 5.1. CENTER FOR ECOLOGICAL ARCHAEOLOGY AND ITS INSTITUTIONAL FOCUS
>
> The Center for Ecological Archaeology (CEA) at Texas A&M University was an example of an archaeological research organization with a strong institutional focus. As the name indicates, this university-based contracting organization took an ecological approach to archaeology. The CEA sought out CRM and grant-funded projects within its main research region (broadly, the southern and eastern parts of Texas) that played to its strengths. Among its staff archaeologists were specialists in geoarchaeology and paleobotany. CEA also worked on a consulting basis with leading experts in soils, geology, zooarchaeology, and other ecological fields.
>
> In the early 1990s, a CEA project took place within the giant foundation trench that was being dug to anchor a dam intended to create a municipal water supply reservoir near San Antonio. CEA archaeologists sampled a series of hunter-gatherer occupations up to twelve meters below the modern floodplain surface. As huge earth-moving machines worked their way downward, the archaeologists cordoned off relatively small areas so they could safely and quickly examine layer after layer of thin occupational lenses often separated by flood deposits. Over several years, the CEA's multidisciplinary research team was able to reveal the interwoven climatic, environmental, and cultural relationships of a ten-thousand-year record of human ecology. Through this project and many smaller ones, the CEA developed a strong research focus, one that gave it a competitive edge because of its proven ability to develop and follow through on scientific-based questions linked to ecology.

other career. It is difficult to design effective research unless you have an abiding interest in the outcome of your work. One of the big reasons lots of CRM work sucks is because the work is done only as a technical service instead of as a research effort. Most serious archaeologists develop many different research interests as their career progresses. Often these are unanticipated outgrowths of research opportunities. For example, one of us (Black) has become a minor expert on earth ovens after working at many sites where such features are common. This personal research interest has led to a number of projects where earth/rock ovens have been a central focus, as well as a series of workshops, presentations, and publications focused on experimental approaches to understanding earth ovens and other hot-rock cooking techniques. This focus has helped define highly targeted research questions regarding technology, subsistence, and site structure that continue to inform ongoing research projects. Such personal research interests are great sources of research questions because you are posing genuine problems from the strength of your prior experience and knowledge. Sustained interest in a topic helps you ask increasingly more productive questions.

- *Fads.* Fads play a big role in archaeological research, sometimes to the good, often not. Archaeological fads often begin when some big-name researcher publishes an article or book that attracts attention. If Lew Binford wrote an article about polyendemic resource procurement, the phrase (and the acronym PRP) would soon find its way into research proposals across the country, first in academia and then in CRM, regardless of whether it made any sense. Too often people latch onto buzzwords and leap on the latest intellectual bandwagon but neglect the real work that is required of substantive research. We've all seen new fads flash on the scene, make the quoting circles, and then fade away.

  The reason fads are fads, though, is that people are interested in them, and that interest does count in the world of CRM archaeology as it does in academia. The interest cuts two ways. One reason that fads fade so quickly is that while the white hot heat is on, lots of researchers are focused squarely on the target. Very quickly papers and reports start popping up that document every wart and pimple revealed by the emperor's new clothes. If you decide to go with a fad, make sure you follow it closely enough to keep current. Nothing is quite as foolish as putting all your research eggs in a faded basket whose bottom was ripped out by a lead article in *American Antiquity* two years ago.

- *Informed knowledge.* Perhaps the best source of research questions is an informed knowledge of what is known, what is worth knowing, and what is knowable in a given circumstance. This goes right back to your training and basic background knowledge as well as your career-long need to keep current. If you are well informed about the state of archaeological knowledge in the geographic and topical areas of interest, you'll see the gaps, holes, and questions worth pursuing in order to advance our understanding of the archaeological record. It starts with your background and builds with your current, up-to-date understanding of the regional record and the wider field of anthropological archaeology. When you throw in your personal and institutional research interests, you can't help but find interesting, exciting, fun research questions around which to focus your work.

Developing research questions is where you get the chance to exercise your archaeological imagination to best effect. The kinds of questions you pose at the beginning of the project will determine to a great extent how many of your colleagues will actually read your final report, how much interest will be generated in your work, and how many beers you will be fed to explain what you did and why. But if all we had to do was think up neat questions, we'd all be selling insurance by now. A good research question is only halfway there—to make a contribution, you need to have a shot at formulating an answer. You need questions for which you can reasonably expect to collect relevant data. So, as soon as you pose what you think is a good research question, you need to evaluate it before you commit yourself to action.

THINK THROUGH YOUR RESEARCH QUESTION— EXCAVATE IN YOUR MIND'S EYE

The value of your research question is relative to its cost and probability of success. Once you have a question in mind, simple or complex, you need to measure it against these acid tests:

- *What data do I need to address this question?* No cheating here. You need to think about this explicitly from an informed perspective. If your question is focused on seasonal resource scheduling, it's not enough to say, "I need preserved plant materials." You need the right kinds of plant materials from the right contexts. You will need data on associated artifacts and features, not to mention reliable dating to provide the cultural and chronological contexts that give meaning to all archaeological research questions. Think about all the data you're going to need. Then ask:
- *Is it likely that data needed to address this question can be obtained?* A useful research question in your region might be, "Were domesticated plants introduced in Late Archaic times, prior to heavy reliance on agriculture?" However, the reality may be that owing to the prevailing preservation conditions at the sites under investigation, you probably won't encounter the materials you need to address that question unless you see a way to infer incipient agriculture from the range of artifacts and features that you will actually encounter. Such indirect evidence may not provide an adequate basis to resolve your research question. So the first cut is

whether you, in your best archaeological judgment, believe that your field efforts will likely encounter the kinds of materials and contexts that will produce the data you need to attack your question. If you are confident that the data will be there to be gleaned, you must then ask:

- *Can I collect the data I need under the proposed scope of work?* Questions focused on materials recovered from deeply buried sites won't be of much use for a reconnaissance survey or limited testing program. At an equally practical level, your project budget may not cover the personnel to process and analyze flotation samples for large volumes of dirt or to cover a survey area at the level of resolution you'd need to produce data relevant to your questions. Logistics—time, place, money—control what you do and how you do it, to a large extent. The most elegant research question targeted at exactly the needed (and existing) data is not worth squat if you can't make the logistics work. Your questions need to point at data that you will be able to effectively collect and analyze within your project constraints. This leads to the final test:
- *Can I analyze and report the data in ways that effectively address the question?* It's not enough just to have the data in hand: You must have a workable plan to analyze and report the data so that it informs your research question. The old adage that you can figure out what it means back in the lab is a vestige of archaeology's heritage as a long-term academic endeavor. Tenured professors may have the luxury of taking their sweet time and of hauling lots of stuff back to the university without any concrete plan for completing the analysis. In the CRM world, a wealth of rich data doesn't help you much if you lack the skills, budget, or time to analyze and report it.

To take a personal example, when we excavated the Higgins site, an Archaic hunter-gatherer campsite, we used then state-of-the-art total data stations (TDSs) and collected incredible three-dimensional provenience data on more than thirty thousand artifacts. It was, and is, an awesome data set, rich with possibility. But because we didn't have the foresight to carefully consider how we would analyze and report this data, it was eventually collapsed into ten-centimeter-thick analytical units for comparison and analysis. It was cool, but it did not go very far toward addressing most of our research questions. We had not budgeted enough time to figure how to derive the data we'd hoped to use to deduce the subtle vertical patterning within the site. This experience highlights what is one of the most critical tests: All the sites, all the data in the world don't advance research much if it's beyond your capabilities to analyze and report them effectively.

Look at each one of your research questions with these tests in mind. A thorough job of thinking through your research questions will save you heartache, time, and money down the road. Once again, keep in mind that the process of thinking through your research questions is ongoing as the project unfolds. As new information, new constraints, and new opportunities become known, you will often need to rethink your initial research questions and run through the acid tests again. You may well find that your once-promising research question must bite the dust. Fine, but do yourself a favor and take a few minutes and write down a succinct and honest justification for why you abandoned your research question. The same goes if you have come up with a new question—think it through and write it down.

Armed with a thorough scope of work and a good set of research questions, you are ready to focus a little closer and figure out how these problems can be incorporated within a workable strategy or research plan.

6
DEVELOPING YOUR RESEARCH STRATEGY

Your research design is part of a larger strategy, a strategy that joins the work you have to do, the resources you have at your disposal, and the research you hope to accomplish. To be effective, this strategy needs to be two-pronged: archaeological and operational. Your archaeological strategy addresses the data collection, sampling, and analytical methods you plan to use to accomplish your research goals. Just as important (and often ignored), your operational strategy needs to address the logistical, organizational, and political needs of your project.

THINK FIRST, DIG LATER— ARCHAEOLOGICAL STRATEGY

Developing a successful archaeological strategy is hard, so you have to *think before you dig*. The larger and more complex the research project, the more thinking ahead you must do. Your job is to clearly understand what it is you are trying to accomplish, within all the contexts you're working in—the natural and cultural landscapes, the regulations, and the sponsor's time constraints. While the concept of thinking before you dig sounds obvious, archaeologists are notorious for arriving in the field without much of a game plan. Oh, sure, the typical archaeological project today follows some sort of research design, but far too often the research strategy boils down to little more than "Let's dig some holes and hope we find something interesting"—archaeology by default.

The mind-set that there is a limited and proven set of correct or standard ways to do things relieves the archaeologist from the responsibility of thinking. Making strategic decisions means explicitly examining the standards and alternatives and stating explicitly from the outset what you will be doing. For others to judge the work you do, the data you collect, and the conclusions you reach, they need to understand how you did the work. Your strategy is the way you define the work you'll do, for yourself and your peers.

Several important elements comprise your archaeological strategy. First you need to focus on the data you need or hope to collect to address your research questions. If your questions have to do with seasonality, you may want to collect floral or faunal data. What kind of materials will you target to get the data you need? Linked inexorably to the data are the methods you'll use to collect your materials and make your observations. If you are targeting plant remains, will you use flotation to collect the light fraction for sorting? Do you need to use a chemical process to break the bonds of a heavy clay soil? How will that affect delicate plant materials and the suitability of surviving plant remains for radiocarbon dating? A big part of any archaeological strategy must address your approach to sampling. Will your float samples be taken from features only? How will that affect the analysis? What size constant-volume sample would provide a reliable background sample? You need to look at how you'll manage your data in the field and how materials will be processed back in the lab. If you're using a total data station (TDS) to collect provenience records, do you have a system in place to ensure against bad shots? Do you have lab procedures in place so that all your samples will be processed in a standard manner? And last, consider how you'll ultimately analyze the data. Will you be trying to compare relative frequencies of materials across or between sites? If so, how will you ensure that the sample sizes are equivalent?

These elements of your archaeological strategy are interdependent and causally linked. Remember, your job isn't to find the perfect strategy but rather to settle on one that will work and, most critically, to make your strategy explicit. Spell out what you're going to do and how you're going to do it. When you've done this, you've made a contribution to the archaeological record, regardless of what you find in the field. By presenting an explicit research strategy, you're providing the archaeological community with a record of what worked and what didn't, and probably some good insights into why. This is the information we all need if we're going to learn and move our field forward.

DEFINING YOUR DATA

First things first: The first thing you need to do is explicitly define the data you want to collect to address your research questions. But what do we mean by data? Data aren't stuff, not dart points, not rim sherds, not even a nice postmold. *Data are systematic observations.* Data may be a properly completed unit level record form, a provenience record for an artifact, or a field interpretation of an artifact type. No data exist until an observation is recorded. The distinction between things and data is basic and critical. The problem is that when we think about things as data, it's all too easy (and common) to point to a pile of things and say, "There's X, Y, Z; anyone can see that," when, in fact, hardly anyone else will see that pile of things the same way that you do. When we recognize that data are observations, we have to be more explicit. "We can tell this is X, because we can see Y and Z." When we're explicit in the way we define our data, the data we collect are more reliable and easier to analyze. And it's only when we're explicit that our peers can make the best use of our data, our work, and our analysis.

One of the first questions you need to ask yourself when you focus on the data for your project is "What is the useful level of resolution for each class of data?" You may want to map the density of artifacts within the excavated areas to identify functional components of a site. If you excavate by ten-centimeter levels on a 1 × 1–meter grid and collect artifacts from the screen, the level of resolution for your map will be one meter horizontally and ten centimeters vertically. If the patterns you're looking for are larger than 1 × 1 × 0.1 meter, then you're probably in good shape. If the patterns are smaller than that, your data are ill suited for this type of analysis. Focusing on the different data sets you hope to collect and defining them explicitly will help remove needless ambiguities from your archaeological strategy.

DATA AND METHODS

It's easy to see how data and methods are linked in a causal chain that follows right through to your analysis. Your methods define and limit the data you can collect. For instance, if you screen the soil from your site through quarter-inch mesh, you'll find it difficult to develop a reliable microdebitage data set. Not only do your methods constrain your data, but the data you collect also set limits on the type and character of the analyses you can perform. What you learn from your work is a direct product of the methodological choices you make right at the beginning.

As you look at a set of data, you should be focusing as well on the methods you'll use to collect it. Define your methods explicitly. If you want to collect constant-volume samples to develop a data set on microdebitage distribution, where are you going to collect them, how are you going to collect them, and what size will they be? How will their locations be measured? Is the depth measured from the top? Bottom? How will their locations be recorded? How will the samples be processed? How will the fractions be analyzed? How will the data from the fraction analysis be matched up with the field observations? It's not as simple as "Take one hundred cubic centimeters of soil, and count the flakes." Every decision you make on the methods you'll use to collect your data will—not *might*, not *could*—*will* affect the data you collect.

Archaeological field methods are the subject of many books and college courses as well as several of the other volumes of The Archaeologist's Toolkit. Wherever you work in archaeology, you will quickly learn the preferred and expected field methods. Often you will get the distinct impression that there is only one right way to do a given task. Nothing exemplifies this like the square hole—the one-meter square that is the basic provenience unit for so much archaeological work. While there are many reasons why square holes make good sense in many different circumstances, there is a wide world of archaeology beyond the square hole.

Methods should be chosen to fit the logistical circumstances and *to yield the targeted data,* not strictly because of expectation or habit—not archaeology by default. The standard, default methods may well be appropriate for your project. They provide consistency and comparability when properly executed. But even if you choose to rest your butt in the well-worn comfort of the "usual" methods, you need to spell out exactly what that means. Just as people have different perceptions of data, "standard" methods seldom are. The small differences—where the depth of a level is measured, what is collected from the screen, the acceptable error in elevation shots—can loom large when data sets are compared.

By carefully defining your methods (and your data), you're building the framework to ensure that the observations that make up your data are being made consistently. This helps you and the people collecting your data. It helps others understand what the data you collect represent. And it provides the background that other researchers need to evaluate your analysis and write-up. Your ultimate goal isn't to create data sets that you alone can analyze and publish. Your goal is to add to our knowledge of the archaeological record. To do that, you need to work to create *reliable* data sets that other researchers can use. The only way to make your data reliable (just like it says—so others can count on it) is to document the data and the methods you used to collect them.

SAMPLING STRATEGIES

Tangled up in data and methods is sampling. We won't go into formal sampling methods here, as statistical methods are beyond the scope of this book. Our favorite published discussion of sampling strategy in archaeology is Kent Flannery's book *The Early Mesoamerican Village* (1976). This humorous yet serious account highlights theoretical and practical aspects of sampling on regional and site levels. You will also find discussions of sampling strategies in the other Archaeologist's Toolkit volumes.

Beyond formal sampling techniques, you should recognize that all archaeological research uses sampling strategies. It is impossible to completely survey, excavate, or study the entire universe. When we collect materials to study or make observations to add to our research data, we are making sampling decisions. When you decide what survey interval to use in a pedestrian survey or reconnaissance, you're making an explicit sampling decision. Your research strategy needs to address explicitly whether this sample will yield the data you need to address your question. If it can, then it's a good strategy. Sampling strategies are part and parcel of the methods you'll use to collect your data.

One of the aspects of sampling that we often ignore is what to do with materials that won't provide data to illuminate the stated research problems. Do you collect all those snail shells on the screen even though you need only a few? Archaeologists have the bad habit of trying to collect everything just because it is there and might someday be useful to someone for something. Materials collected fortuitously and off-hand observations can provide interesting sidelights to your research and can sometimes provide valuable avenues of analysis in and of themselves. But because of tight budgets and strained resources, we also have to consider the costs—in time, money, and space—to process and curate those materials. Like most knotty problems, there is no easy answer; it's something we have to deal with on a case-by-case basis. You should develop an effective sampling strategy that focuses on the data you know you need and preserves enough samples of the rest of the encountered material record so that other kinds of problems can be investigated later.

LAB PROCESSING AND DATA MANAGEMENT

It's easy to ignore the lab processing and data management aspects of your work until the day your bags of materials are stacked on the lab shelf. But the methods you use to collect your data don't stop at the lab door. You should be just as explicit in defining how the materials from the field will be processed and analyzed as you would be in defining how to take a soil sample. Lab folks often are better organized than their field counterparts, so if you have a regular lab staff, they probably already have a set of documentation outlining their standard procedures. The goal, again, is to produce a reliable data set. When you're knee deep in the analysis, you (and anyone else who hopes to use it) must know how the data you're using were generated. The recipe is simple: Explicitly define how your materials will be processed and how your observations will be made.

The data, the observations that are taken in the field and the lab, must be organized and accessible if you are going to use them effectively. Data management is at the core of effectively analyzing your expensive and explicit data. When we think of data management, we usually think of computer databases. But data management isn't just computers. Data management means organizing your data so that they are accessible and safe. Photocopying pages from your field notebook every week and storing them in a binder in the lab is a good way to protect that data from loss—and it's data management. Field forms stuffed into an artifact box with muddy soil samples won't yield the best-quality data, and your data management plan has an obvious hole. Your strategy needs to include how you will move data from the field to the lab to analysis—all the data. You need to plan for problems and develop data management strategies to avoid data loss (e.g., if it's raining, put the field forms in a plastic zip-top bag).

There probably aren't many archaeological projects on the planet today where computers aren't being used. But remember, there is a price to pay. Computers don't really make anything easier, simpler, or less expensive. They require capital investment, power, training, and support. What computers can do is allow us to work with larger and more complex data sets, to do more complex work. But they don't do it by themselves. If your strategy is to use computers to help manage your data, you need to pay the price up front. You need to consider your data, your methods, your sampling strategies, your lab processing, and your analytical needs when putting together your database. You need to ask and answer questions like these: Will data be input in the lab? In the field? Both? How will data be merged? How will data entry be verified?

How can data be reused (for artifact tags, accession cards, etc.)? And the big one: How will we want to retrieve these data during the analysis? Having a well-considered (and, again, documented) data management strategy is a key part of ensuring that the data you need will be there when you need them.

ANTICIPATING THE ANALYSIS

Once you've considered your data and planned your methodological attack, you need to take it a step farther. How will you analyze these data to inform you on your research questions? It's all too easy (and common) to run out to the field, grub up a bunch of stuff, take a pile of field notes, and say to ourselves, "We'll figure it out in the lab." Well, if the data don't match up with the questions you're trying to ask, or if you can't retrieve the data you need, you're screwed, plain and simple. The entire process of developing an archaeological strategy is recursive—you look from your data back to your questions, and from your methods to your data, and from your sampling strategies to your methods, and finally you should consider how your analysis will illuminate your research questions. And from there you look back again at your data, methods, and sampling.

As with the rest of your archaeological strategy, you need to be explicit in defining your analysis. Recognize from the outset that your approach may change, but consider how you would proceed today if you had the data you plan to recover in hand. Just as defining your methods will help you maintain control in data collection, documenting how you plan to analyze the data can provide a jump start to that part of the work and helps keep the focus on the research questions at hand.

FINAL STEPS

The last step in planning your archaeological strategy is to critically review your questions, data, methodology, and analysis. Get someone else—preferably an experienced critical thinker—to read it over and critique it (better now than when it's too late). Where are the holes? What are the hang-ups? You may need to reconsider parts of your questions, which will change aspects of your data and analysis. Stepping through this recursive process in the planning stage pays benefits throughout the course of your research.

Everything costs, in time, money, and energy. Every archaeological research project has limits. You can't make informed, strategic decisions unless and until you take a hard-nosed look at the costs and benefits of each component of your developing research design. Thinking first also means weighing the costs versus the benefits of everything you do.

MAKING TOUGH CHOICES—OPERATIONAL STRATEGY

In the CRM world, you have to make tough choices to balance your research needs against the pressures of time, money, logistics, politics, and bureaucracy. Too often archaeologists are unwilling or unable to make the tough decisions that must be made when, as always, budgets aren't big enough and field seasons aren't quite long enough to do everything we'd like. Without careful planning, these pressures can wedge you into a "dig and figure" mode: You dig stuff up and hope to figure it out later.

As you develop your research strategy, real-world considerations should loom large. Like it or not, you're operating within bureaucratic, logistical, and financial contexts that control what you do and when you do it. A common approach to these complications is to pretend that they don't exist and rail against them when they insistently worm their way into your daily routine. It can be satisfying to ignore repeated calls from your sponsor's accounting department—until the bills come due, and no check is forthcoming. Dodging calls from that regulator about an overdue report makes good bar talk until they put the kibosh on your permit for that new job. You've also got a budget, and a schedule, and people not showing up for work, and broken-down trucks all to make your life a lot more interesting than you'd really like. Getting a handle on all of these things is the other half of your research strategy—it's your strategy for getting things done, your operational strategy.

WORKING WITH BUREAUCRACIES

As a CRM archaeologist, you often have the joy of working with one, or two, or even three or more unique and interesting bureaucracies. You'll almost certainly be dealing with your SHPO reviewers and regulators. If your sponsor is a government agency, you'll have to deal with it and its environmental folks. Whether your organization is large or small, you also have your own bureaucracy to deal with. When we talk about bureaucracy, we're talking about people, people who are trying—for the most part, trying their best—to implement the policies of their organizations. A lot of times, inane bureaucratic BS seems just as inane to the folks who are wielding the bureaucratic stick.

There's one very simple strategy for working with organizations: just deal with them as people (see sidebar 6.1). Find out who you'll be dealing with at each organization. If you don't know them, give them a phone call, introduce yourself, and explain what you'll be working on. If at all possible, arrange to come by and meet them in person. There's no big agenda, just a quick hello and a reminder about the project you'll both be involved in. When you can put human faces to a faceless organization, it will make dealing with bureaucracies much more pleasant. Conversely, the bureaucrats will see you, too, as a person, rather than as an anonymous cog in another annoying project wheel.

PERMITS

Another aspect of working with bureaucracies is dealing with permits. Most government-mandated research requires some sort of permit from a state, federal, or tribal authority. Permits may have very specific, legally binding requirements that can greatly impact your research project. These requirements range from prescribed qualifications for the permit holder to the type of work allowed, the area where the work will be performed, and the time frame within which the work must be completed. As soon as the project is contracted, contact your SHPO and any other agencies involved to get moving on the permits.

In some cases, you may be working under the aegis of an agency's permit or memorandum of agreement (MOA). Some federal agencies such as the Corps of Engineers (COE) negotiate MOAs and programmatic agreements with SHPOs to cover the routine archaeological work they anticipate doing on the properties they manage. In those cases, you're relieved of the responsibility for permitting the project yourself.

In many cases, though, you will be responsible for getting the project permitted and ensuring that the terms of the permit are met. Once you know the project area, your SHPO office can tell you what the permit requirements will be for the project. This is serious business. Permits are generally issued to an individual, and if your project

> **6.1. KEEPING THE PEACE: PEOPLE IN YOUR OPERATIONAL STRATEGY**
>
> It's important to have a strategy for dealing with all of the players in the CRM field. Each of these folks has an interest in your work, and keeping them happy—or at least quiescent—can make the difference between living in harmony and living hell.
>
> **Sponsors**
>
> Naturally, you will need to plan for interaction with the folks who are sponsoring your research and handing out the bucks. They may have no real interest in the work, or they may find it fascinating and want to be closely involved. A healthy professional relationship is your best strategy. Keep your sponsors happy by keeping them informed and by understanding and doing your best to accommodate their schedules and concerns while maintaining your credibility as an ethical archaeologist. Boot-licking client servers are respected by no one. It's much better to earn their respect by doing your job well and paying attention to the details that matter to them.
>
> **Regulators**
>
> You ignore your regulators at your own peril. How you maintain lines of communication depends on the context of your particular project. It could be as simple as making a phone call once a month or as complex as submitting detailed progress reports. The most consistently successful strategy is to develop a healthy professional relationship with your regulators based on mutual respect and cooperation. An antagonistic relationship is always counterproductive and not very bright on your part.

fails to fulfill the requirements of the permit, it is your professional life that is on the line. Read and understand the requirements of the permit, and if you have any questions at all, be sure to ask before it's too late.

As your project progresses, stay in touch with the folks who issued your permit. If you see that you won't be able to make a required deadline, contact the permitting office as soon possible. If you make the effort to be honest and forthright with your permitting agencies, you can avoid a lot of needless hassle. Remember, though, rules are rules and being honest and up-front won't necessarily cut you any slack if you've really screwed up. Most permitting agencies try to be even-handed, and part of that is doling out the bad with the good. That's why you need to take permits seriously.

Research Team

Building a cohesive and effective research team requires more than keeping each team member informed about his or her particular piece of the action. Make sure all the members of your team know the big picture and have a chance to follow the project as it unfolds. Give the lab staff a chance to visit and take part in the fieldwork so they will gain a much better appreciation of the material they are processing. Specialized analysts do a better job when they know the overall objectives of the research and when they develop a good feel for the site and its environs. It's up to you to plot a strategy that creates a successful team.

Peers

Your fellow archaeologists have a legitimate stake in your work because we share common professional goals and depend on one another for information. Therefore, part of your strategy should be to keep your peers informed through conference presentations, progress reports, well-written and accessible final reports, and peer-reviewed journal articles. When possible, invite your archaeological colleagues to visit your field project—show them what you're doing and ask for their observations.

The Public

This is the most difficult operational arena for many archaeologists because we are not trained in communicating and marketing our work to a wide audience. Increasingly, however, archaeologists are learning to develop innovative strategies for involving and engaging the public (see Toolkit, volume 7). Through open houses, volunteer programs, public displays, TV shows, educational videos, websites, public lectures, media events, and popular accounts, archaeologists must ensure our own survival by making our work relevant to those who we ultimately serve.

LOGISTICS

This is where the rubber meets the road, the nuts and bolts of getting people to the field, getting data to the lab, making sure everybody gets paid on time and the outhouse is limed. As you plot your research strategy, you need to look sharply at the time lines, budget, and staff and use your experience and judgment to see whether you can actually do all the work you've planned. Don't forget to figure in weather days, mechanical breakdown, crew loss, and administrative blowouts. Your scope of work outlines what must be done; here you need to really focus on what *can* be done. If you don't feel that you've got the experience you need to make these judgments, call on a colleague with more notches in her belt.

Staffing—having the right people at the right time—can make or break a project. If you're planning to staff the field crew with graduate or undergraduate students on summer break, your five-month field season is going to be problematic. It's more than just having enough bodies on hand. Most successful research projects are accomplished by well-matched teams of people with the right skills and the right attitude. This comes back to the importance of thinking and planning ahead. You have to put together your budget long before you start hiring people. If you have tried to cut costs by planning to pay cheap wages, you will probably end up with an inexperienced crew or one made up of disgruntled, low-achiever types no one else will hire. This is especially true when the CRM economy is in full swing and jobs are plentiful. Supply and demand.

The best way to make sure that you've crossed your logistical t's is to use a checklist. Setting up a checklist of what you need and when you need it is also a good way to focus on equipment, personnel, and supplies while you're writing a proposal or a budget (see appendix D, "The Logistics Checklist"). Using a good checklist is an easy way to help ensure that you've got what you need when you need it. Logistics is more than staffing and supplies, though. Your strategy has to include the administrative end of the logistics train. You need to have a plan for making sure that the paychecks come on time and that per diem is there when the crew is hungry. It may be as simple as wrangling some help from the office administrative assistant, or it may mean creating another checklist of things you need to do at the end of each week. Putting an effective strategy in place boils down to defining what the tasks are (write it down!) and making sure that everyone knows what needs to be done.

Logistics isn't fun and it doesn't seem like archaeology, but working out the how, when, and where is essential to the success of any research project. The graveyard of wrecked archaeological careers is littered with the corpses of the logistically impaired.

THE BUDGET

The budget is an evil, immutable, unyielding force grinding slowly and painfully over your grand plans. Once you've set a budget, you need to develop a strategy for executing your research within it. It's critical to develop a realistic budget up front. But even the most finely crafted budget can fall victim to unexpected problems, crew disasters, and rising prices. This is where the decisions get hard.

The first thing you should do with your budget is to place your analysis and write-up money off limits. When you're in the field, there will be all kinds of unexpected expenses, and it will be tempting to say, "We're not finding any charcoal, so we won't really need those extra five carbon dates." But there are unexpected expenses in the lab, analysis, and write-up as well. Don't touch that money until its time comes.

Next, list all the things you can jettison in each phase of the project if money gets tight. The strategy here is to plan for a kind of financial triage, to choose the parts of your work that you can allow to die so that the project can live. As you do this, you need to balance your research needs against their costs in real dollars. If your research questions are focused on seasonality, you probably can't afford to eliminate your botanical analyst. But you may be able to forego excavation of additional units where the soils don't seem to support the kind of preservation you're interested in.

Last, if the budget is impossibly tight or you have a major disaster, contact your SHPO and your sponsor immediately. We can't emphasize enough the value of close communication. Everyone is truly interested in getting the work completed. Calling your sponsor to report that you're going to run out of time before the work is finished won't be fun, but it can head off total disaster. Going to the sponsor or the SHPO to report that you're out of money shouldn't be a deceitful ploy to get more funding. It's critical that you be completely honest and aboveboard. In a recent case in our area, a contractor found that its crew would not be able to complete its work within the existing budget (which, stupidly, covered only the fieldwork). The sponsor refused to provide any more funds, so the contractor simply walked away from the site, leaving half-completed excavation units open and unprotected. This was a disaster for everyone involved, most of all for the archaeology. When you see a budget problem on the horizon, you need to have a strategy to deal with it rationally, to fulfill your obligation to the sponsor and, most of all, to the archaeological record.

THE SCHEDULE

No matter how much money you have, no matter the skill of the crew, the length of the day, or the speed of your trowel, you will not have enough time to do all the things you want to do. Your time is constrained by the budget, but you'll also have other pressures eating away at the time you have. If you have a crew of students, the opening day of classes will likely see a sharp drop-off in production. If you work in the Southwest or the Great White North (somewhere up past Oklahoma, we hear), the high heat of summer or the icy blasts of winter can limit your days in the field. Or it could be monsoons, or black flies, or a construction schedule—whatever it is, time is immutable.

You need to manage your time just like your budget. Your schedule should allow for weather delays and other problems. Look at your schedule and anticipate problems. Prioritize your work with regard to time as well as money, and be ready with the phone numbers of your sponsor and the SHPO. Your goal is to balance your research needs against the time you have available to do the work. Again, communication is the key. If you see a problem looming, talk it over with all the folks involved. There may be solutions you haven't yet seen.

Your time strategy, like your budget strategy, is to maintain the greatest measure of control you can in the face of unanticipated problems. You can't anticipate everything, but you can be ready for problems by focusing now on the tough choices you may need to make later and hammering out a realistic schedule.

INVOLVING THE PUBLIC

An often overlooked aspect of most CRM work is planning to involve the public. Since they pay the bills and generally support the laws that sponsor most of our research, we need to spend some energy and time of our own to acknowledge and encourage their interest. Some large CRM projects have public information aspects built in by the sponsor—maybe an informational brochure will be produced or a museum exhibit will be developed. More likely there is no plan nor a budget to engage the public in the work you're doing.

Sad fact first: Most of us archaeologists can't write our way out of a paper bag. Years of churning out CRM reports have imbued most of us with the writing style of a stale bagel. Your strategy for sharing information about the work you're doing and getting the public engaged in your work should involve working with people who can and do write for public consumption. Sometimes a project sponsor requires production of a brochure or other popular treatment that they plan to distribute to the public. Often these materials are written in turgid, dumbed-down archaeo-jargon and please nobody, so they languish in boxes somewhere and don't accomplish the worthy goal of giving something back to the public. The archaeologists aren't happy, and the sponsors aren't happy. The intention is good, though, so when presented with the opportunity to prepare a popular product, swallow your pride and farm the job out. Find someone who has experience reaching a general audience, a journalist who writes for the audience you intend to reach, a writer with credits in periodicals that reach your audience, a videographer with experience in documentaries. These things aren't cheap, so if your sponsor wants to tack on $500 to produce a "quick brochure," tactfully decline. Neither of you will be happy with the result.

You don't have to be James Michener or Brian Fagan, though, to reach a general audience with good, solid information about your archaeological project. People in the communities where you're working are interested in what you're doing there. With no more complex skills than being able to answer questions about your work intelligently, you can get the good word out about what archaeologists do and the value of the archaeological record.

Before you embark on any plans for including the public, be sure you run them past your sponsor. Some sponsors will jump at the chance to look good, but some have deeply ingrained institutional paranoia. Work with those you can.

The first step is to find out whether your sponsor or client has a public relations or communications department responsible for generating good press. If they do, contact them and give them the scoop on your planned work. You can help them write a short press release that they'll send to all the newspapers, radio stations, and TV stations in the area where you'll be working. A local angle is gold for these folks, so you might even be contacted by reporters for interviews, photos, or maybe a live radio spot.

If your sponsor doesn't have a PR department, you can do it yourself. Write a one-page summary of the work you're doing that explains why it is interesting and important. Keep it simple. Include information about scheduling and how to contact your project. Send copies to the local papers, radio stations, and TV stations. If even one small-town paper runs a piece on your project, more people probably will read that article than your final report—a depressing thought but one that should spur you to action.

Another way to reach the public is through presentations to community groups. Giving presentations is a great way to focus your ideas about the work you do, and the more presentations you give, the easier—and more fun—it gets. If you're going to be in the field for a month or two, contact local community organizations like the

Optimists or Lions Club. They are always looking for people to give thirty-minute presentations at their monthly meetings, and a live archaeologist is quite a catch. Talk about both the local area and the big picture, adding the human element as much as possible—those are the things the audience will find most interesting.

Another excellent and fun way to include the public is with an open house. Try to schedule it on a weekend so as many people as possible can attend, and use the local media, schools, and community organizations to get the word out. Plan for a crowd, and do your best to have your crew working—people love to see archaeologists actually doing things rather than just pointing at square holes and jawing. People are fascinated with archaeology, and the opportunity to see what it's like in action will almost always draw a crowd. A few years ago in New Mexico we drew many more than four hundred people one cold, drizzly Saturday morning to look at some open units, see the flakes and pot sherds that had been collected, and ask questions about what we were doing. Volume 7 in the Toolkit series has even more information.

CONCLUSION

Your research strategy is the sum of two parts: the archaeological strategy and the operational strategy. You need both to make the most of your opportunity to engage in archaeological research. It may be that you write out parts of your strategy, or you might just think through them and get a game plan straight in your head. It's a matter of degree—a small three-day survey doesn't demand the same level of planning that a six-month excavation does. However you conceptualize your research strategy, if you work through the elements discussed here, you've got your research design right at hand.

7
THE WRITTEN RESEARCH DESIGN

Your written research design can take many forms—a revised scope of work, a proposal, a work plan, a formal research design, or part of a letter or a contract outlining what you have agreed to do. Usually it's part of the package you prepare for your sponsor to win a contract (or funding agency to win a grant). Occasionally it's a stand-alone document written later, after the contract is signed, to satisfy the funding, regulating, or permitting agencies. Whatever its form, most written research designs tend to focus on the problem orientation and the methodological aspects of research.

A good research design acknowledges the contexts of the work, addresses the concerns of the various players, spells out the scope of the work, outlines the approach to be taken, describes the research problems and the targeted data, and links these to appropriate field and analytical methods (see sidebar 7.1). When a research design is created solely to win a contract or grant, it's little more than a pretense—a slick pile of verbiage filed away and forgotten once the work begins. A well-done research design is a dynamic tool used by all the players involved to focus the field work, analysis, reporting, and evaluation of the project.

> **7.1. YOUR WRITTEN RESEARCH DESIGN CHECKLIST**
>
> 1. Before you finalize your research design, show it to an experienced colleague and members of your research team, and ask for feedback. Sometimes the SHPO will agree to review your draft document. Getting feedback takes time, so plan ahead and don't put it off to the last minute.
> 2. Treat your research design just like any other professional document. Check the spelling, have it edited, make certain all the text references are matched by those in the "references cited," format it consistently, and use headings and subheadings to improve the organization.
> 3. Acid-test your plan by trying to reread it from the perspective of your audience. What do they expect? Look for BS, hyperbole, naiveté, and other common faults.
> 4. Don't confuse jargon for clear thinking or scholarship. Plain English is the way to go. Make it easy to read and follow.
> 5. When you think you are done, go back over your scope of work one last time, and make sure your research design responds appropriately.
> 6. Make certain that you credit other people's ideas, particularly those in print. Plagiarism and sloppy scholarship are sure signs of incompetence.
> 7. Avoid the dazzler approach of using fancy paper, special binders, and gratuitous color bar graphs. The sharp reader will recognize these ploys for what they are: smoke and mirrors intended to dress up a feeble document. Dazzle them instead with your coherent ideas and pragmatic attention to necessary detail.
> 8. Don't promise the sky—you cannot do it all. Focus, focus, focus, and make your research count by homing in on important but narrow problems that you can realistically address. Write a doable plan.

WHOM ARE YOU WRITING FOR?

As always, knowing your audience allows you to communicate more effectively. In the CRM world, five audiences may be reading and (hopefully) using your research design:

- *Clients or sponsors.* These are usually nonarchaeologists who are mainly interested in the time and cost aspects of your work.
- *Contract administrators and regulators.* These are archaeologists working for your sponsors and in the SHPO office. Their main interest is in making sure that the various state and federal laws (NHPA, ARPA, NAGPRA, etc.) are followed and that the archaeological work is up to snuff (a very relative term in the wide world of CRM archaeology).

- *The archaeologists and specialists who will be part of your team.* Your crew and consultants will use your research design to help understand their role and to focus what they do.
- *Your peers.* Other archaeologists in your area may be keenly interested in how your strategy and results can inform their own research, or they may just want to kibitz.
- *Most important, yourself.* The act of writing down what you plan to do is one of the most effective ways of focusing your thoughts. Things that seem obvious and simple when floating around in the thick gruel of your brain show all their warts and flaws when laid out in black and white. Writing your research design allows you to hone your ideas and your half-formed thoughts into a concrete and effective plan for wringing knowledge from the sparse and challenging archaeological record.

Keep these various audiences in mind as you write. Tell a sponsor that you are planning to "seek subsistence data on the Early Archaic," and he may shrug and assume that you know what you are doing. Tell a SHPO reviewer the same thing, and she might say, "What the heck do you mean? Don't you realize that Early Archaic sites in this region almost never have any preserved organic remains?" The archaeologically knowledgeable people who have control over your work will look at your research design to determine whether you are asking appropriate questions, proposing appropriate methods, seeking appropriate data, and planning a feasible chunk of work for the circumstance. If they are competent and doing their jobs, they will read your document with a critical eye. Knowing your readers allows you to do a better job of communicating your ideas in ways they will understand.

An all-too-often neglected use of a research design is to inform those who are on your research team about the game plan. If you have gone to the trouble of writing a research design, why not let your crew members and consultants read it as soon as they come on board? This is common sense—people who know the plan are in a much better position to be able to contribute. On one large survey project we survived, the field director told the crew that we were there only to serve as his "hands and eyes" and would be informed about the research strategy only "on a need-to-know basis." Bad for morale, bad for research. The most productive research projects are those in which the crew members and consultants are actively informed and work as an integrated team. By circulating your research design to those you work with, your crew members and consultants become research team members and colleagues working toward a common goal.

Your archaeological peers are likely to be your toughest critics and may well read your research design years after the fact. There is plenty of nit-picking criticism, petty bickering, and spiteful quarreling in archaeology. You can't avoid it; all you can do is try to minimize it by doing the best work possible. Part of that is learning to take criticism for what it's worth and to use it to make your work even better (see sidebar 7.2).

This takes us back to the most important audience—you. Your research design really isn't for SHPO Joe, or Sponsor Sue, or Professor Punwit down at the university. It's a tool you create to help you do your work. You're the one with the most to lose—or gain. A thorough and well-conceived research design will improve your chances of winning the contract, help you create a productive research atmosphere, and serve as your blueprint for success from start to finish.

7.2. MORE CRITICISM, PLEASE!

An essential and often unrecognized skill in archaeology is learning to deal with criticism. Most of us have an automatic, defensive reaction when we're criticized. But think for a moment about how rare it is to get feedback of any kind. Without feedback, the only guiding voice you have is the all-too-familiar sound of your own opinion.

Exchanging ideas and critiquing work is absolutely essential in a productive research environment. This is why all leading journals are peer reviewed. The review process in CRM sometimes involves agency review of draft reports. But when criticism happens only after the fact, there's usually not much you can do but shrug your shoulders. That's why it is so important to get feedback—criticism—early and throughout the research process. Good, focused criticism can help you concentrate on what you're doing and help find flaws or unrealized research potentials. It's not enough to graciously accept criticism; seek it out and use it as a whetstone to keep your research sharp.

Positive strokes are easy enough to accept, but harsh criticism often really hurts. If you let it, it can destroy your self-confidence, and no matter how self-assured you are, disapproving evaluations are upsetting. That's why you need a strategy to turn criticism to your own purposes:

1. Evaluate the critic's motives and abilities. Is he just a jerk with a personal grudge? Maybe she is just too ignorant to recognize the merit of your work (this one's always handy). But maybe the critics just don't know how to frame criticism in constructive terms. Don't lash back at critics, no matter how mean

PROJECTS BIG AND SMALL

When we're talking about archaeological projects and research designs, size does matter a great deal. So does purpose. A research design for a small project intended to answer basic compliance questions—"Are sites present?" and "Are they potentially significant?"—is a much different beast from a research design for a multiyear mitigation project or an ongoing academic research program. For small CRM projects, the formal document will probably be little more than a revised scope of work, coupled with a budget and a contract or a letter authorizing you to do the work. In contrast, the larger projects will almost certainly require much more thorough documentation and justification of your research plan. Despite such real-world differences, there is still a commonality here: An informed research perspective is equally critical to the success of projects big and small.

---

spirited or harsh they seem; it makes you feel good only for a minute. If the criticism is truly nasty, intended only to make the insecure critic feel superior, dismiss it—it's your critic's problem, not yours.

2. Mine the criticism for valid or useful points. The most stinging critiques usually identify something you missed or ignored. Make a list of the things you can use—actually write them down. There is always something you can use in every critique, even if it is just to avoid sending any more draft manuscripts to that jerk. As you work through the critique try to remember the critic's point of view. If she is a pottery expert and seems to harp continually over what you think are typological minutiae, she could simply be looking at it from her particular perspective—or else your pottery analysis really does stink.

3. Acknowledge the effort of the critics. They've taken the time to read or think about your work and have reacted to it, and you stand to benefit. Write a short note to the critics, thank them for their time, and offer yourself to review their work in the future. If the critique is part of a formal review process, note every comment and suggestion and respond to each one. If you can't accommodate a suggestion, tell the reviewer or sponsor why not.

4. Seek out criticism. Your best critics are often the colleagues and associates whose work you respect. Ask them to critique your work, formally or informally. It's easy: Just ask them what they think, and take what they give you to heart. None of us has a corner on the archaeology market, and the way we really learn is by exchanging our ideas. We really mean it when we say, "More criticism, please"!

Most small projects don't require elaborate justification, a detailed methodology, or a big research question, but they do require an informed research perspective and a clear understanding of what will be done. Big projects will always require a lot more in the way of a written research design. Many projects fall in between these extremes—they will require more than a simple scope of work and a letter and a less-than-sixty-page treatise. We can't tell you how much detail you will need. You have to decide based on your experience and the requirements of the particular research project you are designing. But no matter the size or scope of your research design, it should hit five key areas: (1) scope of work and legal rationale, (2) problem definition, (3) data definition, (4) methods and techniques, and (5) analysis.

CRITICAL COMPONENTS OF YOUR WRITTEN RESEARCH DESIGN

The components presented here are the basis for any well-developed research design. How you put them together, how extensive they are, and how the document itself is presented are simply variations on a theme. Your sponsor or SHPO may have a preferred form, so be sure you find that out before you start to write. Become a student of research designs—keep a file of research designs you've glommed onto from other projects, successful and unsuccessful. These can help you get a sense of what is expected in your area, but remember, this is your research design, so think for yourself.

SCOPE OF WORK AND LEGAL RATIONALE

Your scope of work should appear near the beginning of any research design. Start by repeating your marching orders. Your research design has to respond to and correspond with the scope of work. If the scope does not already contain language spelling out the legal and regulatory basis for the work, be sure to discuss this up front as well. This shows that you know what you are doing, and your sponsors and regulators will be comforted by the proper legal lingo. It's often easier to get your plan approved by using familiar words. It is also very important that you do understand the statutory requirements because, like it or not, you are responding to government mandate. These introductory parts set the stage for your proposed research.

PROBLEM DEFINITION

Next, your research design should explain your research context and identify your research questions. One way you can do this is to begin by reviewing the *relevant* research background—not a boilerplate background section with a tedious chronology of every piddling project and published interpretation bearing on your research area, time period, or site type but a succinct, focused discussion of the state of knowledge about your research problem. The goal is to demonstrate that you are on top of what has already been learned or postulated concerning the research area or topic. This isn't accomplished by compiling long lists of citations but by highlighting the most important previous research and discussing the key points and critical data sets that are part of the written archaeological record (and sometimes the recent work that is not yet published). Don't just list and cite the work, but demonstrate in a cogent fashion that you have read and absorbed the importance of what your predecessors have done. The existing on-the-shelf archaeological record is a critical part of your research context.

Sometimes it is appropriate to have a section called "Theoretical (or "Conceptual") Approach" to explain where your research is coming from and where you intend it to go. Under such headings one often finds a confusing and tedious recitation of buzzwords, dense quotes, famous names, and rambling discussions of overused concepts. We've written several just like that. Make yours stand out by cutting through the fog and stating clearly and succinctly the essence of the explanation or model you are inspired by and working within. Most useful theoretical approaches have been explained well in the leading journals and academic press books. Focus on the essence. Don't parrot your sources; just cite the key references, and state the gist of the ideas. Boiling it down will win you much more respect than a lengthy and dense discussion.

In many circumstances you don't even have to discuss your theoretical approach per se. If your research is basic and aimed at addressing compliance and management issues, cut to the chase and focus on your research questions. Whatever references to established theoretical perspectives you need can be mentioned in your

"Research Background" section and in your statement of the problem. The high-church theorists need to belabor the assumptions, the history of ideas, and the abstract understandings that these views encompass as well as point us in new directions from time to time. But we rank-and-file members of the "archaeological proletariat," as Patty Jo Watson put it, have to spend more time fishing than playing with the bait.

As we said in chapter 4, research problems, hypotheses, and questions are simply different names for the same thing. One can split hairs, quote philosophers, and insist on a narrow view of how to structure scientific research, but we don't think it matters much. What does matter is that you can answer the question "What is my research intended to accomplish?" This is your problem, your hypothesis, your question. It may be little more than a basic "find and evaluate," or it may be a much more complex set of ideas. Fit your problem to the situation. Explain enough of the research context to justify why you are choosing to focus on this topic. State your research question(s) clearly.

One way to start is to lay out a central problem and then identify the parts you hope to solve. You can call these parts research questions. Or you can identify these as smaller research problems. You can start with a hypothesis—a proposed explanation of the archaeological patterning you will be investigating—and then list your expectations: what you think you will find and what you think it will mean as a test of your hypothesis. No matter how you structure your approach, the more explicit you are about what you are trying to learn, about your underlying assumptions, and about how your findings could inform your starting point, the better.

Some of the most common faults in research designs (including all of those that we have written) are that the problems are too broad, the questions are too numerous, or the hypotheses are too difficult to test. Ambition and idealism generally are good things, but they must be tempered by reality. While you do want to aim high, you also want to give yourself a realistic chance to hit the target. If your stated problem is too broad, do your best to pare it down while you are writing your research design, instead of waiting until the crucible of the field exacts its inevitable toll.

Most worthwhile research projects end up identifying more questions, narrower problems, and more sophisticated hypotheses than they started with. That is really the point of doing problem-oriented research. Starting with the best questions you can pose and focusing as narrowly and precisely as you can, you end up learning something worth knowing. In the long view one of your most important goals is to learn enough so that you (and your colleagues) can do a better job on the next research opportunity.

DATA DEFINITION

After you have explained your problem, identify the data you will need to address it. Think carefully about the data you need and can collect. You won't be able to collect all the data you need. And you'll probably collect data you won't need for your problem. Justify your data definitions by telling the reader how and why these data will (or won't) inform your problem. Justify your decisions to treat different material classes differentially by telling the reader how you've balanced the costs against the benefits. If you are going to exclude certain classes of data (as you must always do), be upfront about it. For instance, you might reason that the sherds smaller than a half inch (those that would fall through a half-inch mesh screen) are not worth bothering with because you can't learn anything from them that can't be learned from a sample of larger sherds.

You have an ethical obligation to record basic data even if these data will not inform your research questions. This means different things for different kinds of sites and different regions. You will have to meet certain standards, expectations that your colleagues, reviewers, and sponsors have established. Question these when you can and try to justify your deviations from the norm, but you can't plow through a pit house just to see whether there is a burial in its floor. You have an obligation to do a thorough job of recording the features you find and sampling the potentially meaningful deposits you encounter. You should have a plan to address these data. Explain this plan in your research design.

That said, there are circumstances in which you may decide to sacrifice basic data and normal standards to salvage something that you judge to be more important. For example, it may be justifiable to use heavy machinery to scrape off the upper deposits of a site to allow you to carefully investigate an earlier, better-preserved, and more poorly known component. Such decisions can be controversial and must be made in concert with the regulators and perhaps other parties. Such sacrifices are often necessary because you cannot do it all. By defining your critical data, you are also defining the data that are less critical in your particular circumstance.

METHODS AND TECHNIQUES

Data and methods are two sides of the same coin. To understand the data, you and your reader must understand the methods used. Most field methods are selected from a well-established array of proven techniques. You don't have to explain how you will set up a datum, grid the site, and label the bags. You do have to explain your overall methodological strategy and outline how you plan to cover the terrain on a survey, dig holes in a site, or sample a midden. If you intend to break tradition and do something methodologically innovative, take special care to elaborate and explain what you intend to do and why.

The marriage of data and methods is sometimes called your *sampling strategy*. It often helps to acknowledge that sampling pervades all archaeological research, so be explicit about your sampling methods and rationale. Why have you chosen to dig shovel tests every two hundred meters on the terrace margins? Once again, this requires your informed judgment—elaborate when you need to and be as succinct as you can. Avoid painting yourself into a no-win corner; build in contingency plans that show how your sampling strategy might change in response to field realities.

ANALYSIS

The more complex and involved your research is, the more critical it is that you discuss how you plan to analyze the data. Sometimes this section is called your "Analytical Approach." This section should flow logically from your research questions and the nature of the data; explain how you will use the data you will collect to address your questions. For basic questions, your planned analysis may be straightforward; describe and document what you found, determine how old it is, and link it to the known archaeological universe. More sophisticated questions and larger research projects will require you to develop your analytical approach in more detail. Think ahead and plan your analysis so you can budget your time and analytical expenses and so you can ensure that you collect the data your analysis will require. This step is often skipped or glossed over, usually to the detriment of the project. By including an analytical plan, you demonstrate the thoroughness of your research design and inspire confidence in your ability to complete the project within the budget.

ADDITIONAL CONSIDERATIONS

SPECIALISTS AND SPECIALIZED STUDIES

Multidisciplinary projects require close coordination, and this coordination should begin in the planning process. If you plan to use specialists or perform specialized studies, you should address these needs in each relevant section of your research design. For example, you may need to hire a historian to search for scattered records that pertain to the construction of structures once present in your study area. Mention the data you hope to collect in the data definition section, and describe the methods to be used to collect those data. In the analysis section, discuss your specific analytical approach to the data. Specialists may be at another institution and you may never meet face-to-face, so they need a well-crafted research design to understand how their work fits into your overall approach. By anticipating and discussing the specialists' roles in your analysis, your research design will pave the way toward a successful collaboration.

ANTICIPATING COMPLICATIONS

As you assemble your document, do your best to anticipate and discuss the normal kinds of complications that may arise: inclement weather; the hoped-for data fail to materialize; there is not enough water to do water screening; you encounter human remains. Describe briefly your contingency plan for dealing with these difficulties. By doing this explicitly, your research design will show your sponsors that you've got a foot in the real world, and it will provide a handy "plan B" to fall back on when things threaten to fall apart.

If you encounter human remains, you will probably have to follow a specific protocol that varies greatly depending on your source of funding, the ownership of the land, the age and cultural affiliation of the burial, and the region of the country you are in (see sidebar 7.3). In many circumstances, you will need to consult with appropriate tribal authorities and consider their concerns. In some cases, you may

> ### 7.3. NAGPRA
>
> Some of the most contentious issues and greatest confusions in modern American archaeology revolve around the Native American Graves Protection and Repatriation Act (NAGPRA) of 1990. This law essentially says that human graves on federal and tribal lands are to be protected and that Native American descendants should decide the fate of the bones, grave goods, and sacred objects of their ancestors that are encountered or held by the federal government and any institution that receives federal money. The law and its regulations also set up a repatriation process. NAGPRA basically tries to balance the very real concerns and legitimate interests of Native Americans in controlling their cultural patrimony with the scientific community's interests in studying and preserving these materials. For most working CRM folks and academic field archaeologists as well, NAGPRA can impact your work in two ways.
>
> First, if you're doing archaeology on federal or tribal lands, you need to ensure that the appropriate tribes or Native Hawaiian organizations have been consulted, usually before the work begins. It is the sponsoring organization's responsibility to consult, but don't leave it up to them. You should be part of the consultation process. The consultation should spell out what happens if any human remains or burial goods are uncovered as part of your work. You may need to stop work and call in a tribal representative. Or you may need to handle burials and grave goods according to a specific protocol. The key is to follow the law

have to quit digging or avoid any human remains entirely. In others, you may need to have a physical anthropologist examine the remains in the field and then immediately turn them over to a Native American group for reburial. If you don't know the protocol in your area, consult with your SHPO or sponsoring agency. Thinking ahead is far better than dealing with an ugly and costly crisis. Outline your plan for such contingencies in your research design.

RESPONDING TO THE SCOPE OF WORK

Your research design should be written and organized in ways that are directly responsive to the scope of work. Make sure that your research plan fits within the scope and that you have satisfied the concerns and needs of your sponsor and your regulators. This may mean being certain that you include certain language or certain sections that are expected or required. This is particularly critical if your research design is part of your proposal for a competitive bid. RFPs that and all appropriate procedures. NAGPRA is the law, and the law says that federally recognized tribes have a legitimate interest in the remains of their ancestors.

The second way you may find yourself under the NAGPRA umbrella is when human remains or grave goods are encountered in the course of a federally funded or permitted archaeological project that is not located on federal or tribal land. State laws and procedures will generally apply, even on private property, but NAGPRA will control the disposition of Native American human remains and certain objects if they are held by a federal agency or federally funded institution. Know your state burial laws as well as NAGPRA. What you have to do will again depend on where you are and who is sponsoring and permitting your work. Your best bet is to contact your SHPO and federal sponsor and follow their advice.

For the most part, NAGPRA works reasonably well. Most of the archaeological community seems to be taking NAGPRA in stride and complying with the law. But it's complicated and contentious, particularly when it comes to the numerous "culturally unaffiliated" remains that cannot be clearly linked to a living tribe, like virtually all of those that are thousands of years old. Over a decade after NAGPRA's enactment, the final regulations for treatment of these remains still have not been approved.

Handling NAGPRA-related situations requires consultation, compromise, and the willingness to work within established protocol. Accommodating NAGPRA and making productive use of its challenges and opportunities in both the field and the lab is part of your successful research strategy.

---

are written by archaeological bureaucrats may mandate that you follow a particular document format, keep to a certain length, address certain problems, or even follow a particular methodology. You may find such directives maddening and even nonsensical, but you will probably have a better chance of winning the contract if you bite your tongue and follow the guidelines. "Nonresponsive" proposals may be summarily dismissed, even if your objections are reasonable. If you do choose to buck the system and propose an alternative approach, explain your reasoning carefully and noncontentiously. Sometimes it's better to negotiate such deviations after you have the contract in hand.

Responding to the scope of work is really a matter of thinking about who will be reading your research design and anticipating their needs, concerns, and likely responses. Your sponsor or client may not care about your research plan except insofar as it affects the schedule or the budget. Your regulators may not care about scheduling and budgets but may be quite picky about your proposed methods. There is no precise formula for success—you have to create the document that does what is needed in your situation. By carefully reading and rereading the scope of work and making sure your research design is consistent with the scope, you can save yourself lots of unneeded hassle. Don't forget that your peers may be keenly interested in your research design for a variety of reasons. Don't be paranoid, but do try to look at your document from the vantage point of the devil's advocate. How will it read to your professional colleagues?

CONCLUSION

Your written research design spells out just a part of what you will have to do. Most of your operational strategy won't be part of the written document. But it is no less a part of your plan, so don't let it go unheeded. If you've prepared a logistics checklist, grab that now, print a clean copy of your research design, and put those two documents together in a binder or a folder, or paste them in the back of your field book. Just as your logistics checklist is a guide to the mundane, everyday crap that has to be done, your research design is a guide to the legal, regulatory, and intellectual things that need to be done. One won't do you much good without the other.

8
PUTTING YOUR RESEARCH DESIGN TO WORK

Your research design is done, you've got the contract or grant in hand, and now it's time to do the archaeology you've designed. Your research design is a tool, and if you leave it hidden in your toolbox while you're working, it won't help you.

The process of designing your research has forced you to think first. Your research design provides a very real research context that helps you collect data and make observations with the confidence of knowing how this information will help you answer your research questions. You've already considered many of the decisions you will have to make in the field and laid out a reasonable course of action—your archaeological and operational strategies. You've already done the work in your mind, by going through the process of designing your research. With a written research design in your hands, you and your crew have a map to success.

Yet, the unexpected will always arise. When new things pop up that you're not prepared for, the first tool you should grab is your research design. How does the new situation affect the data you're collecting? Is this new data? How does it fit into the analytical model? What changes will it mean to other parts of the data collection and analysis effort? Handling change intelligently is one of the hardest tasks in any endeavor, and your research design provides a structure to handle—and incorporate—those changes in your overall research plan.

Even after the work is done, the analysis is complete, and the report is almost in the bag, your research design is still at work. The last, and most important step—for you—is using your research design to honestly evaluate your work. Looking back on what worked and what didn't, assessing what you would do differently and how you could do better, is how you become a better researcher and archaeologist. Your research design provides not only a structure for your research but a structure for your professional growth.

IMPLEMENTING YOUR RESEARCH DESIGN

Your research design does a very simple thing: It communicates. It codifies the things you hope to learn from the archaeological record; the kinds of data you expect to collect; the ways you plan to collect, record, and manage that data; and the approaches you'll use to analyze those data. It also communicates those things to the people working with you, your sponsors, and regulators. Archaeological projects, even small ones, are complex undertakings, and your research design helps keep you on track and moving in (hopefully) the right direction. Clear communication is the key to effectively implementing your research design.

SHARING THE DREAM

Once you get the green light on your research project, the first thing you should do with your research design is to share it with others, especially the members of your research team. The element of surprise might be useful in battle, but it can be death for an archaeological project. Everyone involved in your project needs to know your approach, the data you hope to collect, and how you hope to collect it. That's just what your research design does. Suggest that the crew read your research design, and, if possible, set aside time for discussion. Don't forget any specialists who will be involved in your research or analysis. When the members of your research team know the context of the research they are undertaking, they become more active and helpful participants.

One part of active participation that some archaeologists have a hard time dealing with is criticism from the crew. It can be disconcerting to expose your ideas and plans to the folks who work for you. The crew might criticize or laugh at your plans, but so what? This is the real world of research—if the crew sees flaws in your work, you can be confident that your peers will as well. Look at this as an opportunity to make your research design and project better. It could be that you've just done a poor job of communicating, that your ideas aren't clearly explained. There's no better way to clarify your thoughts than to try to explain them to someone who doesn't understand. If you're honest with them, respect their criticisms, and use the opportunity to benefit from their experience, you'll have a team with a common goal. If you're hesitant to accept criticism from the those who work for you, how are you going to feel when the report comes out for everybody to read? Evaluation, criticism, and comparison are standard stock in the CRM trade.

Look for opportunities to share your research design with colleagues and peers, too. If colleagues express an interest in visiting your site, invite them, and send them a copy of your research design before the visit. Sometimes this can seem like asking for a poke in the eye with a sharp stick—expect colleagues to point out problems they see and tell you how they would do things. Differences of opinion are to be expected, but they may also call your attention to alternatives you haven't considered, changes you may want to implement right away. No matter what the tone, your colleagues' first impressions can be very valuable feedback. Hear them out, record their observations and concerns in your notes, and use these later during your analysis and write-up to help you strengthen your arguments and explain your case.

FOLLOWING THROUGH

One of the challenges of fieldwork is following through on what you proposed in the research design. This seems obvious and simple but seldom is. In the field or the lab, it's easy to fall back into familiar default patterns—deeply ingrained habits. So use your research design to assess the work you're doing at each stage of your research. A good approach is to do a quick weekly evaluation. Look at your research design—really look at it, don't rely on your memory—get the thing out and review what you said you'd be doing at this stage. Have you revised your data collection methods because of special soil conditions? Even seemingly trivial changes in your methods could mean that you're not collecting the data you expected. Even if your crew members have read and discussed your plan, they, too, may backslide into old habits. If you're using forms to collect systematic observations in the field and lab, you need to review them and make sure that the observations recorded are what you expected. When forms are filled out in a rote fashion, the real variability often gets hidden behind a wall of convenience. A quick weekly review of those data can help you head off those problems. It might be that you weren't as clear as you needed to be, and a quick explanation could fix the problem.

Review the status of your project, your data, and your research design on a regular schedule. On large projects you might be required to produce a regular monthly status report, which is a great opportunity to review what you're doing and how it fits within your research plan. Even if there's no requirement in your contract, it's a good idea for you to perform a regular status check. And it's not enough to just mentally check off what you're doing. Our perception is wonderfully imprecise, and our memories are often conveniently unreliable. You should actually sit down with your research design, spend an hour or so reviewing the work, and make sure that you're staying on track. And write it down! Just a section in your notes is fine, but by writing down your impressions, you create a valuable record of your progress (or lack thereof) that you can look back at during the analysis and write-up.

Sounds like a good idea, you say, but when do I find the time? You won't, unless you build time into your schedule to handle the chore. Maybe you set aside an hour every Friday morning or chop an hour out of Wednesday afternoon. Review and reflection comprise a critical part of any research process, and, if you build them into your project, you'll be a better researcher for it.

WORKING WITH SPECIALISTS

The specialists and those who provide analytical services instrumental to your research design will need special consideration. Those who take part in the fieldwork will obviously need to be scheduled appropriately. For example, the best time for geoarchaeology may be early in the field season if you need depositional interpretation to guide your work or late in the field season if the work depends on maximum excavation exposure. Don't pass up any opportunity to invite the specialists who will do their work later to come visit you in the field. This exposure (plus reading your research design) will give them the contextual information crucial to the success of many specialized studies. Keep in mind that most specialists have their own research agendas that may be quite different from yours. If you don't make it clear what you want from them, they will often give you what they perceive as important instead of what you need. For example, faunal experts who aren't archaeologists may concentrate on identifying species and elements and may not make systematic observations about evidence of modification (butchering marks, burning, gnawing, etc.). With careful coordination, you will be able to integrate your specialists into your research.

WHEN THE PLAN HITS THE FAN

No matter how well you plan, your research design will start to fall apart shortly after you hit the field. Hopefully, you'll just have small problems that nibble away at the edges of your work, like fire-cracked rocks that are too large to be weighed on the field scale you intended to use or aerial photographs that don't show the roads you had counted on using to locate your survey quadrants. The more you count on a perfect field season with everything falling into place just at the right moment, the more likely you'll end up scrambling to finish something that only vaguely resembles what you set out to do. Big or small, problems will occur, and to keep your sanity and your research design intact, you need to plan for failure.

Planning for failure doesn't mean having a contingency plan for every possible disaster. It means approaching problems the same way you approach your research: by thinking first. When you encounter unexpected circumstances, take a step back. Think. If you can use something from your plan to help solve the problem, great. But remember that reality takes precedence. Always. It's not what we want or hope or expect that counts, it's what's there. When you encounter a major logistical problem or an unanticipated archaeological circumstance, the key is to react to it intelligently.

Your research design is a tool for decision making, but only if you believe in it. That is, if you have a well-focused problem and good research questions, do your best to stay on track. When unanticipated wrinkles appear, look to your research design to help you make decisions. How can you accommodate the new situation and still address all or most of your research questions? Remember that failure is not the only thing you'll have to prepare for. Sometimes unexpected success can cause even more problems. You're digging along and suddenly an unexpected and time-consuming feature pops up. Halfway through your survey you realize that your team is finding twice as many sites as you predicted and time is running out. Your excavation of an eighteenth-century house turns up a previously unrecognized prehistoric component rich with promise. What do you do: stick with your game plan, or go for the newfound glory?

We don't often think of archaeology as an experimental science, but it is. Each time we take to the field and drop our shovel into the dirt, we're experimenting. We have an expectation of what we might find and how we're going to deal with it. When we find something unexpected, it's not because we've failed; our experiment is simply showing us new information. Our job, then, is to pause and rethink our strategy to incorporate this new circumstance.

Stop. Think. Think again. You are right back to making tough decisions and weighing costs against benefits as you consider revising your strategy based on the new circumstances. You have three choices: Stick with your original plan, modify your plan to accommodate the new situation, or scrap your plan entirely and come up with a new one. You should consider the latter option only in truly extraordinary circumstances. If you ditch your original plan, you'll need to create a new research design, and you will almost certainly need approval from the powers that be; otherwise, you won't be fulfilling the contract or the terms of your permit. If you abandon your plan and start making ad hoc reactionary decisions on your own without a comprehensive strategy and without approval, you are begging for trouble and are right back to archaeology by default.

Sometimes you may decide to stick with your game plan, keeping your original focus and ignoring the new opportunity. You know you can't do everything, and sometimes staying on course may be the most reasonable alternative, particularly if your sponsor isn't willing to give you the extra time or funding needed to do more than you originally planned. But often you will want to shift your priorities and take advantage of the unexpected.

As long as you approach these things intelligently, handling change in the field isn't too difficult. It's the same as if you had found some new information when you were working on your research design. Look at the new circumstance; see how it impacts your research design. Start by asking yourself these questions:

- How does this change the data I expected to collect?
- What are these new or changed data?
- Do I need to modify my methods to collect the data?
- How can I analyze these data to inform my research questions?
- Do I need to revise or create new research questions to accommodate these data?
- How will this impact the rest of the project in time and money?

As you review these questions, develop a plan for addressing the change, and write it down. You can start this process in your notes, but plan to get the changes down more formally as quickly as possible. This is the research design process all over, but instead of starting from scratch, you've got the work you've already done in hand, and you're just adding a new wrinkle. Of course, you've got a crew standing around waiting, a sponsor anxious to get moving, and three weeks left in your schedule, so you can't spend a lot of time digesting this new information. Using the questions listed earlier, specify—in writing—what the problem or potential is and how you propose to handle it.

As you document your changes, remember that your research design is a record of your approach to the work. Think of the research design as intellectual field notes. Although you don't want to recreate your research design as each change or problem comes along, you do want to document the process you went through to get to the final stage. If you change your methods in the field, add a section to your research design explaining the change and why it was made. Don't simply replace the original information with your revision. That's dishonest. You need to know why you made the changes, and so do your readers.

If you're revising your approach, you must work with your sponsor and regulators to make them aware of any changes to your plan and make sure they concur. If the revised plan is more costly in time or money, you obviously must get the OK before preceding. Often the easiest way to gain their approval is to invite them to visit you in the field and see the new circumstances for themselves. Hand them a copy of your proposed changes, explain your case, accommodate their suggestions, and try to get everybody on board the new and improved plan.

Verbal agreements aren't enough—you need to get this stuff in writing. Write a quick memo or letter reviewing the meeting and who was there, noting that "On XX date we agreed that. . . ." Make sure you spell out the proposed changes to your research design, and fax or mail it to all of the involved parties. Communication, again, is the key. In tense or potentially contentious circumstances, you may want to ask for signed acknowledgment. If contractual obligations change, you may need a formal contract amendment. This annoying paperwork will seem like manna from heaven if contract disputes arise later and you have a paper trail supporting your version of events.

Documenting any changes you make also keeps your crew and specialists on the same page. Whenever changes are made in the field, it's critical that everyone who will be affected, from the field tech in the bottom of the trench to the plant specialist four states away, knows and understands what the change is and how it will affect the data to be collected and analyzed. The quickest, most effective way to do that is to hand them a couple of sheets of paper, let them read it, and talk about what it means. It's simple and it works, and if you don't do it, you may screw yourself.

Documenting these changes is most important to you. When you're knee deep in the analysis or write-up, the last thing you need is to confront ill-remembered problems in the data. When you find six bags of stuff that you vaguely remember were pulled from some hole, somewhere, for some reason, the cold pit in the bottom of your stomach will quickly disappear if you can turn to your notes and jog your memory.

IN FROM THE COLD

Often the time you really have to face reality isn't in the field but when you get back to the lab. You should have been revisiting your research design at each step of the process, but let's be realistic. You may not have recognized (or been able or willing to recognize) that the data you needed to address your research questions just weren't there. Once you're back in the lab, trying to fit the data into your analysis, it may quickly become apparent that there is no way you can address half of your research questions. What do you do—scrap the research design and hope that no one else will go back and read it?

No, this is just another problem to overcome. Don't panic. Use your research design. Look at your research questions in light of the data you've actually collected. Where the data can be applied to your proposed questions, do it. You may not have the whole answer, you may not have all the data, but just say that and move on. If some of your data don't match what you proposed, document it. Write it down, right now. Describe the data and the methods, as accurately as possible. Look at the data in terms of your current questions; there may be places where the data can be applied. If there doesn't seem to be a fit between your data and your research questions, you may need to propose an alternative set of questions or problems. If you do, write them down, and consult with your sponsor and regulators. Remember, revising your research design means adding new information and explaining your changes, not providing a revisionist view of your work. Once you've made your revisions, get those folks to sign off on the changes.

THE LAB STAFF CAN READ, TOO

In the happy event that you return from the field with the targeted data and your research questions still more or less intact, use your research design to prioritize the lab processing and focus your analysis. As with your field crew and specialists, your first step in the lab should be to provide all of the staff with copies of your research design. Spend some time with them discussing the fieldwork completed, the lab work to come, and the looming analysis. Don't let default routines and ad hoc decisions prevent you from reaching your research objectives. The same process that applies in the field also applies in the lab. If you encounter problems, approach them with research design in hand, systematically, and document any changes you might need to make. Keep focused and keep using your research design to guide the work. By sticking with your game plan, you might even be able to save enough time to follow "bonus" research leads that turn up in the course of your work. But keep your priorities straight—follow through on your research design.

ORGANIZING YOUR REPORT

Your research design (as revised) provides you with a logical and compelling organizational structure for your report. Screw the tedious boilerplate and mindless detail—you have the perfect justification to prepare your report in creative and unique ways. Start by laying out your research design, and then structure the report to follow your design to its logical conclusion. If you've followed your research plan, you have sets of data that address specific questions. Lay out your report just like that. When you do problem-oriented research, when you use a research design to guide your work, what you have to report is fundamentally different from the standard CRM fare. You do have the usual compliance crap to regurgitate—so many acres, so many shovel tests. But more important and interesting, you have a set of assertions and the data needed to test them, so you can create compelling arguments about aspects of the archaeological record. Do what you must to satisfy the regulators and the sponsor, but then look back at your research questions and lay out your data, methods, and analysis, and then address the research questions you posed. You've learned something, something about the archaeological record, something about archaeological methods, something about floodproofing your excavation block. Give that to the reader, to your colleagues.

EVALUATING YOUR RESEARCH DESIGN

As the analysis and writing phase draws to a close, the final step of any thorough research program is to evaluate the research design. What worked and what didn't? Which of the grandly stated ideas were you able to effectively test or address with the data in hand? What happened with the other ideas, the ones you could not effectively address? Did all of your methods and analytical strategies pay off? Faced with a similar circumstance, how could you improve on your research design?

Sadly, many of us fail to truly evaluate and honestly judge the effectiveness of our research designs. We tend to scrap the ideas that our data don't speak to and focus on the research questions we can address. Fine enough, but we also tend to skip the crucial step of sitting down and admitting to ourselves and others what worked and what didn't. We gloss over (or completely ignore) our mistakes as if admitting error were tantamount to admitting professional incompetence. Failing to learn from our mistakes, or at least failing to share what we have learned, we doom ourselves and others to repeat our mistakes. We forget that archaeological research is a cumulative and cyclical process and that the point of making mistakes should be to learn from them.

Your research design is a tool for learning about both the archaeological record and the process of doing archaeology. Sections labeled "Interpretations" or "Conclusion" in archaeological reports say something about what we think we have learned. Don't stop there. You should also specifically discuss the effectiveness of your research design. Which lines of evidence and research strategies were helpful? Proudly tout the ideas and strategies that worked, but keep in mind that they only tell a part of the overall story. Equally important are the ideas and strategies that didn't work. By discussing these shortfalls, you (and others as well) can learn from your mistakes.

It's natural to try to avoid talking about what we might see as our failures. But no project is perfect, and the picture presented by many CRM reports—where it seems the work went smoothly (except for weather) and the analysis was flawless—is sheer fantasy. We all know that all kinds of problems crop up, that none of our data sets is without flaws, and that it's difficult to usefully analyze archaeological data. Learning about the archaeological record means sharing information, good, bad, and indifferent.

Consider adding a section to your report called "Evaluating the Research Design." Work through your research design, point by point. Look at your research questions, your methods, the data you hoped to collect, and your analysis. Ask yourself (and answer) these questions about each part of the research design:

- Did this work?
- Did we encounter problems with this in the field, lab, or analysis?
- Did we revise it?
- If it was revised, did that fix the problem?
- How could this be made better or more effective?

The goal here isn't self-flagellation, but be as honest as you can. Don't apologize or make excuses. The goal is to evaluate each section dispassionately. If you've been documenting changes to your research design as you've gone along, point out how and why things changed. Remember, your research design is really a process. Evaluating it completes the cycle.

There are no perfect research designs. Understanding what worked and what didn't will help you do a better job next time. Owning up to half-baked ideas and ineffective strategies is one of the basic obligations of an ethical researcher. It will also help other archaeologists understand the work you did and hopefully avoid some of the traps you found. It will enhance your reputation as forthright researcher. It's your obligation to learn something from the time and money that your archaeological research project has swallowed. Learning what not to do next time is as critical as learning what worked. Report the truth.

NOW IT'S UP TO YOU

Excellent and meaningful archaeological research is being accomplished every day by archaeologists all across the country. There's also much that is overpriced, mindless, and mercenary being done under the banner of "archaeology." The path you'll take is up to you.

The road to accomplishing excellent research in archaeology—CRM or academic—is anything but easy. The money and the opportunities are there in the CRM world, but the pace is often relentless. When and where you work will be determined by engineering concerns, bureaucratic decisions, market factors, political winds, and construction schedules. Your research has to focus where and when these larger forces say, and you have to make it fit within the framework of federal, state, and local preservation laws and regulations. You'll share archaeological decision making with government archaeologists who work as contract managers, reviewers, and regulators and with many interested parties who aren't archaeologists. There will be compromises at every turn, and some of them will turn you inside out. None of us survives for long in the research world with all of our academic ideals intact. Reality always takes precedence.

It's no wonder that the academic world disparages CRM archaeology. It's damn hard to do good research with all of these pressures and constraints. When your research arena is defined by others, you must be opportunistic, persistent, and resourceful. No one likes to think and work with people hanging over her shoulder, yet the whole archaeological bureaucracy is designed around just that concept, and you can't begrudge the need for accountability in publicly funded work. The upshot is that CRM archaeologists simply don't have control over big chunks of their research world.

In spite of this, CRM archaeologists are making substantive contributions to the archaeological record, day in and day out. One of the things that CRM archaeology does is force us to look outside our usual box. What, for instance, would we know about upland sites without CRM archaeology? If the archaeologists were doing the choosing, we'd all be down in the deep sandy loam up to our armpits in stratigraphy. But because the engineers who route the pipelines, highways, and power grids choose our work locations, we've been looking at these thin, shallow, and interestingly different sites for the last thirty years. Just the fact that CRM projects have forced us to deal with such "marginal" (and numerous) sites on the landscape is an important contribution to archaeology. Being told where to work isn't all bad.

It's hard, though, to do interesting research consistently when you're faced with a nonstop stream of small surveys and testing jobs. It's a formidable challenge to you as a researcher, a much greater challenge than excavating a well-stratified site. Your professional objective is the same: to make a useful and interesting contribution to the archaeological record. The logistical and practical problems are immense: short projects, small areas, and few truly significant archaeologically deposits. The retreat into archaeology by default is easy to understand. The same projects, the same analyses, the same reports. It's a hard trap to escape, even when you're determined.

The worst part of the bittersweet reward for enduring the trap is that you become someone who can be counted on to accomplish the technical routine. Not creative, but reliable. The work gets done and the report goes in on time. You won't want for work, but no one will want to read your reports, not even you. The only punishable crimes are to badly overspend your budget, seriously inconvenience the construction schedule, or miss your final due date. The real problem with CRM archaeology is that the easiest path leads to a dreary, soulless technical service.

It doesn't have to be that way. By choosing to do archaeology by design, by carefully thinking about the archaeology, by doing problem-driven research while meeting your compliance obligations, you can make your work exciting, interesting, and meaningful. You have to take a long view of your research. It's not what you find that counts; it's how the data you collect, the methods you use, and the approach you take give you and your colleagues a leg up on the next one. It's also having a research perspective, a plan to carry it out, and the gumption to report what you really did and learned.

The only consistent route to doing interesting, effective, problem-driven research is through the process of designing your research. It's a proven method that we archaeologists and social scientists borrowed from the hard sciences. For all the preaching and proclaiming we've done in these pages, the process itself is pretty simple. First, you have to recognize and grasp the opportunity. Next, you have to have a research perspective, something you care about and are interested in studying. Then, given the work at hand, think about questions you can ask and imagine the data you can collect and how it will help you address those questions. The rest is window dressing, creative (or dull) variations on a theme. The key, though, is simple: Think about your research perspective, problems, and data, and do archaeology by design.

This book is meant to be a guide to helping you learn how to create successful strategies for achieving substantive archaeological research. Our views are shaped by many painful mistakes made and witnessed in the real world of archaeology. It's up to you to glean what you can from these pages and adapt it to the actual circumstances you find yourself in. You will make many of the same mistakes we made. Despite knowing in theory what you should do, in practice you will find yourself caught with not enough time or money or willpower to do the right thing.

Your attitude, skills, and aptitude will determine whether you accomplish archaeological research of consequence during your career.

If you pursue opportunities with eyes wide open and heart and mind set on doing research, your career in archaeology will count for something more than a road to retirement.

Your written research design is often an important part of the process, but most of what it will take will go unwritten. It will take knowledge (much of it acquired on your own), relevant skills, and on-the-job training. It will take a good command of all the interrelated contexts of your work, from politics to the physical and cultural landscape to logistics. It will take a willingness to make the most of thankless situations. A keen sense of humor helps—archaeological research is a surreal world at times, and you'd better be able to appreciate irony and absurdity (it's either that or pulling your hair out). It will take interpersonal skills, from ego stroking to consensus building to learning how to say, "You're fired." It will take tempering your ideals with a heavy dose of pragmatism—you may have to sacrifice large chunks of the in-the-ground archaeological record in order to learn and to save its most significant parts. It will take many, many tough decisions. And on and on. It's all part of the tangled process of designing archaeological research.

The need to be purposeful in your research cannot be overemphasized. In most projects, you won't have the luxury or mandate to protect all sites or take your sweet time doing everything that can be done. Faced with finite resources and ticking time, you must be purposeful and focus your research.

Doing archaeology by design gives you and your organization a competitive advantage. When you focus your research on specific problems of consequence, you accomplish more of substance for a given amount of money. Archaeology by default is expensive even if you are merely going through the motions, doing your fieldwork in the usual way, and collecting, processing, reporting, and curating forever whatever you happen to find. The costs mount up in such an unfocused approach, and the payoff is always uncertain. When you focus your work, you can spend the money more productively because you know what you're after and what you can do without. It's just smart business to concentrate on achieving something worthwhile. Smart business, smart archaeology—and that is the secret to archaeology by design.

Now, it's up to you.

APPENDIX A

GROUPS YOU SHOULD JOIN

The results of your archaeological work join the body of work that we all draw on when we try to understand the archaeological record. You are part of a profession, a discipline, a cohort of archaeologists who contribute to that body of work. The more active a role you take in thinking, talking, and writing about archaeology, the better an archaeologist you will be. To do these things, you need to interact with other archaeologists, folks who will listen to what you have to say and challenge you and your ideas, folks who can inspire you, and folks who you can inspire. Anthropology has a word for such interacting groups—*societies*.

Society for American Archaeology (SAA)
900 Second Street NE, #12
Washington, D.C. 20002-3557
Phone: (202) 789-8200
Fax: (202) 789-0284
headquarters@saa.org
www.saa.org

Every archaeologist in North America should join the Society for American Archaeology, the largest and most influential national group. In addition to publishing *American Antiquity*, *Latin American Antiquity*, and the *SAA Archaeological Record*, the SAA provides access to a variety of resources and a national forum for presenting ideas. There is no better way to get a sense of where your work fits into the larger body of archaeology than at the SAA's annual meeting each spring. Although the SAA has been traditionally dominated by academic archaeologists, CRM archaeologists are assuming greater roles at every level of the organization. This shift is reflected in the content of the the *SAA Archaeological Record* and in the SAA's backing of ROPA, the Register of Professional Archaeologists.

Register of Professional Archaeologists (ROPA)
5024-R Campbell Blvd.
Baltimore, Maryland 21236
Phone: (410) 933-3486
www.rpanet.org

ROPA is a listing of professional archaeologists in the United States who have agreed to abide by an explicit code of conduct and standards of research performance; who hold a graduate degree in archaeology, anthropology, or other germane discipline; and who have substantial practical experience. ROPA was established in 1998, jointly sponsored by the SAA, the Society for Historical Archaeology (SHA), the Archaeological Institute of America (AIA), and the Society of Professional Archaeologists (SOPA). ROPA succeeds SOPA, a group that tried with only modest success to promote professional accreditation since 1976. The joint sponsorship has already resulted in a much larger membership as archaeologists around the country, academic and CRM, are recognizing the advantages of becoming a registered professional archaeologist (RPA). The hope is that a majority of professional archaeologists in the United States will join ROPA and provide our profession with the means to police itself effectively as others do. Do your part—earn your ROPA credentials by gaining the experience and graduate degree you'll need for career advancement.

If you have an area of specialization within archaeology, you should also join appropriate professional organizations such as these:

Society of Historical Archaeology (SHA)
PO Box 30446
Tucson, Arizona 85751-0446
www.sha.org

American Association of Physical Anthropologists (AAPA)
AAPA Membership
Box 1897
Lawrence, Kansas 66044-8897
physanth.org

Society of Archaeological Sciences
c/o Felicia R. Beardsley
Department of Anthropology
University of California–Riverside
Riverside, California 92521-0418
Phone: (909) 787-5524
Fax: (909) 787-5409
E-mail: beardsley@qnet.com
www.wisc.edu/larch/sas/sas.htm

Regional societies and conferences also should form an important part of your archaeological world. Here are a few, but check with your SHPO or professors for the ones in your area.

Plains Anthropological Society
Southeastern Archaeological Conference
Pecos Conference
Mid-Atlantic Archaeological Conference
Mogollon Conference
Caddo Conference

As for groups specializing in CRM, most states now have professional archaeological councils made up mainly of CRM archaeologists. These groups provide important opportunities for professional networking. Your SHPO or the SAA's Council of Councils can give you the right contacts.

The American Cultural Resources Association (ACRA) functions as a trade organization and has become an important forum and lobby for the CRM industry. It also has an excellent Listserv, ACRA-L, and a good website with many useful links.

American Cultural Resources Association (ACRA)
Thomas R. Wheaton
Executive Director, ACRA
6150 East Ponce de Leon Avenue
Stone Mountain, Georgia 30083
www.acra-crm.org

Every CRM archaeologist should be concerned about the preservation of the archaeological record and public support for archaeology. You can play a direct role by participating in state and local archaeological societies and associations. Some societies are strictly collector oriented. But many strong, successful societies engage professionals and avocational archaeologists as well as students and other enthusiasts. Those groups that welcome professional participation often have cadres of avocational archaeologists as knowledgeable of their local areas as any archaeologist alive and eager to share their insights. They can be invaluable sources of information, inspiration, support, and trained help, but it only works when it's a two-way partnership. The societies look to professionals for leadership, training, and knowledge, which we should be glad to provide.

You can't afford the cost in time and money to join and participate in all the archaeology-related groups you might want to, so prioritize them and stay involved with a few. Your membership dues, conference trips, and other expenses can be deducted from your taxes as professional business expenses.

APPENDIX B

JOURNALS YOU SHOULD READ

Dozens of archaeological journals, newsletters, and magazines can keep you up-to-date. You can't afford to subscribe to all of them, and even if you could, you'd never have enough time to read all of them regularly. But don't let this stop you from trying because keeping abreast of the field is essential to being a well-informed professional. Here are some strategies that might work for you.

Subscribe to as many archaeology-related serial publications as you can reasonably afford. This is your livelihood, and subscription costs are tax-deductible. The journals are much easier to refer to if you have them on your bookshelf. Some are free. You can gain access to the others by borrowing them from colleagues or spending a few hours in the library from time to time. Increasingly, journals and magazines are putting abstracts or entire contents online. Browse their websites when you can. For paper journals, scan the table of contents of each issue, and put a sticky note on all the articles you should read. If the journal is not yours, scan the articles of interest and copy the most important ones (five or ten cents a page is cheap for reference material). Keep a stack of flagged journals and copied articles in your favorite reading places, and read the articles as you have a chance. Try not to let the stack build up too high. Set aside a couple of quiet reading lunches each week. Two articles a week is more than a hundred a year. Keeping current with your field is a discipline you need to develop and nurture.

Here is a list of archaeological and related journals of international, national, and regional scope; you'll have to find the state and other regional ones for yourself.

American Anthropologist—flagship journal of the American Anthropological Association
American Antiquity—flagship journal of the Society for American Archaeology; a must
American Archaeology—quarterly from the Archaeological Conservancy
Archaeology—popular articles on sexy archaeology worldwide
Common Ground—The federal archaeology viewpoint quarterly from the National Park Service (NPS)
CRM—another overdesigned rah-rah rag produced by the NPS, but free
Current Anthropology—in-depth academic articles and commentary
Geoarchaeology—for the subdiscipline
Historical Archaeology—flagship journal of the Society for Historical Archaeology
Human Ecology—interdisciplinary academic journal
International Journal of Historical Archaeology—academic
Journal of Anthropological Archaeology—a theory-heavy academic
Journal of Archaeological Method and Theory—academic topical syntheses
Journal of Archaeological Research—academic topical reviews
Journal of Archaeological Science—archaeometry, published in the United Kingdom
Journal of Field Archaeology—academic focusing on methods and results
Journal of Material Culture—interdisciplinary academic journal
The Kiva—regional journal focusing on the American Southwest
Latin American Antiquity—second journal of the Society for American Archaeology
Lithic Technology—specialist journal
Mammoth Trumpet—Paleo-Indian research newsletter with in-depth interviews
Midcontinental Journal of Archaeology—regional journal
North American Archaeologist—not as impressive as its title, but improving
Plains Anthropologist—regional journal
Public Archaeology—new journal produced in the United Kingdom
Public Archaeology Review—small journal focused on CRM archaeology
Quaternary Research—interdisciplinary academic journal emphasizing paleoenvironments

APPENDIX C
THE ABCs OF CRM

Modern archaeology is rife with acronyms. You can't avoid them, so you might as well learn them. Here are a few of the most common ones:

ACHP—Advisory Council on Historic Preservation
ACRA—American Cultural Resources Association
AHPA—Archeological and Historical Preservation Act (of 1974), a.k.a. Moss-Bennett
APE—area of potential effect
ARPA—Archeological Resources Protection Act (of 1979)
BLM—Bureau of Land Management
COE—U.S. Army Corps of Engineers (also USACE)
CRM—Cultural Resource Management
EO 11593—Executive Order 11593 (of 1972)
MOA—memorandum of agreement
MOU—memorandum of understanding
NAGPRA—Native American Graves Protection and Repatriation Act (of 1990)
NCSHPO—National Conference of State Historic Preservation Officers
NEH—National Endowment for the Humanities
NEPA—National Environmental Policy Act (of 1969)
NHPA—National Historic Preservation Act (of 1966)
NPS—National Park Service
NRHP—National Register of Historic Places
NSF—National Science Foundation

PMOA—programmatic memorandum of agreement
RFP—request for proposals
ROPA—Register of Professional Archaeologists
RPA—registered professional archaeologist (accredited by ROPA)
SAA—Society for American Archaeology
SHA—Society for Historical Archaeology
SHPO—State Historic Preservation Office (and officer); pronounced "ship-oh"
TCP—traditional cultural property
THPO—Tribal Historic Preservation Office (and officer); pronounced "tip-oh"

APPENDIX D

THE LOGISTICS CHECKLIST

Logistics can kill a project. If you're fifteen miles and forty-five minutes from town when you use your last collection bag, or the batteries for your TDS aren't charged, or someone forgets to gas the truck, it's not just you who will be twiddling thumbs. It's expensive to field even a small crew, and every hour they sit on their hands because something they need isn't available is eight or ten hours you've lost in the lab or the analysis or the write-up. Good logistics doesn't necessarily mean good archeology, but bad logistics can kill any hope you have to finish a project on time and within budget.

Luckily for us administratively challenged folks, there is a miracle tool that can virtually ensure that you never run out of bags, leave a battery uncharged, or sit by the side of a lonely highway: the checklist. Checklists are simple and powerful tools, and most of us don't use them effectively or enough. In the press of fieldwork, it's all to easy too forget one or two simple things. A checklist lets you place your brain on cruise control to focus on those things for just as long as it take to run through the list. You don't have to worry about remembering things—the checklist remembers for you.

As with any kind of tool, building an effective checklist takes a bit of work. Review the equipment and supplies you're going to need in the field or lab and list everything: shovels, trowels, screens, transits, cameras—anything and everything you might need to do your proposed work. This comprehensive equipment list is what you'll use to build your checklist, so make sure anything you might need is listed.

The next step is to do the same thing with supplies, those things you're going to use up. Write down all of the supplies you can imagine needing, everything from pencils to protractors to toilet paper. You can estimate quantities if you want, but the critical thing is to get it all down on paper.

The last and most often ignored step is operations. Make a list of each thing that needs to be done every day before you go to the field and after you return. This includes tasks like loading the equipment into the trucks, making sure there is gas for the pump, and recharging the batteries for the TDS. Write down everything and anything you can think of and when it needs to be done.

In the planning stage, these lists can help in budgeting. In the field, take the complete list and edit it down to reflect the real-world needs. A lot of stuff you planned for may be superfluous, and you probably forgot some stuff. Combine your equipment, supplies, and chores lists on a single sheet of paper if at all possible. Print the lists on colored paper so you can find them easily. You might want to have separate setup and shutdown lists for each day. It's incredibly anal, but it works, and it will save you time and heartache.

Save your lists from each project. As you work on your proposal for the next one, you've got a good head start on the equipment and supplies you'll need for your new project.

Here are three sample lists from the Higgins project, a CRM project we conducted several years ago in central Texas.

Equipment Checklist

Equipment	Number Needed	On Hand?
Excavation Equipment		
Flat-bladed shovels	8	
Round shovels	8	
Entrenching tools	8	
Pick mattock	2	
Rake	1	
Wheelbarrows	6	
Plumb bobs	2	
Folding saws	4	
Pointed trowels	12	
Square trowels	12	
Line levels	12	
Knee pads	12	
Saw horses	10	
25-m tape measures	4	
100-m tape measures	1	
10-m tape measures	12	
Field scale	1	
5-pound hammer	1	
Claw hammer	1	
Hand saw	1	
Electric drill	1	
Nails/screws/staples	Box	
1/4-inch screens	6	
1/2-inch screens	2	
1/8-inch screens	2	
Folding tables (lab also)	6	
Folding chairs (lab also)	12	
5-gallon buckets (plastic)	12	
3-gallon buckets (metal)	12	
Dustpans	12	
Whisk brooms	12	
Corner pins	100	
Set out pins	25	
Survey/Mapping Equipment		
Optical (4 screw) transit	1	
Transit tripod	1	
Ranging pole	1	
Stadia rod	1	
TDS	1	
TDS tripod	1	
3-centimeter TDS targets	2	
1-centimeter TDS targets (peanuts)	2	

Continued

Equipment Checklist (*continued*)

Equipment	Number Needed	On Hand?
Extendable target poles	4	
50-centimeter target poles	2	
TDS quick charger	1	
TDS overnight charger	1	
TDS battery	2	
SDR-33 data collector	1	
SDR-33 field case	1	
SDR-33 serial cable	1	
SDR-33 Hirose to DB9 connector	1	
Brunton compass	1	
Datum markers	25	
TDS/SDR manuals	1	
Transit/TDS field book	1	
Computer Equipment		
Laptop computers	3	
Desktop computer	1	
Dot matrix printer	2	
100-foot extension cords	1	
25-foot extension cords	3	
6-outlet surge suppressors	4	
Mice	4	
Mouse pads	4	
Printer cables	2	
DB9F-DB25F serial cables	2	
Null modem cable—DB9F-DB9M	2	
Modem	1	
Boot disk (DOS6.22)	4	
Field Lab/Field Office Equipment		
Calipers	2	
Stapler	4	
Metal rulers	4	
3-hole punch	2	
Tape dispenser	2	
Lab scale	1	
3-hole 1.5-inch binders	12	
File boxes	3	
Hanging file folders	100	
Camera	4	
Camera tripod	2	
Camera case	4	
Photo log	2	
Pencil sharpener	2	
Hand lenses	2	
Labeling pens	6	

Supplies—verify inventory in field box daily	
Pin flags	100
String	2 rolls
Burlap bags	20
Large paper bags	100
Small paper bags	100
Big Ziploc bags	100
Medium Ziploc bags	100
Small Ziploc bags	400
Pin fed 3 × 5 cards	200
Pin fed 8 1/2 × 11-inch computer paper	500 sheets
Printer ribbons	2
#2 pencils	48—2 boxes
Erasers	6
Stick pens—black	12
Sharpies—fine-point	20
Sharpies—medium-point	10
SDR33 batteries—9-volt	6
SDR33 batteries—backup	2
Staples	1 box
Flagging tape—red	6 rolls
Flagging tape—orange	6 rolls
Flagging tape—yellow	6 rolls
Aluminum foil	3 rolls
Toilet paper	6 rolls
Film—color	12 rolls
Film—B&W	12 rolls
File folders	50
Strat forms	100
Sample forms	100
Photo log pages	100

Daily Operations Checklist

**Daily Loading
(equipment not housed at
the site in the field office)—Morning—at field house**
Field office boxes
Artifact/sample boxes
Soil sample bags
Rock sample bags
Water screen buckets
 (5-gallon plastic)
Laptops
TDS
Extra TDS battery
SDR33
Cameras
Water
Vehicle fuel

Daily cleanup—At site—shut down
All flagged artifacts shot
 in with TDS
Field equipment in trailer
Tarp open units
Laptops to field house
Tables under trailer
Electrical power disconnected,
 cords in trailer
TDS broken down—to field
 house
Cameras to field house
Inventory field office supplies
Collected artifact numbers
 verified
Collected sample numbers
 verified
Artifacts and samples to field
 house

Daily Check-in—At field house/lab—daily
Check artifacts and samples in
 to field lab
Reconcile provenience
 problems, etc., from
 previous day's artifacts and
 samples
Download SDR33

Continued

140

Daily Operations Checklist (*continued*)

Daily Check-in—At field house/lab—daily (*continued*)
Create Daily.dat backup file
　on floppy
Delete SDR33 jobs
Verify status of SDR33 backup
　batteries (change if low)
Create updated strat list for lab
Create updated provenience
　list for lab
Vacuum field computers
Vacuum field printer
Update lab database from
　field computers
Back up lab computer
　database
Place TDS battery on charger
Install charged TDS battery
　on TDS

REFERENCES

Binford, Lewis R.
    1983   *In Pursuit of the Past*. Thames and Hudson, New York. Most approachable book from America's leading and most controversial archaeological thinker.

Black, Stephen L.
    1993   Nailing the Coffin Shut on the Traditional Approach to Prehistoric Archeology in Texas: An Epitaph and Inquiry into the Afterlife. *CRM News and Views* 5(1):16–19. Critique of archaeology by default as practiced mainly by CRM archaeologists.

Black, Stephen L., Kevin Jolly, Charles D. Frederick, Jason R. Lucas, James W. Karbula, Paul R. Takac, and Daniel R. Potter
    1998   *Investigations and Experimentation at the Higgins Site (41BX184—Module 3)*, 2 volumes, Studies in Archeology 27. Texas Archeological Research Laboratory, University of Texas at Austin. An atypical CRM report in which Black and Jolly attempt to practice what they preach.

Blanton, Dennis B.
    1995   The Case for CRM Training in Academic Institutions. *SAA Bulletin* 13(4):40–41.

Butler, William B.
    1987   Significance and Other Frustrations in the CRM Process. *American Antiquity* 52(4):820–29.

Carnett, Carol
    1991   *Legal Background of Archeological Resources Protection*. Technical Brief No. 11. U.S. Department of the Interior, National Park Service. Technical speak by a lawyer.

Cheek, Annetta L.
  1991 Protection of Archaeological Resources on Public Lands: History of the Archaeological Resources Protection Act. In George S. Smith and John E. Ehrenhard, eds., *Protecting the Past*, CRC Press, Boca Raton, Florida. Reviews ARPA.

Ebert, James I.
  1992 *Distributional Archaeology*. University of New Mexico Press, Albuquerque. Theoretical take on the landscape approach.

Flannery, Kent V., ed.
  1976 *The Early Mesoamerican Village*. Academic Press, New York. Classic study useful for its discussion of sampling strategies.

Friedman, Janet L.
  1996 The Business of Archaeology: Planning the Work of Cultural Resource Compliance. *SAA Bulletin* 14(5):22–24.

Green, Ernestene L., ed.
  1984 *Ethics and Values in Archaeology*. Free Press, New York. Edited volume with useful discussions of ethics in CRM archaeology.

Green, William, and John F. Doershuk
  1998 Cultural Resource Management and American Archaeology. *Journal of Archaeological Research* 6(2):121–67. Useful review of subject and literature.

Jameson, John H. Jr., John E. Ehrenhard, and Wilfred M. Husted
  1990 *Federal Archeological Contracting: Utilizing the Competitive Procurement Process*. Archeological Assistance Program, Technical Brief 7. Bureaucratic process.

Johnson, Ronald W., and Michael G. Schene
  1987 *Cultural Resource Management*. Robert E. Krieger Publishing, Malabar, Florida.

Kerber, Jordan E., ed.
  1994 *Cultural Resource Management: Archaeological Research, Preservation Planning, and Public Education in the Northeastern United States*. Bergin and Garvey, Westport, Connecticut. Useful examples of how CRM has affected the archaeology of one region.

King, Thomas F.
  1998 *Cultural Resource Laws and Practice: An Introductory Guide*. AltaMira Press, Walnut Creek, California. Must-read for any budding CRM archaeologist, written by the principal architect of federal CRM laws.

2000 *Federal Planning and Historic Places: The Section 106 Process.* AltaMira Press, Walnut Creek, California. Another practical guidebook to CRM.

Lyon, Edwin A.
1996 *A New Deal for Southeastern Archaeology.* University of Alabama Press, Tuscaloosa. History of early years of government funded archaeology.

McGimsey, Charles R. III
1985 This, Too, Will Pass: Moss-Bennett in Perspective. *American Antiquity* 50(2):326–31. Historical perspective on AHPA and NHPA.
1998 Headwaters: How the Postwar Boom Boosted Archeology. *Common Ground* 3(2/3):16–21. Historical essay on formative years of federally funded archaeology.

Mueller, James W., ed.
1979 *Sampling in Archaeology.* University of Arizona Press, Tucson. Useful edited volume reflecting emphasis on the subject in 1960s and 1970s.

Neumann, Loretta
1991 The Politics of Archaeology and Historical Preservation: How Our Laws Are Really Made. In George S. Smith and John E. Ehrenhard, eds., *Protecting the Past,* pp. 41–46. CRC Press, Boca Raton, Florida. Title says it all; author was the SAA's lobbyist.

Preucel, Robert W., ed.
1991 *Processual and Postprocessual Archaeologists: Multiple Ways of Knowing the Past.* Occasional Paper 10, Center for Archaeological Investigations, Southern Illinois University at Carbondale. Papers from archaeologists of many different theoretical perspectives, some of them—papers and perspectives—quite dense.

Raab, L. Mark, Timothy C. Klinger, Michael B. Schiffer, and Albert C. Goodyear
1980 Clients, Contracts, and Profits: Conflicts in Public Archaeology. *American Anthropologist* 82(3):539–55. Early discussion of tension created by archaeology as a private business.

Renfrew, Colin, and Paul Bahn
2000 *Archaeology: Theories, Methods, and Practice.* 3d edition. Thames and Hudson, New York. Introductory textbook from a British perspective with useful explanations of contrasting theoretical approaches.

Rossignol, Jacqueline, and LuAnn Wandsnider, eds.
    1992   *Space, Time, and Archaeological Landscapes*. Plenum Press, New York. Method and theory of landscape approaches.

Schuldenrein, Joseph
    1995   The Care and Feeding of Archaeologists: A Plea for Pragmatic Training in the 21st Century. *SAA Bulletin* 13(3):22–24.
    1998   The Changing Career Paths and the Training of Professional Archaeologists: Observations from the Barnard College Forum—Part I. *SAA Bulletin* 16(1):31–33.
    1999   The Changing Career Paths and the Training of Professional Archaeologists: Observations from the Barnard College Forum—Part II. *SAA Bulletin* 17(1):26–29.

Simpson, Kay
    1999   Business 101: The Mysteries of Billable Hours, Overhead, and Contracts. *ACRA Edition* 5(6):1–5. Excellent overview of how the costs of CRM projects are calculated and billed from a business perspective.

Snyder, David
    1995   Cinderella's Choice: The Emerging Role of the State Historic Preservation Office in Cultural Resource Management. *SAA Bulletin* 13(5):19–21. SHPO 101.

Thomas, David Hurst
    1997   *Archaeology*. 3d ed, Harcourt Brace College, Fort Worth, Texas. Approachable introductory textbook from an American perspective with good examples of basic concepts.

Wandsnider, LuAnn
    1996   Describing and Comparing Archaeological Spatial Structures. *Journal of Archaeological Method and Theory* 3(4):319–84. Useful theoretical and methodological discussion of site structure.

Waters, Michael R.
    1992   *Principles of Geoarchaeology: A North American Perspective*. University of Arizona Press, Tucson. Good introductory text on subject.

Watson, Patty Jo, Steven A. LeBlanc, and Charles L. Redman
    1984   *Archaeological Explanation: The Scientific Method in Archaeology*. Columbia University Press, New York. Theoretical primer from a processualist perspective.

Welch, James M.
   1993   From Archaeological Field Crew to Business Administration. *Practicing Anthropology* 15(1):9–11. Personal view of a CRM career and training.

Willey, Gordon R., and Jeremy A. Sabloff
   1993   *A History of American Archaeology*. 3d ed. W. H. Freeman, New York. Latest edition of this standard reference includes some discussion of the impact of CRM on American archaeology.

Zeder, Melinda A.
   1997   *The American Archaeologist—A Profile*. AltaMira Press, Walnut Creek, California. Results of survey of SAA members.

INDEX

Note: Throughout the index, *CRM* is used in place of *cultural resource management.*

AAPA. *See* American Association of Physical Anthropologists
academic archaeology: history of, 7–8, 15; politics of, 28. *See also* research, academic
ACRA. *See* American Cultural Resources Association
Advisory Council on Historic Preservation, 29, 32, 33
alternative mitigation project, 48
American Association of Physical Anthropologists (AAPA), 128
American Cultural Resources Association (ACRA), 12, 32, 58, 129
American Cultural Resources Association e-mail list (ACRA-L), 32, 129
anthropological archaeology, 20
APE. *See* area of potential effect
The Archaeological Record, 32
archaeology: classical, 5–6; culture history, 5, 8, 13, 15, 70; by default, 15–16, 86, 126; developing personal, 73–74; ethnoarchaeology, 24–25, 27; European, 16; geoarchaeology, 24, 25; salvage, 7–8; scientific, 8, 69–70; university-based contract, 8, 10, 77. *See also* academic archaeology; CRM
archaeometric dating, 22
area of potential effect (APE), 6, 7, 56

Belize. *See* Colha Project
bidding/underbidding, 54–55, 60–61, 62, 63, 110–11
Binford, Lewis, 8, 78
Black, Stephen L., 36, 38, 40, 77–78
BLM. *See* Bureau of Land Management
budget, for CRM, 7, 44
building, as cultural resource, 9
Bureau of Land Management (BLM), 10, 75–76

Center for Ecological Archeology (CEA), 77
classical archaeology, 5–6
COE. *See* Corps of Engineers
Colha Project, 37–38

compliance, 8, 100; concerning Native Americans, 31, 109–10, 111; local ordinance, 31–32; section 106 of NHPA, 9, 29–30, 32–33, 47, 49, 56, 71
conservation ethic, 10
consulting firm, 10, 71
context, of research design, 3
contract, for CRM, 7, 39, 45
contract administrator, involving in research design, 100
contract archaeology: university-based, 8, 10. See also CRM
Corps of Engineers (COE), 10–11, 75, 76, 91
CRM: as archaeology-as-business, 71–72; bad reputation of, 10, 11–12; compliance in, 5, 28–29; cost of, 12, 13–14, 60–61, 62, 63; history of, 8–11; interdisciplinary approach to, 49; overview of, 6–7; public obligation of, 12–13, 14–15
cultural ecology, 25–26, 69
cultural resource, classifications of, 9
cultural resource management. See CRM
culture, definition of, 20
culture history archaeology, 5, 8, 13, 15, 70

data: defining, 3–4, 107–8. See also research strategy, CRM
data recovery, 47
dating, 13, 22
Department of Transportation, 76
district, as cultural resource, 9

Earthwatch, 37
ecology, 26
ethics, 50, 51, 58, 95, 107, 123
ethnoarchaeology, 24–25, 27
ethnography, 27
ethnohistory, 27
ethnology, 27
European archaeology, 16

fad, in research, 78
Federal Highway Administration. See Wurzbach Project
field notes, 50, 51
Flannery, Kent, 1, 8, 87
Forest Service, 10, 76
foundation, 28

geoarchaeology, 24, 25, 47
geomorphology, 47
government agencies, involving in research design, 75–76
government archaeologists, 7, 8; growth in numbers of, 10–11; role in scope of work, 71, 110–11
granting agency, 28

Hester, Thomas R., 36–38
Higgins Experiment, 41–42, 80
human ecology. See cultural ecology
hypothesis, definition of, 106

institutional focus research question, 76
intensive testing, 47

Jolly, Kevin, 40
journal, professional, 23–24, 25, 32, 131–32

landscape archaeology, 72–73
letter report, 51
literature survey/background research, 47
local preservation officer, 32

media, involving in research strategy, 97
memorandum of agreement (MOA), 33, 91

INDEX

NAGPRA. *See* Native American Graves Protection and Repatriation Act
National Endowment for the Humanities, 37
National Environmental Protection Act (NEPA), 8
National Geographic Society, 6, 37
National Historic Preservation Act (NHPA), section 106 of, 9, 29–30, 32–33, 47, 49, 56, 71
National Park Service, 8, 29
National Register of Historic Places (NRHP), 9, 32–33, 40, 48, 49
National Science Foundation (NSF), 5, 28, 41
Native American Graves Protection and Repatriation Act (NAGPRA), 109–10, 111
NEPA. *See* National Environmental Protection Act
networking, 33
New Archaeology, 8
NHPA. *See* National Historic Preservation Act
NRHP. *See* National Register of Historic Places
NSF. *See* National Science Foundation

object, as cultural resource, 9
Occupational Safety and Health Administration (OSHA), 61, 64–65
open house, 98
organization, professional, 23, 32–33; local, 129–30; national, 127–29

peers, 101, 102; and research design, 115; and research strategy, 93
physical anthropology, 20
Plains Anthropological Society, 129
processualist, 69
proposal, for CRM, 39, 44–45
public: CRM obligation to, 12–13, 14–15, 50; involving in research, 93, 96–98

radiocarbon dating, 13
reconnaissance project, 47
regional research design, 3, 75–76
Register of Professional Archaeologists (ROPA), 12, 128
registered professional archaeologist (RPA), 58, 128
regulator, 90, 92, 100, 104, 110, 119
regulatory cycle, 58
report series, 52
request for proposals (RFP), 2, 7, 30, 32, 43–44
research, academic: common elements of, 38; design in, 5–6, 44–45, 54; difference from CRM, 11, 35, 36, 41, 42; sample project, 36–39. *See also* academic archaeology
research, CRM: default methodology in, 15–16, 86, 126; difference from academic, 11, 35, 36, 41, 42; pressures/constraints on, 123–24; sample project in, 39–42
research design, CRM, 17, 125; common faults in, 106; elements of, 3–4, 53–54; evaluating, 122–23; tools for, 19; uses for, 12–13. *See also* toolkit building, basis of
research design, CRM, components of written: analysis, 108–9; data definition, 107–8; methods/techniques, 108; problem definition, 105–7; scope of work/legal rationale, 104–5. *See also* research design, CRM, written
research design, CRM, implementing: dealing with criticism of, 114–15; responding

to change in, 117–20; by reviewing work, 115–16; role of specialist in, 116–17, 119–20; role of team/colleagues in, 114–15
research design, CRM, lab processing, 4; focusing on analysis, 121; organizing report, 121; revising design, 120
research design, CRM, written: affect of project size on, 103–4; anticipating complications in, 109–10; audience for, 100–102; checklist for, 100; dealing with criticism of, 102–3; form of, 99; responding to scope of work in, 110–12; role of specialists/specialized studies in, 101, 109, 114. See also research design, CRM, components of written
research process, CRM, overview, 53–54; contract, 45–46; project, 46–49; proposal, 44–45; request for proposals, 43–44; research report, 39, 49–52; scope of work, 44
research project phase, CRM: permutations in, 47–49; of stereotypical project, 46–47
research question, CRM, 3, 67–69; creating, 74–75; testing cost/success of, 79–81
research question, CRM, source of: informed knowledge, 78; institutional, 76, 77; personal interest, 76–78; regional research design, 75–76
research report, CRM, 39, 49–52, 121
research strategy, CRM: data analyzation plan, 89; data management, 88–89; data-method relationship, 85–87; defining data, 3, 85; lab processing, 4, 88; preliminary planning, 83–84; review, 89–90; sampling, 87
research strategy, operational, 90; budget, 94–95; involving public in, 93, 96–98; logistics, 93–94; logistics checklist, 135–40; permits, 91–92; schedule, 95–96; working with bureaucracies, 90–91, 92, 93
research team, CRM: for landscape archaeology, 72–73; qualifications for, 44; research design and, 101–9, 114, 119–20; research strategy and, 93, 94
Reservoir Salvage Act, 8
revised scope of work, for CRM, 44
RFP. See request for proposals
ROPA. See Register of Professional Archaeologists
RPA. See registered professional archaeologist

SAA. See Society for American Archaeology
safety regulations, 61, 64–65
salvage archaeology, 7–8
sampling strategy, 87, 108
schedule, for CRM, 44, 59–60, 95–96
scientific archaeology, 8, 69–70
scope of work, for CRM, 44, 53; aligning with research question, 80; defining physical work location, 56–57; defining work to be performed, 57–59; estimating cost to fulfill, 60–61, 62, 63; expertise source for, 55; logistics in, 61, 64–65; primary areas of focus, 55–56; revised, 44; role of government archaeologist in, 71, 110–11; scheduling, 59–60; sponsor-prepared, 30, 54–55; work standards for, 59; in written research design, 104–5, 110–12

section 106 of NHPA, 9, 29–30, 32–33, 47, 49, 56, 71
Shafer, Harry J., 36–38
SHPO. *See* state historic preservation office/officer
site: as cultural resource, 9; evaluation of, 46–47; significance of, 47, 48
Smithsonian River Basin surveys, 7–8
Society for American Archaeology (SAA), 32, 127–28
Society for Archaeological Sciences, 129
Society for Historical Archaeology (SHA), 128
Society of Professional Archaeologists (SOPA), 128
sociocultural anthropology, 20, 27
SOPA. *See* Society of Professional Archaeologists
specialist: for landscape archaeology, 72–73; for research design, 101, 109, 114, 116–17, 119–20
sponsor: private, 30, 31; public, 28–29, 30, 54, 90–91; research design and, 100, 119; research strategy and, 90, 92, 95, 97; scope of work and, 54–55, 56, 57, 59, 60, 63, 104, 110; value of knowledge about, 30, 31
square hole archaeology, 15–16
staffing, 94
state historic preservation office/officer (SHPO): duties of, 29–32; regional research design and, 3, 75, 76; research strategy and, 91, 95; as source of regional expertise, 26; as source of scope of work expertise, 55
structure, as cultural resource, 9

survey/cultural resource inventory, 46

TARL. *See* Wurzbach Project
technology, for data management, 88–89
testing/site evaluation, 46–47
Texas Archaeological Research Laboratory (TARL). *See* Wurzbach Project
Texas Department of Transportation (TxDOT). *See* Wurzbach Project
Texas Historical Commission (THC). *See* Wurzbach Project
THPO. *See* Tribal Historic Preservation Office/officer
toolkit building, basis of: anthropology/archaeology background, 20–21; keep knowledge current, 23–24; method/theory background, 21–22; multidisciplinary knowledge, 22–23; political skill, 29–31; regional expertise, 24–27
Tribal Historic Preservation Office/officer (THPO), 31–32
TxDOT. *See* Wurzbach Project

university-based contract archaeology, 8, 10, 77
U.S. Army Corps of Engineers. *See* Corps of Engineers
USDA Forest Service. *See* Forest Service

Walker Ranch Historic District. *See* Wurzbach Project
Watson, Patty Jo, 106
Wurzbach Project, 40–42

ABOUT THE AUTHORS AND SERIES EDITORS

Stephen L. Black is a research associate and editor of www.TexasBeyondHistory.net at the Texas Archeological Research Laboratory at the University of Texas at Austin. He has been involved with CRM archaeology since 1975 and has done shovel testing, project administration, and almost everything in between. Steve has a Ph.D. from Harvard, an M.A. from the University of Texas at San Antonio, and a B.A. from the University of Texas at Austin, all focusing on anthropological archaeology. Most of his CRM research has centered on hunter-gatherer archaeology. Steve has been fortunate enough to be able to design and carry out mitigation projects at four significant hunter-gatherer sites in south and central Texas. He has also directed the analysis and reporting of five excavation projects that were begun by others. Through such experiences, he has developed an attitude about accomplishing meaningful research in the real world of archaeology. Steve's other research interests include experimental archaeology, hot rock technology, field methodology, information technology, and public archaeology. When not doing archaeology, he enjoys sailing, fishing, traveling, writing, experimenting with fermentation science, and having fun without his clothes on.

Kevin Jolly is a research fellow of the Texas Archeological Research Laboratory at the University of Texas at Austin and the vice president of technology at RW3 Technologies, Inc. In the late 1970s and early 1980s, he was a contract archaeologist working in Texas, New Mexico, and the southeastern United States. Starting in 1993, along with Steve Black and Dan Potter, Kevin helped put together a series of innovative (controversial) archaeological projects including the Higgins Experiment. To support the data collection efforts of these projects,

Kevin designed and built an archaeological data management system integrating total data stations and field computers. In the late 1990s, he directed the Historical Sites Atlas Project for the Texas Historical Commission, developing a comprehensive GIS database of Texas archaeological and historic sites. His research interests are archaeological data design, data collection, and data management.

❀

Larry J. Zimmerman is the head of the Archaeology Department of the Minnesota Historical Society. He served as an adjunct professor of anthropology and visiting professor of American Indian and Native Studies at the University of Iowa from 1996 to 2002 and as chair of the American Indian and Native Studies Program from 1998 to 2001. He earned his Ph.D. in anthropology at the University of Kansas in 1976. Teaching at the University of South Dakota for twenty-two years, he left there in 1996 as Distinguished Regents Professor of Anthropology.

While in South Dakota, he developed a major CRM program and the University of South Dakota Archaeology Laboratory, where he is still a research associate. He was named the University of South Dakota Student Association Teacher of the Year in 1980, given the Burlington Northern Foundation Faculty Achievement Award for Outstanding Teaching in 1986, and granted the Burlington Northern Faculty Achievement Award for Research in 1990. He was selected by Sigma Xi, the Scientific Research Society, as a national lecturer from 1991 to 1993, and he served as executive secretary of the World Archaeological Congress from 1990 to 1994. He has published more than three hundred articles, CRM reports, and reviews and is the author, editor, or coeditor of fifteen books, including *Native North America* (with Brian Molyneaux, University of Oklahoma Press, 2000) and *Indians and Anthropologists: Vine Deloria, Jr., and the Critique of Anthropology* (with Tom Biolsi, University of Arizona Press, 1997). He has served as the editor of *Plains Anthropologist* and the *World Archaeological Bulletin* and as the associate editor of *American Antiquity*. He has done archaeology in the Great Plains of the United States and in Mexico, England, Venezuela, and Australia. He has also worked closely with a wide range of American Indian nations and groups.

William Green is Director of the Logan Museum of Anthropology and Adjunct Professor of Anthropology at Beloit College, Beloit, Wisconsin. He has been active in archaeology since 1970. Having grown up on the south side of Chicago, he attributes his interest in archaeology and anthropology to the allure of the exotic (i.e., rural) and a driving urge to learn the unwritten past, abetted by the opportunities available at the city's museums and universities. His first field work was on the Mississippi River bluffs in western Illinois. Although he also worked in Israel and England, he returned to Illinois for several years of survey and excavation. His interests in settlement patterns, ceramics, and archaeobotany developed there. He received his Master's degree from the University of Wisconsin–Madison and then served as Wisconsin SHPO staff archaeologist for eight years. After obtaining his Ph.D. from UW–Madison in 1987, he served as State Archaeologist of Iowa from 1988 to 2001, directing statewide research and service programs including burial site protection, geographic information, publications, contract services, public outreach, and curation. His main research interests focus on the development and spread of native agriculture. He has served as editor of the *Midcontinental Journal of Archaeology* and *The Wisconsin Archeologist*, has published articles in *American Antiquity, Journal of Archaeological Research*, and other journals, and has received grants and contracts from the National Science Foundation, National Park Service, Iowa Humanities Board, and many other agencies and organizations.